Graphene-Rubber Nanocomposites

Since the Nobel Prize for the discovery of graphene was presented in 2010, graphene has been frequently leveraged for different applications. Owing to the strategic importance of elastomer-based products in different segments, graphene and its derivatives are often added to different elastomers to improve their properties. The book, *Graphene-Rubber Nanocomposites: Fundamentals to Applications* provides a comprehensive and innovative account of graphene-rubber composites.

Features:
- Provides an up-to-date information and research on graphene-rubber nanocomposites
- Presents a detailed account of the different niche applications ranging from sensors, flexible electronics to thermal, and EMI shielding materials
- Offers a comprehensive know-how on the structure-property relationship of graphene-rubber nanocomposites
- Covers the characterization of graphene-based elastomeric composition
- Delivers a comprehensive understanding of the structure of the graphene, including its chemical modification for usage in elastomer composites

This book will be a valuable resource for graduate-level students, researchers, and professionals working in the fields of materials science, polymer science, nanoscience and technology, rubber technology, chemical engineering, and composite materials.

Graphene-Rubber Nanocomposites
Fundamentals to Applications

Edited by
Titash Mondal and Anil K. Bhowmick

CRC Press
Taylor & Francis Group
Boca Raton London New York

CRC Press is an imprint of the
Taylor & Francis Group, an **informa** business

First edition published 2023
by CRC Press
6000 Broken Sound Parkway NW, Suite 300, Boca Raton, FL 33487-2742

and by CRC Press
4 Park Square, Milton Park, Abingdon, Oxon, OX14 4RN

CRC Press is an imprint of Taylor & Francis Group, LLC

ISBN: 978-1-032-05977-8 (hbk)
ISBN: 978-1-032-06045-3 (pbk)
ISBN: 978-1-003-20044-4 (ebk)

DOI: 10.1201/9781003200444

Typeset in Times
by codeMantra

Ayansh Mondal & Piya Mondal
Tapati Mondal & Tapan Mondal
Asmit Bhowmick
Kundakali Bhowmick

Contents

Preface

In the era of the connected world, the use of nanomaterials for various applications ranging from healthcare devices to intelligent tires is on the rise. Different nanomaterials ranging from carbon nanotubes, graphene, silica, etc., are being used by scientists across the globe. In the current scenario, such nanomaterials are beyond the academic world and are being widely adopted by different industries. Among the different nanomaterials, graphene's contribution as a potential filler in various segments has been peerless. Graphitic material has been a subject of chemical research since the 18th century when the first intercalated compound of graphite was reported. A mixture of sulfuric acid and nitric acid was used for the exfoliation of graphitic layers. Eminent scientists like Brodie took interest in determining the molecular weight of graphitic material. He used strong oxidants like potassium perchlorate for his reaction recipe. Serendipitously, he landed up with a new class of compound known as graphite oxide. Following that, there has been a series of experiments by many stalwarts like Staudenmaier, Hummers, and Offeman. In the 19th century, Brodie's method of oxidation was coupled with an additional step of reduction by Boehm. However, lately, in the 20th century, the gold rush of graphene began, when it was first observed to exist practically as a two-dimensional crystal. A micromechanical exfoliation technique was developed. Ever since, there has been exponential research related to this carbon nanomaterial. Thanks to its exhilarating properties like superior mechanical properties, ability to transport electrons ballistically, and excellent thermal conductivity, graphene is often termed as a wonder material and thereby has become an ultimate choice for many researchers. A common thread wherein the functional benefits of graphene have been significantly explored is in the domain of rubber science and technology. Rubber has always been a strategically important material for its robust utility in various fields ranging from aerospace to healthcare. However, in most cases, it can be observed that the usage of unfilled rubber in practical application is far and few. Hence, it is compounded with a variety of fillers, including graphene. Although we see multiple research articles and patents that capture the progress of the research related to graphene-rubber nanocomposites, a comprehensive package consolidating the fundamental understanding of various aspects of graphene-rubber nanocomposites, their fabrication, and their specific applications is missing. It can be reasonably inferred that there is a pressing need to have a comprehensive information package for scientists across the globe. This gave us an impetus to edit this book, pen down our thoughts from the decade-long learning we had from the work we did in the graphene-rubber nanocomposite, and consolidate the learning from various experts in this domain.

The book is divided into 18 chapters. The first chapter provides the prelude to the wonder material, graphene. The second chapter gives valuable insight into synthesizing graphene from conventional sources and their characterization techniques, while the third chapter discusses the synthesis of graphene from non-conventional sources and their characterization. In most cases, the functionalization of graphene is critical for achieving good dispersion in the rubber matrix. Hence, the fourth chapter is dedicated to the different functionalization strategies involved in graphene and graphite. The functionalized graphene once mixed with the rubber is bound to impact the mechanical properties of the composite. Hence, it is critical to understand the structure-property relationship. It is also essential to comprehend the different characterization techniques to decipher the causality for such improvement in the properties. Chapters 5–8 discuss these crucial parameters. In modern days, computational science and technology and machine learning are becoming essential tools to predict various material properties without performing real experiments in the laboratory. Chapters 9 and 10 provide insight into different computational tools used to predict the performance of graphene-rubber nanocomposites. Chapters 11–18 are application-specific and provide a comprehensive idea about the different processing routes to develop such compositions. Chapter 11 captures a very important property of graphene. Owing to its plate-like morphology, it is effectively used as a barrier material for rubber nanocomposite. Similarly, in the field of thermoplastic elastomer, the usage of graphene for multifaceted applications is possible. Chapter 12 provides an insight into the role of graphene in thermoplastic elastomer nanocomposites. Chapter 13 provides an account of the specific usage of graphene in tire tread composition and how graphene can improve the durability of the tire is provided. Graphene-rubber nanocomposites also find applications in the field of flexible electronics and Chapter 14 provides insight into the role of graphene in developing flexible electronic material. Chapter 15 talks about the development of thermally conductive graphene rubber composite, while Chapter 16 furnishes an overview of the development of Graphene-Elastomer Composite for Energy Storage Applications. Chapter 17 captures the usage of graphene-rubber nanocomposites for biomedical applications, while Chapter 18 discusses the graphene-rubber nanocomposites for electromagnetic shielding applications. It is worth mentioning that the opinions, comments, permission for images from publisher(s), and representation made in the various chapters are from the authors from the respective chapters. The editors don't hold any responsibilities in case of any discrepancies noted. The contribution from the different authors for this book is appreciated, and the editors would like to thank them for their input and valuable time.

Dr. Mondal is grateful to his colleagues at the Rubber Technology Centre, IIT Kharagpur, research collaborators, and funding agencies. He would like to especially thank his mentor, Professor Anil K. Bhowmick, and Professor Ramanan Krishnamoorti from the University of Houston, USA. Dr. Mondal would like to thank his team of Ph.D and master's students: Simran Sharma, Muthamil Selvan T, Aparna Guchait, Prama Adhya, Ajay Haridas, Sayan Chakraborty and Arpita Kundu for their help. Professor Bhowmick is grateful to the University of Houston, especially to Professor Renu Khator, Chancellor and President, Professor Joe Tedesco, Dean, Cullen College of Engineering, Professor T. J. (Lakis) Mountziaris, Department Chair, Department of Chemical and Biomolecular Engineering, Professor Michael P. Harold, Former Department Chair, Department of Chemical and Biomolecular Engineering, Professor A. Karim, Professor R. Krishnamoorti, and Professor M. Robertson, all from the Department of Chemical and Biomolecular Engineering for their support and cooperation. We would also like to thank Gabrielle Vernachio and Allison Shatkins for their help. Last but not least, Dr. Mondal would like to thank Mr. Tapan Mondal, Ms. Tapati Mondal, Ms. Piya Mondal, and Ayansh Mondal for their encouragement and support to complete the book and publish it. Professor Bhowmick would like to thank Dr. Kundakali Bhowmick and Dr. Asmit Bhowmick for their sacrifice for the publication of the book.

Titash Mondal
Indian Institute of Technology Kharagpur, India
Anil K. Bhowmick
The University of Houston, USA
Formerly with Indian Institute of Technology Kharagpur, India

Editors

Dr. Titash Mondal is an Assistant Professor at the Rubber Technology Centre, IIT Kharagpur, India. Dr. Mondal received his Master of Technology degree in 2011 in Rubber Technology from the Indian Institute of Technology Kharagpur, India, and his Ph.D. degree in 2015 jointly from the Indian Institute of Technology of Patna, India, and University of Houston, USA. From 2016 to 2020, he worked as a scientist in Momentive Performance Materials. Since 2020, he has been working as an Assistant Professor at the Indian Institute of Technology Kharagpur, India. His main research interests are nanomaterials and polymer nanocomposites, flexible and printed electronics, adhesive science and technology, sustainable materials, and silicone science and technology. Dr. Mondal got multiple publications in the international journal of high repute, four patents, and a few book chapters. Dr. Mondal has been working in the field of graphene and its composite for the last decade. Dr. Mondal has won himself multiple awards. Some of the notable awards include Modi Rubber Prize 2011, IIT Kharagpur and R&D Innovator of the Year 2017, Momentive Performance Materials. Currently, Dr. Mondal is supervising seven Ph.D. and three Master's students. He is in the reviewer panel of multiple international journals and is also the member of the American Chemical Society.

Professor Anil K. Bhowmick is currently a Research Professor at the Department of Chemical and Biomolecular Engineering at the University of Houston and a former Professor of Eminence, IIT Kharagpur, India. He was previously associated with the University of Akron, Ohio, USA, London School of Polymer Technology, London, and Tokyo Institute of Technology, Japan. His main research interests are nanomaterials and polymer nanocomposites, thermoplastic elastomers and polymer blends, failure and degradation of polymers, sustainable materials, rubber-filler interaction, adhesion and adhesives, polymer modification, and rubber technology. He has more than 600 peer reviewed international publications in these fields, 35 book chapters, and seven books. He was the 2002 winner of the Chemistry of Thermoplastic Elastomers award, the 1997 winner of the George Stafford Whitby award of the Rubber Division, American Chemical Society, 2019 SPE Fred E. Schwab Education Award of the Society of Plastics Engineers for innovative research and teaching, and 2001 K.M. Philip award of the All India Rubber Industries Association for outstanding contribution to the growth and development of rubber industries in India. He has received the 2021 J. L. White Innovation award from the Polymer Processing Society and is the recipient of the 2022 Melvin Mooney Distinguished Technology award of the Rubber Division, American Chemical Society. He was also awarded Trila Tyre times award 2019, Distinguished alumni award 2020, NOCIL award 1991, JSPS award 1990, Commonwealth award 1990, MRF award 1989, and Stanton-Redcroft ITAS award 1989. He is on the Editorial Board of the *Journal of Applied Polymer Science* (USA), *Journal of Materials Science* (USA), *Polymers and Polymer Composites* (UK), *Rubber Chemistry and Technology* (USA), *Polymers for Advanced Technology* (Europe), *Nano-Micro Letters* (UK), *Processes* (Germany), J*ournal of Rubber Research* (Malaysia), and *Natural Rubber Research* (India). He holds 21 patents, including three US, three Japanese, and one German patents. He has guided 50 Ph.D. students and a few students are presently working for Ph.D. under his guidance. He is a Fellow of Indian National Academy of Engineering, Indian National Science Academy and W.B. Academy of Science and Technology. He was an INAE Chair Professor in India and the Founder Director of Indian Institute of Technology Patna.

Contributors

S. Anandhan
Department of Metallurgical and
    Materials Engineering
National Institute of Technology
    Karnataka
Mangaluru, India

Aswathy T. R.
Rubber Technology Centre
IIT Kharagpur
Kharagpur, India

Mohammad Javad Azizli
Department of Chemistry and Chemical
    Engineering
Islamic Azad University
Rasht, Iran

Ajish Babu
Department of Metallurgical and
    Materials Engineering,
Indian Institute of Technology Patna
    Bihar
Bihta, India

Mohammad Barghamadi
Department of Rubber
Iran Polymer and Petrochemical
    Institute
Tehran, Iran

Madhab Bera
Department of Polymer and Process
    Engineering,
Indian Institute of Technology Roorkee
Roorkee, India

Surita Bhatia
Department of Chemistry
SUNY Stony Brook University
Stony Brook, New York

Asit Baran Bhattacharya
Rubber Technology Centre
IIT Kharagpur
Kharagpur, India

Anil K. Bhowmick
Formerly with Rubber Technology Centre
IIT Kharagpur
Kharagpur, India
and
Department of Chemical and
    Biomolecular Engineering
The University of Houston
Houston, Texas

Jagannath Chanda
Department of Advance Material
    Research and Characterization
    (AMRC),
Hari Shankar Singhania Elastomer and
    Tire Research Institute (HASETRI)
Mysuru, India

Santanu Chattopadhyay
Rubber Technology Centre
IIT Kharagpur
Kharagpur, India

Narayan Chandra Das
Rubber Technology Centre
IIT Kharagpur
Kharagpur, India

Deepthi Anna David
Department of Applied Chemistry,
    Cochin University of Science and
    Technology (CUSAT)
Cochin, India

Pritam V. Dhawale
Department of Polymer and Surface
    Engineering,
Institute of Chemical Technology
Mumbai, India

Jinu Jacob George
Department of Polymer Science and
    Rubber Technology,
Cochin University of Science and
    Technology (CUSAT)
Cochin, India

Sanjoy Kumar Ghorai
Rubber Technology Centre
IIT Kharagpur
Kharagpur, India

Ranjan Ghosal
Birla Carbon India Private Limited
Panvel Taluka, Mumbai, India

Akash Ghosh
Rubber Technology Centre
IIT Kharagpur
Kharagpur, India

Bikshan Ghosh
Rubber Technology Centre
IIT Kharagpur
Kharagpur, India

Prasenjit Ghosh
Department of Advance Material
    Research and Characterization
    (AMRC),
Hari Shankar Singhania Elastomer and
    Tire Research Institute (HASETRI)
Mysuru, India

Saikat Das Gupta
Department of Advance Material
    Research and Characterization
    (AMRC),
Hari Shankar Singhania Elastomer and
    Tire Research Institute (HASETRI)
Mysuru, India

Mario Hofmann
Department of Physics
National Taiwan University
Taipei, Taiwan

Ya-Ping Hsieh
Institute of Atomic and Molecular
    Sciences
Academia Sinica
Taipei, Taiwan

Shalmali Hui
Department of Chemistry
Hijli College affiliated to Vidyasagar
    University
West Midnapore, India

Saptarshi Kar
Birla Carbon India Private Limited
Panvel Taluka, Mumbai, India

Mohammed Khalifa
Kompetenzzentrum Holz GmbH, Wood
    K plus
Linz, Austria

Kedar Kirane
Department of Mechanical Engineering
Stony Brook University
Stony Brook, New York

Challa V. Kumar
Department of Chemistry, Department
    of Molecular and Cell Biology and
    the Institute of Material Science
University of Connecticut
Storrs, Connecticut

Leonardo Dantas Machado
Departamento de Física,
Universidade Federal do Rio Grande do
    Norte
Natal, Brazil

Pradip K. Maji
Department of Polymer and Process
    Engineering,
Indian Institute of Technology Roorkee
Roorkee, India

Masoud Mokhtary
Department of Chemistry and Chemical
 Engineering
Islamic Azad University
Rasht, Iran

Soumyadeep Mondal
Department of Mechanical Engineering
National Institute of Technology
 Durgapur
Durgapur, India

Titash Mondal
Rubber Technology Centre
IIT Kharagpur
Kharagpur, India

Rabindra Mukhopadhyay
Department of Advance Material
 Research and Characterization
 (AMRC),
Hari Shankar Singhania Elastomer and
 Tire Research Institute (HASETRI)
Mysuru, India

Selvakumar Murugesan
Department of Metallurgical and
 Materials Engineering
National Institute of Technology
 Karnataka
Mangaluru, India

Kinsuk Naskar
Rubber Technology Centre
IIT Kharagpur
Kharagpur, India

Susmita Naskar
Faculty of Engineering and Physical
 Sciences,
University of Southampton
Southampton, United Kingdom

Dinh-Tuan Nguyen
Institute of Atomic and Molecular
 Sciences
Academia Sinica
Taipei, Taiwan

Peter Samora Owuor
Department of Materials Science and
 NanoEngineering,
Rice University
Houston, Texas

Pramod Rakt Patel
Department of Mechanical Engineering,
 Dr. BR Ambedkar National Institute
 of Technology
Jalandhar, India

Ajith Pattammattel
National Synchrotron Light Source II
Brookhaven National Laboratory
Upton, New York

Sourav Paul
Rubber Technology Centre
IIT Kharagpur
Kharagpur, India

Prasanth Raghavan
Department of Polymer Science and
 Rubber Technology,
Cochin University of Science and
 Technology (CUSAT)
Cochin, India

Katayoon Rezaeeparto
Research Institute of Petroleum Industry
Tehran, Iran

Amrita Roy
Rubber Technology Centre
IIT Kharagpur
Kharagpur, India

Meryem Samancı
Department of Chemical Engineering,
 Atatürk University
Erzurum, Turkey

A. M. Shanmugharaj
Centre for Energy and Alternative Fuels
Vels Institute of Science, Technology
    and Advanced Studies
Chennai, India

Simran Sharma
Rubber Technology Centre
IIT Kharagpur
Kharagpur, India

Sumit Sharma
Department of Mechanical Engineering,
    Dr. BR Ambedkar National Institute
    of Technology
Jalandhar, India

Kishor Balasaheb Shingare
Department of Aerospace Engineering
Indian Institute of Technology Bombay
Mumbai, India

Vijay Kumar Thakur
Biorefining and Advanced Materials
    Research Centre
Edinburgh, United Kingdom

Ayşe Bayrakçeken Yurtcan
Department of Chemical Engineering,
    Department of Nanoscience and
    Nanoengineering,
Atatürk University
Erzurum, Turkey

1 Introduction to Graphene

Ajith Pattammattel
Brookhaven National Laboratory

Challa V. Kumar
University of Connecticut

CONTENTS

1.1 INTRODUCTION

Graphene is a single carbon-atom thick nanomaterial with considerable implications, one of the thinnest stable materials known (Kumar and Pattammattel 2017) and graphene thickness is 100,000 times less than that of ordinary paper (Novoselov et al. 2004). Yet, it is one of the strongest and highest conductivity materials as well. Graphene is dubbed as the material of the millennium due to a vast array of its awesome properties, which stimulated both theoretical and applied research. After its successful isolation in 2004 (Novoselov et al. 2004), research on two-dimensional (2D) nanomaterials that are similar in structure to that of graphene has grown exponentially (Tan et al. 2017). Thus, the discovery of graphene itself has inadvertently catalyzed the growth of 2D materials science. This chapter aims to provide a brief introduction to the family of graphene materials.

DOI: 10.1201/9781003200444-1

1.2 HISTORY OF GRAPHENE

Graphene, consisting of a 2D nanosheet of hexagonally arranged sp^2-hybridized carbon atoms, is the most studied and debated material over the last two decades. Graphene was the talk of the town after characterizing the electronic properties of a sample of single-layer graphene in 2004 by Geim and Novoselov, Nobel Laureate Physicists at the University of Manchester (Novoselov et al. 2004). But was it the only time scientists thought about isolating or synthesizing graphene? History tells us otherwise! One of the studies by Sir Benjamin Collins Brodie at the University of Oxford in 1859, reported the synthesis of oxidized graphene (today known as graphene oxide, GO) from graphite (Figure 1.1) (Brodie 1859). Brodie's 'graphite carbonic acid' is a single layer oxidized carbon sheet. Interestingly, the chemical reduction of GO was also reported in the same work, suggesting one of the first attempts to make graphene. Over 100 years later, Ruess and Vogut reported the transmission electron microscopy (TEM) images of multi-layer GO (Ruess et al. 1948). In 1962, the electron micrographs of monolayer graphene were also reported (Boehm et al. 1962).

Several theoretical calculations in 1990s predicted that self-standing 2D crystals are unstable (Chodos et al. 2009, Geim et al. 2010). Therefore, the existence of graphene was questionable except for growing them on metallic supports (Forbeaux et al. 1998). But, with the famous 'Scotch-tape' method, Geim and Novoselov proved that single-layer atomically thin carbon sheets are thermodynamically stable. They reported the superior electronic properties that led to calling graphene the 'wonder material.' Since 2004, graphene research has grown exponentially, resulting in thousands of publications and patents on this subject. To date, hundreds of stable 2D nanosheets comprising of organic and inorganic compounds that resemble the topography of the graphene sheet are known, and several of them found applications in energy storage, catalysis, biomedicine and so on (Tan et al. 2017). However, several promises and claims first made on graphene are yet to be proved but the wonders of graphene discoveries do not seem to cease even after two decades of its re-discovery.

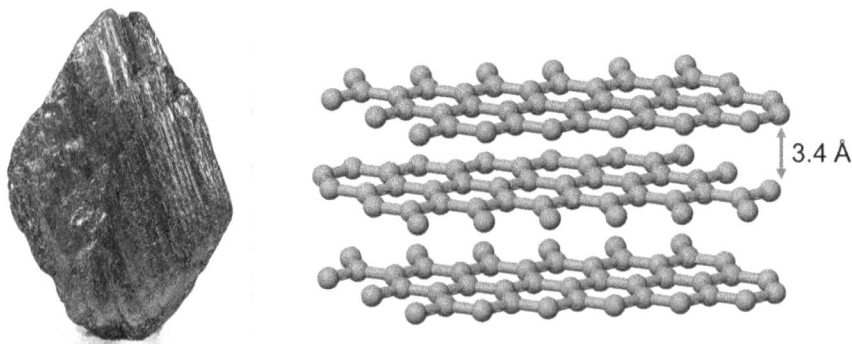

3.4 Å

FIGURE 1.1 Graphite crystal and stacking of carbon layers. (Reprinted with permission from Wikipedia Commons Image, Credit: Rob Lavinsky, iRocks.com and Benjah-bmm27.)

1.3 THE STRUCTURE AND NOMENCLATURE

A single graphene sheet represents a network of sp^2-hybridized carbon atoms. The adjacent carbon atoms in the hexagon structure are bonded with a sigma bond comprising of two sp^2 orbitals, while a π-bond is formed from the remaining p-orbital of each carbon atom (Figure 1.1). The extensive 2D-network of sigma bonds of high bond energies provides the material with high mechanical strength, flexibility, and unusual physical properties. The p$_z$-orbitals of the carbons provide a continuous network of π-electrons which play a vital role in defining the unique optoelectronic and mechanical properties of graphene and graphene-family materials. Extensive delocalization of the network of π-bonds endows graphene with unusual optoelectronic properties, including high thermal and electrical conductivities (Sang et al. 2019).

Graphene is a zero-bandgap semiconductor. The highest occupied molecular orbital of graphene is fully occupied, and the lowest unoccupied molecular orbital is completely empty. The conduction and valence bands meet at the Dirac points and the band structure is discussed later in more detail. Thus, long distance electron transport along the surface of graphene can be visualized from its electronic structure based on the sp^2 hybridization of the carbon atoms present in graphene (Figure 1.2).

When sp^3-carbons are present in place of the sp^2-carbons in the graphene framework, which is often due to the inclusion of another atom such as oxygen or due to defects in its chemical structure, a different terminology is used (Figure 1.3). But several studies in the literature incorrectly term GO and reduced graphene oxide containing sp^3-carbons as 'graphene'. Therefore, one should carefully evaluate the graphene literature to understand the system identified correctly. Graphene derived from graphite through exfoliation methods may have a distribution in number of layers present in each flake, but in a suspension, they are usually present as 'few-layer graphene', a general term used when the number of graphene layers in a flake, on

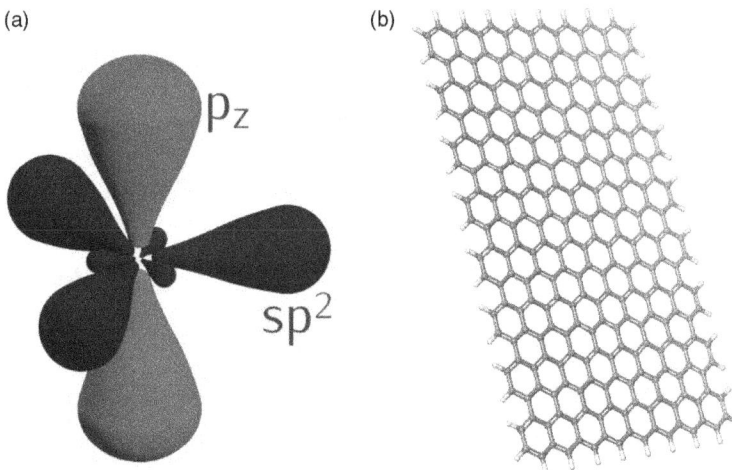

FIGURE 1.2 (a) The sp^2 and p$_z$ orbitals of carbons that make up graphene nanosheet, arranged in a hexagonal honeycomb shape. (Wikipedia Commons Image, Credit: Ponor.). (b) Structure of single layer graphene.

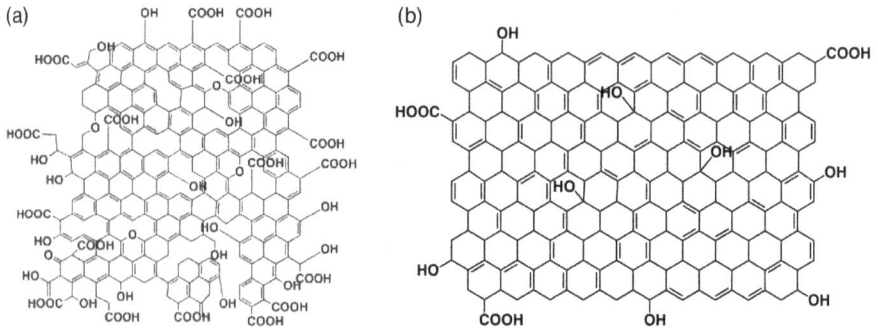

FIGURE 1.3 Representative structures of (a) graphene oxide and (b) reduced graphene oxide.

average, is below five. With a controlled growth method, graphene of 1-, 2- or 3-layers can be precisely prepared, and the resulting material is named accordingly as mono-, bi- or tri-layer graphene.

1.4 OTHER 2D NANOMATERIALS

Since the beginning of the graphene revolution, the hunt for other stable 2D nanolayers began. One of the first members in this class was hexagonal boron nitride (*h*-BN), also called 'white graphene' due to its color (Figure 1.4a) (Stehle et al. 2015). This single-atom-thick 2D material has alternating B and N atoms, and the unoccupied p_z orbital of boron conjugates with the fully occupied p_z orbital of nitrogen atoms, providing a unique network of molecular orbitals, spanning the entire sheet (Caldwell et al. 2019). In comparison to graphene, 2D layers composed of B, P, Sb and Bi are also discovered recently (Gusmão et al. 2017, Ji et al. 2016, Li et al. 2014, Mannix et al. 2015). Transition metal dichalcogenides (TMDs) such as layered MoS_2, WS_2, WSe_2 or FeS_2 belong to a major class of 2D materials that are being studied in detail for numerous practical applications, as well as theoretical studies (Figure 1.4b) (Kaur et al. 2020, Mak et al. 2010, Manzeli et al. 2017, Puglia et al. 2021). To date, taken together, hundreds of stable 2D nanolayered materials are known and scientists can generate nanolayers of materials from the corresponding bulk crystals via a variety

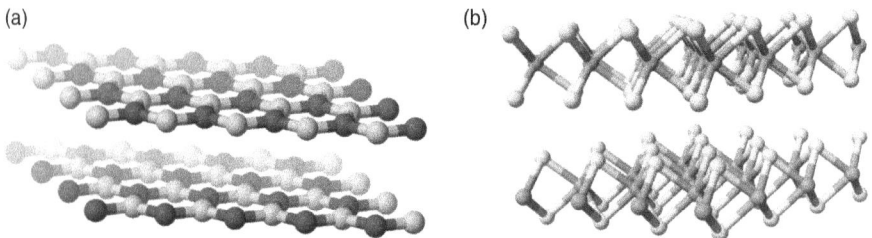

FIGURE 1.4 Structures of some inorganic 2D materials: (a) h-BN and (b) general structure of TMDs (X in yellow). (Reprinted with permission from Wikipedia Commons Images, Credit: Benjah-bmm27.)

of exfoliation methods (Oaki 2021, Puglia et al. 2021). But their foci are mostly centered around their applications in a variety of fields including drug delivery and environmental sensing (Malhotra et al. 2022). One may expect to see many more 2D materials in functional materials and devices in the future.

1.5 SYNTHETIC ROUTES TO GRAPHENE AND ITS ANALOGUES

Graphene family members and inorganic 2D layers can be prepared from top-down or bottom-up approaches. The major consideration in adopting one synthetic protocol over another is the intended application and purity. Generally, structurally homogenous and high purity graphene is needed for electronic applications that is achieved using chemical vapor deposition method (Zhang et al. 2013). But graphene derived from the scotch tape method is also useful for electronic devices (Novoselov et al. 2004).

Solvent exfoliation methods are very economical and scalable to make graphene from bulk graphite crystals/flakes (Nicolosi et al. 2013). Exfoliation of graphene is viable in organic solvents as well as aqueous media (Kumar et al. 2017). In most solvents, graphene has to be stabilized against re-stacking and aggregation using the so-called exfoliation agents. In aqueous medium, surfactants, proteins, DNA, polysaccharides and many small molecules can be used as exfoliation agents (Chabot et al. 2013, Khan et al. 2010, Pattammattel et al. 2015). Ultrasonication, (turbulence-assisted) shearing and simple bench top stirring are found to be effective to delaminate graphite to few-layer graphene in the presence of a suitable exfoliating agent (Ciesielski et al. 2014, Yi et al. 2015). These solvochemical methods are useful to prepare composite materials of graphene with polymers, biomolecules and other inorganic materials. Figure 1.5 summarizes the synthesis of graphene by both these

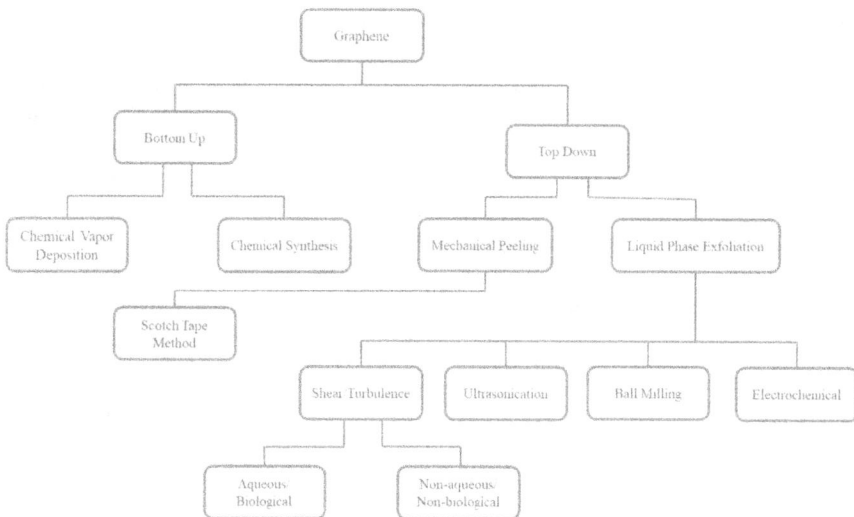

FIGURE 1.5 Major synthetic approaches to making graphene.

approaches. There are myriads of conditions suitable for graphene preparation in solvent systems. And the choice of the medium merely depends on the post-synthetic application and solvent compatibility of the material(s) used for *in situ* composite preparation (Du et al. 2012).

1.6 CHARACTERIZATION OF GRAPHENE

Characterizing graphene and its composites is very important to understand the structure-property relationships. The defects, degree of oxidation, and functionalization can profoundly affect the material and optoelectronic properties of graphene. Multi-modal bulk and surface sensitive techniques are employed to characterize graphene. Characterization of graphene aims to determine the average number of graphene layers in a flake (or thickness), lateral size, the degree and type of defects and stoichiometry. These properties are deduced by using a variety of spectroscopy and microscopy techniques. Raman spectroscopy (micro/bulk) (Ferrari et al. 2013), X-ray photoelectron spectroscopy (XPS) (Yang et al. 2009), and TEM (Meyer 2014) are the most frequently used techniques for graphene characterization. Raman spectroscopy is considered as a 'fingerprint' technique for graphene, and spatially resolved Raman spectroscopy is useful to characterize graphene in pure form as well as in heterogenous environments such as in a composite.

The characteristic Raman modes indicate the number of layers, size and degree of defects, type of defects, functionalization and so on. Empirical formulas are also derived from calculating the lateral size and number of layers of the sheets from Raman spectra (Varrla et al. 2014). Figure 1.6a shows a representative Raman spectrum of exfoliated graphene and that of graphite. The C–C in-plane stretching vibration at ~1,580 cm^{-1} is common for sp^2 carbon allotropes and has similar characteristics for graphene and graphite. But the D band at ~1,350 cm^{-1} is activated by defects and disorders in the layers. So, it is well pronounced in graphene. Therefore, the ratio of intensities of D and G peaks (I_D/I_G) is an indicator of the degree of defects in graphene that may be introduced during the synthesis. The first overtone of the D band called the 2D band is an indicator of number of layers present in the graphene. The intensity of 2D band increases as the number of layers decreases and resonates at a lower wavenumber. However, the 2D band is indistinguishable from graphite, if the graphene sample has more than five layers. Graphene oxide and its derivatives have significantly different Raman peaks when compared to those of the pristine graphene. So, Raman spectroscopy is vital to distinguish graphene from graphene oxide derivatives, although some literature incorrectly term them the same.

XPS measures the binding energy of electrons in the carbon or oxygen atom in graphene. With high energy resolution XPS map, the amounts of both sp^2 and sp^3 carbons can be quantitated, indicating the extent of defect or oxidation of graphene. Table 1.1 shows the characteristic binding energies of possible sp^2 and sp^3 carbons in graphene derivatives. Since XPS is a surface sensitive technique (penetration up to 2 µm), composite materials must be sectioned before data collection, and other carbon bonds present in the matrix can have an effect on the quantitation. Compared to Raman spectroscopy, XPS is not a high-throughput technique that requires an ultra-high vacuum.

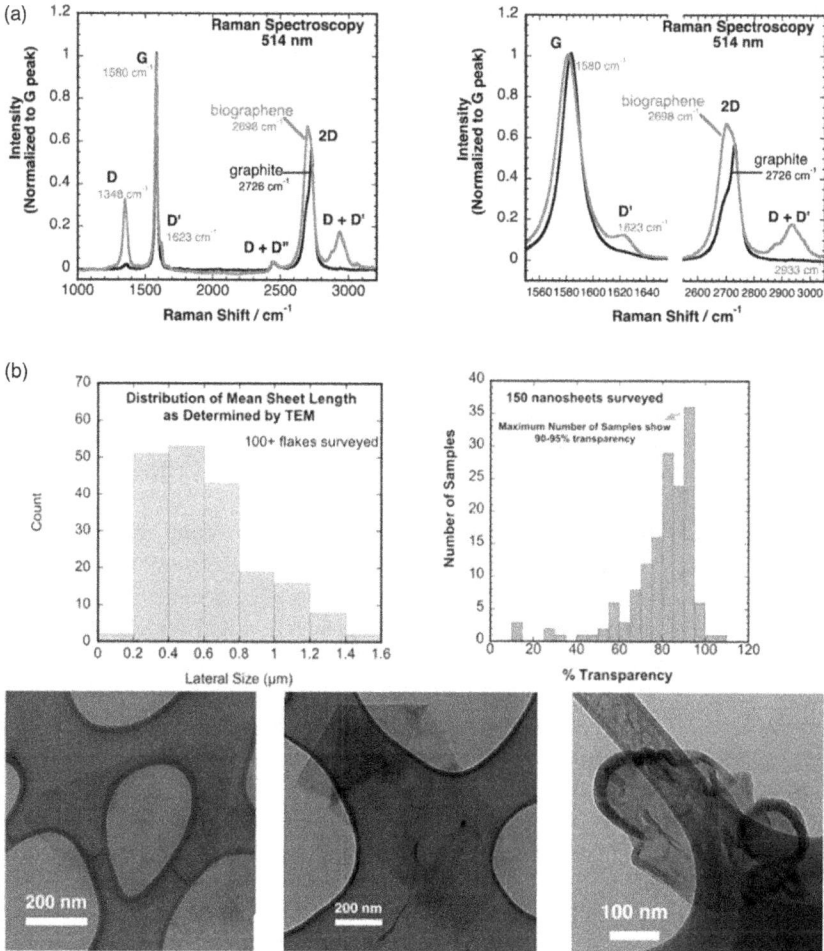

FIGURE 1.6 (a) Raman spectrum of exfoliated graphene in reference to that of graphite. (b) Representative TEM images and estimates of electron transparency and lateral length of exfoliated graphene. (Reprinted with permission from the American Chemical Society.)

TABLE 1.1

Binding Energies of Carbon Bonds in Graphene Derivatives

Bond	~Binding Energy, C1s (eV)	Bond	~Binding Energy, C1s (eV)
C–C	284.8	O=C–O	288.5
C=C	283.4	C–H	285.3
C–O	285.5	C–N	283.7 (sp^{3}N)
O–C–O	286.5	C=N	286.8
C=O	289	π–π* resonance	291.2

Source: Database (2021), **Cooper et al.** (2014), **Scardamaglia et al.** (2014), and **Susi et al.** (2015).

Other spectroscopy methods such as absorption spectroscopy, FT-IR, Surface Plasmon Resonance (SPR) and Zeta potential measurements are combined with Raman spectroscopy or XPS to understand the characteristics of graphene samples. Depending on the sample conditions, these techniques are frequently used in the literature.

Electron, X-ray and atomic force microscopy (AFM) techniques are used to image graphene layers (Figure 1.6b). TEM is a direct imaging technique with an atomic level resolution that can be used to compute the lateral size and roughly inspect the number of layers. And with a selected area diffraction pattern, the number of layers in graphene sheets can be confirmed from electron diffraction peaks in TEM studies. But for statistical validity, hundreds of sheets should be imaged to estimate the lateral size, thickness (electron transparency) and the number of layers of graphene using TEM (Pattammattel et al. 2015). In addition, the folding of graphene sheets, surface functionalization and other morphology details can be characterized with high-resolution TEM imaging. However, composite materials must be sectioned down to sub-micron thickness to achieve electron penetration.

AFM and surface probe microscopy (SPM) reveal the atomic scale details and nanostructure of graphene surface. Accurate thickness measurements (number of layers) can be performed with AFM with a few angstrom accuracies. In addition, lateral size, surface functionalization, surface orientation and folding can also be estimated from AFM measurements (Figure 1.7). AFM is a surface-sensitive technique to image the topology of the sheets and it is not very useful for bulk graphene composites without sectioning. Using contact mode AFM, the electronic properties of graphene sheets can also be examined. AFM can also be coupled with Raman spectroscopy to understand the defects in a spatially resolved manner. The possibility of liquid cell AFM techniques further expands its usefulness in composite characterization. Readers may refer to the other Chapters in this book as well for characterization of these materials in detail.

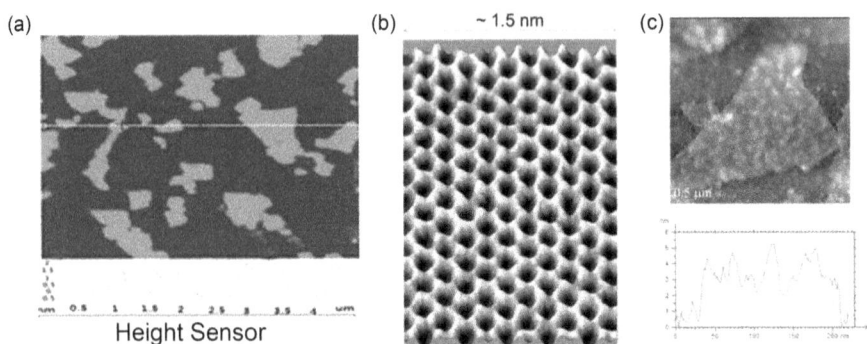

FIGURE 1.7 (a) AFM image showing thickness and morphology of graphene oxide nanolayers. (Modified image from Yu et al. 2016.) (b) Scanning probe microscopy image showing honeycomb atomic structure of graphene. (Reprinted with permission from Wikipedia Commons Image, Credit: U.S. Army Materiel Command.) (c) Morphology of protein functionalized graphene revealed by liquid phase AFM. (Reprinted with permission from the American Chemical Society; Pattammattel et al. 2013.)

1.7 GRAPHENE COMPOSITES

The high surface-to-volume ratio of graphene along with the superior material properties is ideal to prepare hybrid composite materials for mechanical and optoelectronic applications (Terrones et al. 2011). Graphene is hybridized with polymers, biologics, metals, nanomaterials and other carbon allotropes for functional applications. Figure 1.8 summarizes important routes to graphene composites. Graphene composite can be prepared by utilizing the covalent and/or noncovalent interactions. Clearly, oxidized or functionalized graphene is required for covalent mode, and the literature is heavily focused on GO derivatives. The noncovalent mode of interactions aims to use the p–p and van der Waals interactions. The following sections briefly discuss major classes of graphene composites.

1.7.1 GRAPHENE 2D HETEROSTRUCTURES

Van der Waals heterostructures of graphene are realized by hybridizing with other 2D nanolayers. Inorganic nanolayers can be hybridized with graphene by controlled growth, chemical processing in solutions, or co-exfoliation (Puglia et al. 2021). Using the controlled growth method, graphene and the inorganic layers may be in a vertical stack or they can be bonded in the graphene plane (Figure 1.9). Application-focused graphene heterostructures with h-BN, MoS$_2$ and other transition metal dichalcogenides are demonstrated over the last decade. Stable graphene heterostructures are easily prepared by co-exfoliation methods but careful characterization is required to determine the way the composite layers are arranged (Puglia et al. 2021). For example, one could make alternating single layers of a composite of graphene and the dichalcogenide (1:1) or 2:2 or 2:1 etc., Thus, controlling the composition is not sufficient, one has to also control how the composites are arranged in 3D. The direct band gap in some inorganic nanolayer materials can be combined with the unique mechanical strength, conductivity and stability of graphene for applications. MoS$_2$-based graphene composites are also useful for metal-ion batteries and Li-S batteries (Huang et al. 2021, Puglia et al. 2021).

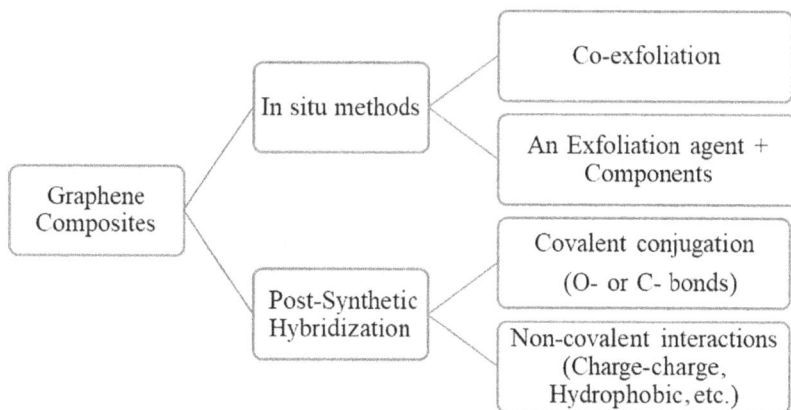

FIGURE 1.8 A general scheme for routes to graphene composites.

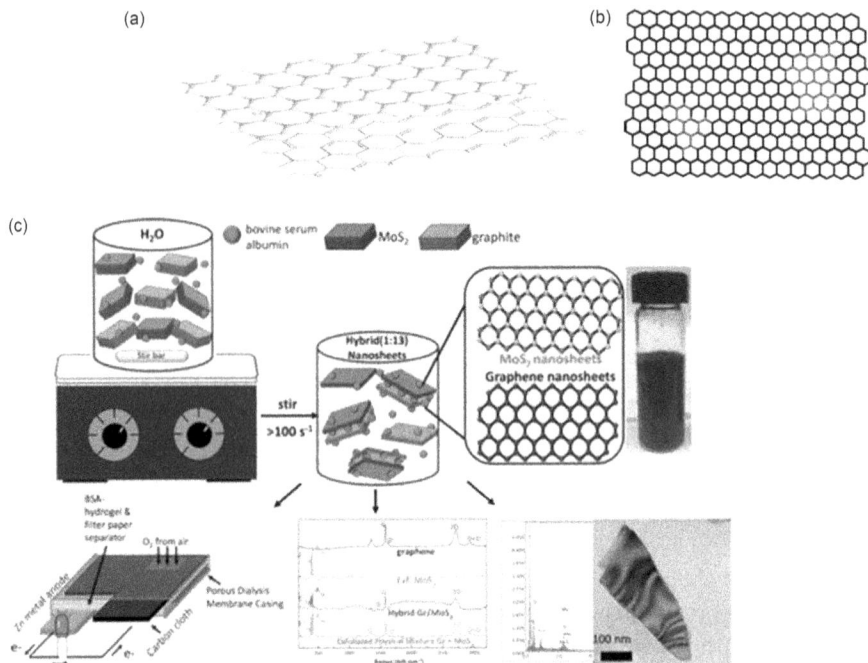

FIGURE 1.9 (a) Vertically stacked graphene heterostructure with inorganic layers. (b) In plane heterostructures where inorganic counterpart is inserted within the same 2D layer. (c) Graphene-MoS$_2$ stacked hybrids prepared by co-exfoliation in the presence of bovine serum albumin as an exfoliating agent (Puglia et al. 2021). (Reprinted with permission from the American Chemical Society.)

1.7.2 Graphene Polymer Composites

Natural and synthetic polymers can be hybridized to harness the electronic and mechanical properties of graphene with flexibility and strength of polymers. For example, 3D printing of graphene was demonstrated by compounding with visco-elastic polymers (Figure 1.10) (Jakus et al. 2015, Ponnamma et al. 2021). The driving forces at the interface of the polymers and graphene can be hydrophobic, π–π stacking and/or covalent conjugation, depending on the polymer. For example, phenyl groups in polystyrene form strong π–π interactions with graphene (Hong et al. 2020). On the other hand, GO composites exploit electrostatic and hydrogen bonding interactions with the polymers, in addition to the π–π stacking. For example, polyamines such as polyethylenimine adsorb strongly to GO layers and composites can be prepared in water (Jiang et al. 2020). Such composites find applications in biology and water treatment (Jiang et al. 2020). Compared to graphene, making GO requires using toxic reagents, and composite preparations are more economically viable and greener than GO itself. Graphene composites' preparations starting from (i) graphite (or other 2D material precursors), (ii) polymers (or monomer + reagents) and (iii) exfoliation agents (*in situ*) are highly viable in an industrial setting (Das et al. 2011). And several polymers are known exfoliators of graphene, so (i) and (ii) are sufficient in many instances.

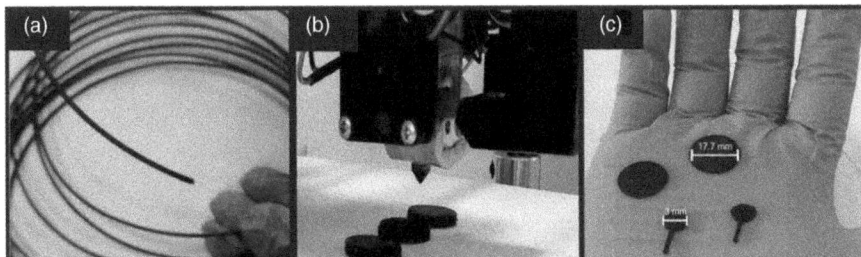

FIGURE 1.10 Scheme for 3D printing graphene polymer composites in different shapes. (Reprinted with permission from Springer Nature; Foster et al. 2017.)

1.7.3 GRAPHENE BIOCOMPOSITES

Biomolecules such as proteins, DNA and polysaccharides can be incorporated into graphene for applications in biomedicine and diagnosis (Xu et al. 2010, Zhou et al. 2014, Zore et al. 2015). Depending on the graphene surface, surface functional groups of the biomolecules, solvent, etc., several modes of interactions are possible at the nanolayer-biomolecule interface (Figure 1.11a). Unlike the polymer composites, the synthesis of biocomposites is usually done under aqueous conditions and pristine graphene has low solubility and stability in water. But proteins and polysaccharides are good exfoliators and stabilizers of graphene in water and this is due to favorable hydrophobic interactions of these biomolecules with the π-system of graphene (Pattammattel et al. 2015). Graphene biocomposites can be prepared directly from graphite and biomolecules on multi-gram scale (Figure 1.11b) and the biographene preparation is stable at 37 °C over months. However, most of the literature on biocomposites is centered around GO since these applications focus on harnessing the high surface-volume ratio of GO rather than the electronic properties of graphene. However, the nanotoxicity of graphene derivatives is an important consideration or concern in use for real-life applications of the composites. Also, GO preparations often contain metallic impurities and small nanosheet fragments, affecting the composites' biological applications (Pattammattel et al. 2015).

1.8 NANOTOXICITY OF GRAPHENE

Nanotechnology, in general, faces the question: Can nanoparticles have adverse effects on humans and the environment? (Fard et al. 2015, Nel et al. 2013). Thus, graphene's nanotoxicity and environmental fate is an intensively researched area. Most biological applications may result in reactions of graphene with biomolecules and these reactivities depend on graphene surface chemistry, functionalization and composite composition. Both in vitro and in vivo models were used to study the acute and chronic responses of graphene in cell and animal models (Seabra et al. 2014). Discussion of nanotoxicity of graphene is beyond the scope of this chapter. But a general trend is that coating the high-energy surface of graphene with benign proteins reduces its reactivity, and smaller biographene sheets show higher cell viability than larger sheets. The higher toxicity of smaller sheets may be because of their ease of

(a)

(b)

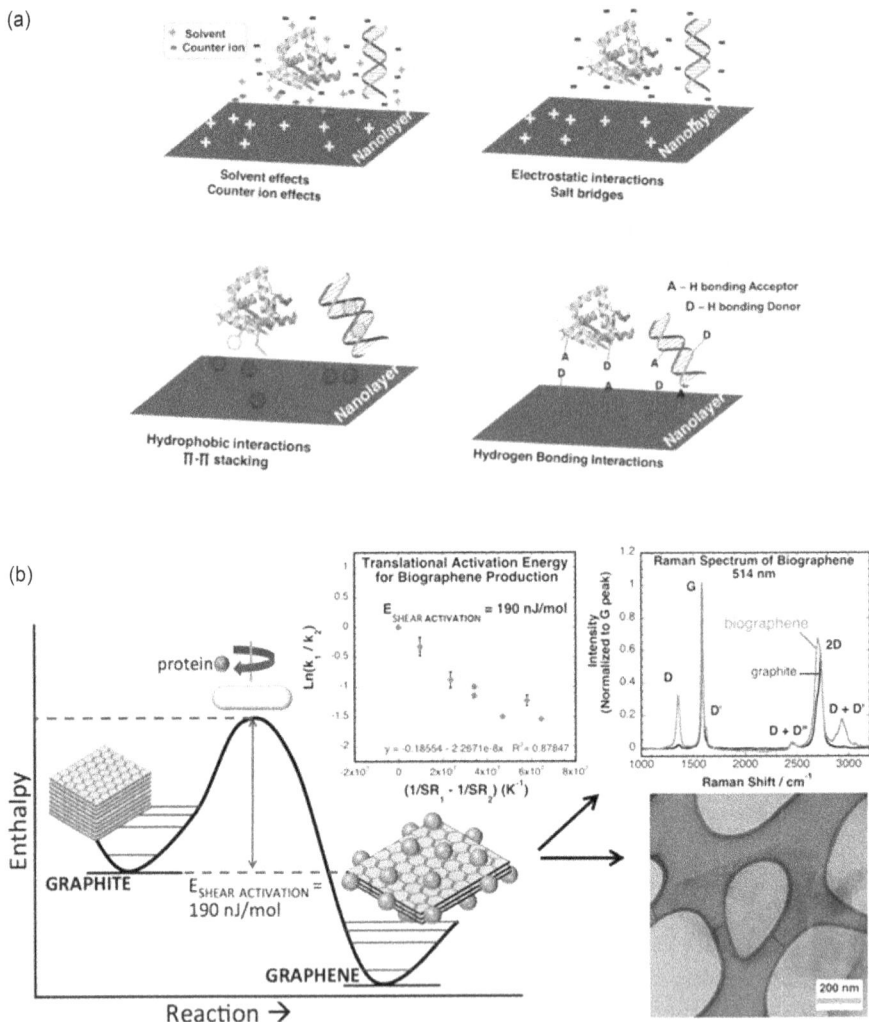

FIGURE 1.11 (a) Scheme showing the potential interactions at the interface of biomolecules and nanolayers. (b) Biographene produced by *in situ* exfoliation of graphene in the presence of proteins. (Reprinted with permission from American Chemical Society; Puglia et al. 2021.)

penetration of the cell membrane when compared to the much larger sheets, on the dimensions of individual cells (Figure 1.12) (Pattammattel et al. 2017). Most of the studies identify the oxidative stress as a major mechanism of graphene toxicity inside the cells (Zhou et al. 2012). Several methodologies may be adopted to mask the toxicity of graphene such as biofunctionalization and polymer modification. However, when graphene is used in composite preparations, gram or kilogram scale or higher amounts of materials are used. So, it is important to access both workplace and environmental toxicities of graphene composites before their deployment in the environment (Morfeld et al. 2012).

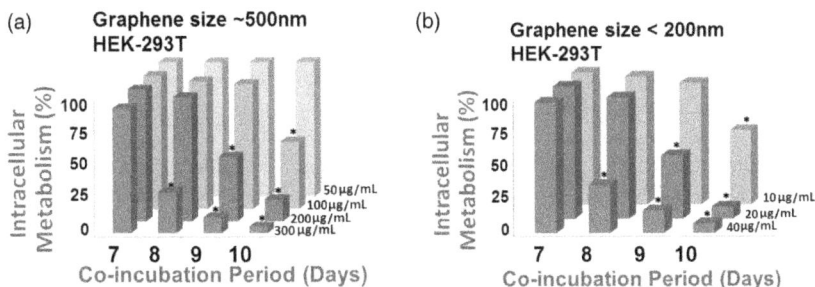

FIGURE 1.12   Size-dependent graphene toxicity assay measuring the relative metabolic activity of (a) human embryonic kidney cells (HEK293T) and (b) HEK cells. Smaller size graphene sheets show toxicity at lower doses than larger sheets. (Reprinted with permission from American Chemical Society; Pattammattel et al. 2017.)

1.9   CONCLUSIONS

Without doubt, graphene is the most discussed and studied material in the 21st century, so far. The invention of graphene opened the path for hundreds of material systems. But graphene is still the 'shining star' in the family of 2D materials. Recent inventions of magic graphene unraveled a new class of material systems where tuning the electronic properties can be achieved by twisting bilayer graphene at specific angles (Cao et al. 2018). Several graphene products also made it into the market. For example, pristine graphene substrates may be purchasable for spectroscopy and microscopy experiments. Graphene inks for electronic applications and for making 3D graphene are commercially available from a variety of vendors. Graphene composites are used in several real-life products such as shoes, light bulbs and headphones (Kong et al. 2019). We hope to see several 'wonders' that might happen with graphene.

REFERENCES

Boehm, Hans-Peter, A Clauss, GO Fischer, and U Hofmann. 1962. "Das adsorptionsverhalten sehr dünner kohlenstoff-folien." *Zeitschrift für anorganische und allgemeine Chemie* 316 (3–4):119–127.

Brodie, Benjamin Collins. 1859. "XIII. On the atomic weight of graphite." *Philosophical transactions of the Royal Society of London* 149:249–259.

Caldwell, Joshua D, Igor Aharonovich, Guillaume Cassabois, James H Edgar, Bernard Gil, and DN Basov. 2019. "Photonics with hexagonal boron nitride." *Nature Reviews Materials* 4 (8):552–567.

Cao, Yuan, Valla Fatemi, Shiang Fang, Kenji Watanabe, Takashi Taniguchi, Efthimios Kaxiras, and Pablo Jarillo-Herrero. 2018. "Unconventional superconductivity in magic-angle graphene superlattices." *Nature* 556 (7699):43–50.

Chabot, Victor, Brian Kim, Brent Sloper, Costas Tzoganakis, and Aiping Yu. 2013. "High yield production and purification of few layer graphene by Gum Arabic assisted physical sonication." *Scientific Reports* 3 (1):1–7.

Chodos, Alan, J Ouellette, and E Tretkoff. 2009. "This month in physics history." *American Physical Society News* 18 (4):5–7.

Ciesielski, Artur, and Paolo Samori. 2014. "Graphene via sonication assisted liquid-phase exfoliation." *Chemical Society Reviews* 43 (1):381–398.

Cooper, Adam J, Neil R Wilson, Ian A Kinloch, and Robert AW Dryfe. 2014. "Single stage electrochemical exfoliation method for the production of few-layer graphene via intercalation of tetraalkylammonium cations." *Carbon* 66:340–350.

Das, Sriya, Ahmed S Wajid, John L Shelburne, Yen-Chih Liao, and Micah J Green. 2011. "Localized in situ polymerization on graphene surfaces for stabilized graphene dispersions." *ACS Applied Materials & Interfaces* 3 (6):1844–1851.

Database, ThermoFisher. 2021. Thermo Fisher XPS Database https://www.thermofisher.com/us/en/home/materials-science/learning-center/periodic-table.html.

Du, Jinhong, and Hui-Ming Cheng. 2012. "The fabrication, properties, and uses of graphene/polymer composites." *Macromolecular Chemistry and Physics* 213 (10–11):1060–1077.

Fard, Javad Khalili, Samira Jafari, and Mohammad Ali Eghbal. 2015. "A review of molecular mechanisms involved in toxicity of nanoparticles." *Advanced pharmaceutical bulletin* 5 (4):447.

Ferrari, Andrea C, and Denis M Basko. 2013. "Raman spectroscopy as a versatile tool for studying the properties of graphene." *Nature nanotechnology* 8 (4):235–246.

Forbeaux, I, J-M Themlin, and J-M Debever. 1998. "Heteroepitaxial graphite on 6 H− SiC (0001): Interface formation through conduction-band electronic structure." *Physical Review B* 58 (24):16396.

Foster, Christopher W, Michael P Down, Yan Zhang, Xiaobo Ji, Samuel J Rowley-Neale, Graham C Smith, Peter J Kelly, and Craig E Banks. 2017. "3D printed graphene based energy storage devices." *Scientific Reports* 7 (1):1–11.

Geim, Andre K, and Konstantin S Novoselov. 2010. "The rise of graphene." In Peter Rodgers (ed.), *Nanoscience and Technology: A Collection of Reviews from Nature Journals*, 11–19. Singapore: World Scientific.

Gusmão, Rui, Zdeněk Sofer, Daniel Bouša, and Martin Pumera. 2017. "Pnictogen (As, Sb, Bi) nanosheets for electrochemical applications are produced by shear exfoliation using kitchen blenders." *Angewandte Chemie* 129 (46):14609–14614.

Hong, Yongming, Senyang Bao, Xiang Xiang, and Xinping Wang. 2020. "Concentration-dominated orientation of phenyl groups at the polystyrene/graphene interface." *ACS Macro Letters* 9 (6):889–894.

Huang, Shaozhuan, Zhouhao Wang, Yew Von Lim, Ye Wang, Yan Li, Daohong Zhang, and Hui Ying Yang. 2021. "Recent advances in heterostructure engineering for lithium–sulfur batteries." *Advanced Energy Materials* 11 (10):2003689.

Jakus, Adam E, Ethan B Secor, Alexandra L Rutz, Sumanas W Jordan, Mark C Hersam, and Ramille N Shah. 2015. "Three-dimensional printing of high-content graphene scaffolds for electronic and biomedical applications." *ACS Nano* 9 (4):4636–4648.

Ji, Jianping, Xiufeng Song, Jizi Liu, Zhong Yan, Chengxue Huo, Shengli Zhang, Meng Su, Lei Liao, Wenhui Wang, and Zhenhua Ni. 2016. "Two-dimensional antimonene single crystals grown by van der Waals epitaxy." *Nature Communications* 7 (1):1–9.

Jiang, Xiangqiong, Guihua Ruan, Yipeng Huang, Zhengyi Chen, Huamei Yuan, and Fuyou Du. 2020. "Assembly and application advancement of organic-functionalized graphene-based materials: A review." *Journal of Separation Science* 43 (8):1544–1557.

Kaur, Harneet, Ruiyuan Tian, Ahin Roy, Mark McCrystall, Dominik Valter Horvath, Guillermo Lozano Onrubia, Ross Smith, Manuel Ruether, Aideen Griffin, and Claudia Backes. 2020. "Production of Quasi-2D platelets of nonlayered iron pyrite (FeS$_2$) by liquid-phase exfoliation for high performance battery electrodes." *ACS Nano* 14 (10):13418–13432.

Khan, Umar, Arlene O'Neill, Mustafa Lotya, Sukanta De, and Jonathan N Coleman. 2010. "High-concentration solvent exfoliation of graphene." *Small* 6 (7):864–871.

Kong, Wei, Hyun Kum, Sang-Hoon Bae, Jaewoo Shim, Hyunseok Kim, Lingping Kong, Yuan Meng, Kejia Wang, Chansoo Kim, and Jeehwan Kim. 2019. "Path towards graphene commercialization from lab to market." *Nature Nanotechnology* 14 (10):927–938.

Kumar, Challa Vijaya, and Pattammattel, Ajith. 2017. *Introduction to Graphene: Chemical and Biochemical Applications.* Amsterdam: Elsevier.

Li, Likai, Yijun Yu, Guo Jun Ye, Qingqin Ge, Xuedong Ou, Hua Wu, Donglai Feng, Xian Hui Chen, and Yuanbo Zhang. 2014. "Black phosphorus field-effect transistors." *Nature Nanotechnology* 9 (5):372–377.

Mak, Kin Fai, Changgu Lee, James Hone, Jie Shan, and Tony F Heinz. 2010. "Atomically thin MoS_2: A new direct-gap semiconductor." *Physical Review Letters* 105 (13):136805.

Malhotra, Mansi, Megan Puglia, Ankarao Kalluri, Dina Chowdhury, and Challa V. Kumar. "Adsorption of metal ions on graphene sheet for applications in environmental sensing and wastewater treatment." *Sensors and Actuators Reports* 4: 100077.

Mannix, Andrew J, Xiang-Feng Zhou, Brian Kiraly, Joshua D Wood, Diego Alducin, Benjamin D Myers, Xiaolong Liu, Brandon L Fisher, Ulises Santiago, and Jeffrey R Guest. 2015. "Synthesis of borophenes: Anisotropic, two-dimensional boron polymorphs." *Science* 350 (6267):1513–1516.

Manzeli, Sajedeh, Dmitry Ovchinnikov, Diego Pasquier, Oleg V Yazyev, and Andras Kis. 2017. "2D transition metal dichalcogenides." *Nature Reviews Materials* 2 (8):1–15.

Meyer, JC. 2014. "Transmission electron microscopy (TEM) of graphene." In Viera Skákalová and Alan B. Kaiser (eds.), *Graphene*, 101–123. Amsterdam: Elsevier.

Morfeld, Peter, Robert J McCunney, Len Levy, and Ishrat S Chaudhuri. 2012. "Inappropriate exposure data and misleading calculations invalidate the estimates of health risk for airborne titanium dioxide and carbon black nanoparticle exposures in the workplace." *Environmental Science and Pollution Research International* 19 (4):1326.

Nel, Andre, Yuliang Zhao, and Lutz Mädler. 2013. "Environmental health and safety considerations for nanotechnology." *Accounts of Chemical Research* 46 (3):605–606.

Nicolosi, Valeria, Manish Chhowalla, Mercouri G Kanatzidis, Michael S Strano, and Jonathan N Coleman. 2013. "Liquid exfoliation of layered materials." *Science* 340 (6139):1226419.

Novoselov, Kostya S, Andre K Geim, Sergei V Morozov, De-eng Jiang, Yanshui Zhang, Sergey V Dubonos, Irina V Grigorieva, and Alexandr A Firsov. 2004. "Electric field effect in atomically thin carbon films." *Science* 306 (5696):666–669.

Oaki, Yuya. 2021. "Exfoliation chemistry of soft layered materials toward tailored 2D materials." *Chemistry Letters* 50 (2):305–315.

Pattammattel, Ajith, and Challa Vijaya Kumar. 2015. "Kitchen chemistry 101: multigram production of high quality biographene in a blender with edible proteins." *Advanced Functional Materials* 25 (45):7088–7098.

Pattammattel, Ajith, Christina L Williams, Paritosh Pande, William G Tsui, Ashis K Basu, and Challa Vijaya Kumar. 2015. "Biological relevance of oxidative debris present in as-prepared graphene oxide." *RSC Advances* 5 (73):59364–59372.

Pattammattel, Ajith, Megan Puglia, Subhrakanti Chakraborty, Inoka K Deshapriya, Prabir K Dutta, and Challa V Kumar. 2013. "Tuning the activities and structures of enzymes bound to graphene oxide with a protein glue." *Langmuir* 29 (50):15643–15654.

Pattammattel, Ajith, Paritosh Pande, Deepa Kuttappan, Megan Puglia, Ashis K Basu, Mary Anne Amalaradjou, and Challa V Kumar. 2017. "Controlling the graphene–bio interface: Dispersions in animal sera for enhanced stability and reduced toxicity." *Langmuir* 33 (49):14184–14194.

Ponnamma, Deepalekshmi, Yifei Yin, Nisa Salim, Jyotishkumar Parameswaranpillai, Sabu Thomas, and Nishar Hameed. 2021. "Recent progress and multifunctional applications of 3D printed graphene nanocomposites." *Composites Part B: Engineering* 204:108493.

Puglia, Megan K, Mansi Malhotra, Ajitha Chivukula, and Challa V Kumar. 2021. "'Simple-Stir' Heterolayered MoS_2/graphene nanosheets for Zn–Air batteries." *ACS Applied Nano Materials* 4 (10):10389–10398.

Ruess, G, and F Vogt. 1948. "Höchstlamellarer Kohlenstoff aus Graphitoxyhydroxyd." *Monatshefte für Chemie und verwandte Teile anderer Wissenschaften* 78 (3):222–242.

Sang, Mingyu, Jongwoon Shin, Kiho Kim, and Ki Jun Yu. 2019. "Electronic and thermal properties of graphene and recent advances in graphene based electronics applications." *Nanomaterials* 9 (3):374.

Scardamaglia, M, B Aleman, M Amati, C Ewels, P Pochet, N Reckinger, J-F Colomer, T Skaltsas, N Tagmatarchis, and R Snyders. 2014. "Nitrogen implantation of suspended graphene flakes: Annealing effects and selectivity of sp^2 nitrogen species." *Carbon* 73:371–381.

Seabra, Amedea B, Amauri J Paula, Renata de Lima, Oswaldo L Alves, and Nelson Durán. 2014. "Nanotoxicity of graphene and graphene oxide." *Chemical Research in Toxicology* 27 (2):159–168.

Stehle, Yijing, Harry M Meyer III, Raymond R Unocic, Michelle Kidder, Georgios Polizos, Panos G Datskos, Roderick Jackson, Sergei N Smirnov, and Ivan V Vlassiouk. 2015. "Synthesis of hexagonal boron nitride monolayer: Control of nucleation and crystal morphology." *Chemistry of Materials* 27 (23):8041–8047.

Susi, Toma, Thomas Pichler, and Paola Ayala. 2015. "X-ray photoelectron spectroscopy of graphitic carbon nanomaterials doped with heteroatoms." *Beilstein Journal of Nanotechnology* 6 (1):177–192.

Tan, Chaoliang, Xiehong Cao, Xue-Jun Wu, Qiyuan He, Jian Yang, Xiao Zhang, Junze Chen, Wei Zhao, Shikui Han, and Gwang-Hyeon Nam. 2017. "Recent advances in ultrathin two-dimensional nanomaterials." *Chemical Reviews* 117 (9):6225–6331.

Terrones, Mauricio, Olga Martín, María González, Javier Pozuelo, Berna Serrano, Juan C Cabanelas, Sofía M Vega-Díaz, and Juan Baselga. 2011. "Interphases in graphene polymer-based nanocomposites: Achievements and challenges." *Advanced Materials* 23 (44):5302–5310. doi:10.1002/adma.201102036.

Varrla, Eswaraiah, Keith R Paton, Claudia Backes, Andrew Harvey, Ronan J Smith, Joe McCauley, and Jonathan N Coleman. 2014. "Turbulence-assisted shear exfoliation of graphene using household detergent and a kitchen blender." *Nanoscale* 6 (20):11810–11819.

Xu, Yuxi, Qiong Wu, Yiqing Sun, Hua Bai, and Gaoquan Shi. 2010. "Three-dimensional self-assembly of graphene oxide and DNA into multifunctional hydrogels." *ACS Nano* 4 (12):7358–7362.

Yang, Dongxing, Aruna Velamakanni, Gülay Bozoklu, Sungjin Park, Meryl Stoller, Richard D Piner, Sasha Stankovich, Inhwa Jung, Daniel A Field, and Carl A Ventrice Jr. 2009. "Chemical analysis of graphene oxide films after heat and chemical treatments by X-ray photoelectron and Micro-Raman spectroscopy." *Carbon* 47 (1):145–152.

Yi, Min, and Zhigang Shen. 2015. "A review on mechanical exfoliation for the scalable production of graphene." *Journal of Materials Chemistry A* 3 (22):11700–11715.

Yu, Huitao, Bangwen Zhang, Chaoke Bulin, Ruihong Li, and Ruiguang Xing. 2016. "High-efficient synthesis of graphene oxide based on improved hummers method." *Scientific Reports* 6 (1):1–7.

Zhang, YI, Luyao Zhang, and Chongwu Zhou. 2013. "Review of chemical vapor deposition of graphene and related applications." *Accounts of Chemical Research* 46 (10):2329–2339.

Zhou, Hejiang, Kai Zhao, Wei Li, Na Yang, Ying Liu, Chunying Chen, and Taotao Wei. 2012. "The interactions between pristine graphene and macrophages and the production of cytokines/chemokines via TLR-and NF-κB-related signaling pathways." *Biomaterials* 33 (29):6933–6942.

Zhou, Wei, Wei Li, Ying Xie, Lei Wang, Kai Pan, Guohui Tian, Mingxia Li, Guofeng Wang, Yang Qu, and Honggang Fu. 2014. "Fabrication of noncovalently functionalized brick-like β-cyclodextrins/graphene composite dispersions with favorable stability." *RSC Advances* 4 (6):2813–2819.

Zore, Omkar V, Ajith Pattammattel, Shailaja Gnanaguru, Challa V Kumar, and Rajeswari M Kasi. 2015. "Bienzyme–polymer–graphene oxide quaternary hybrid biocatalysts: efficient substrate channeling under chemically and thermally denaturing conditions." *ACS Catalysis* 5 (9):4979–4988.

2 Graphene Synthesis and Characterization for Graphene Nanocomposites

Dinh-Tuan Nguyen
Academia Sinica

Mario Hofmann
National Taiwan University

Ya-Ping Hsieh
Academia Sinica

CONTENTS

DOI: 10.1201/9781003200444-2

2.1   INTRODUCTION

While only the seventeenth most abundant element in the Earth's crust, carbon is central to our planet's marvelous chemical and biological diversity. Carbon atoms' ability to link up covalently with many other elements and themselves in innumerable configurations gives rise to million substances, from the simplest compounds to macromolecules such as proteins and DNA, as well as chemical reactions that, among other things, enable and sustain life.

In the nanomaterial world, carbon also plays a dominant role. Among the first nanostructures discovered are buckminsterfullerene C_{60} (Kroto et al. 1985) and carbon nanotubes (Iijima 1991) which at that time generated a tremendous amount of interest that propelled the nascent field of nanotechnology forward. Shortly afterward, scientists found that many different types of fullerenes (C_{60}, C_{70}, C_{72}, C_{84}, ...) can be produced, while carbon nanotubes exist in various single-walled and multi-walled configurations. If carbon bonds are the backbone of all organic molecules, what is the basic building block of all these carbon-based nanomaterials?

As Mildred and Gene Dresselhaus showed in the early 1990s (Saito et al. 1992), carbon nanotubes can be formed by rolling a single-layer sheet of graphite, and the electronic properties of nanotube can be derived from that of the unrolled sheet itself (Figure 2.1). Similarly, the cage-like structure of fullerene can be conceptualized as a result of wrapping the same graphite nanosheet (Georgakilas et al. 2015). This sheet is called graphene, defined by IUPAC as "a single carbon layer of the graphite structure, describing its nature by analogy to a polycyclic aromatic hydrocarbon of quasi infinite size" (Fitzer et al. 1995). While graphene had been synthesized and observed

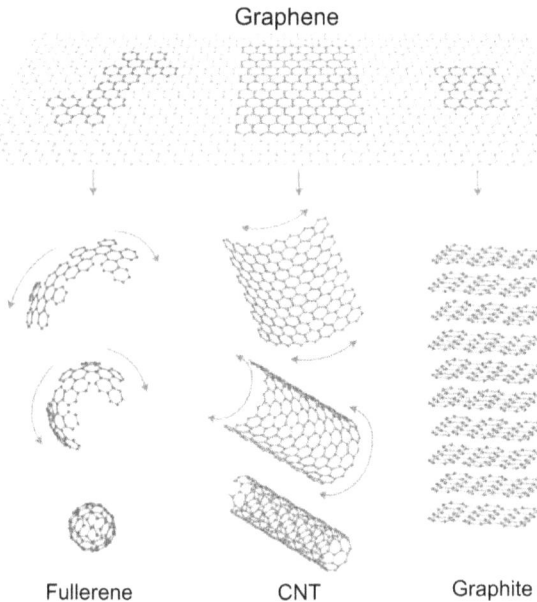

Graphene

Fullerene   CNT   Graphite

FIGURE 2.1   Graphene as the basis of other carbon allotropes. (Adapted from Geim and Novoselov 2009, with permission.)

by several investigating groups as early as 1960s (Dreyer, Ruoff, and Bielawski 2010; Geim 2012), it is a 2004 study by A. Geim and K. Novoselov (Novoselov et al. 2004) that made the breakthrough in synthesizing the 2D material. The study, published in *Science* magazine, was extremely influential because not only did it explicitly identify a new material but it also reported the gigantic carrier mobility of graphene and its unique zero-gap band structure. Subsequent studies revealed a host of other peculiar properties including tremendous mechanical strength (Lee et al. 2008), anomalous quantum Hall effect (Zhang et al. 2005), frequency-independent optical absorption (Nair et al. 2008), among others. That started a "gold rush" of research to explore the ways to synthesize graphene and exploit its properties for practical implications. Geim and Novoselov were awarded Nobel Prize in Physics in 2010, followed by a media sensation outside scientific circles in which graphene was dubbed as the "wonder material" (Larousserie 2013). A race soon started between nations and companies to commercialize graphene, which by 2017 has generated over 53,000 patents worldwide (Yang, Yu, and Liu 2018).

Much like the carbon skeleton in combination with functional groups enables limitless structural richness of organic molecules, multifarious nanostructures can be derived from the simple honeycomb plane of graphene, each with their own unique properties. Aside from the aforementioned nanotube and fullerene, a graphene nanosheet can also be cut into 1D graphene nanoribbons or further sliced into graphene nanoplatelets. The dangling bonds at the edges of graphene and imperfections at the basal plane allow the attachment of various functional groups, which alter pristine graphene's properties and equip it with new functionalities. Moreover, many of graphene's desirable properties can also be found in a structure consisting of a few graphitic layers. This type of 2D material, commonly called few-layer graphene, further extends the range of carbon nanomaterials.

As a result, the graphene-related family of materials has grown quickly over the last decade, with the meaning encompassed the ubiquitous term "graphene" that extends beyond the original IUPAC definition. For a particular application, often only some aspects of graphene's properties are needed, and compromises can be made in others to meet cost and scalability requirements.

Various synthesis routes have been invented to produce diverse forms of graphene nanomaterials. This book chapter aims to review graphene synthesis methods, with a focus on three main types of graphene: single-layer polycrystalline graphene film, noted for superior electronic properties; few-layer monocrystalline graphene nanoflakes, possessing high mechanical strength; and reduced graphene oxide, produced in large quantities but inferior in quality to the previous two. The second part of the chapter will introduce several common techniques to characterize graphene's properties with emphasis on simplicity and versatility. The authors hope to equip the experimentalists who have not previously worked in graphene research with a blueprint and basic tools to produce and utilize graphene in their research projects.

2.2 GRAPHENE SYNTHESIS

Graphene preparation methods can be classified into two main approaches: top-down exfoliation and bottom-up synthesis.

2.2.1 TOP-DOWN SYNTHESIS

2.2.1.1 Mechanical Exfoliation

The famed "scotch-tape method", invented by A. Geim and K. Novoselov, falls into the category of top-down synthesis whose principle is based on the exfoliation of van der Waals-based layered structure from bulk material. The graphite core of a pencil, for example, contains millions of sp^2 carbon layers which are stacked with a bonding energy of merely ~30 meV/atom (Benedict et al. 1998), and therefore, are easily delaminated when a moderate force is applied by human hands, tracing a dark line on the paper. The thickness of delaminated fragments varies vastly, and through meticulous isolation and observation, some monolayer, single-crystalline flakes can be revealed among them (Novoselov et al. 2004).

As historic as the discovery was, scientists quickly realized that this method and its variants (Jayasena and Subbiah 2011; Huang et al. 2015) are not scalable; as a result, the high-quality graphene nanoflakes produced this way are only useful for fundamental studies. Many schemes have been invented to convert external mechanical energy to delamination more efficiently.

One method is ball-milling (He et al. 2009; Zhao et al. 2010; Knieke et al. 2010) in which nanoflakes are produced by grinding graphite powder with hard balls inside a rotating cylinder. Since its principal driving force is imparted through the balls' vertical impact on powder flakes, the method produces highly fragmented graphene nanoplatelets (<100 nm in size and few nm in thickness). The ground powder containing nanoflakes can then be dispersed in a stabilizing solvent. The main advantage of ball milling is that other substances can be ground at the same time with graphite powder to prepare edge-functionalized graphene (León et al. 2011; Jeon et al. 2012; Yan et al. 2012) and graphene nanocomposites (Wu et al. 2011; Guo and Chen 2014; Lv, Wang, and Miura 2017) in a one-pot process. The downsides of ball milling include the damage to the carbon lattice brought about by high-speed collisions, and the low exfoliation efficiency that requires tens of hours to prepare few-layer graphene. Contamination is another concern: Not only the edges of newly created platelets react with environmental agents but the product also contains metallic impurities of the milling balls (Chua et al. 2016). Several innovations have been made to partially ameliorate these negative aspects, including combining with plasma radiation (Lin et al. 2017), reducing the mill's rotating speed (<500 rpm), and adding a small amount of exfoliating liquid (Zhao et al. 2010). One of the best quality graphene prepared by ball-milling was reported by Del Rio-Castillo et al. (2014), who employed slow wet ball milling in carbon nanofiber instead of graphite powder. It achieved single-layer graphene nanosheets, though at the cost of a more complicated process to produce the precursor material. A recently reported method does away with balls altogether, instead using a grinding roller with an intermediate agent to better translate pressure into shear force in order to achieve high efficiency (Zhang et al. 2019).

The exfoliating liquid in wet ball milling translates kinetic force of the balls into shear force applied to the powder flakes, thus enhancing shear-induced delamination and reducing fragmentation. This principle is utilized to the fullest extent in liquid phase exfoliation (LPE), where exfoliation occurs directly in dispersing solvents (aqueous or organic). Exfoliation is introduced through flow turbulence, as in

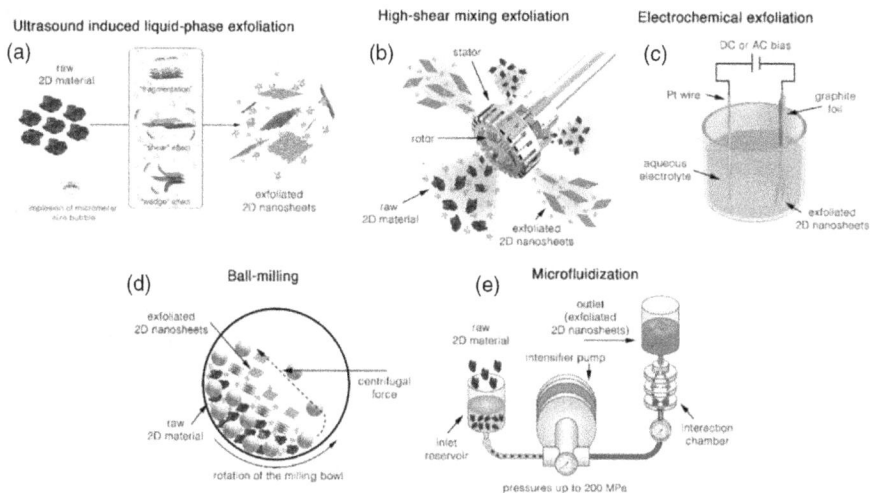

FIGURE 2.2 Most common exfoliation methods for graphene productions. (Adapted from Witomska et al. 2019 with permission.)

high shear exfoliation (Paton et al. 2014), or by high pressure, as in microfluidization (Karagiannidis et al. 2017), or by ultrasonic vibration as in tip sonication (Khan et al. 2011) and bath sonication (Hernandez et al. 2008) (Figure 2.2). Each of these methods has its merits and limitations, as discussed below.

In sonication-assisted exfoliation, the exfoliation is triggered by bubbles that are generated by ultrasonic vibration of the liquid, which intercalate between the graphitic layers (Suslick and Flannigan 2008). The burst of the bubbles generates enough energy to overcome van der Waals attraction between layers and peel them off, a process repeated many times to produce few-layer graphene nanosheets (several hundred to over 1 micron in size). Ultrasonic baths are a common fixture in many laboratories, providing an accessible way of graphene production. However, the exfoliation yield of bath sonication is low, so extended time (from 8 hours to several days) is needed. This can be improved by using tip sonication instead, where the sonication energy is highly focused on the liquid region near the vibration source. The drawback of tip sonication is that the high energy density causes some defects in the basal plane of carbon lattice (Bracamonte et al. 2014), much like the case of ball milling.

Experimentalists whose aim is upscaling graphene production using sonication face a dilemma: as the volume increases, energy density in bath sonication decreases, bringing down the exfoliation efficiency; meanwhile, tip sonication's localized shear is largely independent of volume, but produces lower quality graphene.

High shear exfoliation circumvents this problem by using flow turbulence and collisions, instead of ultrasonication-generated bubbles, inside a confined region to apply shear on graphite flakes. The method's main apparatus is a homogenizer, which has a rotor spinning at high speed enclosed by a cage (stator) perforated with holes for liquid flow. The solution containing graphite powder is sucked into the gap between the rotor and the stator, which is about 0.1 mm large, where the flakes are subjected to high shear rate ($>10^4 s^{-1}$) and delamination (Paton et al. 2014). The nanosheet yield of shear exfoliation is significantly better than that of sonication: under the same conditions (solvent, graphite concentration, etc.), sonication produces up to three times more few-layer graphene than sonication (Lund et al. 2021). Moreover, turbulent flow can be harnessed with much simpler setup: exfoliation of graphene with a kitchen blender and detergent as solvent has been demonstrated (Varrla et al. 2014).

Flow confinement is also the principle behind other methods to achieve the shear rate. One method is jet cavitation (Shen, Li, et al. 2011) in which the graphite solution is repeatedly pumped to a nozzle with cavities at a pressure of ~20 MPa. A more sophisticated approach is microfluidization (Karagiannidis et al. 2017; Paton et al. 2017), where microchannels generate much higher pressure (>200 MPa) and achieve similar few-layer graphene yield (~4%) to high shear exfoliation.

The common feature of all liquid phase exfoliation methods is that the shear is not homogenously applied on the graphite flakes, resulting in a polydisperse solution containing nanosheets with a wide range of sizes and thicknesses. As a result, exfoliation needs to be followed by a size selection process, in the form of either vacuum filtration or centrifugation, to separate the few-layer portions of nanoflakes from thicker ones (Backes et al. 2016).

2.2.1.2   Chemical Exfoliation

Unlike mechanical exfoliation, which directly exfoliates graphite by mechanical impact, chemical exfoliation first transforms graphite into a graphite intercalation compound or graphite oxide that represents an expanded form of graphite which can be easily exfoliated before converting the compound back to graphene.

The related chemistry knowledge has been around a long time: the first demonstration of graphite intercalation was reported in 1840 (Schafhaeutl 1840), while B. Brodie first described a method to synthesize graphite oxide (GO) in 1859 (Brodie 1859). For graphite oxide, Brodie's method is still used today, however, most recent studies use a modified version of Hummers' method which is more efficient and environmentally friendlier (Hummers and Offeman 1958; Botas et al. 2013). The technique involves graphite oxidation by sulfuric acid with the help of catalysts (e.g., $KMnO_4$ and $NaNO_3$) and purification by moderate heating in water and H_2O_2. The product is a powder compound with 20%–40% oxygen content in various functional groups both at the edge and the plane of the graphitic lattice. The powder can be reduced in various ways to eliminate the functional groups and restore sp^2 bonding of carbon atoms. The most common way is the chemical reduction by hydrazine (Stankovich et al. 2007; Wang et al. 2009) or other deoxygenating agents (Smith et al. 2019), but the involved reactions are often slow or toxic. Some notable alternative reduction methods employ pyrolysis (McAllister et al. 2007), electrolysis (Guo et al. 2009), microwave (Zhu et al. 2010), laser annealing (Sokolov, Shepperd, and Orlando 2010), or photocatalysis (Williams, Seger, and Kamat 2008). Laser-induced GO reduction in particular is interesting since its local treatment can be used for patterning graphene nanosheets (Zhou et al. 2010).

Overall, the main virtue of this "oxidation- reduction" synthesis route is that it can produce large amounts (up to 100% efficiency) of large nanosheets (tens of microns in size). The tradeoff is the inferior quality of the end product compared to material produced by mechanical exfoliation. To date, none of the reduction methods can completely eliminate the functional groups from graphene oxide (best results have C/O ratio in the range of 16:1–20:1). Some of the functional groups (e.g., epoxy) are more difficult to remove than others (Sengupta et al. 2018), and the chemical variety of functional groups also leads to irreproducible results in many cases (Zhong, Tian, et al. 2015). The process itself introduces further damages to the lattice, from unhealed lattice points (pentagonal and heptagonal sites) to nanopores (Jankovský et al. 2014). As a result, the product has low thermal and electrical conductivity and is correspondingly termed "reduced graphene oxide" (RGO) instead of "graphene" to denote the differences in structural and transport properties. Finally, the chemical processes raise some serious environmental and health concerns (Ambrosi et al. 2013; Zhong, Tian, et al. 2015).

Similar to the oxidation-reduction route, the process of solvothermal synthesis also involves two main steps, first expanding graphite in a solvent and then exfoliating it at high temperatures. Early attempts employed inorganic acids to intercalate graphite (Li, Wang, et al. 2008; Qian et al. 2009), though recently, non-toxic agents such as salt (Niu et al. 2013) and oxalic acid (Bai et al. 2019) have been found suitable for the purpose. Heating in forming gas or inert gas results in highly porous structures, which require mild sonication to yield graphene nanosheets or graphene nanoribbons (Li, Wang, et al. 2008).

Electrochemical exfoliation is yet another approach able to mass produce graphene from graphite (Liu et al. 2008; Lu et al. 2009). When an electric bias is applied to a graphite electrode within an electrolytic solution, alkaline ions intercalate between graphitic layers near the exposed electrode surface, and hydrolysis-generated gas bubbles expand the layer spacing further until the layers extricate themselves from the electrode (Hofmann et al. 2015). Electrochemically exfoliated nanosheets represent an intermediate condition between LPE graphene and reduced graphene oxide: the nanosheets are large (like RGO), their plane defectiveness and conducting properties are inferior to LPE graphene's but offer an improvement over RGO's (Wang et al. 2011). Electrochemical exfoliation is very fast compared to other methods and suitable for applications requiring an aqueous solution.

2.2.2 BOTTOM-UP SYNTHESIS

Contrary to the top-down approach, the bottom-up route aims to build graphene atom-by-atom from suitable precursors. There are three main methods classified as bottom-up: chemical vapor deposition (CVD), epitaxial growth, and organic synthesis.

2.2.2.1 Chemical Vapor Deposition (CVD)

CVD relies on the decomposition of organic molecules at high temperatures with the help of a catalyst. Considering the decomposition of the simplest organic molecule, methane, this reaction can be expressed as follows:

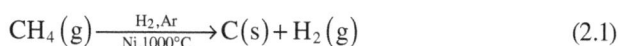

$$CH_4\left(g\right) \xrightarrow[\text{Ni,1000°C}]{\text{H}_2,\text{Ar}} C\left(s\right) + H_2\left(g\right) \qquad (2.1)$$

Carbon atoms diffuse into bulk nickel and then migrate to the surface upon cooling (due to decreased diffusivity at a lower temperature) and, provided that the surface is smooth, they link up in the honeycomb pattern of graphene. Etching away the metal allows transferring it to any other substrate.

The phenomenon of carbon condensation on catalytic surfaces has been known for some time, and indeed, the occurrence of ultrathin carbon layers on Ni, Pt, Co, Ru, etc. have been reported back in the 1960s (Morgan and Somorjai 1968; Blakely, Kim, and Potter 1970). However, the process does not guarantee monolayer growth: the amount of diffused carbon must be precisely controlled by fine-tuning processing parameters; otherwise, additional layers would be formed. Moreover, initial decomposition (often called "carbon seeding") can happen at different sites simultaneously, which generate various graphene grains with different orientations. These grains eventually merge to become a large polycrystalline film. The grain boundaries and other types of defects (point defects, pores, wrinkles, etc.) formed during growth and the required transfer step severely hamper transport properties which represent the core appeal of graphene. Nonetheless, CVD growth of graphene has matured over the past decade by employing expertise gained from the thin film manufacturing in the semiconductor industry and carbon nanotube production.

For example, replacing nickel catalyst with copper facilitates monolayer growth (Li et al. 2009): copper has an extremely low carbon diffusivity, so gas decomposition and carbon condensation happen almost exclusively at the surface, and the area covered by the first layer of graphene is passivated, stopping further deposition at the place (Hsieh, Hofmann, and Kong 2014) (Figure 2.3). In other words, CVD growth on Cu is a self-limited process (Chin et al. 2017). It has also been found that the pre-treatments of substrate, such as electropolishing and annealing (Luo et al. 2011; Nguyen et al. 2019), are essential to achieve a smooth surface with minimal defects. Growth of large-scale, single crystal copper substrate (Brown et al. 2014; Xu et al. 2017) has been attempted to suppress carbon seeding and reduce graphene grain boundaries. Moreover, processing parameters such as air flow and growth duration

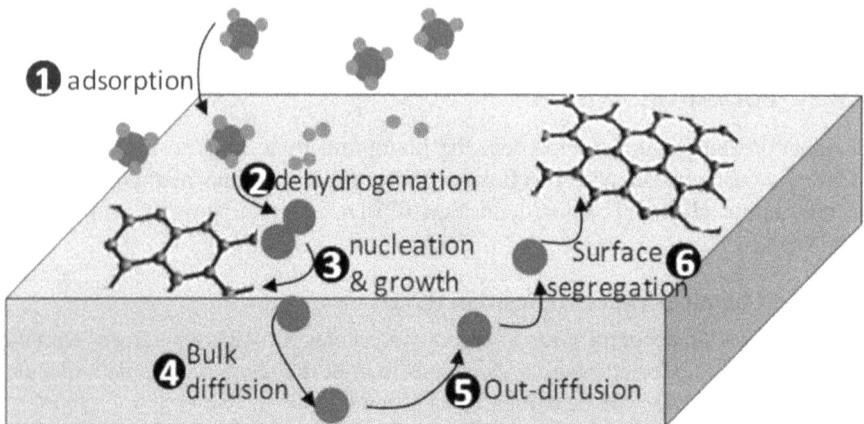

FIGURE 2.3 Chemical vapor deposition of graphene on copper via nucleation on copper and surface segregation on nickel.

have been optimized to control the subtle kinetics of CVD. Novel setup geometries (Chen et al. 2013; Hsieh, Hofmann, and Kong 2014) have also been explored for the same purpose.

Overall, with those optimizations, while not reaching the level of mechanically exfoliated graphene in terms of mechanical and electrical performance, CVD graphene's quality is superior to that of LPE graphene and RGO, and competitive enough for early-stage commercialization. The main advantage of CVD is that it is highly scalable; the size of graphene film, in theory, is only limited by the physical dimension of substrate and vacuum chamber, which can be further multiplied via roll-to-roll or stacking schemes (Bae et al. 2010; Hsieh et al. 2016). CVD can also work with many types of feedstocks, with demonstrations of graphene growth from food waste and plastics (Ruan et al. 2011; Sharma et al. 2014) hinting at a promising way to recycle waste. However, the main obstacle to widespread adoption of CVD graphene is cost. The strict requirements for vacuum, temperature, and environmental control necessitate expensive apparatuses and enormous energy consumption which add to the cost of production (Zhong, Tian, et al. 2015). There have been attempts to remediate this, such as plasma-assisted CVD at lower temperature requirement (Kim et al. 2011), but often at the cost of reduced quality.

2.2.2.2 Non-catalytic Epitaxial Growth

At ultrahigh vacuum and elevated temperature, carbon atoms can be directly generated and form a 2D network even on non-catalytic substrates. The most well-known example of this type of graphene growth is the sublimation of hexagonal SiC. De Heer et al. reported that at $1,200°C$ and 10^{-10} torr, Si atoms volatilize and desorb from the lattice, leaving carbon layers on the surface (Berger et al. 2004) in the form of graphene micrograins. Later wafer-scale growth was realized by using high pressure argon instead of vacuum to restrict the sublimation rate (Emtsev et al. 2009). This on-site graphitization eliminates the need for transfer, but the uneven Si sublimation rate results in poor thickness homogeneity (Borovikov and Zangwill 2009). Wrinkles also inevitably occur during cooling due to the mismatched thermal coefficients of SiC and graphene (Vecchio et al. 2011).

Another method is molecular beam epitaxy (MBE), which uses even higher temperatures ($\sim1,500°C$ to decompose graphite and enables epitaxial growth on various substrates (Park et al. 2010). Overall, the stringent requirements and the lack of control over growth quality in epitaxial growth, in spite of efforts to remediate them (Suemitsu et al. 2014; Ben Jabra et al. 2021), mean that epitaxial growth is only suitable for niche applications in microelectronics (Beshkova, Hultman, and Yakimova 2016).

2.2.2.3 Organic Synthesis

Since graphene can also be conceptualized as a supramolecular structure of covalently bonded carbon atoms, scientists have turned to the wealth of accumulated knowledge in organic chemistry for synthesizing graphene at ambient conditions. Multiple successful reactions have been discovered, which commonly involve polymerization of polycyclic aromatic hydrocarbon (PAH molecules) via cycloaddition or cyclodehydrogenation of their polymer precursors (Cai et al. 2010; Narita et al. 2014). The obtained sp^2-dominant carbon networks are often called "nanographene" for their minuscule size, with the largest product being 12–13 nm graphene nanoribbons (Yang et al. 2008; Narita et al. 2015). As such, they are not suitable for large scale production for applications requiring large size. Another approach is ion/electron irradiation-induced cross-linking of self-assembled molecules (SAM), which produce rather large (0.3 mm²) carbon nanomembranes. The nanomembrane can then be converted to graphene through pyrolysis (Turchanin et al. 2009), but its crystallinity is rather poor. Readers may refer to Chapter 3 for graphene synthesis from non-conventional sources.

2.3 GRAPHENE CHARACTERIZATION

As mentioned in the introduction of this chapter, the unabated enthusiasm in graphene research originates from its many exceptional properties and exciting effects. Its versatility as a platform for various functionalized 2D structures elicits even more study interests. More than 15 years after the first report of massless Dirac fermions in graphene (Novoselov, Geim, et al. 2005), graphene continues to surprise scientists with new attributes in unexpected areas such as superconductivity or ferroelectrics (Cao et al. 2018; Zheng et al. 2020). Therefore, the characterization of all graphene's properties could not be fully covered within the scope of this section. Instead, the authors focus on basic properties most commonly encountered or utilized by scientists working on graphene nanocomposites. We also put an emphasis on the introduction of characterization techniques that are likely to be most accessible.

2.3.1 Size and Thickness

As with other nanoobjects, graphene's shape, size, and thickness are often the most important factors that control fundamental aspects of its properties. Aside from the archetypal single-layer graphene nanosheets, bi-layer and tri-layer graphene have also been sought after due to their tunable bandgap and other extraordinary properties (Zhang et al. 2009; Lui et al. 2011). As the number of layers (N) increases, the material's properties transit from graphene to graphite. However, few-layer graphene ($N < 10$) still has high electrical conductivity and mechanical strength that distinguish it from graphite (Partoens and Peeters 2006) and is often cheaper to produce than single-layer graphene. Particular shapes (nanoribbon, twisted sheets, etc.) also open up new functionalities for graphene devices.

The most straightforward way to characterize graphene morphology is transmission electron microscopy (TEM) observation. Relative TEM contrast on a substrate has been used in early observations of carbon ultrathin films (Boehm et al. 1962), however, the contrast highly depends on focusing conditions. Instead, evaluating the number of layers can be reliably done by counting the folding lines of graphene flakes (Horiuchi et al. 2004). For samples with multiple layers or inhomogeneous thickness, atomic force microscope (AFM) can be used to determine the thickness at the edge and the roughness of graphene samples (Schniepp et al. 2006). However, due to sample-tip interactions and the fact that graphene layers are often folded or wrinkled, AFM may overestimate the thickness of few-layer graphene and require calibration (instead of a simple division of nominal thickness by graphene layer spacing) to ascertain the number of layers (Nemes-Incze et al. 2008). High-resolution TEM and AFM equipment are expensive and their analysis is often time-consuming especially for a batch of samples with thickness variations. In the case of TEM, the high-energy electrons involved can sometimes damage the samples.

To alleviate these challenges, many optical techniques have been explored for graphene characterization. A peculiar property of monolayer graphene is that it absorbs 2.3% of incident light, almost independent of the wavelength (Nair et al. 2008). For multilayer graphene, optical transmittance is only weakly affected by van der Waal interlayer interactions, and Janssen et al. found that N can be calculated from transmittance with a rather simple relation (Zhu, Yuan, and Janssen 2014):

$$T(\lambda) = \left(1 + \frac{C(\lambda)\pi\alpha N}{2}\right)^{-2} \qquad (2.2)$$

where $T(\lambda)$ is the transmittance at wavelength λ, α is the fine structure constant, and C is a correction coefficient.

It is further found that for $\lambda = 550\,\text{nm}$, $C = 1.13$ and is independent of stacking sequence. Based on this observation, the number of layers and the thickness can be measured from a 550 nm laser's transmittance value when using suspended graphene or a transparent support, such as glass.

For the measurement of samples on a solid substrate, optical microscopy can be used to visualize and determine graphene's thickness. In the milestone 2004 paper, Geim and Novoselov noticed that while graphene itself is transparent, its thickness can alter the optical path of incident light on a SiO_2-coated Si wafer and result in the variation of interference color (Novoselov et al. 2004). However, this is only observable in optical microscope with a precisely controlled oxide layer thickness of 300 nm (Novoselov, Jiang, et al. 2005), and the contrast varies with lighting conditions. Later this interference contrast-based method was made more robust by examining the whole reflectance spectrum and by employing confocal microscopy (Ni et al. 2007; Jung et al. 2007).

For better contrast and spatial resolution, Rayleigh imaging uses a laser light source instead of white light to scan the graphene nanosheet's surface (Casiraghi et al. 2007). The method can also be extended to epitaxial graphene on SiC (Ivanov et al. 2014). Nonetheless, the difficulty of depositing a SiO_2 layer as spacer with the required small margin of error – variations in the 5% range can render graphene invisible (Geim and Novoselov 2007) – makes this method impractical in some cases.

In contrast to reflectance-based microscopy and imaging techniques, Raman spectroscopy relies on the inelastic scattering of incident light from the sample. Though this signal is significantly smaller, it is the result of interaction between the incident photons and the material's phonons (molecular vibrations), and therefore, can reveal rich information about the material's structure and many other properties (Figure 2.4). Over the years, Raman spectroscopy has proven to be one of the most versatile tools for carbon nanomaterial metrology (Saito et al. 2011). A prominent feature in graphene's Raman spectra is the 2D band (sometimes called G' band), which occurs near $2,700\,\text{cm}^{-1}$. The band originates from resonance of incident photon and

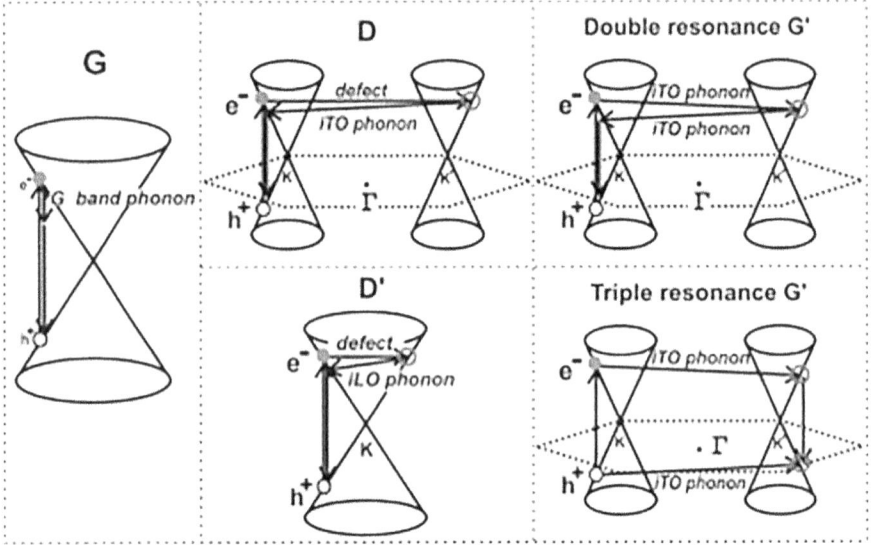

FIGURE 2.4 Raman scattering process of prominent feature in graphene's Raman spectra. (Reproduced from Malard et al. 2009 with permission.)

two phonons near the K point of the lattice's electronic structure and constitutes a single peak for single-layer graphene. In bilayer graphene, energy dispersion splits each phonon branch into two (and progressively more as further layers are added), which creates more phonons that meet the resonant conditions (Malard et al. 2009). As a result, the 2D band widens as the number of layers increases, and deconvolution of the 2D band shape can reveal the number of layers (Figure 2.5). Unfortunately, the shape of the 2D band is also sensitive to other material's properties, such as defectiveness (discussed later), so reliable numeration of graphene layers can only be done in highly crystalline graphene and requires careful calibration.

Another major feature of graphene's Raman spectra is the G band, which is due to a single photon/phonon resonance at the Γ point ("G" stands for "graphite"). The number of phonons available scales with the number of layers N, but laser absorption also gradually increases with film thickness. Therefore, the intensity of G band increases at first but later decreases as N increases. The ratios of G band intensity over 2D band intensity (I_G/I_{2D}) (Reina et al. 2009) or Si intensity (I_G/I_{Si}) (Li et al. 2015) have been used to identify N. However, the nonlinear behavior and the variation of other factors (2D band is sensitive of sample quality, while Si peak is sensitive of excitation wavelength and focusing condition) mean that this method is only suitable for relatively low N.

There are other Raman features that can also be used for thickness determination. For instance, the frequency of the shear mode (also called C mode) is strongly dependent on interlayer coupling (Tan et al. 2012):

$$\text{Pos}(C)_N = \frac{1}{\sqrt{2}\pi c}\sqrt{\frac{\alpha}{\mu}\left(1+\cos\left(\pi/N\right)\right)} \qquad (2.3)$$

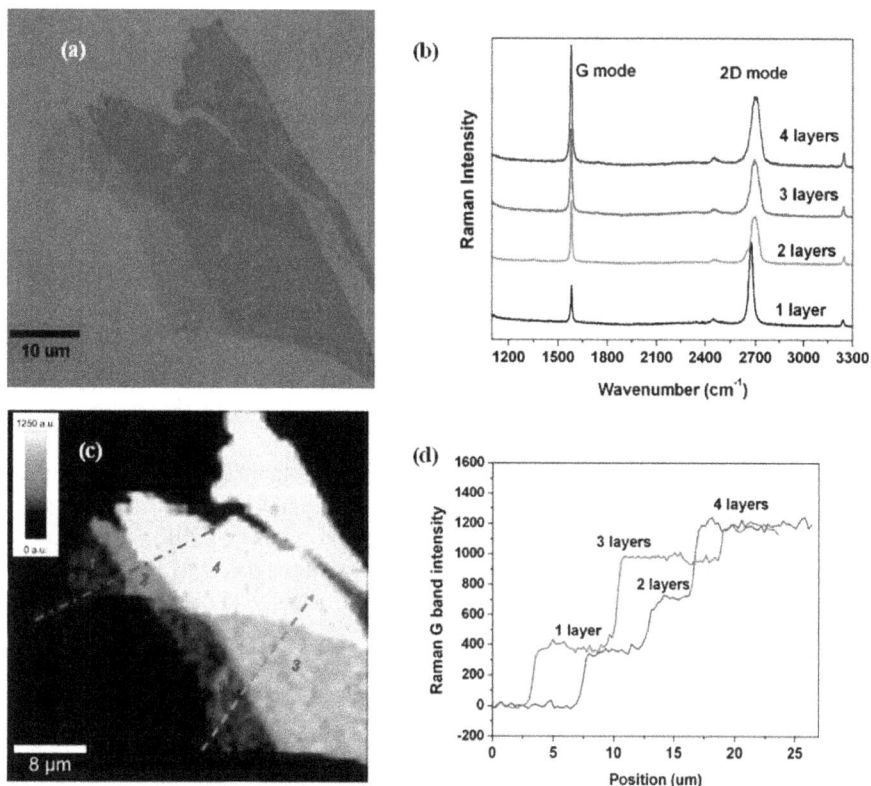

FIGURE 2.5 Determination of graphene layers via optical contrast (a) and Raman spectra (b) and mapping (c, d). (Reproduced from Ni et al. 2007 with permission.)

where α and μ are graphene's interlayer coupling strength and planar density, respectively.

The intensity of the C mode is relatively large, but it is a low-energy interaction appearing at ~40 cm^{-1}, a region overlapped by the much stronger Rayleigh scattering and challenging to measure without expensive optical components. More accessible are combination modes (1,690–2,150 cm^{-1}), including M band at ~1,750 cm^{-1} (Cong et al. 2011; Rao et al. 2011), and N modes at ~1,510 cm^{-1} (Herziger, May, and Maultzsch 2012), whose positions are also dependent on layer stacking, but the signal of these modes is relatively weak. Overall, Raman spectra can be utilized in various ways to determine N on any insulating or semiconducting substrates, and several modes can be simultaneously analyzed (Silva et al. 2020) or combined with many characterization techniques (Bayle et al. 2018; Casiraghi et al. 2007) to improve accuracy/confidence.

A common limitation of the above techniques is that, even if the data analysis can be automated, the process is still unsuitable for high-throughput metrology, since the incident light needs to be focused on a micron-size spot of the sample, a process that needs to be repeated many times for mapping a large-size sample.

Fluorescence quenching microscopy can image areas of hundred microns large and determine the area's thickness by exploiting graphene's strong fluorescence quenching effect. As a broadband light absorber, graphene suppresses photoluminescence of a nearby fluorescence-active molecule via Förster resonant energy transfer (Swathi and Sebastian 2008). Hence, by spin-coating a dye-containing polymer layer over the graphene, graphene layers appear as dark areas in fluorescent images (Kyle et al. 2011; Stöhr et al. 2012). Even diluted graphene solution can be characterized by sandwiching the solution mixed with dye between two glass slides (Kim et al. 2010). Unfortunately, the contrast is quickly saturated as N increases, rendering multilayer graphene ($N > 4$) difficult.

For LPE-produced graphene nanosheets, which have a wide range of thickness size, sample quality is often evaluated by representative measures of an ensemble like the average number of layers $\langle N \rangle$ and average size $\langle L \rangle$. Dynamic light scattering is a common technique for such statistical size measurements of nanoparticles and microscopic objects (Berne and Pecora 2000), but its principle is based on assumptions of spherical particle, which are not valid for 2D materials; hence, the acquired data cannot be directly translated to nanosheet size (Lotya et al. 2013). Backes et al. correlated the size of graphene nanosheets as determined by TEM with their optical absorption and Raman spectra and established formulae to calculate the sheets' dimensions from spectral characteristics:

$$\langle N \rangle = 35.7 \times \frac{\varepsilon_{550}}{\varepsilon_{325}} - 14.8 \tag{2.4}$$

where ε_{550} and ε_{325} are extinction intensity at 550 nm and 325 nm, respectively; and

$$\langle L \rangle = \frac{0.094}{\left(\dfrac{I_D}{I_G}\right)_{graphene} - \left(\dfrac{I_D}{I_G}\right)_{graphite}} \tag{2.5}$$

where I_D and I_G is the intensity of D band and G band in Raman spectra.

This method allows a convenient way to assess nanosheet quality; however, it is not clear how the experiment-based parameters might vary in different exfoliating conditions.

2.3.2 Defectiveness

No crystal is perfect, and defectiveness is another important fundamental property that affects the performance of composites and devices made of graphene. From another perspective, defects can add new functionalities and provide a way to control existing ones and are thus desirable in some cases. Defects exist in many forms, from non-hexagonal cells, doping atoms to functional groups, to pores and grain boundaries. Thus, information on impurity concentration and location (i.e., at the edge or basal plane) is required to evaluate the character of graphene materials.

HR-TEM and scanning tunnel microscopy (STM) are tools to reveal crystal defects with atomic resolution, but they are not efficient in evaluating statistical distributions of impurities. Instead, the most commonly used tool for this task is

Raman spectroscopy. The D band in Raman spectra of graphitic material is so named because its occurrence requires the presence of defects. The D band peak originates from intervalley Raman scattering that involves inelastic scattering by a defect, and it is largely absent in highly crystalline graphene samples.

Tuinstra and Koenig, who first discovered the D band (Tuinstra and Koenig 1970), established the intensity ratio of D and G bands (I_D/I_G) as a measure for the defectiveness of carbonous materials, which was later quantitatively correlated to graphene's planar defect concentration by Cançado et al. (2011):

$$n_D = \frac{(1.8 \pm 0.5) \times 10^{22}}{\lambda_L^4}\left(\frac{I_D}{I_G}\right), \tag{2.6}$$

where n_D is the defect concentration and λ_L is the excitation laser wavelength. This relation, however, only holds true for low levels of defects. For highly defected samples such as RGO or halogenated graphene, D band intensity decreases as defect concentration increases due to the limited number of non-defective sites that provide the phonons matching the resonant condition (Lucchese et al. 2010).

For large-scale graphene films (as prepared by CVD), impurities may not be as significant in limiting the material's mechanical and electrical performance as larger planar defects (grain boundaries) and bulk defects (cracks, pores). Film-induced frustrated etching (FIFE) is a convenient tool to characterize those defects (Hofmann et al. 2012) (Figure 2.6). In CVD-graphene transfer, etching is one step where the copper foil on which graphene is grown is removed in an etchant solution, leaving graphene (with a coated polymer support) on the liquid surface. FIFE inverts this by applying etchant on the side of graphene instead, so that the etching happens selectively at these defective sites where etchant can permeate graphene and reach copper. The color change on the copper "canvass" allows facile detection of defect distribution and density.

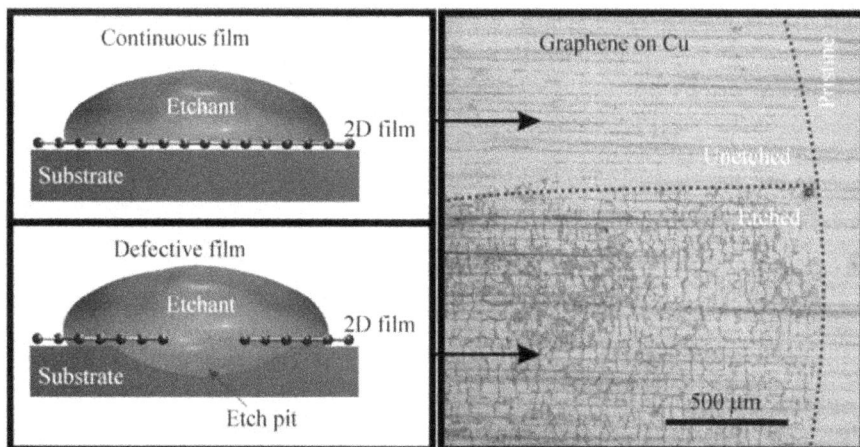

FIGURE 2.6 Schematic (a) and example (b) of defect visualization by film-induced frustrated etching. (Reproduced from Hofmann et al. 2012 with permission.)

For the study of the structure of defects, many tools exist, such as Fourier transform infrared spectroscopy (FTIR), thermal gravimetric analysis (TGA), X-ray diffraction (XRD), X-ray photoelectron spectroscopy (XPS), mass spectroscopy, and nuclear magnetic resonance (NMR) spectroscopy. The latter two have the highest sensitivity, but require expensive equipment and large sample amounts and are therefore limited to elucidating doping nature or functional groups in bulk prepared RGO (Pontiroli et al. 2014; Li et al. 2020).

FTIR and XRD are probably the most accessible tools, and with suitable specialized techniques such as attenuated total reflectance (ATR-FTIR) and small-angle X-ray scattering (SAXS), they are sensitive enough to probe microscopic graphene samples (Yan et al. 2014; Wang et al. 2016). XRD can reveal the degree of short-range order of the lattice and the interlayer spacing which can help distinguish pristine graphene ($d \cong 0.14$ nm), expanded RGO ($d \geq 0.4$ nm), and fully intercalated graphene oxide ($d \geq 0.8$ nm). Meanwhile, molecular vibrations in pristine graphene are not infrared-inactive, but functional groups emit infrared signals, which in practice can be conveniently looked up in correlation tables. However, FTIR (and XPS) quantitative analysis is difficult due to sample contamination and, more importantly, the lack of comprehensive understanding of graphene's IR behavior (Zhang et al. 2015).

Raman spectroscopy can also help to identify the type of defects. The D' band in Raman spectra of graphene located at ~1,600 cm^{-1} is also defect-induced, but unlike the D band, it is produced from low-energy intravalley scattering, which is more likely to involve smaller scattering centers compared to the D band (Malard et al. 2009). The relatively weak signal and proximity to a much stronger G band means that this feature is often overlooked. However, for low defect-density samples, the signal is fairly distinct, and as Eckmann et al. (2012) pointed out, $I_D/I_{D'}$ can indicate whether the defects represent vacancy types or sp^3 functional groups.

The location of defects can also be investigated spectroscopically. The D' band is more likely formed from defective edges than other types of defects, and an $I_D/I_{D'}$ ratio ≈ 3 was found to indicate that edge defects are dominant, whereas the ratio is higher if defects reside in the basal plane (Eckmann et al. 2012). Besides, the locality of defects can also be deduced from the scaling relationship between the level of defectiveness (represented by I_D/I_G) and the average nanosheet length, although this process is very time-consuming (Khan et al. 2010). Moreover, in a testament to the sensitivity of Raman spectroscopy in studying subtle lattice properties, I_D/I_G can be used to determine the chirality of graphene edges, i.e., if they are of zigzag or armchair type, an information that can otherwise only be gathered from highly specialized tools like STM (Cançado et al. 2004; You et al. 2008). A combination of scanning probe microscopy and Raman scattering is called tip-enhanced Raman spectroscopy (TERS), which allows the probing of microscopic defects with a high spatial resolution (Mignuzzi et al. 2015).

2.3.3 TRANSPORT PROPERTIES

One of the most important findings in early graphene research was its record carrier mobility which was experimentally confirmed to be as high as 200,000 cm^{2}V^{-1}s^{-1} on suspended, mechanically cleaved graphene flakes (Bolotin et al. 2008). In this impressive demonstration of Dirac quasiparticles with zero effective mass, electrons in graphene can move ballistically hundreds nanometer at sizeable fractions of the speed of light without scattering. Unfortunately, carrier mobility was found to be sensitively affected by defects and substrate interactions which have to be controlled to make graphene suitable for next-generation electronics material (Lin, Huang, and Duan 2019).

Two main methods are employed to determine graphene's carrier mobility and carrier concentration – field-effect measurement and Hall-effect measurement. Field-effect measurement requires fabricating a field-effect transistor (FET) from the graphene sample (Lemme et al. 2007), whose I-V characteristics allow mobility extraction (from transconductance, which is the slope of I_d/V_d, and V_g). The math involved in this method is straightforward, however, the result underestimates the real mobility values as it does not take into account the considerable contact resistance (Zhong, Zhang, et al. 2015), which can be corrected by fitting the whole I-V curve (Kim, Nah, et al. 2009). Moreover, graphene's FET-measured mobility is influenced by the device's physical dimensions (Chen and Appenzeller 2008) and substrate effects (Ando 2006). The high applied bias means the leakage is another concern, and it is found that higher mobility is achieved by using an improved insulator layer like hexagonal boron nitride instead of SiO_2 (Dean et al. 2010). The Hall effect measurement, on the other hand, exploits the Lorentz force that acts on charge carriers when moving perpendicular to a magnetic field, which generates a detectable voltage between the sides of the device. Early studies prepared the sample by etching graphene into a Hall bar shape (Zhang et al. 2005; Ponomarenko et al. 2009), but more recent studies often employed a much simpler van der Pauw geometry with four contacts (Kim, Zhao, et al. 2009; Shen, Wu, et al. 2011). Since the technique decouples current and voltage measurement, the effect of contact resistance is virtually eliminated and the measurement can be carried out on any arbitrary insulating substrate (Tedesco et al. 2009; Fanton et al. 2011). On the other hand, Hall effect mobility is more temperature dependent than field-effect mobility (Falkovsky 2007).

Despite the attractiveness of ultrahigh mobility, perfect graphene per se is not suitable for field effect transistors in digital electronics: the cone-shape valence conduction band indicates an extremely low carrier density near the Dirac points, and the absence of a bandgap results in a low on-off ratio and graphene devices cannot be switched off like normal transistors (Schwierz 2010). Therefore, many schemes have been investigated to open a band-gap in graphene (Wehling et al. 2008; Zhang et al. 2009; Ritter and Lyding 2009); the most convenient is chemical doping (Liu, Liu, and Zhu 2011). Moreover, some impurities are inevitably introduced during graphene preparation, so graphene samples are naturally slightly doped. Therefore, understanding graphene doping characteristics is crucial for tuning the material's electronic properties.

Aside from the standard electrical methods, optical spectroscopy can also be employed to conveniently characterize graphene doping and other photonic properties simultaneously. For example, the presence of a band gap leads to Pauli blocking of interband transitions with an energy lower than $2|\varepsilon_F|$, which manifests as a drop of absorption at the energy threshold, which usually falls in the infrared range. After considering broadening effects (thermal fluctuations, carrier lifetime), Fermi energy and carrier concentration can be determined from absorption spectra (Li, Henriksen, et al. 2008). Doping also has effects on many Raman spectra features, such as increased D band intensity, G band sharpening and frequency shift (Pisana et al. 2007), decreased I_{2D}/I_G (Berciaud et al. 2010). It should be noted that the intensity of different Raman features varies differently in response to excitation energy and strain, while the frequency shifts are sensitive to the environment, so a comparison is only valid for similar samples measured under identical conditions. Moreover, since doping is closely related to defects, caution is advised in using Raman-based techniques on samples with a high level of defectiveness (Bruna et al. 2014).

For graphene-based nanocomposites and thin film applications, electric performance is influenced not only by the intrinsic quality of graphene material (i.e., scattering-limited ballistic transport within the lattice) but also by the contact resistance between graphene nanosheets or interface with other materials. As such, when comparing macroscopic quantities like resistance and conductivity, the morphology of the film has to be considered (Yao et al. 2020).

2.3.4 MECHANICAL PROPERTIES

A second important property of graphene is its high mechanical strength. With Young's modulus over 1 TPa and tensile strength of 130 GPa (Lee et al. 2008), it is the strongest known material and has the potential to revolutionize many application areas, from aerospace (Siochi 2014) to construction (Yang et al. 2017) and from biomedical engineering (Reina et al. 2017) to birth control (Iliut et al. 2016). Its mechanical strength must be put into the context of its nanoscale thickness, though, and in most cases, is best exploited for making membranes or reinforcing composites. Moreover, the diverse forms of graphene perform drastically different and the elastic modulus in best CVD films can nearly reach the theoretical value of 1 TPa (Lee et al. 2013), whereas introduction of wrinkles may reduce its performance by six times (Ruiz-Vargas et al. 2011). Moreover, Young's modulus drops to ~250 GPa for single-layer RGO (Gómez-Navarro, Burghard, and Kern 2008), while few-layer graphene oxide films may only achieve as low as 30 GPa (Wang et al. 2018). Therefore, it is important to understand the mechanical properties of specific graphene materials and adjust the preparation method depending on the required scale and performance.

Lee et al. first reported a study of graphene's mechanical properties using nanoindentation (Lee et al. 2008). This technique involves depositing graphene onto a SiO_2/Si substrate with circular cavities of micron size, prepared by photolithography and reactive ion etching. An AFM tip is then used to indent the suspended graphene at the center and the force placement is recorded. The elastic response of graphene, under that condition, is nonlinear and described by

$$F = \sigma \pi a d + E q^3 a d^3 \tag{2.7}$$

where σ and E are 2D strain and Young's modulus of graphene, respectively, a is the crater's diameter, d is the displacement (deflection height h normalized by a), while q is a function of graphene's Poisson ratio and found to be $q \approx 1.02$.

By fitting the above model, both the strain and elastic modulus of graphene can be extracted. The accuracy of the measurement is affected by intrinsic rippling of graphene that causes sagging (Fasolino, Los, and Katsnelson 2007; Bao et al. 2009). The concentrated loading at the nanotip generates widely varying results in polycrystalline graphene (such as CVD graphene) depending on the tip's proximity to a grain boundary (Sha et al. 2014).

An alternative characterization method is bulge testing, where pressure is applied at one side of the membrane, causing it to bulge. The pressure-deflection behavior is modeled as biaxial stress-strain on a spherical cap according to (Vlassak and Nix 1992):

$$P = C_1 \frac{\sigma_0}{a^2} h + C_2 \frac{Et}{(1-v)a^4} h^3 \tag{2.8}$$

where C_1 and C_2 are geometry-dependent factors, $C_1 = 4$ and $C_2 = 8/3$ for circular membrane, σ_0 is the residual stress, t is the film thickness, and v is Poisson's ratio of graphene.

A variant of the bulge test is the blister test, where the applied pressure on the membrane is increased until the membrane starts to detach from the substrate, which reveals the adhesion strength of graphene to the substrate (Jensen 1991).

The deflection can be measured by *in situ* AFM (Wang et al. 2017) or light interference profilometry (Nicholl et al. 2015) (see Figure 2.7). The pressure is most commonly applied by sequentially increasing and decreasing the pressure in a chamber (Bunch et al. 2008) or with a hydraulic pump (Kulkarni et al. 2010). The former has a more uniform stress profile, but it assumes gas impermeability of the membrane, which may not hold true for some of the more defected graphene materials. The membrane can also be electrostatically pressurized, either by a separated electrical contact (Nicholl et al. 2015) or an STM tip (Uder et al. 2018). This electrostatic tuning permits fine control of applied pressure, but can only induce low-amplitude deflection and is therefore unsuitable for characterizing the high-stress region.

Compared to nanoindentation and bulge testing, uniaxial tensile testing is more accurate in principle, but sample preparation is difficult. It involves clamping a suspended graphene sample with nanotips and then actuating a clamping tip to apply the tension on the film (Polyzos et al. 2015). For fracture characteristics, a small crack is generated on the lattice, for example, by an ion beam, and the breaking point is monitored under SEM or TEM (Zhang et al. 2014). Recent innovations in graphene transfer allow the loading of graphene samples into micromechanical testers (Cao et al. 2020), and even measurement on centimeter-scale samples can be realized (Wang et al. 2018).

At a microscopic level, strain leads to the stretching of bond length in molecules and it can also be studied with spectroscopic tools. Under uniaxial strain, the *G* band is split into two peaks, and both *G* band and 2D band graphene show strain-dependent peak shift (Mohiuddin et al. 2009). Thus, Young's modulus and other mechanical properties of graphene can be deduced under variable deformation solely from Raman spectroscopy (Metten et al. 2014). However, the interaction of phonons and deformation depend on the strain direction and is therefore sensitive to details of the experimental setup (Cheng et al. 2011). Moreover, it is not always easy to disentangle

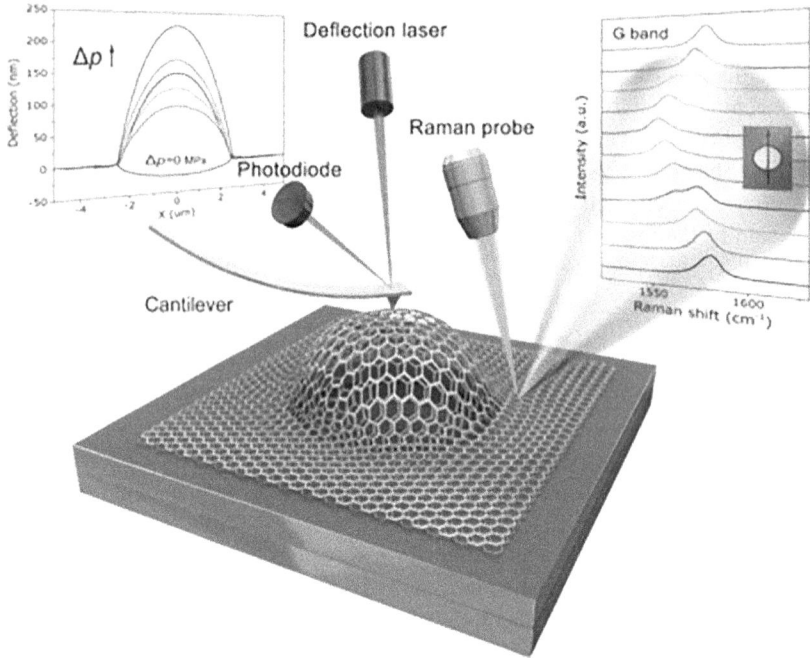

FIGURE 2.7 A bulge testing setup used in combination with Raman spectroscopy for graphene mechanical testing. (Reproduced from Wang et al. 2017 with permission.)

strain from other effects and Raman spectroscopy is best used for quick determination of strain or to complement other testing techniques.

2.4 CONCLUSIONS

In conclusion, graphene exhibits impressive intrinsic properties such as mechanical strength, uniform chemical structure, and carrier mobility. The ability to adjust these properties over a wide range turns graphene into the precursor for an exciting class of nanomaterials. In this chapter, we have introduced different synthesis routes to produce different types of graphene with specific advantages in the parameter space of performance and cost. Due to the large differences in such nanographene's quality, morphology, and electronic structure, their metrology is of great importance toward future applications. We have given an overview of powerful characterization methods that permit the evaluation of nanographene under particular perspectives and at different length scales.

REFERENCES

Ambrosi, Adriano, Gwendeline K. S. Wong, Richard D. Webster, Zdeněk Sofer, and Martin Pumera. 2013. "Carcinogenic organic residual compounds readsorbed on thermally reduced graphene materials are released at low temperature." *Chemistry – A European Journal* 19 (43):14446–14450. doi:10.1002/chem.201302413.

Ando, Tsuneya. 2006. "Screening effect and impurity scattering in monolayer graphene." *Journal of the Physical Society of Japan* 75 (7):074716. doi:10.1143/JPSJ.75.074716.

Backes, Claudia, Beata M. Szydłowska, Andrew Harvey, Shengjun Yuan, Victor Vega-Mayoral, Ben R. Davies, Pei-liang Zhao, Damien Hanlon, Elton J. G. Santos, Mikhail I. Katsnelson, Werner J. Blau, Christoph Gadermaier, and Jonathan N. Coleman. 2016. "Production of highly monolayer enriched dispersions of liquid-exfoliated nanosheets by liquid cascade centrifugation." *ACS Nano* 10 (1):1589–1601. doi:10.1021/acsnano.5b07228.

Bae, Sukang, Hyeongkeun Kim, Youngbin Lee, Xiangfan Xu, Jae-Sung Park, Yi Zheng, Jayakumar Balakrishnan, Tian Lei, Hye Ri Kim, Young Il Song, Young-Jin Kim, Kwang S. Kim, Barbaros Özyilmaz, Jong-Hyun Ahn, Byung Hee Hong, and Sumio Iijima. 2010. "Roll-to-roll production of 30-inch graphene films for transparent electrodes." *Nature Nanotechnology* 5 (8):574–578. doi:10.1038/nnano.2010.132.

Bai, Miaomiao, Wei Wu, Lingna Liu, Jianfeng Chen, Xiangrong Ma, and Yu Meng. 2019. "NaCl and oxalic acid–assisted solvothermal exfoliation of edge-oxidized graphite to produce organic graphene dispersion for transparent conductive film application." *Journal of Nanoparticle Research* 21 (7):145. doi:10.1007/s11051-019-4574-6.

Bao, Wenzhong, Feng Miao, Zhen Chen, Hang Zhang, Wanyoung Jang, Chris Dames, and Chun Ning Lau. 2009. "Controlled ripple texturing of suspended graphene and ultrathin graphite membranes." *Nature Nanotechnology* 4 (9):562–566. doi:10.1038/nnano.2009.191.

Bayle, Maxime, Nicolas Reckinger, Alexandre Felten, Périne Landois, Ophélie Lancry, Bertrand Dutertre, Jean-François Colomer, Ahmed-Azmi Zahab, Luc Henrard, Jean-Louis Sauvajol, and Matthieu Paillet. 2018. "Determining the number of layers in few-layer graphene by combining Raman spectroscopy and optical contrast." *Journal of Raman Spectroscopy* 49 (1):36–45. doi:10.1002/jrs.5279.

Ben Jabra, Zouhour, Isabelle Berbezier, Adrien Michon, Mathieu Koudia, Elie Assaf, Antoine Ronda, Paola Castrucci, Maurizio De Crescenzi, Holger Vach, and Mathieu Abel. 2021. "Hydrogen-mediated CVD epitaxy of graphene on SiC: Implications for microelectronic applications." *ACS Applied Nano Materials* 4 (5):4462–4473. doi:10.1021/acsanm.1c00082.

Benedict, Lorin X., Nasreen G. Chopra, Marvin L. Cohen, A. Zettl, Steven G. Louie, and Vincent H. Crespi. 1998. "Microscopic determination of the interlayer binding energy in graphite." *Chemical Physics Letters* 286 (5):490–496. doi:10.1016/S0009-2614(97)01466-8.

Berciaud, Stéphane, Melinda Y. Han, Kin Fai Mak, Louis E. Brus, Philip Kim, and Tony F. Heinz. 2010. "Electron and optical phonon temperatures in electrically biased graphene." *Physical Review Letters* 104 (22):227401. doi:10.1103/PhysRevLett.104.227401.

Berger, Claire, Zhimin Song, Tianbo Li, Xuebin Li, Asmerom Y. Ogbazghi, Rui Feng, Zhenting Dai, Alexei N. Marchenkov, Edward H. Conrad, Phillip N. First, and Walt A. de Heer. 2004. "Ultrathin epitaxial graphite: 2D electron gas properties and a route toward graphene-based nanoelectronics." *The Journal of Physical Chemistry B* 108 (52):19912–19916. doi:10.1021/jp040650f.

Berne, Bruce J, and Robert Pecora. 2000. *Dynamic Light Scattering: With Applications to Chemistry, Biology, and Physics.* Massachusetts: Courier Corporation.

Beshkova, M., L. Hultman, and R. Yakimova. 2016. "Device applications of epitaxial graphene on silicon carbide." *Vacuum* 128:186–197. doi:10.1016/j.vacuum.2016.03.027.

Blakely, J. M., J. S. Kim, and H. C. Potter. 1970. "Segregation of carbon to the (100) surface of nickel." *Journal of Applied Physics* 41 (6):2693–2697. doi:10.1063/1.1659283.

Boehm, H. P., A. Clauss, G. O. Fischer, and U. Hofmann. 1962. "Das Adsorptionsverhalten sehr dünner Kohlenstoff-Folien." *Zeitschrift für anorganische und allgemeine Chemie* 316 (3–4):119–127. doi:10.1002/zaac.19623160303.

Bolotin, K. I., K. J. Sikes, Z. Jiang, M. Klima, G. Fudenberg, J. Hone, P. Kim, and H. L. Stormer. 2008. "Ultrahigh electron mobility in suspended graphene." *Solid State Communications* 146 (9):351–355. doi:10.1016/j.ssc.2008.02.024.

Borovikov, Valery, and Andrew Zangwill. 2009. "Step-edge instability during epitaxial growth of graphene from SiC(0001)." *Physical Review B* 80 (12):121406. doi:10.1103/PhysRevB.80.121406.

Botas, Cristina, Patricia Álvarez, Patricia Blanco, Marcos Granda, Clara Blanco, Ricardo Santamaría, Laura J. Romasanta, Raquel Verdejo, Miguel A. López-Manchado, and Rosa Menéndez. 2013. "Graphene materials with different structures prepared from the same graphite by the Hummers and Brodie methods." *Carbon* 65:156–164. doi:10.1016/j.carbon.2013.08.009.

Bracamonte, M. V., G. I. Lacconi, S. E. Urreta, and L. E. F. Foa Torres. 2014. "On the nature of defects in liquid-phase exfoliated graphene." *The Journal of Physical Chemistry C* 118 (28):15455–15459. doi:10.1021/jp501930a.

Brodie, Benjamin Collins. 1859. "XIII. On the atomic weight of graphite." *Philosophical Transactions of the Royal Society of London* 149:249–259. doi:10.1098/rstl.1859.0013.

Brown, Lola, Edward B. Lochocki, José Avila, Cheol-Joo Kim, Yui Ogawa, Robin W. Havener, Dong-Ki Kim, Eric J. Monkman, Daniel E. Shai, Haofei I. Wei, Mark P. Levendorf, María Asensio, Kyle M. Shen, and Jiwoong Park. 2014. "Polycrystalline graphene with single crystalline electronic structure." *Nano Letters* 14 (10):5706–5711. doi:10.1021/nl502445j.

Bruna, Matteo, Anna K. Ott, Mari Ijäs, Duhee Yoon, Ugo Sassi, and Andrea C. Ferrari. 2014. "Doping dependence of the Raman spectrum of defected graphene." *ACS Nano* 8 (7):7432–7441. doi:10.1021/nn502676g.

Bunch, J. Scott, Scott S. Verbridge, Jonathan S. Alden, Arend M. van der Zande, Jeevak M. Parpia, Harold G. Craighead, and Paul L. McEuen. 2008. "Impermeable atomic membranes from graphene sheets." *Nano Letters* 8 (8):2458–2462. doi:10.1021/nl801457b.

Cai, Jinming, Pascal Ruffieux, Rached Jaafar, Marco Bieri, Thomas Braun, Stephan Blankenburg, Matthias Muoth, Ari P. Seitsonen, Moussa Saleh, Xinliang Feng, Klaus Müllen, and Roman Fasel. 2010. "Atomically precise bottom-up fabrication of graphene nanoribbons." *Nature* 466 (7305):470–473. doi:10.1038/nature09211.

Cançado, L. G., A. Jorio, E. H. Martins Ferreira, F. Stavale, C. A. Achete, R. B. Capaz, M. V. O. Moutinho, A. Lombardo, T. S. Kulmala, and A. C. Ferrari. 2011. "Quantifying defects in graphene via Raman spectroscopy at different excitation energies." *Nano Letters* 11 (8):3190–3196. doi:10.1021/nl201432g.

Cançado, L. G., M. A. Pimenta, B. R. A. Neves, M. S. S. Dantas, and A. Jorio. 2004. "Influence of the atomic structure on the Raman spectra of graphite edges." *Physical Review Letters* 93 (24):247401. doi:10.1103/PhysRevLett.93.247401.

Cao, Ke, Shizhe Feng, Ying Han, Libo Gao, Thuc Hue Ly, Zhiping Xu, and Yang Lu. 2020. "Elastic straining of free-standing monolayer graphene." *Nature Communications* 11 (1):284. doi:10.1038/s41467-019-14130-0.

Cao, Yuan, Valla Fatemi, Shiang Fang, Kenji Watanabe, Takashi Taniguchi, Efthimios Kaxiras, and Pablo Jarillo-Herrero. 2018. "Unconventional superconductivity in magic-angle graphene superlattices." *Nature* 556 (7699):43–50. doi:10.1038/nature26160.

Casiraghi, C., A. Hartschuh, E. Lidorikis, H. Qian, H. Harutyunyan, T. Gokus, K. S. Novoselov, and A. C. Ferrari. 2007. "Rayleigh imaging of graphene and graphene layers." *Nano Letters* 7 (9):2711–2717. doi:10.1021/nl071168m.

Chen, Shanshan, Hengxing Ji, Harry Chou, Qiongyu Li, Hongyang Li, Ji Won Suk, Richard Piner, Lei Liao, Weiwei Cai, and Rodney S. Ruoff. 2013. "Millimeter-size single-crystal graphene by suppressing evaporative loss of cu during low pressure chemical vapor deposition." *Advanced Materials* 25 (14):2062–2065. doi:10.1002/adma.201204000.

Chen, Z., and J. Appenzeller. 2008. "Mobility extraction and quantum capacitance impact in high performance graphene field-effect transistor devices." *2008 IEEE International Electron Devices Meeting*, 15–17 Dec. 2008.

Cheng, Y. C., Z. Y. Zhu, G. S. Huang, and U. Schwingenschlögl. 2011. "Grüneisen parameter of the *G* mode of strained monolayer graphene." *Physical Review B* 83 (11):115449. doi:10.1103/PhysRevB.83.115449.

Chin, H. T., C. H. Shih, Y. P. Hsieh, C. C. Ting, J. N. Aoh, and M. Hofmann. 2017. "How does graphene grow on complex 3D morphologies?" *Physical Chemistry Chemical Physics* 19 (34):23357–23361. doi:10.1039/C7CP03207B.

Chua, Chun Kiang, Zdeněk Sofer, Bahareh Khezri, Richard D. Webster, and Martin Pumera. 2016. "Ball-milled sulfur-doped graphene materials contain metallic impurities originating from ball-milling apparatus: their influence on the catalytic properties." *Physical Chemistry Chemical Physics* 18 (27):17875–17880. doi:10.1039/C6CP03004A.

Cong, Chunxiao, Ting Yu, Riichiro Saito, Gene F. Dresselhaus, and Mildred S. Dresselhaus. 2011. "Second-order overtone and combination Raman modes of graphene layers in the range of 1690−2150 cm^{-1}." *ACS Nano* 5 (3):1600–1605. doi:10.1021/nn200010m.

Dean, C. R., A. F. Young, I. Meric, C. Lee, L. Wang, S. Sorgenfrei, K. Watanabe, T. Taniguchi, P. Kim, K. L. Shepard, and J. Hone. 2010. "Boron nitride substrates for high-quality graphene electronics." *Nature Nanotechnology* 5 (10):722–726. doi:10.1038/nnano.2010.172.

Del Rio-Castillo, Antonio Esaú, César Merino, Enrique Díez-Barra, and Ester Vázquez. 2014. "Selective suspension of single layer graphene mechanochemically exfoliated from carbon nanofibres." *Nano Research* 7 (7):963–972. doi:10.1007/s12274-014-0457-4.

Dreyer, Daniel R., Rodney S. Ruoff, and Christopher W. Bielawski. 2010. "From conception to realization: An historial account of graphene and some perspectives for its future." *Angewandte Chemie International Edition* 49 (49):9336–9344. doi:10.1002/anie.201003024.

Eckmann, Axel, Alexandre Felten, Artem Mishchenko, Liam Britnell, Ralph Krupke, Kostya S. Novoselov, and Cinzia Casiraghi. 2012. "Probing the nature of defects in graphene by Raman spectroscopy." *Nano Letters* 12 (8):3925–3930. doi:10.1021/nl300901a.

Emtsev, Konstantin V., Aaron Bostwick, Karsten Horn, Johannes Jobst, Gary L. Kellogg, Lothar Ley, Jessica L. McChesney, Taisuke Ohta, Sergey A. Reshanov, Jonas Röhrl, Eli Rotenberg, Andreas K. Schmid, Daniel Waldmann, Heiko B. Weber, and Thomas Seyller. 2009. "Towards wafer-size graphene layers by atmospheric pressure graphitization of silicon carbide." *Nature Materials* 8 (3):203–207. doi:10.1038/nmat2382.

Falkovsky, L. A. 2007. "Unusual field and temperature dependence of the hall effect in graphene." *Physical Review B* 75 (3):033409. doi:10.1103/PhysRevB.75.033409.

Fanton, M. A., J. A. Robinson, C. Puls, Y. Liu, M. J. Hollander, B. E. Weiland, M. LaBella, K. Trumbull, R. Kasarda, C. Howsare, J. Stitt, and D. W. Snyder. 2011. "Characterization of graphene films and transistors grown on sapphire by metal-free chemical vapor deposition." *ACS Nano* 5 (10):8062–8069. doi:10.1021/nn202643t.

Fasolino, A., J. H. Los, and M. I. Katsnelson. 2007. "Intrinsic ripples in graphene." *Nature Materials* 6 (11):858–861. doi:10.1038/nmat2011.

Fitzer, E., K.-H. Kochling, H. P. Boehm, and H. Marsh. 1995. "Recommended terminology for the description of carbon as a solid (IUPAC Recommendations 1995)." *Pure and Applied Chemistry* 67 (3):473–506. doi:10.1351/pac199567030473.

Geim, A. K. 2012. "Graphene prehistory." *Physica Scripta* T146:014003. doi:10.1088/0031-8949/2012/t146/014003.

Geim, A. K., and K. S. Novoselov. 2007. "The rise of graphene." *Nature Materials* 6 (3):183–191. doi:10.1038/nmat1849.

Georgakilas, Vasilios, Jason A. Perman, Jiri Tucek, and Radek Zboril. 2015. "Broad family of carbon nanoallotropes: Classification, chemistry, and applications of fullerenes, carbon dots, nanotubes, graphene, nanodiamonds, and combined superstructures." *Chemical Reviews* 115 (11):4744–4822. doi:10.1021/cr500304f.

Gómez-Navarro, Cristina, Marko Burghard, and Klaus Kern. 2008. "Elastic properties of chemically derived single graphene sheets." *Nano Letters* 8 (7):2045–2049. doi:10.1021/nl801384y.

Guo, Hui-Lin, Xian-Fei Wang, Qing-Yun Qian, Feng-Bin Wang, and Xing-Hua Xia. 2009. "A green approach to the synthesis of graphene nanosheets." *ACS Nano* 3 (9):2653–2659. doi:10.1021/nn900227d.

Guo, Wenman, and Guohua Chen. 2014. "Fabrication of graphene/epoxy resin composites with much enhanced thermal conductivity via ball milling technique." *Journal of Applied Polymer Science* 131 (15). doi:10.1002/app.40565.

He, T., J. L. Li, L. J. Wang, J. J. Zhu, and W. Jiang. 2009. "Preparation and consolidation of alumina/graphene composite powders." *Materials Transactions* 50 (4):749–751. doi:10.2320/matertrans.MRA2008458.

Hernandez, Yenny, Valeria Nicolosi, Mustafa Lotya, Fiona M. Blighe, Zhenyu Sun, Sukanta De, I. T. McGovern, Brendan Holland, Michele Byrne, Yurii K. Gun'Ko, John J. Boland, Peter Niraj, Georg Duesberg, Satheesh Krishnamurthy, Robbie Goodhue, John Hutchison, Vittorio Scardaci, Andrea C. Ferrari, and Jonathan N. Coleman. 2008. "High-yield production of graphene by liquid-phase exfoliation of graphite." *Nature Nanotechnology* 3 (9):563–568. doi:10.1038/nnano.2008.215.

Herziger, Felix, Patrick May, and Janina Maultzsch. 2012. "Layer-number determination in graphene by out-of-plane phonons." *Physical Review B* 85 (23):235447. doi:10.1103/PhysRevB.85.235447.

Hofmann, Mario, Wan-Yu Chiang, Dinh-Tuan Nguyen, and Ya-Ping Hsieh. 2015. "Controlling the properties of graphene produced by electrochemical exfoliation." *Nanotechnology* 26 (33):335607. doi:10.1088/0957-4484/26/33/335607.

Hofmann, Mario, Yong Cheol Shin, Ya-Ping Hsieh, Mildred S. Dresselhaus, and Jing Kong. 2012. "A facile tool for the characterization of two-dimensional materials grown by chemical vapor deposition." *Nano Research* 5 (7):504–511. doi:10.1007/s12274-012-0227-0.

Horiuchi, Shigeo, Takuya Gotou, Masahiro Fujiwara, Toru Asaka, Tadahiro Yokosawa, and Yoshio Matsui. 2004. "Single graphene sheet detected in a carbon nanofilm." *Applied Physics Letters* 84 (13):2403–2405. doi:10.1063/1.1689746.

Hsieh, Ya-Ping, Mario Hofmann, and Jing Kong. 2014. "Promoter-assisted chemical vapor deposition of graphene." *Carbon* 67:417–423. doi:10.1016/j.carbon.2013.10.013.

Hsieh, Ya-Ping, Ching-Hua Shih, Yi-Jing Chiu, and Mario Hofmann. 2016. "High-throughput graphene synthesis in gapless stacks." *Chemistry of Materials* 28 (1):40–43. doi:10.1021/acs.chemmater.5b04007.

Huang, Yuan, Eli Sutter, Norman N. Shi, Jiabao Zheng, Tianzhong Yang, Dirk Englund, Hong-Jun Gao, and Peter Sutter. 2015. "Reliable exfoliation of large-area high-quality flakes of graphene and other two-dimensional materials." *ACS Nano* 9 (11):10612–10620. doi:10.1021/acsnano.5b04258.

Hummers, William S., and Richard E. Offeman. 1958. "Preparation of graphitic oxide." *Journal of the American Chemical Society* 80 (6):1339–1339. doi:10.1021/ja01539a017.

Iijima, Sumio. 1991. "Helical microtubules of graphitic carbon." *Nature* 354 (6348):56–58. doi:10.1038/354056a0.

Iliut, Maria, Claudio Silva, Scott Herrick, Mark McGlothlin, and Aravind Vijayaraghavan. 2016. "Graphene and water-based elastomers thin-film composites by dip-moulding." *Carbon* 106:228–232. doi:10.1016/j.carbon.2016.05.032.

Ivanov, Ivan G., Jawad Ul Hassan, Tihomir Iakimov, Alexei A. Zakharov, Rositsa Yakimova, and Erik Janzén. 2014. "Layer-number determination in graphene on SiC by reflectance mapping." *Carbon* 77:492–500. doi:10.1016/j.carbon.2014.05.054.

Jankovský, Ondřej, Štěpánka Hrdličková Kučková, Martin Pumera, Petr Šimek, David Sedmidubský, and Zdeněk Sofer. 2014. "Carbon fragments are ripped off from graphite oxide sheets during their thermal reduction." *New Journal of Chemistry* 38 (12):5700–5705. doi:10.1039/C4NJ00871E.

Jayasena, Buddhika, and Sathyan Subbiah. 2011. "A novel mechanical cleavage method for synthesizing few-layer graphenes." *Nanoscale Research Letters* 6 (1):95. doi:10.1186/1556-276X-6-95.

Jensen, Henrik Myhre. 1991. "The blister test for interface toughness measurement." *Engineering Fracture Mechanics* 40 (3):475–486. doi:10.1016/0013-7944(91)90144-P.

Jeon, In-Yup, Yeon-Ran Shin, Gyung-Joo Sohn, Hyun-Jung Choi, Seo-Yoon Bae, Javeed Mahmood, Sun-Min Jung, Jeong-Min Seo, Min-Jung Kim, Dong Wook Chang, Liming Dai, and Jong-Beom Baek. 2012. "Edge-carboxylated graphene nanosheets via ball milling." *Proceedings of the National Academy of Sciences* 109 (15):5588. doi:10.1073/pnas.1116897109.

Jung, Inhwa, Matthew Pelton, Richard Piner, Dmitriy A. Dikin, Sasha Stankovich, Supinda Watcharotone, Martina Hausner, and Rodney S. Ruoff. 2007. "Simple approach for high-contrast optical imaging and characterization of graphene-based sheets." *Nano Letters* 7 (12):3569–3575. doi:10.1021/nl0714177.

Karagiannidis, Panagiotis G., Stephen A. Hodge, Lucia Lombardi, Flavia Tomarchio, Nicolas Decorde, Silvia Milana, Ilya Goykhman, Yang Su, Steven V. Mesite, Duncan N. Johnstone, Rowan K. Leary, Paul A. Midgley, Nicola M. Pugno, Felice Torrisi, and Andrea C. Ferrari. 2017. "Microfluidization of graphite and formulation of graphene-based conductive inks." *ACS Nano* 11 (3):2742–2755. doi:10.1021/acsnano.6b07735.

Khan, Umar, Arlene O'Neill, Mustafa Lotya, Sukanta De, and Jonathan N. Coleman. 2010. "High-concentration solvent exfoliation of graphene." *Small* 6 (7):864–871. doi:10.1002/smll.200902066.

Khan, Umar, Harshit Porwal, Arlene O'Neill, Khalid Nawaz, Peter May, and Jonathan N. Coleman. 2011. "Solvent-exfoliated graphene at extremely high concentration." *Langmuir* 27 (15):9077–9082. doi:10.1021/la201797h.

Kim, Jaemyung, Laura J. Cote, Franklin Kim, and Jiaxing Huang. 2010. "Visualizing graphene based sheets by fluorescence quenching microscopy." *Journal of the American Chemical Society* 132 (1):260–267. doi:10.1021/ja906730d.

Kim, Jaeho, Masatou Ishihara, Yoshinori Koga, Kazuo Tsugawa, Masataka Hasegawa, and Sumio Iijima. 2011. "Low-temperature synthesis of large-area graphene-based transparent conductive films using surface wave plasma chemical vapor deposition." *Applied Physics Letters* 98 (9):091502. doi:10.1063/1.3561747.

Kim, Keun Soo, Yue Zhao, Houk Jang, Sang Yoon Lee, Jong Min Kim, Kwang S. Kim, Jong-Hyun Ahn, Philip Kim, Jae-Young Choi, and Byung Hee Hong. 2009. "Large-scale pattern growth of graphene films for stretchable transparent electrodes." *Nature* 457 (7230):706–710. doi:10.1038/nature07719.

Kim, Seyoung, Junghyo Nah, Insun Jo, Davood Shahrjerdi, Luigi Colombo, Zhen Yao, Emanuel Tutuc, and Sanjay K. Banerjee. 2009. "Realization of a high mobility dual-gated graphene field-effect transistor with Al_2O_3 dielectric." *Applied Physics Letters* 94 (6):062107. doi:10.1063/1.3077021.

Knieke, Catharina, Angela Berger, Michael Voigt, Robin N. Klupp Taylor, Jonas Röhrl, and Wolfgang Peukert. 2010. "Scalable production of graphene sheets by mechanical delamination." *Carbon* 48 (11):3196–3204. doi:10.1016/j.carbon.2010.05.003.

Kroto, H. W., J. R. Heath, S. C. O'Brien, R. F. Curl, and R. E. Smalley. 1985. "C60: Buckminsterfullerene." *Nature* 318 (6042):162–163. doi:10.1038/318162a0.

Kulkarni, Dhaval D., Ikjun Choi, Srikanth S. Singamaneni, and Vladimir V. Tsukruk. 2010. "Graphene oxide–polyelectrolyte nanomembranes." *ACS Nano* 4 (8):4667–4676. doi:10.1021/nn101204d.

Kyle, Jennifer Reiber, Ali Guvenc, Wei Wang, Maziar Ghazinejad, Jian Lin, Shirui Guo, Cengiz S. Ozkan, and Mihrimah Ozkan. 2011. "Centimeter-scale high-resolution metrology of entire CVD-grown graphene sheets." *Small* 7 (18):2599–2606. doi:10.1002/smll.201100263.

Larousserie, David. 2013. "Graphene – The new wonder material." *The Guardian.*

Lee, Changgu, Xiaoding Wei, Jeffrey W. Kysar, and James Hone. 2008. "Measurement of the Elastic Properties and Intrinsic Strength of Monolayer Graphene." *Science* 321 (5887):385. doi:10.1126/science.1157996.

Lee, Gwan-Hyoung, Ryan C. Cooper, Sung Joo An, Sunwoo Lee, Arend van der Zande, Nicholas Petrone, Alexandra G. Hammerberg, Changgu Lee, Bryan Crawford, Warren Oliver, Jeffrey W. Kysar, and James Hone. 2013. "High-strength chemical-vapor–deposited graphene and grain boundaries." *Science* 340 (6136):1073. doi:10.1126/science.1235126.

Lemme, M. C., T. J. Echtermeyer, M. Baus, and H. Kurz. 2007. "A graphene field-effect device." *IEEE Electron Device Letters* 28 (4):282–284. doi:10.1109/LED.2007.891668.

León, Verónica, Mildred Quintana, M. Antonia Herrero, Jose L. G. Fierro, Antonio de la Hoz, Maurizio Prato, and Ester Vázquez. 2011. "Few-layer graphenes from ball-milling of graphite with melamine." *Chemical Communications* 47 (39):10936–10938. doi:10.1039/C1CC14595A.

Li, Xiao-Li, Xiao-Fen Qiao, Wen-Peng Han, Yan Lu, Qing-Hai Tan, Xue-Lu Liu, and Ping-Heng Tan. 2015. "Layer number identification of intrinsic and defective multilayered graphenes up to 100 layers by the Raman mode intensity from substrates." *Nanoscale* 7 (17):8135–8141. doi:10.1039/C5NR01514F.

Li, Xiaolin, Xinran Wang, Li Zhang, Sangwon Lee, and Hongjie Dai. 2008. "Chemically derived, ultrasmooth graphene nanoribbon semiconductors." *Science* 319 (5867):1229. doi:10.1126/science.1150878.

Li, Xuesong, Weiwei Cai, Jinho An, Seyoung Kim, Junghyo Nah, Dongxing Yang, Richard Piner, Aruna Velamakanni, Inhwa Jung, Emanuel Tutuc, Sanjay K. Banerjee, Luigi Colombo, and Rodney S. Ruoff. 2009. "Large-area synthesis of high-quality and uniform graphene films on copper foils." *Science* 324 (5932):1312. doi:10.1126/science.1171245.

Li, Z. Q., E. A. Henriksen, Z. Jiang, Z. Hao, M. C. Martin, P. Kim, H. L. Stormer, and D. N. Basov. 2008. "Dirac charge dynamics in graphene by infrared spectroscopy." *Nature Physics* 4 (7):532–535. doi:10.1038/nphys989.

Li, Ziqing, Wenxiu He, Xixin Wang, Xiaoliu Wang, Min Song, and Jianling Zhao. 2020. "N/S dual-doped graphene with high defect density for enhanced supercapacitor properties." *International Journal of Hydrogen Energy* 45 (1):112–122. doi:10.1016/j.ijhydene.2019.10.196.

Lin, Cheng, Lingli Yang, Liuzhang Ouyang, Jiangwen Liu, Hui Wang, and Min Zhu. 2017. "A new method for few-layer graphene preparation via plasma-assisted ball milling." *Journal of Alloys and Compounds* 728:578–584. doi:10.1016/j.jallcom.2017.09.056.

Lin, Zhaoyang, Yu Huang, and Xiangfeng Duan. 2019. "Van der Waals thin-film electronics." *Nature Electronics* 2 (9):378–388. doi:10.1038/s41928-019-0301-7.

Liu, Hongtao, Yunqi Liu, and Daoben Zhu. 2011. "Chemical doping of graphene." *Journal of Materials Chemistry* 21 (10):3335–3345. doi:10.1039/C0JM02922J.

Liu, Na, Fang Luo, Haoxi Wu, Yinghui Liu, Chao Zhang, and Ji Chen. 2008. "One-step ionic-liquid-assisted electrochemical synthesis of ionic-liquid-functionalized graphene sheets directly from graphite." *Advanced Functional Materials* 18 (10):1518–1525. doi:10.1002/adfm.200700797.

Lotya, Mustafa, Aliaksandra Rakovich, John F. Donegan, and Jonathan N. Coleman. 2013. "Measuring the lateral size of liquid-exfoliated nanosheets with dynamic light scattering." *Nanotechnology* 24 (26):265703. doi:10.1088/0957-4484/24/26/265703.

Lu, Jiong, Jia-Xiang Yang, Junzhong Wang, Ailian Lim, Shuai Wang, and Kian Ping Loh. 2009. "One-pot synthesis of fluorescent carbon nanoribbons, nanoparticles, and graphene by the exfoliation of graphite in ionic liquids." *ACS Nano* 3 (8):2367–2375. doi:10.1021/nn900546b.

Lucchese, M. M., F. Stavale, E. H. Martins Ferreira, C. Vilani, M. V. O. Moutinho, Rodrigo B. Capaz, C. A. Achete, and A. Jorio. 2010. "Quantifying ion-induced defects and Raman relaxation length in graphene." *Carbon* 48 (5):1592–1597. doi:10.1016/j.carbon.2009.12.057.

Lui, Chun Hung, Zhiqiang Li, Kin Fai Mak, Emmanuele Cappelluti, and Tony F. Heinz. 2011. "Observation of an electrically tunable band gap in trilayer graphene." *Nature Physics* 7 (12):944–947.

Lund, Sara, Jussi Kauppila, Saara Sirkiä, Jenny Palosaari, Olav Eklund, Rose-Marie Latonen, Jan-Henrik Smått, Jouko Peltonen, and Tom Lindfors. 2021. "Fast high-shear exfoliation of natural flake graphite with temperature control and high yield." *Carbon* 174:123–131. doi:10.1016/j.carbon.2020.11.094.

Luo, Zhengtang, Ye Lu, Daniel W. Singer, Matthew E. Berck, Luke A. Somers, Brett R. Goldsmith, and A. T. Charlie Johnson. 2011. "Effect of substrate roughness and feedstock concentration on growth of wafer-scale graphene at atmospheric pressure." *Chemistry of Materials* 23 (6):1441–1447. doi:10.1021/cm1028854.

Lv, J. L., Z. Q. Wang, and H. Miura. 2017. "The effects of ball milling on microstructures of graphene/Ni composites and their catalytic activity for hydrogen evolution reaction." *Materials Letters* 206:124–127. doi:10.1016/j.matlet.2017.07.001.

Malard, L. M., M. A. Pimenta, G. Dresselhaus, and M. S. Dresselhaus. 2009. "Raman spectroscopy in graphene." *Physics Reports* 473 (5):51–87. doi:10.1016/j.physrep.2009.02.003.

McAllister, Michael J., Je-Luen Li, Douglas H. Adamson, Hannes C. Schniepp, Ahmed A. Abdala, Jun Liu, Margarita Herrera-Alonso, David L. Milius, Roberto Car, Robert K. Prud'homme, and Ilhan A. Aksay. 2007. "Single sheet functionalized graphene by oxidation and thermal expansion of graphite." *Chemistry of Materials* 19 (18):4396–4404. doi:10.1021/cm0630800.

Metten, Dominik, François Federspiel, Michelangelo Romeo, and Stéphane Berciaud. 2014. "All-optical blister test of suspended graphene using micro-Raman spectroscopy." *Physical Review Applied* 2 (5):054008. doi:10.1103/PhysRevApplied.2.054008.

Mignuzzi, Sandro, Naresh Kumar, Barry Brennan, Ian S. Gilmore, David Richards, Andrew J. Pollard, and Debdulal Roy. 2015. "Probing individual point defects in graphene via near-field Raman scattering." *Nanoscale* 7 (46):19413–19418. doi:10.1039/C5NR04664E.

Mohiuddin, T. M. G., A. Lombardo, R. R. Nair, A. Bonetti, G. Savini, R. Jalil, N. Bonini, D. M. Basko, C. Galiotis, N. Marzari, K. S. Novoselov, A. K. Geim, and A. C. Ferrari. 2009. "Uniaxial strain in graphene by Raman spectroscopy: G peak splitting, Grüneisen parameters, and sample orientation." *Physical Review B* 79 (20):205433. doi:10.1103/PhysRevB.79.205433.

Morgan, A. E., and G. A. Somorjai. 1968. "Low energy electron diffraction studies of gas adsorption on the platinum (100) single crystal surface." *Surface Science* 12 (3):405–425. doi:10.1016/0039-6028(68)90089-7.

Nair, R. R., P. Blake, A. N. Grigorenko, K. S. Novoselov, T. J. Booth, T. Stauber, N. M. R. Peres, and A. K. Geim. 2008. "Fine structure constant defines visual transparency of graphene." *Science* 320 (5881):1308. doi:10.1126/science.1156965.

Narita, Akimitsu, Xinliang Feng, Yenny Hernandez, Søren A. Jensen, Mischa Bonn, Huafeng Yang, Ivan A. Verzhbitskiy, Cinzia Casiraghi, Michael Ryan Hansen, Amelie H. R. Koch, George Fytas, Oleksandr Ivasenko, Bing Li, Kunal S. Mali, Tatyana Balandina,

Sankarapillai Mahesh, Steven De Feyter, and Klaus Müllen. 2014. "Synthesis of structurally well-defined and liquid-phase-processable graphene nanoribbons." *Nature Chemistry* 6 (2):126–132. doi:10.1038/nchem.1819.

Narita, Akimitsu, Xiao-Ye Wang, Xinliang Feng, and Klaus Müllen. 2015. "New advances in nanographene chemistry." *Chemical Society Reviews* 44 (18):6616–6643. doi:10.1039/C5CS00183H.

Nemes-Incze, P., Z. Osváth, K. Kamarás, and L. P. Biró. 2008. "Anomalies in thickness measurements of graphene and few layer graphite crystals by tapping mode atomic force microscopy." *Carbon* 46 (11):1435–1442. doi:10.1016/j.carbon.2008.06.022.

Nguyen, Dinh-Tuan, Wan-Yu Chiang, Yen-Hsun Su, Mario Hofmann, and Ya-Ping Hsieh. 2019. "Solid-diffusion-facilitated cleaning of copper foil improves the quality of CVD graphene." *Scientific Reports* 9 (1):257. doi:10.1038/s41598-018-36390-4.

Ni, Z. H., H. M. Wang, J. Kasim, H. M. Fan, T. Yu, Y. H. Wu, Y. P. Feng, and Z. X. Shen. 2007. "Graphene thickness determination using reflection and contrast spectroscopy." *Nano Letters* 7 (9):2758–2763. doi:10.1021/nl071254m.

Nicholl, Ryan J. T., Hiram J. Conley, Nickolay V. Lavrik, Ivan Vlassiouk, Yevgeniy S. Puzyrev, Vijayashree Parsi Sreenivas, Sokrates T. Pantelides, and Kirill I. Bolotin. 2015. "The effect of intrinsic crumpling on the mechanics of free-standing graphene." *Nature Communications* 6 (1):8789. doi:10.1038/ncomms9789.

Niu, Liyong, Mingjian Li, Xiaoming Tao, Zhuang Xie, Xuechang Zhou, Arun P. A. Raju, Robert J. Young, and Zijian Zheng. 2013. "Salt-assisted direct exfoliation of graphite into high-quality, large-size, few-layer graphene sheets." *Nanoscale* 5 (16):7202–7208. doi:10.1039/C3NR02173D.

Novoselov, K. S., A. K. Geim, S. V. Morozov, D. Jiang, M. I. Katsnelson, I. V. Grigorieva, S. V. Dubonos, and A. A. Firsov. 2005. "Two-dimensional gas of massless Dirac fermions in graphene." *Nature* 438 (7065):197–200. doi:10.1038/nature04233.

Novoselov, K. S., A. K. Geim, S. V. Morozov, D. Jiang, Y. Zhang, S. V. Dubonos, I. V. Grigorieva, and A. A. Firsov. 2004. "Electric field effect in atomically thin carbon films." *Science* 306 (5696):666. doi:10.1126/science.1102896.

Novoselov, K. S., D. Jiang, F. Schedin, T. J. Booth, V. V. Khotkevich, S. V. Morozov, and A. K. Geim. 2005. "Two-dimensional atomic crystals." *Proceedings of the National Academy of Sciences of the United States of America* 102 (30):10451. doi:10.1073/pnas.0502848102.

Park, Jeongho, William C. Mitchel, Lawrence Grazulis, Howard E. Smith, Kurt G. Eyink, John J. Boeckl, David H. Tomich, Shanee D. Pacley, and John E. Hoelscher. 2010. "Epitaxial graphene growth by carbon molecular beam epitaxy (CMBE)." *Advanced Materials* 22 (37):4140–4145. doi:10.1002/adma.201000756.

Partoens, B., and F. M. Peeters. 2006. "From graphene to graphite: Electronic structure around the *K* point." *Physical Review B* 74 (7):075404. doi:10.1103/PhysRevB.74.075404.

Paton, Keith R., Eswaraiah Varrla, Claudia Backes, Ronan J. Smith, Umar Khan, Arlene O'Neill, Conor Boland, Mustafa Lotya, Oana M. Istrate, Paul King, Tom Higgins, Sebastian Barwich, Peter May, Pawel Puczkarski, Iftikhar Ahmed, Matthias Moebius, Henrik Pettersson, Edmund Long, João Coelho, Sean E. O'Brien, Eva K. McGuire, Beatriz Mendoza Sanchez, Georg S. Duesberg, Niall McEvoy, Timothy J. Pennycook, Clive Downing, Alison Crossley, Valeria Nicolosi, and Jonathan N. Coleman. 2014. "Scalable production of large quantities of defect-free few-layer graphene by shear exfoliation in liquids." *Nature Materials* 13 (6):624–630. doi:10.1038/nmat3944.

Paton, Keith R., James Anderson, Andrew J. Pollard, and Toby Sainsbury. 2017. "Production of few-layer graphene by microfluidization." *Materials Research Express* 4 (2):025604. doi:10.1088/2053-1591/aa5b24.

Pisana, Simone, Michele Lazzeri, Cinzia Casiraghi, Kostya S. Novoselov, A. K. Geim, Andrea C. Ferrari, and Francesco Mauri. 2007. "Breakdown of the adiabatic Born–Oppenheimer approximation in graphene." *Nature Materials* 6 (3):198–201. doi:10.1038/nmat1846.

Polyzos, Ioannis, Massimiliano Bianchi, Laura Rizzi, Emmanuel N. Koukaras, John Parthenios, Konstantinos Papagelis, Roman Sordan, and Costas Galiotis. 2015. "Suspended monolayer graphene under true uniaxial deformation." *Nanoscale* 7 (30):13033–13042. doi:10.1039/C5NR03072B.

Ponomarenko, L. A., R. Yang, T. M. Mohiuddin, M. I. Katsnelson, K. S. Novoselov, S. V. Morozov, A. A. Zhukov, F. Schedin, E. W. Hill, and A. K. Geim. 2009. "Effect of a high-κ environment on charge carrier mobility in graphene." *Physical Review Letters* 102 (20):206603. doi:10.1103/PhysRevLett.102.206603.

Pontiroli, Daniele, Matteo Aramini, Mattia Gaboardi, Marcello Mazzani, Samuele Sanna, Filippo Caracciolo, Pietro Carretta, Chiara Cavallari, Stephane Rols, Roberta Tatti, Lucrezia Aversa, Roberto Verucchi, and Mauro Riccò. 2014. "Tracking the hydrogen motion in defective graphene." *The Journal of Physical Chemistry C* 118 (13):7110–7116. doi:10.1021/jp408339m.

Qian, Wen, Rui Hao, Yanglong Hou, Yuan Tian, Chengmin Shen, Hongjun Gao, and Xuelei Liang. 2009. "Solvothermal-assisted exfoliation process to produce graphene with high yield and high quality." *Nano Research* 2 (9):706–712. doi:10.1007/s12274-009-9074-z.

Rao, Rahul, Ramakrishna Podila, Ryuichi Tsuchikawa, Jyoti Katoch, Derek Tishler, Apparao M. Rao, and Masa Ishigami. 2011. "Effects of layer stacking on the combination Raman modes in graphene." *ACS Nano* 5 (3):1594–1599. doi:10.1021/nn1031017.

Reina, Alfonso, Xiaoting Jia, John Ho, Daniel Nezich, Hyungbin Son, Vladimir Bulovic, Mildred S. Dresselhaus, and Jing Kong. 2009. "Large area, few-layer graphene films on arbitrary substrates by chemical vapor deposition." *Nano Letters* 9 (1):30–35. doi:10.1021/nl801827v.

Reina, Giacomo, José Miguel González-Domínguez, Alejandro Criado, Ester Vázquez, Alberto Bianco, and Maurizio Prato. 2017. "Promises, facts and challenges for graphene in biomedical applications." *Chemical Society Reviews* 46 (15):4400–4416. doi:10.1039/C7CS00363C.

Ritter, Kyle A., and Joseph W. Lyding. 2009. "The influence of edge structure on the electronic properties of graphene quantum dots and nanoribbons." *Nature Materials* 8 (3):235–242. doi:10.1038/nmat2378.

Ruan, Gedeng, Zhengzong Sun, Zhiwei Peng, and James M. Tour. 2011. "Growth of graphene from food, insects, and waste." *ACS Nano* 5 (9):7601–7607.

Ruiz-Vargas, Carlos S., Houlong L. Zhuang, Pinshane Y. Huang, Arend M. van der Zande, Shivank Garg, Paul L. McEuen, David A. Muller, Richard G. Hennig, and Jiwoong Park. 2011. "Softened elastic response and unzipping in chemical vapor deposition graphene membranes." *Nano Letters* 11 (6):2259–2263. doi:10.1021/nl200429f.

Saito, R., M. Hofmann, G. Dresselhaus, A. Jorio, and M. S. Dresselhaus. 2011. "Raman spectroscopy of graphene and carbon nanotubes." *Advances in Physics* 60 (3):413–550. doi:10.1080/00018732.2011.582251.

Saito, Riichiro, Mitsutaka Fujita, G. Dresselhaus, and M. S. Dresselhaus. 1992. "Electronic structure of graphene tubules based on C_{60}." *Physical Review B* 46 (3):1804–1811. doi:10.1103/PhysRevB.46.1804.

Schafhaeutl, C. 1840. "Ueber die Verbindungen des Kohlenstoffes mit Silicium, Eisen und anderen Metallen, welche die verschiedenen Gallungen von Roheisen, Stahl und Schmiedeeisen bilden." *Journal für Praktische Chemie* 21 (1):129–157. doi:10.1002/prac.18400210117.

Schniepp, Hannes C., Je-Luen Li, Michael J. McAllister, Hiroaki Sai, Margarita Herrera-Alonso, Douglas H. Adamson, Robert K. Prud'homme, Roberto Car, Dudley A. Saville, and Ilhan A. Aksay. 2006. "Functionalized single graphene sheets derived from splitting graphite oxide." *The Journal of Physical Chemistry B* 110 (17):8535–8539. doi:10.1021/jp060936f.

Schwierz, Frank. 2010. "Graphene transistors." *Nature Nanotechnology* 5 (7):487–496. doi:10.1038/nnano.2010.89.

Sengupta, Iman, Samarshi Chakraborty, Monikangkana Talukdar, Surjya K. Pal, and Sudipto Chakraborty. 2018. "Thermal reduction of graphene oxide: How temperature influences purity." *Journal of Materials Research* 33 (23):4113–4122. doi:10.1557/jmr.2018.338.

Sha, Z. D., Q. Wan, Q. X. Pei, S. S. Quek, Z. S. Liu, Y. W. Zhang, and V. B. Shenoy. 2014. "On the failure load and mechanism of polycrystalline graphene by nanoindentation." *Scientific Reports* 4 (1):7437. doi:10.1038/srep07437.

Sharma, Subash, Golap Kalita, Ryo Hirano, Sachin M. Shinde, Remi Papon, Hajime Ohtani, and Masaki Tanemura. 2014. "Synthesis of graphene crystals from solid waste plastic by chemical vapor deposition." *Carbon* 72:66–73. doi:10.1016/j.carbon.2014.01.051.

Shen, Tian, Wei Wu, Qingkai Yu, Curt A. Richter, Randolph Elmquist, David Newell, and Yong P. Chen. 2011. "Quantum Hall effect on centimeter scale chemical vapor deposited graphene films." *Applied Physics Letters* 99 (23):232110. doi:10.1063/1.3663972.

Shen, Zhigang, Jinzhi Li, Min Yi, Xiaojing Zhang, and Shulin Ma. 2011. "Preparation of graphene by jet cavitation." *Nanotechnology* 22 (36):365306. doi:10.1088/0957-4484/22/36/365306.

Silva, Diego L., João Luiz E. Campos, Thales F. D. Fernandes, Jeronimo N. Rocha, Lucas R. P. Machado, Eder M. Soares, Douglas R. Miquita, Hudson Miranda, Cassiano Rabelo, Omar P. Vilela Neto, Ado Jorio, and Luiz Gustavo Cançado. 2020. "Raman spectroscopy analysis of number of layers in mass-produced graphene flakes." *Carbon* 161:181–189. doi:10.1016/j.carbon.2020.01.050.

Siochi, Emilie J. 2014. "Graphene in the sky and beyond." *Nature Nanotechnology* 9 (10):745–747. doi:10.1038/nnano.2014.231.

Smith, Andrew T., Anna Marie LaChance, Songshan Zeng, Bin Liu, and Luyi Sun. 2019. "Synthesis, properties, and applications of graphene oxide/reduced graphene oxide and their nanocomposites." *Nano Materials Science* 1 (1):31–47. doi:10.1016/j.nanoms.2019.02.004.

Sokolov, Denis A., Kristin R. Shepperd, and Thomas M. Orlando. 2010. "Formation of graphene features from direct laser-induced reduction of graphite oxide." *The Journal of Physical Chemistry Letters* 1 (18):2633–2636. doi:10.1021/jz100790y.

Stankovich, Sasha, Dmitriy A. Dikin, Richard D. Piner, Kevin A. Kohlhaas, Alfred Kleinhammes, Yuanyuan Jia, Yue Wu, SonBinh T. Nguyen, and Rodney S. Ruoff. 2007. "Synthesis of graphene-based nanosheets via chemical reduction of exfoliated graphite oxide." *Carbon* 45 (7):1558–1565. doi:10.1016/j.carbon.2007.02.034.

Stöhr, Rainer J., Roman Kolesov, Kangwei Xia, Rolf Reuter, Jan Meijer, Gennady Logvenov, and Jörg Wrachtrup. 2012. "Super-resolution fluorescence quenching microscopy of graphene." *ACS Nano* 6 (10):9175–9181. doi:10.1021/nn303510p.

Suemitsu, Maki, Sai Jiao, Hirokazu Fukidome, Yasunori Tateno, Isao Makabe, and Takashi Nakabayashi. 2014. "Epitaxial graphene formation on 3C-SiC/Si thin films." *Journal of Physics D: Applied Physics* 47 (9):094016. doi:10.1088/0022-3727/47/9/094016.

Suslick, Kenneth S., and David J. Flannigan. 2008. "Inside a collapsing bubble: sonoluminescence and the conditions during cavitation." *Annual Review of Physical Chemistry* 59 (1):659–683. doi:10.1146/annurev.physchem.59.032607.093739.

Swathi, R. S., and K. L. Sebastian. 2008. "Resonance energy transfer from a dye molecule to graphene." *The Journal of Chemical Physics* 129 (5):054703. doi:10.1063/1.2956498.

Tan, P. H., W. P. Han, W. J. Zhao, Z. H. Wu, K. Chang, H. Wang, Y. F. Wang, N. Bonini, N. Marzari, N. Pugno, G. Savini, A. Lombardo, and A. C. Ferrari. 2012. "The shear mode of multilayer graphene." *Nature Materials* 11 (4):294–300. doi:10.1038/nmat3245.

Tedesco, J. L., B. L. VanMil, R. L. Myers-Ward, J. M. McCrate, S. A. Kitt, P. M. Campbell, G. G. Jernigan, J. C. Culbertson, C. R. Eddy, and D. K. Gaskill. 2009. "Hall effect mobility of epitaxial graphene grown on silicon carbide." *Applied Physics Letters* 95 (12):122102. doi:10.1063/1.3224887.

Tuinstra, F., and J. L. Koenig. 1970. "Raman spectrum of graphite." *The Journal of Chemical Physics* 53 (3):1126–1130. doi:10.1063/1.1674108.

Turchanin, Andrey, André Beyer, Christoph T. Nottbohm, Xianghui Zhang, Rainer Stosch, Alla Sologubenko, Joachim Mayer, Peter Hinze, Thomas Weimann, and Armin Gölzhäuser. 2009. "One nanometer thin carbon nanosheets with tunable conductivity and stiffness." *Advanced Materials* 21 (12):1233–1237. doi:10.1002/adma.200803078.

Uder, Bernd, Haibin Gao, Peter Kunnas, Niels de Jonge, and Uwe Hartmann. 2018. "Low-force spectroscopy on graphene membranes by scanning tunneling microscopy." *Nanoscale* 10 (4):2148–2153. doi:10.1039/C7NR07300C.

Varrla, Eswaraiah, Keith R. Paton, Claudia Backes, Andrew Harvey, Ronan J. Smith, Joe McCauley, and Jonathan N. Coleman. 2014. "Turbulence-assisted shear exfoliation of graphene using household detergent and a kitchen blender." *Nanoscale* 6 (20):11810–11819. doi:10.1039/C4NR03560G.

Vecchio, Carmelo, Sushant Sonde, Corrado Bongiorno, Martin Rambach, Rositza Yakimova, Vito Raineri, and Filippo Giannazzo. 2011. "Nanoscale structural characterization of epitaxial graphene grown on off-axis 4H-SiC (0001)." *Nanoscale Research Letters* 6 (1):269. doi:10.1186/1556-276X-6-269.

Vlassak, J. J., and W. D. Nix. 1992. "A new bulge test technique for the determination of Young's modulus and Poisson's ratio of thin films." *Journal of Materials Research* 7 (12):3242–3249. doi:10.1557/JMR.1992.3242.

Wang, Bin, Da Luo, Zhancheng Li, Youngwoo Kwon, Meihui Wang, Min Goo, Sunghwan Jin, Ming Huang, Yongtao Shen, Haofei Shi, Feng Ding, and Rodney S. Ruoff. 2018. "Camphor-enabled transfer and mechanical testing of centimeter-scale ultrathin films." *Advanced Materials* 30 (28):1800888. doi:10.1002/adma.201800888.

Wang, Guorui, Zhaohe Dai, Yanlei Wang, PingHeng Tan, Luqi Liu, Zhiping Xu, Yueguang Wei, Rui Huang, and Zhong Zhang. 2017. "Measuring interlayer shear stress in bilayer graphene." *Physical Review Letters* 119 (3):036101. doi:10.1103/PhysRevLett.119.036101.

Wang, Hailiang, Joshua Tucker Robinson, Xiaolin Li, and Hongjie Dai. 2009. "Solvothermal reduction of chemically exfoliated graphene sheets." *Journal of the American Chemical Society* 131 (29):9910–9911. doi:10.1021/ja904251p.

Wang, Junzhong, Kiran Kumar Manga, Qiaoliang Bao, and Kian Ping Loh. 2011. "High-yield synthesis of few-layer graphene flakes through electrochemical expansion of graphite in propylene carbonate electrolyte." *Journal of the American Chemical Society* 133 (23):8888–8891. doi:10.1021/ja203725d.

Wang, Xu, Weimiao Wang, Yang Liu, Mengmeng Ren, Huining Xiao, and Xiangyang Liu. 2016. "Characterization of conformation and locations of C–F bonds in graphene derivative by polarized ATR-FTIR." *Analytical Chemistry* 88 (7):3926–3934. doi:10.1021/acs.analchem.6b00115.

Wehling, T. O., K. S. Novoselov, S. V. Morozov, E. E. Vdovin, M. I. Katsnelson, A. K. Geim, and A. I. Lichtenstein. 2008. "Molecular doping of graphene." *Nano Letters* 8 (1):173–177. doi:10.1021/nl072364w.

Williams, Graeme, Brian Seger, and Prashant V. Kamat. 2008. "TiO_2-graphene nanocomposites. UV-assisted photocatalytic reduction of graphene oxide." *ACS Nano* 2 (7):1487–1491. doi:10.1021/nn800251f.

Witomska, Samanta, Tim Leydecker, Artur Ciesielski, and Paolo Samorì. 2019. "Production and patterning of liquid phase–exfoliated 2D sheets for applications in optoelectronics." *Advanced Functional Materials* 29 (22):1901126. doi:10.1002/adfm.201901126.

Wu, Hang, Weifeng Zhao, Huawen Hu, and Guohua Chen. 2011. "One-step in situball milling synthesis of polymer-functionalized graphene nanocomposites." *Journal of Materials Chemistry* 21 (24):8626–8632. doi:10.1039/C1JM10819K.

Xu, Xiaozhi, Zhihong Zhang, Jichen Dong, Ding Yi, Jingjing Niu, Muhong Wu, Li Lin, Rongkang Yin, Mingqiang Li, Jingyuan Zhou, Shaoxin Wang, Junliang Sun, Xiaojie Duan, Peng Gao, Ying Jiang, Xiaosong Wu, Hailin Peng, Rodney S. Ruoff, Zhongfan Liu, Dapeng Yu, Enge Wang, Feng Ding, and Kaihui Liu. 2017. "Ultrafast epitaxial growth of metre-sized single-crystal graphene on industrial Cu foil." *Science Bulletin* 62 (15):1074–1080. doi:10.1016/j.scib.2017.07.005.

Yan, Lu, Mimi Lin, Chao Zeng, Zhi Chen, Shu Zhang, Xinmei Zhao, Aiguo Wu, Yaping Wang, Liming Dai, Jia Qu, Mingming Guo, and Yong Liu. 2012. "Electroactive and biocompatible hydroxyl- functionalized graphene by ball milling." *Journal of Materials Chemistry* 22 (17):8367–8371. doi:10.1039/C2JM30961K.

Yan, Ning, Giovanna Buonocore, Marino Lavorgna, Saulius Kaciulis, Santosh Kiran Balijepalli, Yanhu Zhan, Hesheng Xia, and Luigi Ambrosio. 2014. "The role of reduced graphene oxide on chemical, mechanical and barrier properties of natural rubber composites." *Composites Science and Technology* 102:74–81. doi:10.1016/j.compscitech.2014.07.021.

Yang, Haibin, Hongzhi Cui, Waiching Tang, Zongjin Li, Ningxu Han, and Feng Xing. 2017. "A critical review on research progress of graphene/cement based composites." *Composites Part A: Applied Science and Manufacturing* 102:273–296. doi:10.1016/j.compositesa.2017.07.019.

Yang, Xi, Xiang Yu, and Xin Liu. 2018. "Obtaining a sustainable competitive advantage from patent information: A patent analysis of the graphene industry." *Sustainability* 10 (12):4800.

Yang, Xiaoyin, Xi Dou, Ali Rouhanipour, Linjie Zhi, Hans Joachim Räder, and Klaus Müllen. 2008. "Two-dimensional graphene nanoribbons." *Journal of the American Chemical Society* 130 (13):4216–4217. doi:10.1021/ja710234t.

Yao, Heming, Ya-Ping Hsieh, Jing Kong, and Mario Hofmann. 2020. "Modelling electrical conduction in nanostructure assemblies through complex networks." *Nature Materials* 19 (7):745–751. doi:10.1038/s41563-020-0664-1.

You, Yu-Meng, Zhen-Hua Ni, Ting Yu, and Ze-Xiang Shen. 2008. "Edge chirality determination of graphene by Raman spectroscopy." *Applied Physics Letters* 93 (16):163112. doi:10.1063/1.3005599.

Zhang, Chi, Junyang Tan, Yikun Pan, Xingke Cai, Xiaolong Zou, Hui-Ming Cheng, and Bilu Liu. 2019. "Mass production of 2D materials by intermediate-assisted grinding exfoliation." *National Science Review* 7 (2):324–332. doi:10.1093/nsr/nwz156.

Zhang, Cui, Daniel M. Dabbs, Li-Min Liu, Ilhan A. Aksay, Roberto Car, and Annabella Selloni. 2015. "Combined effects of functional groups, lattice defects, and edges in the infrared spectra of graphene oxide." *The Journal of Physical Chemistry C* 119 (32):18167–18176. doi:10.1021/acs.jpcc.5b02727.

Zhang, Peng, Lulu Ma, Feifei Fan, Zhi Zeng, Cheng Peng, Phillip E. Loya, Zheng Liu, Yongji Gong, Jiangnan Zhang, Xingxiang Zhang, Pulickel M. Ajayan, Ting Zhu, and Jun Lou. 2014. "Fracture toughness of graphene." *Nature Communications* 5 (1):3782. doi:10.1038/ncomms4782.

Zhang, Yuanbo, Yan-Wen Tan, Horst L. Stormer, and Philip Kim. 2005. "Experimental observation of the quantum Hall effect and Berry's phase in graphene." *Nature* 438 (7065):201–204. doi:10.1038/nature04235.

Zhang, Yuanbo, Tsung-Ta Tang, Caglar Girit, Zhao Hao, Michael C. Martin, Alex Zettl, Michael F. Crommie, Y. Ron Shen, and Feng Wang. 2009. "Direct observation of a widely tunable bandgap in bilayer graphene." *Nature* 459 (7248):820–823. doi:10.1038/nature08105.

Zhao, Weifeng, Ming Fang, Furong Wu, Hang Wu, Liwei Wang, and Guohua Chen. 2010. "Preparation of graphene by exfoliation of graphite using wet ball milling." *Journal of Materials Chemistry* 20 (28):5817–5819. doi:10.1039/C0JM01354D.

Zheng, Zhiren, Qiong Ma, Zhen Bi, Sergio de la Barrera, Ming-Hao Liu, Nannan Mao, Yang Zhang, Natasha Kiper, Kenji Watanabe, Takashi Taniguchi, Jing Kong, William A. Tisdale, Ray Ashoori, Nuh Gedik, Liang Fu, Su-Yang Xu, and Pablo Jarillo-Herrero. 2020. "Unconventional ferroelectricity in moiré heterostructures." *Nature* 588 (7836):71–76. doi:10.1038/s41586-020-2970-9.

Zhong, Hua, Zhiyong Zhang, Haitao Xu, Chenguang Qiu, and Lian-Mao Peng. 2015. "Comparison of mobility extraction methods based on field-effect measurements for graphene." *AIP Advances* 5 (5):057136. doi:10.1063/1.4921400.

Zhong, Yu Lin, Zhiming Tian, George P. Simon, and Dan Li. 2015. "Scalable production of graphene via wet chemistry: Progress and challenges." *Materials Today* 18 (2):73–78. doi:10.1016/j.mattod.2014.08.019.

Zhou, Yong, Qiaoliang Bao, Binni Varghese, Lena Ai Ling Tang, Chow Khim Tan, Chorng-Haur Sow, and Kian Ping Loh. 2010. "Microstructuring of graphene oxide nanosheets using direct laser writing." *Advanced Materials* 22 (1):67–71. doi:10.1002/adma.200901942.

Zhu, Shou-En, Shengjun Yuan, and G. C. A. M. Janssen. 2014. "Optical transmittance of multilayer graphene." *EPL (Europhysics Letters)* 108 (1):17007. doi:10.1209/0295-5075/108/17007.

Zhu, Yanwu, Shanthi Murali, Meryl D. Stoller, Aruna Velamakanni, Richard D. Piner, and Rodney S. Ruoff. 2010. "Microwave assisted exfoliation and reduction of graphite oxide for ultracapacitors." *Carbon* 48 (7):2118–2122. doi:10.1016/j.carbon.2010.02.001.

3 Synthesis and Characterization of Graphene from Non-Conventional Precursors

Amrita Roy, Titash Mondal, and Kinsuk Naskar
IIT Kharagpur

Anil K. Bhowmick
The University of Houston

CONTENTS

DOI: 10.1201/9781003200444-3

3.1 INTRODUCTION

The science of materials takes a turn as they reach down to a size of 100 nm or lower and thermodynamics of a material change when a material becomes two-dimensional from conventional three dimensions. By definition, a nanomaterial is a material with at least one dimension less than or equal to a hundred nanometers, but curiosity has led scientists toward further down to a level where materials are two-dimensional as graphene, one-dimensional as nanotubes or nanowires, or even zero-dimensional as nanodots or quantum dots (Maynard 2011). At this level, materials behave differently than at the macro-level (Lecloux 2015). For an ideal two-dimensional flat surface, the radius of curvature is essentially infinite, and hence, the surface energy is also ideally infinite. The cohesive energy of the atoms on a flat surface is equal for all the atoms and also very different than that in bulk materials. The reactivity of such flat surfaces is also different, and the surface bonds often behave as 'dangling bonds'. These bonds can be referred to as 'relaxed' or 'free' to react and they lead to rearrangements or reconstruction on the surface (McArthur 2006). This reactivity often causes the flat surfaces to lose their very flat nature and they tend to form connections to similar surfaces through van der Waals interaction. Graphene is a typical example of a two-dimensional flat surface and the three-dimensional form of graphene, in which many flat graphene surfaces are joined together by weak van der Waals force, is known as 'graphite' (Novoselov 2011).

The lamellar structure of graphite was known since the middle of the 19th century, but the term 'graphene' was introduced in 1986, ending with 'ene' as in polycyclic aromatic hydrocarbon. Separation of a single graphene layer was not possible until 2004 when Geim and Novoselov (2009) extracted graphene through a micromechanical cleavage technique with a scotch tape and transferred it on a silicon wafer. Since its discovery, graphene is defined as one atom thick single layer of a hexagonally packed honeycomb-shaped lattice with carbon atoms at each corner of the hexagon (Novoselov et al. 2007, 2012).

All the synthetic routes to any nanomaterial can substantially be categorized into two fundamental approaches, top-down and bottom-up. Any technique that involves controlled breakdown of bulk materials down to a nanoscale is considered the top-down approach. This approach involves micromechanical or sometimes chemical cleavage (Castro Neto et al. 2009, Kostarelos and Novoselov 2014). Classic examples of top-down techniques are dry and wet ball milling, ultra-sonication, or electrochemical reactions (Choi et al. 2010a). Bottom-up techniques are rather intricate and they involve atom by atom fabrication of a nanomaterial. Chemical vapor deposition (CVD) is an archetypal example of a bottom-up technique (Mattevi, Kim, and Chhowalla 2011). Other typical bottom-up methods are electrodeposition, sol-gel method, hydrothermal, solvothermal, reverse micelle route, pyrolysis, etc. (Rao et al. 2009). The top-down and bottom-up approaches toward graphene synthesis are illustrated in Figure 3.1.

In a typical CVD process, a gaseous precursor or a mixture of precursor gases is passed through a chamber at very high temperature and high vacuum over a substrate, and the material is fabricated atom by atom over the substrate (Muñoz and Gómez-Aleixandre 2013a, Li, Colombo, and Ruoff 2016). In the consequent steps,

Bulk Graphite	Exfoliated Graphite	Graphene	Clusters/condensed product	Atoms

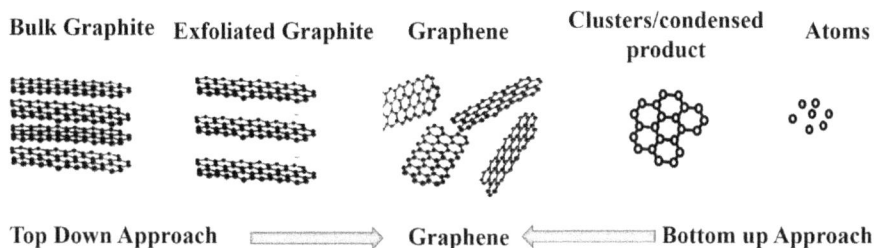

Top Down Approach ⟹ Graphene ⟸ Bottom up Approach

FIGURE 3.1   Different approaches toward the synthesis of graphene.

the nanomaterial is isolated and transferred to desired media (Fang et al. 2015, Deokar et al. 2015). The major advantage of this process is that the nucleation of the nanomaterial can be controlled and customized (Habib et al. 2018). The nanomaterial can be grown in an epitaxial manner in a single crystal format and the materials grown through CVD possess maximum purity (Yan, Peng, and Tour 2014).

A CVD process is essentially a system of high-temperature pyrolysis techniques in a controlled atmosphere with widely varying sets of conditions, instrumentations, and precursors (Xin and Li 2018). Though in a conventional CVD technique, high temperature and high vacuum are the prime reaction conditions, in several other methods, atmospheric pressure or low pressure, lower temperature ranges, and varying gas flow are employed (Carlsson and Martin 2010). Techniques have been developed to introduce plasma, microwave, and catalysts to tailor the synthetic procedures that satisfy specific desires with the product (Yan et al. 2013).

The traditional CVD synthesis of graphene employs H_2/CH_4 gas mixture as precursor and usual substrates are Ni or Cu (Muñoz and Gómez-Aleixandre 2013). The grain sizes and grain boundaries of the substrates are often modified to tailor the number of layers in graphene (Ning et al. 2017). Though these hydrocarbon precursors were proficient in producing high-quality graphene, the questions of sustainability and environment compatibility remained. As these precursors were petroleum-derived and not abundant in nature, large-scale production for widespread industrial applications was not very feasible. A search for alternative sources was necessary, and over the decade, many potential alternative carbon sources have surfaced and proved their potential in producing competing quality graphene. Several non-conventional sources of solid carbon have paved their way into the precinct of graphene synthesis, replacing the potentially hazardous flammable hydrocarbon gas mixtures in the last decade (Raghavan, Thangavel, and Venugopal 2017). Many of these non-conventional sources do not require a high vacuum, or even in some cases a very high temperature. These precursors often do not need catalysts or hazardous reducing gas like hydrogen (Yan et al. 2020). Precursors like wheat straw, rice husk, or reed grass are abundant in nature; these sources are green and reproducible at ease. Many waste materials, biowaste and industrial waste, waste cooking oil, plastics, and coal tar have been successfully converted into graphene, which is a probable answer to the world's disquieting waste-management issue. Common polymers are competent sources of solid carbon which under high vacuum and at very high temperatures can form graphene (Tiwari et al. 2012). Some edible materials such as cookies and

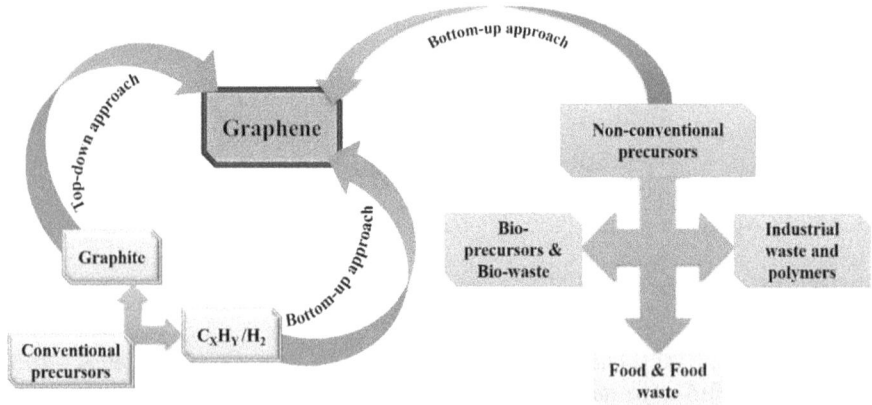

FIGURE 3.2   Different categories of precursors for the synthesis of graphene.

chocolates are also rich sources of solid carbon for the synthesis of graphene. The use of grass, insects, and biological wastes such as dog feces has given a dramatic turn to graphene synthesis (Ruan et al. 2011). This chapter emphasizes the non-conventional precursors for graphene synthesis as illustrated in Figure 3.2.

Doping in graphene became a very attractive trajectory to venture into, as micro and nanoelectromechanical systems aka 'mems and nems' made their debut in the field of electronics. Graphene itself is a zero-band gap semiconductor with its valence and conduction bands meeting at the Fermi level (Guo et al. 2010). With the 'zero band gap', graphene can easily be doped with charge carriers, electrons, or holes without any notable loss in charge carrier mobility (Nicolle et al. 2011). Experiments with the adsorption of NH_3 have resulted in the introduction of electrons (Guo et al. 2010). Adsorption of H_2O or NO_2 created holes in the graphene lattice (Gierz et al. 2008). A new trend has arrived for developing synthetic routes for graphene that is already doped during the synthesis, and many non-conventional sources, such as coir pith, liquid organics, and certain polymers, have fruitfully served the purpose (Karuppannan et al. 2019, Mondal, Bhowmick, and Krishnamoorti 2015).

3.2   SYNTHESIS OF GRAPHENE

3.2.1   SYNTHESIS OF GRAPHENE FROM BIOWASTE AND OTHER BIOMATERIALS

Biowaste has been a favored choice as a rich and abundant carbon source since extreme exploitation of fossil fuels started threatening the future of human civilization and has been widely studied for the production of alternate fuels. There has always been a scope for biowaste to be utilized for the synthesis of high-performance carbon forms like activated carbon or graphene or carbon nanotubes to minimize the use of petroleum-derived materials (Choi et al. 2010b).

Ruan et al. found some unique biowaste to work with, dog feces, cockroach leg, and grass. In the first step, dog feces and grass were dehydrated in a vacuum oven at 102 Torr pressure and 65°C temperature for 10 hours, and in the next step, the materials were kept on a copper (Cu) foil inside a quartz boat and annealed at 1,050°C for 15 minutes (Ruan et al. 2011). A gas mixture containing argon (Ar) and hydrogen (H_2) was passed continuously at the rate of 500 cm³ STP min⁻¹ and 100 cm³ STP min⁻¹ respectively. A constant pressure of 9.3 Torr was maintained throughout the process. Fast cooling of the sample was done under Ar/H_2 flow and then a 100-nm-thick polymethyl methacrylate (PMMA) film was deposited underneath the Cu foil and the Cu was etched away with an acidic $CuSO_4$ solution. A schematic diagram of their work is shown in Figure 3.3.

In another study, mango peel was used as the raw material for synthesizing graphene in the plasma-enhanced chemical vapor deposition (PECVD) method (Shah et al. 2018). Mango peel contains chemical elements like cellulose, hemicellulose, lignin, and pectin, some of which have been used in high-temperature pyrolysis processes to obtain activated carbons or graphitic structures. Mango peels were dried and ground to a particle size of about 1 mm and 10 g of it was evenly distributed on Cu substrate and inserted into the vacuum reactor having 0.2 Torr. pressure and

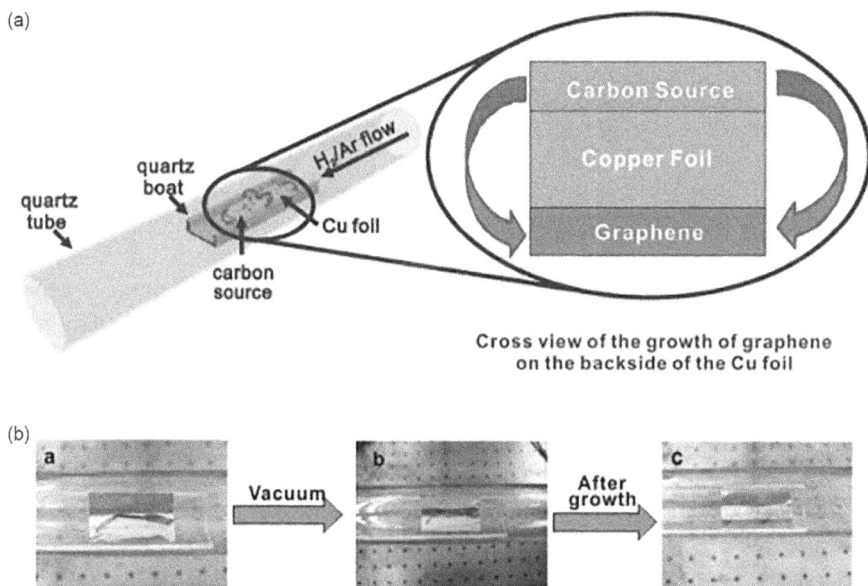

FIGURE 3.3   (a) Diagram of the experimental apparatus for the growth of graphene from food, insects, or waste in a tube furnace. On the top left, the Cu foil with the carbon source contained in a quartz boat is placed at the hot zone of a tube furnace. The growth is performed at 1,050°C under low pressure with H_2/Ar gas flow. On the right is a cross-view that represents the formation of pristine graphene on the backside of the Cu substrate. (b) Growth of graphene from a cockroach leg. (a) One roach leg on top of the Cu foil. (b) Roach leg under vacuum. (c) Residue from the roach leg after annealing at 1,050°C for 15 minutes. The pristine graphene grew on the bottom side of the Cu film (not shown). (Adapted with permission from Ruan, Gedeng, Zhengzong Sun, Zhiwei Peng, and James M. Tour. 2011. "Growth of Graphene from Food, Insects, and Waste." *ACS Nano* 5 (9): 7601–7. Copyright © 2011 American Chemical Society.)

750°C temperature with a ramp rate of 25°C min^{-1} along with a gas flow of 10% H_2/ Ar at a flow rate of 50 sccm. A temperature of 750°C was maintained for 15 minutes and was exposed to a 300 W plasma for 5, 10, 30, and 60 minutes. Samples derived after 5 and 10 minutes of plasma exposure were graphenes, but not monolayer. After 30 minutes of plasma exposure, the formation of single-layer graphene started and 60 minutes of plasma exposure yielded ideally monolayer graphene.

Reed is a tall plant from the grass family which grows rapidly in wet areas and thus can serve well as a renewable source of solid carbon. Reed was cut, dried, and milled to a fine powder with 74-µm particle size, and was treated with a solution of sorbitol and heated in the presence of copper (Cu), nickel (Ni), and magnesium (Mg) salts which acted as a catalyst for the graphitization reaction (Rahbar et al. 2019). The mass ratio of these inorganic salts Ni-Cu/Mg was 40-10/50. The catalyst-soaked reed was dried and subjected to in situ catalytic chemical vapor deposition (CCVD) at a temperature of 800°C attained by a ramp of 10°C min^{-1}, held at the constant temperature for an hour with a steady flow of H_2.

In one of the several graphene synthesis routes reported by Kalita et al., botanically derived camphor was described as a potential raw material for the synthesis of graphene (Kalita et al. 2010). The authors used layered substrates for the nucleation, Si wafers coated with 200-nm-thick SiO_2 and 500-nm-thick polycrystalline Nickel, and a setup of two adjoining furnaces was engaged for the reaction. In the first furnace, camphor was evaporated at 200°C temperature and the vapor was carried by a flow of Ar to the second furnace where pyrolysis took place at 800°C temperature. The products were few layers to multilayer graphene with the number of layers ranging from 4 to 13.

In a unique study, a volatile extract from *Melaleuca alternifolia*, commonly recognized as the tea tree plant, was used in a radiofrequency capacitively coupled PECVD method (Jacob et al. 2015). The extract was obtained by distillation of the tea-tree plant leaves which contain a complex mixture of aromatic hydrocarbons like terpinenes and cineole. This volatile extract was vaporized at 64°C temperature and 15.7 Torr pressure and fed into the reactor; it started forming graphene within few seconds of plasma exposure. In this process, a very large area of graphene was synthesized within a few minutes.

Rice husk was chemically activated by the treatment with KOH and subsequently annealed at 1,123 K temperature for 2 hours, and single- to multilayer large corrugated graphene sheets were produced (Muramatsu et al. 2014). Similarly, dried wheat straw was treated with KOH and then pyrolysis was carried out for 3 hours at 800°C temperature while N_2 was passed at 150 cm^3 min^{-1} for chemical activation. After the removal of excess KOH by washing the pyrolyzed product with 0.5 M HCl solution, thermal treatment at 2,600°C was done for 5 minutes in a graphite furnace under Ar atmosphere. In another study, dried camphor leaves were pyrolyzed at 1,200°C temperature for 4 minutes to obtain a porous graphene material (Shams et al. 2015).

Lignin was refined from sugarcane bagasse and pyrolyzed in a tube furnace at a very high temperature up to 1,200°C under an N_2 atmosphere for 3 hours. The pyrolyzed product was then subjected to hydrothermal carbonization and the resultant product was graphene nanocrystal. These nanocrystals contained 3–5 layers of graphene on average (Tang et al. 2018).

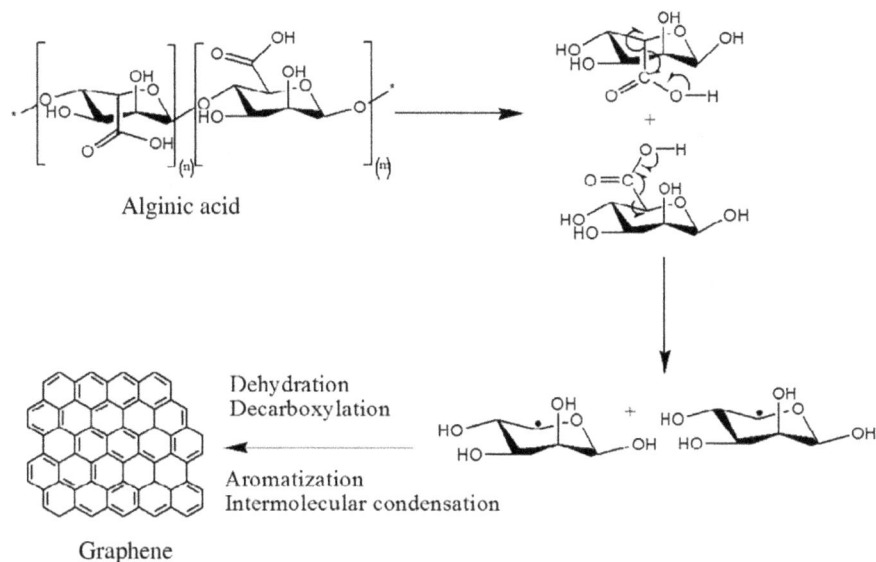

FIGURE 3.4 Plausible pyrolysis reaction mechanism for the synthesis of graphene from alginic acid. (Adapted with permission from Roy, Amrita, Saptarshi Kar, Ranjan Ghosal, Kinsuk Naskar, and Anil K. Bhowmick. 2021. "Facile Synthesis and Characterization of Few-Layer Multifunctional Graphene from Sustainable Precursors by Controlled Pyrolysis, Understanding of the Graphitization Pathway, and Its Potential Application in Polymer Nanocomposites." *ACS Omega* 6 (3): 1809–22. Copyright © 2021 American Chemical Society.)

In recent research, biologically derived chemical precursors, tannic acid and alginic acid, along with dried green tea leaves were taken and pyrolyzed under N_2 atmosphere at 1,100°C temperature for 3 hours. All three precursors generated few layers of graphene sheets, with a few micron lateral sizes. Tannic acid and green tea yielded trilayer graphene on an average and the graphene obtained from alginic acid was mostly bilayer in nature (Roy et al. 2021). A plausible mechanism behind the synthesis of graphene from alginic acid is given in Figure 3.4.

3.2.2 Synthesis of Graphene from Food and Food Waste

Waste chicken fat was put through a low-pressure chemical vapor deposition (LPCVD) process and re-solidified on a Cu substrate as graphene (Rosmi et al. 2016). Chicken fat and skin were heated at 200°C temperature to separate the oil and this extract was laid on a 20-μm-thick previously annealed Cu substrate. The pressure inside the reactor was pumped down to 2 Pa for 10 minutes followed by raising the temperature to 1,080°C in 120 minutes while pure H_2 was passed at a flow rate of 100 sccm. Ar/H_2 flow with a molar ratio of 98:2 was injected and the sample placed on a magnetic boat was inserted into the reaction chamber. The process was continued at 1,080°C temperature and 150 Pa pressure for 60 minutes. The graphene so obtained was transferred onto a PMMA film by a standard delamination process. The polycrystalline Cu substrate used in the study was annealed and re-solidified at 1,100°C under H_2 atmosphere to remove the oxide residues and to increase the grain size, which played a key role in the formation of monolayer graphene. The smoothness of the Cu substrate was reflected in the smoothness of the graphene sheets which were surprisingly uniform and less wrinkled compared to many other regular graphene samples.

In the same work by Ruan et al., food items like cookies and chocolate were also subjected to a similar CVD method devoid of the dehydration step in the process (Ruan et al. 2011). In the gas mixture employed in this work, H_2 acts as a reducing agent for the elimination of the carbon residue and Ar works as a carrier. The graphene sample obtained from the cookie was analyzed under a high-resolution transmission electron microscope and was essentially found to be monolayer graphene.

Graphene was grown on Ni substrate from refined cooking palm oil in a single-zone tube furnace at 800°C, 900°C, and 1,000°C with a constant 200 sccm flow of Ar (Rahman, Mahmood, and Hashim 2014).

Employing waste palm-based cooking oil was also a fecund idea that yielded graphene nanoplatelets (Azam, Mudtalib, and Seman 2018). Palm oil commonly contains mostly triglycerides, fatty acid esters, glycerol, and traces of mono- and diglycerides. During cooking, lengths of the carbon chains in the complex fatty acid esters reduce and changes in the chemical structure eventuate from the water released from food and exposure to oxygen and heat. This waste cooking oil was subjected to a CVD process at a pressure range of 50 m Torr to 1 Torr with continuous Ar flow at three different temperatures, 850°C, 875°C, and 900°C, for 5 minutes. The resultants were few layers of graphene nanoplatelets with smooth and semi-transparent surfaces.

Edible sugar and gelatin are two cheap, reliable, and green precursors for graphene synthesis (Ruiz-Hitzky et al. 2011). The chief chemical component of edible sugar is sucrose, which undergoes graphitization at 750°C temperature. Sucrose provides the advantage for easy handling and processing, and for this reason, it has been an ideal choice for the synthesis of graphene-based hybrid materials. The sugar solution was mixed with river sand and pyrolyzed at 750°C under N_2 atmosphere to prepare graphene-sand composite which exhibited excellent performance in water purification (Gupta et al. 2012a). Sucrose, in the form of molasses, was mixed with carbon black and pyrolyzed at 750°C temperature in a tube furnace under continuous nitrogen flow. The resultant graphene-carbon black hybrid was a high-performance carbonaceous nanomaterial suitable for versatile applications, such as in energy storage, or, in rubber nanocomposites, as a reinforcing filler (Roy et al. 2022). A scheme for the synthesis of graphene-carbon black hybrid from the sucrose-carbon mixture is given in Figure 3.5.

3.2.3 SYNTHESIS OF GRAPHENE FROM INDUSTRIAL WASTE

In their pioneering work, Ruan et al. (2011) also used waste plastic Petri-dishes along with food and insect, a solid carbon source, for the synthesis of graphene and thus

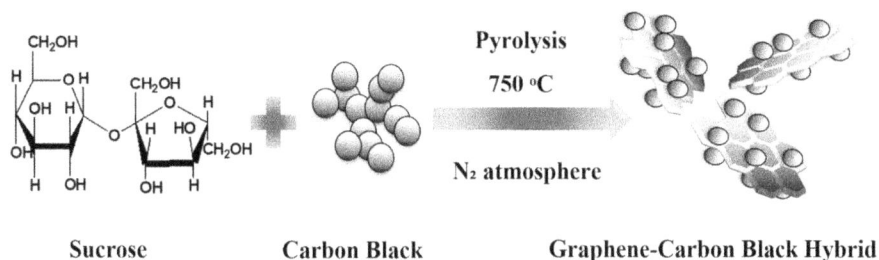

FIGURE 3.5 (a) Scheme of synthesis of graphene–carbon black hybrid from sucrose (molasses) by pyrolysis.

opened a pathway that addressed the worldwide concern raised by an enormous amount of industrial and polymer wastes accumulated every day.

In another outstanding work, PMMA thin film was converted into high-quality monolayer graphene on a metallic surface acting as the catalyst. PMMA thin film was coated on Cu surface which under a reductive/carrier gas flow of H_2/Ar within a temperature range of 800°C–1,000°C under low-pressure conditions resulted in uniform monolayer graphene (Sun et al. 2010a).

Plastic food cover and the fibrous waste from palm oil industries were also subjected to a traditional CVD method to obtain graphene (Tahir et al. 2020). The ratio of reductive (H_2) and carrier (Ar) gases was varied to optimize the conditions of graphene synthesis and it was observed that with only argon, the optimum growth time and temperature were 90 minutes and 1,020°C, while with only hydrogen the optimum growth conditions were 30 minutes and 1,000°C temperature.

Inexpensive polymer sources such as polystyrene (PS), polyacrylonitrile (PAN), and PMMA were dissolved in solvents like chloroform or dimethylformamide (DMF) and spin-coated on SiO_2/Si substrate to obtain thin films of 10 nm thickness. Traditional carrier/reductive gas flow of Ar at 50 sccm and H_2 at 10 sccm under vacuum of 4 Torr was applied to these polymer films for 1 min at 1,000°C temperature to acquire high-quality tri-layer graphene (Byun et al. 2011).

Waste plastic from fruit packaging was employed in an atmospheric pressure chemical vapor deposition (APCVD) method to derive large single-crystal bilayer graphene (Sharma et al. 2014). A split two furnace setup was dedicated to this experiment. The polymer waste was kept on the first furnace at 500°C and the annealed Cu substrate with enlarged grain size was kept on the second at 1,020°C with a mixture gas flow of Ar/H_2 with a flow rate of 98/2.5 sccm.

Another industrial waste, coal tar pitch, was dissolved in quinoline and spin-coated on SiO_2/Si wafer and the coal tar layer was subsequently capped with Ni. Ni layer prevented the coal tar from vaporizing at elevated temperature as well as catalyzed the growth of graphene. Annealing of this sample was done at 900°C–1,100°C at 330 mTorr pressure for 4 minutes with Ar/H_2 with the flow. After graphene was grown, the Ni layer was chemically removed. A similar experiment was also carried out with a Cu capping at a lower temperature of 1,000°C to obtain high-quality graphene (Seo et al. 2015).

3.2.4 IN SITU SYNTHESIS OF DOPED GRAPHENE

Crystal sugar and urea in a mass ratio of 1:60 were dissolved in warm water and then water was evaporated to obtain a dry and homogeneous mixture of sugar and urea. This mixture was subjected to pyrolysis under 60 sccm flow of Ar. The experiment was carried out at three different temperatures, 800°C, 900°C, and 1,000°C, though the results inferred that the pyrolysis temperature had no effect on the morphology of the graphene prepared (Pan et al. 2013).

Pyrolysis of chitosan film at 600°C–800°C temperature under inert atmosphere also yields nitrogen-doped single- to few-layer graphenes without the presence of any catalytic component (Primo et al. 2012).

A sophisticated technique of tailoring a polymer molecule to derive doped graphene was developed by Mondal et al. (2015). A heteroatom polymer, polyurethane acrylate, was synthesized from acrylic acid and bisphenol A glycidyl ether. Another heteroatom polymer, acrylonitrile-butadiene rubber (NBR) was also selected as another precursor. Two non-heteroatom polymers, such as styrene-butadiene rubber (SBR) and natural rubber (NR) were also subjected to a similar CVD process. These polymers were dissolved in tetrahydrofuran (THF) at 5% concentration and spin-coated on the Cu surface before the CVD process. Ar/H$_2$ gas flow at 50/10 sccm was passed over the precursors at two different temperatures, 800°C and 1,000°C, at 4 Torr pressure for 15 minutes. The heteroatom polymers resulted in high-quality nitrogen-doped graphenes, while the other two polymers yielded undoped graphene. A schematic representation of the process is shown in Figure 3.6.

Nitrogen and sulfur co-doped graphene was prepared from a single precursor, coir pith (Karuppannan et al. 2019). Coir pith was carbonized at 400°C temperature for 3 hours in a tube furnace. Followed by this carbonization, the coir pith was ground with water in a ball mill at 600 rpm speed and air atmosphere. Thiourea was added to the resultant paste and wet grinding was continued for another 30 minutes. The coir pith-thiourea paste was then dried and graphitized at 850°C temperature for an hour under the Ar atmosphere. Graphene thus produced comprised of 78.2% carbon, 18.5% nitrogen, and 3.2% sulfur and displayed a wrinkled and porous structure.

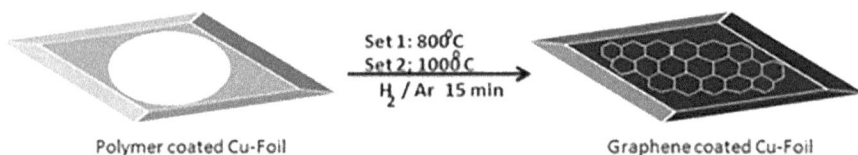

Set 1: 800°C
Set 2: 1000°C
H$_2$ / Ar 15 min

Polymer coated Cu-Foil   Graphene coated Cu-Foil

FIGURE 3.6 Fabrication of nitrogen-doped/undoped graphene using a CVD reactor. (Adapted with permission from Mondal, Titash, Anil K. Bhowmick, and Ramanan Krishnamoorti. 2015. "Controlled Synthesis of Nitrogen-Doped Graphene from a Heteroatom Polymer and Its Mechanism of Formation." *Chemistry of Materials* 27 (3): 716–25. Copyright © 2015 American Chemical Society.)

3.3 CHARACTERIZATION OF GRAPHENE

3.3.1 X-RAY DIFFRACTION (XRD)

The crystallinity in graphene two-dimensional crystal is different from the crystallinity of three-dimensional graphite. Though graphene has a hexagonal lattice structure, its unit cell comprises only two atoms A and B, whereas graphite consists of a hexagonal lattice with four atoms in its unit cell noted as A, A', B, B' as shown in Figure 3.7a and b. There are two possible configurations for a graphite crystal, such as Bernal, in which the positions of A atom and B atom in adjacent layers exactly superpose with each other, and turbostratic, in which positions of A and B atoms in adjacent layers differ by a definite angle θ as graphically represented in Figure 3.7c and d (Garlow et al. 2016).

The preferred orientation in graphite and graphene is (002), which gives a strong diffraction pattern at $2\theta = 25°–26°$, but other diffraction peaks are also observed along [100], [101], and [102] directions in some samples if the growth is not epitaxial, but turbostratic. Notably, graphene synthesized from bio-precursors without a transition metal substrate is often turbostratic, while copper or nickel substrate with specific grain sizes promotes epitaxial growth. Highly oriented graphite gives one sharp peak for the (002) plane at $2\theta = 25°$ (Reference: JCPDS file: ICDD-023-0064). Multilayer graphene gives a sharp peak at $2\theta = 26°$ for the same, and the peak is

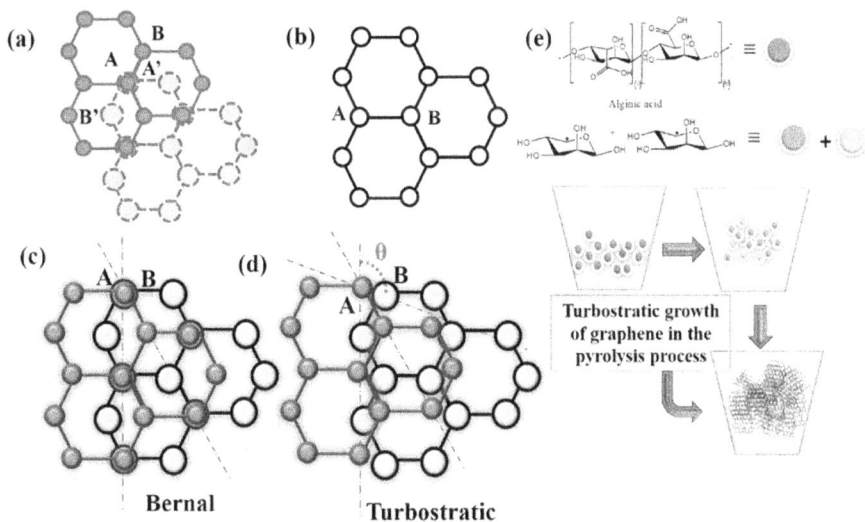

FIGURE 3.7 (a) Unit cell in graphite crystal, (b) unit cell in graphene crystal, (c) Bernal arrangement in bilayer graphene, (d) turbostratic arrangement in bilayer graphene, and (e) schematic diagram for the synthesis of turbostratic graphene from alginic acid (Adapted with permission from Roy, Amrita, Saptarshi Kar, Ranjan Ghosal, Kinsuk Naskar, and Anil K. Bhowmick. 2021. "Facile Synthesis and Characterization of Few-Layer Multifunctional Graphene from Sustainable Precursors by Controlled Pyrolysis, Understanding of the Graphitization Pathway, and Its Potential Application in Polymer Nanocomposites." *ACS Omega* 6 (3): 1809–22. Copyright © 2021 American Chemical Society.)

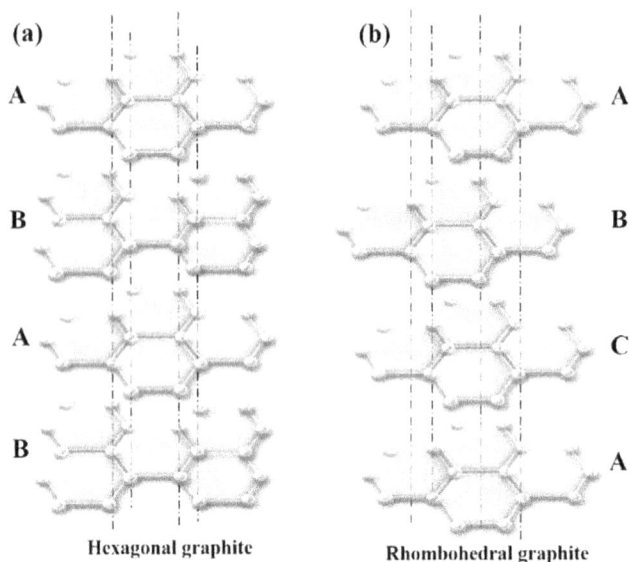

FIGURE 3.8 (a) Hexagonal packing of graphite and (b) rhombohedral packing of graphite. (Adapted with permission from Roy, Amrita, Saptarshi Kar, Ranjan Ghosal, Kinsuk Naskar, and Anil K. Bhowmick. 2021. "Facile Synthesis and Characterization of Few-Layer Multifunctional Graphene from Sustainable Precursors by Controlled Pyrolysis, Understanding of the Graphitization Pathway, and Its Potential Application in Polymer Nanocomposites." *ACS Omega* 6 (3): 1809–22. Copyright © 2021 American Chemical Society.)

gradually widened as the structure breaks down to nanometers and for the broadening interlayer distance (Clark et al. 2013). Turbostratic graphene often displays diffraction patterns from (100), (101), and (102) planes at 2θ values of 42°, 43°, and 49° respectively. The turbostratic growth of graphene from alginic acid is pictorially demonstrated in Figure 3.7e.

The phenomenon of peak broadening in graphene can be attributed to two different causes, microcrystalline broadening and stacking fault broadening. A graphite crystal can have two different stacking configurations; the layers can either be stacked in hexagonal (ABAB) pattern or in rhombohedral (ABCA) pattern as shown in Figure 3.8. In the case of graphene, stacking faults or defects are generated because of few number of layers and structural defects that occur from synthetic procedures. These broadening effects are often considered as parts of the full-width-half-maximum, FWHM, of an X-ray Diffraction (XRD) peak, β, and denoted as β_c, the microcrystalline broadening and β_p, the stacking fault broadening (Zhang and Wang 2017).

d spacing of graphene (002) plane is calculated from Bragg's law of X-ray diffraction as stated in the following equation:

$$n\lambda = 2d\sin\theta \tag{3.1}$$

where θ is the diffraction angle, n is the degree of diffraction, and λ is the wavelength of Cu Kα radiation (0.154 nm).

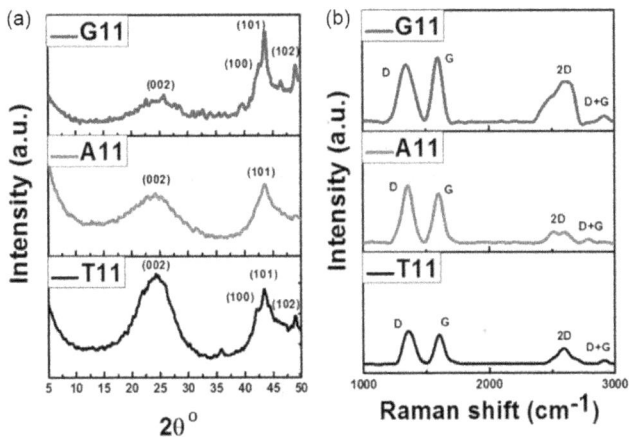

FIGURE 3.9   (a) XRD plot for T11, A11, and G11 and (b) Raman spectra for T11, A11, and G11. (Adapted with permission from Roy, Amrita, Saptarshi Kar, Ranjan Ghosal, Kinsuk Naskar, and Anil K. Bhowmick. 2021. "Facile Synthesis and Characterization of Few-Layer Multifunctional Graphene from Sustainable Precursors by Controlled Pyrolysis, Understanding of the Graphitization Pathway, and Its Potential Application in Polymer Nanocomposites." *ACS Omega* 6 (3): 1809–22. Copyright © 2021 American Chemical Society.)

The thickness of the layers is calculated by the Scherrer equation (Eq. 3.2) using the full-width-half-maxima (FWHM) values of the (002) diffraction peak. τ is the size of the crystal inversely related to β, the FWHM in radians, and cos of the diffraction angle. K is the shape factor that was considered to be unity for the calculations.

$$\tau = K\lambda/\beta\cos\theta \qquad (3.2)$$

The graphene samples derived from the pyrolysis of alginic acid, tannic acid, and green tea at 1,100°C are denoted as A11, T11, and G11 respectively and their characteristic XRD patterns are shown in Figure 3.9a.

3.3.2   RAMAN SPECTROSCOPY AND RAMAN IMAGING

Graphite often shows two characteristic Raman peaks, a G peak (graphitic peak) at 1,580 cm^{-1} which is generally sharp and highly intense, and a 2D peak at 2,700 cm^{-1} (Kudin et al. 2008). Graphene shows three characteristic peaks, D at 1,350 cm^{-1}, G at 1,580 cm^{-1}, and 2D at 2,700 cm^{-1} respectively (Kaniyoor and Ramaprabhu 2012). D peak or diamondoid peak is absent in most samples of pure highly oriented graphite, as the D peak is a direct consequence of the defects present in the graphitic sample, but defects are induced in graphene, especially in graphenes grown through CVD or pyrolysis process (Tuinstra and Koenig 1970). Sometimes, another defect-induced peak is also present often merged with the G peak, known as the D' peak at 1,650 cm^{-1}. In specific cases, $D+G$ peaks are also visible near 2,900 cm^{-1} in graphene occasionally (Das, Chakraborty, and Sood 2008). Raman spectra for the graphene samples derived from alginic acid (A11), tannic acid (T11), and green tea (T11) are shown in Figure 3.9b.

It is noteworthy that the D and G peaks of these samples are not discrete, rather they are joined together, most possibly owing to the presence of a peak at 1,480 cm^{-1}, enshrouded by the highly intense D and G bands, which is a direct contribution of the amorphous carbon present in the samples as pyrolysis by-products. This is a common occurrence in graphene and doped graphene samples synthesized from non-conventional sources as the separation of amorphous carbon from graphene is often unfeasible. A Gaussian-Lorentzian fitting was deployed to the Raman spectra obtained from different polymers such as polyurethane and SBR at 1,000°C, referred to PU10 and SBR10, and confirmed the presence of amorphous carbon in the graphene samples as illustrated in Figure 3.10.

The number of layers in graphene can also be calculated from Raman spectra with the help of the following formula:

$$W_G = 1581.6 + 11 / \left(\left(1 + n^\wedge 1.6 \right) \right) \tag{3.3}$$

where W_G is the value of Raman shift in cm^{-1} and n is the number of layers in the graphene sample (Fesenko et al. 2015).

The ratio of the intensity of the D peak to the G peak is very significant as it relates to the structural defects in graphene. Higher intensity of the G band compared to the D band indicates a low degree of the defect (Pimenta et al. 2007). The I_D/I_G ratio is expected to increase as the flakes of graphene become smaller. The $I_D/I_{D'}$ value is also important to get an idea about the type of defects that are introduced to the layers

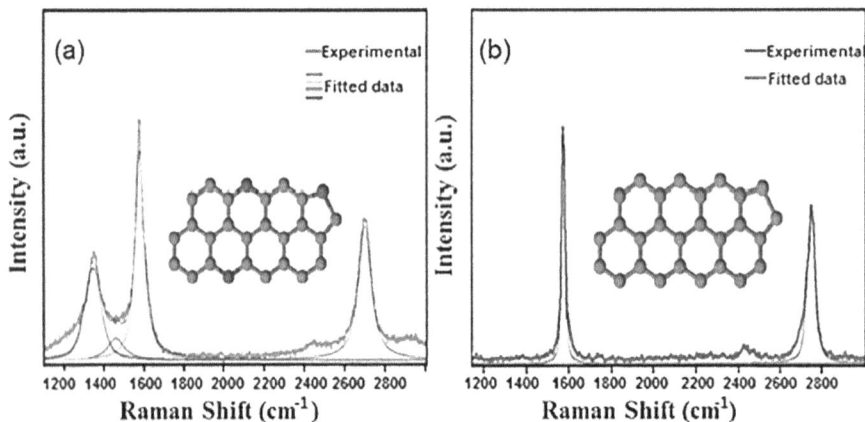

FIGURE 3.10 Raman spectra of (a) PU10 and (b) SBR10. Experimentally observed data were correlated with the theoretically fitted data. A Gaussian–Lorentzian fitting was adopted for the present fitting. The experimental data reported here are the average of three data sets collected using 514.5 nm Ar+ as the laser source. (Adapted with permission from Mondal, Titash, Anil K. Bhowmick, and Ramanan Krishnamoorti. 2015. "Controlled Synthesis of Nitrogen-Doped Graphene from a Heteroatom Polymer and Its Mechanism of Formation." *Chemistry of Materials* 27 (3): 716–25. Copyright © 2015 American Chemical Society.)

(Paton 2014). The value lies near 13 for sp³ type defect and it is approximately equal to 7 when vacancy defects or basal plane defects are introduced. This ratio becomes near to 3.5 or lowers when only edge defects or boundary defects are present. This condition ensures the formation of defect-free high-quality graphene (Parvez et al. 2014).

The 2D band is very crucial for graphene as its nature concerns the number of layers present in graphene (Malard et al. 2009). 2D is a sharp and intense peak in the case of highly oriented pyrolytic graphite (HOPG), and when analytically deconvoluted, it is found that there are highly intense peaks at higher Raman shift with a smaller peak at lower Raman shift merged. 2D peak is flattened and appears as a band in the cases of single- and few-layer graphenes; a critical transition takes place when the number of layers is four when this 2D band is deconvoluted into three peaks with slightly higher intensity at higher Raman shift. Typically, trilayer graphene displays a 2D band which can be deconvoluted to at least six peaks, and bilayer graphene, at least four peaks. Splitting of the 2D band into several Gaussian peaks is a phenomenon occurring from the double resonance Raman processes accompanied with a phonon-electron scattering in graphene (Ferrari et al. 2006). Single-layer graphene gives a single 2D signal at a slightly lower value of Raman shift. From the calculation of the number of layers in graphenes derived from alginic acid (A11), tannic acid (T11), and green tea (G11) from the deconvolution of 2D band, it was inferred that A11 was essentially bilayer, while T11 and G11 were trilayer graphenes as displayed in Figure 3.11a–c.

Under a Raman microscope, point by point Raman spectra are taken and the desired area of the sample is scanned and the image is often displayed in grayscale or with imposed false colors. From this Raman mapping technique, the number of layers in a specific graphene sample is determined (Ni et al. 2008).

3.3.3 THERMOGRAVIMETRIC ANALYSIS

Thermogravimetric analysis is another key characterization for graphene which determines the purity of graphene. Graphite and graphene are thermally stable at very high temperatures, even beyond 1,200°C. Any instance of weight loss at temperatures lower than 900°C is generally considered as a result of the degradation of the impurities in graphene. Weight loss at 100°C occurs when moisture is present in the sample. Slight weight loss at higher temperatures, such as 800°C–900°C, is a probable effect of degradation of the amorphous carbon and release of carbon dioxide. The percentage of weight loss can be used for the determination of the percentage of amorphous carbon or other impurities present in a graphene sample. The TGA weight loss versus temperature plots for A11, T11, and G11 are illustrated in Figure 3.11d, which indicates the presence of a small quantity of amorphous carbon in these graphenes.

3.3.4 X-RAY PHOTOELECTRON SPECTROSCOPY (XPS)

Though graphene produced from conventional hydrocarbon precursors through the CVD technique is almost always chemically pure, graphenes produced from non-conventional sources often contain chemical impurities. Surface atomic contents

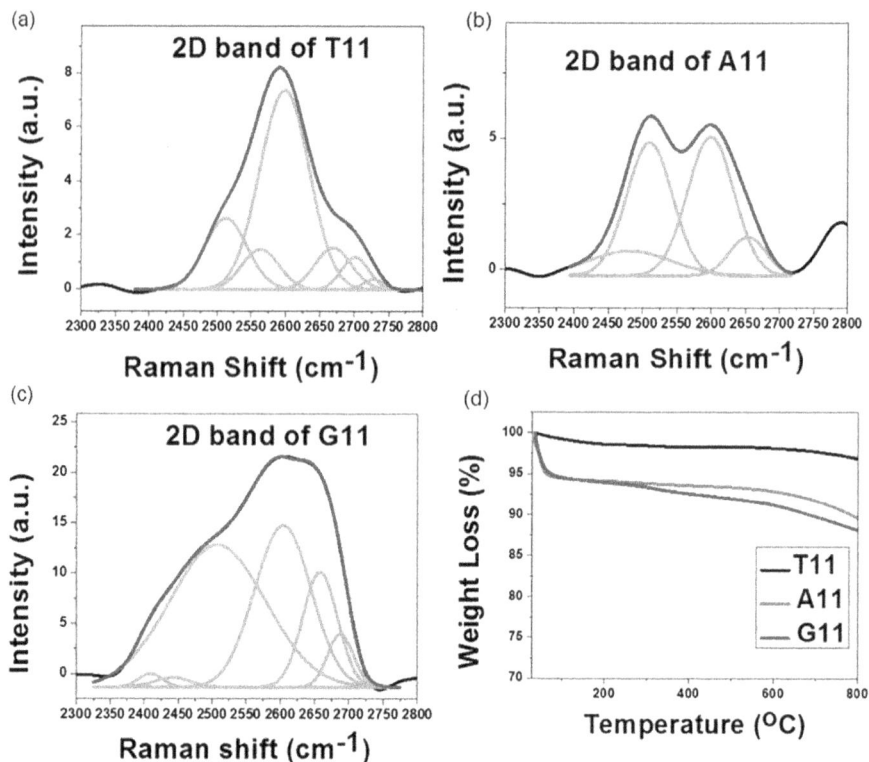

FIGURE 3.11 Deconvolution of Raman 2D band of (a) T11, (b) A11, and (c) G11. (d) TGA plot of T11, A11, and G11. (Adapted with permission from Roy, Amrita, Saptarshi Kar, Ranjan Ghosal, Kinsuk Naskar, and Anil K. Bhowmick. 2021. "Facile Synthesis and Characterization of Few-Layer Multifunctional Graphene from Sustainable Precursors by Controlled Pyrolysis, Understanding of the Graphitization Pathway, and Its Potential Application in Polymer Nanocomposites." *ACS Omega* 6 (3): 1809–22. Copyright © 2021 American Chemical Society.)

in a specific graphene sample are ideally determined by X-ray Photoelectron Spectroscopy (XPS) analysis (Siokou et al. 2011). Typically, any graphene sample comprises mostly carbon and traces of oxygen, and in case of doping, nitrogen, sulfur, or any other dopant molecule. Graphene oxide usually contains higher amounts of surface oxygen, usually ranging between 20% and 30%, while reduced graphene oxide might exhibit a lower but widely varied content of oxygen depending upon the synthetic route (Stobinski et al. 2014). XPS of the graphene samples grown from tannic acid, alginic acid, and green tea exhibited high carbon content, such as 91.28%, 92.41%, and 92.75% respectively, and as an obvious consequence, displayed quite low oxygen content such as 8.26%, 6.79%, and 6.81%. XPS analysis also gives insight into the quantification of sp^3 defects in a specific sample of graphene. Sp^2 carbon contents in graphenes obtained from tannic acid, alginic acid, and green tea were

68.6%, 72.5%, and 75.2% indicating the presence of low content of sp³ carbon, such as 13.3%, 11.5%, and 11.6% respectively (Roy et al. 2021).

3.3.5 FIELD EMISSION SCANNING ELECTRON MICROSCOPY (FESEM) AND ENERGY DISPERSIVE X-RAY (EDX) ANALYSIS

Field Emission Scanning Electron Microscopy (FESEM) is a direct and precise technique to investigate the morphology of graphene. Often under FESEM, wrinkled structures of graphene are found (Azam, Mudtalib, and Seman 2018; Rahbar et al. 2019). The lateral sizes of these wrinkled graphene sheets are also derivable from FESEM images which vary from a few hundred nanometers to few microns. Crucial features, such as the presence of pores, pore size, and pore size distribution in graphene are also observable under FESEM (Kalita et al. 2010). FESEM can also be utilized as an indispensable tool to optimize the conditions for graphene nucleation and growth as illustrated in Figure 3.12, where Mondal, Bhowmick, and Krishnamoorti recorded FESEM images at two different temperatures at which polyurethane, NR, and SBR were annealed.

FESEM is often coupled with Energy Dispersive X-Ray (EDX) analysis which enables elemental mapping over a particular graphene sheet and thus ascertains the elemental purity of carbon (Tahir et al. 2020; Muramatsu et al. 2014).

3.3.6 HIGH-RESOLUTION TRANSMISSION ELECTRON MICROSCOPY (HRTEM) AND SAED SELECTED AREA ELECTRON DIFFRACTION (SAED)

High-Resolution Transmission Electron Microscopy (HRTEM) is the most exclusive technique to visualize graphene, magnifying it from nanoscale to the resolution level of human eyes. Translucent graphene sheets are visible under HRTEM. Layered

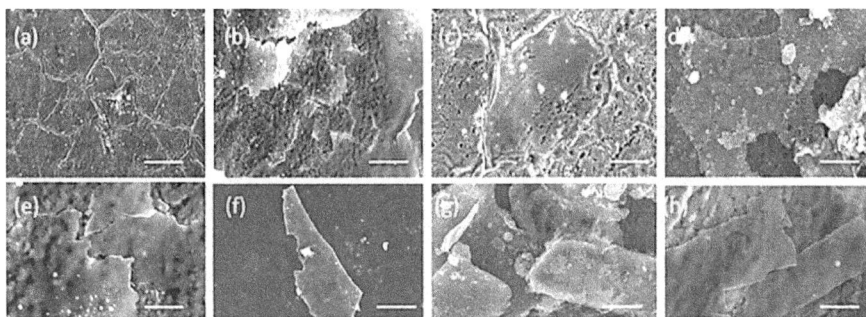

FIGURE 3.12 Selective area SEM micrograph of isolated product on a copper substrate. (a–d) Products obtained from NBR, PU, NR, and SBR at 800°C and (e–h) products obtained from NBR, PU, NR, and SBR at 1,000°C. The scale bar corresponds to 2 μm, and the micrographs were recorded at 10 kV. (Adapted with permission from Mondal, Titash, Anil K. Bhowmick, and Ramanan Krishnamoorti. 2015. "Controlled Synthesis of Nitrogen-Doped Graphene from a Heteroatom Polymer and Its Mechanism of Formation." *Chemistry of Materials* 27 (3): 716–25. Copyright © 2015 American Chemical Society.)

edges in few or multilayer graphenes are only observed under HRTEM from which the number of layers can be directly counted (Guoxiu et al. 2008).

HRTEM instruments also provide the facility of placing the nanometrically thin samples parallel to the electron beam which results in electron diffraction from the crystalline structures, generating a series of bright spots, each indicating a satisfied diffraction pattern. Typically, each A-B type unit cell in graphene gives a diffraction spot, and as a cumulative effect, a 2D graphene crystal gives six dots in Selected Area Electron Diffraction (SAED), stipulating sixfold symmetry in graphene (Sun et al. 2010b; Fang et al. 2015; Sun et al. 2010a; Guoxiu et al. 2008). In a polycrystalline graphene sample, multiple rings are obtained in SAED instead of clear dots. HRTEM images of A11, G11, and T11 and their corresponding SAED patterns are disclosed in Figure 3.13.

Moir fringes in overlapping graphene sheets are seen in magnified HRTEM images (Habib et al. 2018; Liu, Liu, and Zhu 2011; N'Diaye et al. 2008). HRTEM also allows to envisage atomic fringes at extreme magnification levels, such as 2–5 nm, and profiling of the fringes leads to an accurate determination of the d-spacing of graphene crystals. HRTEM is the ultimate microscopic solution for viewing the crystallographic structure of a material and the representation is more realistic than any other microscopic technique developed to date. The atomic fringes of graphene are synthesized from the pyrolysis of molasses, and the subsequent determination of d-spacing is exemplified pictorially in Figure 3.14.

3.3.7 Atomic Force Microscopy (AFM)

Atomic Force Microscopy (AFM) is an explicit technique that not only comprehends the morphology, topography, structure, and lateral size of a graphene sheet but also accurately measures the thickness of a graphene sheet and hence the number of layers in a specific graphene sample (Pérez, Bajales, and Lacconi 2019; Shams et al. 2015; Ferrari et al. 2006). Sophisticated AFM instruments are even capable of scanning a sample within a 1 nm^2 area where the interaction between the scanning probe and the carbon atoms in graphene lies within picometer ranges and originate images of atomic contrast. These images are particularly advantageous in studying the atomic-level defects in graphene or doped graphene samples (Boneschanscher et al. 2012). Thickness and the average number of layers in A11, T11, and G11, i.e., graphenes derived from alginic acid, tannic acid, and green tea, were assessed from the AFM; thickness of the samples ranged between 1.2 and 1.5 nm and the number of layers was 2–3, as manifested in Figure 3.15.

3.4 PROPERTIES AND APPLICATION OF GRAPHENE PREPARED FROM NON-CONVENTIONAL SOURCES

Graphene synthesized from non-conventional sources is designed to have equal or better electronic and electrochemical properties, gas adsorption, improved tribological behavior, and various other ameliorated functions compared to conventional systems. A contrast between the properties of graphene obtained from conventional and non-conventional precursors is given in Table 3.1. Biomass-derived graphene sample

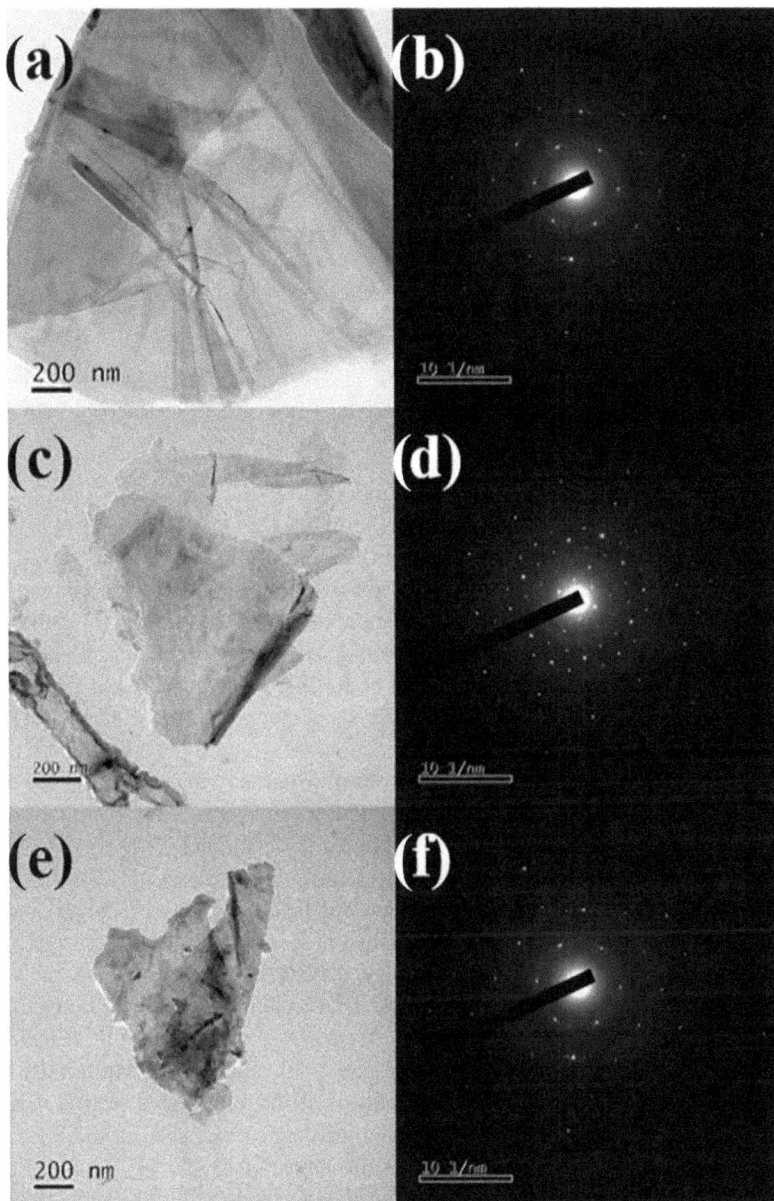

FIGURE 3.13 (a) HRTEM image of T11, (b) SAED pattern of T11, (c) HRTEM image of A11, (d) SAED pattern of A11, (e) HRTEM image of G11, and (f) SAED pattern of G11. (Adapted with permission from Roy, Amrita, Saptarshi Kar, Ranjan Ghosal, Kinsuk Naskar, and Anil K. Bhowmick. 2021. "Facile Synthesis and Characterization of Few-Layer Multifunctional Graphene from Sustainable Precursors by Controlled Pyrolysis, Understanding of the Graphitization Pathway, and Its Potential Application in Polymer Nanocomposites." *ACS Omega* 6 (3): 1809–22. Copyright © 2021 American Chemical Society.)

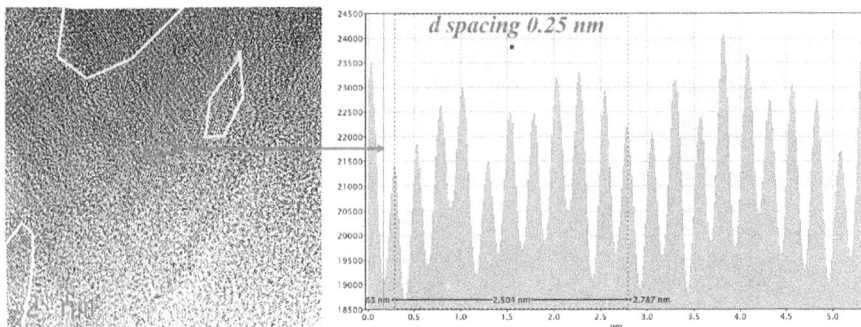

FIGURE 3.14   (a) Atomic fringes of graphene observed under HRTEM (regions marked with yellow border); (b) d spacing profile determined from the region marked with red, d spacing of graphene is 0.25 nm.

was effectively used in volatile organic compound (VOC) adsorption with gasoline working capacity of the activated graphene being 65.2 g/$l_{adsorbent}$ (Rahbar et al. 2019) which was superior to the toluene adsorption reported for graphene synthesized from methane (Wang et al. 2019). Enhanced slipping friction and wear resistance were reported by graphene coatings with a coefficient of friction lower than 0.1(Tahir et al. 2020) which was comparable with an anisotropic coefficient of friction, measured for conventional CVD grown graphene (Long et al. 2017). Transparent electrodes were prepared with graphene synthesized by pyrolysis of camphor leaves and optimization between the transparency and sheet resistance was thoroughly conducted. The optimum value of sheet resistance 203 Ω sq^{-1} with transparency in terms of 63.5% transmission was reported (Shams et al. 2015). This performance was commensurable with that of a graphene sample synthesized in the conventional CVD process, which displayed 230 Ω sq^{-1} at 72% transparency (Gomez De Arco et al. 2010). A resistive random access memory (RRAM) device was fashioned with graphene obtained from tea-tree oil extract and strong memristive property over 400 on/off cycles was reported (Jacob et al. 2015). On the contrary, decent resistive endurance and retention property was reported only up to a hundred switching cycles for conventional graphene (Yang et al. 2013). Rice-husk-derived graphene exhibited high hydrogen adsorption capacity and fair value of electrochemical capacitance (Muramatsu et al. 2014). Graphene from alginic acid precursor was utilized in the fabrication of a rubber latex-based supercapacitor thin film with high capacitance such as 137 F g^{-1} (Roy et al. 2021). However, slightly higher specific capacitance, 203.4 F g^{-1}, has been reported for a graphene-poly(vinylidene fluoride) electrode for graphene synthesized from a traditional precursor, methane (Zhou et al. 2015). A maximum capacitance of 247 F g^{-1} was reported for nitrogen and sulfur dual-doped graphene with a maximum power density of 3790 W/Kg (Karuppannan et al. 2019). A graphene-sand composite synthesized from sugar and river sand displayed excellent dye absorbing capacity, 55 mg g^{-1} (Gupta et al. 2012b). Standard graphene synthesized from plasma-enhanced CVD of methane was reported to show the highest electrocatalytic oxygen reduction at a current density of 6.3 mA cm^{-2} at 0.91 V (Wu et al. 2020). In comparison, an improved oxygen reduction electrocatalytic property with a kinetic current density

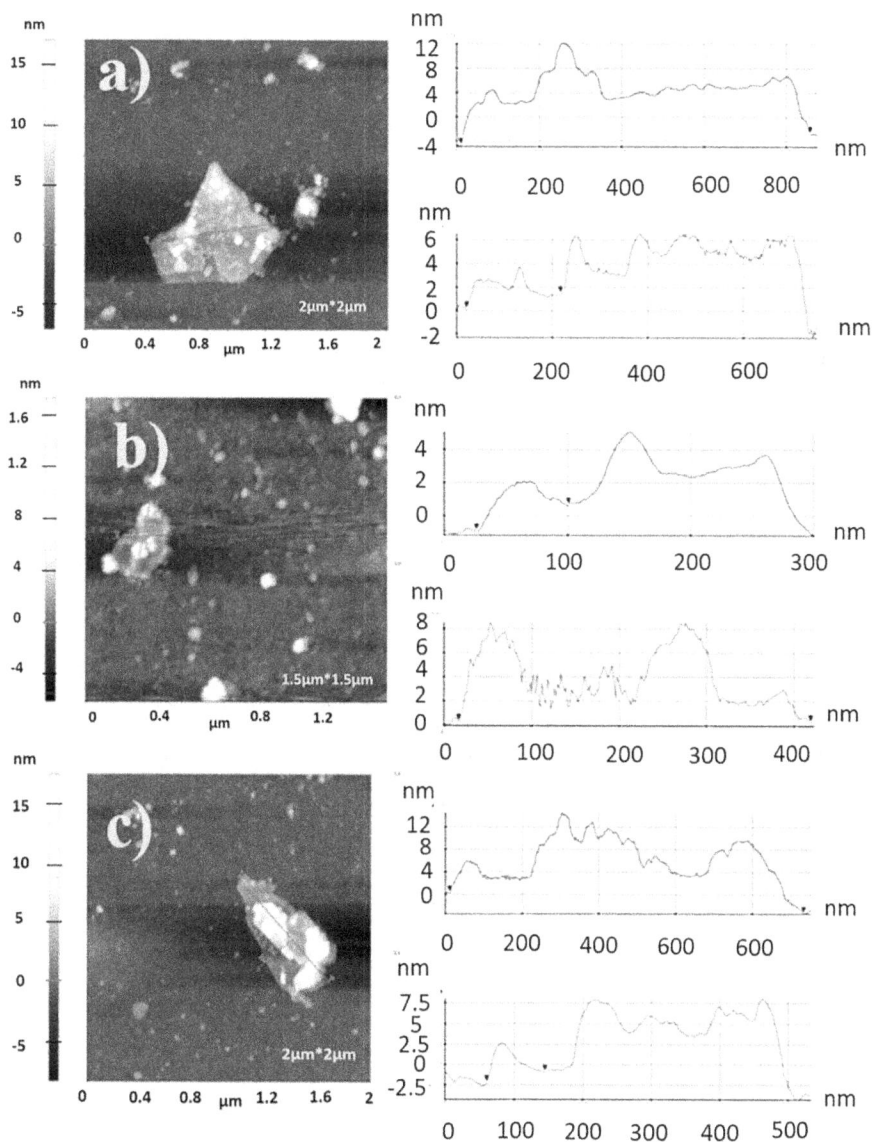

FIGURE 3.15 AFM height images and profiles for (a) T11, (b) G11, and (c) A11. (Adapted with permission from Roy, Amrita, Saptarshi Kar, Ranjan Ghosal, Kinsuk Naskar, and Anil K. Bhowmick. 2021. "Facile Synthesis and Characterization of Few-Layer Multifunctional Graphene from Sustainable Precursors by Controlled Pyrolysis, Understanding of the Graphitization Pathway, and Its Potential Application in Polymer Nanocomposites." *ACS Omega* 6 (3): 1809–22. Copyright © 2021 American Chemical Society.)

TABLE 3.1

Comparison of Properties between Graphene from Conventional and Non-Conventional Precursors

Property	Graphene from Conventional Precursor	Graphene from Non-Conventional Precursor
d spacing (nm)	0.337–0.344 (Garlow et al. 2016)	0.336 (Chen et al. 2016)
I_D/I_G value	0.17–0.37 (Garlow et al. 2016)	0.37 (Chen et al. 2016)
Carbon content	86.9 (Deokar et al. 2015)	98.88 (Chen et al. 2016)
Sheet resistance of transparent electrode	230 Ω sq^{-1} at 72% transmission (Gomez De Arco et al. 2010)	203 Ω sq^{-1} at 63.5% transmission (Shams et al. 2015)
Resistive endurance	Over 100 cycles (Yang et al. 2013)	Over 400 cycles (Jacob et al. 2015)
Capacitance	203.4 F g^{-1} (Zhou et al. 2015)	137 F g^{-1} (Roy et al. 2021)

as high as 21.33 mA cm^{-2} at -0.25 V (vs Ag/AgCl) in O_2-saturated 0.1 M KOH electrolyte solution was obtained from the nitrogen-doped graphene (Pan et al. 2013). Compared to applications in energy area, the technology is still in the nascent stage in the field of rubber composites.

3.5 CONCLUSIONS

From insects to specially tailored polymers, drifting the focus from traditional precursors toward the neoteric state-of-the-art precursors has made graphene research even more interesting than it used to be. The utilization of biowaste and industrial waste products for the synthesis of this high-performance nano-carbon opens up a wide possibility to address the rising concern for the environment and establish sustainability. Copious applications of these graphenes in energy devices show the silver lining around the issue of cost-effective and non-petroleum-derived energy for the future. When the quality of graphene synthesized from non-conventional precursors is considered, these graphenes are capable of competing with the ones produced from hydrocarbons, with the additional advantage of cost-effectiveness and renewability. Though some of the techniques discussed herein report a high level of sophistication in the experimental setup, such as a plasma or high vacuum and high temperature requirements, the production of high quality, uniform, non-wrinkled, large-area monolayer graphene compensates the complicacy of these techniques. A thorough review of the astonishingly widespread properties and applications of such graphene samples encourages further dedication and anchored scientific concentration toward the field with the hope of graphene being established as the greatest component of future technology.

REFERENCES

Azam, Mohd Asyadi, Nur Elina Shahrina Amza Abd Mudtalib, and Raja Noor Amalina Raja Seman. 2018. "Synthesis of Graphene Nanoplatelets from Palm-Based Waste Chicken Frying Oil Carbon Feedstock by Using Catalytic Chemical Vapour Deposition." *Materials Today Communications* 15: 81–87. doi:10.1016/j.mtcomm.2018.02.019.

Boneschanscher, Mark P., Joost Van Der Lit, Zhixiang Sun, Ingmar Swart, Peter Liljeroth, and Daniël Vanmaekelbergh. 2012. "Quantitative Atomic Resolution Force Imaging on Epitaxial Graphene with Reactive and Nonreactive AFM Probes." *ACS Nano* 6 (11): 10216–21. doi:10.1021/nn3040155.

Byun, Sun Jung, Hyunseob Lim, Ga Young Shin, Tae Hee Han, Sang Ho Oh, Jong Hyun Ahn, Hee Cheul Choi, and Tae Woo Lee. 2011. "Graphenes Converted from Polymers." *Journal of Physical Chemistry Letters* 2 (5): 493–97. doi:10.1021/jz200001g.

Carlsson, Jan Otto, and Peter M. Martin. 2010. *Chemical Vapor Deposition. Handbook of Deposition Technologies for Films and Coatings*. 3rd Ed. Elsevier Ltd. doi:10.1016/B978-0-8155-2031-3.00007-7.

Castro Neto, Antonio Helio de, Francisco Guinea, Nuno Peres, Konstantin Novoselov, and Andre Geim. 2009. "The Electronic Properties of Graphene." *Reviews of Modern Physics* 81 (1): 109–62. doi:10.1103/RevModPhys.81.109.

Chen, Feng, Juan Yang, Tao Bai, Bo Long, and Xiangyang Zhou. 2016. "Facile Synthesis of Few-Layer Graphene from Biomass Waste and Its Application in Lithium Ion Batteries." *Journal of Electroanalytical Chemistry* 768: 18–26. doi:10.1016/j.jelechem.2016.02.035.

Choi, Wonbong, Indranil Lahiri, Raghunandan Seelaboyina, and Yong Soo Kang. 2010a. "Synthesis of Graphene and Its Applications: A Review." *Critical Reviews in Solid State and Materials Sciences* 35 (1): 52–71. doi:10.1080/10408430903505036.

Clark, S. M., Ki Joon Jeon, Jing Yin Chen, and Choong Shik Yoo. 2013. "Few-Layer Graphene under High Pressure: Raman and X-Ray Diffraction Studies." *Solid State Communications* 154 (1): 15–18. doi:10.1016/j.ssc.2012.10.002.

Das, Anindya, Biswanath Chakraborty, and Ajay Kumar Sood. 2008. "Raman Spectroscopy of Graphene on Different Substrates and Influence of Defects." *Bulletin of Materials Science* 31 (3): 579–84. doi:10.1007/s12034-008-0090-5.

Deokar, Geetanjali, Jose Avila, Ivy Razado-Colombo, Jean Louis Codron, Christophe Boyaval, Elisabeth Galopin, Maria Carmen Asensio, and Dominique Vignaud. 2015. "Towards High Quality CVD Graphene Growth and Transfer." *Carbon* 89: 82–92. doi:10.1016/j.carbon.2015.03.017.

Fang, Wenjing, Allen L. Hsu, Yi Song, and Jing Kong. 2015. "A Review of Large-Area Bilayer Graphene Synthesis by Chemical Vapor Deposition." *Nanoscale* 7 (48): 20335–51. doi:10.1039/c5nr04756k.

Ferrari, Andrea C., Jannik Meyer, Vittorio Scardaci, Cinzia Casiraghi, Michele Lazzeri, Francesco Mauri, S. Piscanec, et al. 2006. "Raman Spectrum of Graphene and Graphene Layers." *Physical Review Letters* 97 (18): 1–4. doi:10.1103/PhysRevLett.97.187401.

Fesenko, Olena, Galyna Dovbeshko, Andrej Dementjev, Renata Karpicz, Tommi Kaplas, and Yuri Svirko. 2015. "Graphene-Enhanced Raman Spectroscopy of Thymine Adsorbed on Single-Layer Graphene." *Nanoscale Research Letters* 10 (1): 1–7. doi:10.1186/s11671-015-0869-4.

Garlow, Joseph A., Lawrence K. Barrett, Lijun Wu, Kim Kisslinger, Yimei Zhu, and Javier F. Pulecio. 2016. "Large-Area Growth of Turbostratic Graphene on Ni(111) via Physical Vapor Deposition." *Scientific Reports* 6 (January): 1–11. doi:10.1038/srep19804.

Geim, Andrew, and Konstantin Novoselov. 2009. "The Rise of Graphene." *Nanoscience and Technology: A Collection of Reviews from Nature Journals*, 11–19. doi:10.1142/9789814287005_0002.

Gierz, Isabella, Christian Riedl, Ulrich Starke, Christian R. Ast, and Klaus Kern. 2008. "Atomic Hole Doping of Graphene." *Nano Letters* 8 (12): 4603–7. doi:10.1021/nl802996s.

Gomez De Arco, Lewis, Yi Zhang, Cody W. Schlenker, Koungmin Ryu, Mark E. Thompson, and Chongwu Zhou. 2010. "Continuous, Highly Flexible, and Transparent Graphene Films by Chemical Vapor Deposition for Organic Photovoltaics." *ACS Nano* 4 (5): 2865–73. doi:10.1021/nn901587x.

Guo, Beidou, Qian Liu, Erdan Chen, Chen Zhu, Liang Fang, and Jian Ru Gong. 2010. "Controllable N-doping of graphene." *Nano Letters* 10: 4975–80. https://doi.org/10.1021/nl103079j

Guoxiu, Wang, Yang Juan, Park Jinsoo, Gou Xinglong, Wang Bei, Liu Hao, and Yao Jane. 2008. "Facile Synthesis and Characterization of Graphene Nanosheets." *Journal of Physical Chemistry C* 112 (22): 8192–95. doi:10.1021/jp710931h.

Gupta, Soujit Sen, Theruvakkattil Sreenivasan Sreeprasad, Shihabudheen Mundampra Maliyekkal, Sarit Kumar Das, and Thalappil Pradeep. 2012a. "Graphene from Sugar and Its Application in Water Purification." *ACS Applied Materials and Interfaces* 4 (8): 4156–63. doi:10.1021/am300889u.

Habib, Mohammad Rezwan, Tao Liang, Xuegong Yu, Xiaodong Pi, Yingchun Liu, and Mingsheng Xu. 2018. "A Review of Theoretical Study of Graphene Chemical Vapor Deposition Synthesis on Metals: Nucleation, Growth, and the Role of Hydrogen and Oxygen." *Reports on Progress in Physics* 81 (3). doi:10.1088/1361-6633/aa9bbf.

Jacob, Mohan V., Rajdeep S. Rawat, Bo Ouyang, Kateryna Bazaka, D. Sakthi Kumar, Dai Taguchi, Mitsumasa Iwamoto, Ram Neupane, and Oomman K. Varghese. 2015. "Catalyst-Free Plasma Enhanced Growth of Graphene from Sustainable Sources." *Nano Letters* 15 (9): 5702–8. doi:10.1021/acs.nanolett.5b01363.

Kalita, Golap, Matsushima Masahiro, Hideo Uchida, Koichi Wakita, and Masayoshi Umeno. 2010. "Few Layers of Graphene as Transparent Electrode from Botanical Derivative Camphor." *Materials Letters* 64 (20): 2180–83. doi:10.1016/j.matlet.2010.07.005.

Kaniyoor, Adarsh, and Sundara Ramaprabhu. 2012. "A Raman Spectroscopic Investigation of Graphite Oxide Derived Graphene." *AIP Advances* 2 (3). doi:10.1063/1.4756995.

Karuppannan, Mohanraju, Youngkwang Kim, Yung Eun Sung, and Oh Joong Kwon. 2019. "Nitrogen and Sulfur Co-Doped Graphene-like Carbon Sheets Derived from Coir Pith Bio-Waste for Symmetric Supercapacitor Applications." *Journal of Applied Electrochemistry* 49 (1): 57–66. doi:10.1007/s10800-018-1276-1.

Kostarelos, Kostas, and Kostya S. Novoselov. 2014. "Graphene Devices for Life." *Nature Nanotechnology* 9 (10): 744–45. doi:10.1038/nnano.2014.224.

Kudin, Konstantin N., Bulent Ozbas, Hannes C. Schniepp, Robert K. Prud'homme, Ilhan A. Aksay, and Roberto Car. 2008. "Raman Spectra of Graphite Oxide and Functionalized Graphene Sheets." *Nano Letters* 8 (1): 36–41. doi:10.1021/nl071822y.

Lecloux, André J. 2015. "Discussion about the Use of the Volume-Specific Surface Area (VSSA) as Criteria to Identify Nanomaterials According to the EU Definition: First Part: Theoretical Approach." *Journal of Nanoparticle Research* 17 (11): 1–18. doi:10.1007/s11051-015-3239-3.

Li, Xuesong, Luigi Colombo, and Rodney S. Ruoff. 2016. "Synthesis of Graphene Films on Copper Foils by Chemical Vapor Deposition." *Advanced Materials* 28 (29): 6247–52. doi:10.1002/adma.201504760.

Liu, Hongtao, Yunqi Liu, and Daoben Zhu. 2011. "Chemical Doping of Graphene." *Journal of Materials Chemistry* 21 (10): 3335–45. doi:10.1039/c0jm02922j.

Long, Fei, Poya Yasaei, Wentao Yao, Amin Salehi-Khojin, and Reza Shahbazian-Yassar. 2017. "Anisotropic Friction of Wrinkled Graphene Grown by Chemical Vapor Deposition." *ACS Applied Materials and Interfaces* 9 (24): 20922–27. doi:10.1021/acsami.7b00711.

Malard, Leandro M., Marcos Assuncao Pimenta, George Dresselhaus, and Mildred Spiewak Dresselhaus. 2009. "Raman Spectroscopy in Graphene." *Physics Reports* 473 (5–6): 51–87. doi:10.1016/j.physrep.2009.02.003.

Mattevi, Cecilia, Hokwon Kim, and Manish Chhowalla. 2011. "A Review of Chemical Vapour Deposition of Graphene on Copper." *J. Mater. Chem.* 21 (10): 3324–34. doi:10.1039/C0JM02126A.

Maynard, Andrew D. 2011. "Don't Define Nanomaterials." *Nature* 475: 31.

McArthur, Sally L. 2006. "Thin Films of Vanadium Oxide Grown on Vanadium Metal." *Surface and Interface Analysis* 38 (c): 1380–85. doi:10.1002/sia.

Mondal, Titash, Anil K. Bhowmick, and Ramanan Krishnamoorti. 2015. "Controlled Synthesis of Nitrogen-Doped Graphene from a Heteroatom Polymer and Its Mechanism of Formation." *Chemistry of Materials* 27 (3): 716–25. doi:10.1021/cm503303s.

Muñoz, Roberto, and Cristina Gómez-Aleixandre. 2013a. "Review of CVD Synthesis of Graphene." *Chemical Vapor Deposition* 19 (10–12): 297–322. doi:10.1002/cvde.201300051.

Muramatsu, Hiroyuki, Yoong Ahm Kim, Kap Seung Yang, Rodolfo Cruz-Silva, Ikumi Toda, Takumi Yamada, Mauricio Terrones, Morinobu Endo, Takuya Hayashi, and Hidetoshi Saitoh. 2014. "Rice Husk-Derived Graphene with Nano-Sized Domains and Clean Edges." *Small* 10 (14): 2766–70. doi:10.1002/smll.201400017.

N'Diaye, Alpha T., Johann Coraux, Tim N. Plasa, Carsten Busse, and Thomas Michely. 2008. "Structure of Epitaxial Graphene on Ir(111)." *New Journal of Physics* 10 (111). doi:10.1088/1367-2630/10/4/043033.

Ni, Zhenhua, Yingying Wang, Ting Yu, and Zexiang Shen. 2008. "Raman Spectroscopy and Imaging of Graphene." *Nano Research* 1 (4): 273–91. doi:10.1007/s12274-008-8036-1.

Nicolle, Jimmy, Denis MacHon, Philippe Poncharal, Olivier Pierre-Louis, and Alfonso San-Miguel. 2011. "Pressure-Mediated Doping in Graphene." *Nano Letters* 11 (9): 3564–68. doi:10.1021/nl201243c.

Ning, Jing, Dong Wang, Yang Chai, Xin Feng, Meishan Mu, Lixin Guo, Jincheng Zhang, and Yue Hao. 2017. "Review on Mechanism of Directly Fabricating Wafer-Scale Graphene on Dielectric Substrates by Chemical Vapor Deposition." *Nanotechnology* 28 (28). doi:10.1088/1361-6528/aa6c08.

Novoselov, Konstantin 2011. "Nobel Lecture: Graphene: Materials in the Flatland." *Reviews of Modern Physics* 83 (3): 837–49. doi:10.1103/RevModPhys.83.837.

Novoselov, Konstantin, Vladimir Fal'Ko, Luigi Colombo, P. R. Gellert, Matthias Georg Schwab, and Kwanpyo Kim. 2012. "A Roadmap for Graphene." *Nature* 490 (7419): 192–200. doi:10.1038/nature11458.

Novoselov, Konstantin, Sergei Morozov, Tariq Mohammed Ghulam Mohiuddin, Leonid Ponomarenko, Daniel Cunha Elias, Rong Yang, I. I. Barbolina, et al. 2007. "Electronic Properties of Graphene." *Physica Status Solidi (B) Basic Research* 244 (11): 4106–11. doi:10.1002/pssb.200776208.

Pan, Fuping, Jutao Jin, Xiaogang Fu, Qiao Liu, and Junyan Zhang. 2013. "Advanced Oxygen Reduction Electrocatalyst Based on Nitrogen-Doped Graphene Derived from Edible Sugar and Urea." *ACS Applied Materials and Interfaces* 5 (21): 11108–14. doi:10.1021/am403340f.

Parvez, Khaled, Zhong-shuai Wu, Rongjin Li, Xianjie Liu, and Robert Graf. 2014. "Exfoliation of Graphite into Graphene in Aqueous Solutions of Inorganic Salts." *Journal of the American Chemical Society* 136 (16): 6083–91. doi:10.1021/ja5017156.

Paton, Keith R. 2014. "Scalable Production of Large Quantities of Defect-Free Few-Layer Graphene by Shear Exfoliation in Liquids." *Nature Materials* 13 (6): 624–30. doi:10.1038/nmat3944.

Pérez, Luis A., Noelia Bajales, and Gabriela I. Lacconi. 2019. "Raman Spectroscopy Coupled with AFM Scan Head: A Versatile Combination for Tailoring Graphene Oxide/ Reduced Graphene Oxide Hybrid Materials." *Applied Surface Science* 495 (April): 143539. doi:10.1016/j.apsusc.2019.143539.

Pimenta, M. A., G. Dresselhaus, M. S. Dresselhaus, L. G. Cançado, A. Jorio, and R. Saito. 2007. "Studying Disorder in Graphite-Based Systems by Raman Spectroscopy." *Physical Chemistry Chemical Physics* 9 (11): 1276–91. doi:10.1039/b613962k.

Primo, Ana, Pedro Atienzar, Emilio Sanchez, José María Delgado, and Hermenegildo García. 2012. "From Biomass Wastes to Large-Area, High-Quality, N-Doped Graphene: Catalyst-Free Carbonization of Chitosan Coatings on Arbitrary Substrates." *Chemical Communications* 48 (74): 9254–56. doi:10.1039/c2cc34978g.

Raghavan, Nivea, Sakthivel Thangavel, and Gunasekaran Venugopal. 2017. "A Short Review on Preparation of Graphene from Waste and Bioprecursors." *Applied Materials Today* 7: 246–54. doi:10.1016/j.apmt.2017.04.005.

Rahbar Shamskar, Kobra, Alimorad Rashidi, Parviz Aberoomand Azar, Mohammad Yousefi, and Sahar Baniyaghoob. 2019. "Synthesis of Graphene by in Situ Catalytic Chemical Vapor Deposition of Reed as a Carbon Source for VOC Adsorption." *Environmental Science and Pollution Research* 26 (4): 3643–50. doi:10.1007/s11356-018-3799-8.

Rahman, Shaharin Fadzli Abd, Mohamad Rusop Mahmood, and Abdul Manaf Hashim. 2014. "Growth of Graphene on Nickel Using a Natural Carbon Source by Thermal Chemical Vapor Deposition." *Sains Malaysiana* 43 (8): 1205–11.

Rao, C. N.R., Kanishka Biswas, K. S. Subrahmanyam, and A. Govindaraj. 2009. "Graphene, the New Nanocarbon." *Journal of Materials Chemistry* 19 (17): 2457–69. doi:10.1039/b815239j.

Rosmi, Mohamad Saufi, Sachin M. Shinde, Nor Dalila Abd Rahman, Amutha Thangaraja, Subash Sharma, Kamal P. Sharma, Yazid Yaakob, et al. 2016. "Synthesis of Uniform Monolayer Graphene on Re-Solidified Copper from Waste Chicken Fat by Low Pressure Chemical Vapor Deposition." *Materials Research Bulletin* 83: 573–80. doi:10.1016/j.materresbull.2016.07.010.

Roy, Amrita, Saptarshi Kar, Ranjan Ghosal, Rabindra Mukhopadhyay, Kinsuk Naskar, and Anil K. Bhowmick. 2022. "Synthesis and Characterization of Graphene Sheets Decorated with Carbon Black by Direct Pyrolysis of a Molasses-Carbon Clack Mixture as a Potential Versatile Filler for Rubber". *Rubber Chemistry and Technology* (in press); doi:10.5254/rct.21.79928

Roy, Amrita, Saptarshi Kar, Ranjan Ghosal, Kinsuk Naskar, and Anil K. Bhowmick. 2021. "Facile Synthesis and Characterization of Few-Layer Multifunctional Graphene from Sustainable Precursors by Controlled Pyrolysis, Understanding of the Graphitization Pathway, and Its Potential Application in Polymer Nanocomposites." *ACS Omega* 6 (3): 1809–22. doi:10.1021/acsomega.0c03550.

Ruan, Gedeng, Zhengzong Sun, Zhiwei Peng, and James M. Tour. 2011. "Growth of Graphene from Food, Insects, and Waste." *ACS Nano* 5 (9): 7601–7. doi:10.1021/nn202625c.

Ruiz-Hitzky, Eduardo, Margarita Darder, Francisco M. Fernandes, Ezzouhra Zatile, Francisco Javier Palomares, and Pilar Aranda. 2011. "Supported Graphene from Natural Resources: Easy Preparation and Applications." *Advanced Materials* 23 (44): 5250–55. doi:10.1002/adma.201101988.

Seo, Hong Kyu, Tae Sik Kim, Chibeom Park, Wentao Xu, Kangkyun Baek, Sang Hoon Bae, Jong Hyun Ahn, Kimoon Kim, Hee Cheul Choi, and Tae Woo Lee. 2015. "Value-Added Synthesis of Graphene: Recycling Industrial Carbon Waste into Electrodes for High-Performance Electronic Devices." *Scientific Reports* 5 (October): 1–10. doi:10.1038/srep16710.

Shah, Javishk, Janneth Lopez-Mercado, M. Guadalupe Carreon, Armando Lopez-Miranda, and Maria L. Carreon. 2018. "Plasma Synthesis of Graphene from Mango Peel." *ACS Omega* 3 (1): 455–63. doi:10.1021/acsomega.7b01825.

Shams, S. Saqib, Li Sheng Zhang, Renhao Hu, Ruoyu Zhang, and Jin Zhu. 2015. "Synthesis of Graphene from Biomass: A Green Chemistry Approach." *Materials Letters* 161: 476–79. doi:10.1016/j.matlet.2015.09.022.

Sharma, Subash, Golap Kalita, Ryo Hirano, Sachin M. Shinde, Remi Papon, Hajime Ohtani, and Masaki Tanemura. 2014. "Synthesis of Graphene Crystals from Solid Waste Plastic by Chemical Vapor Deposition." *Carbon* 72: 66–73. doi:10.1016/j.carbon.2014.01.051.

Siokou, A., F. Ravani, S. Karakalos, O. Frank, M. Kalbac, and C. Galiotis. 2011. "Surface Refinement and Electronic Properties of Graphene Layers Grown on Copper Substrate: An XPS, UPS and EELS Study." *Applied Surface Science* 257 (23): 9785–90. doi:10.1016/j.apsusc.2011.06.017.

Stobinski, L., B. Lesiak, A. Malolepszy, M. Mazurkiewicz, B. Mierzwa, J. Zemek, P. Jiricek, and I. Bieloshapka. 2014. "Graphene Oxide and Reduced Graphene Oxide Studied by the XRD, TEM and Electron Spectroscopy Methods." *Journal of Electron Spectroscopy and Related Phenomena* 195: 145–54. doi:10.1016/j.elspec.2014.07.003.

Sun, Zhengzong, Zheng Yan, Jun Yao, Elvira Beitler, Yu Zhu, and James M. Tour. 2010a. "Growth of Graphene from Solid Carbon Sources." *Nature* 468 (7323): 549–52. doi:10.1038/nature09579.

Tahir, Noor Ayuma Mat, Mohd Fadzli Bin Abdollah, Noreffendy Tamaldin, Mohd Rody Bin Mohamad Zin, and Hilmi Amiruddin. 2020. "Optimisation of Graphene Grown from Solid Waste Using CVD Method." *International Journal of Advanced Manufacturing Technology* 106 (1–2): 211–18. doi:10.1007/s00170-019-04585-2.

Tang, Pei Duo, Qi Shi Du, Da Peng Li, Jun Dai, Yan Ming Li, Fang Li Du, Si Yu Long, Neng Zhong Xie, Qing Yan Wang, and Ri Bo Huang. 2018. "Fabrication and Characterization of Graphene Microcrystal Prepared from Lignin Refined from Sugarcane Bagasse." *Nanomaterials* 8 (8). doi:10.3390/nano8080565.

Tiwari, Rajanish N., M. Ishihara, Jitendra N. Tiwari, and Masamichi Yoshimura. 2012. "Transformation of Polymer to Graphene Films at Partially Low Temperature." *Polymer Chemistry* 3 (10): 2712. doi:10.1039/c2py20448g.

Tuinstra, F., and J. L. Koenig. 1970. "Raman Spectrum of Graphite." *The Journal of Chemical Physics* 53 (3): 1126–30. doi:10.1063/1.1674108.

Wang, Yaling, Zehui Li, Cheng Tang, Haixia Ren, Qiang Zhang, Mo Xue, Jin Xiong, et al. 2019. "Few-Layered Mesoporous Graphene for High-Performance Toluene Adsorption and Regeneration." *Environmental Science: Nano* 6 (10): 3113–22. doi:10.1039/c9en00608g.

Wu, Zhiheng, Yongshang Zhang, Lu Li, Yige Zhao, Yonglong Shen, Shaobin Wang, and Guosheng Shao. 2020. "Nitrogen-Doped Vertical Graphene Nanosheets by High-Flux Plasma Enhanced Chemical Vapor Deposition as Efficient Oxygen Reduction Catalysts for Zn-Air Batteries." *Journal of Materials Chemistry A* 8 (44): 23248–56. doi:10.1039/d0ta07633c.

Xin, Hao, and Wei Li. 2018. "A Review on High Throughput Roll-to-Roll Manufacturing of Chemical Vapor Deposition Graphene." *Applied Physics Reviews* 5 (3). doi:10.1063/1.5035295.

Yan, Kai, Lei Fu, Hailin Peng, and Zhongfan Liu. 2013. "Designed CVD Growth of Graphene via Process Engineering." *Accounts of Chemical Research* 46 (10): 2263–74. doi:10.1021/ar400057n.

Yan, Yuxin, Fathima Zahra Nashath, Sharon Chen, Sivakumar Manickam, Siew Shee Lim, Haitao Zhao, Edward Lester, Tao Wu, and Cheng Heng Pang. 2020. "Synthesis of Graphene: Potential Carbon Precursors and Approaches." *Nanotechnology Reviews* 9 (1): 1284–1314. doi:10.1515/ntrev-2020-0100.

Yan, Zheng, Zhiwei Peng, and James M. Tour. 2014. "Chemical Vapor Deposition of Graphene Single Crystals." *Accounts of Chemical Research* 47 (4): 1327–37. doi:10.1021/ar4003043.

Yang, Po Kang, Wen Yuan Chang, Po Yuan Teng, Shuo Fang Jeng, Su Jien Lin, Po Wen Chiu, and Jr Hau He. 2013. "Fully Transparent Resistive Memory Employing Graphene Electrodes for Eliminating Undesired Surface Effects." *Proceedings of the IEEE* 101 (7): 1732–39. doi:10.1109/JPROC.2013.2260112.

Zhang, Zhuo, and Qi Wang. 2017. "The New Method of XRD Measurement of the Degree of Disorder for Anode Coke Material." *Crystals* 7 (1): 1–10. doi:10.3390/cryst7010005.

Zhou, Shuang, Junling Xu, Yubin Xiao, Ni Zhao, and Ching Ping Wong. 2015. "Low-Temperature Ni Particle-Templated Chemical Vapor Deposition Growth of Curved Graphene for Supercapacitor Applications." *Nano Energy* 13: 458–66. doi:10.1016/j.nanoen.2015.03.010.

4 Functionalization of Graphite and Graphene

Akash Ghosh and Simran Sharma
IIT Kharagpur

Anil K. Bhowmick
The University of Houston

Titash Mondal
IIT Kharagpur

CONTENTS

DOI: 10.1201/9781003200444-4

4.1   INTRODUCTION

Over the last two decades, research related to the carbon-based nanomaterials for different applications ranging from healthcare, automotive section, aerospace industries are on the rise. Carbon nanomaterials can be classified based on the dimensionality. For instance, fullerene is considered to be zero-dimensional, while carbon nanotubes (CNTs) are one-dimensional and graphene is considered to be the two-dimensional material. It is worth mentioning that graphene is considered to be the mother of all carbon nanomaterials. Among different classes of carbon nanomaterials that are used, graphene is getting significant attention from the researchers across the globe. The exhilarating properties demonstrated by graphene are considered as a matter of keen area of research in the past two decades. The discovery of atomically thin two-dimensional graphene sheets from graphite by Novoselov and Geim has led to a revolution in graphene research. A scotch tape technique was used to isolate the graphene from the highly oriented pyrolytic graphite block. The isolated material demonstrated Young's modulus of 1 TPa, inherent thermal conductivity (3,000 W mK^{-1}), and electron mobility (2.0×10^5cm^{2}V^{-1}s^{-1}). Further, the material is also impermeable to gases and demonstrated high optical absorption (Bolotin et al. 2008; Nair et al. 2008). This makes graphene a versatile material useful for wide range of applications including sensors (Vuorinen et al. 2016; Eswaraiah, Balasubramaniam, and Ramaprabhu 2011; Feng et al. 2012), photodetectors (Liu et al. 2015; Chang et al. 2017), and supercapacitors (Yoo et al. 2011). Owing to its multifunctional properties, nowadays, graphene finds its application beyond the electronic segments as well. There are multiple reports, wherein graphene or its hybrid has been utilized as filler for the development of tire composition (Mondal et al. 2018).

As shown in Figure 4.1a, graphene demonstrates a hexagonal lattice. Ideally, the lattice system of graphene cannot be classified as a Bravais lattice. Typically, the lattice system of graphene is considered as a fusion of two triangular lattices wherein the one individual carbon atom is linked to three other neighboring carbon atoms with carbon single bond carbon distance of 0.142 nm. This forms the two-dimensional crystal, which is of *P6mm* plane group. The structural rigidity is attributed to the formation of sigma bonds, due to the interaction of three half-filled valence orbitals, thereby leaving a half-filled orbital in an uncoupled state. These uncoupled orbitals result in the formation of a pi-cloud as shown in Figure 4.1b.

These out of plane p_z orbitals (pi clouds) play a critical role toward controlling the processability of the filler further. Owing to the presence of the pi-cloud, the graphene or graphitic fillers demonstrate an inert behavior. It does not possess affinity toward many solvents. Dispersibility in solvent is essential to enhance its attributes and makes it an ideal choice for practical applications. Thus, it can be unequivocally suggested that functionalization of graphene and graphitic fillers is needed. The functionalization of graphene is an emerging and essential area of research to use this wonderful material in real-time applications.

In quest of understanding the functionalization of graphene and graphitic materials, it is important to understand the structure of the graphene and graphitic material. The contribution of thermodynamics and kinetics factors toward modification of graphene is critical to understand. Ideally, functionalization of graphene can be

(a) (b)

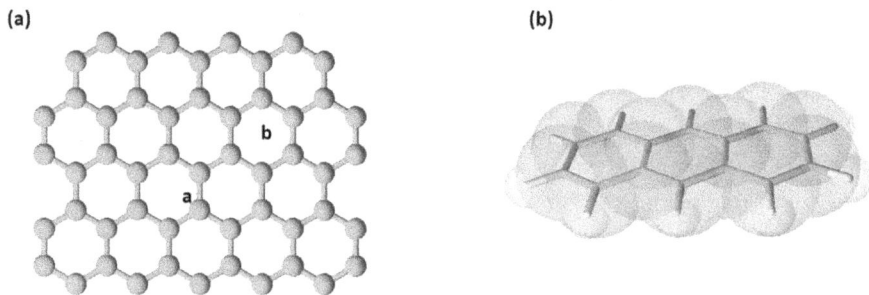

FIGURE 4.1 (a) A fragment of a graphene lattice with two triangular sublattices marked as *a* and *b*. (b) Representative example of the out-of-plane p_z orbitals forming delocalized π cloud above and below the basal plane of graphene.

targeted at the basal plane or at the edges of the material. However, modification at the basal plane of the graphene is tricky. Modification at the basal plane results in the formation of high energy radicals. Thermodynamically, the formation of such radicals is highly unfavorable. In terms of the kinetic purview, change in hybridization of modified carbon (post-modification) introduces geometric constraints. Hence, modification of graphene at the basal plane is challenging task, except under specific conditions. One of such specific case is the defect specific modification of the graphene. Graphene and graphitic materials demonstrate topological isolated defects in their ring, commonly referred to as the Stone–Wales defect. In an ideal situation, graphene and graphitic materials demonstrate a perfect array of hexagonal ring system. However, in the case of Stone–Wales defect, one carbon-carbon bond of the ring is flipped by 90° (participation of four hexagonal rings) and results in the formation of two heptagonal and two pentagonal rings in the system as shown in Figure 4.2a. The other common type of defect pertains to the missing atom in the ring. Vacancy of atoms in the framework leads to Jahn–Teller distortion at those sites. This results in the saturation of three dangling bonds over the vacant site, while the other bond remains further apart due to the introduction of the constraint in the geometry. This results in the formation of five- and nine-membered ring in the skeleton as shown in Figure 4.2b. Defect specific modification of graphene and its analogs is possible and will be discussed in detail in the later section of the chapter. In contrary to the basal plane modification, edge specific modifications can be an easier route for modification of the graphene and its analogs. This is so because the C=C bonds at the edges are more strained and surface modifications favor the conversion of sp² hybridized carbon of graphene to take up a more relaxed sp³ hybridized form.

In terms of the nature of the functionalization, it can be further classified via the non-covalent and the covalent modification route. The non-covalent route involves the interaction of π systems which are helpful in applications such as protein, organic supramolecules, and protein-DNA stabilization (Georgakilas et al. 2012). For instance, ionic liquids have been used to non-covalently wrap the graphene oxide for the development of polymer composite (Damlin et al. 2015). Similarly, biomaterials like chitosan and dextran have been also utilized to non-covalently modify graphene (Xie et al. 2016). The covalent route involves the formation of covalent

(a)   (b)

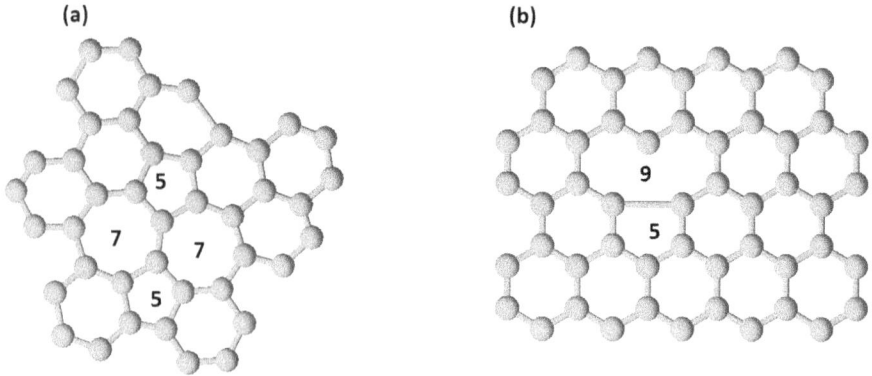

FIGURE 4.2 (a) Stones–Wales defect [5+7] in graphene. (b) Defect due to Jahn–Teller distortion resulting in [9+5] in graphene.

bonds between the graphene and its modifier. However, this chapter will deal mostly with covalent functionalization, which is more challenging and has more opportunity. Covalent functionalization of graphene can be done by small organic molecules or by the macromolecules. Further, covalent functionalization works on either C–C or C–N bond formations that are more stable. It is worth mentioning that most of the functionalization of graphene starts by using graphene oxide as the precursor. Eminent scientists like Brodie took interest in determining the molecular weight of graphitic material. He used strong oxidants like potassium perchlorate for his reaction recipe. Serendipitously, he landed up with a new class of compound known as graphite oxide (Brodie 1983). Following that, there has been a series of experiment by many stalwarts like Staudenmaier (1898) and Hummers and Offeman (1958). Not only the graphene oxide serves as a gateway for further modification, but as material itself, graphene oxide can be utilized as a filler in multiple polymer and elastomer composite. For instance, Mondal et al. utilized the graphene oxide as a filler to improve their thermal properties of acrylate-based polymers (Mondal, Chandra and Bhowmick 2016). Treatment with mixed acid (H_2SO_4 and HNO_3) and subsequent heating at higher temperature causes most effective modification of natural graphite as compared to other reagents (George, Bandyopadhyay and Bhowmick 2008). The modified graphene was used to prepare ethylene vinyl acetate based composites which exhibited superior thermal and mechanical properties. Similarly, the readers are suggested to read seminal review work on graphene-elastomer composite, wherein the utility of graphene oxide as a filler for polymer and rubber nanocomposite is captured by Mondal et al. (2017).

Hitherto, it can be noted that covalent functionalization of graphene can lead to a broad spectrum of graphene-based fillers with tunable properties. Furthermore, covalent functionalization is a more frequently used technique to functionalize graphene, and therefore, this chapter focuses on various approaches that can be used for the same (Criado et al. 2015; Englert et al. 2011). Different types of functionalization of graphene are shown in Figure 4.3. In the remaining section of the chapters, details about different functionalization of graphene will be captured.

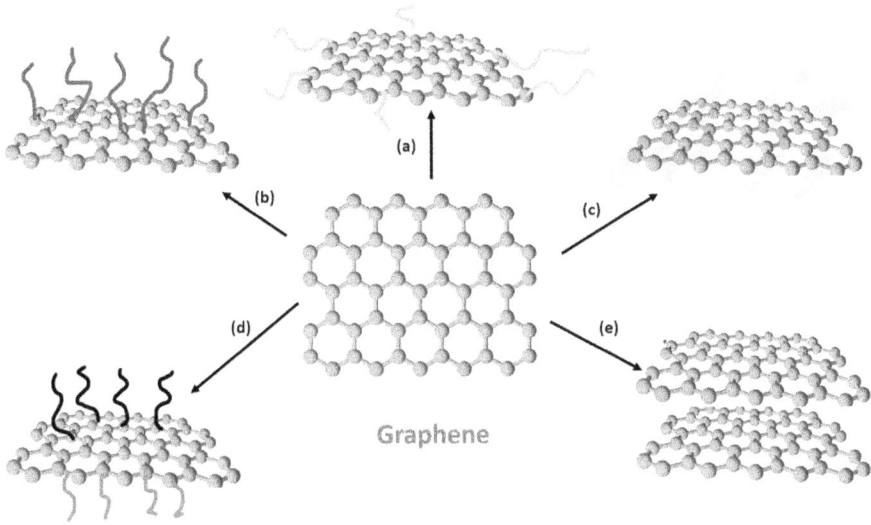

FIGURE 4.3 Various functionalization routes for graphene: (a) via the edges, (b) on the basal plane, (c) non-covalent route, (d) asymmetrically modified basal plane, and (e) self-assembling of modified graphitic sheets.

4.2 COVALENT FUNCTIONALIZATION OF GRAPHENE AND ITS ANALOGS VIA CARBON-CARBON BOND FORMATION (SMALL MOLECULES)

4.2.1 FUNCTIONALIZATION VIA DIAZOTIZATION CHEMISTRY

Diazonium chemistry has a significant contribution in the covalent functionalization of material. It emphasizes the carbon-carbon bond formation using diazonium salt for the tuning of graphitic-like materials—the functionalization reaction is believed to be governed by the free radical mechanism. The generic functionalization of graphene via the diazonium chemistry is shown in Figure 4.4. Liu et al. reported that aniline was functionalized on graphene nanosheet using diazonium coupling, further reduced by hydrazine. The functional attachment was confirmed by X-ray photoelectron spectroscopy (XPS), Fourier Transform Infrared (FTIR) Spectroscopy, and Raman Spectroscopy. The diazonium salt releases a molecule of nitrogen gas and produces a highly reactive aryl radical using any thermal, photochemical, or electrochemical stimuli. As the radical gets covalently attached with the graphitic material, the hybridization of graphitic carbon changes from sp^2 to sp^3. Consequently, the aromatic system will get disrupted, and a change is observed in the electronic behavior (Liu et al. 2014).

In the same fashion, Sinitskii et al. (2010) experimentally exhibited that the conductivity of functionalized graphene can be tuned by controlling the reaction. Controlled functionalization reaction introduces an energy bandgap in the graphene. Bekyarova et al. reported that 4-nitrobenzenediazonium tetrafluoroborate was attached to the epitaxial graphene on the SiC substrate. The material was characterized by XPS

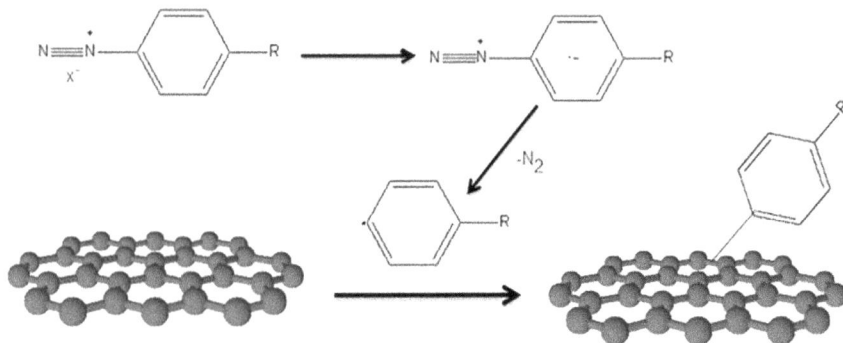

FIGURE 4.4 Schematic perspective on the functionalization of graphene via diazotization chemistry.

and FTIR Spectroscopy analysis (Bekyarova et al. 2012). Sun et al. used top-down synthesized graphene and thoroughly investigated the reactivity of nitrophenyl with it. Single-layer and multi-layer graphenes were prepared on a silicon wafer from the graphite surface and annealed, further reacted with a diazonium salt. The top and edge of the graphene sheet were found to be more reactive than the core (Sun et al. 2016).

Sinitskii et al. (2010) reported an oxidative unzipping of CNT to generate nitro-phenyl grafted graphene through a diazonium reaction. The sample was analyzed with the help of AFM, SEM, and XPS to propose a kinetic model explaining the conductivity of the modified graphene. Greenwood et al. used 4-nitrobenzenedia-zonium tetrafluoroborate and 3,5-bis-tert-butylbenzenediazonium chloride to alter the surface of graphene. The addition of diazonium moeity is considered to follow an electron transfer process from graphene to diazonium to produce the aryl radical (Greenwood et al. 2015).

Various other groups like nitro, cyano, chloro, fluoro, bromo, methoxy, and car-boxy can be placed in the para position of diazonium salt to impart new characteris-tics. These substituted diazonium salts have a versatile role in improving graphene's dispersibility in the solvents. Initially, the dispersion of nanostructured carbon was a challenge, and surfactant used to assist in the dispersion. The modified graphene with functional group shows a good dispersion in various solvents. Sun et al. used diazonium salt to link bromophenyl on a graphene sheet. The edge and defect sites were explicitly functionalized (Sun et al. 2016). Similarly, Hirsch et al. utilized p-sulfonylphenyldiazonium chloride and p-t-butylphenyldiazonium tetrafluoroborate to modify graphene covalently. This functionalized graphene shows the solubility of about 27 µg mL^{-1} in chloroform (Englert et al. 2012). The carboxyl terminated dia-zonium salt is famous for its post-functionalized usage. Reactions like esterification and amidation can attach the organic group with the graphene. Xu and co-worker synthesized carboxylated phenyl covalently linked with graphene and subjected to post-esterification with perfluoro-1-octanol. The functionalized graphene was solu-ble in water, but it was found to be dispersed in o-chlorobenzene. Moreover, perfluo-roalkyl introduces a band gap in the system which can be used for the photovoltaic

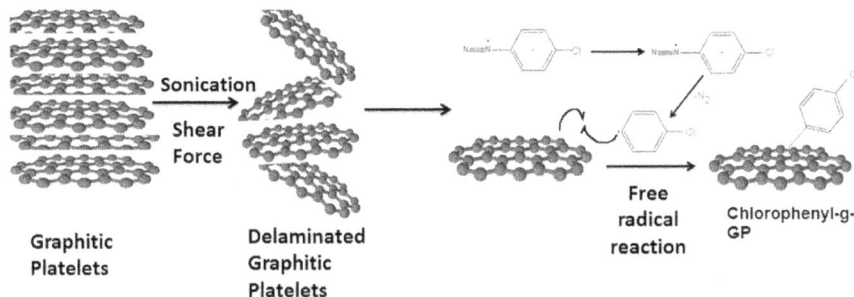

FIGURE 4.5 Modification of graphene using 4-chloroaniline as the starting material for diazotization process.

system (Ye et al. 2012). On a similar note, Mondal, Bhowmick, and Krishnamoorti investigated the antimicrobial activity of chlorophenyl functionalized graphene. Incidentally, this antimicrobial efficacy of the modified graphene (using diazotization chemistry) was reported for the first time by our group and was in the list of first few reports on antimicrobial graphene in the literature. Graphene nanoplatelets were functionalized with chlorophenyl using a diazonium salt reaction and is shown in Figure 4.5 (Mondal, Bhowmick, and Krishnamoorti 2012). The antimicrobial efficacy of these engineered nanomaterials was noted against the *Escherichia coli* and *Staphylococcus aureus* bacteria. The antimicrobial activity of the engineered graphene particles was demonstrated from the turbidity and the zone of inhibition analysis. Based on the experimental window, it was noted that the chlorophenyl modified graphene demonstrated superior antimicrobial properties compared to the neat graphene and graphene oxide. This is attributed to the presence of chlorine atoms in the modified nanomaterials. Typically, chlorines are well reported for their biocidal activities.

Kotal and Bhowmick (2013) developed a novel chemical approach to design a three-dimensional porous network of reduced graphene oxide (GO)/multiwalled carbon nanotube (MWCNT) hybrid using the diazo chemistry. The process involved simultaneous functionalization, reduction, and stitching of GO by p-phenylenediamine and subsequent diazotization followed by C–C coupling with MWCNT. The chemical hybrid exhibited good electrical conductivity (210.48 S m^{-1}), promising specific capacitance (277 F g^{-1}) even at high current density (10 A g^{-1}), remarkable energy density (21.32 W h kg^{-1}) especially at high power density (3.13 kW kg^{-1}), outstanding cyclability (89%) even after 2,000 cycles, and good dye adsorption capacity (245 mg g^{-1}) for crystal violet.

4.2.2 FUNCTIONALIZATION VIA DIELS–ALDER REACTION

The Diels–Alder reaction is renowned and widely used in organic chemistry to form carbon-carbon bonds between dienophile alkene and a conjugated diene. This reaction proceeds through the [4+2] cycloaddition pathway to yield a six-membered cyclic compound and successfully applied on various carbon nanomaterials (Chang and Liu. 2009; Wang et al. 2005; Delgado et al. 2004; Munirasu et al. 2010). Graphene being a polyaromatic system can resemble both dienophile and diene depending on the counter reactant. Sarkar et al. investigated the dual character of graphene in Diels–Alder cycloaddition reaction. Graphene acts as diene when reacting with tetra-cyanoethylene (TCNE) and maleic anhydride, whereas reaction with 2,3-dimethoxy-1,3-butadiene (DMBD) or 9-methyl anthracene resembles dienophile like behavior (Sarkar et al. 2011). This reaction has unique features like thermal reversibility and rapid reaction kinetics involving catalyst-free reaction. Pristine graphene can be regenerated using thermal treatment and can regain its electronic properties. The reaction scheme is shown in Figure 4.6.

The retro reaction of tetra functionalized graphene in the p-Xylene has been carried out at 100°C and verified by the Raman spectroscopy and XPS studies. Cao et al. studied different reactive sites of graphene nanosheets using various functional groups via Diels–Alder cycloaddition. They reported that edge functionalization was positively achieved using all four reactants, but the inner part remained unhampered (Cao et al. 2013). Li et al. reported a facile and rapid functionalized graphene using dihydronapthalene as *cis*-diene. The hydroxyl group on cis-diene decorated over the graphene enhanced the compatibility of dispersion in the polymer matrices like PMMA (Li et al. 2016). Chen et al. reported functional graphene and polyvinyl alcohol (PVA)-based nanocomposite showing high thermal conductivity. The graphene was initially modified with acrylic acid via Diels–Alder reaction, and other PVA was grafted onto the surface using the free carboxyl group (Chen et al. 2020). Polymer capped with cyclopentadienyl can be used as a diene, binding the polymer chain onto the graphene. Similar chemistry has been used in the case of cyclopentadienyl terminated polyethylene glycol (PEG) grafted graphene via Diels–Alder one-step reaction at 80°C (Yuan et al. 2012). Using the facile technique, atom transfer radical polymerization (ATRP) initiator can be attached over the graphene and then polymer growth can occur over the graphitic surface. Munirasu et al. (2010) used furfuryl group and maleic anhydride derivatives containing the molecule to decorate the graphitic surface (Zydziak, Yameen, and Barner-Kowollik 2013). Huang and his group used the Diels–Alder reaction to crosslink the polymer/graphene nanocomposite and utilized its distinct feature to exhibit self-healing properties.

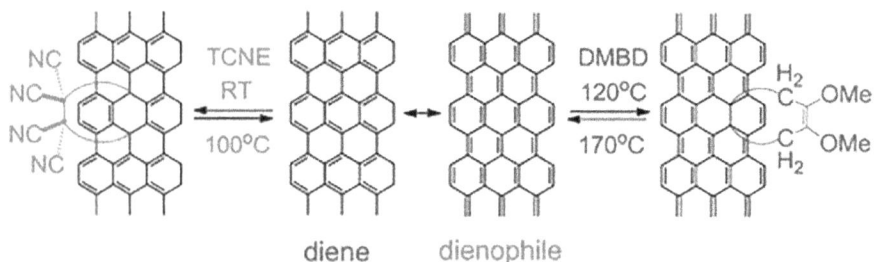

FIGURE 4.6 Schematics of the Diels–Alder reaction on graphene, wherein the graphene can act both like diene and dienophile. (Adapted with permission from Sarkar et al. 2011. "Diels–Alder Chemistry of Graphite and Graphene: Graphene as Diene and Dienophile." *Journal of the American Chemical Society* 133 (10): 3324–27. Copyright 2011 American Chemical Society.)

The crosslinking site healed using multiple stimuli like heat, infrared light and cleaved in a controlled manner (Li et al. 2019). Seo, Jeon, and Baek (2013) reported a solid-state Diels–Alder reaction of graphite in the presence of maleic anhydride or maleimide via ball milling process. This resulted in an edge functionalized graphene nanoplate. The same group also reported a solvent-free Diels–Alder reaction that helped synthesize functionalized graphene nanosheets from graphite. Graphite was sealed in an inert atmosphere with maleimide or maleic anhydride at 160°C for 12 hours to synthesize a surface decorated graphene nanomaterial (Seo and Baek 2014). Zhong et al. performed functionalization of graphene using aryne cycloaddition. The 1,2-(trimethylsilyl) phenyltriate and Cesium fluoride were used to generate the aryne intermediate (Zhong et al. 2010).

4.2.3 FUNCTIONALIZATION VIA REACTION WITH CARBENE

Carbene is one of the most reactive organic intermediates capable of forming C–C bonds with graphene. This intermediate mainly undergoes an insertion reaction with C–H, X–H bond (where X=O, S, N), and cyclopropanation reaction (cycloaddition type [1+2]) with C=C to give a three-membered cycloalkane. As graphene has the basis of sp^2 carbon and some sp^3 carbon present in the edge or defect, it gives an opening as an interactive site for carbene to form a functionalized graphene. Highly reactive carbene is mainly produced from the haloform and cyclic diazoalkane precursor. The chloroform in the concentrated alkaline medium forms in situ dichlorocarbene. This electron-deficient carbene can be transferred to graphene nanoplates and form a three-membered ring by the [1+2] addition reaction. Sainsbury et al. (2016) exhibited covalent functionalization of graphene with dibromocarbene in additive fashion and tuned the electronic properties. The intermediate generated from the bromoform in situ and can be modified with nucleophilic substitution or reductive coupling as shown in Figure 4.7.

Diazirine is another broadly used precursor for carbene generation. However, the synthesis is time-consuming and less productive, limiting its outreach, unlike nitrene. In addition to this, the diazoalkane is unstable and inapplicable in the organic synthesis. Diazirine is a three-membered cyclic compound that is relatively stable in the family. The thermal or radiative treatment of diazirine instantly produces electron-deficient carbene and a nitrogen molecule, similar to azide. 3-Aryl-3(trifluoromethyl)-diazirine derivatives are best known for carbene production. The molecular structure does not give scope for intermolecular rearrangement, leading to reduced yield. Primarily, 3-aryl-3(trifluoromethyl)-diazirine was used as the photoaffinity labels for the proteins. It was also used in surface modification and synthesis of functional materials. Ismaili et al. (2011) used 3-aryl-3(trifluoromethyl)-diazirine as an organic linker to immobilize gold nanoparticles on the surface of reduced graphene oxide. They used alkane thiol to functionalize gold nanoparticles initially, and a few of these are bound to diazirine moiety with thio-alkoxyl linkage. Gold modified particles were irradiated in UV light to form a Gold–Graphene hybrid system. The system showed a significant change in the absorbance for Au-modified graphene, and a peak of Au was observed in the XRD. Many groups have monitored the effect of different carbene and modification position on electronic properties using density functional

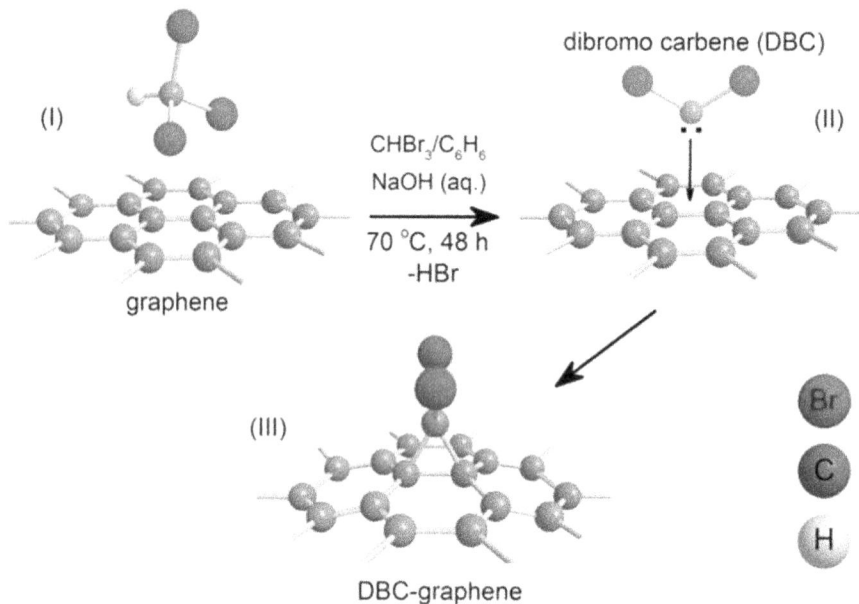

FIGURE 4.7 Schematics of the carbene addition reaction on graphene. (Adapted with permission from Sainsbury et al. 2011. "Covalent Carbene Functionalization of Graphene: Toward Chemical Band-Gap Manipulation. *ACS Applied Materials & Interfaces*, 8(7), 4870–77, 2011. Copyright 2016 American Chemical Society.)

theory. Petrushenko investigated the impact of cycloaddition [1+2] on various sites of graphene at the center, edge, and corner. The addition to the center generates a morphological distortion. However, other carbons at the corner and edge align with the six-membered ring to have a similar planar structure. A shift of both HOMO and LUMO has also been observed depending on the position of functionalization (Petrushenko 2014). Majidi et al. used different carbene groups (CR_2 with R = Cl, H, CN, CH_3, and NO_2) to tailor the electronic band gap. To conclude, functionalization of CCl_2 and CH_2 changes the energy gap depending on the concentration. However, groups like CR_2 with R = CN, CH_3, NO_2 behold a semiconductor property (Majidi and Rabczuk 2020).

4.3 COVALENT FUNCTIONALIZATION OF GRAPHENE AND ITS ANALOGS VIA CARBON-CARBON BOND FORMATION (MACROMOLECULES)

The modification of graphene and its analogs can be extended to the macromolecules as well. However, it is worth mentioning that tethering of the macromolecular chains can be primarily achieved using two techniques, namely *grafting from* and *grafting to* technique. In the next section of the chapter, these two methodologies will be discussed in detail.

4.3.1 GRAFTING FROM TECHNIQUE

In this technique, surface graphitic material is modified with the initiator moiety and associated with the monomer. These initiators are attached with hydroxyl and carboxyl groups of graphene oxide or covalently modified with a small organic moiety to generate the desired functionality for initiator attachment. Initiator molecule gets anchored onto the basal plane and edges of graphene and initiates the polymerization reaction. The growth of the polymer chain initiates from the graphene surface without any steric hindrance. These make polymer growth more feasible and controllable. Generally, controlled radical polymerization (CRP) has been dominantly used for the synthesis of polymer with narrow molecular weight distribution. Other factors like grafting density, grafting site, functionality, and polymer thickness can easily be controlled using CRP. ATRP and reversible addition fragment transfer (RAFT) are highly considered CRP for preparing polymer brushes on graphene in situ. However, ATRP technique is mainly preferred over RAFT as a narrow polydispersed system can be grown within a controlled manner. Different block copolymers can be synthesized using the technique on the surface of graphitic material.

Lee et al. have performed experiments to graft various monomers on the graphene surface. Initially, 2-Bromo-2-methylpropionyl bromide has been decorated on the surface. Further, the chain length was manipulated with monomer concentration. They investigated the variation of the structure with the monomer and the initiator concentration. Grafted polymer brush's molecular weight was estimated by cleaving the chain from the graphene surface using a saponification reaction. The grafting density increased by increasing the initiator content. Similarly, increasing the monomer concentration amplifies the polymeric brush's molecular weight and chain length (Lee et al. 2010).

Fang et al. used the ATRP method for grafting the polystyrene chain onto the graphitic surface using a 2-Bromobutyl Bromide initiator (Fang et al. 2010).The pristine graphitic material was functionalized with 2(4-aminophenyl)ethanol via diazonium chemistry followed by initiator binding. This site makes the initiator immobilized over the graphene for the polymerization reaction. The resulting nanocomposite exhibits multi-fold increments in mechanical and thermal conductivity. Increasing glass transition temperature of nanocomposite with the grafting density has also been reported (Eskandari et al. 2019).

Nowadays, the polymeric brush has gained massive attraction for its versatile utility in biomedical applications. The polymeric brush can be produced by controlling the grafting density and polymer chain length. Ohno et al. used novel amine-based ATRP initiator 2-((3-((2-bromo-2-methylpropanoyl)oxy)propyl)thio)ethylamine hydrochloride for graphene oxide to form polymer brush (Ohno, Zhao and Nishina 2019). The amine group of initiators reacted with the surface epoxy group of graphene oxide followed by the chain growth of different types of monomer. This polymeric brush grafted graphene oxide exhibits a liquid crystal-forming behavior in both ionic and non-ionic solvents. Steenecker et al. reported direct functionalization of graphene oxide using photopolymerization with styrene (Steenackers et al. 2011). They used CVD grown graphene on Copper foil filled with various monomers like styrene, methyl methacrylate (MMA), N,N-dimethylaminoethyl methacrylate

(MAEMA), methacrylatoethyl trimethyl ammonium chloride (METAC), and 4-vinyl pyridine (4VP) under UV light (350 nm) for 16 hrs. A patterned polymer brush of only styrene has been found to form on CVD graphene on the existing defect sites of the basal plane. Further grafting can be done using electron beam carbon deposition on the surface.

4.3.2 GRAFTING TO TECHNIQUE

Polymer chains are initially synthesized and tethered to graphitic material using amide, ester linkage, or click chemistry in this grafting technique. By using the esterification reaction, Salavagione et al. attached the PVA with the Graphene oxide in the presence N, N′-dicyclohexylcarbodiimide (DCC) and 4-dimethyl aminopyridine (DMAP) catalyst (Salavagione, Gomez and Martinez 2009). Another experiment by the same group was performed using graphene oxide treatment with $SOCl_2$ to form acid chloride on the surface that further reacted with the PVA to form an ester link between hydroxyl and chloride group. Similarly, Li et al. reported poly(piperazine spirocyclic pentaerythritol bisphosphonate) (PPSPB), which was functionalized with GO via esterification reaction. This GO-PPSB was reduced and used to prepare nanocomposite with ethylene-vinyl acetate copolymer and that exhibits good thermal stability and flame retardancy (Huang et al. 2012).

Liu et al. successfully prepared the PEG functionalized nano graphene oxide. The PEG amine stars and graphene oxide undergo an amidation reaction in the presence of carbodiimide. This nanocomposite was utilized for an anti-cancer drug delivery (Liu et al. 2008). Similarly, Jin et al. used PEG diamine to graft the graphene oxide via amide bond forming. They investigated the effect of the free amine group on thermostability and trypsin activity of various serine proteases. An azido terminated polymer can be grafted on alkyne-derived graphene (Jin et al. 2012). Pan et al. grafted poly(N-isopropyl acrylamide) on graphene for drug delivery application. The azide terminated polymer was synthesized using an azide-based chain transfer agent by the RAFT method. This retained the azide molecule in the polymer chain which helped in covalently linking with the graphene surface. Polymers like azide terminated Polystyrene, Polymethylmethacrylate, and Poly (4-vinyl pyridine) were used to graft onto the surface and the structure and solubility in various systems (Pan et al. 2011) were studied. Hitherto, we can note that multiple attempts were made to study the grafting to methodology. However, the major caveat associated with the method is to graft high molecular polymer chains onto graphene. In quest of addressing the same, Mondal et al. were the first to report a synthetic way to graft high molecular weight poly(caprolactone) onto anisotropic graphene using grafting to methodology as shown in Figure 4.8. The molecular weight of the model polymer was 80,000. Typically, the steric hindrance between the polymer chains and the functional groups of graphene results in challenges in the grafting process. However, the authors have meticulously modified the graphene with toluene diisocyanate. The isocyanate group demonstrates a linear shape and it prefers to stay out of the plane. This made the polymer chain easily approachable to the reaction site. The polymer brushes generated were stable (more than 7 months) and were thoroughly characterized by neutron scattering experiments. Additionally, the graphene-assisted polycaprolactone (PCL)

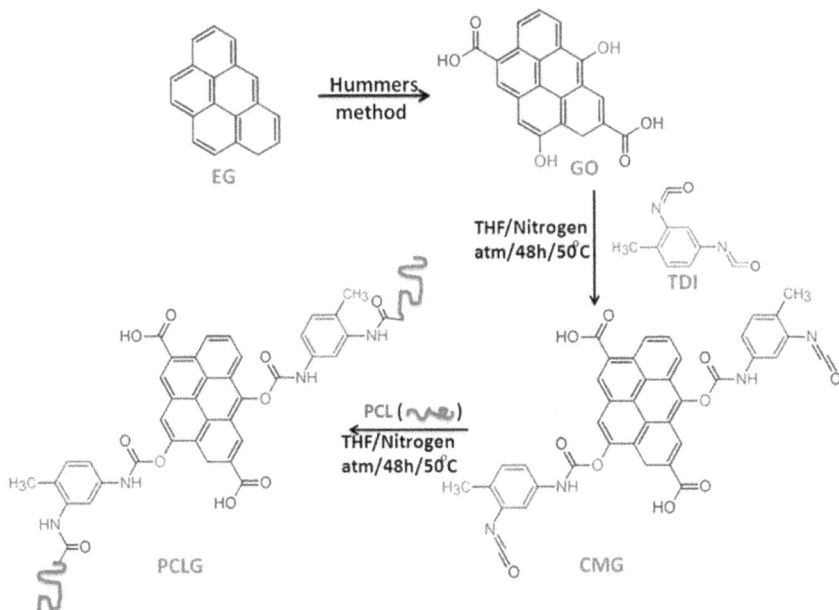

FIGURE 4.8 Schematics of the development of PCL brush over graphene via the grafting to technique.

exhibited superior thermal stability with an accelerated crystallization, which is used in solution-based processing and fabrication of graphitic nanocomposite (Mondal et al. 2016).

4.4 COVALENT FUNCTIONALIZATION OF GRAPHENE AND ITS ANALOGS VIA CARBON-NITROGEN BOND FORMATION

In the previous section, we have seen the modification of graphene via the formation of carbon bonds. The next most important atom which is significantly leveraged to modify the graphene is nitrogen atom. In the next part of the chapter, we will discuss different modifications of graphene using nitrogen.

4.4.1 FUNCTIONALIZATION WITH AMINE

Amine-based graphene functionalization seeks attention in various fields like polymer solar cell, sensor, drug delivery, and energy storage. The epoxy and carboxylic groups were mainly used to react with the amine for surface anchoring. A similar strategy was utilized by Mondal et al. to modify the surface of the graphene oxide with blocked amine. 1-Methyl imidazole-based ionic liquid was used as the modifier for the graphene oxide. The ionic liquid modified graphene was further utilized in polyurethane-based foam composition to generate pores with uniformity (Mondal, Basak, and Bhowmick 2017). Bourlinos et al. used different amines and amino acids to modify the surface of graphene oxide. The epoxy group on the graphene oxide under-goes the substitution and nucleophilic reaction to modify the properties (Bourlinos et al. 2003). Using microwaves, Caliman et al. (2018) proposed a direct method to prepare amine functionalized graphene oxide. They produced four different amine functionalized graphenes by the use of dibenzyl amine, p-phenylenediamine, diiso-propylamine, and piperidine. It was reported that the diisopropylamine and piperi-dine functionalized graphenes exhibit a better life cycle and specific capacitance of 290 F g^{-1} leading to its usage in the field of supercapacitors. Aguilar-Bolados and his co-workers reported the reductive amination of graphene oxide using the Leuckart reaction. Ammonium formate is used to reduce the carbonyl group present on the graphene oxide surface with simultaneous addition of amine (Aguilar-Bolados et al. 2017). Jeyaseelan et al. (2021) anchored the ethylenediamine molecule on graphene oxide for fluoride removal application.

Valentini et al. used butylamine modified graphene for polymer solar cell applica-tion. Graphene was initially treated with plasma in CF_4 gas. Further, it was dispersed in butylamine liquid for 1 hour of ultrasonication. Functionalized material acts as a hole acceptor in polymer-based photovoltaic applications (Valentini et al. 2010). Rabchinskii et al. reported a scalable method for synthesizing aminated graphene using liquid-phase treatment via a two-step reaction. In mild condition, graphene oxide was treated with hydrobromic acid and ammonia solution. The efficiency of amination can be enhanced by increasing the bromine content in the bromination step. This method can be further utilized for the post-modification of an amine group like 3-chlorobenzoyl chloride that was anchored to the aminated graphene via amida-tion reaction (Rabchinskii et al. 2020). The Hoffman rearrangement can also be used for preparing amino-functionalized graphene as reported by Zhang et al. Modified graphene oxide has been used to reinforce polyacrylate films and exhibit superior mechanical properties and better dispersion in the matrix (Zhang et al. 2016).

4.4.2 FUNCTIONALIZATION WITH NITRENE

Nitrene-based chemical modification of carbon nanomaterial is well known for syn-thesizing functional and smart materials. Organic azides are a significant source for nitrene generation, which initiates the C–N bond formation. This azide group liber-ates a molecule of nitrogen gas and a highly active nitrene in the presence of heat or light. However, there are plenty of azides in the market; but aromatic azide has attracted the scientific community for its various isomeric forms. A singlet phenyl azide gets isomerized to benzazirine and further to cyclic ketone imine. Both the singlet and the triplet do not react with alkene or alkane.

The nitrene attaches with a C=C bond to form a three-membered aziridine ring over the substrate, known as [1+2] cycloaddition reaction. The organic part of azide plays a vital role in decorating the graphene with added properties. For example, azide with a group like carboxy, hydroxyl, bromo, and amine has been used to form modified graphene with the respective functional group. This modified graphene can anchor to metal nanoparticles, producing magnetic and conductive polymer nanocomposite reported by Gao et al (2009, 2010). Nitrene linkage was identified in functionalized graphene by observing an increasing intensity of the ratio D/G band in Raman Spectra. It depicted the transformation of hybridization of graphene carbon from sp^2 to sp^3. Choi et al. used azido-trimethyl silane to tailor the band-gap of epitaxial graphene in vacuum condition. Thermal treatment has been done to link the trimethyl silane, and a bandgap of 0.66 eV was observed in functionalized graphene. The material was stable during the thermal treatment at 250°C for 5 minutes. However, bandgap diminished and material regained metallic characteristics on annealing at 850°C for 5 minutes (Choi et al. 2009).

Xu et al. (2012) have reported graphene with tetraphenyl ethylene (TPE) azide for fabricating nanodevices. The TPE azirine is well known for its aggregation-induced emissive property. The fluorescent spectra of TPE azide in THF can be observed by observing the maximum intensity that is found around 480 nm in the emission band. The aggregation of the TPE molecule decreased with the dilution and confirmed by the downfall of fluorescence emission. However, functionalized graphene with TPE does not behave analogously. Fluorescent emission of modified graphene around 480 nm quenched out due to the intermolecular interaction between TPE and graphene (Ji et al. 2020).

Polymer grafting to the graphitic nanosheet can be done via nitrene chemistry. Polymer must possess an azide moiety in the chain for this type of grafting. Synthesis of phenylaniline graphene was reported by Strom et al. (2010) using exfoliated graphene flakes. They performed the same reaction using protected phenylaniline to avoid the peptide linkage. The higher extent of functionalization was found in protected phenylaniline (73% wt loss) than the other (68% wt loss). The percentage of weight loss has been correlated with the extent of functionalization using TGA data. Similarly, Xu et al. reported the attachment of alkyl azides in the side chain of polyacetylene, and grafting of graphene to polyacetylene occurred through nitrene addition. The graphene-based polymeric material shows good dispersion in organic solvent with improved luminescence properties (Xu et al. 2011).

There are some other azides, like iodine azide, which react differently. Devadoss and Chidsey (2007) used iodine azide to decorate the graphitic surface. Unlike alkyl azide, it gets directly attached to the graphene surface and substitutes alkyl azide. A similar strategy was utilized by Mondal et al. to directly graft iodine azide on the surface of graphene as shown in Figure 4.9 (Mondal, Bhowmick, and Krishnamoorti 2014b). The azide modified surface was further ethynyl fluorophenyl via the click chemistry. This resulted in the generation of stress onto the graphene and thereby the band gap of the graphene was found to be tuned. The extent of stress generated on the graphene was calculated by leveraging mathematical tensors. A biaxial compressive

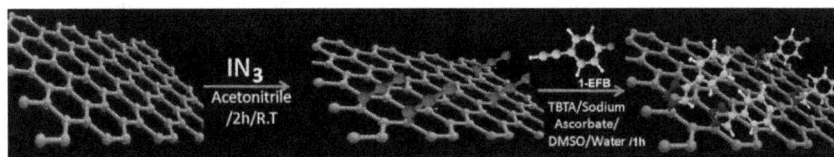

FIGURE 4.9 Schematics of the modification of graphene surface using iodine azide (first product) followed by click chemistry to yield the triazole ring containing product (second product).

stress of 1.60 GPa was exerted on the graphene sheet by the modifier. Owing to the generation of the stress, the band gap of the graphene was affected. From a semi-metallic character, the modified material demonstrated a semiconductor behavior.

4.5 OTHER CLASSES OF MODIFIED GRAPHENE

4.5.1 NITROGEN-DOPED GRAPHENE

Nitrogen atom impurities can be imparted in the graphene framework by substitutional doping. This class of material is known as nitrogen-doped graphene. Synthesis of nitrogen-doped graphene can be classified into two classes: direct synthesis and post-synthesis. In the direct method, nitrogenous gas is annealed with carbon-containing gas using the chemical vapor deposition (CVD) technique. Precursor gas at high temperature splits into small fragments. These fragments collide and rearrange to form nitrogen-doped graphene using a metal catalyst.

Wei et al. used a mixture of ammonia and methane gases to fabricate nitrogen-doped graphene on the copper catalyst in the CVD chamber (Wei et al. 2009). Similarly, Reddy et al. annealed acetonitrile in a CVD reactor to generate nitrogen-doped graphene successfully. In the CVD technique, the atomic proportion of carbon and nitrogen can easily be manipulated by changing the nitrogen and carbon precursor ratio (Reddy et al. 2010). In the same fashion, Jin et al. used pyridine precursors to be annealed in the presence of copper (Jin et al. 2011). Deng et al. synthesized nitrogen-doped graphene using the solvothermal technique. Lithium nitride and carbon tetrachloride were initially mixed and heated in solvothermal conditions. This technique can be used as a direct method in preparing N-doped graphene (Deng et al. 2011).

In the post-treatment method, graphitic material is heated with nitrogen precursor at an elevated temperature. Guo et al. heated graphene in ammonia gas at $1,100°C$ to produce nitrogen-doped graphene (Guo et al. 2010). A similar type of methodology was used by Geng et al. (2011), but they reported that nitrogen-doped graphene could be synthesized at lower temperatures ranging from $800°C$ to $900°C$. Liu et al. used graphene oxide and aniline as nitrogen sources for generating nitrogen-doped graphene. The obtained graphene was annealed around $700°C–1,000°C$ (Guo et al. 2020a).

Apart from using the traditional techniques for developing nitrogen-doped graphene by incorporating gaseous precursors, our team has smartly designed nitrogen containing polymer and further subjected these polymers to appropriate conditions in CVD reactor to yield nitrogen-doped graphene (Mondal, Bhowmick and Krishnamoorti. 2015). This was incidentally the first report on developing the nitrogen doped graphene from polymeric precursors. The authors successfully established the mechanistic route for the formation of nitrogen-doped graphene via validation using mass spectroscopy. It was observed that aromatic rings containing polymers demonstrated an ease in the formation of graphene or nitrogen-doped graphene. Polyurethane acrylate (containing aromatic ring) in its structure was used as the model polymer for the reaction condition. A series of free radical reactions at higher temperature yielded the nitrogen-doped graphene as shown in Figure 4.10.

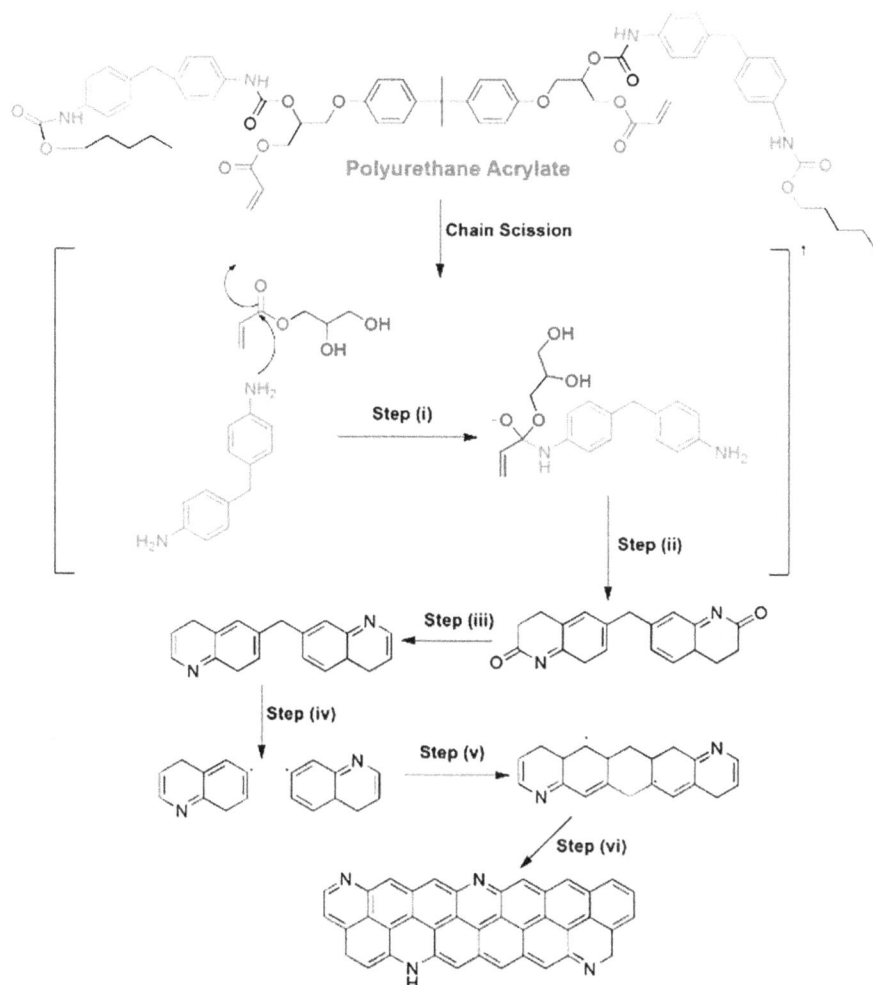

FIGURE 4.10 Schematics of the formation of nitrogen-doped graphene. (Adapted with permission from Mondal, Bhowmick and Krishnamoorti. 2015. "Controlled Synthesis of Nitrogen-Doped Graphene from a Heteroatom Polymer and Its Mechanism of Formation. *Chemistry of Materials*, 27(3), 716–25. Copyright 2015 American Chemical Society.)

4.5.2 Carboxylated Graphene

Carboxyl functionalized graphene can be used for the biomedical or dye absorption application. Already there are carboxyl groups present over the graphene surface; however, their density is less. Deliberate addition of carboxyl group over the graphene opens up a large domain of applications like biosensor, microfilter, and dye extractors (Mondal, Bhowmick, and Krishnamoorti 2013; Zhao et al. 2017). Park et al. used chloroacetic acid in an alkaline medium to transform the hydroxyl and epoxy groups into carboxyl groups. Modified graphene was further treated with manganese nitrate in a hydrothermal medium to give Carbonylated-graphene–Mn_2O_3. This material can be used as a molecular sieve and supercapacitor (Park 2014). Similarly, many researchers have reported different carboxylic functional graphenes using chloroacetic acid and alkaline solution (Ma et al. 2019; Ziółkowski, Górski, and Malinowska 2017). Zhao et al. reported the preparation of carboxylated graphene oxide sponge using the former method for dye remover application. The efficiency of dye removal was found to be double after carboxylation of GO (Zhao et al. 2017). Guo et al. suggest that a mild alkaline medium is preferred for functionalizing graphene oxide. In excess alkaline medium, graphene oxide gets partially reduced, and the efficiency of functionalization reaction gets reduced (Guo et al. 2020b). Ryzhkov et al. used peroxide and potassium hydroxide to oxidize graphene oxide into carboxylated graphene oxide. XPS spectra confirmed that 10.9% carboxyl group addition (Ryzhkov et al. 2020). Mondal et al. reported a novel technique to prepare carboxyl group decorated graphene using organolithium reagent for dye adsorbing material. Butyl Lithium plays a dual role as proton scavenger as well as a nucleophile. The nucleophilic reaction is quite difficult to achieve, but it becomes feasible with the help of organolithium. Butyl group is successfully bound on the graphene surface due to the high nucleophilicity of BuLi, which acts as a divider for further agglomeration. Prepared bifunctional graphene can be purged with carbon dioxide gas, resulting in a carboxyl group on the proton abstraction site (Mondal, Bhowmick and Krishnamoorti. 2013). This was a unique method to prepare the carboxylic group containing graphene. Such a finding was reported for the first time, wherein organolithium chemistry was successfully leveraged to develop the functional graphene. Further, the same strategy was utilized by the same group to graft silyl polymer onto the bifunctional graphene to generate electrically conducting instant adhesive (Mondal, Bhowmick and Krishnamoorti 2014c). The developed composite demonstrated an instant adhesive behavior. Under inert conditions, it remains as a liquid. Upon application, it forms bonds with the substrates. This extended the applicability of such materials as an alternative to solder material used in the electronic industry. The versatility of the silane chemistry was also explored later on. Industrially important polyoligomeric silsesquioxane (POSS) was utilized as a modifier for the bi-functional graphene (Mondal, Bhowmick and Krishnamoorti 2014a). Liu et al. reported a supercapacitor nanocomposite material based on carboxylated graphene oxide and polyaniline. Initially, graphene oxide was sonicated in a concentrated HBr solution to catalyze the ring-opening reaction. After transforming all the epoxide groups into hydroxyls, the system was treated with oxalic acid to perform an esterification reaction. Polyaniline was pre-synthesized and grafted to the free carboxylic group of oxalic acid (Liu et al. 2012). Yuan et al. used glycine amino acid to functionalize the graphene oxide for desalination application. Prepared functionalized graphene oxide exhibited a better hydrophilicity, permeability, and salt rejection capability. Jeon et al. successfully reported cost-efficient and mass production of edge carboxylated graphene (Jeon et al. 2012). Pristine graphene was grinded with dry ice in a small ball mill resulting in synthesizing edge carboxylated graphene. This functionalized graphene can be used to produce graphene via thermal decarboxylation which shows electrical conductivity of about 1,214 S cm^{-1}.

4.5.3 Fluorographene

In the last decade, fluorinated polymers have gained massive recognition because of their high stability toward hydrolysis, thermolysis, and most of the solvents. These exceptional properties motivated the scientific community to work more on fluorinated graphene. Preparation of fluorographene can be done by two ways: (i) Top-Down Approach (Mechanical or chemical Exfoliation) and (ii) Bottom-Up Approach (Fluorination)

Using top-down methodology, fluorinated graphene can be prepared by mechanical and chemical treatment of bulk graphitic fluoride. However, the mechanical exfoliation is carried out in drastic condition resulting in poor quality of monolayered fluorographene. Bourlinos et al. prepared stoichiometric graphene fluoride from graphitic fluoride via mechanical exfoliation technique. Highly oriented pyrolytic graphene and fluorine at 600°C in 1 atm pressure produced graphitic fluoride (Bourlinos et al. 2003). Owing to chemical exfoliation, solvent plays a crucial role in the weakening of Van der Waals interaction between the graphitic layer. The solvents like sulfolane, dimethylformamide (DMF), and isopropyl alcohol are used to delaminate the graphene layers. Hence, sonication of bulk graphitic fluoride in the presence of these chemicals results in a colloidal solution of fluorographene. Bulusheva et al. (2014) performed a thermal exfoliation of graphite with BrF_3 in liquid Br_2 to develop a fluorographene with limited fluorine concentration (~1.4%). In another paper, Sysoev et al. used BrF_3 as a fluorine source for preparing fluorinated graphene for gas sensors via ultrasonication method. The product was chemically stable and less interactive with other gas molecules (Sysoev et al. 2017).

Fluorination is another widely used method to prepare fluorographene. Baraket et al. were first to report the fluorination of CVD-based graphene on Cu foil. Graphene was exposed to XeF_2 in an inert environment at 30°C to form single-sided fluorinated graphene (Baraket et al. 2010). Nair et al. followed the same protocol to prepare the fluorographene using SiO_2/Si substrate. The results suggested double-sided fluorination of graphene with a stoichiometry of C_1F_1. The obtained material was stable and inert in the air up to 400°C like Teflon. On fluorination, the graphitic sp^2 carbons transform into sp^3 states in fluorinated graphene that results in a drastic change in electrical and optical properties. Modified graphene becomes an electrical insulator with its resistance value in the order of 10^{12} Ω. Apart from XeF_2, fluorine gas is another fluorinating agent which can be used to functionalize graphene at different conditions (Nair et al. 2010). Mazanek et al. reported mass production of fluorinated graphene oxide using fluorine gas in an autoclave. Stoichiometric composition was found to be around C1F1.05 that exhibited high hydrophobicity and controlled luminescence properties (Mazánek et al. 2015). Jankovský et al. (2014) utilized fluorine gas to prepare fluorinated graphene oxide and further reduced it to fluorographene. The fluorographene was water soluble with tunable luminescence property.

Plasma-mediated fluorination is another strategy to synthesize fluorographene in a controlled manner. Common precursors like CF_4, SF_6, SF_4, and MoF_6 were used as a fluorinating agent to produce the corresponding ions. Wang et al. reported the synthesis of fluorographene using CF_4 plasma in a CVD reactor. They focused on the effect of grain boundary and structural defects on the rate of fluorination and other morphological characteristics (Wang, Wang and Zhu 2014). Baraket et al. utilized SF_6/Ar for reversible functionalization of graphene via plasma treatment. Photochemical method is an eco-friendly and cost-efficient technique for producing fluorographene (Baraket et al. 2010). Lee et al. employed fluoropolymer-coated graphene under laser irradiation to generate highly reactive fluorine that binds with the graphitic framework. Usage of solid fluoropolymer as a precursor results in fluorinated graphene with the overall composition of CF_4 (Lee et al. 2012).

Amount of fluorine manipulates entire properties of fluorographene such as electrical resistivity, hydrophobicity, and luminescence properties. Sudeep et al. reported low fluorine-containing graphene to generate a partial positive charge on the graphene surface, significantly preventing thromboembolism. It can also be used as a drug delivery carrier due to its lipophilicity. Moreover, fluorinated graphene can be used in catalysis, ultrasound imaging, and sensing applications. All these properties can be tuned by manipulating the composition of fluorographene. Hence, the controlled fluorination and defluorination of graphene come into special attention for various applications (Sudeep et al. 2015).

Defluorination of graphene can be performed through electron beam radiation, hydrothermal, and ultrasonication. Withers et al. reported that fluorine concentration was decreased by exposing fluorographene under electron beam after keeping it in benzene (Withers et al. 2011). Tadi et al. used the ultrasonication technique to defluorinate FGO in THF. Being a polar solvent, THF interacts with the C–F bond, eliminating the fluorine upon prolonged sonication (Tadi et al. 2017). Similarly, N-methyl-2-pyrrolidinone (NMP) was used for defluorination via ultrasonication reported by Wang et al. (2016). Thermal defluorination occurs around 400°C–600°C resulting in decomposition of C–F bonds from fluorinated graphene. Graphene can be regenerated from fluorographene by treating it with hydrazine vapor at 100°C–200°C.

4.6 CONCLUSIONS

Diazonium salt, Diels–Alder reaction, and Carbene are major routes for graphene functionalization with small molecules via C–C bond formation. Moreover, macromolecular covalent functionalization of graphene is also possible by using techniques such as 'grafting to' and 'grafting from'. While diazonium salt functionalization is based on free radical mechanism, graphene's behavior as diene and dienophile promotes cycloaddition in the Diels–Alder reaction. Functional groups such as bromo, chloro, methoxy, and carboxy can be added to graphene rings via diazonium salt, whereas cycloaddition of another diene or dienophile to graphene structure can happen via Diels–Alder reaction. Carbene is a highly reactive and interesting material for functionalizing graphene to alter its electronic properties via C–C bond formation. However, the attachment of macromolecules to graphene sheets isn't possible via these routes, and hence, techniques such as 'grafting-to' and 'grafting-from' are useful. While 'grafting-from' techniques involve in situ formation of macromolecules on its structure, 'grafting-to' technique works more on the attachment of macromolecules to graphene via. various linkages. Another major technique to functionalize graphene involves C–N bond formation. Generally, amines and nitrenes are responsible for such functionalization. While amines interact with amino groups to the graphene system, Nitrene interaction is based on azide linkages. Azides can be used to anchor various functional groups such as carboxy, methoxy, bromo, and chloro on graphene's surface. Not only has this but azides can also be used to graft macromolecules to the graphene surface. Apart from these approaches, nitrogen-doped graphene, carboxylated graphene, and fluorographene are other functionalized graphene classes that are readily making their space in research and in the market.

REFERENCES

Aguilar-Bolados, Héctor, Daniela Vargas-Astudillo, Mehrdad Yazdani-Pedram, Gabriela Acosta-Villavicencio, Pablo Fuentealba, Ahirton Contreras-Cid, Raquel Verdejo, and Miguel A. López-Manchado. 2017. "Facile and Scalable One-Step Method for Amination of Graphene Using Leuckart Reaction." *Chemistry of Materials* 29 (16): 6698–6705. doi:10.1021/acs.chemmater.7b01438.

Baraket, M., S. G. Walton, E. H. Lock, J. T. Robinson, and F. K. Perkins. 2010. "The Functionalization of Graphene Using Electron-Beam Generated Plasmas." *Applied Physics Letters* 96 (23): 94–97. doi:10.1063/1.3436556.

Bekyarova, E., S. Sarkar, S. Niyogi, M. E. Itkis, and R. C. Haddon. 2012. "Advances in the Chemical Modification of Epitaxial Graphene." *Journal of Physics D: Applied Physics* 45 (15): 154009. IOP Publishing. doi:10.1088/0022-3727/45/15/154009.

Bolotin, K. I., K. J. Sikes, Z. Jiang, M. Klima, G. Fudenberg, J. Hone, P. Kim, and H. L. Stormer. 2008. "Ultrahigh Electron Mobility in Suspended Graphene." *Solid State Communications* 146 (9–10): 351–55. doi:10.1016/j.ssc.2008.02.024.

Bourlinos, Athanasios B., Dimitrios Gournis, Dimitrios Petridis, Tamás Szabó, Anna Szeri, and Imre Dékány. 2003. "Graphite Oxide: Chemical Reduction to Graphite and Surface Modification with Primary Aliphatic Amines and Amino Acids." *Langmuir* 19 (15): 6050–55. doi:10.1021/la026525h.

Brodie, B. C. 1983. "On Tha Atomic Weight of Graphite." *Philosophical Transactions of the Royal Society B: Biological Sciences* 303 (1113): 1–62. http://rstb.royalsocietypublishing.org/cgi/doi/10.1098/rstb.1983.0080.

Bulusheva, L. G., V. A. Tur, E. O. Fedorovskaya, I. P. Asanov, D. Pontiroli, M. Riccò, and A. V. Okotrub. 2014. "Structure and Supercapacitor Performance of Graphene Materials Obtained from Brominated and Fluorinated Graphites." *Carbon* 78: 137–46. doi:10.1016/j.carbon.2014.06.061.

Caliman, C. C., A. F. Mesquita, D. F. Cipriano, J. C.C. Freitas, A. A.C. Cotta, W. A.A. Macedo, and A. O. Porto. 2018. "One-Pot Synthesis of Amine-Functionalized Graphene Oxide by Microwave-Assisted Reactions: An Outstanding Alternative for Supporting Materials in Supercapacitors." *RSC Advances* 8 (11): 6136–45. Royal Society of Chemistry. doi:10.1039/c7ra13514a.

Cao, Yang, Sílvia Osuna, Yong Liang, Robert C. Haddon, and K. N. Houk. 2013. "Diels–Alder Reactions of Graphene: Computational Predictions of Products and Sites of Reaction." *Journal of the American Chemical Society* 135 (46): 17643–49. doi:10.1021/ja410225u.

Chang, Chia Ming, and Ying Ling Liu. 2009. "Functionalization of Multi-Walled Carbon Nanotubes with Furan and Maleimide Compounds through Diels–Alder Cycloaddition." *Carbon* 47 (13). Elsevier Ltd: 3041–49. doi:10.1016/j.carbon.2009.06.058.

Chang, Po-Han, Yi-Chen Tsai, Shin-Wei Shen, Shang-Yi Liu, Kuo-You Huang, Chia-Shuo Li, Hei-Ping Chang, and Chih-I Wu. 2017. "Highly Sensitive Graphene–Semiconducting Polymer Hybrid Photodetectors with Millisecond Response Time." *ACS Photonics* 4: 2335–44. doi:10.1021/acsphotonics.7b00626.

Chen, Weilong, Kun Wu, Qi Liu, and Mangeng Lu. 2020. "Functionalization of Graphite via Diels–Alder Reaction to Fabricate Poly (Vinyl Alcohol) Composite with Enhanced Thermal Conductivity." *Polymer* 186: 122075. doi:10.1016/j.polymer.2019.122075.

Choi, Junghun, Ki-jeong Kim, Bongsoo Kim, Hangil Lee, and Sehun Kim. 2009. "Covalent Functionalization of Epitaxial Graphene by Azidotrimethylsilane." *The Journal of Physical Chemistry C* 113 (22): 9433–35. American Chemical Society. doi:10.1021/jp9010444.

Criado, Alejandro, Michele Melchionna, Silvia Marchesan, and Maurizio Prato. 2015. "The Covalent Functionalization of Graphene on Substrates." *Angewandte Chemie - International Edition* 54 (37): 10734–50. doi:10.1002/anie.201501473.

Damlin, Pia, Milla Suominen, Markku Heinonen, and Carita Kvarnström. 2015. "Non-Covalent Modification of Graphene Sheets in PEDOT Composite Materials by Ionic Liquids." *Carbon* 93: 533–43. doi:10.1016/j.carbon.2015.05.055.

Delgado, Juan L., Pilar de la Cruz, Fernando Langa, Antonio Urbina, Juan Casado, and Juan T. López Navarrete. 2004. "Microwave-Assisted Sidewall Functionalization of Single-Wall Carbon Nanotubes by Diels–Alder Cycloaddition." *Chemical Communications* 4 (15): 1734–35. doi:10.1039/b402375g.

Deng, Dehui, Xiulian Pan, Liang Yu, Yi Cui, Yeping Jiang, Jing Qi, Wei-Xue Li, et al. 2011. "Toward N-Doped Graphene via Solvothermal Synthesis." *Chemistry of Materials* 23 (5). American Chemical Society: 1188–93. doi:10.1021/cm102666r.

Devadoss, Anando, and Christopher E. D. Chidsey. 2007. "Azide-Modified Graphitic Surfaces for Covalent Attachment of Alkyne-Terminated Molecules by 'Click' Chemistry." *Journal of the American Chemical Society* 129 (17): 5370–71. American Chemical Society. doi:10.1021/ja071291f.

Englert, Jan M, Christoph Dotzer, Guang Yang, Martin Schmid, Christian Papp, J. Michael Gottfried, Hans-Peter Steinrück, Erdmann Spiecker, Frank Hauke, and Andreas Hirsch. 2011. "Covalent Bulk Functionalization of Graphene." *Nature Chemistry* 3: 279–86. doi:10.1038/NCHEM.1010.

Englert, Jan M., Kathrin C. Knirsch, Christoph Dotzer, Benjamin Butz, Frank Hauke, Erdmann Spiecker, and Andreas Hirsch. 2012. "Functionalization of Graphene by Electrophilic Alkylation of Reduced Graphite." *Chemical Communications* 48 (41): 5025–27. The Royal Society of Chemistry. doi:10.1039/C2CC31181J.

Eskandari, Parvaneh, Zahra Abousalman-Rezvani, Hossein Roghani-Mamaqani, Mehdi Salami-Kalajahi, and Hanieh Mardani. 2019. "Polymer Grafting on Graphene Layers by Controlled Radical Polymerization." *Advances in Colloid and Interface Science* 273: 102021. Elsevier B.V.. doi:10.1016/j.cis.2019.102021.

Eswaraiah, Varrla, Krishnan Balasubramaniam, and Sundara Ramaprabhu. 2011. "Functionalized Graphene Reinforced Thermoplastic Nanocomposites as Strain Sensors in Structural Health Monitoring." *Journal of Materials Chemistry* 21: 12626. doi:10.1039/c1jm12302e.

Fang, Ming, Kaigang Wang, Hongbin Lu, Yuliang Yang, and Steven Nutt. 2010. "Single-Layer Graphene Nanosheets with Controlled Grafting of Polymer Chains." *Journal of Materials Chemistry* 20 (10): 1982–92. The Royal Society of Chemistry. doi:10.1039/B919078C.

Feng, L., L. Wu, J. Wang, J. Ren, X. Qu, D. Miyoshi, and N. Sugimoto. 2012. "Detection of a Prognostic Indicator in Early-Stage Cancer Using Functionalized Graphene-Based Peptide Sensors." *Adv. Mater* 24 (1): 125–31. doi:10.1002/adma.201103205.

Gao, Limin, Tsu-Wei Chou, Erik T. Thostenson, Ajay Godara, Zuoguang Zhang, and Luca Mezzo. 2010. "Highly Conductive Polymer Composites Based on Controlled Agglomeration of Carbon Nanotubes." *Carbon* 48 (9): 2649–51. doi:10.1016/j.carbon.2010.03.027.

Gao, Chao, Hongkun He, Li Zhou, Xing Zheng, and Yu Zhang. 2009. "Scalable Functional Group Engineering of Carbon Nanotubes by Improved One-Step Nitrene Chemistry." *Chemistry of Materials* 21 (2): 360–70. American Chemical Society. doi:10.1021/cm802704c.

Geng, Dongsheng, Ying Chen, Yougui Chen, Yongliang Li, Ruying Li, Xueliang Sun, Siyu Ye, and Shanna Knights. 2011. "High Oxygen-Reduction Activity and Durability of Nitrogen-Doped Graphene." *Energy & Environmental Science* 4 (3): 760–64. The Royal Society of Chemistry. doi:10.1039/C0EE00326C.

Georgakilas, Vasilios, Michal Otyepka, Athanasios B. Bourlinos, Vimlesh Chandra, Namdong Kim, K. Christian Kemp, Pavel Hobza, Radek Zboril, and Kwang S. Kim. 2012. "Functionalization of Graphene: Covalent and Non-Covalent Approaches, Derivatives and Applications." *Chemical Reviews* 112 (11): 6156–6214. doi:10.1021/cr3000412.

George Jinu Jacob, Abhijit Bandyopadhyay, and Anil K. Bhowmick. 2008 "New Generation Layered Nanocomposites Derived from Ethylene-Co-Vinyl Acetate and Naturally Occurring Graphite" *Journal of Applied Polymer Science*, 108, 1603–1616

Greenwood, John, Thanh Hai Phan, Yasuhiko Fujita, Zhi Li, Oleksandr Ivasenko, Willem Vanderlinden, Hans Van Gorp, et al. 2015. "Covalent Modification of Graphene and Graphite Using Diazonium Chemistry: Tunable Grafting and Nanomanipulation." *ACS Nano* 9 (5): 5520–35. American Chemical Society. doi:10.1021/acsnano.5b01580.

Guo, Beidou, Qian Liu, Erdan Chen, Hewei Zhu, Liang Fang, and Jian Ru Gong. 2010. "Controllable N-Doping of Graphene." *Nano Letters* 10 (12): 4975–80. American Chemical Society. doi:10.1021/nl103079j.

Guo, Jiahao, Songlin Zhang, Mingxun Zheng, Jing Tang, Lei Liu, Junming Chen, and Xuchun Wang. 2020a. "Graphitic-N-Rich N-Doped Graphene as a High Performance Catalyst for Oxygen Reduction Reaction in Alkaline Solution." *International Journal of Hydrogen Energy* 45 (56): 32402–12. doi:10.1016/j.ijhydene.2020.08.210.

Guo, Shi, Jésus Raya, Dingkun Ji, Yuta Nishina, Cécilia Ménard-Moyon, and Alberto Bianco. 2020b. "Is Carboxylation an Efficient Method for Graphene Oxide Functionalization?" *Nanoscale Advances* 2 (9): 4085–92. doi:10.1039/d0na00561d.

Huang, Guobo, Suqing Chen, Shouwan Tang, and Jianrong Gao. 2012. "A Novel Intumescent Flame Retardant-Functionalized Graphene: Nanocomposite Synthesis, Characterization, and Flammability Properties." *Materials Chemistry and Physics* 135 (2): 938–47. doi:10.1016/j.matchemphys.2012.05.082.

Hummers, William S, and Richard E. Offeman. 1958. "Preparation of Graphitic Oxide." *Journal of the American Chemical Society* 80 (6): 1339. American Chemical Society. doi:10.1021/ja01539a017.

Ismaili, Hossein, Dongsheng Geng, Andy Xueliang Sun, Trissa Trisevgeni Kantzas, and Mark S. Workentin. 2011. "Light-Activated Covalent Formation of Gold Nanoparticle-Graphene and Gold Nanoparticle-Glass Composites." *Langmuir* 27 (21): 13261–68. doi:10.1021/la202815g.

Jankovský, Ondřej, Petr Šimek, David Sedmidubský, Stanislava Matějková, Zbyněk Janoušek, Filip Šembera, Martin Pumera, and Zdeněk Sofer. 2014. "Water-Soluble Highly Fluorinated Graphite Oxide." *RSC Advances* 4 (3): 1378–87. doi:10.1039/c3ra45183f.

Jeon, In Yup, Yeon Ran Shin, Gyung Joo Sohn, Hyun Jung Choi, Seo Yoon Bae, Javeed Mahmood, Sun Min Jung, et al. 2012. "Edge-Carboxylated Graphene Nanosheets via Ball Milling." *Proceedings of the National Academy of Sciences of the United States of America* 109 (15): 5588–93. doi:10.1073/pnas.1116897109.

Jeyaseelan, Antonysamy, Ayman A. Ghfar, Mu. Naushad, and Natrayasamy Viswanathan. 2021. "Design and Synthesis of Amine Functionalized Graphene Oxide for Enhanced Fluoride Removal." *Journal of Environmental Chemical Engineering* 9 (4): 105384. doi:10.1016/j.jece.2021.105384.

Ji, Si-Yuan, Wei Zhao, Hang Gao, Jian-Bin Pan, Cong-Hui Xu, Yi-Wu Quan, Jing-Juan Xu, and Hong-Yuan Chen. 2020. "Highly Efficient Aggregation-Induced Electrochemiluminescence of Polyfluorene Derivative Nanoparticles Containing Tetraphenylethylene." *IScience* 23 (1): 100774. doi:10.1016/j.isci.2019.100774.

Jin, Liling, Kai Yang, Kai Yao, Shuai Zhang, Huiquan Tao, Shuit-Tong Lee, Zhuang Liu, and Rui Peng. 2012. "Functionalized Graphene Oxide in Enzyme Engineering: A Selective Modulator for Enzyme Activity and Thermostability." *ACS Nano* 6 (6): 4864–75. American Chemical Society. doi:10.1021/nn300217z.

Jin, Zhong, Jun Yao, Carter Kittrell, and James M. Tour. 2011. "Large-Scale Growth and Characterizations of Nitrogen-Doped Monolayer Graphene Sheets." *ACS Nano* 5 (5): 4112–17. American Chemical Society. doi:10.1021/nn200766e.

Kotal, Moumita, and Anil K. Bhowmick. 2013. "Multifunctional Hybrid Materials Based on Carbon Nanotube Chemically Bonded to Reduced Graphene Oxide" *Journal of Physical Chemistry C* 117, 25865–25875. doi:10.1021/jp4097265.

Lee, Sun Hwa, Daniel R. Dreyer, Jinho An, Aruna Velamakanni, Richard D. Piner, Sungjin Park, Yanwu Zhu, Sang Ouk Kim, Christopher W. Bielawski, and Rodney S. Ruoff. 2010. "Polymer Brushes via Controlled, Surface-Initiated Atom Transfer Radical Polymerization (ATRP) from Graphene Oxide." *Macromolecular Rapid Communications* 31 (3). John Wiley & Sons, Ltd: 281–88. doi:10.1002/marc.200900641.

Lee, Wi Hyoung, Ji Won Suk, Harry Chou, Jongho Lee, Yufeng Hao, Yaping Wu, Richard Piner, Deji Akinwande, Kwang S. Kim, and Rodney S. Ruoff. 2012. "Selective-Area Fluorination of Graphene with Fluoropolymer and Laser Irradiation." *Nano Letters* 12 (5): 2374–78. doi:10.1021/nl300346j.

Li, Guanghao, Peishuang Xiao, Shengyue Hou, and Yi Huang. 2019. "Rapid and Efficient Polymer/Graphene Based Multichannel Self-Healing Material via Diels–Alder Reaction." *Carbon* 147: 398–407. Elsevier Ltd. doi:10.1016/j.carbon.2019.03.021.

Li, Jing, Meng Li, Li Zhou, Shuang Yan Lang, Hai Yan Lu, Dong Wang, Chuan Feng Chen, and Li Jun Wan. 2016. "Click and Patterned Functionalization of Graphene by Diels–Alder Reaction." *Journal of the American Chemical Society* 138 (24): 7448–51. doi:10.1021/jacs.6b02209.

Liu, Muchun, Yuexin Duan, Yan Wang, and Yan Zhao. 2014. "Diazonium Functionalization of Graphene Nanosheets and Impact Response of Aniline Modified Graphene/Bismaleimide Nanocomposites." *Materials & Design* 53: 466–74. doi:10.1016/j.matdes.2013.07.027.

Liu, Yan, Ruijie Deng, Zan Wang, and Hongtao Liu. 2012. "Carboxyl-Functionalized Graphene Oxide–Polyaniline Composite as a Promising Supercapacitor Material." *Journal of Materials Chemistry* 22 (27): 13619–24. The Royal Society of Chemistry. doi:10.1039/C2JM32479B.

Liu, Zhaoyang, Khaled Parvez, Rongjin Li, Renhao Dong, Xinliang Feng, and Klaus Müllen. 2015. "Transparent Conductive Electrodes from Graphene/PEDOT:PSS Hybrid Inks for Ultrathin Organic Photodetectors." *Advanced Materials* 27 (4): 669–75. doi:10.1002/adma.201403826.

Liu, Zhuang, Joshua T. Robinson, Xiaoming Sun, and Hongjie Dai. 2008. "PEGylated Nanographene Oxide for Delivery of Water-Insoluble Cancer Drugs." *Journal of the American Chemical Society* 130 (33): 10876–77. American Chemical Society. doi:10.1021/ja803688x.

Ma, Fuqiu, Jinru Nian, Changlong Bi, Ming Yang, Chunhong Zhang, Lijia Liu, Hongxing Dong, Mingxin Zhu, and Boran Dong. 2019. "Preparation of Carboxylated Graphene Oxide for Enhanced Adsorption of U(VI)." *Journal of Solid State Chemistry* 277 (May). Elsevier Ltd: 9–16. doi:10.1016/j.jssc.2019.05.042.

Majidi, Roya, and Timon Rabczuk. 2020. "Tailoring the Band Gap of A2-Graphyne through Functionalization with Carbene Groups: A Density Functional Theory Study." *Chemical Papers* 74 (10): 3581–87. doi:10.1007/s11696-020-01195-1.

Mazánek, Vlastimil, Ondřej Jankovský, Jan Luxa, David Sedmidubský, Zbyněk Janoušek, Filip Šembera, Martin Mikulics, and Zdeněk Sofer. 2015. "Tuning of Fluorine Content in Graphene: Towards Large-Scale Production of Stoichiometric Fluorographene." *Nanoscale* 7 (32): 13646–55. doi:10.1039/c5nr03243a.

Mondal, Titash, Anil K. Bhowmick, and Ramanan Krishnamoorti. 2014a. "Butyl Lithium Assisted Direct Grafting of Polyoligomeric Silsesquioxane onto Graphene." *RSC Advances* 4 (17): 8649–56. The Royal Society of Chemistry. doi:10.1039/C3RA47373B.

Mondal, Titash, Anil K. Bhowmick, and Ramanan Krishnamoorti. 2014b. "Stress Generation and Tailoring of Electronic Properties of Expanded Graphite by Click Chemistry." *ACS Applied Materials & Interfaces* 6 (10). American Chemical Society: 7244–53. doi:10.1021/am500471q.

Mondal, Titash, Anil K. Bhowmick, and Ramanan Krishnamoorti. 2014c. "Conducting Instant Adhesives by Grafting of Silane Polymer onto Expanded Graphite." *ACS Applied Materials & Interfaces* 6 (18): 16097–105. American Chemical Society. doi:10.1021/am5040472.

Mondal, Titash, Anil K. Bhowmick, and Ramanan Krishnamoorti. 2015. "Controlled Synthesis of Nitrogen-Doped Graphene from a Heteroatom Polymer and Its Mechanism of Formation." *Chemistry of Materials* 27 (3): 716–25. doi:10.1021/cm503303s.

Mondal, Titash, Anil K. Bhowmick, Ranjan Ghosal, and Rabindra Mukhopadhyay. 2017. "Graphene-Based Elastomer Nanocomposites: Functionalization Techniques, Morphology, and Physical Properties BT – Designing of Elastomer Nanocomposites: From Theory to Applications." In, edited by Klaus Werner Stöckelhuber, Amit Das, and Manfred Klüppel, 267–318. Cham: Springer International Publishing. doi:10.1007/12_2016_5.

Mondal, Titash, Anil K. Bhowmick, Ranjan Ghosal, and Rabindra Mukhopadhyay. 2018. "Expanded Graphite as an Agent towards Controlling the Dispersion of Carbon Black in Poly (Styrene –Co-Butadiene) Matrix: An Effective Strategy towards the Development of High Performance Multifunctional Composite." *Polymer* 146: 31–41. doi:10.1016/j.polymer.2018.05.031.

Mondal, Titash, Anil K. Bhowmick, and Ramanan Krishnamoorti. 2012. "Chlorophenyl Pendant Decorated Graphene Sheet as a Potential Antimicrobial Agent: Synthesis and Characterization." *Journal of Materials Chemistry* 22 (42): 22481–87. doi:10.1039/c2jm33398h.

Mondal, Titash, Anil K. Bhowmick, and Ramanan Krishnamoorti. 2013. "Synthesis and Characterization of Bi-Functionalized Graphene and Expanded Graphite Using n-Butyl Lithium and Their Use for Efficient Water Soluble Dye Adsorption." *Journal of Materials Chemistry A* 1 (28): 8144–53. doi:10.1039/c3ta11212h.

Mondal, Titash, Rana Ashkar, Paul Butler, Anil K. Bhowmick, and Ramanan Krishnamoorti. 2016. "Graphene Nanocomposites with High Molecular Weight Poly(ε-Caprolactone) Grafts: Controlled Synthesis and Accelerated Crystallization." *ACS Macro Letters* 5 (3): 278–82. doi:10.1021/acsmacrolett.5b00930.

Mondal, Titash, Suman Basak, and Anil K. Bhowmick. 2017. "Ionic Liquid Modification of Graphene Oxide and Its Role towards Controlling the Porosity, and Mechanical Robustness of Polyurethane Foam." *Polymer* 127: 106–18. doi:10.1016/j.polymer.2017.08.054.

Mondal, Titash, Varunesh Chandra, and Anil K. Bhowmick. 2016. "Unique Method to Improve the Thermal Properties of Bisphenol a Tetraacrylate by Graphite Oxide Induced Space Confinement." *RSC Advances* 6 (106): 104483–90. The Royal Society of Chemistry. doi:10.1039/C6RA22252H.

Munirasu, Selvaraj, Julio Albuerne, Adriana Boschetti-de-Fierro, and Volker Abetz. 2010. "Functionalization of Carbon Materials Using the Diels–Alder Reaction A." *Macromolecular Rapid Communications* 31 (6): 574–79. doi:10.1002/marc.200900751.

Nair, R. R., P. Blake, A. N. Grigorenko, K. S. Novoselov, T. J. Booth, T. Stauber, N. M.R. Peres, and A. K. Geim. 2008. "Fine Structure Constant Defines Visual Transparency of Graphene." *Science* 320 (5881): 1308. doi:10.1126/science.1156965.

Nair, Rahul R., Wencai Ren, Rashid Jalil, Ibtsam Riaz, Vasyl G. Kravets, Liam Britnell, Peter Blake, et al. 2010. "Fluorographene: A Two-Dimensional Counterpart of Teflon." *Small* 6 (24): 2877–84. doi:10.1002/smll.201001555.

Ohno, Kohji, Chenzhou Zhao, and Yuta Nishina. 2019. "Polymer-Brush-Decorated Graphene Oxide: Precision Synthesis and Liquid-Crystal Formation." Research-article. *Langmuir* 35: 10900–909. American Chemical Society. doi:10.1021/acs.langmuir.9b01747.

Pan, Yongzheng, Hongqian Bao, Nanda Gopal Sahoo, Tongfei Wu, and Lin Li. 2011. "Water-Soluble Poly(N-Isopropylacrylamide)–Graphene Sheets Synthesized via Click Chemistry for Drug Delivery." *Advanced Functional Materials* 21 (14): 2754–63. John Wiley & Sons, Ltd. doi:10.1002/adfm.201100078.

Park, Kyeong Won. 2014. "Carboxylated Graphene Oxide-Mn$_2$O$_3$ Nanorod Composites for Their Electrochemical Characteristics." *Journal of Materials Chemistry A* 2 (12): 4292–98. doi:10.1039/c3ta14223j.

Petrushenko, Igor K. 2014. "[2+1] Cycloaddition of Dichlorocarbene to Finite-Size Graphene Sheets: DFT Study." *Monatshefte Fur Chemie* 145 (6): 891–96. doi:10.1007/s00706-014-1181-1.

Rabchinskii, Maxim K., Sergei A. Ryzhkov, Demid A. Kirilenko, Nikolay V. Ulin, Marina V. Baidakova, Vladimir V. Shnitov, Sergei I. Pavlov, et al. 2020. "From Graphene Oxide towards Aminated Graphene: Facile Synthesis, Its Structure and Electronic Properties." *Scientific Reports* 10 (1): 1–12. Springer US. doi:10.1038/s41598-020-63935-3.

Reddy, Arava Leela Mohana, Anchal Srivastava, Sanketh R. Gowda, Hemtej Gullapalli, Madan Dubey, and Pulickel M. Ajayan. 2010. "Synthesis of Nitrogen-Doped Graphene Films For Lithium Battery Application." *ACS Nano* 4 (11): 6337–42. American Chemical Society. doi:10.1021/nn101926g.

Ryzhkov, S. A., M. K. Rabchinskii, V. V. Shnitov, M. V. Baidakova, S. I. Pavlov, D. A. Kirilenko, and P. N. Brunkov. 2020. "On the Synthesis of the Carboxylated Graphene via Graphene Oxide Liquid-Phase Modification with Alkaline Solutions." *Journal of Physics: Conference Series* 1695 (1). doi:10.1088/1742-6596/1695/1/012008.

Sainsbury, Toby, Melissa Passarelli, Mira Naftaly, Sam Gnaniah, Steve J. Spencer, and Andrew J. Pollard. 2016. "Covalent Carbene Functionalization of Graphene: Toward Chemical Band-Gap Manipulation." *ACS Applied Materials and Interfaces* 8 (7): 4870–77. doi:10.1021/acsami.5b10525.

Salavagione, Horacio J., Marián A. Gómez, and Gerardo Martínez. 2009. "Polymeric Modification of Graphene through Esterification of Graphite Oxide and Poly(Vinyl Alcohol)." *Macromolecules* 42 (17): 6331–34. American Chemical Society. doi:10.1021/ma900845w.

Sarkar, Santanu, Elena Bekyarova, Sandip Niyogi, and Robert C. Haddon. 2011. "Diels–Alder Chemistry of Graphite and Graphene: Graphene as Diene and Dienophile." *Journal of the American Chemical Society* 133 (10): 3324–27. doi:10.1021/ja200118b.

Seo, Jeong Min, and Jong Beom Baek. 2014. "A Solvent-Free Diels–Alder Reaction of Graphite into Functionalized Graphene Nanosheets." *Chemical Communications* 50 (93): 14651–53. Royal Society of Chemistry. doi:10.1039/c4cc07173e.

Seo, Jeong-Min, In-Yup Jeon, and Jong-Beom Baek. 2013. "Mechanochemically Driven Solid-State Diels–Alder Reaction of Graphite into Graphene Nanoplatelets." *Chemical Science* 4 (11): 4273–77. The Royal Society of Chemistry. doi:10.1039/C3SC51546J.

Sinitskii, Alexander, Ayrat Dimiev, David A. Corley, Alexandra A. Fursina, Dmitry V. Kosynkin, and James M. Tour. 2010. "Kinetics of Diazonium Functionalization of Chemically Converted Graphene Nanoribbons." *ACS Nano* 4 (4): 1949–54. doi:10.1021/nn901899j.

Staudenmaier, L. 1898. "Verfahren Zur Darstellung Der Graphitsäure." *Berichte Der Deutschen Chemischen Gesellschaft* 31 (2). John Wiley & Sons, Ltd: 1481–87. doi:10.1002/cber.18980310237.

Steenackers, Marin, Alexander M. Gigler, Ning Zhang, Frank Deubel, Max Seifert, Lucas H. Hess, Candy Haley, et al. 2011. "Polymer Brushes on Graphene," 10490–98.

Strom, T. Amanda, Eoghan P. Dillon, Christopher E. Hamilton, and Andrew R. Barron. 2010. "Nitrene Addition to Exfoliated Graphene: A One-Step Route to Highly Functionalized Graphene." *Chemical Communications* 46 (23): 4097–99. The Royal Society of Chemistry. doi:10.1039/C001488E.

Sudeep, P. M., S. Vinayasree, P. Mohanan, P. M. Ajayan, T. N. Narayanan, and M. R. Anantharaman. 2015. "Fluorinated Graphene Oxide for Enhanced S and X-Band Microwave Absorption." *Applied Physics Letters* 106 (22): 1–6. doi:10.1063/1.4922209.

Sun, Zhiyao, Dechao Guo, Shuhong Wang, Cheng Wang, Yingjian Yu, Dongge Ma, Rongrong Zheng, and Pengfei Yan. 2016. "Efficient Covalent Modification of Graphene by Diazo Chemistry." *RSC Advances* 6 (70): 65422–25. doi:10.1039/c6ra09963g.

Sysoev, Vitalii I., Alexander V. Okotrub, Igor P. Asanov, Pavel N. Gevko, and Lyubov G. Bulusheva. 2017. "Advantage of Graphene Fluorination Instead of Oxygenation for Restorable Adsorption of Gaseous Ammonia and Nitrogen Dioxide." *Carbon* 118 (2): 225–32. Elsevier Ltd. doi:10.1016/j.carbon.2017.03.026.

Tadi, Kiran Kumar, Santosh Kumar Bikkarolla, Kapil Bhorkar, Shubhadeep Pal, Narayan Kunchur, Indulekha N., Sruthi Radhakrishnan, Ravi K. Biroju, and Tharangattu N. Narayanan. 2017. "Defluorination of Fluorographene Oxide via Solvent Interactions." *Particle & Particle Systems Characterization* 34 (5): 1600346. John Wiley & Sons, Ltd. doi:10.1002/ppsc.201600346.

Valentini, Luca, Marta Cardinali, Silvia Bittolo Bon, Diego Bagnis, Raquel Verdejo, Miguel Angel Lopez-Manchado, and Josè M. Kenny. 2010. "Use of Butylamine Modified Graphene Sheets in Polymer Solar Cells." *Journal of Materials Chemistry* 20 (5): 995–1000. doi:10.1039/b919327h.

Vuorinen, Tiina, Juha Niittynen, Timo Kankkunen, Thomas M. Kraft, and Matti Mäntysalo. 2016. "Inkjet-Printed Graphene/PEDOT:PSS Temperature Sensors on a Skin-Conformable Polyurethane Substrate." *Scientific Reports* 6: 35289. Nature Publishing Group. doi:10.1038/srep35289.

Wang, Bei, Junjie Wang, and Jun Zhu. 2014. "Fluorination of Graphene: A Spectroscopic and Microscopic Study." *ACS Nano* 8 (2): 1862–70. doi:10.1021/nn406333f.

Wang, Guan Wu, Zhong Xiu Chen, Yasujiro Murata, and Koichi Komatsu. 2005. "[60]Fullerene Adducts with 9-Substituted Anthracenes: Mechanochemical Preparation and Retro Diels–Alder Reaction." *Tetrahedron* 61 (20): 4851–56. doi:10.1016/j.tet.2005.02.089.

Wang, Xu, Weimiao Wang, Yang Liu, Mengmeng Ren, Huining Xiao, and Xiangyang Liu. 2016. "Controllable Defluorination of Fluorinated Graphene and Weakening of C–F Bonding under the Action of Nucleophilic Dipolar Solvent." *Physical Chemistry Chemical Physics* 18 (4): 3285–93. The Royal Society of Chemistry. doi:10.1039/C5CP06914A.

Wei, Dacheng, Yunqi Liu, Yu Wang, Hongliang Zhang, Liping Huang, and Gui Yu. 2009. "Synthesis of N-Doped Graphene by Chemical Vapor Deposition and Its Electrical Properties." *Nano Letters* 9 (5): 1752–58. American Chemical Society. doi:10.1021/nl803279t.

Withers, Freddie, Thomas H. Bointon, Marc Dubois, Saverio Russo, and Monica F. Craciun. 2011. "Nanopatterning of Fluorinated Graphene by Electron Beam Irradiation." *Nano Letters* 11 (9): 3912–16. American Chemical Society. doi:10.1021/nl2020697.

Xie, Meng, Hailin Lei, Yufeng Zhang, Yuanguo Xu, Song Shen, Yanru Ge, Huaming Li, and Jimin Xie. 2016. "Non-Covalent Modification of Graphene Oxide Nanocomposites with Chitosan/Dextran and Its Application in Drug Delivery." *RSC Advances* 6 (11): 9328–37. The Royal Society of Chemistry. doi:10.1039/C5RA23823D.

Xu, Xiujuan, Qingying Luo, Wei Lv, Yongqiang Dong, Yi Lin, Quanhong Yang, Aiguo Shen, et al. 2011. "Functionalization of Graphene Sheets by Polyacetylene: Convenient Synthesis and Enhanced Emission." *Macromolecular Chemistry and Physics* 212 (8): 768–73. John Wiley & Sons, Ltd. doi:10.1002/macp.201000608.

Xu, Xiujuan, Wei Lv, Jing Huang, Jijun Li, Runli Tang, Jiawei Yan, Quanhong Yang, Jingui Qin, and Zhen Li. 2012. "Functionalization of Graphene by Tetraphenylethylene Using Nitrene Chemistry." *RSC Advances* 2 (18): 7042–47. The Royal Society of Chemistry. doi:10.1039/C2RA20460F.

Ye, Lei, Ting Xiao, Ni Zhao, Haihua Xu, Yubin Xiao, Jianbin Xu, Yuzi Xiong, and Weijian Xu. 2012. "Derivitization of Pristine Graphene for Bulk Heterojunction Polymeric Photovoltaic Devices." *Journal of Materials Chemistry* 22 (33): 16723–27. doi:10.1039/c2jm32729e.

Yoo, Jung Joon, Kaushik Balakrishnan, Jingsong Huang, Vincent Meunier, Bobby G. Sumpter, Anchal Srivastava, Michelle Conway, et al. 2011. "Ultrathin Planar Graphene Supercapacitors." *Nano Letters* 11 (4): 1423–27. American Chemical Society. doi:10.1021/nl200225j.

Yuan, Jinchun, Guohua Chen, Wengui Weng, and Yuanze Xu. 2012. "One-Step Functionalization of Graphene with Cyclopentadienyl-Capped Macromolecules via Diels–Alder 'Click' Chemistry." *Journal of Materials Chemistry* 22 (16): 7929–36. doi:10.1039/c2jm16433g.

Zhang, Wenbo, Jianzhong Ma, Dangge Gao, Yongxiang Zhou, Congmin Li, Jiao Zha, and Jing Zhang. 2016. "Preparation of Amino-Functionalized Graphene Oxide by Hoffman Rearrangement and Its Performances on Polyacrylate Coating Latex." *Progress in Organic Coatings* 94: 9–17. Elsevier B.V. doi:10.1016/j.porgcoat.2016.01.013.

Zhao, Lianqin, Sheng Tao Yang, Shicheng Feng, Qiang Ma, Xiaoling Peng, and Deyi Wu. 2017. "Preparation and Application of Carboxylated Graphene Oxide Sponge in Dye Removal." *International Journal of Environmental Research and Public Health* 14 (11). doi:10.3390/ijerph14111301.

Zhong, Xing, Jun Jin, Shuwen Li, Zhiyong Niu, Wuquan Hu, Rong Li, and Jiantai Ma. 2010. "Aryne Cycloaddition: Highly Efficient Chemical Modification of Graphene." *Chemical Communications* 46 (39): 7340–42. The Royal Society of Chemistry. doi:10.1039/C0CC02389B.

Ziółkowski, Robert, Łukasz Górski, and Elżbieta Malinowska. 2017. "Carboxylated Graphene as a Sensing Material for Electrochemical Uranyl Ion Detection." *Sensors and Actuators, B: Chemical* 238: 540–47. doi:10.1016/j.snb.2016.07.119.

Zydziak, Nicolas, Basit Yameen, and Christopher Barner-Kowollik. 2013. "Diels–Alder Reactions for Carbon Material Synthesis and Surface Functionalization." *Polymer Chemistry* 4 (15): 4072–86. doi:10.1039/c3py00232b.

5 Structure-Property Relationships for the Mechanical Behavior of Rubber-Graphene Nanocomposites

Kedar Kirane and Surita Bhatia
Stony Brook University

CONTENTS

5.1 INTRODUCTION

Rubber is a cross-linked polymer that has found widespread use in many engineering applications, e.g., tires used in automotive and aerospace industry, gaskets, hoses, sealants, insulation materials, electromagnetic shields, and shock absorption. These applications are enabled by the unique mechanical behavior of rubber, primarily its ability to sustain very high, yet reversible deformations before failure (Gent and Walter 2006). For further improvement of the mechanical properties of rubber, and

DOI: 10.1201/9781003200444-5

109

in general polymers/elastomers, micro- or nano-reinforcement has been employed in many cases. In fact, the most common use of rubber in tires includes a micro-reinforcement in the form of micron sized particulate fillers, viz. carbon black, giving the tires the familiar black color. Recent studies have shown that in comparison to micro-fillers, the enhancement in the mechanical properties (such as stiffness, strength, and toughness) can be considerably higher with nano-fillers even for small volume fractions. This is primarily due to the large surface area of the nano-fillers. As a result, a variety of nano-fillers have been explored by researchers such as nano-clays, carbon nanotubes, and graphene, which is the focus of this chapter.

With the discovery of graphene and its first unambiguous production in 2004 (Geim and Novoselov 2007), it has increasingly been considered as a nano-filler for rubber and other polymers. In fact, it has sparked a major revolution in the development of graphene-based nanocomposites which have realized unprecedented enhancements to mechanical and other properties. Graphene is the strongest material ever tested (Lee 2008, Cao 2020) with an intrinsic tensile strength of 130 GPa (with representative engineering tensile strength of ~50 to 60 GPa for large-area freestanding graphene) and Young's modulus close to 1 TPa. It also has very good thermal, electrical, and gas barrier properties compared to many other nano-filler materials (Wang et al. 2017, Balandin et al. 2008, Lee et al. 2008) allowing multi-functional capabilities in its composites. Further, it is a 2D material with atomic thickness in the out-of-plane direction. This allows a high surface area, maximizing the mechanical contact with the surrounding matrix/elastomer, translating into a highly effective reinforcement. It also can be easily functionalized allowing easier and uniform dispersion in polymeric matrices. These characteristics make graphene one of the most ideal nano-filler materials.

The goal of this chapter is to survey key findings with regard to the structure-property relationships (SPRs) for rubber-graphene nanocomposites. The primary focus here is on the mechanical properties, which mainly include the modulus (stiffness), strength, and deformation at failure, which can alternately be understood as the toughness. We note that it is restrictive to seek quantitative SPRs, given the amount of variability from rubber compound to compound as well as processing and fabrication techniques. Therefore, the focus of discussions here is more on the various physical reinforcement mechanisms at play. Understanding these is key to being able to harness them for appropriately designing volume fractions and fabrication techniques for specific applications involving these composites. First reviewed are the salient behaviors of rubber, which is integral to understanding the behavior of its composites. Then discussed are the various strengthening/toughening mechanisms that are brought about by the addition of graphene to rubber, leading to an improvement in the mechanical properties. Next, we discuss the effect of graphene on the kinetics of vulcanization reaction in rubber, which also has a bearing on the mechanical properties. Structural studies are reviewed next which shed further light on the fundamental mechanisms and dynamics in these nanocomposites. This is followed by a brief discussion on various micromechanics-based homogenization techniques for soft nanocomposites (such as rubber/graphene) and their ability to account for the various reinforcement mechanisms. To conclude, we discuss some of the important knowledge gaps and potential areas of research.

5.2 MECHANICAL BEHAVIOR OF RUBBER

One of the salient aspects of the mechanical behavior of rubber is its ability to sustain large and mostly reversible deformations before failure. In fact, natural rubber (NR), when cross-linked, can be stretched to strains over 700% without rupture (Ozbas et al. 2012). In addition, rubber exhibits a variety of interesting mechanical behaviors. For instance, it exhibits viscoelasticity which makes its modulus dependent on strain rate and temperature. It has a very high bulk modulus and is typically treated as virtually incompressible (Poisson's ratio ~0.5). If fillers are incorporated, rubber exhibits the Payne effect which involves a dependence of the viscoelastic storage modulus on the cyclic loading strain amplitude. Filled rubber's stress-strain behavior also exhibits the Mullins effect, which refers to cyclic stress softening. It also exhibits hysteresis during loading and unloading, implying energy dissipation in each load-unload cycle, which is the primary cause behind a tire's rolling resistance. Rubber also exhibits stiffening at high strains due to strain-induced crystallization (SIC). Many of these behaviors can be explained by understanding the molecular structure of rubber (Gent and Walter 2006).

The most intuitive explanation for cross-linked rubber's SPRs involves comparing its structure with spaghetti and meatballs, which is shown schematically in Figure 5.1. The spaghetti represents the long-chain molecules of rubber, intertwined, and tangled up, while the meatballs represent the cross-links. Due to this structure, rubber in its natural (unvulcanized) form behaves almost like a fluid with a high viscosity and is unusable in structural applications such as tires. For this purpose, it needs to be chemically processed at high temperatures or "cured". The curing process consists of engineering cross-links between these long-chain polymer molecules in the form of chemical bonds (meatballs) or cross-links. These cross-links form at certain temperatures and allow creating a loose but permanent 3D network of molecule chains. At the macroscale, this results in an elastic solid, transformed from the original viscous state. In rubber, the chemical bonds are effectively achieved by sulfur cross-links, and the process of rubber curing is also called vulcanization. A wide variety of chemical reagents are used to facilitate/accelerate this process (Barlow 2018).

The resulting molecular structure brings about the various observed effects in rubber's mechanical behavior. For example, Mullins effect involves cyclic stress softening, which results from the slipping or detachment of rubber molecules from the filler surfaces during the cyclic loading (Mullins 1969). A similar effect occurs at small strains and is referred to as the Payne effect caused by a different mechanism

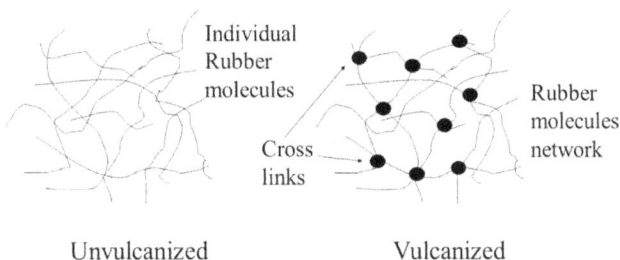

FIGURE 5.1 Schematic of the molecular structure of unvulcanized and vulcanized rubber.

involving the disruption of aggregation of the filler particles (Payne 1963). However, in general, it is difficult to separate the effects of these two mechanisms. Since the original structure of rubber behaves like a viscous fluid, it is not surprising that the filled rubber also exhibits viscoelasticity (Ferry 1980). Rubber also exhibits temperature dependence in its mechanical response, and the effect of rising temperature can often be likened to that of increasing the strain rate (which in practice is termed as time temperature superposition). With this mechanistic understanding, a variety of constitutive models for rubber have been developed that can capture these effects and are used in practice for predicting the mechanical behavior and assisting the design of various engineering applications. The SPRs for rubber/graphene nanocomposites are best interpreted in the context of these behaviors.

5.3 REINFORCEMENT MECHANISMS IN RUBBER-GRAPHENE NANOCOMPOSITES

Research interest has grown exponentially in the recent past in nanocomposites made from rubber and graphene. A host of studies have shown that the addition of graphene notably enhances the mechanical properties of rubber, including the strength, stiffness, and deformation at failure. Here, stiffness refers to resistance to deformation, strength refers to maximum load bearing capacity before failure, and toughness refers to energy dissipation capacity by failure. So, in this sense, the toughness is also indicative of the maximum deformation capability before failure. In this section, we will review the main findings. The reader is also referred to other excellent review articles on this topic (Wang et al. 2017, Stöckelhuber et al. 2017, Mondal et al. 2016, Papageorgiou et al. 2015, Liu et al. 2018, Srivastava and Mishra 2018, Bokozba 2019, Zhang et al. 2019, Mensah et al. 2018, and Bhavitha et al. 2021).

Song et al. (2010) and Zhan et al. (2011) conducted some of the initial investigations on the effect of various nano-fillers on the mechanical properties of NR composites. They considered and compared various nano-fillers such as graphene, graphite, carbon black, and multi-walled carbon nanotubes (MWCNTs). Among these, graphene was found to cause the highest enhancements to the tensile and tear strengths for the same filler volume fractions (e.g., 47% and 50% increase, respectively, at 2% volume fraction in Zhan et al. 2011). Also, for a given filler volume fraction, graphene was found to stiffen NR twice as much as the corresponding MWCNT composite. Similar findings were reported by Li et al. (2017) where graphene nanoplatelets were found to outperform carbon black by nearly three times for the same volume fraction. Since then, many other studies have reported similar or better enhancements from graphene addition for not only NR but also SBR, NBR, and silicone-based rubbers (Wang et al. 2017, Lian et al. 2011, Mao et al. 2013, Varghese et al. 2013). Findings from Wu et al. (2013) are shown in Figure 5.2 as an example.

The superior mechanical reinforcement from graphene has been primarily attributed to the increased interfacial interactions made possible by its large surface area. Of course, the precise enhancement to the properties can be seen to vary from study to study. This is due to a variety of reasons from the fabrication technique and the uniformity of dispersion of the filler, to chosen functionalization. But the key finding is that just by adding a small volume fraction of graphene, a huge change occurs in the mechanical properties.

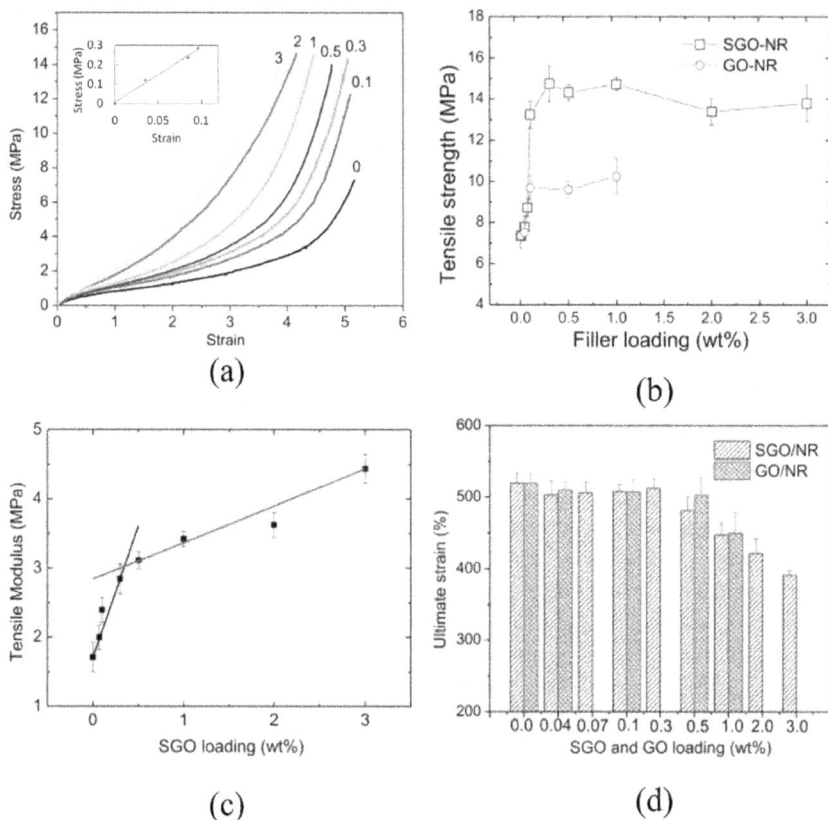

FIGURE 5.2 Enhancement of various mechanical properties vs volume fraction of graphene. (a) Stress-strain response, (b) tensile strength, (c) tensile modulus, and (d) ultimate strain (%). (Reproduced from Wu et al. 2013, with permission from Elsevier.)

In some cases, graphene addition has been observed to achieve the so-called holy grail of nanocomposites, i.e., an increase in both the strength and the toughness. (In general, when one increases the other decreases). For instance, Xing et al. (2014) have reported that adding a low content of GE can remarkably increase the tensile strength and the initial tensile modulus of NR. With the incorporation of as low as 0.5 phr of GE, a 48% increase in the tensile strength and an 80% increase in the initial tensile modulus were achieved without observing a significant decrease in the ultimate strain to failure. And some other studies have reported an increase in the elongation at break, accompanying an increase in the strength and stiffness. For example, Gan et al. (2015) reported a 64% increase in the elongation to break for silicone rubber by addition of 0.4% graphene. They also reported modulus and strength increases of 93% and 67% at 2 wt. % graphene. Similarly, in other works on butyl rubber, Sadasivuni et al. (2013) reported huge increases (over 100% each) in modulus, strength as well as the elongation to break. This is a highly desirable effect because in general any additives or fillers that serve to increase the stiffness and strength also tend to decrease the elongation to break. It is worth noting, however, that not all studies report the same conclusions. There have been some studies that report a marked decrease in the strain to failure (e.g., see Figure 5.2d). This needs to be understood well since increasing the strength at the cost of toughness may not be a desirable outcome for many engineering applications. So, processes that lead to increasing both the strength and the toughness especially need to be paid special attention. These and other observed effects can be explained by exploring the various reinforcement mechanisms at play, which form the basis to understanding the SPRs with the ultimate goal of identifying the best and most effective fabrication techniques and material microstructure designs for rubber graphene composites.

5.3.1 INTERFACIAL INTERACTIONS

In general, in filler-based elastomer composites, the filler elastomer interface is the primary stress transfer location. Naturally, the interfacial interactions are the key factor influencing the reinforcement mechanisms and thereby the property enhancements. The best outcome generally results when the filler dispersion is uniform with minimal aggregation, enabling maximum filler-matrix interface area (Young et al. 2012). Overall, the studies on rubber and graphene offer findings consistent with this understanding. These and other studies have shed light on a variety of reinforcement mechanisms that are specific to graphene/rubber nanocomposites. In addition to the foregoing cited works, many other examples may be listed with NR, styrene butadiene rubber (SBR) as well as synthetic rubber (SR) (Mensah et al. 2015, Yaragalla et al. 2015, Bai et al. 2011). There are mainly two effects related to the interfacial interactions between graphene and rubber, which greatly impact the resulting mechanical properties, viz. (i) interfacial adhesion and (ii) dispersion of graphene. Both are greatly affected by the surface chemistry of graphene. For the best results, it is desirable to have excellent interfacial adhesion and uniform dispersion of graphene to maximize the surface contact area between graphene and rubber.

Zhan et al. (2011) discuss that the mechanisms which lead to a superior performance of graphene are essentially mechanical effects such as interlocking and adhesion, which can be maximized due to the large contact surface (Ramanathan et al. 2008). These effects are present, but comparatively to a lesser degree with other fillers such as MWCNTs which act as 1D fillers and carbon black which acts as a sphere-shaped filler. The atomically thin 2D structure of graphene thus has an advantage over the 1D or 3D shaped fillers. The 2D nature of graphene is favorable also because it appears to develop wrinkles, resulting in molecular scale surface roughness which can contribute to improvement of the mechanical interlocking, stress transfer, and adhesion. Such mechanisms of stress transfer between NR and graphene nanoplatelets can be probed via techniques such as stress-induced Raman band shifts, as demonstrated in Li et al. (2017). Consistent evidence was reported in Xue et al. (2019), who found that a higher surface roughness of graphene translated directly to a more pronounced strengthening and toughening effect for NR. This is shown in Figure 5.3. The findings in Mowes et al. (2014) were consistent with this too and also extended to electrical and thermal properties.

An additional powerful effect of graphene is the increased density of cross-links in the molecular structure. This is again made possible by the high contact area of

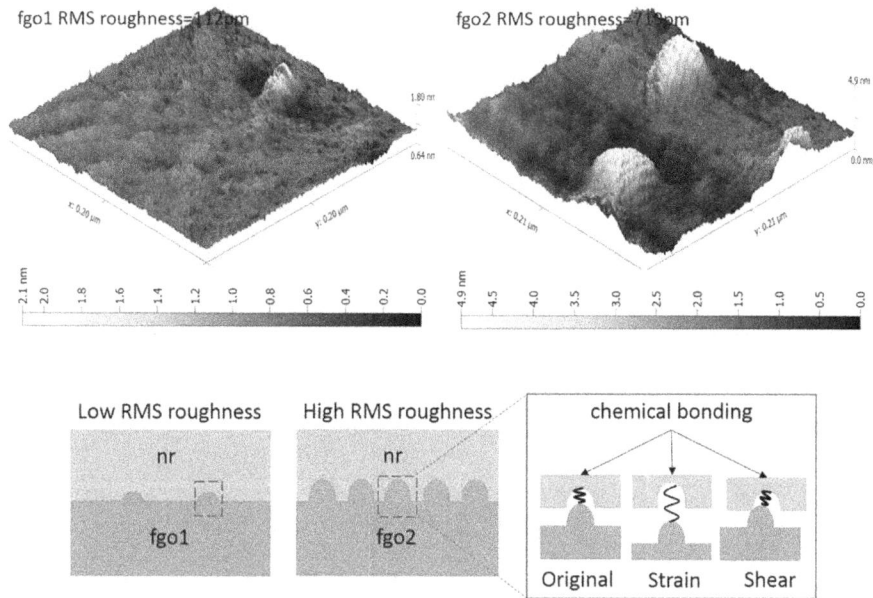

FIGURE 5.3 Figure showing surface roughness of graphene and a schematic of the mechanical interlocking between NR and FGO. (Reproduced from Xue et al. 2019, with permission from Elsevier.).

graphene, offering a higher number of physical cross-linking locations. At a more fundamental level, graphene is believed to alter the dynamics of the molecular chains of rubber, especially so at the rubber-graphene interface, as discussed in Valentin et al. (2010). For example, the chains interacting with graphene undergo considerable changes to their dynamics, which affect the glass transition temperature, as also explained in previous works such as Wang (1998) and Putz et al. (2008). This change to chain dynamics manifests by the formation of a glassy interface layer between the rubber and the filler as elucidated early on by Tsagaropoulos et al. (1996). This effect is known to occur for other fillers too and in fact forms the basis of material models to capture the Payne and Mullins effects (Merabia et al. 2008). The glassy layer can also be viewed as the "interphase" material which is known to develop in other polymer nanocomposites (e.g., Ciprari et al. 2006). It acts like a glue between the filler and the rubber and creates a local gradient in material properties in the vicinity of the filler. The effect of this interphase is amplified for nano-sized fillers due to the considerably enhanced contact surface area, which is the case for graphene, thus explaining the property enhancements.

The addition of graphene as filler also significantly reduces the overall swelling capacity of rubber (Valentin et al. 2010) as shown in Figure 5.4. This is because graphene itself is not prone to swelling, and when bonded strongly to rubber, it locally affects the swelling capacity. This further explains the significantly improved stiffness. According to Valentin et al. (2010), this is another effect of the wrinkled geometry of graphene. Such effects are absent with other fillers such as clays.

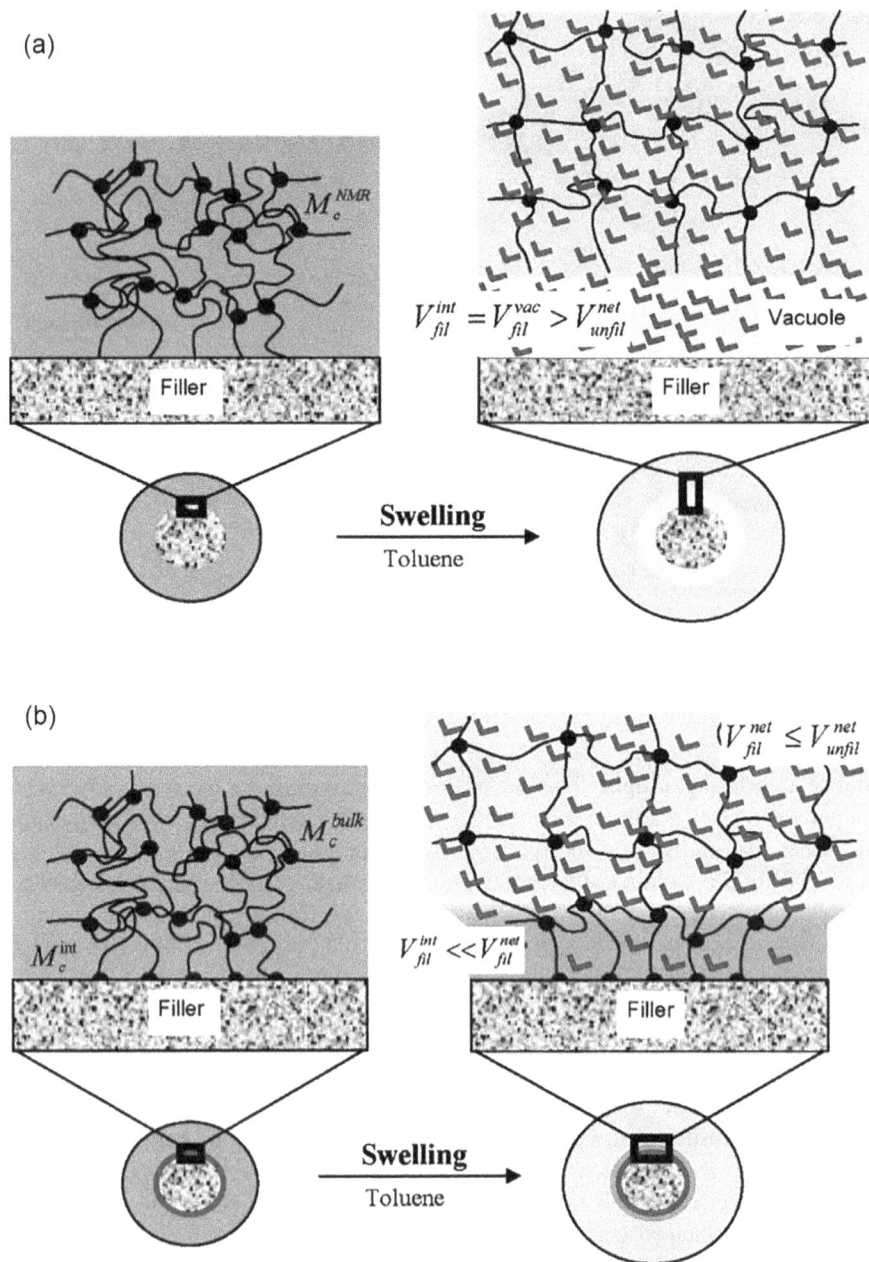

FIGURE 5.4 Schematic showing the swelling behavior of an elastomer composite without (a) and with (b) interactions between rubber-filler interactions. (Adapted with permission from Valentin et al. 2010, Copyright 2010 American Chemical Society.).

Since the interfacial/surface effects are so instrumental, many efforts have been undertaken to understand and modify the surface chemistry of graphene and optimize these interactions. A notable example in this regard is the work by Yang et al. (2016), who developed an innovative rubber graphene interface mediated by covalent bonds of ortho-quinone. Reduced graphene oxide was used as the filler, and such a tailored interface was found to achieve very high reinforcing efficiency and a very good dispersion of the filler. This strategy was found to increase the tensile strength by a staggering 223% at only 1.1 phr graphene. Further, the strong covalent bonds nearly eliminated the frictional slip between the rubber and the filler, resulting in minimal dynamic energy loss during cyclic loads. These energy losses contribute toward hysteretic losses in rubber products, e.g., a tire's rolling resistance. Thus, such alterations of the interface chemistry have the potential to minimize the rolling resistance and allow the development of engineering designs that are environmentally sustainable as well. Another notable example involves the work by Wen et al. (2017) on tailoring the interfacial interactions by means of varying oxidation degrees of GO. Other such examples are discussed in Section 5.4, which focuses on graphene functionalization targeted at suitably altering its surface chemistry.

The interfacial effects have another important consequence in addition to local adhesion and mechanical interlocking and alteration of chain dynamics. It is the effect on the uniformity of the dispersion of the filler, which as is well known is essential to realize the full benefits of adding the filler. For instance, Tang et al. (2014) showed that graphene's surface chemistry governs its dispersion in matrices such as rubber. Their investigation is interesting because it provides an excellent multiscale correlation of molecular scale property to structure to macroscale property. The key features at the molecular scale that govern the performance of these composites include the wettability of the filler by rubber, the interfacial adhesion, and the tendency of the filler to agglomerate or disperse. These can be quantified by the relevant surface energies, which led to the finding of the existence of a critical fraction of CO_x in graphene. When this fraction is below 0.2, graphene dispersion was found to be much easier. On the other hand, at higher fractions, graphene was found to agglomerate, which is not desirable for achieving the enhancements to properties.

5.3.2 FILLER MORPHOLOGY

Beyond the interfacial effects, one of the most important aspects that affects the SPRs of rubber-graphene nanocomposites is the morphology of the filler. The filler morphology can be interpreted at two scales, one being the overall morphology in the entire composite sample, which is related to the distribution/dispersion of the filler, and the other being the specific morphology of the individual filler particles. The latter aspect was touched upon already in the previous subsection when discussing the surface roughness of graphene flakes affecting an enhanced adhesion and interlocking. Here, other aspects of the filler morphology and their effect on the property enhancements in rubber will be discussed.

As mentioned previously, a host of studies have reported notable increases in the stiffness and strength of rubber due to the addition of graphene. However, on the flip side, many studies do report a decrease in the strain to failure. Thus, it appears that the strength and stiffness are increased at the cost of toughness. The strong interfacial interaction between graphene and rubber serves to immobilize the long molecule chains causing them to break at relatively lower strains. This being true, it does appear to be possible to increase all three. A possible explanation for these differences could be found in the work by Potts et al. (2012) according to which the roots could be found in the processing technique.

Rubber graphene composites can be fabricated by a variety of techniques of which two of the most common ones are the solution treatment method or the milling method (Beckert et al. 2014). It is revealed in the work by Potts et al. (2012) that the solution treatment method leads to the formation of a well-connected network of fillers within the interstitial areas between the elastomer particles as shown in Figure 5.5. This is true especially if the filler is reduced graphene oxide (rGO). In addition, the graphene platelets cluster within the interstitial areas between the elastomer particles and prevent the interparticle diffusion of the rubber chains. This creates an effect of binding or cohesion between the elastomer particles. This web-like filler network can significantly stiffen and strengthen the composite. However, since the network itself is inflexible, consisting primarily of graphene, it decreases the overall strain to failure of the composite.

On the other hand, composites formed by the milling technique consist of graphene flakes that are more uniformly dispersed, which do not necessarily connect and form a network. This is because the two-roll milling process serves to break

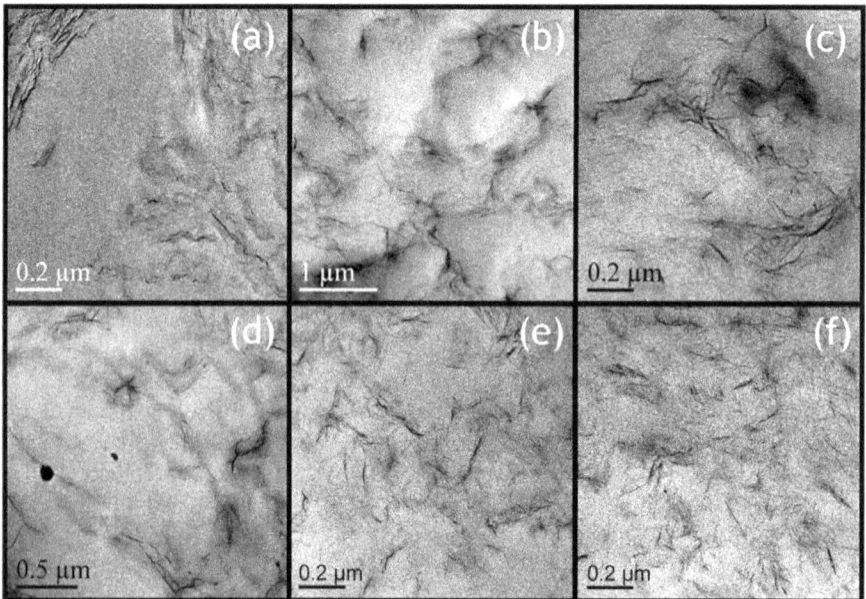

FIGURE 5.5 TEM micrographs of RG-O/NR nanocomposite sections. Images (a) and (b) show the network of dispersed RG-O platelets in the uncured composites, images (c) and (d) show the dispersion in the solution-treated samples, while images (e) and (f) show the morphology of the milled nanocomposites indicating lack of a network type structure. (Adapted with permission from Potts et al. 2012, Copyright 2012 American Chemical Society.)

down the structure of the filler network, as shown in Figure 5.5. This uniform, well-dispersed morphology of the filler allows a relatively easy movement of the long-chain molecules even at high values of deformation. In fact, the presence of graphene can even enhance these movements and lead to an increase in the deformation to failure in addition to an increase in the stiffness and strength. It should be noted though that the increase in the stiffness caused by well-dispersed graphene is lower than that caused by the filler network, as shown in Figure 5.6.

The filler network has a considerable effect on the transport properties as well. The composites fabricated via the solution treatment, i.e., having the filler network, exhibit a remarkably high thermal and electrical conductivity, compared to the nanocomposite fabricated via the milling process, i.e., consisting of well-dispersed graphene flakes. This is to be expected due to the physical connectivity of the 3D graphene/filler network.

A similar investigation was carried out by Yan et al. (2014) by considering composites of NR and reduced graphene oxide (rGO). The fabrication techniques involved (i) latex mixing, coagulation, and static hot-press and (ii) twin roll mixing process. Similar to Potts et al. (2012), here too, the composites exhibited either a 3D filler network with the latex mixing process or a homogeneous dispersion of rGO

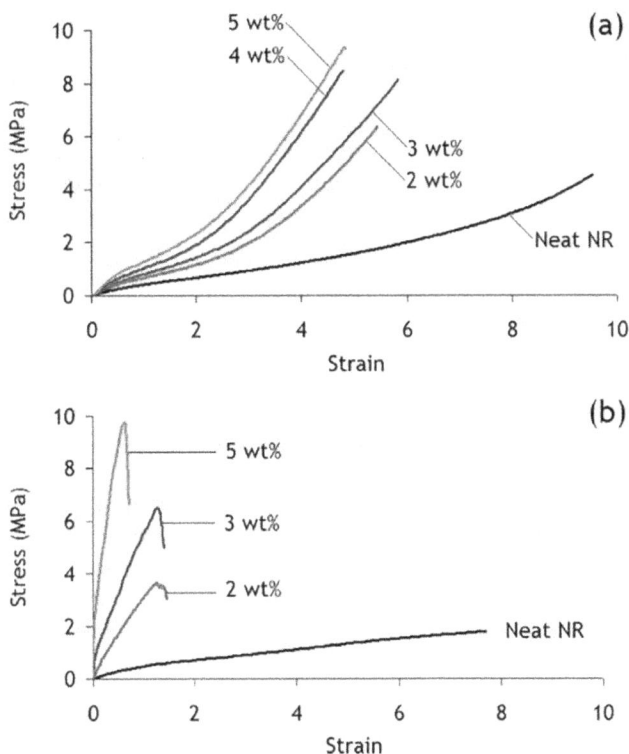

FIGURE 5.6 Representative stress–strain curves of the (a) milled and (b) solution-treated rGO/NR nanocomposites showing marked difference in the overall strain to failure. (Adapted with permission from Potts et al. 2012, Copyright 2012 American Chemical Society.)

platelets with the twin roll mixing process. Their findings on property enhancement were consistent with those from Potts et al. (2012).

Thus, the mechanisms of reinforcement are strongly influenced by the dispersion morphology, which in turn is affected by the fabrication technique. It is notable that this observed difference is only attributable to the overall filler distribution, since in both cases the individual graphene flakes were always well-exfoliated with a high aspect ratio. This understanding is important since the fabrication technique could be tailored depending on the application or the desired property to be enhanced.

The filler network morphology can be characterized by 3D transmission electron microscopy (TEM) as demonstrated in Das et al. (2014). Based on this finding, many other studies have attempted to build and optimize such filler networks to achieve desired properties (He et al. 2015). For example, Lin et al. (2016) discuss a simple pre-construction approach to yield a 3D segregated graphene network in a SBR matrix, aimed at superior electrical conductivity. Likewise in Hao et al. (2020), a similar graphene network structure was constructed by means of liquid-phase redispersion and self-assembly.

The foregoing studies elucidate the effect of the overall filler morphology in terms of forming an extensive 3D network vs. being homogeneously dispersed. However, the morphology of the individual filler particle level is also important. For instance, when graphene nanoflakes are used as fillers, they generally tend to stack and form platelets, which then serve as the filler. Investigations by Schopp et al. (2014) revealed that the fewer the number of graphene nanoflakes within a platelet, the better is the enhancement to the properties. They considered SBR latex as the elastomer and functionalized graphene (FG) as the filler. They fabricated the nanocomposites via dispersion blending with aqueous emulsifier-free FG dispersions, containing predominantly single FG nanosheets. This technique was found to yield a very effective dispersion of graphene, without using any organic solvents. Their finding is likely to hold for other rubbers too, which consist of uniformly dispersed graphene platelets. Additionally, this work also confirmed that overall, graphene outperformed other fillers such as carbon black and carbon nanotubes. Graphene sheet size was also found to matter in Wu et al. (2015). Of course, controlling the sheet size in a composite is a challenge since it is governed by the size of the initial graphite flakes (Zhou and Liu 2010). Despite this, their findings suggest that smaller sheet size of graphene leads to higher reinforcement effects compared to larger sheet sizes.

Overall, it is clear that just like macroscale fiber composites, for nanocomposites, knowing just the reinforcement volume fraction of fibers is not sufficient to assess or determine the overall composite properties. The geometry and distribution of the reinforcement is extremely important. For nanocomposites too, not only does the geometry of each individual graphene flake or platelet matters but also the overall dispersion and morphology of the filler in the composite.

5.3.3 STRAIN-INDUCED CRYSTALLIZATION

As mentioned previously, rubber can sustain very high deformations prior to failure. At high strains, it undergoes a phase transformation, changing from predominantly an amorphous structure to a crystalline one. This phenomenon is referred to as strain-induced crystallization (SIC) and it occurs because the strain, especially uni-axial, serves to align the initially entangled molecular chains, making them ordered (Huneau 2011). SIC occurs if the temperature is always held above the glass transition temperature and if the starting molecular structure is not too irregular. Thus, it brings about a self-reinforcing effect in the mechanical behavior of rubber at higher strains.

The addition of graphene has a significant effect on SIC and its consequences. Interestingly, it appears that the reinforcement mechanism of graphene in rubber can depend on the deformation level due to SIC being an effect appearing only beyond certain strain levels. For example, Potts et al. (2012) showed that at low strains, the reason for enhanced stiffness is the well-known shear lag effect (Gong et al. 2010), which occurs even in macroscale fiber-reinforced composites. However, at higher strain levels, the mechanism alters due to the occurrence of SIC.

Primarily, observations suggest that graphene addition lowers the strain at which the onset of SIC occurs as shown in Figure 5.7. Further, for increasing volume fraction (or loadings) of graphene, a progressively greater decrease occurs in the SIC onset strain. (Potts et al. 2012). Other studies have elucidated possible mechanisms behind this (Ozbas et al. 2012). By investigating the SIC of thermally exfoliated graphite oxide/NR nanocomposites, it was revealed that the presence of graphene platelets (with a large interface area) amplifies the strains locally (as expected), which then promotes early onset of SIC, at least locally. The resulting effect is a pronounced enhancement to the modulus not only due to the high modulus of graphene and cre-ation of a high modulus glassy interface but also due to the promotion of SIC in the surrounding rubber. Notably, other nano-fillers also have similar effects (Trabelsi et al. 2003). Carbon black also serves to decrease the SIC onset strain, but at much higher volume fractions when compared to graphene.

In fact, studies have shown that the cross-linking density does not affect the SIC onset strain, but the addition of filler does (Gent and Zhang 2002, Poompradub et al. 2005). The graphene addition (or in general any filler) leads to localized SIC around the filler. Thus, at moderate overall strains, the filler is now surrounded by crystallized

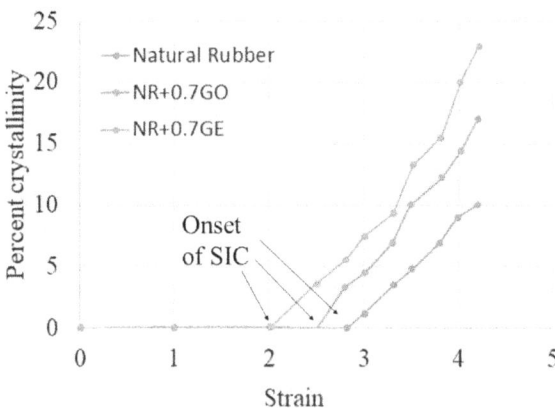

FIGURE 5.7 Percentage crystallinity in rubber showing a shift in the SIC onset strain due to the addition of graphene and graphene oxide. (Data from Li et al. 2013.).

rubber due to the localized strain concentrations. This surrounding crystallized rubber has often been termed as "immobilized rubber" since its stiffness is higher than the amorphous state. The immobilized rubber can also be distinguished energetically since it exhibits slower dynamics than the amorphous counterpart (Li et al. 2013). According to the NMR investigations reported by Rault et al. (2006), the fillers have a heterogeneous strain field around them, consisting of over-strained and thus crystallized chains.

The amount of immobilized rubber can depend on the precise morphology of the graphene nanoflakes and also the surface chemistry and dynamics. Other studies call this locally immobilized rubber as "crystallites" and offer similar explanations for early SIC-induced enhanced stiffness from addition of graphene (Li et al. 2013, Xing et al. 2014). Some researchers have proposed that the volume fraction of the immobilized rubber should in fact be counted as part of the reinforcement or filler volume fraction, since its overall effect on the rubber is similar to that of a filler. In other words, the localized SIC process has an effect similar to increasing the volume fraction of the filler. However, a quantitative relationship for the effective volume fraction to be used in homogenization rules remains elusive. For example, assuming the volume fraction of the filler to scale with the surface area of the filler does not suffice to explain the magnitude of increase in the stiffness. It should also be noted that this alteration of the SIC onset strain can be observed clearly if the graphene flakes are uniformly dispersed. These effects were found to not occur in the solution treated nanocomposites, since the strain to failure decreased so as to not allow the onset of SIC. That said, it is notable that graphene addition decreases the onset strain for SIC much more effectively than other fillers.

5.3.4 TOUGHENING MECHANISMS

Previous discussions pertained mostly to the enhancement of stiffness and strength of rubber-graphene nanocomposites. However, in many applications, their fracturing behavior is also of interest. Graphene addition alters the fracturing behavior of rubber too and brings about variety of toughening mechanisms as shown in Figure 5.8, which lend the nanocomposite a higher energy dissipation capability by fracturing. As is apparent from the SIC related discussion, any local strain concentrations can promote an early onset of SIC. Since flaws and crack tips also involve local stress/strain concentrations, SIC can be expected to occur at crack tips (Bruning et al. 2013, Trabelsi et al. 2002). The local stiffening due to SIC around the crack tip can provide additional resistance to crack growth.

While that is intuitive, studies have shown the possibility of two opposite effects of graphene on the crack growth resistance (Yan et al. 2013). In fact, at lower strains, graphene was found to accelerate crack growth, possibly due to a competition between SIC and cavitation at crack tip. However, at higher overall strains, localized SIC at the crack tips, accelerated due to graphene addition served to retard crack growth. Yan et al. (2013) found that the GE/NR composite exhibited SIC at the crack tip at a strain of 30% which was not observed for the unfilled counterpart.

Similar to other graphene or CNT-based polymer nanocomposites, other mechanisms are expected to be at play in rubber-graphene nanocomposites too as far as toughness enhancement is concerned. These mechanisms need not necessarily lead to enhancement to the stiffness or strength of the material but can contribute to the

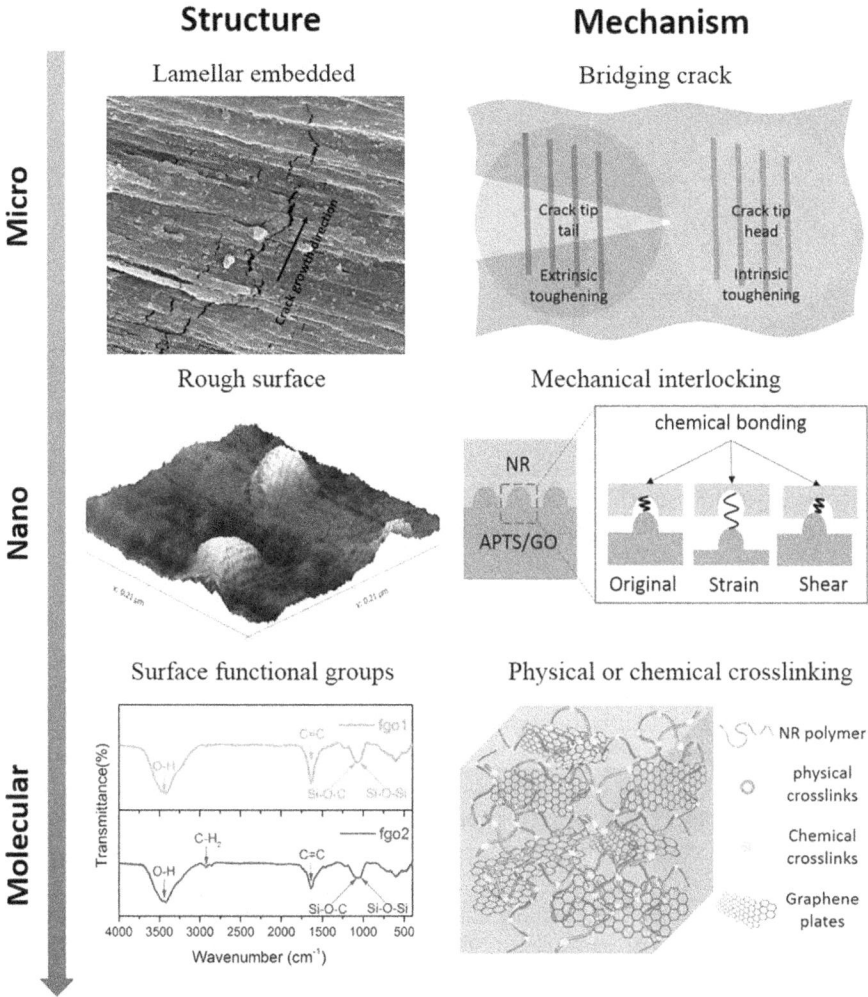

FIGURE 5.8 Multiscale nature of toughening mechanisms of graphene in rubber. (Schematic adapted from Xue et al. 2019, with permission from Elsevier.).

overall energy dissipation during failure thus imparting higher toughness to the material. One such mechanism is reinforcement pullout (Luo et al. 2016), also referred to as the well-known fiber pullout mechanism at the macroscale in long and short fiber-reinforced composites (such as GFRP or CFRPs). It is an important mechanism leading to increased toughness. This is schematically shown in Figure 5.9. When a crack propagates across a region with the fibers (or reinforcement), the fibers essentially bridge the crack faces. So, for further propagation, the fiber must be pulled out of either one or both sides. The frictional dissipation during this process contributes to the enhanced toughness. This mechanism must be operative at molecular scales for nano-cracks in the elastomer. Evidence for nano-crack bridging by graphene flakes has yet to be investigated in rubber. Since the primary source of energy dissipation

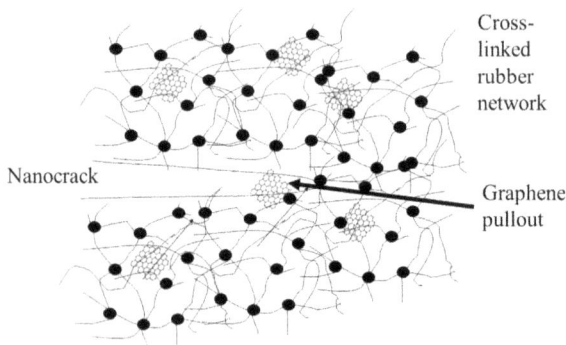

FIGURE 5.9   Schematic of graphene pullout in a rubber nano-crack.

is surface to surface friction, it is expected to be an extremely important effect for graphene which has such high surface area.

Crack deflection (Rafiee et al. 2010) is an important mechanism for toughness enhancement, observed in many composites. Since GE/NR nanocomposites will tend to have a higher volume fraction of crystallized regions, for a given strain, these can promote blocking and branching of nano- and micro-cracks. Although Yan et al. (2013) found qualitative evidence for this to occur in rubber too, a quantitative evaluation of this mechanism has not yet been conducted for rubber.

Partly, the reason for this is the multiscale nature of the failure process which adds to the complexity (see Figure 5.8). Studies have shown that the precise mechanisms that lead to reinforcement and toughening can vary depending on the length scale. For example, Xue et al. 2019 discuss that these mechanisms in APTS edited graphene/carbon black/NR composites at micron, nano, and molecular levels are different. At the molecular scales, the toughening mechanism consists of the physicochemical cross-linking between rubber and graphene surface functional groups after modification. On the nanoscale, the surface roughness of the modified graphene increases, irregular nano-arrays formed on the surface, and static friction between graphene and rubber matrix increases, which ultimately improves the toughness of the composites as discussed earlier here too. On the micron scale, the toughening effect of graphene occurs due to creation of a barrier to micro-cracks forcing them to deflect, thus requiring a higher energy. At higher scales, as mentioned previously, the mechanism of crack bridging can exist. This can be more likely in solution treated nanocomposites which consist of a 3D network of the filler. However, this needs to be investigated further. It should be noted that many of the available studies on toughness enhancements appear to be more qualitative, and there is need for more quantitative evaluations for rubber-graphene nanocomposites. Such studies are far more common for other stiffer polymers such as epoxy.

Despite this, judging from the enhancements to all mechanical properties brought about by graphene, there appears to be an upper bound to the enhancements (e.g., Xing et al. 2014). This is because with increasing amounts of graphene addition, the uniformity of dispersion becomes increasingly difficult, which in turn then leads to suboptimal property enhancements. Thus, there appears to be a sweet spot in terms of graphene volume fractions to be added to rubber, approximately around

1%–2% volume fraction. This is one of the reasons why many studies have considered functionalization of graphene to achieve better dispersion even at higher graphene volume fractions, as discussed next.

5.4   GRAPHENE MODIFICATION AND FUNCTIONALIZATION

Since the interfacial interactions are of primary importance, a large variety of studies have explored modification of graphene to result in improved interactions and thereby enhancements to properties. As explained by Tang et al. (2014), since pure graphene is defect free and inert, it can tend to aggregate severely and not bond properly with polymers. As a result, graphene oxide (GO) and reduced graphene oxide (rGO) have been widely considered as fillers. Studies have suggested that the oxygen-containing groups in GO and rGO not only enhance the interfacial stress transfer with rubber but also assist with better dispersion of the filler, thus operating at both important length scales in terms of the filler morphology. For example, in Bai et al. (2015), rGO was found to work better than GO because the reduction reaction removed a large number of oxygen-containing groups from the surface of GO. This allowed the extent of exfoliation to increase, resulting in improved interfacial stress transfer and mechanical properties. In fact, the higher the degree of reduction of GO, the better was the improvement to properties. Consistent findings were reported by Yaragalla et al. (2015) who showed a dramatic enhancement to modulus (~282%) in a composite of NR and 3 phr rGO. Further enhancements were found to be possible by thermal treatments in Bai et al. (2015) with further findings related to the degree of oxidation in Chen et al. (2018).

Significant research efforts have been devoted to understanding the functionalization of graphene for use as fillers, instead of pure graphene. These are referred to as functionalized graphene sheets or FGS's. The high amenability to being functionalized can be considered as another benefit of graphene. Studies involving epoxy and hydroxyl groups have found great success in this regard (Figure 5.10). A notable example is the work presented by She et al. (2014) who observed that functionalization of GO with epoxy and hydroxyl groups enhanced the interfacial interactions and stress transfer with rubber. As discussed in this work, the hydrogen bonds were found to be the key factor contributing to the enhancements. The epoxidized NR particles showed a tendency to assemble on the GO sheets due to the hydrogen bonds, which prevented close stacking and agglomeration of GO sheets. This enabled a uniform dispersion of the filler causing the improved properties.

On the other hand, vinyl groups have been found to be successful too. For instance, Luo et al. (2016) reported marked improvements in the mechanical properties of graphene/solution-polymerized styrene-butadiene rubber (SSBR) composites, with higher vinyl content. Their molecular dynamics (MD) simulations on graphene pull-out explained these findings, showing increased interfacial interactions and pullout energy with higher vinyl content. Staggering increases to strength (400%–500%) were obtained by Yin et al. (2017) at 5 phr functional GO, by using polyvinylpyrrolidone as the modifier. Other research groups have suggested using silane coupling agents and surfactant treatments as effective functionalization techniques, e.g., works by Wu et al. (2013), Zhang et al. (2016), and Xie et al. (2017), which were found to improve the interfacial interactions between graphene and rubber.

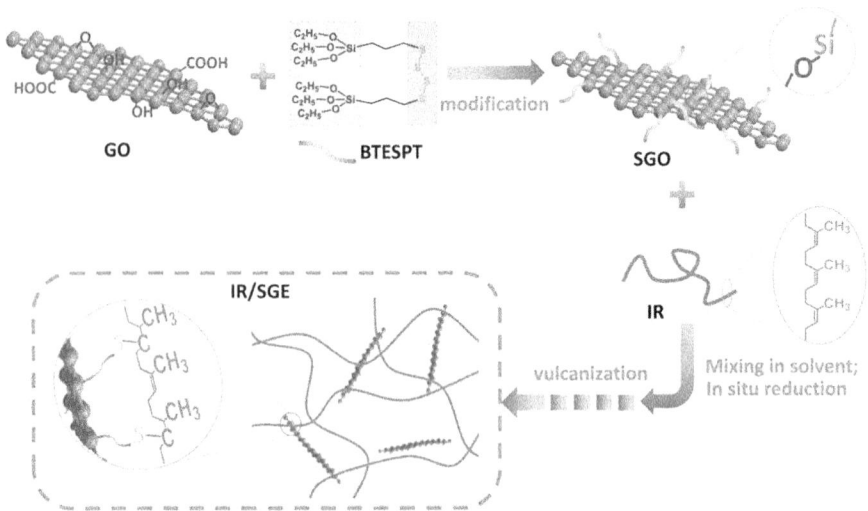

FIGURE 5.10 Schematic showing functionalization of graphene by hydroxyl groups and the preparation of preparation of Isoprene Rubber and surface-modified graphene nanocomposites. (Reproduced from Xie et al. 2017, with permission from Elsevier.)

Examples of other successful approaches involve the use of nano-ZnO by Lin et al. (2015) for graphene and NR composites, carboxylated nitrile-butadiene rubber (XNBR)/graphene hybrid nanocomposites by Liu et al. (2015) and Xue et al. (2017), oleyl amine groups causing enhanced cross-linking with sulfur in Liu et al. (2015), amine functional groups in Jose et al. (2021), and the use of ionic liquids in Yin et al. (2017). The common theme across all these studies is the attempt to alter the surface chemistry of graphene to bring about improved interfacial interactions and therefore achieve better reinforcement of rubber.

5.5   EFFECT OF GRAPHENE ON CURING KINETICS

The foregoing discussions focused on the mechanisms active in rubber graphene composites after they have been fabricated. However, the action of graphene begins sooner than that since graphene also influences the vulcanization reaction of rubber. This is especially true for FG as discussed in Hernandez et al. (2012). This is because FG enhances the interfacial interactions with rubber, and as a result, it accelerates the cross-linking reaction during vulcanization. The enhanced interfacial interactions may be evidenced by the altered timescales of the chain dynamics. This alteration contributes toward improved reinforcement and mechanical properties.

The fact that the addition of fillers affects the vulcanization is well known as discussed in Wu et al. (2013) and some of the references therein. The precise mechanisms for graphene were investigated in detail by Wu et al. (2013) for NR with the sulfur curing system. They discovered a marked effect of graphene on vulcanization process as shown in Figure 5.11. Not only was the induction period of vulcanization decreased but also the reaction rate was increased. However, this was true for lower graphene volume fractions and the trend was reversed at higher graphene

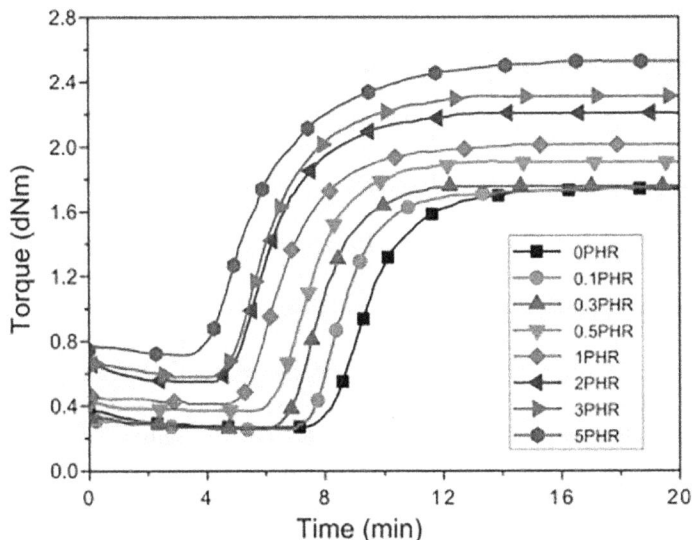

FIGURE 5.11 Effect of graphene on curing curves of various natural rubber and graphene nanocomposites. (Reproduced from Wu et al. 2013b, with permission from Elsevier.).

volume fractions, serving to decrease the reaction rate. So, the optimal cure time decreased for lower graphene loadings but then increased with increased graphene content. On the other hand, the cross-link density monotonically increased with increasing graphene content. Consistent results were reported in Tang et al. (2014) for SBR with dicumyl peroxide.

The mechanisms behind this effect were explained by considering the effects of graphene on the activation energies of the chemical reaction as well as the diffusion process. Further explanations in Xing et al. (2017) on SBR suggest that GO addition generates free radicals upon heating which favors the formation of covalent bonds during curing, leading to a higher cross-link density, which can be evidenced by the slower chain dynamics in the graphene-rubber interface layer. Thus, it is revealed that graphene alters the mechanical behavior of rubber as soon as the vulcanization begins. The reinforcement mechanisms become active well before the curing process is complete. The next section reviews some important findings related to the dynamics of rubber/graphene nanocomposites.

5.6 STRUCTURAL STUDIES

As described above, it is crucial to quantify the rubber graphene nanocomposite morphology in order to fully understand the reinforcement mechanism of the graphene-based filler and mechanical properties of the composite material. One must consider not only the morphology of the graphene-based fillers (e.g., degree of dispersion, formation of clusters or networks) but also the impact that fillers may have on the structure of the rubber matrix (e.g., impacts of the filler on the cross-linked polymer network, mesh size, and inhomogeneity). Changes to either the nanoscale arrangement of filler particles or the structure of the polymeric network can yield dramatic changes to the mechanical properties.

Techniques such as TEM, field emission scanning electron microscopy (FESEM), and atomic force microscopy (AFM) can be very useful for visualizing the structure and can provide qualitative information on the arrangement of graphene-based fillers, how well-dispersed filler particles are, and so on. To complement imaging studies, small angle scattering and wide-angle diffraction approaches can be used to provide quantitative information on nanocomposite structure and aid in connecting structural characteristics to mechanical properties. In rubber graphene composites, it can be difficult to highlight the structure of the filler in a scattering or diffraction experiment, since both the filler and the polymer matrix are carbon-based and will exhibit similar contrast. However, these techniques can often yield information on changes in the structure of the rubber matrix resulting from added graphene-based fillers. Combining TEM with scattering and diffraction studies will then provide a more complete picture of the nanocomposite morphology.

For example, Yan et al. (2014) conducted small-angle X-ray scattering (SAXS) and TEM studies for two different series of composites comprising NR and reduced graphene oxide (rGO), prepared either by a static hot press process or through a two-roll milling process. TEM images showed that these two processing approaches yielded very different filler morphologies. Composites prepared through the static hot press process showed some evidence of areas with an rGO network with a backbone consisting of several platelets, while composites that had been subject to roll milling showed single platelets of rGO dispersed throughout the rubber matrix. The authors refer to the first type of composite as having a "segregated" structure and the second as a "non-segregated" structure, and they hypothesize that the strong shear of the roll milling process disrupted any aggregation or network formation of rGO particles.

These different filler morphologies could not be observed in SAXS, presumably due to similar contrast between the rGO particles and the matrix. However, the presence of polysulfidic species in the rubber matrix, which scatter more strongly than carbon, did provide information on how the matrix morphology was affected by rGO. Neat NR samples with no rGO exhibited scattering similar to a polymer mesh, with scattered intensity, I, varying roughly as a power law of the scattering vector q, with a very broad peak arising from correlations between polysulfidic species (Yan et al. 2014). For both series of nanocomposites, this broad peak disappeared at the highest concentrations of rGO, which the authors interpret as evidence that rGO interferes with the formation of polysulfidic clusters. However, this effect is seen at lower rGO concentrations for the samples with well-dispersed "non-segregated" rGO fillers as compared to the samples with "segregated" rGO clusters or networks. Mechanical measurements of the nanocomposites showed a dramatic increase in the tensile strength with low rGO values for both filler morphologies. At the same loading of rGO, a greater increase in the tensile strength was observed in the samples with "segregated" rGO clusters or networks as compared to the composites with well-dispersed "non-segregated" rGO (Yan et al. 2014). However, it is not clear whether this effect is due to the partial network formed by the rGO filler, the fact that this filler morphology does not appear to interfere as strongly with the formation of polysulfidic clusters in the rubber matrix or some combination of these effects.

Studies such as these highlight the importance of both controlling the filler morphology and understanding the effects of graphene-based filler on the rubber network morphology when developing SPRs for graphene rubber composites. Mohamed et al. (2018) explored a strategy for controlling morphology of graphene fillers through the use of added surfactants. "Graphene-philic" surfactants containing aromatic phenyl functionalities in their headgroups were used in the formulation of composites containing graphene nanoplatelets and NR. FESEM, AFM, high-resolution TEM (HRTEM), and small-angle neutron scattering (SANS) were used to compare the structure of these composites containing phenylated surfactants to the structure of composites containing conventional commercial surfactants such as sodium dodecyl sulfate (SDS) and sodium dodecylbenzene sulfonate (SDBS), as well as to the structure of neat NR. FESEM images show the aggregation of graphene nanoplatelets in composites with SDS and SDBS surfactants, while those with phenylated surfactants show much better dispersion of graphene nanoplatelets and individual graphene nanoplatelet flakes (Mohamed et al. 2018). Although the morphology of the rubber composite was not directly probed with scattering experiments, SANS on aqueous mixtures of the graphene nanoplatelets with surfactants was consistent with dispersions of assemblies containing graphene platelets with a surface layer of surfactants, which promoted stability of the dispersions.

Graphene-based fillers can also impact SIC of polymer matrices and development of anisotropic structures in the matrix. This latter effect is an important consideration, as SIC is often employed to stiffen NRs during processing. There is some evidence that conventional fillers such as carbon black impede crystallization, although there are conflicting results on this point in the literature. Ozbas et al. (2012) studied these effects in NR containing FGS nano-fillers, comparing the results with neat NR and NR with carbon black fillers. Tensile tests showed an enhancement of Young's modulus of NR at low loadings of the graphene sheet filler; however, the elongation at break was found to decrease with graphene sheet loading. Combined SAXS and wide-angle X-ray diffraction (WAXD) were employed to probe the effect of filler particles on atomic-scale ordering of polymer chains within the matrix and the nanoscale alignment of graphene sheets. Two-dimensional WAXD showed that systems with graphene sheet fillers displayed SIC at lower strains of 0.75–1.25 than both composites containing NR with carbon black fillers and neat NR, which showed crystallization at strains of 1.75 and 2.25, respectively (Ozbas et al. 2012). Thus, the presence of graphene sheets seems to promote crystallization as pointed out previously. However, after subjecting the samples to strong elongational deformation, all samples displayed roughly the same degree of crystallinity. The degree of crystallinity was estimated from the 1D WAXD patterns at high strains, where the intensity of all crystalline reflections was found to be 10%–15% of the total intensity for all samples. In other words, although graphene sheet fillers were found to enhance the onset of crystallization, the degree of crystallinity at high strains is not expected to be significantly different for the rubber graphene composites as compared to the carbon black rubber composites and the neat NR composites. SAXS studies performed during tensile testing showed evidence of alignment of graphene sheets parallel to the stretching direction; moreover, the presence of graphene sheets appears to induce stronger alignment of crystallites in the rubber matrix as compared to composites containing NR and carbon black. These studies demonstrate the effectiveness of probing the local structural dynamics between the fillers and the graphene for a better view and understanding of the SPRs.

5.7 MICROMECHANICS, HOMOGENIZATION, AND CONSTITUTIVE MODELS

To analyze larger structures, the homogenized effective properties of the composite need to be used. This is because explicitly modeling the matrix and reinforcement phases is computationally expensive and nearly impossible for composites with nano-sized fillers. For this purpose, considerable research has amassed into determination of the effective properties of composites (mechanical and otherwise). For conventional fiber-reinforced composites, many models exist to predict effective elastic properties (Halpin–Tsai, variational, self-consistent field method, etc.) (Daniel and Ishai 2006). The simplest of these are the well-known Voigt (parallel) and Reuss (series) models providing convenient, closed-form equations for the effective properties. However, generally, these don't work for nanocomposites, primarily because the constant stress or strain assumptions become highly inaccurate, as is evident from the discussions on interfacial effects in the previous sections.

To determine effective properties of nanocomposites, the micromechanics analysis must be centered around considering the effects of nano-inclusions and the perturbations in stress/strain fields. One of the pioneering theoretical developments in this regard is the Eshelby inclusion theory, viz. the Mori–Tanaka method (Li et al. 2011, Raju et al. 2018, Nemat-Nasser and Hori 2013). Numerical approaches also have found success in this regard which involve hierarchal multiscale models where the smallest scale is MD simulations (Li et al. 2011) or hybrid approaches that incorporate Mori–Tanaka analyses as the base scale (Liu and Brinson 2012). For graphene monolayers in an elastic medium, the shear lag theory as well as the Halpin–Tsai model has been applied (Young et al. 2012, Young et al. 2018).

For rubber/graphene nanocomposites, the Guth–Gold equation has been employed commonly to predict the effective modulus (Guth 1945, Potts et al. 2012, Bergstrom and Boyce 1999). It accounts for interactions between neighboring filler particles and is expressed for spherical particles as follows:

$$E_{\mathrm{eff}} = E_m \left(1 + 2.5 V_f + 14.1 V_f^2\right) \tag{5.1}$$

where E_{eff} is the effective modulus of the composite, E_m is the modulus of the matrix (rubber), and V_f is the filler volume fraction. This equation predicts well the dependence of the effective modulus on the filler content. Several variations of this model have been proposed with varying degrees of success. Notable models include those developed in Govindjee and Simo (1992) and Ponte Castaneda (1989). For graphene/rubber nanocomposites too, the Guth–Gold equation has been applied (Potts et al. 2012) with success at low values of strains. Thus, in general, these models appear to work well only for predicting the effective initial modulus in the small, i.e., infinitesimal strain regime.

In order to predict the entire nonlinear stress-strain curve of the composite, finite strain-based homogenization needs to be considered, e.g., Brun et al. (2007), Jimenez and Pellegrino (2012), Bergstrom and Boyce (2000) (and the references therein). The initial developments are due to Hill who extended the infinitesimal strain formulation to finite strains (Hill 1972) which was followed by the effective nonlinear elasticity theory by Ball (1997). Soon after, Ogden (1978) established extremum on strain energies in composites undergoing finite strains by evaluating boundary value problems in rubber reinforced by carbon fillers. Later, Ponte Castaneda (1989) built from Hill's and Ball's works to provide additional refinement of Ogden's lower bound (Reuss) by proposing a second-order derivation that applied the polyconvexity assumption to Hashin–Shtrikman bounds. Later studies applied this second-order homogenization technique to carbon black-filled elastomers (Ponte Castaneda and Tiberio 2000). Especially notable is the second-order homogenization theory developed in Ponte Castaneda (2002). Some other works include the model for mica-reinforced rubber (Li and Yao 2019), the model presented by Nakamura and Leonard (2016), and the application of simple rules of mixture to soft graphene-based sandwich composites (Redzematovic and Kirane 2021).

For finite strains, the concept of strain amplification has been introduced by Mullins and Tobin (1957, 1965), which the Guth–Gold model too incorporates implicitly. Various models that incorporate this idea seem to work well in capturing the stiffness enhancements due to filler addition. For example, Bergstrom and Boyce (1999) proposed a constitutive model based on the amplification of the first strain invariant and presented very interesting micromechanics-based results. Their findings suggest that the effect of filler particles on the equilibrium behavior of elastomers can be accurately predicted using 3D stochastic simulations. So, a successful treatment of the heterogeneous microstructure of the composite is adequate without considering molecular level concepts such as alteration of chain mobility or effective cross-linking density in the elastomeric phase of the material. It remains to be seen if this finding holds true for graphene as well.

It appears that for accurate prediction of the effective behavior for rubber-graphene nanocomposites, models must account for the strong interfacial effects, such as the enhanced SIC around the fillers, which leads to considerable stiffening effects. To this end, models dedicated toward SIC are valuable. These need to be integrated into the constitutive models of overall homogenized behavior. For example, starting from the classical model predicted by Taylor and Darin (1955), for the tensile strength of amorphous crystallizable elastomers, to the recent model by Poompradub et al. (2005) for predicting the strain at the onset of SIC are very useful mechanistic approaches. Also notable is the model developed for carbon black in Fukahori (2003, 2005) which should be extended to graphene. Also noteworthy is the entanglement bound rubber tube (EBT) model that has been used in various studies to predict the effective properties of the composite in the presence of fillers (Heinrich and Vilgis 1993) and has been also applied to graphene recently (Li et al. 2013). These offer great promise toward the development of an SPR-based predictive capability. However, many of the developments are with respect to microscale fillers such as carbon black and their applicability with respect to nano-fillers such as graphene cannot be taken for granted and need to be checked. Thus overall, it appears that the development of dedicated constitutive models for rubber-graphene nanocomposites which build on the specific SPRs require further work.

5.8   OUTLOOK AND CURRENT CHALLENGES

It is thus evident that graphene is arguably one of the best nano-fillers that can be used to enhance the mechanical properties of rubber. Its excellent mechanical properties, high surface area, strong interfacial interactions with rubber, and ability to being functionalized make it an ideal candidate as a nano-filler. This can benefit a wide variety of engineering applications of rubber. While this chapter focuses on the mechanical properties, it is worth noting that graphene enhances other properties too, such as the electrical and thermal conductivity (Araby et al. 2014), as well as gas barrier properties. Thus, it imparts the composite distinct multi-functional capabilities (Lim et al. 2019, Qin et al. 2020, Zhang et al. 2016).

Many developments with regard to rubber-graphene nanocomposites are relatively recent, and there are several challenges regarding their development and application (Yang et al. 2017). The first and foremost is obtaining significant property enhancements at higher graphene volume fractions. Many studies show that good reinforcement and enhancement to properties is primarily obtained for low graphene volume fractions. This is because pure defect-free graphene tends to self-agglomerate, preventing a uniform dispersion. Continuing to pursue effective approaches to alter the surface chemistry of graphene and functionalize graphene are highly valuable in this regard. These should be pursued further to achieve better control over filler dispersion. Further, creative approaches to combine graphene with other fillers also need to be investigated. For example, some researchers (Wang et al. 2018) have found a remarkable enhancement to mechanical properties of SBR with the addition of both CB and rGO. Das et al. (2012), Ponnamma et al. (2013), and Li et al. (2015) have shown success with the hybrid fillers too (graphene and carbon nanotubes). Further, we note that majority of studies so far have investigated the effect of graphene on mechanical properties such as strength, toughness, and stiffness. Further investigations should be dedicated toward understanding the effect of graphene on other mechanical behaviors which are important for engineering applications, such as the Payne effect, Mullins effect, and friction properties and the underlying mechanisms.

Another significant challenge is with regard to modeling and performance prediction of these composites at a continuum level length scales which are relevant to many engineering applications. Many theories employed for the homogenization of rubber composites work only for small strains. Finite strain-based options need to be employed and developed further. Models that account for the actual filler morphology in terms of dispersion are necessary too. It is notable that nearly all the homogenization rules assume a uniform filler dispersion, and therefore, cannot capture the proper effect of fillers when they form a 3D network. A significant challenge is the lack of experimental techniques providing information on the precise nature of the interphase between rubber and graphene. This information is crucial to inform and calibrate models that attempt to predict the effective behavior. For this purpose, an understanding of the SPRs and structural dynamics at various length scales and its translation into predictive models is necessary. The ultimate prize would be a holistic modeling capability that captures mechanistically the various reinforcement mechanisms and predicts the full range of mechanical behaviors of rubber-graphene nanocomposites. In addition, theories and models for the failure behavior of these nanocomposites also need extensive research. Thus, despite various developments, accurate constitutive model development that builds on various SPRs remains to be a rich area of research for rubber-graphene nanocomposites.

ACKNOWLEDGMENT

The authors would like to gratefully acknowledge the support from NSF grant number 1922639.

REFERENCES

Araby, Sherif, Qingshi Meng, Liqun Zhang, Hailan Kang, Peter Majewski, Youhong Tang, and Jun Ma. 2014. "Electrically and thermally conductive elastomer/graphene nanocomposites by solution mixing." *Polymer* 55, 1: 201–210.

Bai, Xin, Chaoying Wan, Yong Zhang, and Yinghao Zhai. 2011. "Reinforcement of hydrogenated carboxylated nitrile–butadiene rubber with exfoliated graphene oxide." *Carbon* 49, 5: 1608–1613.

Bai, Y., Cai, H., Qiu, X., Fang, X., and Zheng, J. 2015. "Effects of graphene reduction degree on thermal oxidative stability of reduced graphene oxide/silicone rubber nanocomposites." *High Performance Polymers* 27, 8: 997–1006.

Balandin, Alexander A., Suchismita Ghosh, Wenzhong Bao, Irene Calizo, Desalegne Teweldebrhan, Feng Miao, and Chun Ning Lau. 2008. "Superior thermal conductivity of single-layer graphene." *Nano Letters* 8, 3: 902–907.

Ball, John M. "Constitutive inequalities and existence theorems in nonlinear elastostatics." *In Nonlinear Analysis and Mechanics: Heriot-Watt Symposium* 1, 4: 187–241. Pitman London, 1977.

Barlow, Fred W. *Rubber Compounding: Principles, Materials, and Techniques.* Boca Raton, FL: CRC Press, 2018.

Beckert, Fabian, Stephanie Trenkle, Ralf Thomann, and Rolf Mülhaupt. 2014. "Mechanochemical route to functionalized graphene and carbon nanofillers for graphene/SBR nanocomposites." *Macromolecular Materials and Engineering* 299, 12: 1513–1520.

Bergstrom, Jorgen S., and Mary C. Boyce. 1999. "Mechanical behavior of particle filled elastomers." *Rubber Chemistry and Technology* 72, 4: 633–656.

Bergström, Jorgen S., and Mary C. Boyce. 2000. "Large strain time-dependent behavior of filled elastomers." *Mechanics of Materials* 32, 11: 627–644.

Bhavitha, K. B., Srinivasarao Yaragalla, C. H. China Satyanarayana, Nandakumar Kalarikkal, and Sabu Thomas. 2021. "Natural rubber/graphene nanocomposites and their applications." In *Graphene Based Biopolymer Nanocomposites*, pp. 203–220. Singapore: Springer.

Bokozba, Liliane. 2019. "Natural rubber nanocomposites: a review." Nanomaterials 9, 1: 12.

Brun, Michele, O. Lopez-Pamies, and P. Ponte Castaneda. 2007. "Homogenization estimates for fiber-reinforced elastomers with periodic microstructures." *International Journal of Solids and Structures* 44, 18–19: 5953–5979.

Brüning, Karsten, Konrad Schneider, Stephan V. Roth, and Gert Heinrich. 2013. "Strain-induced crystallization around a crack tip in natural rubber under dynamic load." *Polymer* 54, 22: 6200–6205.

Cao, K. 2020. "Elastic straining of free-standing monolayer graphene". *Nature Communications* 11, 284: 284

Castañeda, Pedro Ponte, and E. Tiberio. 2000. "A second-order homogenization method in finite elasticity and applications to black-filled elastomers." *Journal of the Mechanics and Physics of Solids* 48, 6–7: 1389–1411.

Castaneda, Pedro Ponte. 2002. "Second-order homogenization estimates for nonlinear composites incorporating field fluctuations: I—theory." *Journal of the Mechanics and Physics of Solids* 50, 4: 737–757.

Chen, Yang, Qing Yin, Xumin Zhang, Hongbing Jia, Qingmin Ji, and Zhaodong Xu. 2018. "Impact of various oxidation degrees of graphene oxide on the performance of styrene–butadiene rubber nanocomposites." *Polymer Engineering & Science* 58, 8: 1409–1418.

Ciprari, Dan, Karl Jacob, and Rina Tannenbaum. 2006. "Characterization of polymer nanocomposite interphase and its impact on mechanical properties." *Macromolecules* 39, 19: 6565–6573.

Daniel, Isaac M., Ori Ishai, *Engineering Mechanics of Composite Materials*. Vol. 1994. New York: Oxford university press, 2006.

Das, Amit, Gaurav R. Kasaliwal, René Jurk, Regine Boldt, Dieter Fischer, Klaus Werner Stöckelhuber, and Gert Heinrich. 2012. "Rubber composites based on graphene nanoplatelets, expanded graphite, carbon nanotubes and their combination: A comparative study." *Composites Science and Technology* 72, 16: 1961–1967.

Das, Amit, Regine Boldt, René Jurk, Dieter Jehnichen, Dieter Fischer, Klaus Werner Stöckelhuber, and Gert Heinrich. 2014. "Nano-scale morphological analysis of graphene–rubber composites using 3D transmission electron microscopy." *Rsc Advances* 4, 18: 9300–9307.

Ferry, John D. *Viscoelastic Properties of Polymers*. New York: John Wiley & Sons, 1980.

Fukahori, Yoshihide. 2003. "The mechanics and mechanism of the carbon black reinforcement of elastomers." Rubber chemistry and technology 76, 2: 548–566.

Fukahori, Yoshihide. 2005. "New progress in the theory and model of carbon black reinforcement of elastomers." *Journal of Applied Polymer Science* 95, 1: 60–67.

Gan, Lu, Songmin Shang, Chun Wah Marcus Yuen, Shou-xiang Jiang, and Nicy Mei Luo. 2015. "Facile preparation of graphene nanoribbon filled silicone rubber nanocomposite with improved thermal and mechanical properties." *Composites Part B: Engineering* 69: 237–242.

Geim, A., Novoselov, K. 2007. The rise of graphene. *Nature Mater* 6, 183–191.

Gent, A. N., and L-Q. Zhang. 2002. "Strain-induced crystallization and strength of rubber." *Rubber Chemistry and Technology* 75, 5: 923–934.

Gent, Alan Neville, and Joseph D. Walter. 2006. "Pneumatic tire." In *US Department of Transportation*. Washington, DC: NHTSA.

Gong, Lei, Ian A. Kinloch, Robert J. Young, Ibtsam Riaz, Rashid Jalil, and Konstantin S. Novoselov. 2010. "Interfacial stress transfer in a graphene monolayer nanocomposite." *Advanced Materials* 22, 24: 2694–2697.

Govindjee, Sanjay, and Juan C. Simo. 1992. "Mullins' effect and the strain amplitude dependence of the storage modulus." *International Journal of Solids and Structures* 29, 14–15: 1737–1751.

Guth, Eugene. 1945. "Theory of filler reinforcement." *Rubber Chemistry and Technology* 18, 3: 596–604.

Hao, Shuai, Jian Wang, Marino Lavorgna, Guoxia Fei, Zhanhua Wang, and Hesheng Xia. 2020. "Constructing 3D graphene network in rubber nanocomposite via liquid-phase redispersion and self-assembly." *ACS Applied Materials & Interfaces* 12, 8: 9682–9692.

He, Canzhong, Xiaodong She, Zheng Peng, Jieping Zhong, Shuangquan Liao, Wei Gong, Jianhe Liao, and Lingxue Kong. 2015. "Graphene networks and their influence on free-volume properties of graphene–epoxidized natural rubber composites with a segregated structure: rheological and positron annihilation studies." *Physical Chemistry Chemical Physics* 17, 18: 12175–12184.

Heinrich, G., and Thomas A. Vilgis. 1993. "Contribution of entanglements to the mechanical properties of carbon black-filled polymer networks." *Macromolecules* 26, 5: 1109–1119.

Hernández, Marianella, María del Mar Bernal, Raquel Verdejo, Tiberio A. Ezquerra, and Miguel A. López-Manchado. 2012. "Overall performance of natural rubber/graphene nanocomposites." *Composites Science and Technology* 73, 2012: 40–46.

Hill, Rodney. 1972. "On constitutive macro-variables for heterogeneous solids at finite strain." *Proceedings of the Royal Society of London. A. Mathematical and Physical Sciences* 326, 1565: 131–147.

Huneau, Bertrand. 2011. "Strain-induced crystallization of natural rubber: a review of x-ray diffraction investigations." *Rubber Chemistry and Technology* 84, 3: 425–452.

Jiménez, Francisco López, and Sergio Pellegrino. 2012. "Constitutive modeling of fiber composites with a soft hyperelastic matrix." *International Journal of Solids and Structures* 49, 3–4: 635–647.

Jose, Jophy, and A. P. Susamma. 2021. "Studies on natural rubber nanocomposites by incorporating amine functionalised graphene oxide." *Plastics, Rubber and Composites* 50, 1: 35–47.

Lee, Changgu. 2008. "Measurement of the elastic properties and intrinsic strength of monolayer graphene". *Science* 321, 385: 385–388.

Lee, Changgu, Xiaoding Wei, Jeffrey W. Kysar, and James Hone. 2008. "Measurement of the elastic properties and intrinsic strength of monolayer graphene." *Science* 321, 5887: 385–388.

Li, Donghai, and Yin Yao. 2019. "An approximate method to predict the mechanical properties of small volume fraction particle-reinforced composites with large deformation matrix." *Acta Mechanica* 230, 9: 3307–3315.

Li, Fayong, Ning Yan, Yanhu Zhan, Guoxia Fei, and Hesheng Xia. 2013. "Probing the reinforcing mechanism of graphene and graphene oxide in natural rubber." *Journal of Applied Polymer Science* 129, 4: 2342–2351.

Li, Hengyi, Lei Yang, Gengsheng Weng, Wang Xing, Jinrong Wu, and Guangsu Huang. 2015. "Toughening rubbers with a hybrid filler network of graphene and carbon nanotubes." *Journal of Materials Chemistry A* 3, 44: 22385–22392.

Li, Suhao, Zheling Li, Timothy L. Burnett, Thomas JA Slater, Teruo Hashimoto, and Robert J. Young. 2017. "Nanocomposites of graphene nanoplatelets in natural rubber: microstructure and mechanisms of reinforcement." *Journal of Materials Science* 52, 16: 9558–9572.

Li, Yaning, Anthony M. Waas, and Ellen M. Arruda. 2011. "A closed-form, hierarchical, multi-interphase model for composites—Derivation, verification and application to nanocomposites." *Journal of the Mechanics and Physics of Solids* 59, 1: 43–63.

Li, Yaning, Anthony M. Waas, and Ellen M. Arruda. 2011. "The effects of the interphase and strain gradients on the elasticity of layer by layer (LBL) polymer/clay nanocomposites." *International Journal of Solids and Structures* 48, 6: 1044–1053.

Lian, Huiqin, Shuxin Li, Kelong Liu, Liangrui Xu, Kuisheng Wang, and Wenli Guo. 2011. "Study on modified graphene/butyl rubber nanocomposites. I. Preparation and characterization." *Polymer Engineering & Science* 51, 11: 2254–2260.

Lim, Lai Peng, Joon Ching Juan, Nay Ming Huang, Leng Kian Goh, Fook Peng Leng, and Yi Yee Loh. 2019. "Enhanced tensile strength and thermal conductivity of natural rubber graphene composite properties via rubber-graphene interaction." *Materials Science and Engineering: B* 246: 112–119.

Lin, Yong, Shuqi Liu, Jian Peng, and Lan Liu. 2016. "Constructing a segregated graphene network in rubber composites towards improved electrically conductive and barrier properties." *Composites Science and Technology* 131: 40–47.

Lin, Yong, Yizhong Chen, Zhikai Zeng, Jiarong Zhu, Yong Wei, Fucheng Li, and Lan Liu. 2015. "Effect of ZnO nanoparticles doped graphene on static and dynamic mechanical properties of natural rubber composites." *Composites Part A: Applied Science and Manufacturing* 70: 35–44.

Liu, Hua, and L. Catherine Brinson. 2006. "A hybrid numerical-analytical method for modeling the viscoelastic properties of polymer nanocomposites.": 758–768.

Liu, Xin, Le-Ying Wang, Li-Fen Zhao, Hai-Feng He, Xiao-Yu Shao, Guan-Biao Fang, Zhen-Gao Wan, and Rong-Chang Zeng. 2018. "Research progress of graphene-based rubber nanocomposites." *Polymer Composites* 39, 4: 1006–1022.

Liu, Xuan, Wenyi Kuang, and Baochun Guo. 2015 "Preparation of rubber/graphene oxide composites with in-situ interfacial design." *Polymer* 56: 553–562.

Luo, Yanlong, Runguo Wang, Suhe Zhao, Yiyi Chen, Huifang Su, Liqun Zhang, Tung W. Chan, and Sizhu Wu. 2016. "Experimental study and molecular dynamics simulation of dynamic properties and interfacial bonding characteristics of graphene/solution-polymerized styrene-butadiene rubber composites." *RSC Advances* 6, 63: 58077–58087.

Mao, Yingyan, Shipeng Wen, Yulong Chen, Fazhong Zhang, Pierre Panine, Tung W. Chan, Liqun Zhang, Yongri Liang, and Li Liu. 2013. "High performance graphene oxide based rubber composites." *Scientific Reports* 3, 1: 1–7.

Mensah, Bismark, Dinesh Kumar, Dong-Kwon Lim, Seung Gyeom Kim, Byeong-Heon Jeong, and Changwoon Nah. 2015. "Preparation and properties of acrylonitrile–butadiene rubber–graphene nanocomposites." *Journal of Applied Polymer Science* 132, 36.

Mensah, Bismark, Kailash Chandra Gupta, Hakhyun Kim, Wonseok Wang, Kwang-Un Jeong, and Changwoon Nah. 2018. "Graphene-reinforced elastomeric nanocomposites: A review." *Polymer Testing* 68: 160–184.

Merabia, Samy, Paul Sotta, and Didier R. Long. 2008. "A microscopic model for the reinforcement and the nonlinear behavior of filled elastomers and thermoplastic elastomers (Payne and Mullins effects)." *Macromolecules* 41, 21: 8252–8266.

Mohamed, Azmi, Tretya Ardyani, Suriani Abu Bakar, Masanobu Sagisaka, Yasushi Umetsu, J. J. Hamon, Bazura Abdul Rahim et al. 2018. "Rational design of aromatic surfactants for graphene/natural rubber latex nanocomposites with enhanced electrical conductivity." *Journal of Colloid and Interface Science* 516: 34–47.

Mondal, Titash, Anil K. Bhowmick, Ranjan Ghosal, and Rabindra Mukhopadhyay. 2016. "Graphene-based elastomer nanocomposites: functionalization techniques, morphology, and physical properties." *Designing of Elastomer Nanocomposites: From Theory to Applications*: 267–318.

Möwes, Markus M., Frank Fleck, and Manfred Klüppel. 2014. "Effect of filler surface activity and morphology on mechanical and dielectric properties of NBR/Graphene nanocomposites." *Rubber Chemistry and Technology* 87, 1: 70–85.

Mullins, Leonard, and N. R. Tobin. 1957. "Theoretical model for the elastic behavior of filler-reinforced vulcanized rubbers." *Rubber Chemistry and Technology* 30, 2: 555–571.

Mullins, Leonard, and N. R. Tobin. 1965. "Stress softening in rubber vulcanizates. Part I. Use of a strain amplification factor to describe the elastic behavior of filler-reinforced vulcanized rubber." *Journal of Applied Polymer Science* 9, 9: 2993–3009.

Mullins, Leonard. 1969. "Softening of rubber by deformation." *Rubber Chemistry and Technology* 42, 1: 339–362.

Nakamura, Toshio, and Marc Leonard. 2016. "Large deformation of particle-filled rubber composites." In *Mechanics of Composite and Multi-functional Materials*, Volume 7, pp. 149–153. Cham: Springer.

Nemat-Nasser, Siavouche, and Muneo Hori. *Micromechanics: Overall Properties of Heterogeneous Materials*. The Netherlands: Elsevier, 2013.

Ogden, R. W. 1978. "Extremum principles in non-linear elasticity and their application to composites—I: theory." *International Journal of Solids and Structures* 14, 4: 265–282.

Ozbas, Bulent, Shigeyuki Toki, Benjamin S. Hsiao, Benjamin Chu, Richard A. Register, Ilhan A. Aksay, Robert K. Prud'homme, and Douglas H. Adamson. 2012. "Strain-induced crystallization and mechanical properties of functionalized graphene sheet-filled natural rubber." *Journal of Polymer Science Part B: Polymer Physics* 50, 10: 718–723.

Papageorgiou, Dimitrios G., Ian A. Kinloch, and Robert J. Young. 2015. "Graphene/elastomer nanocomposites." *Carbon* 95: 460–484.

Payne, A. R. 1963. "Dynamic properties of heat-treated butyl vulcanizates." *Journal of Applied Polymer Science* 7, 3: 873–885.

Ponnamma, Deepalekshmi, Kishor Kumar Sadasivuni, Michael Strankowski, Qipeng Guo, and Sabu Thomas. 2013. "Synergistic effect of multi walled carbon nanotubes and reduced graphene oxides in natural rubber for sensing application." *Soft Matter* 9, 43: 10343–10353.

Ponte Castaneda, P. 1989. "The overall constitutive behaviour of nonlinearly elastic composites." *Proceedings of the Royal Society of London. A. Mathematical and Physical Sciences* 422, 1862: 147–171.

Poompradub, Sirilux, Masatoshi Tosaka, Shinzo Kohjiya, Yuko Ikeda, Shigeyuki Toki, Igors Sics, and Benjamin S. Hsiao. 2005. "Mechanism of strain-induced crystallization in filled and unfilled natural rubber vulcanizates." *Journal of Applied Physics* 97, 10: 103529.

Potts, Jeffrey R., Om Shankar, Ling Du, and Rodney S. Ruoff. 2012. "Processing–morphology–property relationships and composite theory analysis of reduced graphene oxide/natural rubber nanocomposites." *Macromolecules* 45, 15: 6045–6055.

Putz, Karl W., Marc J. Palmeri, Rachel B. Cohn, Rodney Andrews, and L. Catherine Brinson. 2008. "Effect of cross-link density on interphase creation in polymer nanocomposites." *Macromolecules* 41, 18: 6752–6756.

Qin, Hongmei, Chaoran Deng, Shengjun Lu, Yong Yang, Guichao Guan, Zhen Liu, and Qiuhao Yu. 2020. "Enhanced mechanical property, thermal and electrical conductivity of natural rubber/graphene nanosheets nanocomposites." *Polymer Composites* 41, 4: 1299–1309.

Rafiee, Mohammed A., Javad Rafiee, Iti Srivastava, Zhou Wang, Huaihe Song, Zhong-Zhen Yu, and Nikhil Koratkar. 2010. "Fracture and fatigue in graphene nanocomposites." *Small* 6, 2: 179–183.

Raju, Benjamin, S. R. Hiremath, and D. Roy Mahapatra. 2018. "A review of micromechanics based models for effective elastic properties of reinforced polymer matrix composites." *Composite Structures* 204: 607–619.

Ramanathan, T., A. A. Abdala, S. Stankovich, D. A. Dikin, M. Herrera-Alonso, R. D. Piner, D. H. Adamson et al. 2008. "Functionalized graphene sheets for polymer nanocomposites." *Nature nanotechnology* 3, 6: 327–331.

Rault, J., J. Marchal, P. Judeinstein, and P. A. Albouy. 2006. "Stress-induced crystallization and reinforcement in filled natural rubbers: 2H NMR study." *Macromolecules* 39, 24: 8356–8368.

Redzematovic, Mersim, and Kedar Kirane. 2021. "Homogenization of the Mooney-Rivlin coefficients of graphene-based soft sandwich nanocomposites." *Mechanics of Soft Materials* 3, 1: 1–16.

Sadasivuni, Kishor Kumar, Allisson Saiter, Nicolas Gautier, Sabu Thomas, and Yves Grohens. 2013. "Effect of molecular interactions on the performance of poly (isobutylene-co-isoprene)/graphene and clay nanocomposites." *Colloid and Polymer Science* 291, 7: 1729–1740.

Schopp, Stephanie, Ralf Thomann, Karl-Friedrich Ratzsch, Sabrina Kerling, Volker Altstädt, and Rolf Mülhaupt. 2014. "Functionalized graphene and carbon materials as components of styrene-butadiene rubber nanocomposites prepared by aqueous dispersion blending." *Macromolecular Materials and Engineering* 299, 3: 319–329.

She, Xiaodong, Canzhong He, Zheng Peng, and Lingxue Kong. 2014. "Molecular-level dispersion of graphene into epoxidized natural rubber: Morphology, interfacial interaction and mechanical reinforcement." *Polymer* 55, 26: 6803–6810.

Song, Sung-Ho, O. Kwon, Ho-Kyun Jeong, and Yong-Gu Kang. 2010. "Properties of styrene-butadiene rubber nanocomposites reinforced with carbon black, carbon nanotube, graphene, graphite." *Korean Journal of Materials Research* 20, 2: 104–110.

Srivastava, Suneel Kumar, and Yogendra Kumar Mishra. 2018. "Nanocarbon reinforced rubber nanocomposites: detailed insights about mechanical, dynamical mechanical properties, Payne, and Mullin effects." *Nanomaterials* 8, 11: 945.

Stöckelhuber, Klaus Werner, Amit Das, and Manfred Klüppel, eds. *Designing of Elastomer Nanocomposites: From Theory to Applications.* Cham: Springer International Publishing, 2017.

Tang, Mao-zhu, Wang Xing, Jin-rong Wu, Guang-su Huang, Hui Li, and Si-duo Wu. 2014. "Vulcanization kinetics of graphene/styrene butadiene rubber nanocomposites." *Chinese Journal of Polymer Science* 32, 5: 658–666.

Tang, Zhenghai, Liqun Zhang, Wenjiang Feng, Baochun Guo, Fang Liu, and Demin Jia. 2014. "Rational design of graphene surface chemistry for high-performance rubber/graphene composites." *Macromolecules* 47, 24: 8663–8673.

Taylor, G. R., and S. R. Darin. 1955. "The tensile strength of elastomers." *Journal of Polymer Science* 17, 86: 511–525.

Trabelsi, S., P-A. Albouy, and J. Rault. 2002. "Stress-induced crystallization around a crack tip in natural rubber." *Macromolecules* 35, 27: 10054–10061.

Trabelsi, S., P-A. Albouy, and J. Rault. 2003. "Effective local deformation in stretched filled rubber." *Macromolecules* 36, 24: 9093–9099.

Tsagaropoulos, George, Joon-Seop Kim, and Adi Eisenberg. 1996. "Chain mobility restrictions in random ionomers studied by electron spin resonance spectroscopy." *Macromolecules* 29, 6: 2222–2228.

Valentín, J. L., I. Mora-Barrantes, J. Carretero-González, M. A. López-Manchado, P. Sotta, D. R. Long, and K. Saalwachter. 2010. "Novel experimental approach to evaluate filler–elastomer interactions." *Macromolecules* 43, 1: 334–346.

Varghese, Tony V., H. Ajith Kumar, S. Anitha, S. Ratheesh, R. S. Rajeev, and V. Lakshmana Rao. 2013. "Reinforcement of acrylonitrile butadiene rubber using pristine few layer graphene and its hybrid fillers." *Carbon* 61: 476–486.

Wang, Jian, Kaiye Zhang, Qiang Bu, Marino Lavorgna, and Hesheng Xia. 2017. "Graphene-rubber nanocomposites: Preparation, structure, and properties." In *Carbon-related Materials in Recognition of Nobel Lectures by Prof. Akira Suzuki in ICCE*, pp. 175–209. Cham: Springer.

Wang, Jian, Kaiye Zhang, Zhengang Cheng, Marino Lavorgna, and Hesheng Xia. 2018. "Graphene/carbon black/natural rubber composites prepared by a wet compounding and latex mixing process." *Plastics, Rubber and Composites* 47, 9: 398–412.

Wang, Meng-Jiao. 1998. "Effect of polymer-filler and filler-filler interactions on dynamic properties of filled vulcanizates." *Rubber Chemistry and Technology* 71, 3: 520–589.

Wen, Yanwei, Qing Yin, Hongbing Jia, Biao Yin, Xumin Zhang, Pengzhang Liu, Jingyi Wang, Qingmin Ji, and Zhaodong Xu. 2017. "Tailoring rubber-filler interfacial interaction and multifunctional rubber nanocomposites by usage of graphene oxide with different oxidation degrees." *Composites Part B: Engineering* 124: 250–259.

Wu, Jinrong, Guangsu Huang, Hui Li, Siduo Wu, Yufeng Liu, and Jing Zheng. 2013. "Enhanced mechanical and gas barrier properties of rubber nanocomposites with surface functionalized graphene oxide at low content." *Polymer* 54, 7: 1930–1937.

Wu, Jinrong, Wang Xing, Guangsu Huang, Hui Li, Maozhu Tang, Siduo Wu, and Yufeng Liu. 2013. "Vulcanization kinetics of graphene/natural rubber nanocomposites." *Polymer* 54, 13: 3314–3323.

Wu, X., T. F. Lin, Z. H. Tang, B. C. Guo, and G. S. Huang. 2015. "Natural rubber/graphene oxide composites: Effect of sheet size on mechanical properties and strain-induced crystallization behavior." *Express Polymer Letters* 9, 8.

Xie, Zheng-Tian, Xuan Fu, Lai-Yun Wei, Ming-Chao Luo, Yu-Hang Liu, Fang-Wei Ling, Cheng Huang, Guangsu Huang, and Jinrong Wu. 2017. "New evidence disclosed for the engineered strong interfacial interaction of graphene/rubber nanocomposites." *Polymer* 118: 30–39.

Xing, Wang, Hengyi Li, Guangsu Huang, Li-Heng Cai, and Jinrong Wu. 2017. "Graphene oxide induced crosslinking and reinforcement of elastomers." *Composites Science and Technology* 144: 223–229.

Xing, Wang, Jinrong Wu, Guangsu Huang, Hui Li, Maozhu Tang, and Xuan Fu. 2014. "Enhanced mechanical properties of graphene/natural rubber nanocomposites at low content." *Polymer International* 63, 9: 1674–1681.

Xing, Wang, Maozhu Tang, Jinrong Wu, Guangsu Huang, Hui Li, Zhouyue Lei, Xuan Fu, and Hengyi Li. 2014. "Multifunctional properties of graphene/rubber nanocomposites fabricated by a modified latex compounding method." *Composites Science and Technology* 99: 67–74.

Xue, Chen, Hanyang Gao, Yuchen Hu, and Guoxin Hu. 2019. "Hyperelastic characteristics of graphene natural rubber composites and reinforcement and toughening mechanisms at multi-scale." *Composite Structures* 228: 111365.

Xue, Xiaodong, Qing Yin, Hongbing Jia, Xuming Zhang, Yanwei Wen, Qingmin Ji, and Zhaodong Xu. 2017. "Enhancing mechanical and thermal properties of styrene-butadiene rubber/carboxylated acrylonitrile butadiene rubber blend by the usage of graphene oxide with diverse oxidation degrees." *Applied Surface Science* 423: 584–591

Yan, Ning, Giovanna Buonocore, Marino Lavorgna, Saulius Kaciulis, Santosh Kiran Balijepalli, Yanhu Zhan, Hesheng Xia, and Luigi Ambrosio. 2014. "The role of reduced graphene oxide on chemical, mechanical and barrier properties of natural rubber composites." *Composites Science and Technology* 102: 74–81.

Yan, Ning, Hesheng Xia, Yanhu Zhan, and Guoxia Fei. 2013. "New Insights into Fatigue Crack Growth in Graphene-Filled Natural Rubber Composites by Microfocus Hard-X-Ray Beamline Radiation." *Macromolecular Materials and Engineering* 298, 1: 38–44.

Yang, Zhijun, Baochun Guo, and Liqun Zhang. 2017. "Challenge of rubber/graphene composites aiming at real applications." *Rubber Chemistry and Technology* 90, 2: 225–237.

Yang, Zhijun, Jun Liu, Ruijuan Liao, Ganwei Yang, Xiaohui Wu, Zhenghai Tang, Baochun Guo et al. 2016. "Rational design of covalent interfaces for graphene/elastomer nanocomposites." *Composites Science and Technology* 132: 68–75.

Yaragalla, Srinivasarao, A. P. Meera, Nandakumar Kalarikkal, and Sabu Thomas. 2015. "Chemistry associated with natural rubber–graphene nanocomposites and its effect on physical and structural properties." *Industrial Crops and Products* 74: 792–802.

Yin, Biao, Xumin Zhang, Xun Zhang, Jingyi Wang, Yanwei Wen, Hongbing Jia, Qingmin Ji, and Lifeng Ding. 2017. "Ionic liquid functionalized graphene oxide for enhancement of styrene-butadiene rubber nanocomposites." *Polymers for Advanced Technologies* 28, 3: 293–302.

Young, Robert J., Ian A. Kinloch, Lei Gong, and Kostya S. Novoselov. 2012. "The mechanics of graphene nanocomposites: a review." *Composites Science and Technology* 72, 12: 1459–1476.

Young, Robert J., Mufeng Liu, Ian A. Kinloch, Suhao Li, Xin Zhao, Cristina Vallés, and Dimitrios G. Papageorgiou. 2018. "The mechanics of reinforcement of polymers by graphene nanoplatelets." *Composites Science and Technology* 154: 110–116.

Zhan, Yanhu, Jinkui Wu, Hesheng Xia, Ning Yan, Guoxia Fei, and Guiping Yuan. 2011. "Dispersion and exfoliation of graphene in rubber by an ultrasonically-assisted latex mixing and in situ reduction process." *Macromolecular Materials and Engineering* 296, 7: 590–602.

Zhang, Guangwu, Fuzhong Wang, Jing Dai, and Zhixiong Huang. 2016. "Effect of functionalization of graphene nanoplatelets on the mechanical and thermal properties of silicone rubber composites." *Materials* 9, 2: 92.

Zhang, Hao, Wang Xing, Hengyi Li, Zhengtian Xie, Guangsu Huang, and Jinrong Wu. 2019. "Fundamental researches on graphene/rubber nanocomposites." *Advanced Industrial and Engineering Polymer Research* 2, 1: 32–41.

Zhang, Xumin, Jingyi Wang, Hongbing Jia, Shiyu You, Xiaogang Xiong, Lifeng Ding, and
    Zhaodong Xu. 2016. "Multifunctional nanocomposites between natural rubber and poly-
    vinyl pyrrolidone modified graphene." *Composites Part B: Engineering* 84: 121–129.
Zhou, Xufeng, and Zhaoping Liu. 2010. "A scalable, solution-phase processing route to
    graphene oxide and graphene ultralarge sheets." *Chemical Communications* 46, 15:
    2611–2613.

6 Structure-Property Relationship of Graphene-Rubber Nanocomposite

Mohammad Javad Azizli and Masoud Mokhtary
Islamic Azad University

Mohammad Barghamadi
Iran Polymer and Petrochemical Institute

Katayoon Rezaeeparto
Research Institute of Petroleum Industry

CONTENTS

6.1 INTRODUCTION

Nowadays, nanoparticles play a major role in technology. Researchers have paid special attention to two-dimensional graphene sheets due to their special properties such as exceptional electrical conductivity and mechanical properties and high surface area. Graphite is the lightest, thinnest, and strongest material ever discovered. It was

DOI: 10.1201/9781003200444-6

first produced as a thick monoatomic layer by mechanical exfoliation. Since graphene is only 1 atom thick, using graphene as an one- dimensional or atomic scaffold is possible to create effectively intercalated graphene compounds (Aliofkhazraei 2013). Graphene is easily synthesized from graphite as a raw material and then reduced to give graphene oxide (GO) in which the reduction reaction is rarely 100% complete. Many other forms of graphene can be produced using graphene oxide. Thus, reduced graphene oxide (RGO) often contains impurities, such as catalyst residues, when used in the reduction process, and also has some oxidized sites, often in the form of – C=O or –COOH groups. However, many researchers still use RGO in their research and believe that their findings are also applicable for pure graphene. In addition to single plates of graphene or monolayers, graphene has been studied in other forms such as bilayers or multilayers, nanoribbons, and even quantum dots. Electrical conductivity of pure, defect-free graphene is about 3.6×10^8 S m^{-1} or 6X that of Cu or Ag and its thermal conductivity is 2,000–4,000 W m^{-2} K^{-1}. These values are typical for suspended monolayers, and purer samples have higher values.

At present, most composites used are at the front line of materials technology, due to their performance and costs appropriate for highly demanding applications, and have conquered various sectors such as aerospace industry, aeronautics, automotive industry, manufacturing industries, construction, and the marine industry. Also, more use of composite technologies is emerging. This highly competitive market continues to grow, with the major focus on the past to produce materials with adequate strength and high wear resistance. In the long run, the production of significant products with exceptional capacities will be possible with the help of composites, but these materials are often made in a limited series of production or on a small scale. Today, large industries such as aerospace and automotive require large-scale production for their end-use sectors. Therefore, the composite industry must meet the production requirements of the large series that creative industries need (Summerscales et al. 2010). For the composite industry, there are huge opportunities that have grown rapidly thanks to the innovation that new consumer applications have created and built on existing applications.

In the future, growth for various industries would depend on other capabilities. Due to their cost-performance benefits, composites become the choice of various industries. Also, extended developments are predicted in the composite field in the next years, provided that the composite industry can meet the needs of large-scale production to increase its widespread influence in several advanced industries. In combining new materials for automobiles and other means of transportation, weight reduction and improved mechanical performance are key factors. Thus, the excellent quality and lightweight of the composites cause these materials to become the most attractive candidate for industry, specially automotive industry. Composite materials with fiber-reinforced polymer (FRP) are alternatives to metals. Also, the addition of a filler material affects the mechanical properties. Graphene is one of the most important filler materials, in particular for polymer composites. Graphene has low mass density and exceptional mechanical properties, thus it has been broadly used as engineering materials in different challenging applications. The fibers used in the manufacture of FRPs are mainly synthetic fibers such as glass or steel and thermosetting or thermoplastic matrices of petrochemical origin (Das et al. 2021).

Graphene with a special hexagonal-lattice monolayer structure is much more rigid and heavier compared to the carbon nanotube (CNT). The researchers found that by nano-indentation on an independent monolayer graphene, the modulus (E) and the intrinsic strength (σ) of Young at a strain of $\varepsilon = 0.25$ equal to 1.0 TPa and 130 GPa, respectively, are obtained (Scharfenberg et al. 2011).

Das et al. (2021) claimed that composite materials show improved properties compared to metals and polymeric materials which cause the composites to be used as structural components. In this regard, the emerging field of high-performance natural fibers, especially Bast fibers, has been considered. Various Bast fibers include flax, hemp, and jute. As Bast fibers provide comparable or improved tensile strength and stiffness compared to that of glass fiber, they become attractive. Also, these fibers have favorable vibration dampening and no-abrasive properties. Due to their beneficial properties, low prices, legislative engines, and consumer preferences, natural fibers and resins are on the rise. Natural fiber composites have mainly developed. The users of natural fibers, mainly industry sectors, are now engaged in an eco-design approach. Kuilla et al. in their detailed review article on graphene-based polymer nanocomposites have clarified the importance and use of graphene in different host materials. They made a notable comparison of different nanofillers and described their significant applications in detail (Kuilla et al. 2010). Jang and Zhamu (2008) reviewed the processing of graphene nanoplatelets for fabrication of composite materials. Hansma et al. studied the effective manufacture of nanocomposites based on graphene. They optimized the quantity and combination of adhesives and high-strength nanostructures (graphene) required to produce a solid, low-density, lightweight, and damage-resistant composite material (Hansma et al. 2006). effectively. Lv et al. (2021) observed interesting properties for a small number of epoxy-based graphene nanocomposites that make them suitable for the development of thermal interface materials in the electronics industry.

Graphene has a two-dimensional structure with sp^2 hybridization. In graphene, the orbitals involved in the sigma bond are in the plane p_x, p_y, while p_z, which is responsible for the π bond, is perpendicular to the plane. Orbitals of interatomic length ~1.42 Å form an in-plane sigma bond, which strengthens the C–C bond and increases the mechanical strength (Novoselov et al. 2004). Additionally, the π band is half-filled causing zero band gaps between the valence and conductive bands which results in the free movement of electrons making it highly conductive. p_z *orbital* which is perpendicular to the plane makes a weak π bond resulting in weak van der Waals interaction between the monolayers of graphene which allows them to move under very weak shares (Novoselov et al. 2004). In the past, it was believed that two-dimensional structure cannot exist as it is thermally unstable, but Novoselov et al. who invented a single layer of graphene in their laboratory in 2004 proposed that it is a carbon monolayer that is arranged in the shape of honeycomb. The invention of graphene has opened a new window into the field of materials science. Ideal graphene has high conductivity with zero band gap, high thermal conductivity, and good tensile strength at room temperature (Shen et al. 2013). In spite of all these great properties, graphene cannot be used alone, which limits its potential application. So, there is need for incorporating graphene with different materials like functionalized thin films, fibers, and coatings for the better use of graphene at industry level (Kamaraj et al. 2020). Graphene can also be used as a nanofiller-based carbon to enhance mechanical and thermal properties for the development of high-performance polymer-based nanocomposites (Bharadiya et al. 2019). Due to the high electric conductivity and mechanical flexibility, it can substitute the metal conductors. Graphene has these remarkable properties due to its high aspect ratio and single thin layer. Due to its high surface area, graphene agglomerates in contact with the polymer. Therefore, to reduce this agglomeration, graphene needs to be functionalized, which can be done by oxidizing or by reducing agents.

6.2 GRAPHENE-BASED POLYMER COMPOSITE MATERIALS

The development of composite materials has led to the use of new materials, which in turn has reduced costs, increased efficiency, and optimal use of existing natural resources. In addition, more emphasis on green materials has made the production of natural fibers with high strength to weight ratio more important for entering the automotive, construction, and other industries. Despite the availability of various filler materials, one of the most effective filler used is graphene. Due to its low mass density and exception mechanical properties, graphene plays a potential role in enhancing the properties of green polymer-based composites. Challenges of using renewable resources in the production of industrial materials include reliability, flexibility, high efficiency and sustainability, biodiversity, and environmental impacts. The latest advances in natural fibers used in composite materials are (Zimniewska and Wladyka-Przybylak 2016):

- Improving the strength and rigidity to afford with glass fiber;
- finding new application areas for broadening footprints in various industries;
- improving efficiency and rigidity for automotive applications;
- expecting advanced industries to use modern reinforced composites to replace parts that are currently made with other materials or composite forms.

Re-interest in biobased goods has led to concern for the environment. Natural fibers, particularly in automotive engineering, are considered as an environmentally friendly substitute for glass fibers composite. Due to their wide availability, low cost, low density, high-specific mechanical properties, and being eco-friendly, they are increasingly used as reinforcements in polymer matrix composites. Today, eco-friendly green materials have become important as environmental issues increase and oil reserves decrease. Cellulose fibers or textiles seem to be the most suitable means to solve these specific problems. The ability of cellulose fibers to reinforce composite materials is well known, but many changes have been made due to the re-interest in natural fibers to make them equal to or even superior to synthetic fibers. An emerging trend is the implementation of natural fiber-reinforced composites with graphene as a filler material for replacing the synthetic fiber-reinforced composites, taking into account the high-performance level of composite materials in terms of durability, maintenance, and cost effectiveness (Williams and Starke 2003).

Polymer composite materials based on graphene have been studied (Choi et al. 2010). They can include various components such as conducting polymers (CPs) like poly(aniline), poly(3,4-ethylenedioxythiophene) (PEDOT), poly(phenylene sulfide), and poly(diacetylene); epoxies; polyurethanes; polystyrenes (PS); poly(vinyl alcohol) (PVA); poly(ethylene terephthalate) (PET); polycarbonate; poly(vinylidene fluoride) (PVDF); Nafion®, poly(ε-caprolactone); poly(lactic acid); poly(methyl methacrylate) (PMMA); and high-density polyethylene (HDPE).

A recent review, simple and predictable synthesis methods for these composites have been discussed in detail. The proposed potential applications for these graphene-based polymer composites are very diverse, some of which include organic LEDs, solar cells, transparent conductors, energy storage, sensors, biomedical devices, drug delivery, and EMI shielding (EMI-SE) (Choi et al. 2010). Graphene sheets are reinforced with CNTs for greater strength and toughness. This is done in a similar manner to reinforcing concrete with steel rods, the result of which is called "graphene rebar" (Yan et al. 2014). For "graphene rebar", applications have been proposed such as conductive and transparent films and more environmentally durable photovoltaics.

In general, graphene is used as an additive fiber in composite materials and coatings, to increase strength, conductivity, and other properties and to achieve a combination of properties. Examples include graphene-containing printer powders, the replacement of small amounts of carbon fiber with graphene in tennis rackets to increase stiffness and weight, and the addition of functionalized graphene to commercial epoxy resins (Ng et al. 2019, Chandrasekhar 2018). Recently, a patch has been developed that can be used as a wearable electrocardiographic recorder and it is claimed that it will not come off even if submerged in water. In this patch, a specialized and conductive adhesive based on a composite of graphene powder, CNTs, and poly(dimethyl siloxane) (PDMS) is used, which mimics the structure and adhesive strength based on the van der Waals force of the gecko feet (Eisenhaure and Kim 2017).

Elastomers are polymers that exhibit elasticity (McNaught and Wilkinson 1997). Common properties of most elastomers are high elongation, viscoelasticity, and low modulus with weak intermolecular forces. Easy deforming upon elongation and the flexibility before mechanical failure are other properties of elastomers. These properties have led to elastomers being widely used in various industries, health care applications, and so on. The use of natural rubber, a type of biopolymer (Hosler et al. 1999), dates back to the American people, but did not become popular until the research work of Charles Goodyear and Thomas Hancock in 1839 on vulcanization (Heinrich 2011).

At present, polymer industries have been receiving huge industrial and academic research interests due to the increasing demand of high quality and engineering materials, which will cause a remarkable impact in global revenues in the near future (Bhavitha et al. 2021). These materials were used in manufacturing and development of efficient polymeric materials, which have brought revolution in material science and engineering (Azizli et al. 2018, Paul and Robeson 2008, Azizli et al. 2014, Pavlidou and Papaspyrides 2008).

Biopolymer nanocomposites that can be originated from elastomers exhibit excellent improved properties. Nevertheless, the synthetic methods of these nanocomposites are differing from the regular polymer nanocomposites; they should undergo crosslinking by means of vulcanization during preparation. Several research groups have prepared efficient elastomeric nanocomposites and proposed different fillers such as layered silicates, silica nanoparticles, carbon black, multi-walled carbon nanotubes, and organoclay montmorillonite (Bokobza 2007, Azizli et al. 2017, 2019, 2020a, 2020b, Yaragalla et al. 2019, Donnet and Voet 1976, Ponnamma et al. 2014, Hasanzadeh Kermani et al. 2020). The fillers notably affect the final performance of the nanocomposite and each one has its own impact on the end use applications. Importantly, a homogeneous distribution of the fillers in elastomers is needed for enhanced mechanical, thermal, and electrical properties. To obtain this, chemical treatment of nanofillers is required to boost the interfacial interactions between filler and polymer as well as filler and filler (Basu et al. 2014).

Many studies showed that the incorporation of small quantities of graphene into polymers significantly improves their physical performance (Yaragalla et al. 2015). Many research articles have reported graphene-based elastomers, thermosets, and thermoplastics polymer nanocomposites (Yaragalla et al. 2019, Terrones et al. 2011). The orientation and dispersion of the filler inside the polymer can determine the physical performance and other characteristics of the resultant hybrid materials and nanocomposites. Several methods including solution blending and melt mixing are used for the preparation of such composites. However, each method has its own importance concerning the distribution of the filler in polymer (Kim et al. 2010). To produce elastomer-based nanocomposites, a combination of methods is needed as it could not be prepared in a single step. In order to obtain composites with desired properties, a series of materials like crosslinking agents, catalysts, and processing aids are required to process the elastomer mixture. Thus, processing of elastomer composites consists of various stages in which elastomers/polymers are mixed with fillers and then crosslinked via two-roll mill (Sadasivuni et al. 2013).

Currently, much research is being done to improve the mechanical and electrical properties of polymer graphene nanocomposites (Cote et al. 2009, Prasad et al. 2009). At present, to improve the mechanical and electrical properties of polymer-graphene nanocomposites, many researches have been initiated. Significant improvements have been made in the mechanical properties of graphene/PVA composite films, for example, 150% increase in tensile strength for 1.8 vol.% graphene nanosheets, which indicates efficient load transfer between graphene nanosheets and the matrix. Young's modulus of the composite is more than ten times greater than that of IWA (Zhao et al. 2010). With the addition of only 0.9 wt.% PS-modified graphene sheets to PS, the resulting PS composite film shows 57% and 70% increase in Young's modulus and tensile strength, respectively, over PS. In PVDF, sample storage modulus (G') increases with the increase in PMMA-functionalized graphene (MG) content and the highest increase of G' is 124%, which is comparable to that of PVDF-multiwalled nanotube (MWNT) nanocomposites. The stress at break increases significantly from 39 to 100 MPa for 5% (w/w) MG with a 157% increase, and Young's modulus of PVDF increases from 1.44 to 6.07 GPa for 5% MG showing 321% increase

(Layek et al. 2010). The results show that Young's modulus increases by 31% for 0.1 wt.% thermally exfoliated graphene/epoxy, 120% for 1 wt.% chemically reduced graphene/thermoplastic polyurethane, and for 0.25 wt.% graphene/silicone foam that exfoliated thermally by 200% (Liang et al. 2009). Prasad et al. (2009) observed synergistic enhancement of mechanical properties of PVA composite with a mixture of two different carbon nanomaterials. In a recent report, the PVA/GO composite showing a considerable enhancement of mechanical properties (1,019% in G¢, 1,300% in Young's modulus, and 186% in tensile strength) with GO-oriented structure at high concentrations of GO (40–60 wt.%) in the plane of alignment (Putz et al. 2010). Also, the electrical conductivity of the graphene-polymer composite is greatly increased compared to that of pristine polymer. Stankovich et al. (2006a) reported a PS-graphene nanocomposite showing conductivity of 10^{-3} S cm^{-1} at only 1% (v/v) of graphene, which is sufficient for many electrical applications, and exhibits the lowest percolation threshold at room temperature among different carbon nanostructures. In the epoxy resin/graphite nanocomposite, the electrical conductivity is shown to increase by 12 orders. High conductivity is achieved using latex-based technology for PS/graphene composite formation with 1.6 wt.% graphene (Tkalya et al. 2010). In polyester/exfoliated graphite nanocomposites, the electrical percolation data are compared well with the rigidity percolation from rheological measurements (Kim and Macosko 2008). The conductivity of the electrochemically modified graphene (ECG)/PVA composite was 10^{-7} S cm^{-1} for 6 wt.% filler. For graphene/PS composite with 1.2 wt.% graphene, the conductivity was 0.15 S cm^{-1}; for graphene/polycarbonate composite with 2.2 vol.% graphene, it was 0.5 S cm^{-1} (Yoonessi and Gaier 2010); and for RGO/PVDF composites with 4 wt.% RGO, it was 10^{-4} S cm^{-1} (Ansari and Giannelis 2009). Thus, many efforts have been made to increase mechanical properties of the graphene-polymer nanocomposites for household applications and to induce electrical properties in the composites for the electronic/optoelectronic industry applications.

6.3 MELT MIXING/BLENDING

This technique is favored by industry, because it combines low cost and fast production (Dao et al. 2014, Allahbakhsh et al. 2013). In this process, fillers are initially dispersed in the selected polymer matrix by applying a shear force. The main advantage of this method lies in the absence of various organic solvents. There are some problems associated with this technique despite the aforementioned advantages, because high temperature is required to disperse nanofillers inside the elastomer, which can cause degradation of the material. Filler dispersion can be obstructed by the high amounts of filler and polymer viscosity and it requires high shear forces to control the viscosity, which can destroy the graphene layers. Kim et al. (2010) in their investigation of thermally reduced graphene (TRG) filled thermoplastic polyurethane (TPU) composites found that there was a poor dispersion of filler in the polymer matrix with this method (Figure 6.1). They concluded that it is very difficult to achieve the highest homogeneous distribution of the filler by using this method.

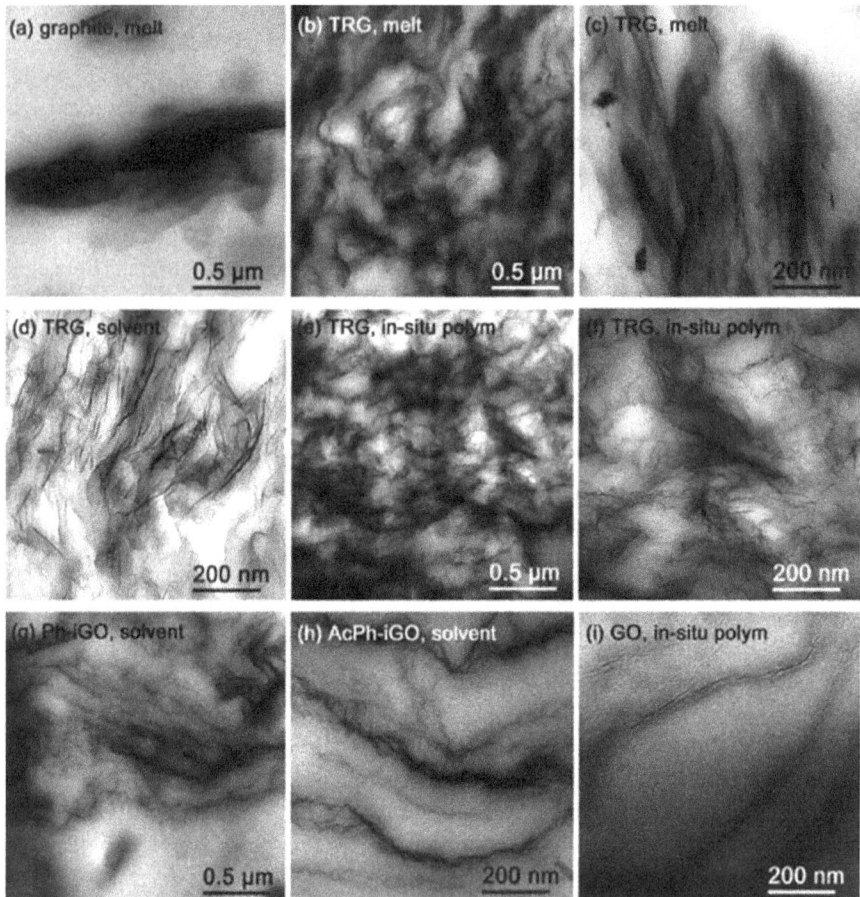

FIGURE 6.1 Dispersion of graphite, graphene, GO, and TRG in polyurethane matrix with respect to comparison of different processing methods. (Kim et al. 2010, with permission.)

6.4 SOLUTION/LATEX BLENDING

Due to the ease of processing graphite derivatives and graphene in various solvents, solution/latex blending is presented as an efficient fabrication technique and is generally used for the preparation of graphene-based polymeric materials (Araby et al. 2014, Wu et al. 2013, Tang et al. 2012). In this process, the mixing of graphene colloidal suspensions with the selected elastomer is carried out in a particular solvent by simple stirring or shear mixing. In order to get better dispersion of graphitic fillers inside the elastomer, sonication is performed. However, agglomeration of graphene layers is possible even at lower loadings owing to the van der Waals forces. The restacking of graphene platelets in the solution mixing process could be prevented using organic surfactants, which have the great tendency to stabilize colloidal nanoparticle suspensions. Latex blending method is an effective technique to produce elastomer or biopolymer graphene based nanocomposites, so the aggregation of the fillers inside biopolymer can be controlled kinetically during processing (Zhan et al. 2012). The elimination or separation of solvents used in the preparation of biopolymer composites is one of the major difficulties involved with this method. Therefore, the latex blending method is limited only for certain applications, where homogeneous distribution of graphene filler could be achieved.

6.5   IN SITU POLYMERIZATION

Graphene/biopolymer or elastomer composites have been synthesized by in situ polymerization methods. In this technique, a high level of graphene dispersion could be achieved in comparison with solution mixing. Initially, the graphene/graphite or functionalized graphene is added to the neat monomers. In this process, the inter-layer distance of graphite greatly increases. Kim et al. (2010) reported the synthesis of TPU/graphene composites via in situ polymerization technique. It is a quite useful method for the fabrication of graphite/graphene-filled elastomer composites, but it has some limitations. The technique is not applicable if the polymerization takes place between the graphitic layers (Shioyama 2000). Furthermore, it required many reagents for the polymerization.

6.6   MECHANICAL PROPERTIES

Incorporation of carbon nanomaterials into elastomers results in elastomer carbon nanocomposites, which exhibit excellent mechanical properties, compared to individual elastomers. As these fillers can enhance the mechanical strength of the elastomer, high stress is required to break the resultant nanocomposite. The physical features of these nanomaterials are very important for tuning the physical performance of such nanocomposites. Mainly, homogeneous dispersion of filler in the matrix could affect the final performance of the elastomer nanocomposites. In fact, agglomeration of nanomaterials or nano reinforcement in the polymer matrix can suppress the mechanical strength of the resultant nanocomposite. Natural rubber-graphene nanocomposites exhibited enhanced mechanical properties that mainly depend on the level of graphene dispersion, regardless of the preparation conditions (Potts et al. 2012). Interfacial interactions of graphene and the polymer could greatly affect the final properties of the composite. There are many reports regarding the effect of nanofillers such as CNTs (Ponnamma et al. 2014), silica (Alexandre et al. 2000), and clays (Favier et al. 1995) on the physical performance of the polymer. The reason for paying more attention to graphene-based composites is that other nanofillers do not show unique properties like graphene.

6.7   TENSILE PROPERTIES

Much research has disclosed regarding the physical performance of graphene-reinforced elastomer composites (Sadasivuni et al. 2013, Dong et al. 2015). Zhan et al. investigated the tensile properties of graphene-reinforced NR composites. They observed significant improvement in modulus and tensile strength of neat NR (Potts et al. 2012). In addition, it was observed that with graphene loading, the neat NR elongation decreases, which indicates the π–π interactions of graphene layers with NR chains. It is believed that cost-effective composites with promising mechanical properties and high efficiency can be made with graphene fillers. Graphene-reinforced elastomeric composites showed considerable enhancement in modulus compared to other graphene/polymer composites, which confirm the strong interfacial interactions among graphene and elastomer chains. In contrast, tensile strength of the elastomers is compromised by the addition of rigid fillers, and in most cases, elastomer-graphene composites show a decline in tensile strength. Potts et al. (2013) studied RGO-reinforced natural rubber composites. They found that tensile performance of the nanocomposites prepared by latex blending method dramatically improved. In another report, they explored the effect of thermally exfoliated graphene oxide (TEGO) on the physical performance of natural rubber with the aid of a two-roll mill. They found that TEGO has significantly improved the tensile properties of natural rubber as shown in Figure 6.2. Transmission electron microscopy (TEM) confirmed the homogeneous dispersion of TEGO in the elastomer, which causes such an improvement in the mechanical performance.

Tensile properties of the materials are listed in Table 6.1. As can be seen, the increment in stiffness is less than the report of Potts et al. (2011), but an improvement in extension/strain was achieved. Overall, considerable improvement in tensile

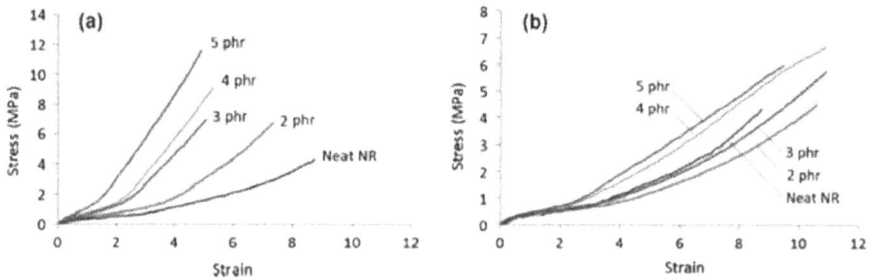

FIGURE 6.2 Tensile properties of graphene and modified graphene-reinforced NR composites: (a) pre-mixed thermally exfoliated graphite oxide with latex (L-TEGO)/NR and (b) TEGO/NR. (Potts et al. 2013, with permission.)

TABLE 6.1

Mechanical Properties of 5 phr of Graphene, Modified Graphene, and Carbon Black-Reinforced Natural Rubber Composites

Filler	E_{100} (MPa)	E_{300} (MPa)	σ_f (MPa)	Strain at Break
Neat NR	0.41	0.84	5.15	8.44
TEGO	0.43	1.28	6.05	8.96
L-TEGO	1.07	5.19	10.90	5.05
RGO	1.59	9.01	10.18	3.19
CB	0.51	1.19	6.28	8.34

TABLE 6.2
Tensile Properties of Graphene, Modified Graphene, Carbon Black, and CNT Reinforced SBR Composites

Type of Filler	E_{300} (MPa)	E_{50} (MPa)	σ_f (MPa)	Maximum Strain at Break
Neat SBR	4.7	0.8	5.1	3.24
TRGO (25 phr)	-	10.9	17.5	0.88
CRGO (25 phr)	12.4	2.8	12.6	2.95
CB (25 phr)	6.7	1.2	8.2	3.65
CNT (25 phr)	-	2.1	9.2	2.60

properties of the composites with graphene can be achieved using a two-roll mill method. It is also evident from Table 6.1 that higher levels of strength and stiffness can be obtained with GO than with carbon black. Graphene oxide (1 phr) reinforced natural rubber composite showed significant improvement in fatigue resistance under cyclic loading (Dong et al. 2015). In fact, best mechanical performance of the composites was obtained by solution blending. However, the use of hazardous solvents making this method less applicable. Also, considerable improvement in tensile properties of graphene/elastomer nanocomposites has been achieved via ecofriendly two-roll mill method, which is industrially viable.

Beckert et al. reported the modified graphene and rGO-reinforced styrene-butadiene rubber (SBR) composites (Beckert et al. 2014). Considerable increase in the tensile properties of SBR was noticed with the addition of both thermally reduced GO (TrGO) and chemically reduced GO (CrGO) (see Table 6.2). As illustrated in this table, the mechanical properties of graphene-based composites compared to those with other nanofillers were improved. In another report, Zhan et al. (2011) showed improvement in the tear strength (50%) and tensile strength (47%) for the natural rubber latex (NRL)/GE (Chemically Reduced Graphene, CRG-2 wt.%) composites. In contrast, the low improvement in tensile properties of NR/MWCNT (multi-walled carbon nanotube) and NR/CB (carbon black) composites was reported. They concluded that strong interfacial interactions among graphene and NR chains and great reinforcing action of graphene are the reasons for their observed high mechanical properties.

6.8   DYNAMIC MECHANICAL PROPERTIES

Dynamic mechanical analysis (DMA) can provide the details about viscous and elastic nature of graphene-reinforced elastomer composites (Sadasivuni et al. 2013, Potts et al. 2013). In general, filler particles improve the storage modulus of a composite at a temperature above the glass transition temperature (T_g) compared to the neat elastomer (Cho et al. 2005). In addition, nanofillers can induce crystallinity in the whole composite, which in turn enhances T_g of the composite, due to the segmental mobility suppression of the elastomer chains around the graphitic filler. Araby et al. (2014) observed significant improvement in the elastic modulus at the rubbery region of graphene-incorporated SBR (Figure 6.3). In another case, notable increase in modulus was recorded by Potts et al. (2012) with GO-reinforced natural rubber

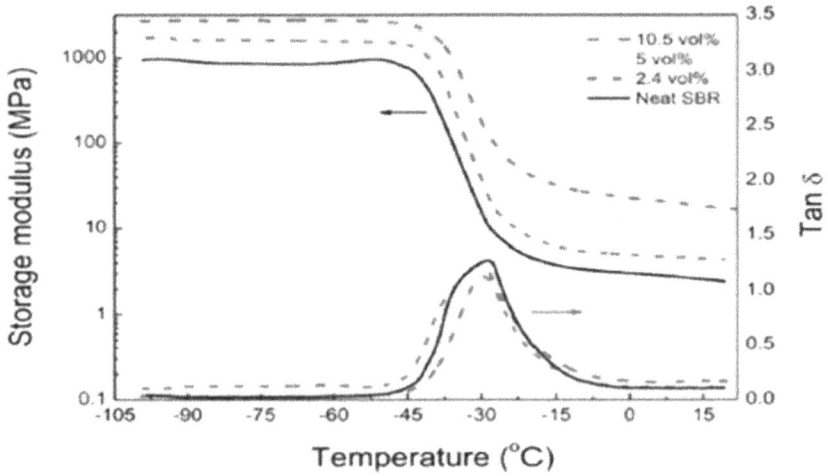

FIGURE 6.3 Dynamic mechanical analysis of SBR composites with respect to graphene loading. (Araby et al. 2014, with permission.)

composites, due to the increased interface region between the fillers and the matrix, which were boosted by the solution blending method. Nanocomposites prepared by solution mixing method showed superior mechanical properties compared to that of nanocomposites produced by melt mixing, two-roll mill methods due to the weak interfacial interactions and agglomeration of nanomaterials. Araby et al. (2014) at another instance revealed the effect of graphene in enhancing the glass transition temperature of SBR composites by virtue of the fact that the elastomer chain mobility suppressed by the graphene. In contrast, a report showed the plasticizing nature of graphene where a decrease in glass transition can be observed in the case of elastomer nanocomposites (Chen et al. 2012).

6.9 PREPARATION OF GRAPHENE POLYMER COMPOSITES

Various preparation methods such as melt mixing, solution mixing, and in situ polymerization are used to synthesize polymer nanocomposites reinforced with graphite fillers. In addition to traditional methods, many other methods are used to make materials that are introduced here. As shown in Figure 6.4a, in conventional strip casting method (Zhou et al. 2012a), a mixture of natural flake graphite powder (NGP) and polyvinyl butyral (PVB) is magnetically stirred in ethanol. After removing the air bubbles by evacuation, the mixture is casted on a plastic film using a blade. The shear force created by the blade causes the orientation of anisotropic graphite particles in the composite. This method was tested with different concentrations of graphite powder from 10 to 95 wt.%, using blades with different heights from 300 to 500 μm. The results showed that in order to create a strong shear force for better orientation of the particles, a narrow blade gap is required and the thermal conductivity as well as the degrees of orientation are affected by the test conditions (Zhou et al. 2012b).

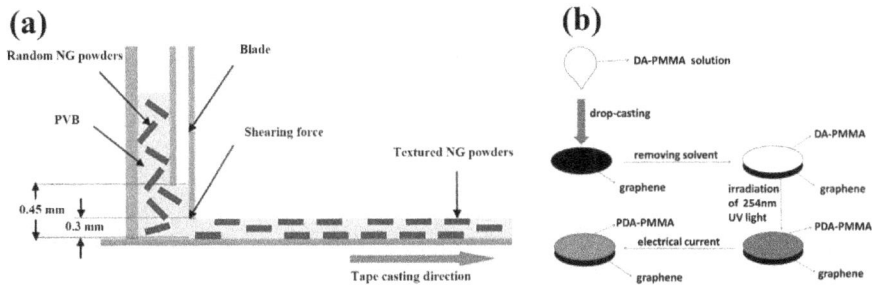

FIGURE 6.4 Schematic of (a) tape-casting process and (b) electrochromic process. (Zhou et al. (2012) and Zhang et al. 2013, with permission.)

In another preparation process, polydiacetylene (PDA)-PMMA/graphene composites were fabricated with the ability to respond by changing the color to an electrical current. In these composites, which are made by electrochromic process (Figure 6.4b), PDA acts as an electrochromic material and graphene provides conductivity. PMMA acts as an inert polymer matrix that improves mechanical properties and the colorimetric phenomenon. In comparison to PDA/graphene blue-red, the phase transition is clearly observed. By changing the graphene content, the critical color transition current can be changed (Zhang et al. 2013). Mixing functionalized GO with epoxy resin by sonication causes the modified GO particles to be transferred from water to epoxy. In another fabrication method, after decanting the water, the mixture is heated to form a dark violet epoxy-modified GO composite and the epoxy is cured by adding a hardener. The volume fraction of functionalized GO in the final composites was calculated to be 1.16 and 2.2 g cm^{-3} (Gudarzi and Sharif 2013).

6.10 PREPARATION OF GRAPHENE RUBBER COMPOSITES

Graphene nanocomposites have recently attracted the attention of researchers and industry experts. In 2020, Azizli et al. produced nanocomposites based on carboxylated acrylonitrile butadiene rubber (XNBR)/EPDM with different amounts of graphene (0, 0.1, 0.3, 0.5, 0.7, and 1 phr). They investigated the effect of graphene on the morphology and curing and mechanical properties of the nanocomposites and the adaptive role of EPDM-grafted maleic anhydride (EPDM-g-MA) compatibilizer. Also, they studied the nanocomposite microstructure with TEM and scanning electron microscopy (SEM). It is found that increasing the graphene content of the composite causes torque, curing time, and scorch time to increase.

The morphological studies of the fracture surface indicated that the presence of EPDM-g-MA causes an improvement of graphene dispersion in XNBR/EPDM matrix and a uniform dispersion of graphene with less aggregates was obtained. In other words, the fracture surface was rough due to the presence of graphene. However, it was observed that the addition of EPDM-g-MA compatibilizer and increase in the amount of graphene caused a uniform dispersion and reduced the dispersed phase size of EPDM in the XNBR matrix. Also, there was a significant enhancement in hardness, tensile strength, fatigue, modulus, and elongation at break of XNBR/EPDM nanocomposite (Table 6.3).

The researchers revealed that the reaction of polar carboxylic functional groups of XNBR with anhydride groups of EPDM-g-MA and the π–π interaction of nonpolar EPDM and C=C groups of graphene were probably the main causes of maximum torque increase with the addition of graphene content (Figure 6.5). This can change the crosslinking density and dispersion of graphene, both of which affect the nano-composite rigidity. Dynamic mechanical thermal analysis (DMTA) results indicate an improvement in loss factor and storage modulus by increasing the amount of gra-phene in XNBR/EPDM compounds (Figure 6.6 and Table 6.3).

In other work of Azizli and coworkers, elastomeric nanocomposites based on NR/EPDM blends are prepared using EPDM-g-MA as a compatibilizer and different amounts of GO (1, 3, 5, 7, and 10 phr) are prepared by melt mixing method (Azizli et al. 2020). The dispersion of GO nanoplatelets in the rubber matrix was verified by both TEM and SEM micrographs. Comparison of Figures 6.7a and 6.7b shows that with the same amount of graphene oxide, the sample containing compatibilizer has a better dispersion and distribution of GO (Figure 6.7).

It is found that in NR/EPDM-g-MA/EPDM blends, the presence of the compati-bilizer and of GO content increase accelerated the curing reaction and reduced t_5 and t_{90} of the nanocomposites. The results indicated that mechanical properties of the blends including tensile strength, modulus, elongation at break, hardness, and fatigue were significantly increased, while the resilience and abrasion resistance were decreased (Table 6.4). The researchers considered the good dispersion of GO in NR/EPDM-g-MA/EPDM blends and the penetration of EPDM and NR chains in the presence of EPDM-g-MA between graphene oxide nanoplatelets as the main rea-sons for the tangible increase in mechanical properties. After ring opening reaction of the anhydride groups of EPDM-g-MA, the hydrogen bond formation between the carboxyl functional groups of XNBR and EPDM-g-MA was facilitated (Figure 6.8).

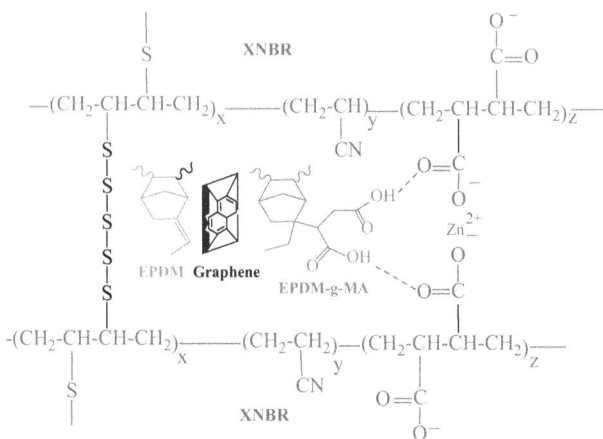

FIGURE 6.5 Most probable chemical interaction of EPDM: XNBR with EPDM-g-MA and graphene with EPDM-g-MA. (Azizli et al. 2020, with permission.)

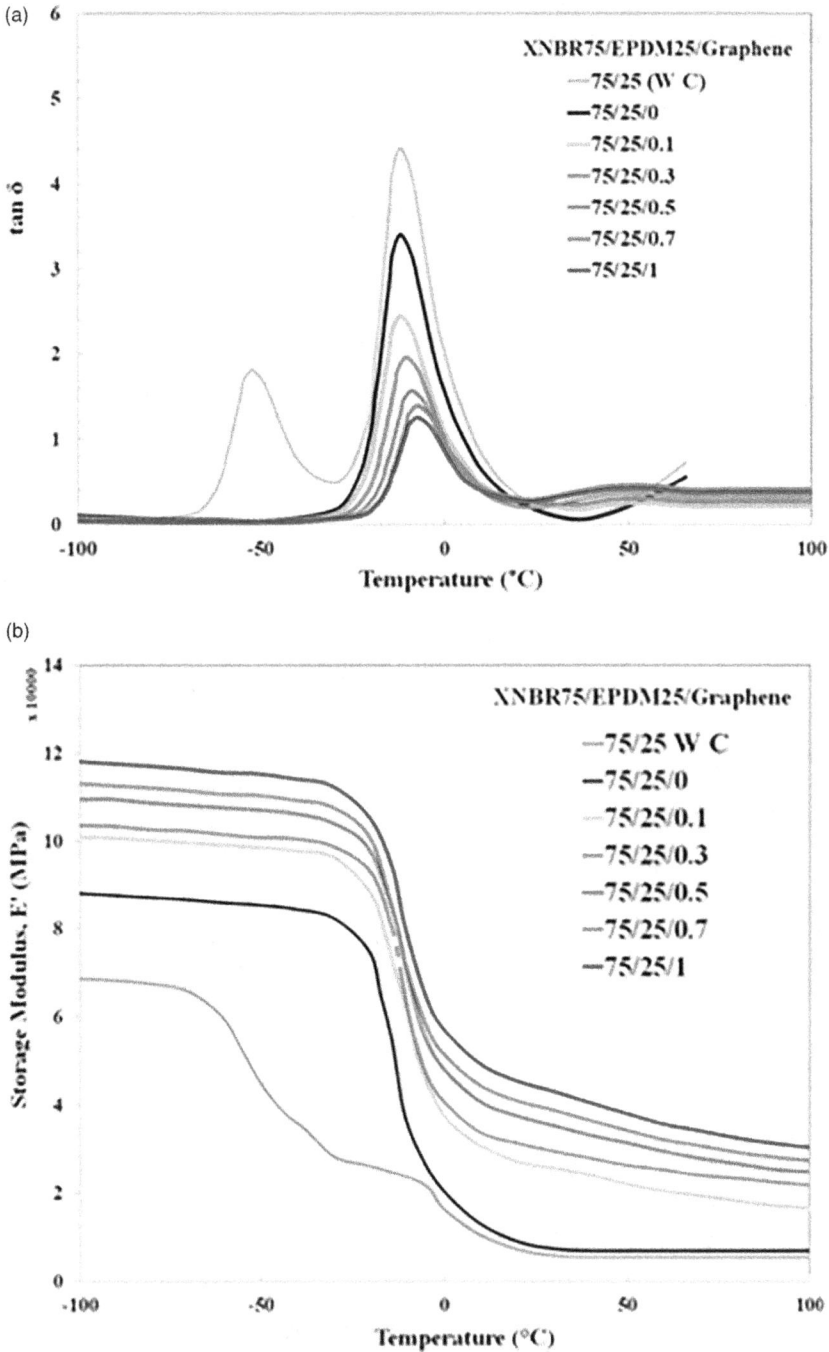

FIGURE 6.6 The result of (a) loss factor and (b) storage modulus of XNBR75/EPDM25 blends with increasing graphene and adding EPDM-*g*-MA. (Azizli et al. 2020, with permission.)

TABLE 6.3

The Result of (DMTA) Parameters of XNBR75/EPDM25 Blend with Increasing Graphene and Adding EPDM-g-MA

Samples	T_g from E' (°C)	T_g from tan δ (°C)	T_i from tan δ (°C)	Height of T_g Peak (-)	Decrease in Peak tan δ (-)	Height of T_i Peak (-)	E' at 25°C (MPa)
75/25	−13.4	−11.6	-	4.44	0	-	2.416
(WC)	−51.4	−49.9	-	2.322	0	-	6.951
75/25/0	−12.1	−10.4	-	3.193	1.255	-	9.121
75/25/0.1	−10.6	−8.9	50.2	2.638	1.811	0.271	10.089
75/25/0.3	−9.01	−6.8	49.1	1.925	2.523	0.313	10.548
75/25/0.5	−7.3	−5.3	48.1	1.704	2.744	0.359	11.411
75/25/0.7	−6.2	−4.1	46.2	1.289	3.159	0.424	11.742
75/25/1	−5.0	−3.9	45.0	1.211	3.237	0.428	11.948

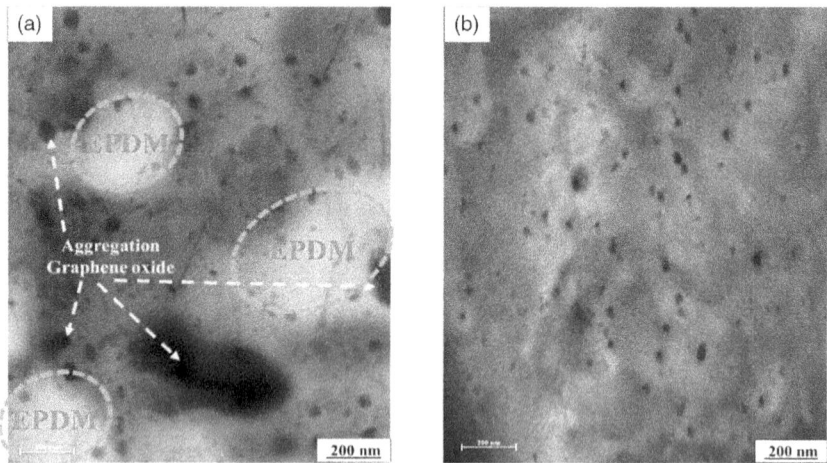

FIGURE 6.7 TEM micrographs of NR75/EPDM25 nanocomposites: (a) with 10 phr graphene oxide and (b) with 10 phr graphene oxide and compatibilizer. (Azizli et al. 2020, with permission.)

SEM and TEM micrographs verified these results. DMTA results and TEM micrographs also indicated that the compatibility of EPDM and NR was due to the addition of EPDM-g-MA which caused better GO dispersion and penetration between the nanoplatelets by EPDM and NR chains.

Theoretical study of the mechanical properties of the nanocomposites was performed using Mori–Tanaka model, which was in good agreement with the experimental data. Furthermore, rheological behavior of the nanocomposites was also assessed by Carreau–Yasuda model and different parameters of the model were calculated.

Figure 6.9 shows the stress-strain diagram of NR/EPDM nanocomposites. As is evident from this figure, as the graphene oxide content increased in the presence of

TABLE 6.4

The Result of Mechanical Properties of NR/EPDM Nanocomposites with the Increasing Graphene Oxide

Sample	Modulus 100% (MPa)	Tensile Strength (MPa)	Elongation (%)	Resilience (%)	Hardness Shore A	Abrasion loss (wt.%)	Fatigue to Failure (Cycles×100)
75/25 (W C)	1.42±0.02	13.47±0.26	640±13	65.32±1.30	66±1.32	41.55±0.83	403±8
75/25/0	1.48±0.02	14.42±0.28	712±14	64.33±1.28	66.5±1.33	39.33±0.78	528±10
75/25/1	1.61±0.03	16.96±0.33	762±15	63.19±1.26	68±1.36	36.21±0.72	731±14
75/25/3	1.82±0.03	21.08±0.42	859±17	62.14±1.24	70±1.40	33.25±0.66	942±18
75/25/5	1.98±0.03	24.43±0.48	931±19	61.28±1.22	72±1.44	30.81±0.61	1,049±21
75/25/7	2.12±0.04	28.19±0.56	977±20	60.12±1.20	74±1.48	28.44±0.56	1,096±22
75/25/10	2.25±0.04	30.02±0.60	1,045±21	59.81±1.19	75.5±1.51	26.12±0.52	1,173±23

EPDM-*g*-MA, tensile strength of the samples increased, which is due to the GO good dispersion in the NR/EPDM matrix.

Azizli et al. (2021) studied nanocomposites based on the styrene-acrylonitrile/carboxylated acrylonitrile butadiene rubber (SAN/XNBR) blend experimentally and theoretically and reported nonlinear elastoplastic behavior. Graphene, graphene

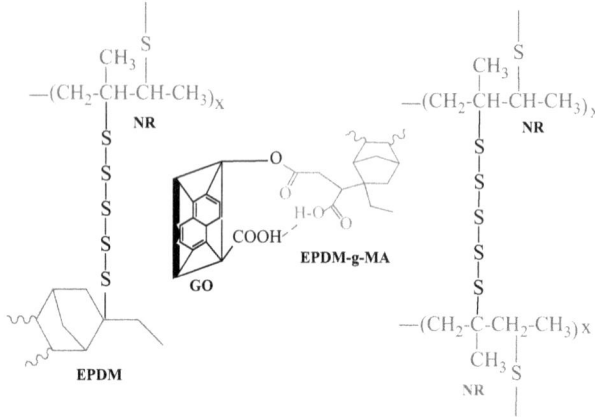

FIGURE 6.8 The possible interaction mechanism for NR/EPDM and EPDM-*g*-MA/graphene oxide nanocomposites. (Azizli et al. 2020, with permission.)

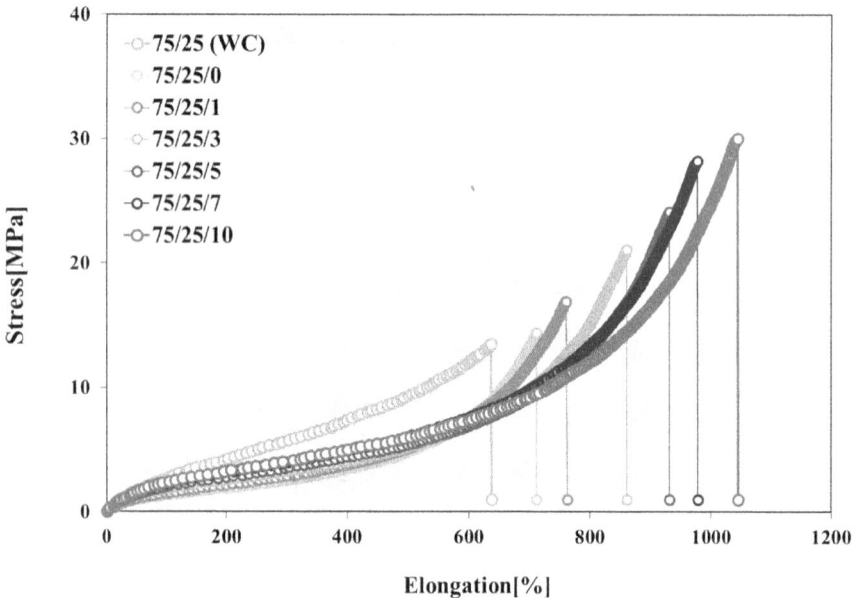

FIGURE 6.9 Stress-strain curves of NR/EPDM nanocomposites by adding EPDM-*g*-MA and increasing graphene oxide. (Azizli et al. 2020, with permission.)

oxide nanoparticles, and glycidyl methacrylate grafted carboxylated acrylonitrile butadiene rubber (XNBR-g-GMA) compatibilizer in the SAN/XNBR blends. This study was focused on the stress-strain behavior modeling of these nanocomposites with respect to the effect of the interfacial interactions created by the compatibilizer. The images of field emission scanning electron microscopy (FESEM) and TEM indicated that the dispersion and localization of the nanoparticles can affected by the presence of compatibilizer. The results of tensile test showed that the presence of the compatibilizer also enhanced mechanical properties of the nanocomposites, in particular elongation at break. The elongation at break of nanocomposites containing compatibilizer and graphene oxide increased about 570% compared to that of nanocomposites without compatibilizer (Figure 6.10). It is claimed that these improvements are due to better dispersion of GO and chemical interaction of the components in the presence of the XNBR-g-GMA (Figure 6.11) that is confirmed by TEM (Figure 6.12). It is found that the results obtained from analyzing the nonlinear elastoplastic behavior of the nanocomposites by Bergstrom–Boyce model showed a proper agreement with the experimental data in the elastic region. Nevertheless, the deviations in the viscoplastic region were specifically close to the breaking elongation region.

Moreover, Azizli et al. (2021) studied the effect of GO on the properties of rubber hybrid nanocomposites based on phenyl-vinyl-methyl-polysiloxane

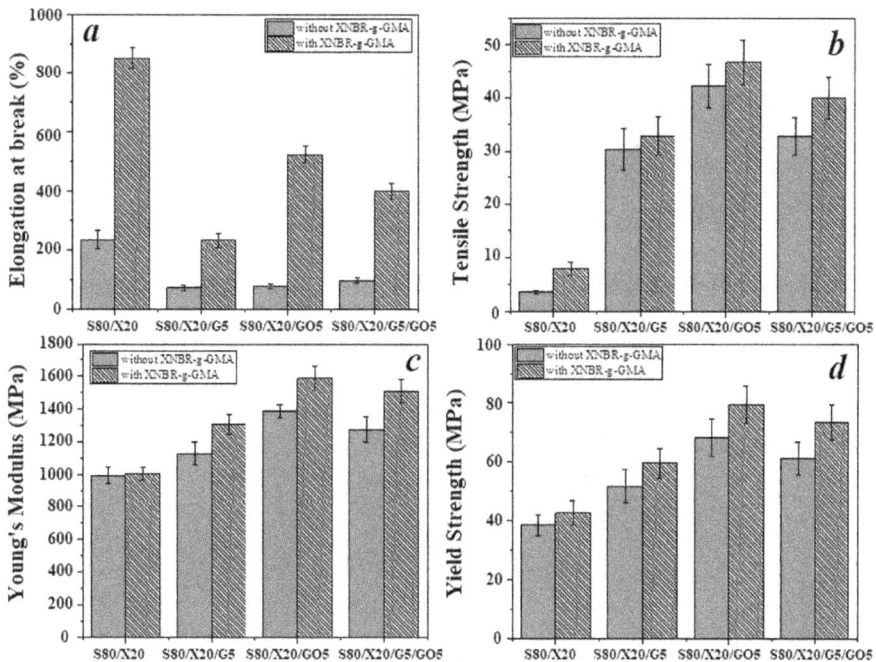

FIGURE 6.10 The result of mechanical properties of SAN/XNBR blends containing graphene oxide and graphene without and with XNBR-g-GMA: (a) elongation at break, (b) tensile strength, and (c) Young's modulus. (Azizli et al. 2021, with permission.)

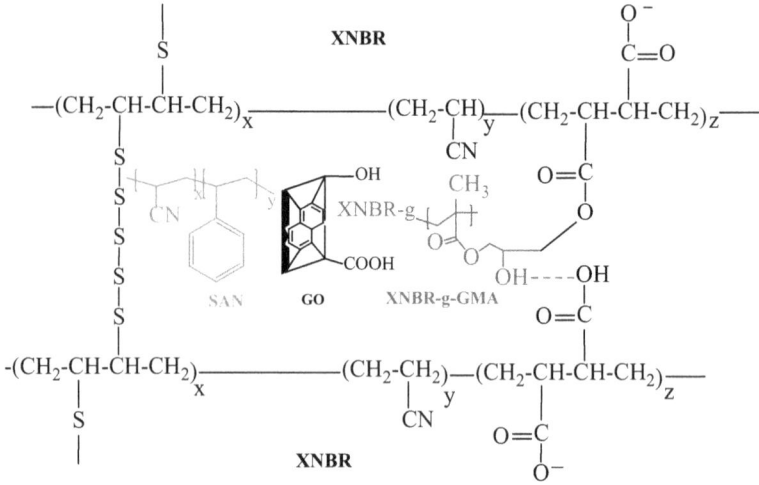

FIGURE 6.11 Most probable chemical interaction of XNBR/XNBR-g-GMA, SAN, and graphene oxide. (Azizli et al. 2021, with permission.)

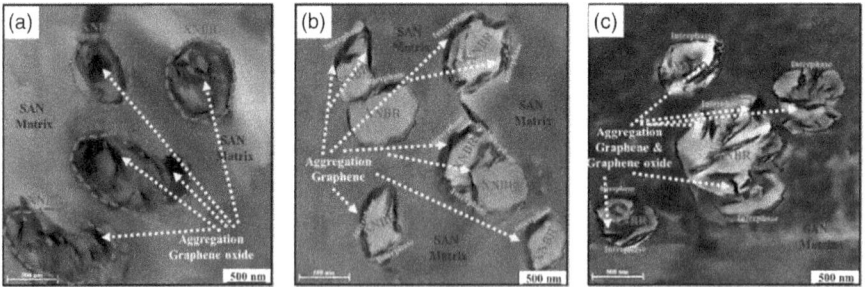

FIGURE 6.12 TEM micrographs of SAN/XNBR composites: (a) S80/X20/GO5, (b) S80/X20/G5, and (c) S80/X20/G2.5/GO2.5. (Azizli et al. 2021, with permission.)

(PVMQ)/XNBR-g-GMA/XNBR. They evaluated the adhesion between PVMQ and XNBR (blend phases) by microscopic observations and swelling test results at the same time. It is concluded that the incorporation of XNBR-g-GMA as a compatibilizer increased the adhesion of GO on PVMQ/XNBR rubber matrix. Also, the scorch time and the curing time decreased by 24% and 26%, respectively, with the increasing GO content to 10 phr; at the same time, the curing rate and the maximum curing torque increased by 27% and 15%, respectively. Tear strength, hardness, and compression set of the samples increased with increasing GO content. The SEM images showed that porosity of the prepared nanocomposites with 5 phr GO reduced from 17.34 to 4.84 μm (Figure 6.13). XNBR dispersed phase size is reduced with the presence of a compatibilizer. This means that the bonds formed between the blend phases are stronger. TEM images also illustrated excellent dispersion of GO nanoplatelets due to the addition of compatibilizer to PVMQ/XNBR rubber matrix

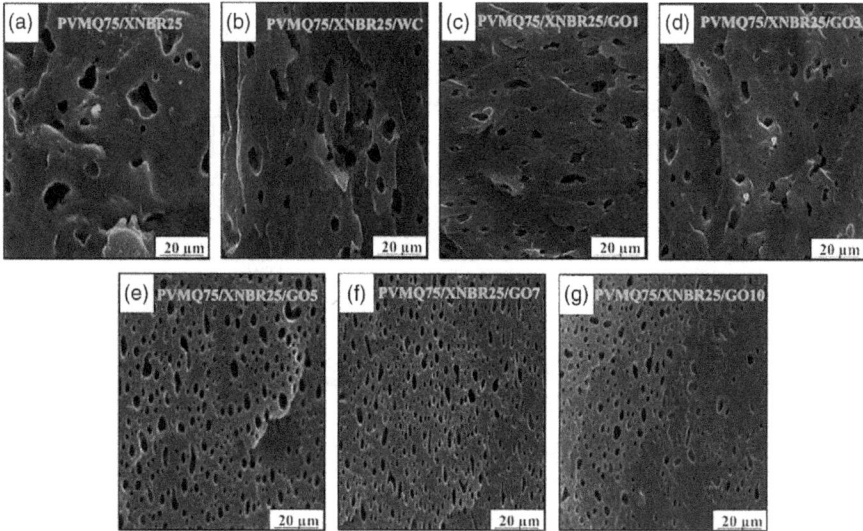

FIGURE 6.13 SEM micrographs of PVMQ75/XNBR25: (a) pure blend, (b) with compatibilizer, (c) 1, (d) 3, (e) 5, (f) 7, and (g) 10 phr of graphene oxide. (Azizli et al. 2021, with permission.)

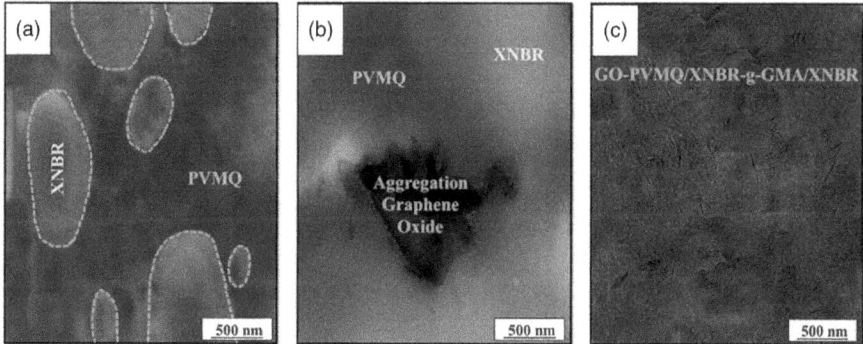

FIGURE 6.14 TEM micrographs of (a) PVMQ75/XNBR25 pure blend, (b) PVMQ75/XNBR25 graphene oxide, and (c) PVMQ75/XNBR25 with XNBR-*g*-GMA and graphene oxide. (Azizli et al. 2021, with permission.)

(Figure 6.14). Mooney–Rivlin and Carreau–Yasuda models were used, respectively, for the prediction of mechanical properties and rheological behavior of the samples. The parameters of the two models were determined theoretically and a good agreement was observed by comparing with the experimental data. It is believed that the presence of hydroxyl and carboxyl functional groups on the surface of GO acts as an accelerator to increase the ring opening rate of XNBR-*g*-GMA (Figure 6.15).

Figure 6.16 displays the stress/strain diagram of PVMQ/XNBR compounds. Due to the possible good dispersion of graphene oxide in the PVMQ/XNBR matrix, the

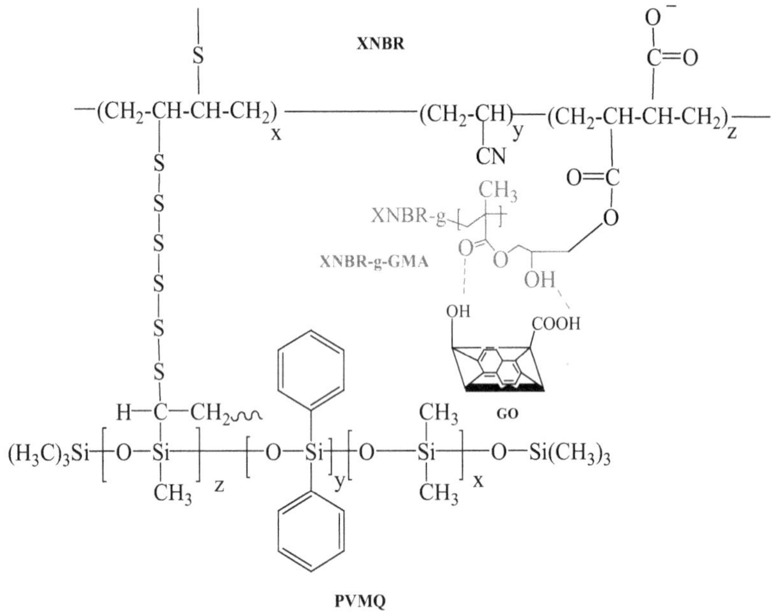

FIGURE 6.15 The possible interaction mechanism between GO and PVMQ/XNBR matrix and XNBR-*g*-GMA compatibilizer. (Azizli et al. 2021, with permission.)

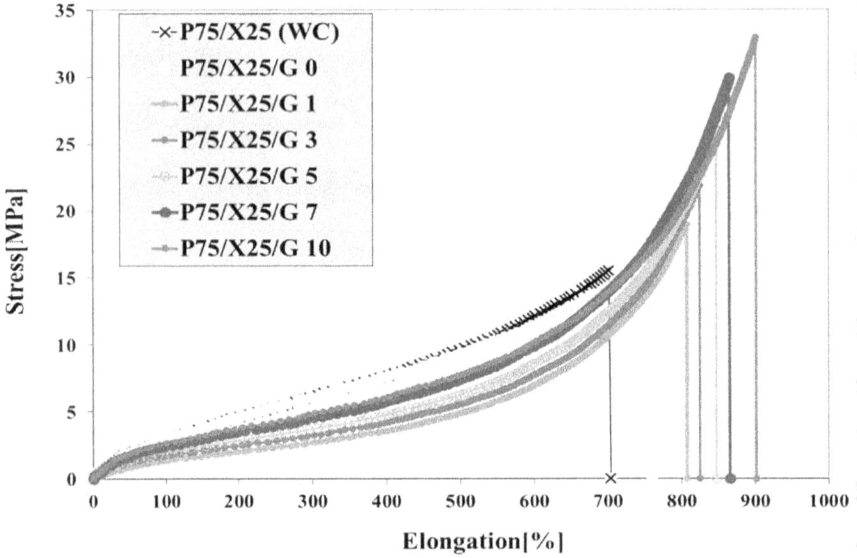

FIGURE 6.16 Stress versus elongation graphs of PVMQ/XNBR blends with incasing of graphene oxide. (Azizli et al. 2021, with permission.)

tensile strength of the samples was increased by adding XNBR-*g*-GMA and increasing the graphene oxide content.

Also using the melt mixing method, elastomeric nanocomposites were prepared on the basis of polyamide 6 (PA6)/CR blends using EPDM-*g*-MA compatibilizer with various amounts of GO (Azizli et al. 2021). The effects of compatibilizer and reinforcement concentration in the PA6/CR blend matrix were investigated using theoretical and experimental analysis. TEM (Figure 6.17) and FESEM (Figure 6.17) micrographs confirmed the dispersion of nanoplatelets in the rubber blend matrix. Modified microstructure of the samples showed considerable effect of EPDM-*g*-MA and GO on CR droplets size reduction in PA6 continuous phase that can be attributed to the good interactions, like hydrogen bonding, of PA6 and GO in the presence of EPDM-*g*-MA compatibilizer (Figure 6.18).

The effects of EPDM-*g*-MA and GO as an effective nucleating agent in PA6-enriched GO/CR (PA6EGO/CR) were approved by the results of differential scanning

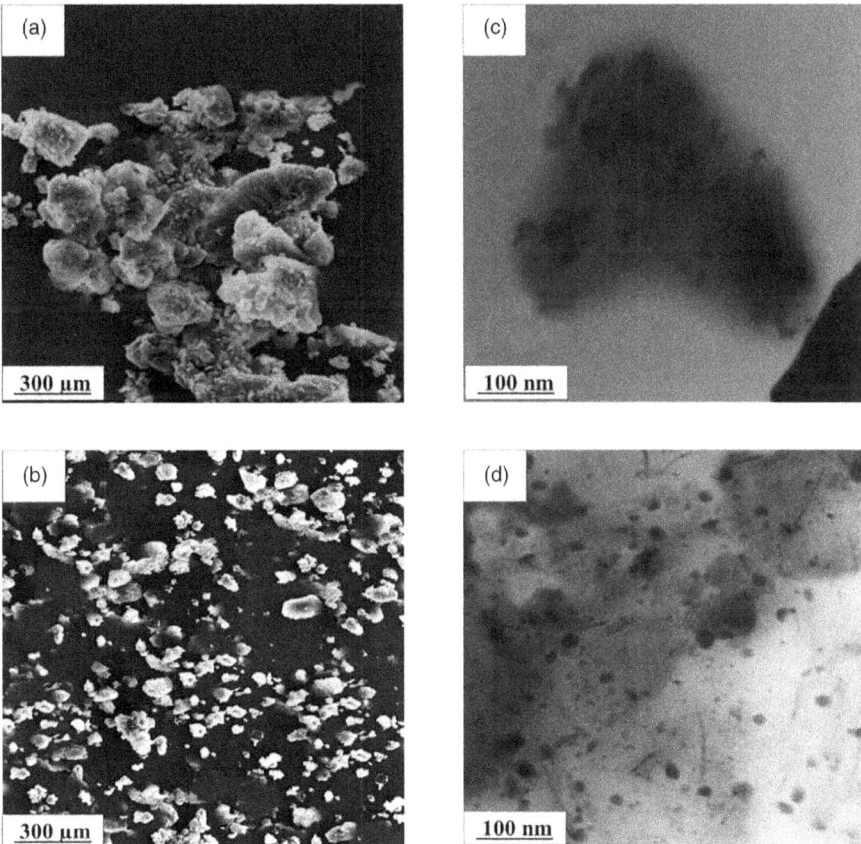

FIGURE 6.17 (a) and (b) FESEM micrographs and (c) and (d) TEM micrographs of PA6/CR nanocomposites with 5 phr GO: (a) and (c) without compatibilizer and (b) and (d) with EPDM-*g*-MA as a compatibilizer. (Azizli et al. 2021, with permission.)

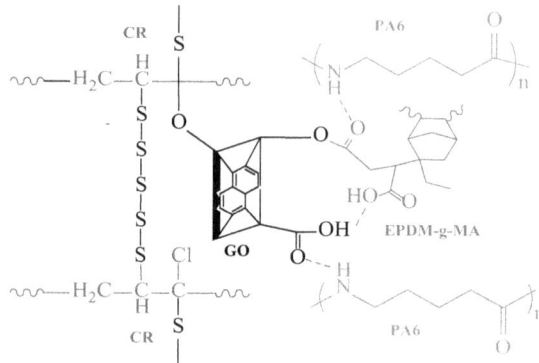

FIGURE 6.18 The possible chemical interactions of GO, CR, EPDM-*g*-MA, and PA6. (Azizli et al. 2021, with permission.)

calorimetry (DSC) and DMTA demonstrated. The results of thermogravimetric analysis (TGA) showed that graphene oxide can cause a higher thermal stability in PA6EGO/CR in the presence of EPDM-*g*-MA as a compatibilizer (Figure 6.19 and Table 6.5).

Investigation of mechanical properties showed that by adding a compatibilizer to the PA6/CR blend, the tensile strength changed from 39.0 MPa to 45.1 MPa, Young's modulus varied from 522.2 to 716.0 MPa, and the elongation at break changed from 246.8% to 222.2%. While the tensile strength increased by 25.2%, Young's modulus increased by 36.6%, and the elongation at break decreased by 20% due to the incorporation of 5 phr GO to the compatibilized blend (Figure 6.20). To predict the effect of different loadings of GO and EPDM-*g*-MA on Young's modulus, Christensen–Lo model was used to analyze the stiffness of PA6EGO/EPDM-*g*-MA/CR with emphasis on the effect of the interphase region. The rheology and creep tests showed a significant effect of EPDM-*g*-MA and GO content on rheological behavior of the nanocomposites.

6.11 CHARACTERIZATION OF POLYMER NANOCOMPOSITES

Morphology, structure, and great properties of graphite fillers and their polymer composites have been confirmed by many characterization techniques. In this section, three more important analysis techniques are explained: X-ray diffraction (XRD) spectroscopy, Fourier transformation infrared (FTIR) spectroscopy, and Raman spectroscopy. XRD pattern of graphite, GO, and their dispersion mode in polyaniline (PANI) is shown in Figure 6.24. The size of GO grain was estimated as 5.1 nm according to the observed peak at $2\theta = 10.04°$. A broad band extending from 15° to 34° was observed for the crystallized PANI (Wei et al. 2013). For PANI composite, the intensity of the peak at 10.04° is affected (Figure 6.21A). Due to the effect of experimental conditions on the exfoliation rate of RGO fillers, XRD spectrum undergoes some variation for PDA composites prepared under different temperature conditions. Figure 6.21B shows the peak centered at 26° that corresponds to the (002) plane of graphite as observed in Ansari and Giannelis (2009). EG illustrates the

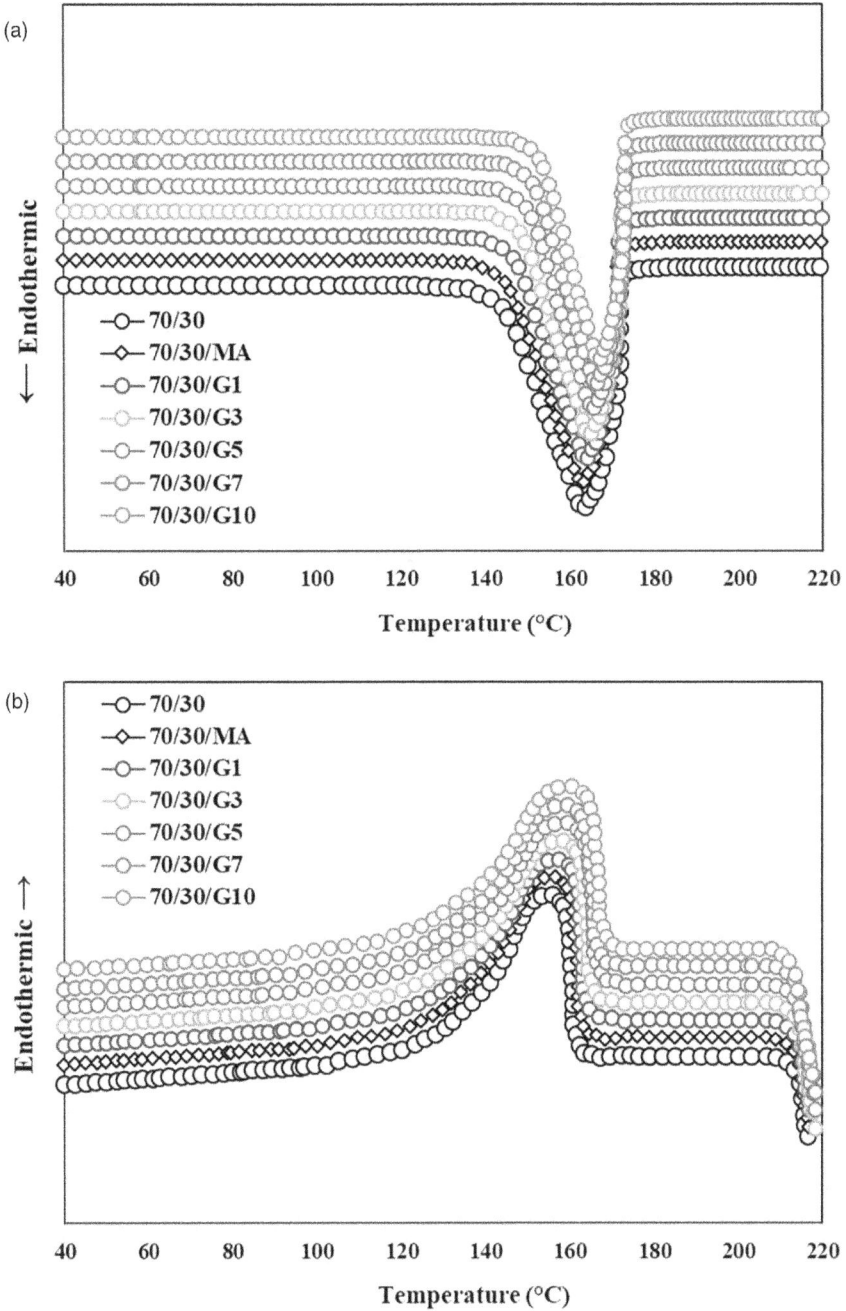

FIGURE 6.19 Non-isothermal DSC thermogram of (a) cooling curves and (b) heating curves of the PA6/CR nanocomposites with the increasing graphene oxide. (Azizli et al. 2021, with permission.)

TABLE 6.5

The Result of Non-isothermal Crystallization and Melting Parameters of the PA6/CR Nanocomposites with the Increasing Graphene Oxide

Sample Code	Initial Crystallization Temp., T_c (°C)	Peak Temp. T_{cp} (°C)	Crystallization Enthalpy ΔH_c (J g⁻¹)	Peak Melting Temp. T_{mp} (°C)	Melting Enthalpy ΔH_m (J g⁻¹)	X_c (%)
70/30	141.8	137.6	51.68	153.4	57.89	20.22
70/30/MA	145.1	141.5	52.22	153.9	56.58	20.01
70/30/G1	146.2	143.4	53.14	154.2	56.46	19.28
70/30/G3	147.1	144.5	54.19	155.3	56.22	18.79
70/30/G5	147.9	144.8	55.21	156.5	56.11	17.74
70/30/G7	148.5	145.3	54.89	156.9	57.02	16.88
70/30/G10	149.2	146.4	55.44	157.2	56.48	16.02

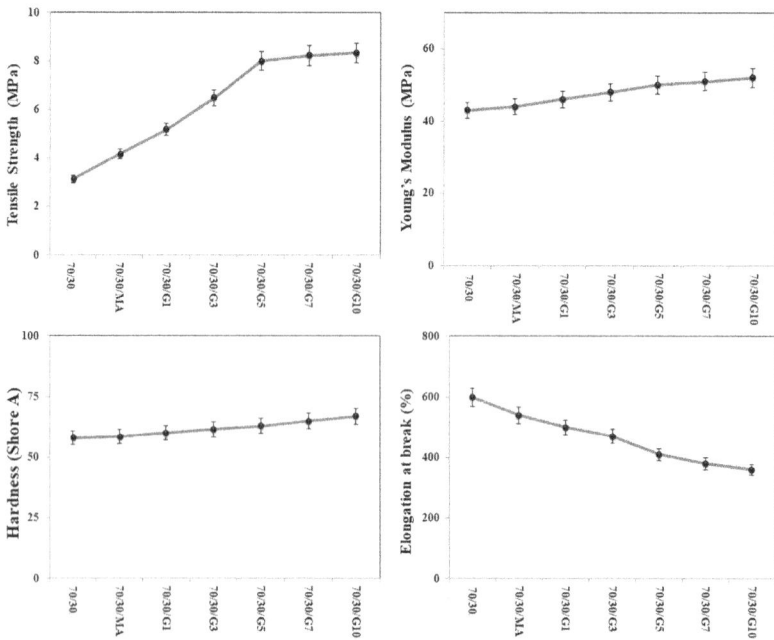

FIGURE 6.20   The results of mechanical properties of PA6/CR nanocomposites by adding EPDM-g-MA and increasing graphene oxide. (Azizli et al. 2021, with permission.)

same peak, that for GO, the diffraction peak is shifted to 9.8°, which corresponds to a d-spacing of 0.9 nm. The weak diffraction maximum of EG compared to graphite is attributed to its exfoliated nature. The d-spacing in GO depends on the preparation

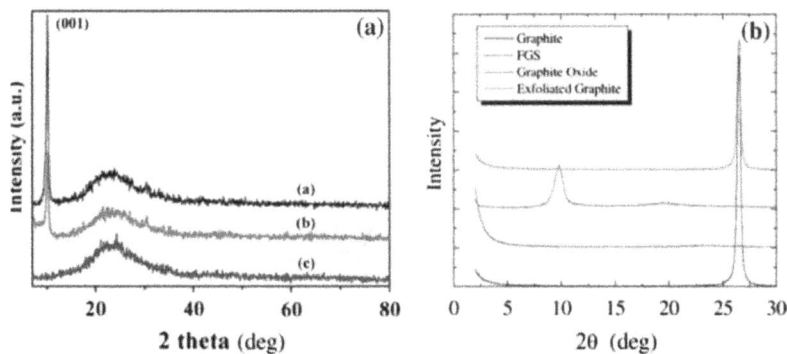

FIGURE 6.21 (A) XRD patterns of (a) GO film, (b) PANI/GO nanocomposite film, and (c) PANI film, Wei et al. (2013). (B) XRD patterns of FGS, GO, and EG. (Ansari et al. 2009, with permission.)

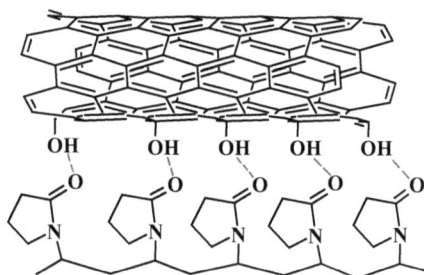

FIGURE 6.22 Structure of interpolymer adduct of PVP with MWNT-OH. (Haghighat and Mokhtary 2017, with permission.)

method and the number of trapped water layers in the structure of the material. No characteristic peak was obtained for functionalized graphene sheets (FGS), which indicates the loss of long-range stacking order in the material. Because pure graphene has no stack, Bragg peaks do not exist at all for graphene.

FTIR is another important characterization technique, which confirms the chemical structure and interactions of the nanocomposites. Haghighat and Mokhtary (2017b) prepared polyvinylpyrrolidone (PVP)/hydroxyl-functionalized multi-walled carbon nanotube and sulfonyl-functionalized multi-walled carbon nanotube nanocomposites in aqueous media (Figure 6.22). The FTIR spectra clearly showed that the hydrogen bonding and ionic interaction take place between MWNT-OH and MWNT-SO$_3$H with PVP, respectively.

Also, Haghighat and Mokhtary (2017a) inserted magnetic decorated carboxylic acid functionalized multi-walled carbon nanotube (MWCNT-Fe$_3$O$_4$) successfully in PVP matrix in the aqueous media. The interaction of MWCNT-Fe$_3$O$_4$ and PVP was confirmed by FTIR spectra. Due to the better distribution of CNTs in the PVP matrix, strong interaction with ~31 cm^{-1} red shift along with good complexation of carbonyl functional group of PVP with MWCNT-Fe$_3$O$_4$ was observed for PVP/MWCNT-Fe$_3$O$_4$ (5% w/w) nanocomposite (Figure 6.23).

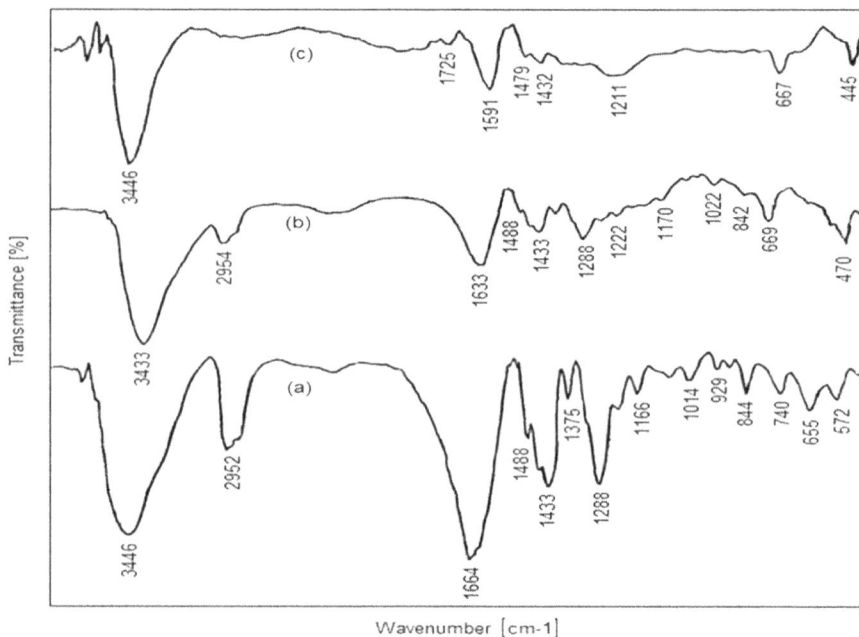

FIGURE 6.23 FTIR spectra of (a) PVP, (b) PVP/MWCNT-Fe$_3$O$_4$ (5% w/w), and (c) MWCNT-Fe$_3$O$_4$. (Haghighat and Mokhtary 2017, with permission.)

The uniform dispersion of MWNT-Fe$_3$O$_4$ in the PVP matrix was shown by SEM image of PVP/MWCNT-Fe$_3$O$_4$ (Figure 6.24) that is directly related to the good interactions between MWNT-Fe$_3$O$_4$ and PVP functional groups. In addition, the dispersion of MWCNT-Fe$_3$O$_4$ in the cross-section of PVP was examined through EDX mapping image, which indicated the homogeneous distribution of MWCNT-Fe$_3$O$_4$ in the PVP matrix of PVP/MWCNT-Fe$_3$O$_4$ (5% w/w) nanocomposite (Figure 6.25).

6.12 DISPERSION OF GRAPHENE

Graphene has a poor dispersion in solvents with low Hildebrand solubility parameters and low boiling points. For graphene dispersion, a solvent with high boiling point and Hildebrand solubility parameter of approximately 23 MPa$^{1/2}$ is the best. N-methylpyrrolidone (NMP) with a solubility parameter of about 23 MPa$^{1/2}$ and a boiling point close to 200°C can form a good and stable dispersion of graphene (O'Neill et al. 2011). Probably, it is the main reason for the difficulty of graphene dispersion in a majority of solvents that are usually used to make polymer composites. Graphite oxide can be dispersed in almost all solvents, except dichloromethane and n-hexane, and to a lesser extent, methanol and oxylene, in just sonicated samples. Nevertheless, many of these dispersions in polar solvents like acetone, ethanol, 1-propanol, dimethyl sulfoxide (DMSO), and pyridine are only stable for a short time and precipitate completely within a few hours or days (Paredes et al. 2008). But the graphite oxide dispersions in four organic solvents, ethylene glycol, NMP, N,N'-dimethylformamide (DMF), and

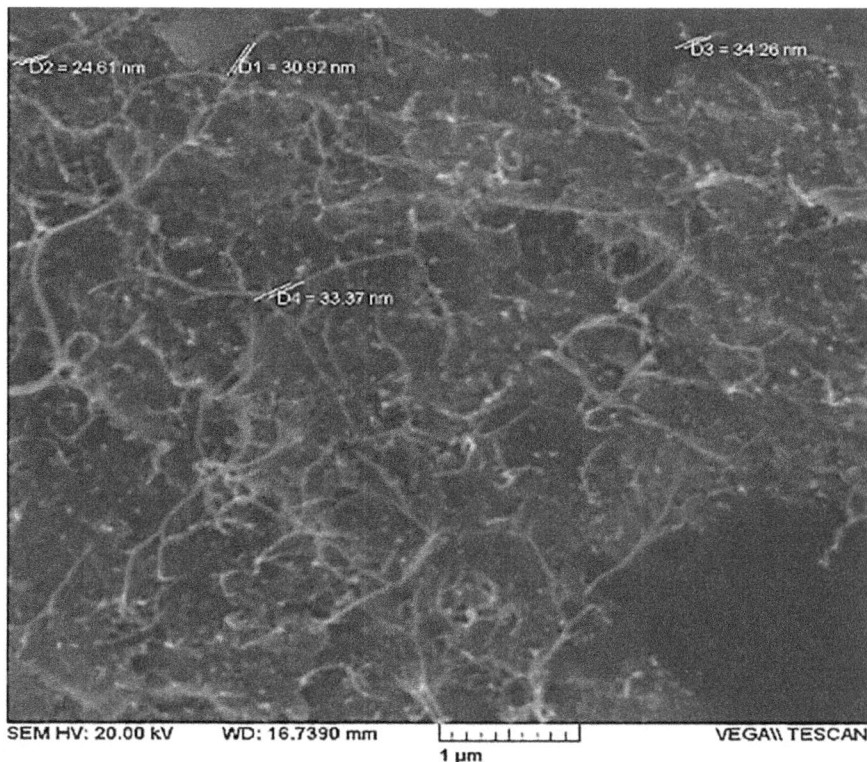

SEM HV: 20.00 kV   WD: 16.7390 mm   [scale bar]   VEGA\\ TESCAN
1 µm

FIGURE 6.24 SEM of PVP/MWNT-Fe$_3$O$_4$ (5% w/w). (Haghighat and Mokhtary 2017, with permission.)

tetrahydrofuran (THF), have long-term stability. Sensitive morphological ordering, fine interface control, and processing facilitate are very basic for producing high-performance polymer nanocomposites that have uniform dispersion of graphene (Putz et al. 2010). To enhance exfoliation of graphene through electrostatic and physical interactions, surfactants and polyelectrolytes are often used (Stankovich et al. 2006b). These methods are based on the creation of noncovalent intermolecular interactions, such as hydrogen bonding or electrostatic interactions, to adhere to molecular systems with 2D graphene or a graphene oxide (GO) monolayers. The solubility of a noncovalently attached conjugated triblock copolymer (PEG-OPE-PEG)/reduced GO (RGO; by π–π interaction) is tested in a series of solvents, such as toluene, THF, chloroform, acetone, DMF, ethanol, methanol, DMSO, and water. It is found that the material is dispersed well in these solvents, resulting in clear and homogeneous solutions (PEG-OPE-rGO, 1.22 mg mL^{-1}) that can exhibit excellent stability of dispersion for months (Qi et al. 2010). Although this graphene sheet exfoliation strategy is relatively easy, it is difficult to control and quantify the nanomaterials produced due to the inherent instability of these supramolecular systems. On the other hand, covalent attachment of small molecules (Kamaras et al. 2003) or polymers to the extended aromatic surfaces of graphene and formation of the resulted composites produced materials that

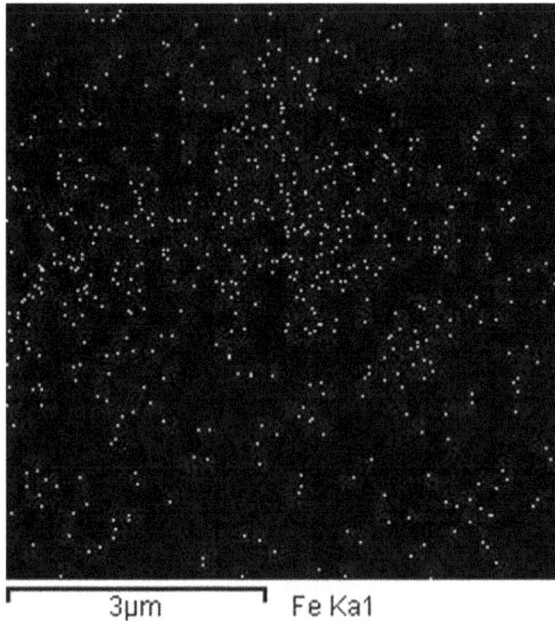

FIGURE 6.25   EDX distributions of MWCNT-Fe_3O_4 in the PVP/MWCNT-Fe_3O_4 (5% w/w) nanocomposite. (Mokhtary et al. 2017, with permission.)

are much more stable under variable external conditions. Also, the ability of creating a band gap in graphene with a simple method of covalent functionalization facilitates the fabrication of graphene-based devices. The sp^2 to sp^3 hybridized carbon transformation introduces a barrier to electron flow by opening a band gap, which allows the generation of insulating and semiconducting regions in graphene. To improve graphene dispersion in nonpolar polymers, organic molecules such as phenyl isocyanate and porphyrin are also attached to graphene surfaces (Fang et al. 2009). An important route to produce water-soluble graphene is sulfonation of GO, so that it can interact supramolecularly with many polymers to form a good dispersion in the composite. But polymer functionalization is necessary to achieve a uniform stable dispersion of graphene and to induce different nanostructures of graphene (Kim et al. 2010, Vickery et al. 2009). Polymer Functionalization is done in threefold manner: (i) the exfoliation can be facilitated by increasing the spacing between individual graphene sheets, (ii) the functional groups of the grafted polymers may assist with the miscibility of modified graphene in the matrix of other polymers through specific interactions yielding a high-performance composite, and (iii) the formation of a band gap due to covalent functionalization of the polymer produces new materials suitable for electronic/optoelectronic devices. There are a few reports in the literature about the surface modification of graphene sheets through polymer grafting. Although both "grafting-to" (Salavagione et al. 2009) and "grafting-from" (Gonçalves et al. 2010) approaches have been successfully applied, for proper modification, the latter is somewhat more advantageous than the former due to lesser steric hindrance.

Monomers enter the gallery of graphene sheets during chain propagation in the grafting-from technique; the interlayer spacing increases by the growing chain and helps to exfoliate the graphene sheets in a thin layer, which is rather difficult using the grafting-to approach. The physicochemical and optoelectronic properties of graphene sheets depend on the nature of the surface modification of graphene; thus, in order to achieve the optimum properties for proper applications, the adjustment of the modification is essential. An ideal and advanced technique for the modification of graphene sheets is surface-initiated (SI) controlled polymerization. Covalently FGS with poly-L-lysine (PLL) are soluble in water and biocompatible, which makes it a novel material showing much promise in biological applications (Shan et al. 2009). Common organic solvents readily dissolve polyacetylene-functionalized graphene, which is useful for the electronics industry. The covalent functionalization of graphene by PMMA, poly(butyl acrylate) (PBA), and PS are very important to produce polymer composites with highly improved mechanical and conducting properties (Fang et al. 2009). Also, the solubility of functionalized graphene is greatly affected by the polymer chain length of the bonded polymer. For example, the shortest polymer chains at the same concentration of PS-grafted graphene in the sample contained the highest amount of GO, producing the darkest solution.

ACKNOWLEDGMENTS

Financial support by Rasht Branch, Islamic Azad University is gratefully acknowledged. Also, the authors thank Wiley publications for allowing the use of some of the figures, diagrams, and data in the following articles: Functionalized graphene sheet—poly(vinylidene fluoride) conductive nanocomposites. Ansari, S., and Giannelis, *J. Polym. Sci. B Polym. Phys.* 2009,47:888. Mechanochemical route to functionalized graphene and carbon nanofillers for graphene/SBR nanocomposites. Beckert, et al., *Macromol. Mater. Eng.* 2014,299:151. Compatibilizer/graphene/carboxylated acrylonitrile butadiene rubber (XNBR)/ ethylene-propylene diene monomer (EPDM) nanocomposites: morphology, compatibility, rheology and mechanical properties. Azizli, et al., *J. App. Polym. Sci.*2020,137:app49331. Compatibility, mechanical and rheological properties of hybrid rubber NR/EPDM-g-MA/EPDM/graphene oxide nanocomposites: Theoretical and experimental analyses. Azizli, et al., *Compos. Commun.* 2020, 22:10044. Analysis and modeling of modified styrene-acrylonitrile/ carboxylated acrylonitrile butadiene rubber nanocomposites filled with graphene and graphene oxide: Interfacial interaction and nonlinear elastoplastic behavior. Azizli, et al., *Polym. Eng. Sci.* 2021,61:2894–2909, in Chapter 6 of this book.

REFERENCES

Alexandre, M., and Dubois, P. 2000. Polymer-layered silicate nanocomposites: Preparation, properties and uses of a new class of materials. *Mater Sci Eng R Rep* 28:1–63.
Aliofkhazraei, M. 2013. *Advances in Graphene Science*. Rijeka: Intech.
Allahbakhsh, A., Mazinani, S., Kalaee, M. R., and Sharif, F. 2013. Cure kinetics and chemorheology of EPDM/graphene oxide nanocomposites. *Thermochim Acta* 563:22–32.
Ansari, S., and Giannelis, E. P. 2009. Functionalized graphene sheet—Poly (vinylidene fluoride) conductive nanocomposites. *J Polym Sci B Polym Phys* 47:888–97.

Araby, S., Meng, Q., Zhang, L., Kang, H., et al. 2014. Electrically and thermally conductive elastomer/graphene nanocomposites by solution mixing. *Polymer* 55:201–10.

Azizli, M. J., Abbasizadeh, S., Hoseini, M., Rezaeinia, S., and Azizli, E. 2017. Influence of blend composition and organic cloisite 15A content in the structure of isobutylene–isoprene rubber/ethylene propylene diene monomer composites for investigation of morphology and mechanical properties. *J Compos Mater* 51:1861–73.

Azizli, M. J., Barghamadi, M., Rezaeeparto, K., et al. 2021. Enhancement of thermal, morphological, and mechanical properties of compatibilized based on PA6-enriched graphene oxide/EPDM-*g*-MA/CR: Graphene oxide and EPDM-*g*-MA compatibilizer role. *J Appl Polym Sci* 138:49901.

Azizli, M. J., Barghamadi, M., Rezaeeparto, K., Mokhtary, M., and Parham, S. 2020. Compatibility, mechanical and rheological properties of hybrid rubber NR/EPDM-*g*-MA/EPDM/graphene oxide nanocomposites: Theoretical and experimental analyses. *Compos Commun* 22:100442.

Azizli, M. J., Barghamadi, M., Rezaeeparto, K., Mokhtary, M., Parham, S., and Darabi, M. J. 2021. Theoretical and experimental analyses of rheological, compatibility and mechanical properties of PVMQ/XNBR-g GMA/XNBR/GO ternary hybrid nanocomposites. *Iran Polym J* 30:1001–18.

Azizli, M. J., Dehaghi, F. M., Nasrollahi, B., et al. 2021. Analysis and modeling of modified styrene–acrylonitrile/carboxylated acrylonitrile butadiene rubber nanocomposites filled with graphene and graphene oxide: Interfacial interaction and nonlinear elastoplastic behavior. *Polym Eng Sci* 61:2894–909.

Azizli, M. J., Khonakdar, H. A., Mokhtary, M., and Goodarzi, V. 2019. Investigating the effect of organoclay montmorillonite and rubber ratio composition on the enhancement compatibility and properties of carboxylated acrylonitrile-butadiene rubber/ethylene-propylene-diene monomer hybrid elastomer nanocomposites. *J. Polym Res* 26:221.

Azizli, M. J., Mokhtary, M., Khonakdar, H. A., and Goodarzi, V. 2020a. Hybrid rubber nanocomposites based on XNBR/EPDM: Select the best dispersion type from different nanofillers in the presence of a compatibilizer. *J Inorg Organomet Polym Mater* 30:2533–50.

Azizli, M. J., Mokhtary, M., Khonakdar, H. A., and Goodarzi, V. 2020b. Compatibilizer/ graphene/carboxylated acrylonitrile butadiene rubber (XNBR)/ethylene propylene diene monomer (EPDM) nanocomposites: Morphology, compatibility, rheology and mechanical properties. *J Appl Polym Sci* 137:app49331.

Azizli, M. J., Naderi, G., Bakhshandeh, G. R., Soltani, S., Askari, F., and Esmizadeh, E. 2014. Improvement in physical and mechanical properties of IIR/CR rubber blend organoclay nanocomposites. *Rubber Chem. Technol* 87:10–20.

Azizli, M. J., Rezaeinia, S., Rezaeeparto, K., Mokhtary, M., and Askari, F. 2020. Enhanced compatibility, morphology, rheological and mechanical properties of carboxylated acrylonitrile butadiene rubber/chloroprene rubber/graphene nanocomposites: Effect of compatibilizer and graphene content. *RSC Adv* 10:11777–90.

Azizli, M. J., Ziaee, M., Rezaeinia, S., et al. 2018. Studying the roles of nanoclay and blend composition on the improved properties of natural rubber/chloroprene composites. *Polym Compos* 39:1562–74.

Basu, D., Das, A., Stöckelhuber, K. W., Wagenknecht, U., and, Heinrich, G. 2014. Advances in layered double hydroxide (LDH)-based elastomer composites. *Prog Polym Sci* 39:594–626.

Beckert, F., Trenkle, S., Thomann, R., and Mülhaupt, R. 2014. Mechanochemical route to functionalized graphene and carbon nanofillers for graphene/SBR nanocomposites. *Macromol Mater Eng* 299:1513–20.

Bharadiya, P. S., Singh, M. K., and Mishra, S. 2019. Influence of graphene oxide on mechanical and hydrophilic properties of epoxy/banana fiber composites. *JOM* 71:838–43.

Bhavitha, K. B., Yaragalla, S., Satyanarayana, C. H. C., Kalarikkal, N., and Thomas, S. 2021. *Natural Rubber/Graphene Nanocomposites and Their Applications, in: Graphene Based Biopolymer Nanocomposites.* Singapore: Springer,.

Bokobza, L. 2007. Multiwall carbon nanotube elastomeric composites: A review. *Polymer* 48:4907–20.

Chandrasekhar, P. 2018. *Conducting Polymers, Fundamentals and Applications.* Switzerland: Springer.

Chen, B., Ma, N., Bai, X., Zhang, H., and Zhang, Y. 2012. Effects of graphene oxide on surface energy, mechanical, damping and thermal properties of ethylene-propylene-diene rubber/petroleum resin blends. *RSC Adv* 2:4683–9.

Cho, D., Lee, S., Yang, G., Fukushima, H., and Drzal, L. T. 2005. Dynamic mechanical and thermal properties of phenylethynyl-terminated polyimide composites reinforced with expanded graphite nanoplatelets. *Macromol Mater Eng* 290:179–87.

Choi, W., Lahiri, I., Seelaboyina, R., and Kang, Y. S. 2010. Synthesis of graphene and its applications: A review. *Criti Rev Solid State Mater Sci* 35:52–71.

Cote, L. J., Cruz-Silva, R., and Huang, J. 2009. Flash reduction and patterning of graphite oxide and its polymer composite. *J Ame Chem Soc* 131:11027–32.

Dao, T. D., Lee, H., and Jeong, H. M. 2014. Alumina-coated graphene nanosheet and its composite of acrylic rubber. *J Colloid Interface Sci* 416:38–43.

Das, P. P., Chaudhary, V., and Mishra, S. 2021. *Emerging Trends in Green Polymer Based Composite Materials: Properties, Fabrication and Applications, in: Graphene Based Biopolymer Nanocomposites.* Singapore: Springer.

Dong, B., Liu, C., Zhang, L., and Wu, Y. 2015. Preparation, fracture, and fatigue of exfoliated graphene oxide/natural rubber composites. *RSC Adv* 5:17140–8.

Donnet, J. B., and Voet, A. 1976. *Carbon Black: Physics, Chemistry, and Elastomer Reinforcement.* New York: M. Dekker.

Eisenhaure, J., and Kim, S. 2017. A review of the state of dry adhesives: Biomimetic structures and the alternative designs they inspire. *Micromachines* 8:125.

Fang, M., Wang, K., Lu, H., Yang, Y., and Nutt, S. 2009. Covalent polymer functionalization of graphene nanosheets and mechanical properties of composites. *J Mater Chem* 19:7098–105.

Favier, V., Chanzy, H., and Cavaillé, J. 1995. Polymer nanocomposites reinforced by cellulose whiskers. *Macromolecules* 28:6365–7.

Gonçalves, G., Marques, P. A. A. P., Barros-Timmons, A., et al. 2010. Graphene oxide modified with PMMA via ATRP as a reinforcement filler. *J Mater Chem* 20:9927–34.

Gudarzi, M. M., and Sharif, F. 2012. Enhancement of dispersion and bonding of graphene-polymer through wet transfer of functionalized graphene oxide. *Express Polym Lett* 6:1017–31.

Haghighat, F., and Mokhtary, M. 2017a. Preparation and characterization of polyvinylpyrrolidone/magnetite decorated carboxylic acid functionalized multiwalled carbon nanotube (PVP/MWCNT-Fe$_3$O$_4$) Nanocomposite. *J Inorg Organomet Polym Mater* 27:779–87.

Haghighat, F., and Mokhtary, M. 2017b. Preparation and characterization of polyvinylpyrrolidone-functionalized multiwalled carbon nanotube (PVP/f-MWNT) nanocomposites. *Polym Plast Technol Eng* 56:794–803.

Hansma, P. K., Turner, P. J., and Ruoff, R. S. 2006. Optimized adhesives for strong, lightweight, damage-resistant, nanocomposite materials: New insights from natural materials. *Nanotechnology* 18:44026.

Hasanzadeh Kermani, H., Mottaghitalab, V., Mokhtary, M., and AlizadehDakhel, A. 2020. Morphological, rheological, and mechanical properties of ethylene propylene diene monomer/carboxylated styrene butadiene rubber/multiwall carbon nanotube nanocomposites. *Int J Polym Anal Char* 25:479–98.

Heinrich, G., 2011. *Advanced Rubber Composites.* Heidelberg: Springer.

Hosler, D., Burkett, S. L., and Tarkanian, M. J. 1999. Prehistoric polymers: Rubber processing in ancient Mesoamerica. *Science* 284:1988–91.

Jang, B. Z., and Zhamu, A. 2008. Processing of nanographene platelets (NGPs) and NGP nanocomposites: A review. *J Mater Sci* 43:5092–101.

Kamaraj, M., Dodson, E. A., and Datta, S. 2020. Effect of graphene on the properties of flax fabric reinforced epoxy composites. *Adv Compos Mater* 29:443–58.

Kamaras, K., Itkis, M. E., Hu, H., Zhao, B., and Haddon, R. C. 2003. Covalent bond formation to a carbon nanotube metal. *Science* 301:1501.

Kim, H., Macosko, C. W. 2008. Morphology and properties of polyester/exfoliated graphite nanocomposites. *Macromolecules* 41:3317–27.

Kim, H., Miura, Y., and Macosko, C. W. 2010. Graphene/polyurethane nanocomposites for improved gas barrier and electrical conductivity. *Chem Mater* 22:3441–50.

Kuilla, T., Bhadra, S., Yao, D., Kim, N. H., Bose, S., and Lee, J. H. 2010. Recent advances in graphene based polymer composites. *Prog Polym Sci* 35:1350–75.

Layek, R. K., Samanta, S., Chatterjee, D. P., and Nandi, A. K. 2010. Physical and mechanical properties of poly (methyl methacrylate)-functionalized graphene/poly (vinylidene fluoride) nanocomposites: Piezoelectric β polymorph formation. *Polymer* 51:5846–56.

Liang, J., Xu, Y., Huang, Y., et al. 2009. Infrared-triggered actuators from graphene-based nanocomposites. *J Phys Chem C* 113:9921–7.

Lv, R., Ren, Y., Guo, H., Bai, S. 2021. Recent progress on thermal conductivity of graphene filled epoxy composites. *Nano Mater Sci* In press.

McNaught, A. D., and Wilkinson, A. 1997. *The Orange Book, IUPAC Compendium of Chemical Terminology*. London, Black Well Scientific Publications.

Ng, L. W. T., Hu, G., Howe, R. C. T., Zhu, X., Yang, Z., Jones, C. G., and Hasan, T. 2019. *Printing of Graphene and Related 2D Materials*. Switzerland: Springer.

Novoselov, K. S., Geim, A. K., Morozov, S. V., et al. 2004. Electric field effect in atomically thin carbon films. *Science* 306:666–9.

O'Neill, A., Khan, U., Nirmalraj, P. N., Boland, J., and Coleman, J. N. 2011. Graphene dispersion and exfoliation in low boiling point solvents. *J Phys Chem C* 115:5422–8.

Paredes, J. I., Villar-Rodil, S., Martínez-Alonso, A., and Tascon, J. M. D. 2008. Graphene oxide dispersions in organic solvents. *Langmuir* 24:10560–4.

Paul, D. R., and Robeson, L. M. 2008. Polymer nanotechnology: Nanocomposites. *Polymer* 49:3187–204.

Pavlidou, S., and Papaspyrides, C. D. 2008. A review on polymer–layered silicate nanocomposites. *Prog Polym Sci* 33:1119–98.

Ponnamma, D., Sadasivuni, K. K., Grohens, Y., Guo, Q., and Thomas, S. 2014. Carbon nanotube based elastomer composites–an approach towards multifunctional materials. *J Mater Chem C* 2:8446–85.

Potts, J. R., Dreyer, D. R., Bielawski, C. W., Ruoff, R. S. 2011. Graphene-based polymer nanocomposites. *Polymer* 52:5–25.

Potts, J. R., Shankar, O., Du, L., and Ruoff, R. S. 2012. Processing–morphology–property relationships and composite theory analysis of reduced graphene oxide/natural rubber nanocomposites. *Macromolecules* 45:6045–55.

Potts, J. R., Shankar, O., Murali, S., Du, L., and Ruoff, R. S. 2013. Latex and two-roll mill processing of thermally-exfoliated graphite oxide/natural rubber nanocomposites. *Compos Sci Technol* 74:166–72.

Prasad, K. E., Das, B., Maitra, U., Ramamurty, U., and Rao, C. N. R. 2009. Extraordinary synergy in the mechanical properties of polymer matrix composites reinforced with 2 nanocarbons. *Proc Natl Acad Sci* 106:13186–9.

Putz, K. W., Compton, O. C., Palmeri, M. J., and Nguyen, S. T. L. C. 2010. High nanofiller content graphene oxide–polymer nanocomposites via vacuum-assisted self-assembly. *Adv Func Mater* 20:3322–9.

Qi, X, Pu, K., Li, H. et al. 2010. Amphiphilic graphene composites. *Angewandte Chemie* 49:9426–9.

Sadasivuni, K. K., Saiter, A., Gautier, N., Thomas, S., and Grohens, Y. 2013. Effect of molecular interactions on the performance of poly (isobutylene-co-isoprene)/graphene and clay nanocomposites. *Colloid Polym Sci* 291:1729–40.

Salavagione, H. J., Gomez, M. A., and Martinez, G. 2009. Polymeric modification of graphene through esterification of graphite oxide and poly (vinyl alcohol). *Macromolecules* 42:6331–4.

Scharfenberg, S., Rocklin, D. Z., Chialvo, C., Weaver, R. L., Goldbart, P. M., and Mason, N. 2011. Probing the mechanical properties of graphene using a corrugated elastic substrate. *Appl Phys Lett* 98:91908.

Shan, C., Yang, H., Han, D., Zhang, Q., Ivaska, A., and Niu, L. 2009. Water-soluble graphene covalently functionalized by biocompatible poly-L-lysine. *Langmuir* 25:12030–3.

Shen, M. Y., Chang, T. Y., Hsieh, T. H. et al. 2013. Mechanical properties and tensile fatigue of graphene nanoplatelets reinforced polymer nanocomposites. *J Nanomater.*

Shioyama, H. 2000. The interactions of two chemical species in the interlayer spacing of graphite. *Synth Met* 114:1–15.

Stankovich, S., Dikin, D. A., and Dommett, G. H. B., et al. 2006a. Graphene-based composite materials. *Nature* 442:282–6.

Stankovich, S., Piner, R. D., Nguyen, S. T., and Ruoff, R. S. 2006b. Synthesis and exfoliation of isocyanate-treated graphene oxide nanoplatelets. *Carbon* 44:3342–7.

Summerscales, J., Dissanayake, N. P. J., Virk, A. S., and Hall, W. 2010. A review of bast fibres and their composites. Part 1–Fibres as reinforcements. *Compo Part A: Appl Sci Manuf* 41:1329–35.

Tang, Z., Wu, X., Guo, B., Zhang, L., and Jia, D. 2012. Preparation of butadiene–styrene–vinyl pyridine rubber–graphene oxide hybrids through co-coagulation process and in situ interface tailoring. *J Mater Chem* 22:7492–501.

Terrones, M., Martín, O., González, M., et al. 2011. Interphases in graphene polymer-based nanocomposites: Achievements and challenges. *Adv Mater* 23:5302–10.

Tkalya, E., Ghislandi, M., Alekseev, A., Koning, C., and Loos, J. 2010. Latex-based concept for the preparation of graphene-based polymer nanocomposites. *J Mater Chem* 20:3035–9.

Vickery, J. L., Patil, A. J., and Mann, S. 2009. Fabrication of graphene–polymer nanocomposites with higher-order three-dimensional architectures. *Adv Mater* 21:2180–4.

Wei, H., Zhu, J., Wu, S., Wei, S., and Guo, Z. 2013. Electrochromic polyaniline/graphite oxide nanocomposites with endured electrochemical energy storage. *Polymer* 54:1820–31.

Williams, J. C., and Starke Jr, E. A. 2003. Progress in structural materials for aerospace systems. *Acta Mater* 51:5775–99.

Wu, S., Tang, Z., Guo, B., Zhang, L., and Jia, D. 2013. Effects of interfacial interaction on chain dynamics of rubber/graphene oxide hybrids: A dielectric relaxation spectroscopy study. *RSC Adv* 3:14549–59.

Yan, Z., Peng, Z., Casillas, G., et al. 2014. Rebar graphene. *ACS Nano* 8:5061–8.

Yaragalla, S., Mishra, R. K., Thomas, S., Kalarikkal, N., and Maria, H. J. 2019. *Carbon-Based Nanofillers and Their Rubber Nanocomposites: Fundamentals and Applications.* Amsterdam: Elsevier.

Yaragalla, S., Sindam, B., Abraham, J., Raju, K. C. J., Kalarikkal, N., and Thomas, S. 2015. Fabrication of graphite-graphene-ionic liquid modified carbon nanotubes filled natural rubber thin films for microwave and energy storage applications. *J Polymr Res* 22:137.

Yoonessi, M., and Gaier, J. R. 2010. Highly conductive multifunctional graphene polycarbonate nanocomposites. *Acs Nano* 4:7211–20.

Zhan, Y., Lavorgna, M., Buonocore, G., and Xia, H. 2012. Enhancing electrical conductivity of rubber composites by constructing interconnected network of self-assembled graphene with latex mixing. *J Mater Chem* 22:10464–8.

Zhan, Y., Wu, J., Xia, H., Yan, N., Fei, G., and Yuan, G. 2011. Dispersion and exfoliation of graphene in rubber by an ultrasonically-assisted latex mixing and in situ reduction process. *Macromol Mater Eng* 296:590–602.

Zhang, W., Xu, H., Chen, Y., Cheng, S., and Fan, L. J. 2013. Polydiacetylene-polymethylmethacrylate/graphene composites as one-shot, visually observable, and semiquantative electrical current sensing materials. *ACS Appl Mater Interf* 5:4603–6.

Zhao, X., Zhang, Q., Chen, D., and Lu, P. 2010. Enhanced mechanical properties of graphene-based poly (vinyl alcohol) composites. *Macromolecules* 43:2357–63.

Zhou, S., Chiang, S., Xu, J., Du, H., Li, B., Xu, C., and Kang, F. 2012a. Modeling the in-plane thermal conductivity of a graphite/polymer composite sheet with a very high content of natural flake graphite. *Carbon* 50:5052–61.

Zhou, S. X., Zhu, Y., Du, H. D., Li, B. H., and Kang, F. Y. 2012b. Preparation of oriented graphite/polymer composite sheets with high thermal conductivities by tape casting. *New Carbon Mater* 27:241–9.

Zimniewska, M., and Wladyka-Przybylak, M. 2016. *Natural fibers for composite applications, in: Fibrous and Textile Materials for Composite Applications.* Singapore: Springer.

7 Dispersion and Characterization of Graphene in Elastomer Composite

Meryem Samancı and Ayşe Bayrakçeken Yurtcan
Atatürk University

CONTENTS

7.1 INTRODUCTION

Rubber is the name given to materials that can change in size when stretched and return to their original state when the deforming stress is removed. It is derived from the South American word 'caoutchouc' meaning 'weeping wood'. Originally, the word Rubber was used for the product obtained from a dense milky liquid called latex found in some plants (Ciesielski 1999). Rubbers are widely used in transportation, medicine, construction, and everyday life due to their excellent durability and flexibility. The most obvious usage area is the automotive industry where the tire and inner tire production takes place. The consumption of rubber, which is a product that human beings inevitably use in their daily life, is increasing day by day. While rubbers of natural origin were used in the beginning, synthetic rubbers began to be produced due to the increasing needs of the rapidly increasing world population, technological developments, and wars. Rubber manufacturing generally consists of three stages: (i) mixing (mills, internal mixing), (ii) forming, and (iii) vulcanizing (Mall, Zhang, and Geng 2010). A brief summary about the commonly used natural and synthetic rubbers is given in Tables 7.1 and 7.2.

In industry, the word 'elastomer' is used synonymously with rubber. Elastomers are the most important commercial polymers with their low weight, corrosion resistance, ease of processing, and economy. Elastomers are obtained after chemical conversion (cross-linking reaction or vulcanization) of rubber materials. Its elasticity is due to

DOI: 10.1201/9781003200444-7

TABLE 7.1
General Properties of Natural Rubbers

Natural Rubbers (General Features)

Natural Sources	Commercial Name	Chemical Composition	Explanation
Hevea brasiliensis (Para rubber)	Indian Standard Natural Rubber (ISNR)	Raw content: Rubber 30%–40%	pH: 6.5–7.1 Density: 0.92–0.98 g cm^{-3}
Parthenium argentatum (Guayule rubber)	Standard Malaysian Rubber (SMR) Standard Indonesian	Proteins 1%–1.5% Resins 1.5%–3.0% Minerals 0.7%–0.9%	Particle size: 0.02–2 μm
Manihot glaziovii (Ceara rubber)	Rubber (SIR) Standard Lanka Rubber	Carbohydrates 0.8%–1.0% Water 55%–60%	Latex technology: Coagulated or dried
Castilla elastica (Panama rubber)	(SLR) Thai Tested Rubber (TTR)	Dry rubber content: cis-l,4-poryisoprene	Centrifuging, creaming,
Ficus elastica (India rubber)	Specified Singapore Rubber (SSR)	(2-methyl-1,3 butadiene) Rubber 60%	evaporation, and electrodecantation
Funtumia elastica (Lagos silk rubber)	Nigerian Standard Rubber (NSR)	Nonrubber constituent+ remaining water 3%–5%	methods Preservative of latex: Ammonia, formalin

Source: Sisanth et al. (2017), Niyogi (2007), and Mathew (2001).

TABLE 7.2
Commonly Used Synthetic Rubbers

Synthetic Rubbers

Commercial Name	Chemical Composition	Explanation
Styrene-Butadiene Rubber (SBR)	Styrene (23%–25%) and 1,3-butadiene (75%–77%) Nitrile butadiene rubber (NBR) (nitrile content of 34%)	Technique: Solution and emulsion polymerization of styrene and butadiene
Polybutadiene Rubber (PBR/PBD)	Polymers containing 90%–98% The homopolymer of butadiene	Technique: Solution polymerization
Polyisoprene rubber (IR)	Isoprene (catalyst:hexane)	Technique: Solution polymerization
Acrylonitrile-Butadiene Rubber (NBR)	1,3-Butadiene and acrylonitrile (25%–50%)	Technique: Emulsion copolymerization
Neoprene Rubber (CR)	2-chloro-1,3 butadiene	Technique: Emulsion polymerization
Ethylene-Propylene Rubber (EPM)	Monomers ethylene and propylene	Technique: Solution copolymerization
Butyl Rubber (IIR)	Isoprene (2%–3%) and isobutylene (97%–98%)	Technique: Cationic copolymerization
Polyacrylate rubbers (ACM)	Acrylic esters (ethyl and methyl acrylate, 95%) and reactive cure site monomer (carboxylic acid or chloroethyl vinyl ether, 5%).	Technique: Solution or emulsion polymerization

(Continued)

TABLE 7.2 (*Continued*)
Commonly Used Synthetic Rubbers

Synthetic Rubbers

Commercial Name	Chemical Composition	Explanation
Chlorobutyl Rubber (CIIR)	Chlorine (1.2 wt.%) in the butyl polymer	Technique: Copolymerization
Chloroprene Rubber (CR)	Also known as Neoprene. The starting material: 2-chloro-1,3-butadiene	Technique: Radical polymerization in aqueous emulsion
Chlorosulfonated polyethylene (CSM)	Also known as Hypalon rubber CSM grade contains 25%–43% by weight of chlorine and 0.8%–1.5% by weight of sulfur	This rubber is a modification of polyethylene; crystalline thermoplastic material to an amorphous elastomer by chlorosulfonation
Polysulfide Rubber (T rubber)	Also known as Thiokol Ethylene dichloride and sodium tetrasulfide	Polysulfide polymers the condensation of sodium polysulfide with dichloroalkanes
Silicone Rubber (SI)	Methyl, phenyl, vinyl, or trifluoropropyl group (dimethyl siloxanes, methyl phenyl siloxanes, methyl vinyl siloxane, fluorosilicone component (40%–90%)	Q rubber is an inorganic polymer, the backbone consists of silicone and oxygen atoms ($-Si-O-$) linkage
Polyurethane rubbers (PUR)	Reaction of the isocyanate group with the hydroxyl group of the alcohol	
Fluorocarbon Elastomers (FKM):	The vinylidene fluoride (least 75%) based elastomers VITON-A is a copolymer of vinylidene fluoride and hexafluoropropylene VITON-B is the terpolymer of vinylidene fluoride, hexafluoropropylene, and tetrafluoroethylene	Technique: Emulsion polymerization
Ethylene-Vinyl Acetate (EVA) Copolymer	Polyethylene, modified by small amount comonomers (vinyl acetate, ã-olefins, etc.)	Technique: Copolymerization
Thermoplastic Elastomers (TPE)	-Thermoplastic Polyurethanes (TPU) -Styrene thermoplastic elastomers -Ethylene–propylene (EP) copolymers -Polyether–polyester TPE -Polyether–polyamide copolymers	Technique: Chemical reaction and blending

Source: Sisanth et al. (2017) and Niyogi (2007).

the ability of the atoms in the cross-linked, long and flexible polymer chain to rotate around single bonds. The individual molecules are separated from each other when the force is applied and the molecules return to their original position when this force is removed (Datta 2001). Due to these properties, it is widely used in the production of elastic bands, tires, gaskets, and hoses (Arguello and Santos 2016). The methods used in making elastomers and polymerization variables (mechanism, catalyst, and process) determine the final micro- and macrostructure of the polymer. In other words, the structure and usage characteristics of the material determine whether the elastomer is suitable for an application. Different processes can be used in the production of elastomers. They are (i) emulsion polymerization (free radical chain polymerization), (ii) bulk polymerization (the pure liquid monomer as reaction medium), (iii) solution polymerization (the monomer is dissolved in an organic solvent), and (iv) suspension polymerization (the monomer is dissolved in an organic solvent) (Datta 2004).

The usage areas of elastomers have increased with the development of industrial activities. This situation has made it inevitable to increase the diversity of elastomers used for different purposes. The development and improvement of the properties of elastomers have been achieved by making elastomer composites. The changes in the elastomer structure of very small amounts of nanofillers materials included in the polymer matrix have been extensively investigated. The most commonly used filling materials are carbon black, silica, clays, boehmite, and carbon nanotubes. Recently, graphene has been widely used as a nanofiller material to produce polymer nanocomposites with new functionalities. Graphene is a single-atom-thick layer of sp^2-bonded carbon atoms in the form of a honeycomb, which is a monolayer of graphite. This structure has enabled it to have excellent physical and chemical properties. Because of these extraordinary properties, graphene can significantly improve the mechanical, thermal, electrical, and barrier properties of the nanocomposite. In this research, nanocomposites formed with graphene-based nanofillers and elastomers were investigated. The distribution of graphene in the elastomer and the morphological and physical changes it causes in the structure were investigated.

7.2   DISPERSION OF GRAPHENE IN ELASTOMER COMPOSITES

In the last 20 years, studies on elastomer nanocomposites have increased considerably. This is due to the need to synthesize elastomers with different properties to meet the demands of many critical applications with the development of polymer science and technology. Very small amounts of nanofillers added to rubber improve the physical and chemical properties of the structure. Fillers are typically used to improve properties such as mechanical, electrical, thermal, gas/liquid barrier resistance, aging, and flame retardancy. In particular, three different fillers are used in the synthesis of rubber nanocomposites. They are carbon fillers (carbon black, carbon fibers, carbon nanotubes, and graphene), inorganic fillers (aluminum oxide, calcium carbonate, halloysite, kaolin, nano clay, silica, talc, titanium oxide, and zinc oxide), and biofillers (cellulose, coir, husk, and wood) (Song 2017). The excellent physical properties (particle size and structure, high surface area and anisotropy (aspect ratio)) and surface properties (presence of reactive/functional groups, number of active sites on their surfaces, wettability, and surface energy) of many nanofiller materials are the most important factors that determine the characteristic properties of the composites (Mall, Zhang, and Geng 2010). Some additional components are used while synthesizing elastomer nanocomposites. These are summarized in Table 7.3. The amount of components used in nanocomposite synthesis is given in volume percent (vol.%), mass percent (wt.%), and phr (parts by weight per hundred parts rubber).

Elastomeric nanocomposites are generally produced by three different techniques. They are (i) solution mixing, (ii) melt mixing, and (iii) in situ polymerization (Mensah et al. 2018). The synthesis steps of the production techniques of elastomeric nanocomposites are summarized in Figure 7.1. Various difficulties are encountered during the production of nanocomposites. The degree of formation of material with desired properties (structure-property relationship) in composite synthesis depends on (i) particle size, which is the most important parameter (surface-area to volume ratio), (ii) rubber-filler interface interaction (adhesion, physical adsorption, chemical adsorption,

---

TABLE 7.3

Components Used in Elastomer Nanocomposite Synthesis

Ingredient	Explanation
Fillers	Carbon fillers, inorganic fillers, biofillers
Curing or vulcanization agent	Zinc oxide, sulfur
Viscosity reducers	Water, trisodium phosphate, sodium dinaphthylmethane disulfonate
Thickening agents	Colloidal silica
Wetting agents	Sulfonated oils
Homogenizing agents	Aromatic hydrocarbon resins
Accelerators	Guanidines, thioureas, thiazoles, sulfenamides, aldehyde/amine condensates, thiurams, dithiocarbamates, xanthates
Activators	The fatty acids (e.g., stearic, palmitic, and lauric acids) and organic activators: polyalcohols (e.g., ethylene glycol) and amino alcohols (e.g., triethanolamine)
Antioxidants	Amine antioxidants and phenolic antioxidants (Hindered phenols, styrenated phenols, amine derivatives such as phenylene diamines, phenyl beta-naphthylamine, ketone-amine condensates)
Softeners	Liquid paraffin, paraffin wax, and stearic acid
Dispersing agents	Ammonium oleate, mixtures of fatty acids, and metal soaps
Emulsifying agents	Soap: The cationic part (ammonia, KOH, or amine), the anionic part (oleic, stearic, or rosin acid)
Stabilizers	Soap and proteins (e.g. casein), sodium salt of cetyl/oleyl sulfate
Pigments	Inorganic pigments: Titanium dioxide, iron oxide, chromium oxide green, and cadmium Organic pigments: Permanent Red 2B, phthalocyanine blue, phthalocyanine green, carbazole violet, and diarylide yellow

Source: Niyogi (2007).

Solution Mixing technique	Melt mixing technique	In-situ polymerization technique
Dispersion of GSD in desired Solvent via prolonged sonication. **Step 1**	Mastication of polymer matrix under high temperature . **Step 1**	Mixing of polymer monomer with GSD Under desired reaction conditions. **Step1**
Dissolving of polymer matrix in desired solvent based on matrix polarity. **Step 2**	Blending of GSD into the polymer matrix. **Step 2**	Swelling the mixture in Step 1. **Step 2**
Mixing of Step1 and Step 2 under stirring to achieve homogenous mixture. **Step 3**	Mixing of curing additives into mixture in Step 2. **Step 3**	Addition of curing ingredient into the mixture in Step 2. **Step 3**
Evaporation of solvent, mixing of cure ingredient into the new mixture and cure	Curing of mixture in Step 3. **Step 4**	Allow for polymerization. **Step 4**

GSD-Polymer Nanocomposites

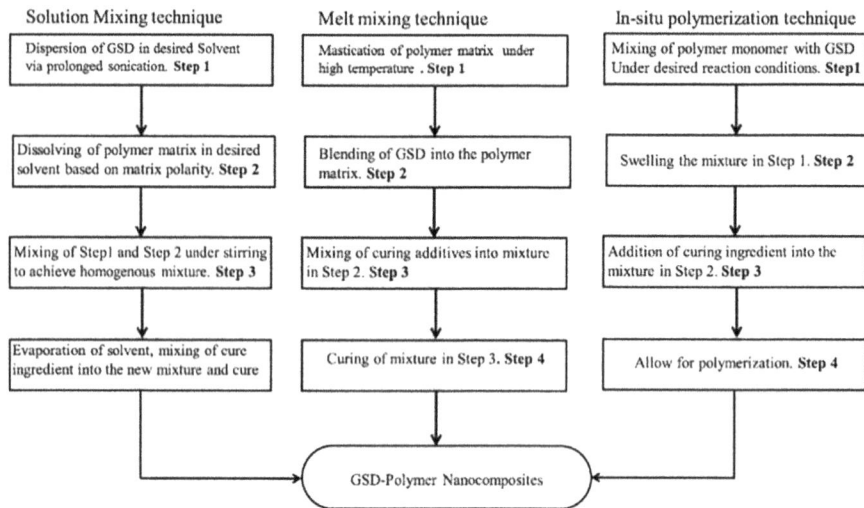

FIGURE 7.1   Common nanofiller mixing methods used in fabrication of graphene oxide nanosheets (GSD)-elastomeric nanocomposites (Mensah et al. 2018).

and mechanical interaction), (iii) filler-filler interface interaction, (iv) compatibility of the shape and structure of the filler with the polymer matrix, (v) optimum filler concentration, and (vi) filler dispersion in the matrix (Alex 2010). The most common of these challenges is to ensure that nanofillers are homogeneously dispersed within the rubber matrix without agglomeration. In general, to overcome this situation, nanofillers are subjected to chemical modification pretreatments. In addition, the filler content added to the rubber should be in optimum proportions. In particular, nanofiller addition at high proportions is expected to increase agglomeration due to increasing clustering rates in the composite structure. In addition, the high filler content increases the weight of the composite and reduces its workability (Stephen and Thomas 2010). During the production stages, situations may occur that adversely affect human health and environmental cleanliness (Mall, Zhang, and Geng 2010).

Graphene (2D material, sp^2 connected carbon atoms) is the only layered form of graphite (3D material) in nature, which is very plentiful in nature. The corner structure of the honeycomb provides its unique electronic, thermal, mechanical, and chemical properties. Single-layered graphene has shown a large surface area of $2,600 \, m^2 g^{-1}$, high charge carriers (electrons and holes) mobility of $230,000 \, cm^2 (Vs)^{-1}$, Young modulus of ~1 TPa, the ultimate strength of 130 GPa, the thermal conductivity of $5,000 \, W \, (mK)^{-1}$, high flexibility, and high electrical conductivity of $6,000 \, S \, cm^{-1}$ (Anwar et al. 2016, Sanna 2014). Graphene is synthesized by several different methods such as mechanical exfoliation, liquid-phase exfoliation, thermal exfoliation, and chemical vapor deposition (CVD) methods as given in the earlier chapters in this book. The desired graphene structure and final properties are highly dependent on the production methods used. Therefore, graphene has been widely used in the development of synthesis of commercial nanocomposites with desired properties. Due to these features, it has been widespread used as nanofiller materials in elastomers recently in the industry. Even with the addition of very low quantities into the elastomer, it changes the characteristics of the composite. Various methods are used for graphene synthesis. The most common methods are the mechanical peeling, chemical flake, chemical synthesis, thermal synthesis, chemical vapor deposition (CVD), and epitaxial growth (Das and Choi 2011). Despite the advantages described above, the graphene also has some disadvantages. They are as follows: It is quite difficult to control the particle size during the synthesis in very large quantities of the graphene (Oliveira et al. 2018). It is difficult to make its composites with polymers and metal oxides due to the tendency to be agglomerate with its hydrophobic structure. In order to overcome these challenges, studies related to modified graphene structures were performed. Therefore, the modified graphenes with the derivatives of the graphene (graphene oxide (GO), reduced graphene oxide (rGO), etc.) replace the pure graphene in potential applications. Modified graphenes usually include some structural defects and functional groups (carboxyl, carbonyl, hydroxyl, and epoxide groups) containing too much oxygen (He et al. 2017). They usually have different hydrophilicity degrees according to the differences in modification transactions. Depending on the hydrophilicity of the modified graphene, it is usually used in the solution mixing or in situ polymerization methods to ensure that the filler is homogenized in the matrix (Zhang and Cho 2018). Tian et al. prepared a nanocomposite of carboxylated acrylonitrile-butadiene rubber (XNBR) and graphene oxide nanosheets (GONS) nanofiller material using ultrasonic-assisted latex mixing process. The schematic representation of the process steps is given in Figure 7.2. In Figure 7.2a and b, the dispersion of GONS in the XNBR polymer matrix is shown schematically. An interaction resulted from hydrogen bonds formed between the carboxylic groups in XNBR and the oxygen-containing functional groups in GONS. Then, the hot pressing and curing processes shown in Figure 7.2c were carried out. Figure 7.2d shows the separated network structure of graphene in the polymer matrix (Tian et al. 2014).

The properties of nanocomposites largely depend on preparation methods and structures of polymer matrices and nanofillers. Aguilar-Bolados et al. synthesized two different graphite oxides using the methods of Hummers and Brodie. Thermal reduction to graphite oxides was performed. Two different oxidation processes caused differences in the structure and morphology of reduced graphene oxides (rGO). Thermal rGO was used as a filler in NR. Thermal rGO-B prepared by Brodie's method showed a more homogeneous distribution in the polymer matrix. This is due to the fact that the thermal rGO-B obtained by Brodie's method consists of more exfoliated regular structures than that obtained by the Hummers method (rGO-H). Therefore, nanocomposites prepared with thermal rGO-B showed more advanced mechanical and electrical properties than nanocomposites prepared with thermal RGO-H (Aguilar-Bolados et al. 2016). Wen et al. synthesized GO at different oxidation degrees with the Hummers method and synthesized elastomer nanocomposites using carboxylated acrylonitrile-butadiene and latex coagulation technique. As the oxidation degree of GO increased, the functional groups in the structure and accordingly the hydrogen bonds increased. Hydrogen bonds improved dispersion by increasing cross-linking due to the interfacial interaction between the polymer matrix and the filler material. As the dispersion improved, the nanocomposite was improved structurally and morphologically compared to pure rubber (Wen et al. 2017). Potts et al. prepared reduced

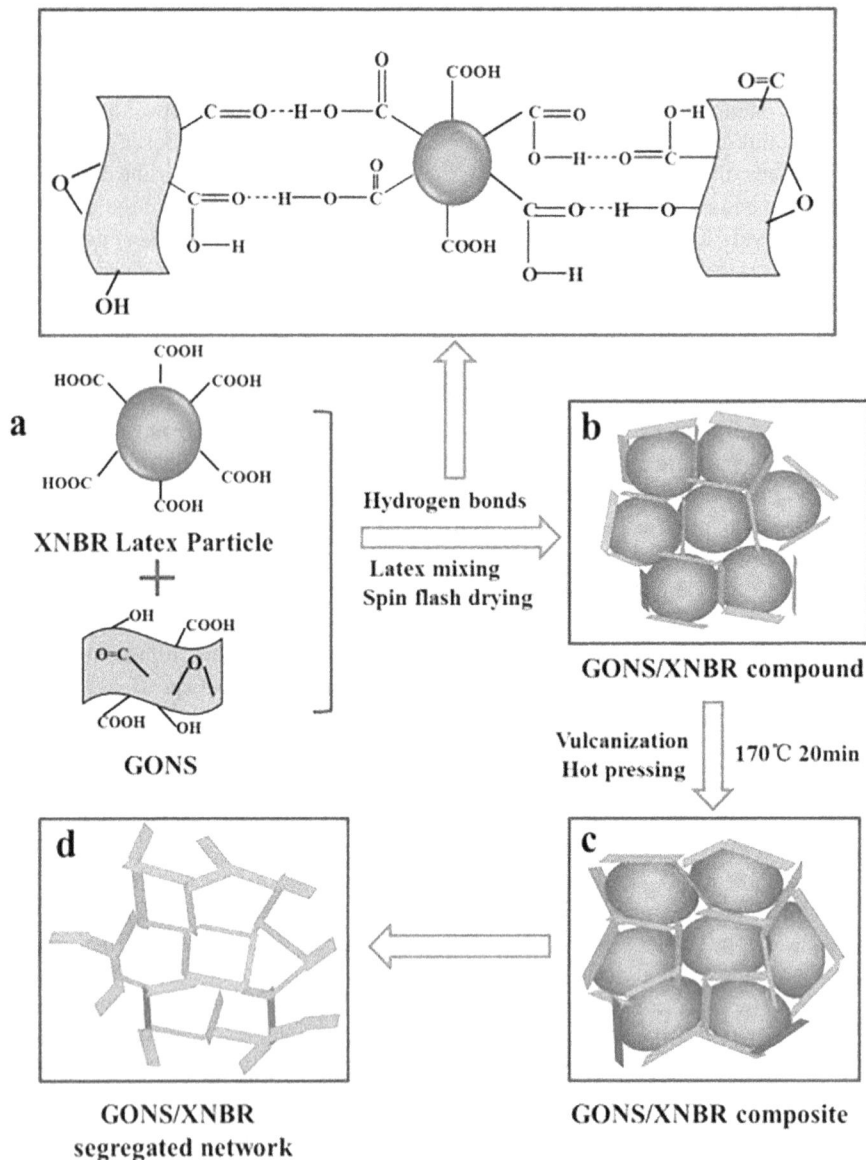

FIGURE 7.2 Schematic representation of the preparation of GONS/XNBR composites with segregated network structure (Tian et al. 2014).

graphene oxide (rG-O)/NR nanocomposites by latex coagulation technique. Then, the curing agents were incorporated into the composite by two different methods: solution treatment or two-roll grinding. The physical properties of the composites have also changed considerably with different curing methods. It was observed that the dispersion quality and the homogeneity of rG-O in the NR matrix were better in two-roll milling. This is explained by the fact that the separated mesh morphology of the composite does not deteriorate in the solution processing technique and disappears in the grinding process. This change in composite morphology in the two-roll milling process resulted in less low-aspect platelets (Potts et al. 2012).

7.3 CHARACTERIZATION OF GRAPHENE/ ELASTOMER COMPOSITES

Characterization techniques such as atomic force microscopy (AFM), wide-angle X-ray diffraction (WAXD), small-angle X-ray scattering (SAXS), scanning electron microscopy (SEM), and transmission electron microscopy (TEM) are generally used as powerful characterization devices for graphene/elastomer nanocomposites. AFM analysis is used to determine the average layer thickness of graphene and graphene-based materials, which are the nanofilling materials to be used. WAXD and SAXS analyses are used to determine the graphene layer numbers and the distance between the layers. The morphology of the nanocomposite is observed by SEM analysis. The size of the graphene dispersed in the nanocomposite is examined by TEM analysis.

The changes sourced by the modifications of graphene in the dimensions of graphene are examined by AFM analysis. Wu et al. synthesized amine-functional reduced graphene oxide (NH_2-RGO) with primary amine (–NH_2) groups attached to the surfaces with toluene-2,4-diisocyanate. AFM observations of rGO and NH_2-rGO were carried out to compare the thicknesses of the graphene nanoplates. The average thickness of the rGO nanoplates was approximately 1.01 nm, while the average thickness of the NH_2-rGO nanosheets was determined as 2.08 nm. rGO and amine-functional rGO are covalently incorporated into a water-based polyurethane (WPU) matrix. They reported that the amine-functional rGO was dispersed very thinly and homogeneously in the WPU matrix due to the amine groups, and this improved the physical properties of the nanocomposite (Wu, Shi, and Zhang 2016). Li et al. performed polyvinyl pyrrolidone (PVP) coating as well as reducing GO to rGO to modify graphene. AFM analysis was performed for GO, GP/PVP, and rGO/PVP. The GO layer is 1 nm, and the GO/PVP and rGO/PVP layers were about 2 nm due to the coating of the PVP on the graphene surfaces. Then, they prepared TPU composites from these modified graphenes. They reported that PVP acts as a dispersant and improves the interface interaction (Li et al. 2015). Chen et al. synthesized GOs of different oxidation degrees using the Hummers method. Then, they produced GO/SBR nanocomposites using the latex coagulation method. AFM analyses were performed to characterize the oxidation degrees of GO. Oxygen-containing functional groups in GO increased with increasing oxidation degrees which reduced the thickness of GO from about 7 to 2 nm (Chen et al. 2018).

Frasca et al. synthesized nanocomposites using different elastomers including Chlorine-Isobutylene-Isoprene Rubber (CIIR), Nitrile-Butadiene Rubber (NBR), Natural Rubber (NR), and Styrene-Butadiene Rubber (SBR) and multilayered (approximately 10 graphene layers) graphene (MLG) as filler material at the same rate (3 phr). Morphological structures of unreinforced elastomers and nanocomposites were compared using SEM images. They reported that the surfaces of unreinforced elastomers were smooth and the surfaces of elastomer nanocomposites became rough and protruding which is due to the strong interfacial interaction between polymer matrices and nanofillers (Frasca et al. 2015).

Xie et al. used a silane coupling reagent (bis-[γ-(triethoxysilyl)propyl]-tetrasulfide (BTESPT)) to modify GO (SGO). The modified GO was mixed with IR by solution blending and in situ reduction was performed to prepare the IR/SGE composite. In Figure 7.3, SEM images of modified GE, TEM, and XRD analyses of polymer nanocomposites (IR/SGE and IR/GE) made with modified and unmodified GOs are given. The composites show a broad amorphous peak at $2\theta = 18.7°$ in contrast to the prominent characteristic peaks of graphene. Sharp diffraction peaks above 30° represent characteristic peaks from curing agents (ZnO and stearic acid). In TEM images, it was observed that the modified GO facilitated homogeneous dispersion in the IR. In addition, aggregations occurred with increasing the loading rate of the filler (Xie et al. 2017).

Xing et al. prepared graphene/NR nanocomposites by a modified latex mixing method with in situ chemical reduction. XRD, SEM, and TEM analyses of graphene/NR nanocomposites were performed. The XRD patterns of the composites showed no characteristic peaks of graphite or GO. This shows that the filling material does not form layers in the NR, on the contrary, it is an amorphous structure in which it is homogeneously dispersed. In the TEM image, they show that the graphene exfoliates

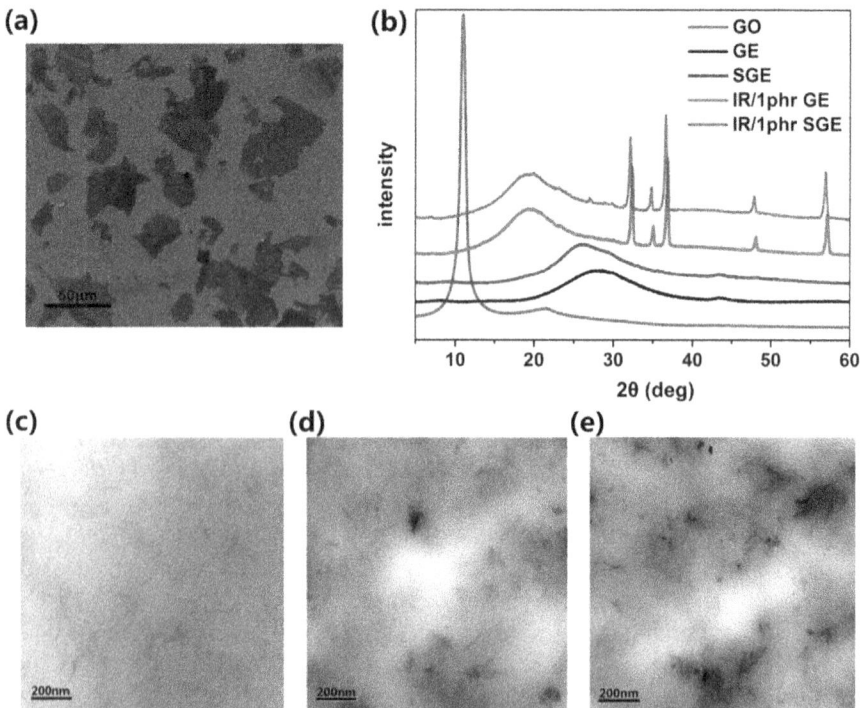

FIGURE 7.3 (a) SEM image of well-exploited SGO; (b) XRD patterns of pure GO, SGE, and their IR composites; TEM image of IR composite with 0.5 phr SGE (c), 2 phr SGE (d), and 2 phr GE (e) (Xie et al. 2017).

in NR, and in the SEM image, the composite structure is rougher than NR. This is explained by the fact that graphene is homogeneously dispersed in NR and the graphene is wrapped with a polymer matrix (Xing, Wu, et al. 2014). Zhang et al. prepared NR nanocomposites reinforced with different ratios of GO (0.5–5 w%) using the latex mixing and casting method. As the proportion of filling material in the composite increased, the tendency of GO to agglomerate and re-stack increased. The TEM images in Figure 7.4 show that GO has a networked morphology within the polymer matrix, and as the filler ratio increases, GO spreads with thicker, multilayered layers that are interconnected (Zhang et al. 2017). Gan et al. opened MWCNTs as nanostrips using strong oxidizers to obtain graphene (GNR). Then, this GNR was dispersed into SR by solution mixing method and nanocomposites were produced. Oxygen-containing functional groups in the GNR structure facilitated the dispersion in the SR. The structure and surface morphology of SR/GNR nanocomposites were investigated using XRD and SEM techniques. From the SEM images, it was observed that the CNT is sheet-shaped and the GNR has different widths. It was observed that the GNR was distributed very well without clustering in the SR. When we look at the XRD spectra, it was seen that the GNR has characteristic peaks at $2\theta = 10.8°$ and $2\theta = 27.4°$, and the SR at $2\theta = 12.2°$. On the other hand, it was observed that the GNR/SR nanocomposites show an intact, weaker, and broader SR-like peak and the GNR peaks almost disappear. This explains that the crystallinity of nanocomposites decreases and turns into an amorphous structure (Gan et al. 2015). Wu et al. achieved coagulation with saturated sodium chloride solution after mixing the aqueous dispersion of GO with NR. Then, in situ reduction of GO with hydrazine hydrate was

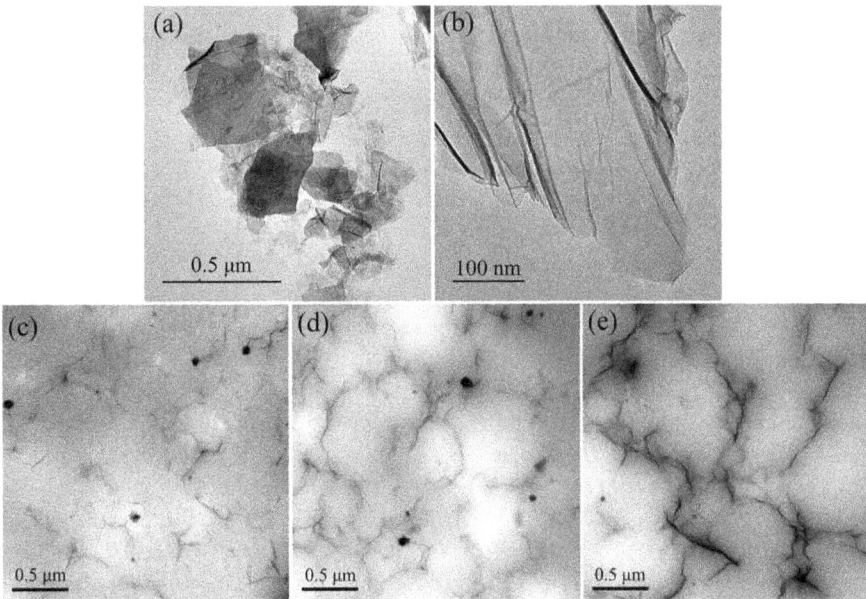

FIGURE 7.4 TEM micrographs of GO sheets (a) and in higher magnification (b), GO/NR nanocomposites with GO loadings of 1% (c), 2% (d), and 5% (e) (Zhang et al. 2017).

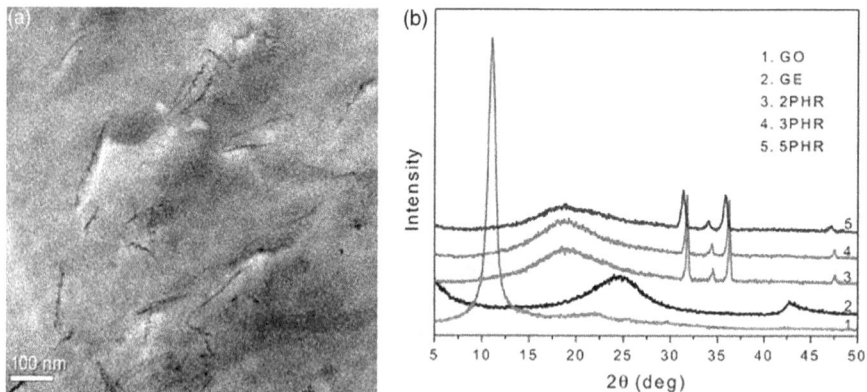

FIGURE 7.5 (a) TEM image of the GE/NR nanocomposite with 1 phr of graphene loading; (b) XRD patterns of graphite oxide, graphene prepared by chemical reduction in aqueous suspension, and GE/NR nanocomposites with graphene loadings of 2, 3, and 5 phr (Wu et al. 2013).

performed. TEM and WAXD analyses of synthesized GE/NR nanocomposites are given in Figures 7.5a and b. When we look at the TEM image of the GE/NR nanocomposite, GE is effectively dispersed throughout the NR matrix when the latex mixing method was used. WAXD patterns of GO, GE, and GE/NR (2, 3, and 5 phr) are shown in Figure 7.5b. GO has a sharp characteristic peak at $2\theta = 11.0°$ and GE has a broad characteristic peak at $2\theta = 24.7°$ which means that the graphene layers are re-stacked after the reduction process of GO. GE/NR nanocomposites show a broad peak at $2\theta = 18.7°$. This shows that the composites have an amorphous structure, and even when the graphene loading is increased from 2 to 5 phr, the graphene plates are homogeneously dispersed in the polymer matrix without agglomeration (Wu et al. 2013).

7.4 PHYSICAL PROPERTIES OF GRAPHENE/ ELASTOMER COMPOSITES

The physical properties of elastomer nanocomposites are the most important parameters in determining the purpose for which they will be used. In the literature studies of graphene/elastomer nanocomposites, the mechanical, thermal, electrical, and gas/liquid barrier properties of the materials were generally investigated. In this chapter, mechanical (tensile strength, elongation at break), thermal conductivity, and electrical conductivity properties of graphene/elastomer composites are summarized. In Tables 7.4–7.6, literature review has been made regarding the changes in physical properties before and after the addition of graphene filler to the elastomer.

One of the most important factors for improving the mechanical properties is that the filling materials dispersed in the form of a network in the polymer matrix perform the best load transfer which is achieved by developing different dispersion techniques. Xie et al. synthesized IR nanocomposites with surface-modified graphene at different phr ratios. At reinforcement ratios greater than 0.5 phr, there is a tendency to decrease in fracture strength sourced by the structural defects due to aggregates formed (Xie

TABLE 7.4

Some Literature Studies on Mechanical Properties of Graphene/Elastomer Nanocomposites

Nanocomposites	Processing	Tensile Strength (MPa)		Elongation at Break (%)		Filler Cont.	References
		Pure Rubber	Composite	Pure Rubber	Composite		
G/M/D-NR	Colloidal dispersion	25.0 ± 1.5	34.9 ± 2.0	869	820	1.5 phr	George et al. (2017)
XNBR/GO 1	Latex co-coagulation	12.47 ± 1.30	16.76 ± 1.40	489 ± 8	452± 17	4 phr	Wen et al. (2017)
XNBR/GO 2	Latex co-coagulation	12.47 ± 1.30	21.52 ± 0.85	489 ± 8	432 ± 9	4 phr	Wen et al. (2017)
XNBR/GO 3	Latex co-coagulation	12.47 ± 1.30	26.76 ± 0.93	489 ± 8	425 ± 6	4 phr	Wen et al. (2017)
NBR/GO	Latex compounding	4.25	8.12	773	606	2 phr	Zhang and Cho (2018)
NBR/MGO	Latex compounding	4.25	9.91	773	438	2 phr	Zhang and Cho (2018)
GO/NR	Latex mixing and casting	1.53 ± 0.49	2.54 ± 0.29	825 ± 38	829 ± 34	0.5 wt.%	Zhang et al. (2017)
GO/NR	Latex mixing and casting	1.53 ± 0.49	7.18 ± 0.18	825 ± 38	688 ± 11	5 wt.%	Zhang et al. (2017)
IR/SGE	Solution blending and in situ reduction	24.1	33.1	-	-	0.5 phr	Xie et al. (2017)
SR/GNR	Solution mixing	0.24	0.37	100	164	0.4 wt.%	Gan et al. (2015)
SR/GNR	Solution mixing	0.24	0.40	100	78	2 wt.%	Gan et al. (2015)
PRGO/NR	Latex hetero-coagulation	17.01 ± 0.69	20.98 ± 0.37	681 ± 13	479 ± 6	3 phr	Zhang et al. (2016)
WPU/FGS	Emulsion mixing	29.6 ± 0.9	40.0 ± 4.6	478 ± 44	590 ± 88	3 phr	Raghu et al. (2008)
WPU/FGS	Emulsion mixing	29.6 ± 0.9	35.8 ± 8.8	478 ± 44	551 ± 185	0.5 phr	Raghu et al. (2008)
CIIR/MLG	Ultrasonically solution mixing	5.28 ± 0.89	5.93 ± 0.29	837 ± 29	868 ± 26	3 phr	Frasca et al. (2015)
NBR/MLG	Ultrasonically solution mixing	7.70 ± 1.23	8.04 ± 0.41	1,083 ± 26	1,065 ± 34	3 phr	Frasca et al. (2015)
NR/MLG	Ultrasonically solution mixing	5.73 ± 0.92	12.50 ± 1.21	635 ± 17	498 ± 21	3 phr	Frasca et al. (2015)

(Continued)

TABLE 7.4 (Continued)
Some Literature Studies on Mechanical Properties of Graphene/Elastomer Nanocomposites

Nanocomposites	Processing	Tensile Strength (MPa)		Elongation at Break (%)		Filler Cont.	References
		Pure Rubber	Composite	Pure Rubber	Composite		
SBR/MLG	Ultrasonically solution mixing	0.91 ± 0.13	4.14 ± 0.60	1,663 ± 284	774 ± 88	3 phr	Frasca et al. (2015)
GO/XNBR	Latex co-coagulation	3.6 ± 0.3	16.0 ± 0.4	504 ± 23	373 ± 13	1.9 vol.%	Kang et al. (2014)
GO-NBR4	Solution-mixing	6.6 ± 0.28	8.1 ± 0.73	559 ± 12	467 ± 22	2 phr	Mensah et al. (2014)
Graphene/NR	Self-assembly latex compounding	12.25 ± 0.29	14.90 ± 0.61	540 ± 21	654 ± 9	0.21 vol.%	Luo et al. (2014)
NR/GE	Latex mixing and in situ reduction	17.1	25.2	579	564	2 wt.%	Zhan et al. (2011)
XNBR/GNS–HDA-2	Solution mixing	2.9	4.63	227	365	2 phr	Manna and Srivastava (2017)
FIL 2.0 (IIrGO/FKM)	Two-roll mixing	6.108	7.395	225	143	2 phr	Moni et al. (2018)

TABLE 7.5

Some Studies on Thermal Conductivity of Graphene/Elastomer Nanocomposites

Nanocomposite	Processing	Thermal Conductivity W (mK)$^{-1}$		Filler Content	References
		Pure Rubber	Composite		
GE/SBR	Modified-latex compounding	0.22	0.27	7 phr	Xing, Tang, et al. (2014)
GO/xNBR	Latex	0.16	0.223	1.5 vol.%	Wang et al. (2013)
RG/L-S(1-2)	Spray drying	0.2043	2.922	15 vol.%	Li et al. (2017)
G/M/D-NR	Emulsion mixing	0.065 (23°C)	0.379 (23°C)	1.5 phr	George et al. (2017)
SBR/GO (SG9)	Latex coagulation	0.161	0.207	4 phr	Chen et al. (2018)
rGO-PU/PSR	Hydrothermal reduction and vacuum impregnation	0.239 (PSR)	0.598	0.9 wt.%	Tao et al. (2021)
PA6 (PGF)	In situ bulk polymerization	0.210	0.847	2 wt.%	Li et al. (2016)
PRGO/NR	Latex heterocoagulation	0.112	0.147	5 phr	Zhang et al. (2016)
GO-IL/SBR	Latex heterocoagulation	0.161	0.188	5 phr	Yin et al. (2017)
rGO/NR	Latex co-coagulation (solution treatment)	0.157	0.219	5 wt.%	Potts et al. (2012)
rGO/NR	Latex co-coagulation (two-roll milling)	0.157	0.188	5 wt.%	Potts et al. (2012)
GnP/SBR	Solution-mixing	0.17	0.48	24 vol.%	Araby et al. (2014)
NR/GE	Latex coagulation	0.1741	0.1963	2 wt.%	Zhan et al. (2011)
rGO-NRL-0.1	Emulsion mixing	0.173 (40°C)	0.236 (40°C)	0.1 phr	Lim et al. (2019)
GO/WPU	Emulsion mixing (high-shear)	0.23	0.33	2 wt.%	Bernard et al. (2020)
rGO/WPU	Emulsion mixing	0.19	0.22	1 wt.%	Wu, Shi, and Zhang (2016)
NH$_2$-rGO/WPU	Emulsion mixing	0.19	0.68	1 wt.%	Wu, Shi, and Zhang (2016)

et al. 2017). Zhang et al., firstly, modified GO with 3-Mercaptopropyltrimethoxysilane (MPTMS) to increase the hydrophilicity. Then, nanocomposite was prepared by dispersing it in NBR using latex compounding method. Compared to unmodified GO, modified GO has stronger interfacial interaction with the polymer matrix through the effect of covalent bonds. This strong interaction between the polymer matrix and the filler material will enable the nanocomposite to effectively transfer stress throughout

TABLE 7.6

Some Studies on Electrical Conductivity of Graphene/Elastomer Nanocomposites

Nanocomposites	Processing	Electrical Conductivity (S cm^{-1})		Filler Content	References
		Pure Rubber	Composite		
G/M/D-NR	Colloidal dispersion	4.56×10^{-6}	7.31×10^{-6}	1.5 phr	George et al. (2017)
PA12/graphene	Melt compounding	2.8×10^{-16}	6.7×10^{-4}	0.3 vol.%	Yan et al. (2012)
MLGS/SBR	Heterocoagulation	4.52×10^{-13} (0.1 wt.%)	4.56×10^{-7}	5 wt.%	Kim et al. (2011)
CRG/NRL	Vacuum-assisted self-assembly	0.04 (10 phr)	1.04	30 phr	Yang et al. (2014)
SBR-TRGO	Solution mixing and coagulation	6.4×10^{-15} (SBR/G)	7.2×10^{-6}	25 phr	Beckert et al. (2014)
Graphene/NR	Self-assembly integrating 1.latex compounding	-	0.0731	4.16 Vol.%	Luo et al. (2014)
rGO/NR	Emulsion mixing	-	0.493	5 vol.%	Dong, Zhang, and Wu (2016)
WPUCL-4	In situ polymerization	8.84×10^{-11}	7.87×10^{-4}	4 phr	Lee et al. (2009)
WPUMG-4	In situ polymerization	1.68×10^{-11}	1.91×10^{-3}	4 phr	Lee et al. (2009)
WPU30F2S6	Colloidal dispersion	6.95×10^{-11}	7.80×10^{-8}	2 phr	Choi et al. (2012)
WPU35F2S6	Colloidal dispersion	6.73×10^{-11}	2.21×10^{-7}	2 phr	Choi et al. (2012)
WPU40F4S12	Colloidal dispersion	7.21×10^{-11}	9.95×10^{-6}	4 phr	Choi et al. (2012)
WPU/FGS	Emulsion mixing	1.34×10^{-10}	2.75×10^{-4}	6 phr	Raghu et al. (2008)
WPU/FGS	Emulsion mixing	1.34×10^{-10}	5.47×10^{-4}	5 phr	Raghu et al. (2008)

the filler network. This improved the mechanical properties of the nanocomposite (Zhang and Cho 2018).

Improving the mechanical performance of graphene/rubber nanocomposites is achieved by the uniform dispersion of graphene in the rubber and the strong interfacial interaction between graphene and rubber. Graphene loading into the rubber matrix should be at optimum rates. For example, as the graphene loading rate increases, the tensile strength and elongation at break decrease due to agglomeration formation and restriction of the movement of polymer chains. Zhang et al. prepared PRGO/NR nanocomposites using the latex heterocoagulation technique with NR and PVP-modified rGO. The tensile strength of PRGO/NR increases by 23% when the PRGO content is increased from 0 to 3 phr. However, when more PRGO is added, the tensile strength decreases due to stress concentration sourced by the increased agglomeration in the polymer matrix (Zhang et al. 2016). Kang et al. prepared GE/NR nanocomposites by direct mechanical mixing method with graphene having high specific surface area obtained by chemical reduction to GO and GO by Hummer method.

The tensile strength reached its highest value at 2 phr GE loading rate. Compared to pure NR, the tensile strength of the GE/NR nanocomposite increased by ~17% (Kang et al. 2017). Xing et al. synthesized graphene/poly(styrene-co-butadiene-costyrene) (GE/SBS) nanocomposites using the modified latex coupling method. Graphene was found to be very well dispersed in the rubber at the molecular level and had a strong interfacial interaction with SBR with this method. The tensile strength increased approximately 11 times with a graphene loading of 7 phr (Xing, Tang, et al. 2014). Graphene/NR nanocomposites were prepared by a modified latex mixing method with in situ chemical reduction by Xing et al. The GE nanolayers were found to be well dispersed and have strong interfacial interaction with NR. Therefore, adding a low GE content can significantly increase the tensile strength and initial tensile modulus of NR. Addition of 0.5 phr graphene into NR resulted in an increase of 48% in tensile strength and 80% in tensile modulus (Xing, Wu, et al. 2014). Table 7.4 presents a literature review regarding the changes in mechanical properties before and after the addition of graphene filler to the elastomer.

One of the properties that are tried to be developed in graphene/elastomer nanocomposites is thermal conductivity. The filling material should be homogeneously dispersed in the polymer matrix and an effective interfacial interaction should occur while improving the thermal conductivity of nanocomposites, because the thermal conduction within the structure is transferred through the interaction of adjacent particles in the form of lattice vibrations (Im and Kim 2012). Phonons are responsible for thermal conduction in amorphous polymers and phonon scattering must be minimized. Filler materials that make strong bonds with the polymer matrix and have low interfacial resistance should be used in order to achieve this. Graphene is one of the ideal nanofiller materials with its high thermal conductivity (5,000 W mK^{-1}) and plate structure that facilitates thermal conduction (Papageorgiou, Kinloch, and Young 2015). Yin et al. primarily functionalized GO with 1-allyl-3-methyl-imidazolium chloride (AMICl), an ionic liquid. Later, nanocomposites were produced using the latex hetero-coagulation technique with GO-IL and SBR. They reported that the thermal conductivity of composite was increased by 34% compared to pure rubber. These results show that there is good interfacial interaction between GO-IL and SBR (Yin et al. 2017). Wen et al. synthesized GO at different oxidation degrees using the Hummers method. Then, GOs were made nanocomposites with carboxylated acrylonitrile-butadiene rubber (XNBR) using the latex co-coagulation technique. As the oxidation degree of GO increased, the degree of hydrophilicity increased. As the degree of hydrophilicity increased, the filling material was finely dispersed in the pure rubber. In this case, the filling material, which is well dispersed in the polymer matrix, increases the thermal conductivity (Wen et al. 2017). Some studies on the thermal conductivity of graphene/elastomer nanocomposites are summarized in Table 7.5.

Electrical conductivity measurements are one of the most important parts of examining the potential properties of graphene/elastomer composites. Elastomer nanocomposites with electrically functional properties are produced by adding electrically conductive fillers to non-conductive or slightly conductive polymer matrices. The main factor in this production is the selection of the ideal filler material. The quality and purity of the filler material is decisive factor in the production of the composite with the desired electrical properties. The ideal dispersion of the filler material provides more paths that require transmission by tunneling with closely spaced filler networks. In some homogenized dispersions, the tunneling condition is prevented due to the coating of the polymer matrix along with the filler network, thus negatively affecting the electrical conductivity. Graphene is an ideal nanofiller material with high surface area and high electrical conductivity due to its natural structure (sp^2 orbitals provide electrical conduction by stacking graphene sheets on top of the plane of the graphene monolayer) (Papageorgiou, Kinloch, and Young 2015). Xing et al. loaded GE into the SBR using the modified latex compounding method. The electrical conductivity of GE/SBR nanocomposites increased significantly as GE loading increased. The electrical conductivity increased until the GE supplementation was 7 phr. This is due to the fact that the reduction to GO is effective and the layers of GE form a conductive network within the polymer matrix (Xing, Tang, et al. 2014). Zhan et al. synthesized GE/NR nanocomposites using different synthesis methods with GE and NR at the same loading rate (1.78 vol.%). They compared the nanocomposites prepared by self-assembly in latex and static hot pressing, and the nanocomposites prepared by traditional methods by taking electrical conductivity measurements. The composite exhibited five times higher electrical conductivity (0.03 S cm^{-1}) than composites made by conventional methods. This increase in electrical conductivity is due to the presence of a conductively separated cross-linked network of GE in the NR polymer matrix prepared by the self-assembly process in latex (Zhan et al. 2012). Some studies on the electrical conductivity of graphene/elastomer nanocomposites are summarized in Table 7.6.

7.5 CONCLUSIONS

The usage areas of elastomers have increased with the increasing consumption due to the rapidly increasing world population and the development of industrial activities. It becomes important to provide diversity in elastomer production with the emergence of different needs requiring different properties. In order to overcome this situation, elastomer nanocomposites started to be produced with different nanofiller materials. Synthesis methods used to produce nanocomposites with desired properties, structural and chemical properties of the filler material are the most important factors. Generally, carbon fillers, inorganic fillers, and biofillers are used depending on the purpose of production. Recently, with the discovery of its superior properties, the use of graphene as a filler material in elastomer nanocomposites is highly preferred. Improving the interfacial interaction of graphene in the polymer matrix provides easy dispersion and homogenization. Modified graphene structures have been developed for this purpose. In this chapter, the dispersion of graphene and graphene-based materials in elastomer nanocomposites was investigated. AFM, SEM, TEM, and XRD characterizations and physical properties (mechanical properties, thermal conductivities, and electrical conductivities) for nanocomposites are summarized. It was observed that the morphological properties of the polymer matrix change, and accordingly, the physical properties are improved and developed, with the incorporation of graphene and modified graphene structures into the elastomer structure, even in very small amounts. Due to these improved properties of graphene/elastomer nanocomposites, it is expected that their use will spread in both traditional and advanced applications in the fields of aviation, tires, automotive components, all kinds of gaskets, hydraulic hoses, shoes, biomedical products, sensing devices, and nanoelectronics.

REFERENCES

Aguilar-Bolados, Héctor, Miguel A Lopez-Manchado, Justo Brasero, F Avilés, and Mehrdad Yazdani-Pedram. 2016. "Effect of the morphology of thermally reduced graphite oxide on the mechanical and electrical properties of natural rubber nanocomposites." *Composites Part B: Engineering* 87:350–356.

Alex, Rosamma. 2010. "Nanofillers in rubber–rubber blends." In *Rubber Nanocomposites: Preparation, Properties, and Applications*, 209–238, (Eds. S. Thomas and R. Stephen). Singapore: John Wiley & Sons.

Anwar, Zanib, Ayesha Kausar, Irum Rafique, and Bakhtiar Muhammad. 2016. "Advances in epoxy/graphene nanoplatelet composite with enhanced physical properties: A review." *Polymer-Plastics Technology and Engineering* 55 (6):643–662.

Araby, Sherif, Qingshi Meng, Liqun Zhang, Hailan Kang, Peter Majewski, Youhong Tang, and Jun Ma. 2014. "Electrically and thermally conductive elastomer/graphene nanocomposites by solution mixing." *Polymer* 55 (1):201–210.

Arguello, JM, and A Santos. 2016. "Hardness and compression resistance of natural rubber and synthetic rubber mixtures." *Journal of Physics: Conference Series* 687 (1):012088.

Beckert, Fabian, Stephanie Trenkle, Ralf Thomann, and Rolf Mülhaupt. 2014. "Mechanochemical route to functionalized graphene and carbon nanofillers for graphene/SBR nanocomposites." *Macromolecular Materials and Engineering* 299 (12):1513–1520.

Bernard, C, DG Goodwin, X Gu, M Celina, M Nyden, D Jacobs, L Sung, and T Nguyen. 2020. "Graphene oxide/waterborne polyurethane nanocoatings: Effects of graphene oxide content on performance properties." *Journal of coatings technology and research* 17 (1):255–269.

Chen, Yang, Qing Yin, Xumin Zhang, Hongbing Jia, Qingmin Ji, and Zhaodong Xu. 2018. "Impact of various oxidation degrees of graphene oxide on the performance of styrene–butadiene rubber nanocomposites." *Polymer Engineering & Science* 58 (8):1409–1418.

Choi, Sang Hyop, Dong Hoon Kim, Anjanapura V Raghu, Kakarla Raghava Reddy, Hyung-Il Lee, Koo Sik Yoon, Han Mo Jeong, and Byung Kyu Kim. 2012. "Properties of graphene/waterborne polyurethane nanocomposites cast from colloidal dispersion mixtures." *Journal of Macromolecular Science, Part B* 51 (1):197–207.

Ciesielski, Andrew. 1999. *An Introduction to Rubber Technology.* Shrewsbury: Rapra Technology Limited.

Das, Santanu, and Wonbong Choi. 2011. "Graphene synthesis." *Graphene: Synthesis and Applications* 3:27–63.

Datta, S. 2001. "Synthetic elastomers." *Rubber Technologist's Handbook* 1:47–74.

Datta, Sudhin. 2004. "Special-purpose elastomers." In *Rubber Compounding*, 83–108, (Ed. B. Rodgers). New York: CRC Taylor & Francis group.

Dong, Bin, Liqun Zhang, and Youping Wu. 2016. "Highly conductive natural rubber–graphene hybrid films prepared by solution casting and in situ reduction for solvent-sensing application." *Journal of Materials Science* 51 (23):10561–10573.

Frasca, Daniele, Dietmar Schulze, Volker Wachtendorf, Christian Huth, and Bernhard Schartel. 2015. "Multifunctional multilayer graphene/elastomer nanocomposites." *European Polymer Journal* 71:99–113.

Gan, Lu, Songmin Shang, Chun Wah Marcus Yuen, Shou-xiang Jiang, and Nicy Mei Luo. 2015. "Facile preparation of graphene nanoribbon filled silicone rubber nanocomposite with improved thermal and mechanical properties." *Composites Part B: Engineering* 69:237–242.

George, Gejo, Suja Bhargavan Sisupal, Teenu Tomy, Bincy Akkoli Pottammal, Alaganandam Kumaran, Vemparthan Suvekbala, Rajmohan Gopimohan, Swaminathan Sivaram, and Lakshminarayanan Ragupathy. 2017. "Thermally conductive thin films derived from defect free graphene-natural rubber latex nanocomposite: Preparation and properties." *Carbon* 119:527–534.

He, Daping, Haolin Tang, Zongkui Kou, Mu Pan, Xueliang Sun, Jiujun Zhang, and Shichun Mu. 2017. "Engineered graphene materials: Synthesis and applications for polymer electrolyte membrane fuel cells." *Advanced Materials* 29 (20):1601741.

Im, Hyungu, and Jooheon Kim. 2012. "Effect of homogeneous Al (OH) 3 covered MWCNT addition on the thermal conductivity of Al_2O_3/epoxy-terminated poly (dimethylsiloxane) composites." *Journal of Materials Science* 47 (16):6025–6033.

Kang, Hailan, Kanghua Zuo, Zhao Wang, Liqun Zhang, Li Liu, and Baochun Guo. 2014. "Using a green method to develop graphene oxide/elastomers nanocomposites with combination of high barrier and mechanical performance." *Composites Science and Technology* 92:1–8.

Kang, Hailan, Yinyin Tang, Lei Yao, Feng Yang, Qinghong Fang, and David Hui. 2017. "Fabrication of graphene/natural rubber nanocomposites with high dynamic properties through convenient mechanical mixing." *Composites Part B: Engineering* 112:1–7.

Kim, Jin Sil, Ju Ho Yun, Il Kim, and Sang Eun Shim. 2011. "Electrical properties of graphene/ SBR nanocomposite prepared by latex heterocoagulation process at room temperature." *Journal of Industrial and Engineering Chemistry* 17 (2):325–330.

Lee, Yu Rok, Anjanapura V Raghu, Han Mo Jeong, and Byung Kyu Kim. 2009. "Properties of waterborne polyurethane/functionalized graphene sheet nanocomposites prepared by an in situ method." *Macromolecular Chemistry and Physics* 210 (15):1247–1254.

Li, Xiaoyu, Hua Deng, Zhen Li, Hao Xiu, Xiaodong Qi, Qin Zhang, Ke Wang, Feng Chen, and Qiang Fu. 2015. "Graphene/thermoplastic polyurethane nanocomposites: Surface modification of graphene through oxidation, polyvinyl pyrrolidone coating and reduction." *Composites Part A: Applied Science and Manufacturing* 68:264–275.

Li, Xuheng, Linbo Shao, Na Song, Liyi Shi, and Peng Ding. 2016. "Enhanced thermal-conductive and anti-dripping properties of polyamide composites by 3D graphene structures at low filler content." *Composites Part A: Applied Science and Manufacturing* 88:305–314.

Li, Ying, Fan Xu, Zaishan Lin, Xianxian Sun, Qingyu Peng, Ye Yuan, Shasha Wang, Zhiyu Yang, Xiaodong He, and Yibin Li. 2017. "Electrically and thermally conductive underwater acoustically absorptive graphene/rubber nanocomposites for multifunctional applications." *Nanoscale* 9 (38):14476–14485.

Lim, Lai Peng, Joon Ching Juan, Nay Ming Huang, Leng Kian Goh, Fook Peng Leng, and Yi Yee Loh. 2019. "Enhanced tensile strength and thermal conductivity of natural rubber graphene composite properties via rubber-graphene interaction." *Materials Science and Engineering: B* 246:112–119.

Luo, Yongyue, Pengfei Zhao, Qi Yang, Dongning He, Lingxue Kong, and Zheng Peng. 2014. "Fabrication of conductive elastic nanocomposites via framing intact interconnected graphene networks." *Composites Science and Technology* 100:143–151.

Mall, Jun, Li-Qun Zhang, and Li Geng. 2010. "Manufacturing techniques of rubber nanocomposites." *Rubber Nanocomposites: Preparation, Properties, and Applications*:21.

Manna, Rakesh, and Suneel Kumar Srivastava. 2017. "Fabrication of functionalized graphene filled carboxylated nitrile rubber nanocomposites as flexible dielectric materials." *Materials Chemistry Frontiers* 1 (4):780–788.

Mathew, NM. 2001. "Natural rubber." In *Rubber Technologist's Handbook. Shawbury, Shrewbury, Shropshire: Rapra Technology Limited*, 11–46, (Eds. Sadhan K. De and Jim R. White). Exeter: Polestar Scientifica.

Mensah, Bismark, Kailash Chandra Gupta, Hakhyun Kim, Wonseok Wang, Kwang-Un Jeong, and Changwoon Nah. 2018. "Graphene-reinforced elastomeric nanocomposites: A review." *Polymer Testing* 68:160–184.

Mensah, Bismark, Sungjin Kim, Sivaram Arepalli, and Changwoon Nah. 2014. "A study of graphene oxide-reinforced rubber nanocomposite." *Journal of Applied Polymer Science* 131 (16):1–9.

Moni, Grace, Anshidha Mayeen, Amalu Mohan, Jinu Jacob George, Sabu Thomas, and Soney C George. 2018. "Ionic liquid functionalised reduced graphene oxide fluoroelastomer nanocomposites with enhanced mechanical, dielectric and viscoelastic properties." *European Polymer Journal* 109:277–287.

Niyogi, Utpal Kumar. 2007. "Natural and synthetic rubber." www.niscair.res.in

Oliveira, Ana Elisa Ferreira, Guilherme Bettio Braga, César Ricardo Teixeira Tarley, and Arnaldo César Pereira. 2018. "Thermally reduced graphene oxide: Synthesis, studies and characterization." *Journal of Materials Science* 53 (17):12005–12015.

Papageorgiou, Dimitrios G, Ian A Kinloch, and Robert J Young. 2015. "Graphene/elastomer nanocomposites." *Carbon* 95:460–484.

Potts, Jeffrey R, Om Shankar, Ling Du, and Rodney S Ruoff. 2012. "Processing–morphology–property relationships and composite theory analysis of reduced graphene oxide/natural rubber nanocomposites." *Macromolecules* 45 (15):6045–6055.

Raghu, Anjanapura V, Yu Rok Lee, Han Mo Jeong, and Cheol Min Shin. 2008. "Preparation and physical properties of waterborne polyurethane/functionalized graphene sheet nanocomposites." *Macromolecular Chemistry and Physics* 209 (24):2487–2493.

Sanna, Roberta. 2014. "Synthesis and characterization of new polymeric materials for advanced applications." (Doctoral dissertation). University of Sassari, Sassari, Italy.

Sisanth, KS, MG Thomas, J Abraham, and S Thomas. 2017. "General introduction to rubber compounding." In *Progress in Rubber Nanocomposites*, 1–39, (Ed. Sabu Thomas). Woodhead Publishing, Elsevier. Duxford, United Kingdom

Song, K. 2017. "Micro-and nano-fillers used in the rubber industry." In *Progress in Rubber Nanocomposites*, Woodhead Publishing Series in Composites Science and Engineering, 41–80. (Eds. S. Thomas and H. J. Maria). Elsevier.

Stephen, Ranimol, and Sabu Thomas. 2010. "Nanocomposites: State of the art, new challenges and opportunities." *Rubber Nanocomposites-Preparation, Properties, and Applications*:1–19. Singapore: John Wiley & Sons (Asia) Pte Ltd.

Tao, Wenjie, Shaohua Zeng, Ying Xu, Wangyan Nie, Yifeng Zhou, Pengbo Qin, Songhua Wu, and Pengpeng Chen. 2021. "3D Graphene–sponge skeleton reinforced polysulfide rubber nanocomposites with improved electrical and thermal conductivity." *Composites Part A: Applied Science and Manufacturing* 143:106293.

Tian, Ming, Jing Zhang, Liqun Zhang, Suting Liu, Xiaoqing Zan, Toshio Nishi, and Nanying Ning. 2014. "Graphene encapsulated rubber latex composites with high dielectric constant, low dielectric loss and low percolation threshold." *Journal of Colloid and Interface Science* 430:249–256.

Wang, Jingyi, Hongbing Jia, Yingying Tang, Dandan Ji, Yi Sun, Xuedong Gong, and Lifeng Ding. 2013. "Enhancements of the mechanical properties and thermal conductivity of carboxylated acrylonitrile butadiene rubber with the addition of graphene oxide." *Journal of Materials Science* 48 (4):1571–1577.

Wen, Yanwei, Qing Yin, Hongbing Jia, Biao Yin, Xumin Zhang, Pengzhang Liu, Jingyi Wang, Qingmin Ji, and Zhaodong Xu. 2017. "Tailoring rubber-filler interfacial interaction and multifunctional rubber nanocomposites by usage of graphene oxide with different oxidation degrees." *Composites Part B: Engineering* 124:250–259.

Wu, Shengli, Tiejun Shi, and Liyuan Zhang. 2016. "Preparation and properties of amine-functionalized reduced graphene oxide/waterborne polyurethane nanocomposites." *High Performance Polymers* 28 (4):453–465.

Wu, Jinrong, Wang Xing, Guangsu Huang, Hui Li, Maozhu Tang, Siduo Wu, and Yufeng Liu. 2013. "Vulcanization kinetics of graphene/natural rubber nanocomposites." *Polymer* 54 (13):3314–3323.

Xie, Zheng-Tian, Xuan Fu, Lai-Yun Wei, Ming-Chao Luo, Yu-Hang Liu, Fang-Wei Ling, Cheng Huang, Guangsu Huang, and Jinrong Wu. 2017. "New evidence disclosed for the engineered strong interfacial interaction of graphene/rubber nanocomposites." *Polymer* 118:30–39.

Xing, Wang, Maozhu Tang, Jinrong Wu, Guangsu Huang, Hui Li, Zhouyue Lei, Xuan Fu, and Hengyi Li. 2014. "Multifunctional properties of graphene/rubber nanocomposites fabricated by a modified latex compounding method." *Composites Science and Technology* 99:67–74.

Xing, Wang, Jinrong Wu, Guangsu Huang, Hui Li, Maozhu Tang, and Xuan Fu. 2014. "Enhanced mechanical properties of graphene/natural rubber nanocomposites at low content." *Polymer International* 63 (9):1674–1681.

Yan, Dong, Hao-Bin Zhang, Yu Jia, Juan Hu, Xian-Yong Qi, Zhong Zhang, and Zhong-Zhen Yu. 2012. "Improved electrical conductivity of polyamide 12/graphene nanocomposites with maleated polyethylene-octene rubber prepared by melt compounding." *ACS Applied Materials & Interfaces* 4 (9):4740–4745.

Yang, Hongsheng, Ping Liu, Tongping Zhang, Yongxin Duan, and Jianming Zhang. 2014. "Fabrication of natural rubber nanocomposites with high graphene contents via vacuum-assisted self-assembly." *RSC Advances* 4 (53):27687–27690.

Yin, Biao, Xumin Zhang, Xun Zhang, Jingyi Wang, Yanwei Wen, Hongbing Jia, Qingmin Ji, and Lifeng Ding. 2017. "Ionic liquid functionalized graphene oxide for enhancement of styrene-butadiene rubber nanocomposites." *Polymers for Advanced Technologies* 28 (3):293–302.

Zhan, Yanhu, Marino Lavorgna, Giovanna Buonocore, and Hesheng Xia. 2012. "Enhancing electrical conductivity of rubber composites by constructing interconnected network of self-assembled graphene with latex mixing." *Journal of Materials Chemistry* 22 (21):10464–10468.

Zhan, Yanhu, Jinkui Wu, Hesheng Xia, Ning Yan, Guoxia Fei, and Guiping Yuan. 2011. "Dispersion and exfoliation of graphene in rubber by an ultrasonically-assisted latex mixing and in situ reduction process." *Macromolecular Materials and Engineering* 296 (7):590–602.

Zhang, Yinhang, and Ur Ryong Cho. 2018. "Enhanced thermo-physical properties of nitrile-butadiene rubber nanocomposites filled with simultaneously reduced and functionalized graphene oxide." *Polymer Composites* 39 (9):3227–3235.

Zhang, Xumin, Jingyi Wang, Hongbing Jia, Shiyu You, Xiaogang Xiong, Lifeng Ding, and Zhaodong Xu. 2016. "Multifunctional nanocomposites between natural rubber and poly-vinyl pyrrolidone modified graphene." *Composites Part B: Engineering* 84:121–129.

Zhang, Chunmei, Tianliang Zhai, Yi Dan, and Lih-Sheng Turng. 2017. "Reinforced natural rubber nanocomposites using graphene oxide as a reinforcing agent and their in situ reduction into highly conductive materials." *Polymer Composites* 38:E199–E207.

8 Graphene-Based Hybrid Fillers as New Reinforcing Agents in Rubber Compounds for the Tire Industry

Bikshan Ghosh and Sourav Paul
IIT Kharagpur

Saptarshi Kar and Ranjan Ghosal
Birla Carbon India Private Limited

Amrita Roy and Titash Mondal
IIT Kharagpur

Anil K. Bhowmick
The University of Houston

CONTENTS

DOI: 10.1201/9781003200444-8

199

8.1   INTRODUCTION

In recent times, polymer nanocomposites have garnered significant attention due to several exceptional properties (Roy, Sengupta, and Bhowmick 2012). Development of elastomeric nanocomposite based on graphite derivatives and graphene is gaining widespread popularity among the community of polymer science and technology due to the unique properties exhibited by these nanofillers (Ajayan 1999). Incorporation of nanofiller in a polymer matrix creates a polymer-filler interface, leading to unique polymer-filler interaction; it plays an important role in controlling properties of the nanocomposites prepared thereof (Mondal, Basak, and Bhowmick 2017). Recently, graphene has become the rising star because of its excellent characteristics of electron mobility, thermal conductivity, mechanical stiffness, strength and elasticity (McCreery 2008).

The unique nanostructure of graphene enables it to possess many desired properties and promising applications (Blake et al. 2008). Graphene is a zero band-gap semi-conductor with exorbitant carrier mobility of up to $200,000\,cm^2\,VS^{-1}$ (Li et al. 2008). The optical transmittance of monolayer graphene can be as high as 98%. This leads graphene to be an excellent electronic material such as field emission transistors (Wang et al. 2009). In fact, the high aspect ratio and strong sp^2 C=C bonds of graphene show excellent mechanical properties (Guoxiu et al. 2008). Young's modulus for a single layer of graphene is reported to be 0.5–1 TPa by atomic force microscopy measurement. Large spring constant of 1–5 Nm^{-1} has also been reported for graphene. High order of stiffness in graphene is attributed to the strong C=C bond dispersed over a broad area (Dreyer et al. 2010). Zhang et al. prepared styrene butadiene rubber (SBR) composites with graphene oxide and found that it demonstrated much better mechanical properties as compared to the carbon black filled samples (Zhang, Wang, and Zhang 2015). Addition of graphene to a polymer matrix often improves several mechanical and functional properties of the composite (George and Bhowmick 2008). However, this observation holds good at a low filler loading. At higher loading of graphene, the enhanced properties displayed by polymer nanocomposite either remain constant or start showing negative effect (Bai et al. 2011). The probable reason might relate to the fact that at higher loading, large filler aggregates are formed inside a polymer matrix leading to phase separation, which results in an inferior performance of composites.

Arduous efforts have been applied to solve the problem of agglomeration and efficient use of nanofillers. Graphite and its derivatives are used along with carbon black (hybrid filler system) and are mixed with an elastomer using the compounding techniques commonly practiced in the industries (two roll mixing) (Zhang, Wang, and Zhang 2015). However, very few literatures are available on the use of graphene-carbon black hybrid filler system (Song, Jeong, and Kang 2010). As tires arguably classify as a high-performance composite product, it is crucial to have an optimum balance of the magic triangle (rolling resistance, traction including ice traction as well as wet traction, and abrasion resistance) without compromising on dry handling for the fabrication of a tire. The present study attempts to address the issue and focuses on studying the reinforcing effect and its mechanism of the graphene-carbon black dual filler system (hybrid filler system) on the polymer matrix.

Natural rubber (NR) and SBR as the polymers forming the matrix and N234 grade carbon black as filler have been chosen for the study. Four different graphenic materials – Graphene oxide (GO), Reduced Graphene Oxide (RGO), Graphene nanoplatelets (GP1 and GP2) – have been taken in this study for complete evaluation in NR and SBR compounds. Pristine nanofillers at high loading do not impart significant properties as evident from our previous studies, and hence, hybrid fillers are used in the present study. We have examined four different aspects, i.e. the effect of nature of graphene in hybrid filler system, the effect of loading of graphene in hybrid filler system, the effect of loading of carbon black keeping graphene level constant and the effect of nature of rubber by using hybrid filler system. The effect of nature of graphene was shown at 45 phr of total loading keeping graphene loading at 5 phr. A reference compound with 45 phr of N234, without graphene, was also prepared. The effect of loading of graphene was shown at 45 phr of total carbon black loading, varying graphene loading at 1, 3, 5 and 7 phr levels. A comparative study between NR and SBR filled compound is also shown here at 45 phr of total loading, keeping graphene loading at 5 phr. The microstructure of the hybrid filler system was probed using transmission electron microscopic facility. In conjugation with the microscopic techniques, the reinforcing nature of the hybrid filler system was analyzed by network formation of carbon black in the presence and absence of graphene in NR/SBR nanocomposite sample. Interestingly, the nanocomposite when examined for its dynamic mechanical properties demonstrated a unique balance of the tire properties. All the above tests were compared and contrasted with carbon black filled NR/SBR (reference material). For comparison, graphene reinforcement in silica filled rubber has been discussed. A mechanism of reinforcement by the hybrid filler has been proposed.

8.2 EXPERIMENTAL SECTION

8.2.1 MATERIALS

NR latex (DRC 60%) was obtained from Harrisons Malayalam Ltd. (Kochi, India). N234 carbon black (surface area-119 m^2g^{-1}, DBP Absorption-125 $cm^3/100g$) was obtained from Birla Carbon (Mumbai, India). GO and RGO were obtained from United nanotech India Pvt. Ltd. (Bangalore, India). GP1 and GP2 were obtained from internal sources of Birla Carbon (USA). The characteristics of the fillers are given in Table 8.1. All other chemicals – Zinc oxide, Stearic acid and 2-mercaptobenzothiazole disulfide (MBTS) – were obtained from Fisher Scientific Ltd. (Massachusetts, USA) and Sulfur was obtained from Loba Chemie (Mumbai, India). The chemicals were used as received.

TABLE 8.1
Physical Characteristics of the Graphene Fillers

Graphene Used	Polarity	Lateral Dimension (μm)	Thickness (nm)	Surface Area (m²g⁻¹)	Bulk Density (g cm⁻³)
GO	Polar	5–10	3–6	110	0.5
RGO	Slightly polar	~10	0.8–2	150	0.4
GP1	Non polar	<2	5–10	750	0.2–0.4
GP2	Non polar	<2	5–10	500	0.2–0.4

8.2.2 METHODOLOGY

In order to establish the effect of loading of carbon black keeping graphene level constant, batches at 30, 45 and 60 phr total filler loading were made at 5 phr replacement by various graphenes in NR systems. To establish the effect of nature of graphene in hybrid filler system, batches were made at 45 phr of total loading keeping graphene replacement level at 5 phr in the rubber systems. For investigation of the effect of loading of graphene in hybrid filler system, batches were made at 45 phr of total filler loading at 1/3/5/7 phr replacement level of graphenes in the NR systems. For SBR, 5 phr graphene was used replacing 5 phr carbon black in 45 phr carbon black filled system.

The mixing of the fillers with the rubbers was performed by a Birla Carbons' proprietary wet process. The final compounding was done in a lab scale open two roll mill (Santosh Engg. Pvt. Ltd. Mumbai, India) at room temperature for 8 minutes. The ASTM formulation for mixing is given in Table 8.2. The compounded samples were molded in a hydraulic compression press (supplied by MonTech, Germany) at a pressure of 5 MPa. The specimens were conditioned at room temperature for 16 hours before carrying out the testing for further characterization. Molding conditions are given in Table 8.3.

8.2.3 CHARACTERIZATION OF DIFFERENT COMPOSITES

Moving die rheometer (MDR) test (ASTM D5289): MDR is a standard instrument for studying the cure characteristics of a rubber compound. This instrumental

TABLE 8.2
Typical Recipe for Mixing (ASTM D3192)

Ingredient	phr
Masterbatch	150
ZnO	5
Stearic acid	3
MBTS	0.6
Sulfur	2.5

TABLE 8.3

Molding Conditions (Including ASTM Specification No. and Parameters)

Mold	Specimen Type	ASTM	Specimen Size (mm) L×B×H, Φ×H	Input wt. Per Sample (gm)	Temp (°C)	Time (min)
Tensile	Square slab	D412	150×50×2	60–61	160	$2 \times t_c$ 90
Abrasion	Cylindrical button	D5963	Φ16×6	2.5–2.75	145	60
Resilience	Cylindrical button	D7121	Φ45×12.2	25–26	145	40
Heat build-up (HBU)	Cylindrical button	D623	Φ17.8×25.4	8–8.5	145	60

set-up was rotorless; the lower die underwent oscillation. The torque generated by the rubber compound was recorded at constant oscillation frequency and amplitude at constant temperature. From the torque-time curve generated by software, scorch time, optimum cure time, cure rate, maximum torque and minimum torque were obtained at 160°C.

Rubber process analyzer (RPA) test (ASTM D5289): The test was done at 160°C temperature and 1.7 Hz frequency for 30 minutes. This test was based on measuring torque or stress while applying strain. In this test, storage modulus (G′) and loss modulus (G″) were obtained in strain sweep (from 0.5% strain to 25% strain), and from G′ values, the Payne effect was measured.

Bound rubber content: The bound rubber content (BdR) is the polymer portion which cannot be extracted by the corresponding solvent due to the adsorption of filler onto the polymer chains. BdR was measured by extracting unbound polymer from toluene solvent after 7 days at room temperature. The solvent was renewed every 24 hours to keep the solvation action constant. Thereafter, the samples were removed from the toluene and dried at 90°C for 24 hours. The samples were weighed before and after the extraction. BdR was calculated according to Eq. 8.1. BdR indicates the extent of polymer-filler interaction.

$$BdR = 100 \times \frac{W_{fg} - W_t \left[m_f / \left(m_f + m_r \right) \right]}{W_t \left[m_r / \left(m_f + m_r \right) \right]} \quad (8.1)$$

where W_{fg} and W_t are the weight of the filler-gel and the sample. m_f and m_r are the weight of the filler and the rubber in the compound. The average value of five measurements is reported.

Hardness test (ASTM D2240): This test was done in SHORE A hardness tester following ASTM D2240 using Wallace hardness tester (Wallace Instruments, Wells Place, England).

Tensile test (ASTM D412): This test was done in a Universal testing machine by INSTRON (USA) and the data were analyzed by Bluehill software. Gauge length was 25 mm in this test. For the test, dumbbell-shaped specimen was cut and fixed in between the grips and tensile force was applied until the specimen broke. 100%,

200% and 300% modulus, tensile strength and elongation at break (%) were calculated from the generated curves. A total of five samples were put to test for reporting each data.

Tear test (ASTM D 624): ANGLE-shaped notched specimen was cut using ASTM Die 'C'. Samples were fixed in between the grips without extensometer in an Instron Machine (USA) Gauge length was 40 mm in this test. Tear force was then applied to the specimen until it broke. Maximum load and tear strength values were then noted down. A total of five samples were tested to report each data.

Rebound resilience (ASTM D7121): This test was done at room temperature in a Zwick resilience tester (Zwick Roell, Ulm, Germany). Initially, an angle of 5° was set, and after fixing the sample in grip, six rebounds were done and final reading was noted as rebound resilience (in %).

Heat build-up (ASTM D623-A): In this test done at 70°C in a Goodrich flexometer, sample was first kept inside the chamber at 70°C for 30 minutes. Specimen was having 17.8 mm diameter and 25.4 mm height. Operating frequency and dynamic stroke range were 30 Hz and 4.45 mm respectively.

Swelling test (ASTM D3616-95): This test was done for swelling index and cross-link density measurement. In this test, 0.4 gm of the sample was cut into square shape and dissolved in 100 mL toluene for duration of 22 hours in dark environment. After 22 hours, the sample was taken out from beaker and swollen weight was taken. Now the sample was kept in an oven at 70°C for 3 hours and the sample was taken away from the oven and deswollen weight was taken. The following parameters were calculated from this test:

$$\text{Swelling index} = \frac{\text{swollen weight}}{\text{initial weight}} \qquad (8.2)$$

The extent of swelling was determined by the solvent power of the liquid and crosslink density. The higher the crosslink density, the lower was the swelling. This relationship is expressed quantitatively by Flory–Rehner equation:

$$1/M_c = -\tfrac{1}{2}\rho V_0 \left[\left(\ln(1 - V_R) + V_R + \mu V_R^2 \right) / \left(V_R^{1/3} - \tfrac{1}{2} V_R \right) \right] \qquad (8.3)$$

The crosslink density is given by $1/2M_C$.

Here, ρ is the density of rubber; μ is the solvent-rubber interaction parameter (normally 0.43 for NR); V_0 is the molecular volume of the solvent (toluene) absorbed; V_R is the volume fraction of the rubber in the swollen material and was calculated using the values of sample weight, swollen weight, deswollen weight and density of sample and solvent. If the crosslink density is more, the swelling is less.

Dynamic mechanical analysis: DMA was carried out at temperature sweep from −30°C to 100°C by using (EPLEXOR 500, GABO, TESTANLAGEN GMBH, Germany). DMA can be used in predicting the performance of a tire at different conditions. Tan δ and E′ values at different temperatures were used to predict the performance of tires. Specifications of DMA test were as follows: Gauge length: 11 mm, Static strain: 10%, Dynamic strain: 5%, Frequency: 11 Hz and Sample thickness: 2 mm.

Transmission electron microscopy: The dispersion of the filler material in the polymer matrix was probed using the FEI transmission electron microscope (TEM) with an operating voltage of 120 kV. The composite samples were prepared using an ultra-cryo-microtomy at −80°C and the microtomed samples were placed over 400 mesh size copper grid. The thickness of the sample was ca. 50 nm.

(Note: For the last sections of this chapter, 8.3.6 and 8.3.7, readers may refer directly to the published papers for details of the measurements. The references are included in the sections.)

8.3 RESULTS AND DISCUSSION

8.3.1 Effect of Loading of Carbon Black Keeping Graphene Level Constant

Figures 8.1–8.3 demonstrate the effect of various graphenes on the rubber properties with respect to carbon black loaded batches. It has been observed that though at 30 and 45 phr loadings the overall improvement in tensile strength was prominent with some graphenes, it deteriorated at 60 phr level. This is due to the fact that at 60 phr filler loading, the fillers act like a diluting agent in the rubber matrix. Thus, due to the dilution effect after a certain loading level (i.e. after 45 phr), the tensile strength decreases, more so with mixed fillers. At 45 phr loading, GP2 exhibited highest tensile strength due to the better reinforcement of GP2 with the rubber matrix. At 30 and 45 phr loadings, GO and GP2 showed much improved modulus than the carbon black, whereas at 60 phr level, the effect was not so prominent in the case of GP2.

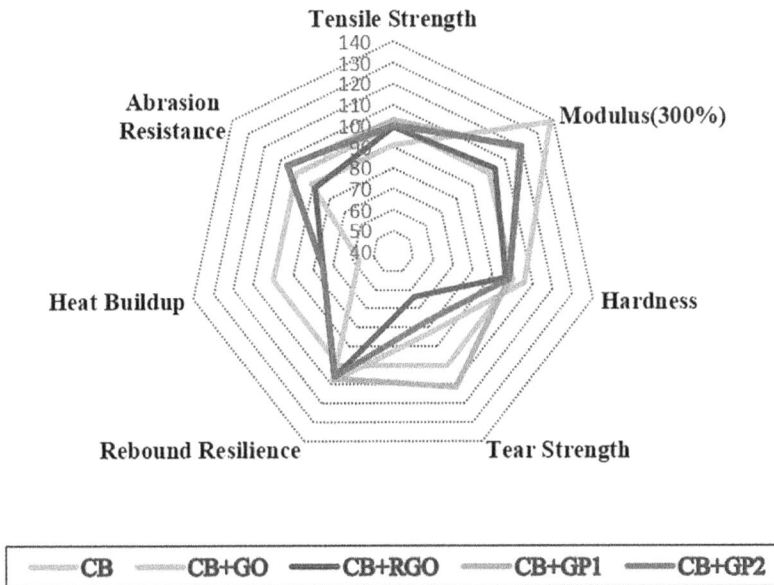

FIGURE 8.1 Properties at 30 phr loading level at 5 phr replacement of graphene.

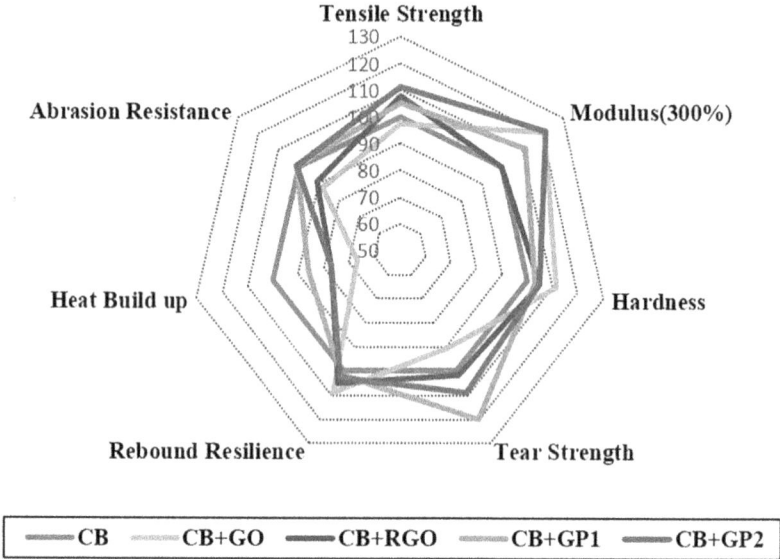

FIGURE 8.2 Properties at 45 phr loading level at 5 phr replacement of graphene.

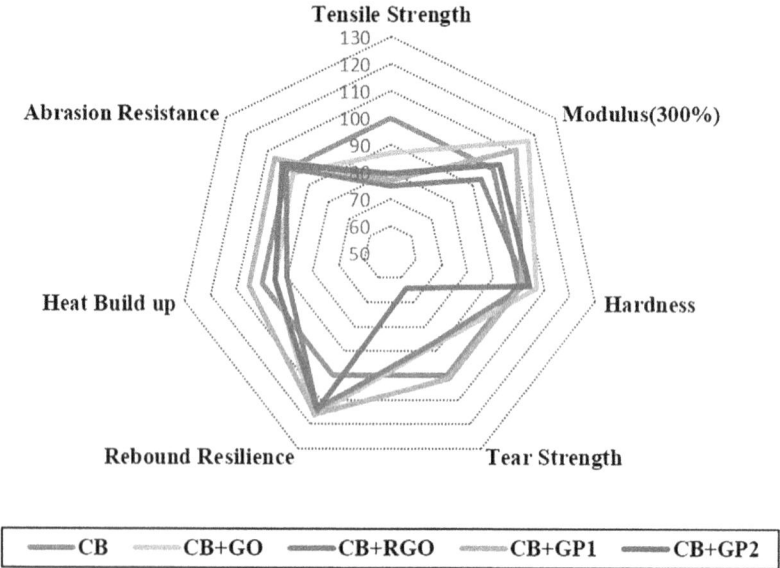

FIGURE 8.3 Properties at 60 phr loading level at 5 phr replacement of graphene.

(Carbon black filled batch has been taken as 100 value in the chart and then percentage improvement of Graphene modified batches with respect to pure carbon black filled batches has been shown.) Overall, GO displayed the highest modulus. Owing to the plate-like structure and larger lateral dimension of GO compared to the

other three graphenes, it has been observed that GO exhibits different types of reinforcement with the rubber, which makes GO compound stiffer. In the case of 30 and
60 phr loadings, tear strength was not improved compared to the reference batches
except GP1 at 30 phr loading. But at 45 phr loading, the tear strength of RGO, GP1
and GP2 was much better than the reference batches.

Resilience is the ratio of energy of indenter after the impact to the energy before
impact, expressed as percentage. It has been shown that for every defined filler loading (30, 45 and 60 phr), the rebound resilience improved. In general, the rebound
resilience was much better for GO and GP1.

In Figures 8.1 and 8.2, the improvement of heat build-up due to the replacement of
carbon black by graphene was very prominent in the case of 30 and 45 phr loadings,
whereas at 60 phr loading (in Figure 8.3), an adverse effect was observed with GP1.
With increasing filler loading, the heat build-up increased probably because of the
increased filler-filler interaction and subsequently the frictional properties.

The process of abrasion involves the removal of small particles leaving behind pits
in the surface followed by the removal of large particles. Detachment of small particles plays an important role in initiating the abrasion, which is related to the localized stresses in the rubber. Since abrasion is clearly a manifestation of mechanical
failure, Schallamach (1968) used tearing energy to describe the rubber wear mechanisms and established an equation in which the wear loss was related to macroscopic
mechanical properties such as tensile strength and elongation at break. It has been
observed that GO and RGO showed inferior abrasion resistance as compared to GP1
and GP2. GO and RGO consist of much larger flakes, which abrade in higher quantity during abrasion testing.

So, in this part of the experiments, an optimum loading level has been established
keeping graphene replacement level at 5 phr. Considering the results of the previous
experiments, it has been seen that at 60 phr loading level, percentage improvement
of properties was lower than that at 30 and 45 phr loadings. This is due to the fact
that some kind of dilution took place at 60 phr loading and the reinforcing effect of
graphene filler was not prominent. For industrial applications, 30 phr loading is low.
Thus, taking all these parameters into consideration, 45 phr loading has been chosen
as the optimized level and considered for further study.

8.3.2 Effect of Nature of Graphene in Hybrid Filler System

From the spider chart in Figure 8.2, performance of each graphene at 5 phr replacement level is understood. Pure carbon black filled batch has been taken as 100 value
in the chart and then percentage improvement of graphene modified batches with
respect to reference carbon black filled batches is shown.

Figure 8.2 demonstrates the higher modulus, tensile strength and abrasion resistance shown by GP2. Higher rebound resilience and hardness were shown by GO,
although GP1 and GP2 mixed compounds were also better on those properties with
respect to the reference CB mixed compound. GP1 displayed very good tear strength
and abrasion resistance. It is evident from Figure 8.2 that GP2 exhibited optimum
improvement of properties when overall performance was considered.

DMA was carried out at temperature sweeps from −30°C to 100°C. DMA could be used to predict the performance of a tire compound under different conditions. tan δ and E′ values at different temperatures are used for such predictions about all of these composites. The results are shown in Table 8.4.

Ice traction, dry traction, wet traction, wet skid resistance and rolling resistance have been predicted from the DMA plots. Much improved ice traction was observed in the graphene modified compounds compared to the carbon black sample. This improvement was much prominent in GP1 and GP2. An improved wet traction was also seen in GP1 and GP2 loaded batches. A correlation between traction, rolling resistance and abrasion loss is shown in Figure 8.4. Generally, on increasing rolling resistance, traction is sacrificed. But here in GP1 and GP2, much improved rolling resistance was observed with improved ice and wet tractions keeping abrasion loss minimum. With these observations, a correlation between these three properties was established.

In order to get further insight, filler-filler interaction was studied using the RPA data. Elastic modulus of the filled rubber is strongly dependent on the deformation and decreases substantially at higher strains, which is caused by filler-filler interaction. This is due to the presence and breakdown of the filler networks on dynamic deformation. The sigmoidal decrease of the storage modulus at zero amplitude (G_o) to the storage modulus at high amplitude plateau (G_∞) was interpreted by Payne due to breakage of the physical bonds between the filler particles.

$$\text{Payne effect} = \left(G_o - G_\infty \right) \tag{8.4}$$

With the increase of structure, filler-filler interaction increases, and hence, there is a rise in the loss of energy in the form of heat. Payne effect values gradually increased from 30, 45 and 60 phr loadings due to increase in total filler loading, as the total filler-filler interaction increased. Interestingly, the values in the case of graphene loaded batches were much lower than the reference batches, i.e. CB loaded batches in each loading level, due to low filler-filler interaction. A representative table is shown in Table 8.5. At higher frequency, Payne effect was lower than that at lower frequency, but the graphene fillers still reduced the Payne effect.

In order to explain the results in the previous section, crosslink density, swelling index and dispersion were investigated. Much higher crosslink density was observed for GO and GP2 with respect to carbon black filled sample (Table 8.6). Hence, lower amount of the solvent was entrapped in between the chains. This resulted in slightly lower swelling index.

Significantly higher IFM dispersion values are shown for GP1 and GP2 loaded batches than the CB loaded batch, which indicates much better reinforcement of GP1 and GP2. As we know that GP1 and GP2 have higher surface area (750 and 500 $m^2 g^{-1}$ respectively), these interact uniformly with the rubber. On the other hand, inferior dispersion than carbon black was observed in GO and RGO due to its high lateral dimension and low surface area (surface area −110 and 150 $m^2 g^{-1}$).

The bound rubber content was much better in graphene filled compounds (Table 8.7). Basically, GP1 and GP2 displayed improved BdR than the other graphenes. Higher BdR contents for all four graphene containing batches than the

TABLE 8.4

Tire Performance Prediction from DMA

Property	Ice Traction	Wet Traction	Dry Traction	Wet Skid Resistance	Rolling Resistance
	Higher Value is Better	Higher Value is Better	Higher Value is Better	Higher Value is Better	Lower Value is Better
Condition	$\tan \delta$ @ $-10°C$	$\tan \delta$ @ $10°C$	$\tan \delta$ @ $30°C$	E' @ $30°C$, MPa	$\tan \delta$ @ $60°C$
NR + CB (45 phr)	0.278	0.131	0.097	19.21	0.087
NR + CB (40 phr) + GO (5 phr)	0.292	0.130	0.098	19.87	0.073
NR + CB (40 phr) + RGO (5 phr)	0.291	0.128	0.086	17.10	0.071
NR + CB (40 phr) + GP1 (5 phr)	0.314	0.140	0.098	17.20	0.078
NR + CB (40 phr) + GP2 (5 phr)	0.316	0.141	0.096	18.10	0.077

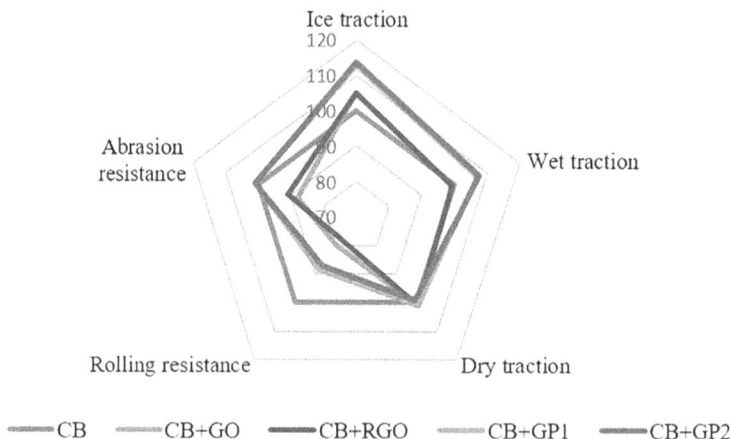

FIGURE 8.4 DMA-assisted tire prediction properties at 45 phr loading level at 5 phr replacement of graphene.

TABLE 8.5
Comparison of Payne Effect Among the 45 phr Loading Batches (Strain Rate 0.5%–25%)

Sample Name	$\Delta G'$ @ 1 Hz	$\Delta G'$ @ 10 Hz
NR + CB	1,498	1,412
NR + CB (40 phr) + GO (5 phr)	1,295	1,199
NR + CB (40 phr) + RGO (5 phr)	1,200	1,027
NR + CB (40 phr) + GP1 (5 phr)	1,343	1,138
NR + CB (40 phr) + GP2(5 phr)	1,297	1,140

TABLE 8.6
Results of Crosslink Density, Swelling Index and IFM Dispersion

Sample Name	Cross-Link Density (mmol cc⁻¹)	Swelling Index	IFM Dispersion (%)
NR + CB (45 phr)	0.92×10^{-3}	2.99	90.7
NR + CB (40 phr) + GO (5 phr)	1.13×10^{-3}	2.79	86.0
NR + CB (40 phr) + RGO (5 phr)	0.88×10^{-3}	3.03	72.4
NR + CB (40 phr) + GP1 (5 phr)	0.93×10^{-3}	2.98	94.0
NR + CB (40 phr) + GP2 (5 phr)	0.99×10^{-3}	2.93	98.0

reference batch indicated better reinforcement in the case of graphenes than the reference batch.

TEM images were obtained by cryo-microtomy of cross-sectional area of vulcanized sheet. A layered structure of GO was observed in Figure 8.5(b), which resulted

TABLE 8.7

Results of Bound Rubber Content (BdR)

Sample Name	Bound Rubber Content (%)
NR Latex + CB (45 phr)	79
NR Latex + CB (40 phr) + GO (5 phr)	80
NR Latex + CB (40 phr) + RGO (5 phr)	81
NR Latex + CB (40 phr) + GP1 (5 phr)	84
NR Latex + CB (40 phr) + GP2 (5 phr)	87

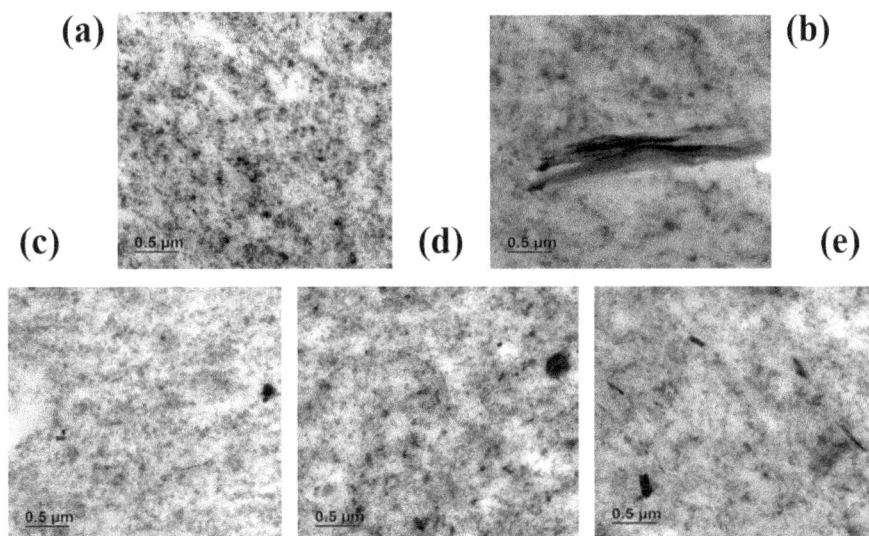

FIGURE 8.5 TEM images of (a) only carbon black, (b) GO, (c) RGO, (d) GP1 and (e) GP2 loaded rubber sample at 5 phr replacement level (Total loading 45 phr).

in inferior dispersion and hence higher aggregate formation. The dispersion as observed in Figure 8.5d and e for GP1 and GP2 modified rubber compounds showed improved dispersibility, as the exfoliation of the two graphenes namely G1 and G2 resulted in much better aqueous dispersion with lesser aggregate formation although the surface area was much larger.

In the schematic diagram (Figure 8.6), it is shown that the mechanism of reinforcement by graphene is completely different from carbon black. For the carbon black, aggregates adsorb onto the surface of rubber droplets, whereas in the case of graphene, they adsorb on the sheets of graphene. Interaction of the rubber chain with carbon black and graphene after coagulation is also shown. The mechanistic aspect of reinforcement by the hybrid is discussed in a later section.

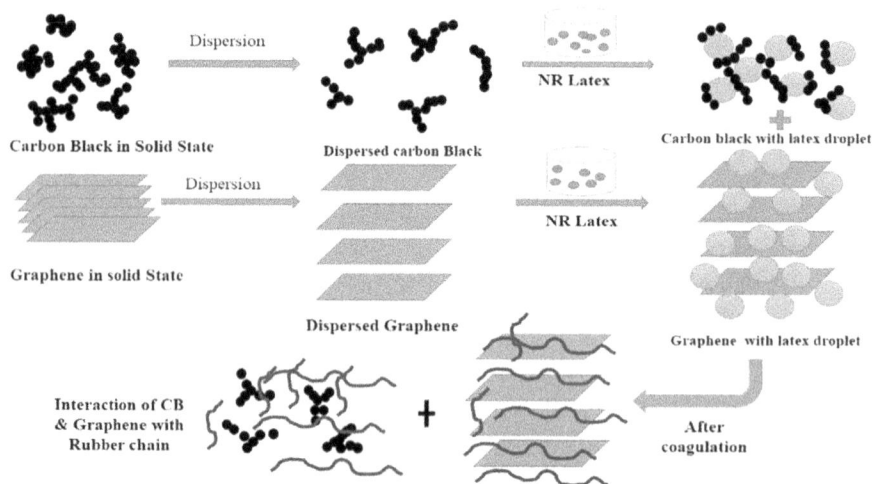

FIGURE 8.6   Schematic diagram of reinforce mechanism of carbon black and graphene into rubber latex.

8.3.3   Effect of Loading of Graphene in Hybrid Filler System

In Figure 8.7a, the values of tensile strength using different graphenic materials at different replacement levels with respect to carbon black loaded batches are reported. It was observed that except GO, all the nanofillers showed improved or similar tensile strength with respect to the reference batch. Also, it was found that GP2 exhibited highest tensile strength at 5 phr replacement level. The improvement of tensile strength was directly related to the better reinforcement of filler in the elastomer matrix. As the exfoliation in aqueous dispersion of GP1 and GP2 is much better than GO and the size of the platelets is favorable, these two graphenic materials interact in a much better way with the rubber matrix, which improves the tensile strength. At replacement level of 7 phr, the values of tensile strength decreased with respect to the reference batch. This is due to the fact that at higher replacement level, there are problems related to dispersibility.

Figure 8.7b displays the improvement of tear strength using various graphenes at specific replacement levels with respect to the carbon black loaded batch. At a higher replacement level, the value of tear strength has a general trend to decrease. Tear strength is found to be of highest magnitude at 1 phr replacement level; thus at lower replacement level, the rubber-filler interaction as well as dispersion of graphene should be more prominent, and hence, the tear strength is much higher than the reference batch. Substantial improvement in tear strength was observed for GP1 and GP2. We found from Table 8.1 that the surface area of GO and RGO is much lower than that of GP1 and GP2. GO and RGO remained as flakes in the matrix and it was easy to cut, whereas GP1 and GP2 having smaller flake size and probably interacted more strongly with the rubber matrix.

In Figure 8.7c, it was found that GO showed highest modulus at all replacement levels. From the MDR results, it was observed that the GO modified compounds

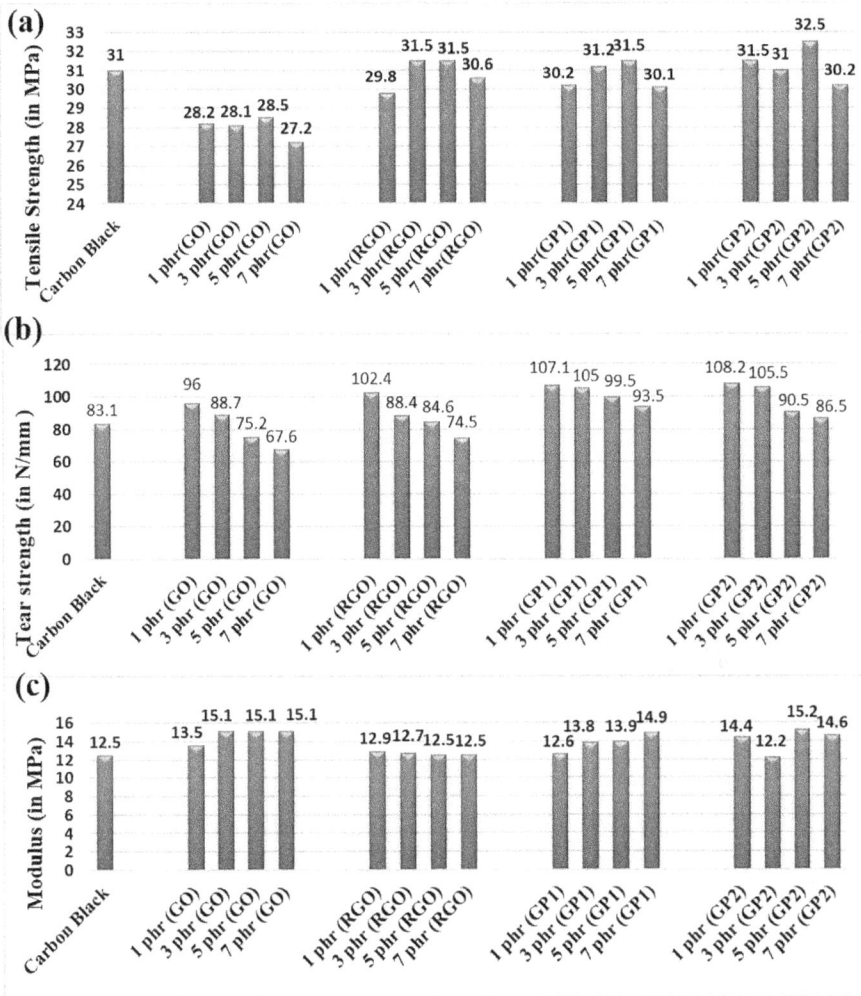

FIGURE 8.7 (a) Tensile strength (in MPa), (b) tear strength (in N mm⁻¹) and (c) modulus (in MPa) at different loadings of graphene.

got stiffer due to the flake-like structure and higher lateral dimension. It was also observed that at 5 phr replacement level, GP2 exhibited the highest modulus among all the four graphenes.

Improvement in rebound resilience using various graphenes at different replacement levels with respect to carbon black loaded batches is shown in Figure 8.8a. In general, much improved resilience values were shown by GO and GP2 at each replacement level.

Lowest heat build-up value (Figure 8.8b) was obtained at 5 phr replacement level. Generally, heat build-up shows a decreasing trend with the increasing amount of non-reinforcing filler. But contrary to the trend, the heat build-up improved along with the

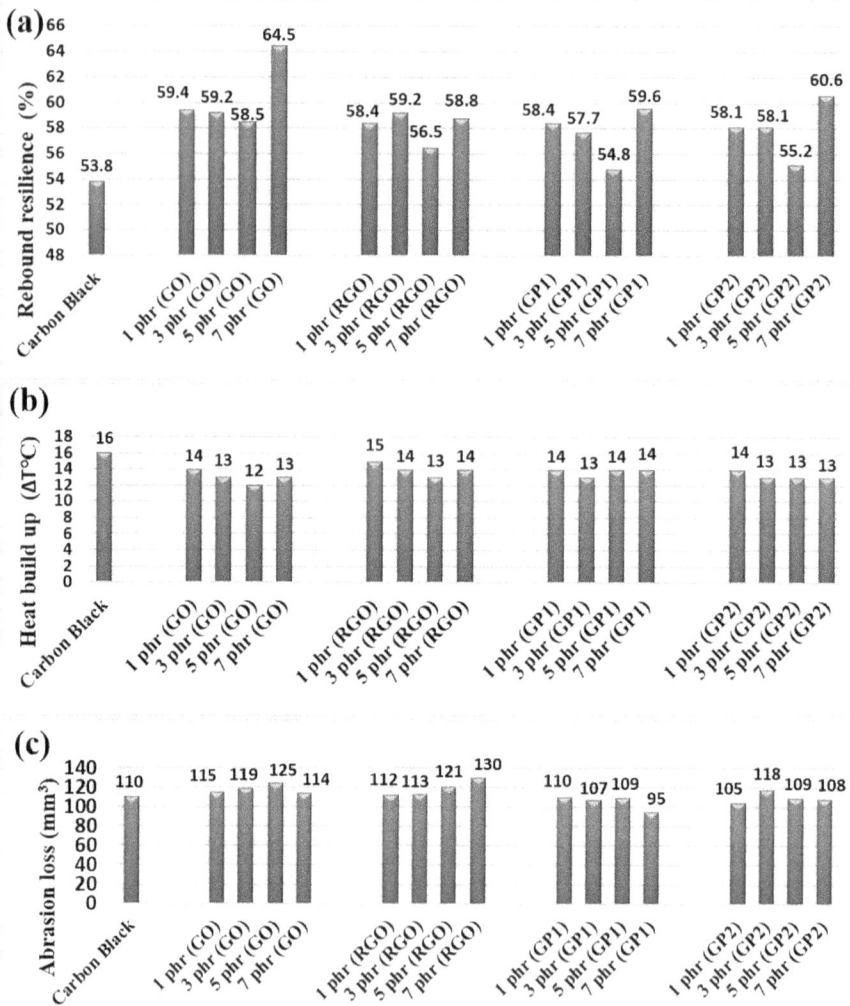

FIGURE 8.8 (a) Rebound resilience (%), (b) heat build-up (ΔT in °C) and (c) abrasion loss (mm³) at different loadings of graphene.

enhancement of modulus and tensile strength using the graphene, where graphene acted as a reinforcing filler. The results can be explained using thermal conductivity. The conductivity of a graphene filled sample was around 0.80 W m⁻¹ K⁻¹, whereas that of the carbon black filled compound was 0.36 W m⁻¹ K⁻¹.

In Figure 8.8c, an improved abrasion resistance performance was observed for GP1 modified compounds. As GP1 and GP2 have very high surface area and better exfoliation in aqueous dispersion, these interact very strongly with the rubber matrix, which is reflected in marginally improved abrasion resistance.

The results showed a clear tangible improvement of mechanical properties, i.e. increased tensile strength, modulus and elongation at break were observed at 5 phr replacement level, whereas tear strength was much improved at 1 and 3 phr replacement level. The heat build-up showed highest improvement at 5 phr replacement level of carbon black.

8.3.4 Graphene Modified SBR Compound

It was noted in the previous section that 45 phr of filler including 5 phr graphene was very effective in NR; further experiments were carried out using SBR at specified carbon black/graphene ratio.

Table 8.8 depicts the performance of different graphenes at 5 phr replacement level at a total loading of 45 phr. Among the three different graphenes, GP2 clearly indicates higher value of modulus, tensile strength and tear strength. The higher the value of the modulus and failure resistance, the greater is the reinforcement effect. In the case of GP1 and GP2, higher value of rebound resilience and heat build-up was observed. Hardness value was higher than the reference compound. Abrasion loss results from the removal of localized particles from an abrading surface, and generally with an increasing order of reinforcement, the abrasion loss should show a decreasing trend. All the three graphenes employed display similar abrasion loss as compared to the reference compound. The results are not in line with the NR compound.

The performance of GP2 with respect to reference compound shows marginal improvement of rolling resistance, ice traction as well as dry handling without much abrasion loss. The performance of GP2 with carbon black has been displayed in the radar plot in Figure 8.9a. Generally, with monotonic increment of the rolling resistance, traction shows a strictly decreasing trend but ice traction shows improvement.

8.3.5 Comparison of Property Change in
Percentage in between NR and SBR

The present study encourages us to make a head-to-head comparison of percentage change in properties between NR and SBR. Figure 8.9b depicts that SBR-Graphene modified compounds showed better improvement of modulus than the NR-Graphene

---

TABLE 8.8

Properties of SBR Modified Graphene Compounds at 45 phr of Total Loading and 5 phr of Graphene Replacement Level

Sample	Modulus @ 300 % (MPa)	Tensile Strength (MPa)	Abrasion Loss (mm³)	Tear Strength (N mm⁻¹)	Hardness (Shore A)	Heat Build Up (°C)	Rebound Resilience (%)
CB	11.6	19.4	163	32.1	60	20	39
CB + GO	10.6	16.1	169	30.3	61	21	40
CB + GP1	12.8	19.8	169	32.9	64	22	40
CB + GP2	15.4	20.7	168	33.9	63	24	41

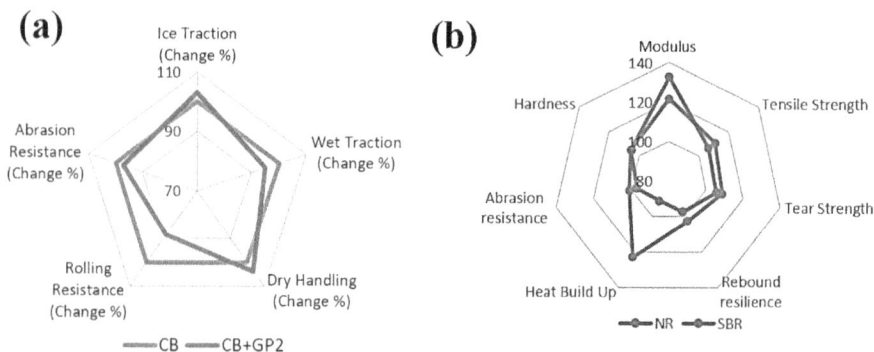

FIGURE 8.9 (a) Tire prediction performance for graphene modified SBR compound and (b) comparison of property change in percentage in between NR and SBR.

modified compounds. However, NR showed to have an edge over SBR on the basis of tensile strength, tear strength and rebound resilience.

It is to be noted that the abrasion resistance and the heat build-up are reported on the basis of percentage change, and hence, the trend indicates that the higher the value the better the performance. Hence, NR demonstrated the improved performance of heat build-up and abrasion resistance over SBR.

8.3.6 GRAPHENE-SILICA HYBRIDS

We have discussed the graphene-carbon black hybrids in the earlier section. Due to the impact of sustainability, silica-based rubber compounds have become very important. In addition, there are a few properties which could be obtained only with silica filled systems. It is quite natural that hybrid fillers have been attempted also in this area. Among many nanofillers, graphene has also been tested as an adjunct filler. In a recent study, Mazumder et al. reported GP2 (500 m²g⁻¹ surface area) from the earlier section in a silica filled tire compound. Table 8.9 reports the moduli, tensile strength, elongation at break at room temperature and at 70°C, hardness, swelling index, DIN abrasion resistance, tan δ values, thermal conductivity and electrical conductivity of the hybrid nanofiller filled compounds at various loadings of graphene. The compound S1 is the control silica filled compound as in reference (Mazumder et al. 2021).

The modulus showed an increasing tendency with higher nanofiller loading, due to more reinforcement, which is due to the high aspect ratio of the nanofillers providing a high interfacial area in the compound. There is an increase in tensile strength. The swelling index (SI) was reduced with increment in nanofiller dosage, indicating a higher crosslinking density and higher filler-polymer interaction. The abrasion loss decreased with an increase in the nanofiller loading due to better reinforcement in the compounds. The LAT100 wet traction properties increase for the graphene hybrid nanofiller compounds. The tear fatigue properties were measured. In the 5 phr graphene loaded compound, the crack growth rate was lower as compared to the control at a particular tearing energy due to surface area, anisotropy of the fillers and

TABLE 8.9

Properties of Rubber Compounds Containing Graphene Nanofiller at Different Levels

Test Parameter	S1	G-1s	G-2s	G-3s	G-5s
Modulus at 100% elongation (MPa)	3.2	3.1	3.5	3.5	3.9
Modulus at 200% elongation (MPa)	8.4	8.7	9.6	9.7	10.7
Tensile strength (MPa)	14.2	13.8	13.9	15.7	16.4
Elongation at break (%)	300	267	262	285	269
Hardness (Shore A)	68	67	68	68	69
Tensile strength at 70°C (MPa)	7.8	-	-	-	10.0
Elongation at break at 70°C (%)	241	-	-	-	215
Swelling Index	1.30	1.28	1.21	1.18	1.15
DIN abrasion loss (mm³)	82	82	80	77	75
tan δ at −25°C	0.61	0.79	0.74	0.60	0.75
tan δ at 0°C	0.45	0.46	0.45	0.44	0.45
tan δ at 30°C	0.22	0.22	0.22	0.22	0.22
tan δ at 70°C	0.13	0.13	0.13	0.13	0.13
Thermal conductivity (W m-K^{-1})	0.223	-	-	-	0.259
Electrical conductivity (S cm^{-1})	$1.2 - 10^{-12}$	-	-	-	5.4×10^{-6}

more tortuous path to the crack propagation. The wet traction as revealed from the tan delta value was higher at all the temperatures, and at 2°C, the traction for G-5s was higher by 33% than that of S1. From Figure 8.10, it is clearly visible that with FSSBR-HD silica (functionalized SBR and highly dispersible silica) and FSSBR-hybrid nanofiller, the magic triangle with the improvement of rolling resistance, wet traction and wear resistance is extended.

8.3.7 MECHANISM OF REINFORCEMENT BY HYBRID FILLER

Mondal et al. (2018) investigated the mechanism of graphite reinforcement in a carbon black filled SBR matrix at different volume fractions of the filler loading. At a similar filler loading, the system containing the graphite-carbon black hybrid filler demonstrated better tensile properties when compared with the carbon black system. For instance, the tensile strength of carbon black loaded- and graphene-carbon black hybrid filler loaded SBR at 25 phr filler loading was increased by 627% and 883% respectively over the pristine material. Similarly, the elongation at break registered an increase by 89% and 134%. The results could be interpreted using bound rubber content, which was higher for the hybrid filler indicating stronger interaction and hence better properties. They then tried to explain the results using molecular models.

It was assumed that due to the topological constraint, the segmental dynamics of the polymer chains are confined in a tube-like environment induced by the surrounding polymers (Heinrich and Vilgis 1993). The modulus due to crosslinks

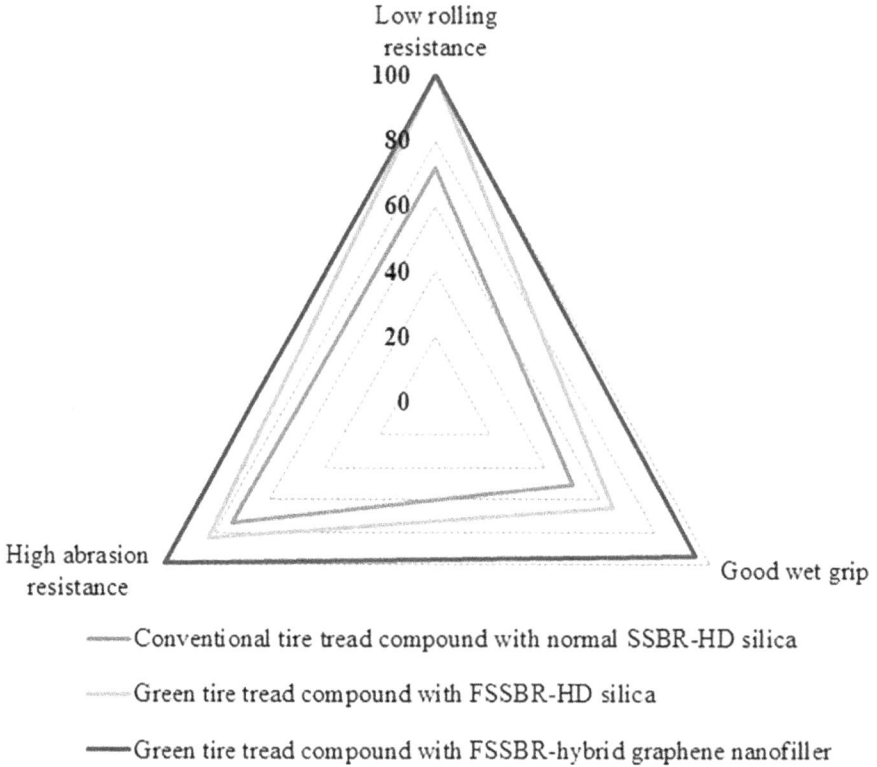

FIGURE 8.10   Magic triangle of graphene filled SBR compounds.

(G_c) and entanglement modulus (G_e) were then calculated using the following equation:

$$\sigma^* = G_c + G_e\, f(\alpha) \tag{8.5}$$

where σ^* is the reduced stress and $f(\alpha)$ is a function given by the following equation:

$$f(\alpha) = \frac{2\left(\alpha^{0.5\beta} - \alpha^{-\beta}\right)}{\beta\left(\alpha^2 - \alpha^{-1}\right)} \tag{8.6}$$

Commonly, the value of β is taken as 1. α value is the intrinsic extension ratio corrected for the amplification factor involved in the case of the high filler containing system (Peddini et al. 2015). The value of G_c and G_e was derived from the C_1 and C_2 values from the Mooney–Rivlin plot (Klüppel and Schramm 2000).

G_c was correlated to the molecular parameters as follows:

$$G_c = A_c V_c k_B T \tag{8.7}$$

where A_C, the microstructural factor, is taken as 0.67, V_c is the crosslink density, k_B and T are the Boltzmann constant and absolute temperature respectively (Heinrich and Klüppel 2008). The topological constraint modulus was expressed using the following equation:

$$G_e = \frac{k_B T \; \eta_s \; (l_s)^2}{4 \; (6)^{0.5} (d_o)^2} \tag{8.8}$$

where l_s is the mean length of Kuhn statistical segment (taken as 1.09 nm for SBR), d_0 is the fluctuation range of a chain segment. n_s, the density of polymer segment, is taken as the ratio of the product of the density of rubber and Avogadro number to that of the molar mass of the statistical segment (M_s, 211 g mol^{-1} for SBR) was 2.56 nm^{-3}. The relationship between the fluctuation range of a chain segment and the average number of statistical segments between successive entanglement (n_e) was given by the following equation:

$$d_0 = l_s n_e^{0.5} \tag{8.9}$$

And the average molecular weight (M_c) of the network chains was expressed in terms of crosslink modulus as follows:

$$M_c = \frac{\rho RT}{G_c} \tag{8.10}$$

where ρ, the density of the rubber, is taken as 0.94 g cc^{-1}, R and T are the gas constant and absolute temperature respectively. The average number of statistical segments (N) can be expressed in terms of M_c as

$$N = \frac{M_c}{M_s} \tag{8.11}$$

Further, the crosslink density (V_c) can be correlated in terms of M_c as

$$V_c = \frac{\rho N_A}{M_c} \tag{8.12}$$

Various parameters obtained at the molecular level from these equations are demonstrated in Table 8.10.

Both G_c and G_e values were also significantly higher for the hybrid system, indicating a crucial role of the plate-like morphology of the graphite in developing physical entanglement inside the rubber matrix as well as the high specific surface area of the graphite in the formation of a large number of chemical crosslinking points. Higher the loading, higher was the value of G$_c$. There was an abrupt enhancement of G_e in the initial stage. M_c, R_c and n_e decreased accordingly with an increase of v_c. Addition of the graphite had a pronounced effect on d_0 and $<R^2_0>$, both of which decreased with the addition of graphene. Hence, it can be concluded that the SBR chains were more space confined in the case of the hybrid filler system over carbon

TABLE 8.10

Molecular Parameters Obtained from Tube Model

Sample	G_c (MPa)	G_e (MPa)	M_c (g/mol)	R_c (nm)	$v_c \times 10^4$ (mol/cc)	N	d_0 (nm)	n_e	$<R^2_0>$ (nm)
SBR	0.56	0.14	4158	4.83	1.36	19.70	2.88	6.98	23.40
25S22	0.76	0.36	3064	4.15	1.84	14.52	1.80	2.72	17.24
25SG22	0.89	0.40	2616	3.83	2.16	12.39	1.70	2.43	14.71
35S22	1.39	0.42	1675	3.06	3.37	7.93	1.66	2.31	9.42
35SG22	1.49	0.45	1563	2.96	3.62	7.40	1.61	2.18	8.79
45S22	1.43	0.47	1628	3.02	3.47	7.71	1.57	2.07	9.15
45SG22	2.14	0.51	1088	2.47	5.20	5.15	1.51	1.91	6.11

Source: Reprinted from Polymer, 146, Mondal et al., Expanded graphite as an agent towards controlling the dispersion of carbon black in poly (styrene –co-butadiene) matrix: An effective strategy towards the development of high performance multifunctional composite, 31–41. Copyright (2018), with permission from Elsevier.

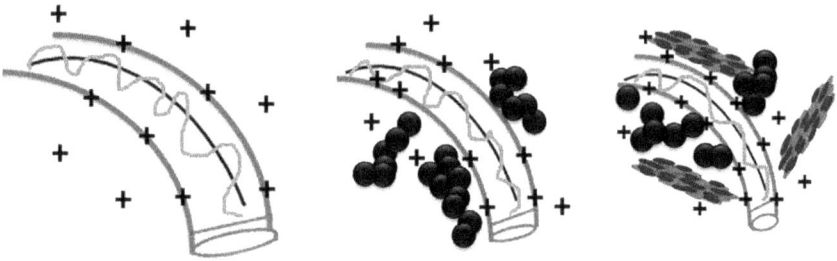

FIGURE 8.11 Representative schematic representation of the tube model for SBR, xS22 and xSG22 respectively. (Reprinted from Polymer, Mondal et al., Expanded graphite as an agent towards controlling the dispersion of carbon black in poly (styrene-co-butadiene) matrix: An effective strategy toward the development of high performance multifunctional composite. Copyright (2018), with permission from Elsevier.)

black filled system at the same loading. This is schematically shown in Figure 8.11. Thus, a better polymer-filler interaction was noted for hybrid filler system due to this typical space confinement effect.

There was another interesting effect on the use of graphite/graphene nanofiller. The edges of the graphene nanofiller were sharp like razor blade. There was a possibility that the graphene fillers might cut the carbon black aggregate to finite required size helping in the dispersion of the carbon black. In order to check this, the TEM of the graphene nanofiller (Figure 8.12) was taken, and AC conductivity as well as dynamic mechanical properties were measured. The increase in the AC conductivity of the hybrid filled system (1.07×10^{-10} S cm^{-1}, at 1 Hz) compared to the carbon black filled sample (5.47×10^{-12} S cm^{-1}, at 1 Hz) supported the formation of percolating network. The storage modulus improvement definitely indicated the higher

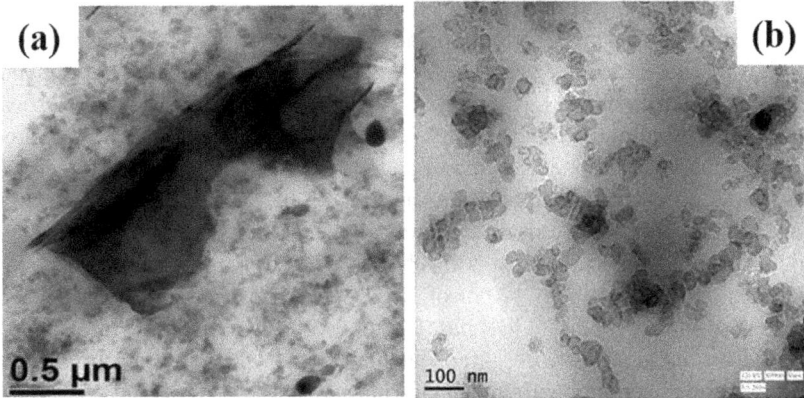

FIGURE 8.12 TEM micrograph of hybrid filled system at 25 phr total loading showing (a) graphene flake in the matrix and (b) percolation network formed by the filler in the matrix. The samples were prepared using cryo-microtomy. (Reprinted from Polymer, Mondal et al., Expanded graphite as an agent towards controlling the dispersion of carbon black in poly (styrene-co-butadiene) matrix: An effective strategy towards the development of high performance multifunctional composite. Copyright (2018), with permission from Elsevier.)

polymer-filler interaction in the case of hybrid filler system. In addition, the graphite platelets might act as a lubricating agent, and in the course of mixing, the graphite platelets might slide off the carbon black structure resulting in better distribution of the carbon black filler in the matrix. These results were further confirmed by Roy et al. who noted that the hybrid filler containing composite in the unvulcanized state demonstrated significant improvement in terms of the physico-mechanical properties such as tensile strength and modulus, compared to the neat carbon black filled system (Roy et al. 2021). Stress relaxation studies indicated that the hybrid system registered minimal decay in the force with time and demonstrated higher gel fraction compared to the carbon black filled SBR at the same loading.

8.4 CONCLUSIONS

In order to get the maximum benefit out of the graphenic materials, the thick graphene flakes need to be broken down into the nano-sized graphene sheets which may be incorporated to the rubber matrix through efficient mixing processes. Previous works indicated the reinforcing ability of the graphenic materials, but many had shown the benefits with respect to the raw rubbers and did not compare with the high-end carbon blacks. In the present study, it was shown that the graphenic materials with suitable dimensions and surface characteristics could produce rubber compound with higher mechanical strength and better application properties such as rolling resistance, traction and abrasion, as compared to N234 filled NR compound, if incorporated efficiently in the rubber matrix through the suitable mixing techniques.

In a typical ASTM D3192 formulation, graphene nano-platelets were found to exhibit superior mechanical and dynamic properties as compared to GO and RGO.

Incorporation of GP2 and GP1 in NR compound had significantly enhanced 300% modulus, tensile strength, tear strength and rebound resilience. The enhancement in properties was attributed to a greater degree of exfoliation of graphene nano-platelets in the rubber matrix as well as its rough surface characteristics. The properties of NR compound containing the graphenic materials were found to reach an optimum at a loading of 5 phr graphenic material and 40 phr carbon black. A decline in tear strength was observed with the increasing loading of graphenes. On the contrary, tensile strength, 300% modulus and rebound resilience were maximum at 5 phr loading of graphenes. The enhancement in rubber properties was also observed in SBR-based formulation containing 40 phr of N234 and 5 phr of graphene nano-platelet. The extent of improvement in SBR-based compound was found to be lower than the NR-based formulation. However, graphene displayed outstanding improvement in a silica filled tire compound with high wet grip and excellent abrasion resistance. The mechanism of reinforcement by graphene was explained using a carbon black filled rubber SBR compound. Addition of the graphite had a pronounced effect on fluctuation range of a chain segment and radius of gyration, both of which decreased with the addition of graphene. In addition, graphene fillers might cut the carbon black aggregate to finite required size helping in the dispersion of the carbon black.

8.5   CONFLICT OF INTEREST

There is no conflict of interest.

8.6   ACKNOWLEDGMENTS

Professor A. K. Bhowmick gratefully acknowledges the funding received from Birla Carbon, Mumbai for the sponsored research work at IIT Kharagpur and the University of Houston for preparation of the manuscript. The authors used the experimental facilities provided at IIT Kharagpur as well as Birla Carbon and hence are grateful to these Organizations. IIT Kharagpur provided fellowship to Bikshan Ghosh, Sourav Pal and Amrita Roy for this work.

REFERENCES

Ajayan, Pulickel M. 1999. "Nanotubes from Carbon." *Chemical Reviews* 99 (7): 1787–99.
Bai, Xin, Chaoying Wan, Yong Zhang, and Yinghao Zhai. 2011. "Reinforcement of Hydrogenated Carboxylated Nitrile-Butadiene Rubber with Exfoliated Graphene Oxide." *Carbon* 49 (5): 1608–13. doi:10.1016/j.carbon.2010.12.043.
Blake, Peter, Paul D. Brimicombe, Rahul R. Nair, Tim J. Booth, Da Jiang, Fred Schedin, Leonid A. Ponomarenko, et al. 2008. "Graphene-Based Liquid Crystal Device." *Nano Letters* 8 (6): 1704–8. doi:10.1021/nl080649i.
Dreyer, Daniel R., Sungjin Park, Christopher W. Bielawski, and Rodney S. Ruoff. 2010. "The Chemistry of Graphene Oxide." *Chemical Society Reviews* 39 (1): 228–40. doi:10.1039/b917103g.
George, Jinu Jacob, and Anil K. Bhowmick. 2008. "Ethylene Vinyl Acetate/Expanded Graphite Nanocomposites by Solution Intercalation: Preparation, Characterization and Properties." *Journal of Materials Science* 43 (2): 702–8. doi:10.1007/s10853-007-2193-6.

Guoxiu, Wang, Yang Juan, Park Jinsoo, Gou Xinglong, Wang Bei, Liu Hao, and Yao Jane. 2008. "Facile Synthesis and Characterization of Graphene Nanosheets." *Journal of Physical Chemistry C* 112 (22): 8192–95. doi:10.1021/jp710931h.

Heinrich, Gert, and Manfred Klüppel. 2008. "Rubber Friction, Tread Deformation and Tire Traction." *Wear* 265 (7–8): 1052–60. doi:10.1016/j.wear.2008.02.016.

Heinrich, G., and T. A. Vilgis. 1993. "Contribution of Entanglements to the Mechanical Properties of Carbon Black Filled Polymer Networks." *Macromolecules* 26 (5): 1109–19.

Klüppel, Manfred, and Joachim Schramm. 2000. "A Generalized Tube Model of Rubber Elasticity and Stress Softening of Filler Reinforced Elastomer Systems." *Macromolecular Theory and Simulations* 9 (9): 742–54. doi:10.1002/1521-3919(20001201)9.

Li, Xiaolin, Xinran Wang, Li Zhang, Sangwon Lee, and Hongjie Dai. 2008. "Chemically Derived, Ultrasmooth Graphene Nanoribbon Semiconductors." *Science* 319 (5867): 1229–32. doi:10.1126/science.1150878.

Mazumder, Amrita, Jagannath Chanda, Sanjay Bhattacharyya, Saikat Dasgupta, Rabindra Mukhopadhyay, and Anil K. Bhowmick. 2021. "Improved Tire Tread Compounds Using Functionalized Styrene Butadiene Rubber-Silica Filler/Hybrid Filler Systems." *Journal of Applied Polymer Science* 138 (42): 1–19. doi:10.1002/app.51236.

McCreery, Richard L. 2008. "Advanced Carbon Electrode Materials for Molecular Electrochemistry." *Chemical Reviews* 108 (7): 2646–87. doi:10.1021/cr068076m.

Mondal, Titash, Anil K. Bhowmick, Ranjan Ghosal, and Rabindra Mukhopadhyay. 2018. "Expanded Graphite as an Agent towards Controlling the Dispersion of Carbon Black in Poly (Styrene –Co-Butadiene) Matrix: An Effective Strategy towards the Development of High Performance Multifunctional Composite." *Polymer* 146: 31–41. doi:10.1016/j. polymer.2018.05.031.

Mondal, Titash, Suman Basak, and Anil K. Bhowmick. 2017. "Ionic Liquid Modification of Graphene Oxide and Its Role towards Controlling the Porosity, and Mechanical Robustness of Polyurethane Foam." *Polymer* 127: 106–18. doi:10.1016/j.polymer. 2017.08.054.

Peddini, S. K., C. P. Bosnyak, N. M. Henderson, C. J. Ellison, and D. R. Paul. 2015. "Nanocomposites from Styrene-Butadiene Rubber (SBR) and Multiwall Carbon Nanotubes (MWCNT) Part 2: Mechanical Properties." *Polymer* 56: 443–51. doi:10.1016/j. polymer.2014.11.006.

Roy, Amrita, Titash Mondal, Saptarshi Kar, Kinsuk Naskar, Ranjan Ghosal, Rabindra Mukhopadhyay, and Anil K. Bhowmick. 2021. "Study of Reinforcement Mechanism and Structural Elucidation of Expanded Graphite-Carbon Black Hybrid Filler-SBR Nanocomposites through Comprehensive Analysis of Mechanical Properties and Small Angle X-Ray Data." *Journal of Applied Polymer Science* 138 (13): 1–13. doi:10.1002/ app.49093.

Roy, Nabarun, Rajatendu Sengupta, and Anil K. Bhowmick. 2012. "Modifications of Carbon for Polymer Composites and Nanocomposites." *Progress in Polymer Science* 37: 781–819. doi:10.1016/j.progpolymsci.2012.02.002.

Schallamach, A. 1968. "Abrasion, Fatigue, and Smearing of Rubber." *Journal of Applied Polymer Science* 12 (2): 281–93. doi:10.1002/app.1968.070120204.

Song, S. H., H. K. Jeong, and Y. G. Kang. 2010. "Preparation and Characterization of Exfoliated Graphite and Its Styrene Butadiene Rubber Nanocomposites." *Journal of Industrial and Engineering Chemistry* 16 (6): 1059–65. doi:10.1016/j.jiec.2010.07.004.

Wang, Guoxiu, Xiaoping Shen, Bei Wang, Jane Yao, and Jinsoo Park. 2009. "Synthesis and Characterisation of Hydrophilic and Organophilic Graphene Nanosheets." *Carbon* 47 (5): 1359–64. doi:10.1016/j.carbon.2009.01.027.

Zhang, Hongmei, Chunwei Wang, and Yong Zhang. 2015. "Preparation and Properties of Styrene-Butadiene Rubber Nanocomposites Blended with Carbon Black-Graphene Hybrid Filler." *Journal of Applied Polymer Science* 132 (3): 1–7. doi:10.1002/APP.41309.

9 Comprehensive Reviews on the Computational Micromechanical Models for Rubber-Graphene Composites

Kishor Balasaheb Shingare
Indian Institute of Technology Bombay

Soumyadeep Mondal
National Institute of Technology Durgapur

Susmita Naskar
University of Southampton

CONTENTS

9.1 INTRODUCTION

Rubber is one of the most widely utilized polymeric matrixes in the industry due to its excellent energy absorption and mechanical characteristics. Natural rubber (NR), polybutadiene rubber (PBR), styrene-butadiene rubber (SBR), isobutylene isoprene rubber (IIR), and poly(styrene-butadiene-styrene) (SBS) rubbers have been thoroughly explored so far. Due to its high elasticity, cracking resistance, and other outstanding mechanical features, NR is one of the most intensively investigated rubbers for practical applications such as tire (Ahmed et al. 2013). In order to improve their multifunctional properties, extensive research is devoted to the introduction of graphene as a modifier to conventional bulk composites. Several researchers utilized derivatives of graphene in a very effective way in the form of graphene nanosheets (GNs), graphene nanoplatelets (GnPs), graphene oxide (GO), or reduced GO into the matrix. These derivatives of graphene demonstrate promising ways toward bulk fabrication due to their cost-effective chemical reduction and oxidation techniques and the use of graphene with excellent electro-thermo-mechanical and gas barrier properties for commercial applications (Zhu et al. 2014; Ji et al. 2016). Xing et al. (2014) performed an experimental study on graphene/natural rubber (GE/NR) nanocomposite and found that, with a small addition of GE (0.5 phr), the tensile strength and initial tensile modulus of NR were increased by 48% and 80%, respectively. Zhan et al. (2011) experimentally found that 2 wt.% incorporation of GE increased the tear strength of pure NR by 50%, which is significantly greater than that of natural rubber/carbon black (NR/CB) and natural rubber/multiwalled carbon nanotube (NR/MWCNT) composites. Wang et al. (2015) performed an experimental investigation on how various grades of GnPs such as GnP-C750 and GnP-5 influence different mechanical and thermal properties of epoxy nanocomposites. Wang et al. (2017) reported a comprehensive analysis of graphene-rubber composite preparation techniques (latex mixing, melt mixing, and solution mixing). They also discussed the benefits and drawbacks of utilizing such composites. Likewise, many researchers (Zhu et al. 2014; Ji et al. 2016) reported such a substantial enhancement in the mechanical characteristics of graphene-based structures in their experimental findings. Now, coming to non-linear responses of rubber, researchers extensively studied hyperelasticity and compared different hyperelastic models for rubber-like materials (Ali, Hosseini, and Sahari 2010; Bergström 2015). Using Mooney–Rivlin (MR) hyperelastic model, Redzematovic and Kirane (2021) examined a non-linear sandwich structured representative volume element (RVE) of soft polymer (hydrogel) embedded with graphene films. Zhang, Yu, and Gu (2017) carried out an experimental and micromechanical study on sepiolite fiber-reinforced rubber composites subjected to large deformation. In the present study, Ogden model is chosen initially to model the hyperelastic matrix due to its high accuracy (Bergström 2015) and available experimental data (Khisaeva and Ostoja-Starzewski 2006) for NRs to validate the results. The pioneering work on this hyperelastic model was carried out by Ogden (Ogden 1972; Ogden, Saccomandi, and Sgura 2004). Before proceeding further, we discuss some analytical and finite element (FE) models to determine the effective properties of advanced composites.

Alfonso et al. (2016) studied Halpin–Tsai (HT) model and its various dependent parameters in particulate composites. Naskar and his coauthors (Naskar, Mukhopadhyay, and Sriramula 2018; Naskar et al. 2017) reported a probabilistic micromechanical analysis and presented a novel concept of stochastic representative volume element (SRVE). Shingare and Kundalwal (2020) and Shingare, Gupta, and Kundalwal (2020) investigated the electromechanical response of graphene-reinforced piezoelectric composite (GRPC) made of piezoelectric graphene sheets embedded in a polymer matrix. They used graphene as a nanofiber in this study and found that size-dependent features including piezoelectricity, flexoelectricity, and surface effect have an impact on the elastic behavior of GRPC nanobeams.

They demonstrated that these effects cannot be ignored at the nanoscale. Afterward, Shingare and Naskar (2021) used different analytical and FE models to determine the elastic and piezoelectric properties of GRPC, including two- and three-phase mechanics of materials (MOM), HT, rules of mixture (ROM), and modified ROM models. They also compared their findings with the existing experimental estimates and found these to be in excellent agreement. These analytical models were also employed by López Jiménez and Pellegrino (2012) in his hyperelastic composite model for validation of the computational findings. Therefore, in this chapter, we studied the computational micromechanical models using different analytical and FE models to study the elastic behavior of rubber-graphene platelets composites (referred to as "GPL/NR nanocomposites"). In this, ROM and HT models are used to validate the results obtained from FE modeling for the proposed GPL/NR nanocomposites, while Ogden model is used to validate the existing experimental findings.

9.2 COMPUTATIONAL MICROMECHANICAL MODELS

9.2.1 CONSTITUTIVE MODELS

In this chapter, micromechanical analysis of two-phase composite with hyperelastic matrix has been performed under small deformation cases. The selection of proper micromechanical theory is a fundamental step for analysis at the microscopic level. In this section, the constitutive equations of the phases and their characteristics have been discussed separately.

Rubber-like materials, having long and flexible molecular structures, generally exhibit large deformation responses. To predict their load-carrying capacity or homogenized properties in a deterministic way, two possible means can be followed: experimental stress-strain responses or an already available analytical hyperelastic model. Hyperelasticity is a non-linear generalization of linear elasticity that can be used to estimate huge strains. By a simple assumption of linear stress-strain relationship, rubber-like hyperelastic material can be approximated as a linear elastic material at small strain zones (Ali, Hosseini, and Sahari 2010). Rubber is also considered to be isotropic in the elastic range and almost incompressible. There exist different hyperelastic predictive models such as Neo-Hookean, MR, Yeoh, Ogden, Extended Tube (ET), and Horgan–Saccomandi, and among these, the two most efficient models are 3 pair Ogden (Ogden 1972) and ET model (Bergström 2015). Due to the availability of experimental material parameters of Ogden model, for NR, we used this model for plotting its non-linear stress-strain curve. Material parameters and stress-strain curves are discussed in Section 9.3.

According to the 3 pair Ogden model, the strain energy potential for a compressible isotropic hyperelastic material can be written as (Khisaeva and Ostoja-Starzewski 2006):

$$U = \sum_{i=1}^{3} \frac{2\mu_i}{m_i^2} \left[\tilde{\lambda}_1^{m_i} + \tilde{\lambda}_2^{m_i} + \tilde{\lambda}_3^{m_i} - 3 \right] + \frac{\alpha}{2} (J-1)^2 \qquad (9.1)$$

where $\tilde{\lambda}_i$ are the principal stretches such that $\tilde{\lambda}_1\tilde{\lambda}_2\tilde{\lambda}_3 = 1$; J is the Jacobian (>0) whose value is 1 for incompressible materials; and μ_i, m_i, and α are the material parameters – modules and its exponent, respectively, whereas α indicates the degree of compressibility of the material which can be given as follows:

$$\alpha = 1,000 \ G\alpha_p\alpha_d \tag{9.2}$$

in which G, α_p, and α_d are the shear modulus, the penalty factor, and the compressibility factor, respectively. The incompressibility of the material can be improved by increasing the value of α_p, and if the penalty factor is equal to 1, the corresponding Poisson's ratio is equal to 0.4995. As the matrix material is assumed to be a linear elastic material, we use tangent modulus determined from the slope of Ogden type stress-strain curve in uniaxial tension at the small strain zone (0–0.05).

In the present work, the derivative of graphene (graphene nanoplatelets as GPL) nanoparticles is used as reinforcement phase and it is modeled as isotropic which follow the traditional linear elasticity law (Hooke's law) which can be given as follows:

$$\sigma = C : \epsilon \tag{9.3}$$

where σ and ϵ are the stress and strain and ':' indicates the inner product; and C is the fourth-order stiffness tensor whose values for isotropic material can be written as (Reddy 2003):

$$C = \frac{E}{(1+v)(1-2v)} \begin{bmatrix} 1-v & v & v & 0 & 0 & 0 \\ v & 1-v & v & 0 & 0 & 0 \\ v & v & 1-v & 0 & 0 & 0 \\ 0 & 0 & 0 & \frac{1-2v}{2} & 0 & 0 \\ 0 & 0 & 0 & 0 & \frac{1-2v}{2} & 0 \\ 0 & 0 & 0 & 0 & 0 & \frac{1-2v}{2} \end{bmatrix} \tag{9.4}$$

From the aforementioned equation, only two independent parameters – E and v – are sufficient for modeling such materials. All the material properties and parameters of both the phases used in the FEM tool are given in the later section.

Therefore, we can now employ different existing analytical homogenization techniques to validate present computational (FEM) results as both the phases come under the isotropic elastic models. The ROM and HT models can be used to predict the transverse stiffness, even in the case of composite with hyperelastic matrix (López Jiménez and Pellegrino 2012). According to ROM, the transverse elastic modulus can be approximated as follows:

$$E_2 = \frac{E_{GPL}E'_m}{\left(1-V_f\right)E_{GPL}+V_fE'_m}; \text{ where } E'_m = \frac{E_{NR}}{1-v_{NR}} \tag{9.5}$$

However, HT model gives more accurate results than the aforementioned ROM, and thus, the transverse stiffness can be written as follows:

$$E_2 = E_{NR}\frac{\left(1+\eta\xi V_f\right)}{1-\eta V_f}; \text{ where } \eta = \frac{E_{GPL}-E_{NR}}{E_{GPL}+\xi E_{NR}} \tag{9.6}$$

Here, V_f and E_{GPL} are the volume fraction of GPL in the matrix and Young modulus of GPL, respectively. ξ $(0<\xi<\infty)$ is one parameter whose value depends on the loading conditions, an aspect ratio (AR) of the reinforcement and can be measured from the curve fitting with experimental observations (Young et al. 2012). E_{NR} denotes the initial tangent modulus of NR determined from its slope of the stress-strain curve. v_{NR} denotes Poisson's ratio which is 0.5 for incompressible material.

9.2.2 GEOMETRY DEFINITION OF RVE

To capture all the heterogeneities at the microscopic level and to predict the equivalent homogeneous macrostructure, it is necessary to consider an RVE for the present rubber matrix composite (RMC). The traditional concept of RVE can also be extended to non-linear analysis/behavior (Wang and Huang 2018). Here, a FE model of RVE is modeled using an effective FE package. In the present study, GPLs are assumed as an isotropic solid cylindrical disk with a diameter (d_{GPL}) and thickness (t_{GPL}), and it is perfectly bonded with the rubber matrix with no agglomerations (Liu and Brinson 2008; López Jiménez and Pellegrino 2012) (shown in Figures 9.1 - 9.2). Here, no interphase is considered between two phases for the sake of brevity. In this FE software environment, the AR of the inclusion is defined as the ratio of thickness to in-plane diameter, i.e., $\frac{t_{GPL}}{d_{GPL}}$. Therefore, we focused on the graphene nanoplatelets/natural rubber (GPL/NR) nanocomposite with two different spatial distributions of GPL within the square RVEs. They are (i) 2D aligned and (ii) 3D random distributions which are shown in Figure 9.3.

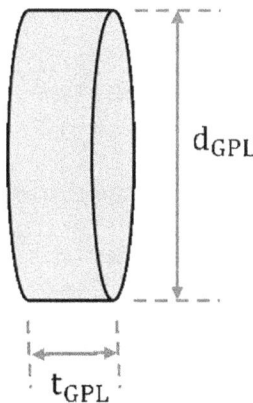

FIGURE 9.1   Disk-shaped GPL nanoplatelet.

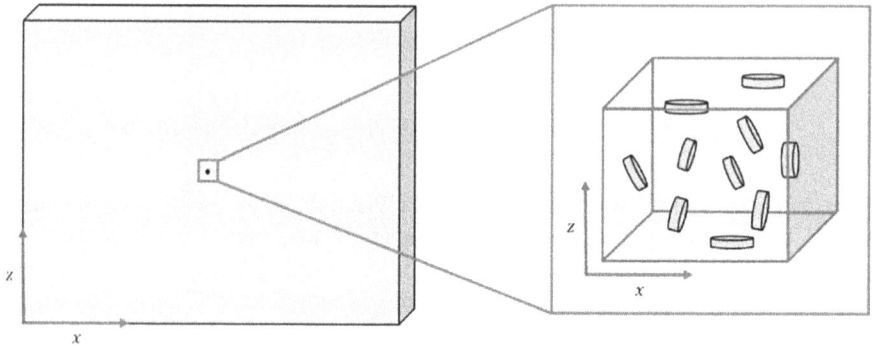

FIGURE 9.2   Definition of RVE of GPL/NR nanocomposite.

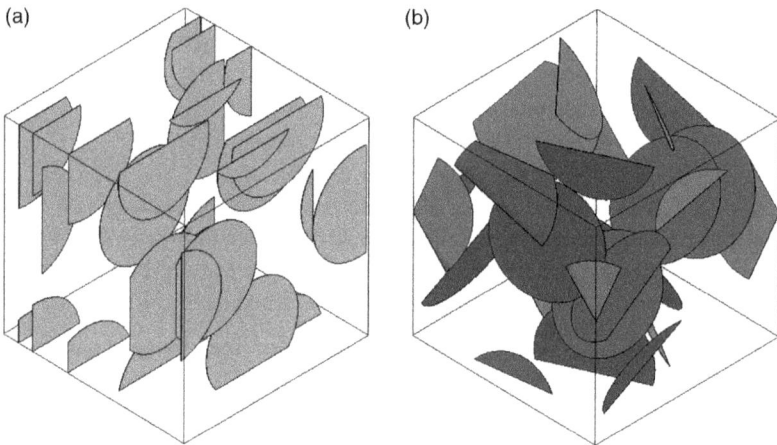

FIGURE 9.3   Distributions of GPL nanoparticles in GPL/NR nanocomposite RVE: (a) 2D aligned and (b) 3D random.

To obtain the positions of each platelet within the volume of RVE in random distribution, an orientation distribution function (ODF) plays a crucial role. The orientation of each platelet can be described with the help of one direction unit vector $(\vec{p})$ in the local coordinate (spherical) system attached with it. This vector is aligned with the symmetry axis for symmetrical inclusions. Figure 9.4 illustrates one such typical GPL and its coordinate system where each platelet is associated with two angles – polar (φ) and azimuth (θ) angles. These angles have a direct effect on the effective stiffness of the material for a particular distribution of inclusions. For this instance, Feng et al. (2017) perfectly analyzed this effect and gave one closed-form analytical expression of the effective stiffness (under uniaxial stretching) which can be written as follows:

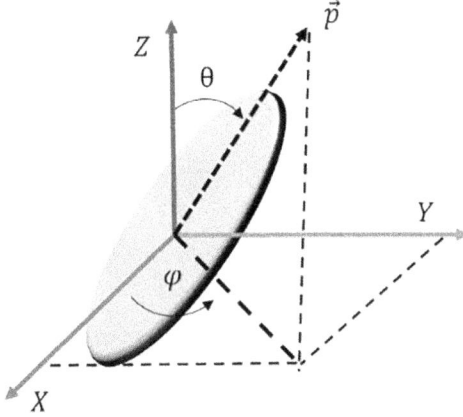

FIGURE 9.4 The spherical local coordinate system of GPL.

$$C_{\text{eff}} = C_{\text{matrix}} + \frac{\displaystyle\int_0^{2\pi}\int_0^{\pi} \rho(\theta,\varphi) V_f \left(C_{\text{GPL}} - C_{\text{matrix}}\right) A \sin\theta \; d\theta d\varphi}{\displaystyle\int_0^{2\pi}\int_0^{\pi} \rho(\theta,\varphi) \sin\theta \; d\theta d\varphi} \tag{9.7}$$

where ODF defines the probability density of the GPL distribution and is represented as $\rho(\theta,\varphi)$. For random distribution, it is equal to unity, signifying a uniform distribution of GPLs along any direction. A is the mechanical concentration tensor of the filler in the global coordinate system. For aligned 2D distribution, the values of θ and φ of each GPLs are 90° and 0°, respectively. In FE tools, the determination of the center coordinates of platelets also depends on other parameters such as the distance between neighboring platelets. In present GPL/NR nanocomposites, GnP-C750 grade is used as GPL inclusions whose diameter and thickness are 0.5 μm and 2 nm, respectively (Wang et al. 2015), and the minimum relative distance between inclusions is taken as 0.05 times the inclusion diameters. The random sequential adsorption algorithm can be employed to generate the random sequence of GPL in RVE (Rintoul and Torquato 1997). The obtained distributions are discussed in Section 9.3.

9.2.3 BOUNDARY CONDITIONS – LOADING CASES

In the present study, periodic geometry and periodic boundary conditions (PBCs) are used to define the 3D RVE. Any inclusion that intersects one of the RVE faces will have its complement placed in the same RVE at regular intervals, resulting in opposite identical faces. This phenomenon is illustrated with the help of a 2D periodic case in Figure 9.5.

For getting effective stiffness, the PBCs are imposed on all the faces of RVE. By this condition, it is assumed that the flux of field variables like displacement is periodic concerning the faces of an RVE. It is more accurate in comparison to the other

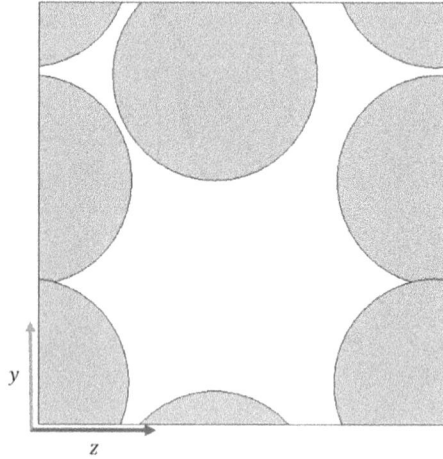

FIGURE 9.5   2D periodic geometry of GPL.

two available boundary condition types (Dirichlet and Mixed) and its convergence rate is also very high as the size of the volume element increases. Also, six different loading states (strain state) are studied. They all are monotonic and quasi-static in nature applied on the RVE within the aforementioned strain range. All six loading conditions applied on the faces of RVE are shown in Figure 9.6a–f.

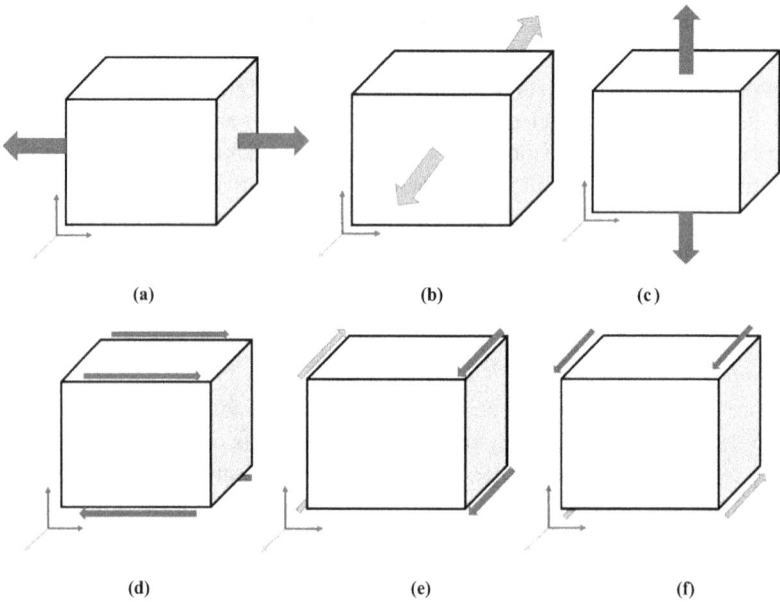

(a)   (b)   (c)

(d)   (e)   (f)

FIGURE 9.6   Representation of uniaxial strain loading along the (a) X-axis, (b) Y-axis, and (c) Z-axis and shear strain loading along the (d) XZ-plane, (e) XY-plane, and (f) YZ-plane.

9.2.4 Finite Element Modeling

In the present study, the second-order conforming tetrahedral mesh elements are used for discretization. Average element size is decided after performing a mesh convergence study which is discussed in a later section. In Figures 9.7 and 9.8, the meshing of both the phases is shown with aligned and random distributions, respectively.

In FE modeling, the RVE generation process is an iterative method. The automatic random RVE generation algorithm along with its meshing is shown in the following flowchart (Figure 9.9).

(a)   (b)

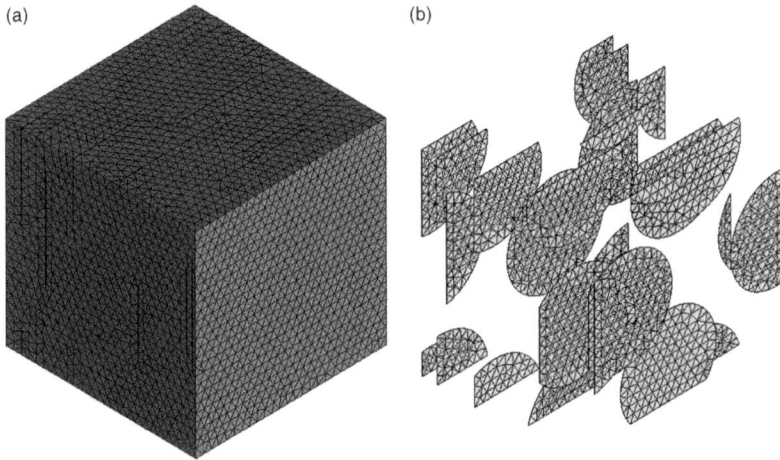

FIGURE 9.7 FE meshing of (a) GPL/NR nanocomposite RVE and (b) aligned GPL (mesh distribution size 0.04).

(a)   (b)

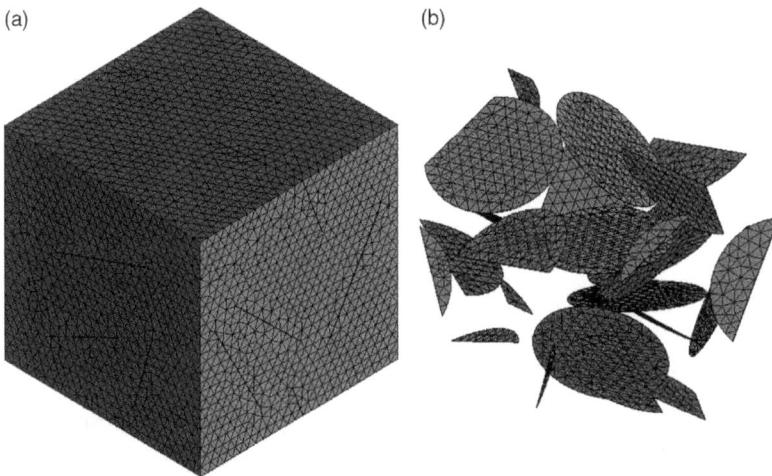

FIGURE 9.8 FE meshing of (a) GPL/NR nanocomposite RVE and (b) random GPL (mesh distribution size 0.04).

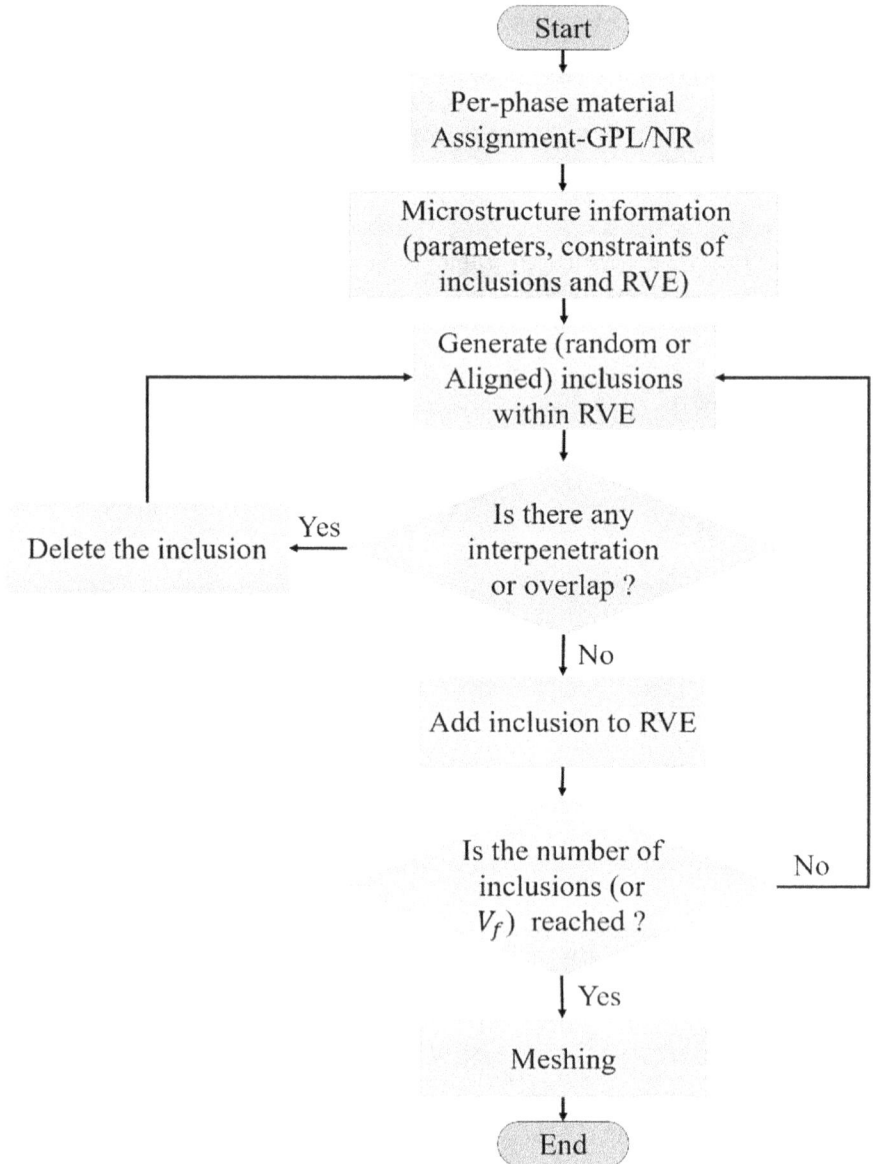

FIGURE 9.9 Flowchart representing the generation of RVE and meshing.

9.3 RESULTS AND DISCUSSION

In this study, the properties of graphene nanoplatelets/NR nanocomposite are investigated with the help of three-dimensional (3D) RVEs. Here, the size of cubic RVE (length, width, and height) is considered with dimensions of 1 μm × 1 μm × 1 μm, i.e., unit volume. Material properties of both reinforcement and matrix phases of GPL/NR nanocomposite are summarized in Table 9.1.

TABLE 9.1
Geometrical and Material Properties of Constituents of GPL/NR Nanocomposites

Elastic and Geometrical Properties of GPL		Hyper-Elastic Properties of NR	
ρ_{GPL}	1,060 kg/m^3	ρ_{NR}	920 kg/m^3
ν_{GPL}	0.186	(m_1, μ_1)	$\left(1.3, 4.095 \times 10^5 \text{ N/m}^2\right)$
d_{GPL}	0.5 µm	(m_2, μ_2)	$\left(5, 0.03 \times 10^5 \text{ N/m}^2\right)$
t_{GPL}	2 nm	(m_3, μ_3)	$\left(-2, 0.1 \times 10^5 \text{ N/m}^2\right)$
E_{GPL}	1,010 GPa	μ_0	$4.225 \times 10^5 \text{ N/m}^2$

Source: Arefi et al. (2018) and Khisaeva and Ostoja-Starzewski (2006).

For getting equivalent elastic properties for NR, we have first plotted its actual stress-strain curve in Figure 9.10a based on its material parameters for the Ogden model. Figure 9.10a shows non-linear relation between the stress and the strain when a uniaxial tension is applied to the present hyperelastic system. As the present work is limited to the small deformation case, we have plotted the aforementioned stress-strain curve for strain values between 0 and 0.05 in Figure 9.10b. From Figure 9.10b, it is observed that at small deformation, non-linear hyperelastic rubber can be modeled as linear material as the nature of the curve is linear and it also validates the assumption mentioned in Section 9.2.1. The value of E_{NR} mentioned in Eqs. 9.5 and 9.6 is then calculated from the slope of the linear curve-fitted line shown in Figure 9.10b. The calculated value of linearized Young's modulus (E_{NR}) is 1.225 MPa.

Figures 9.11 (a, b and c) show the random distribution of the geometrical centers of each GPL inside a square-shaped 3D RVE which is generated using an automatic

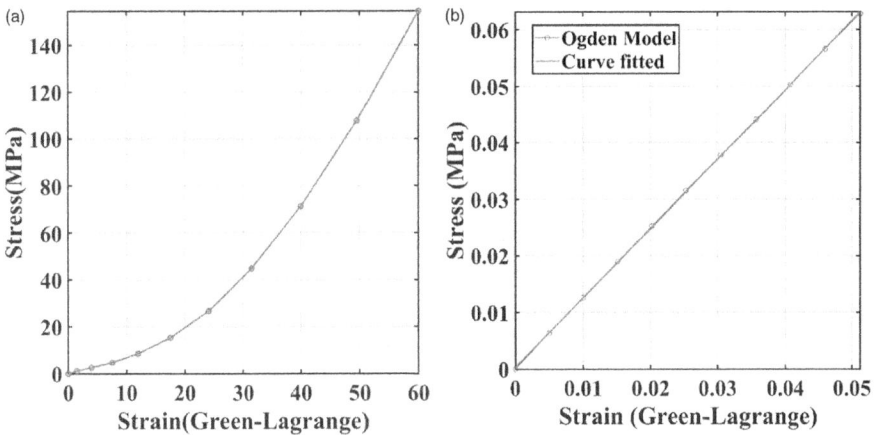

FIGURE 9.10 Uniaxial stress–strain response using Ogden model considering (a) large and (b) small deformations.

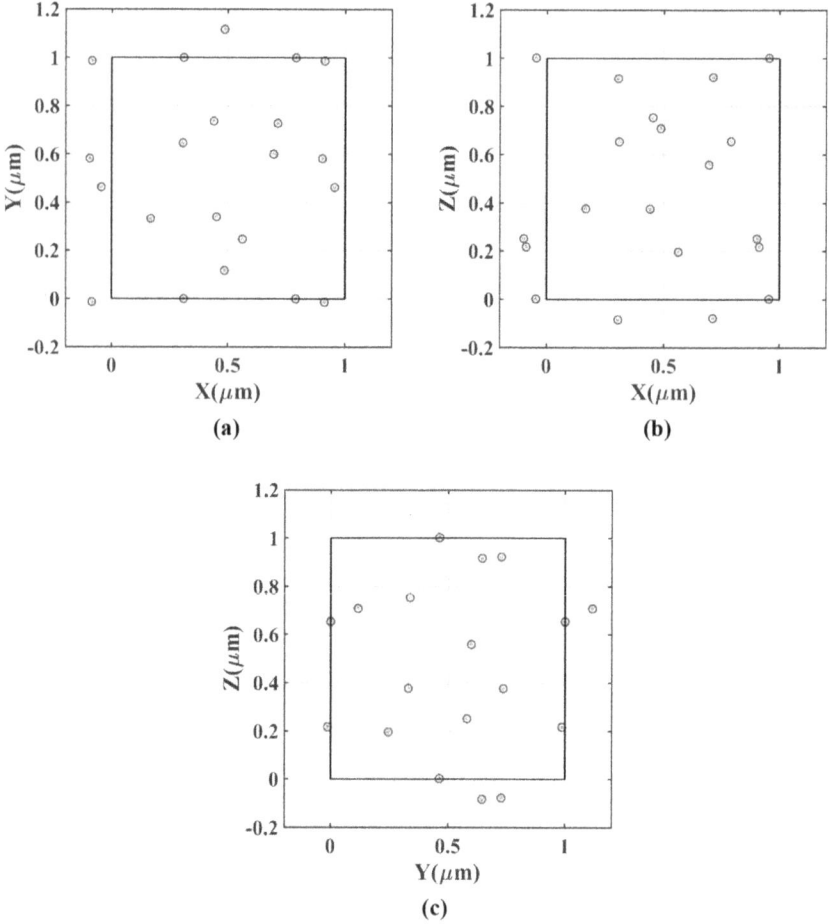

FIGURE 9.11 Distribution of GPL centers in $1\mu m \times 1\mu m \times 1\mu m$ RVE with 0.5% volume fraction in the (a) XY-plane, (b) XZ-plane, and (c) YZ-plane.

random algorithm kernel in the FE tool. Here, solid rectangular lines show the boundary of the RVE on the three principal planes (XY, XZ and YZ). In Fig. 9.11, the positions of GPLs are computed for the volume fraction of 0.5%.

While studying FE analysis, a convergence study is important to carry out for obtaining trustworthy results before proceeding further. Therefore, the convergence study is carried out to investigate the influence of mesh size or the number of elements on the transverse stiffness of the composite. In this, different sizes of meshing elements are considered. The overall range of average element size is taken into consideration in between 0.03 and 0.06. The results of the convergence study with respect to stiffness value and its mesh statistics used in FE analysis are enlisted in Table 9.2. To balance the computational intensiveness and reliability of the results, a mesh size of 0.04 is used in further analysis. There are many physical parameters of GPLs which can influence the elastic properties of the proposed composite such as

TABLE 9.2

Convergence Study of FEM for Transverse Stiffness (E_2) $\left(V_f = 0.5\%\right)$

Element Size	Number of Domain Elements	Number of Nodes	Transverse Stiffness, E_2 (MPa)
0.06	37,782	62,002	2.358
0.05	54,594	89,568	2.616
0.04	92,412	146,440	2.610
0.03	153,305	244,960	2.620

its volume fraction, distribution, AR, agglomeration, and interphase. In this section, only the first three parameters (volume fraction, distribution, and AR) are analyzed. According to FE modeling, the mechanical behavior of composite can be treated as an orthotropic.

Figure 9.12a and b presents the variation of transverse stiffness with respect to the volume fraction of GPL for both distributions. Here, we also used two analytical methods – ROM and HT models – for its validation. It is observed that there exists a large difference between FEM results and ROM predictions. It is because as mentioned in the existing literature (Young et al. 2012), this method usually gives only upper or lower bound of the property, especially in the case of $E_{GPL} \gg E_{NR}$. However, HT model with $\xi = 202$ shows a good agreement with the FEM results of the 2D aligned distribution, while the value of $\xi = 110$ for random distribution gives good predictions. In both the cases, the value of ξ is beyond its usual range (1–2) which is due to the inclusion of incompressibility in the matrix material (López Jiménez and Pellegrino 2012). Here, the little discrepancy (average about ~1.2%) between the HT model and FEM can be explained from the randomness point of view. As the present distribution of GPLs within microstructure is random in nature and its generation

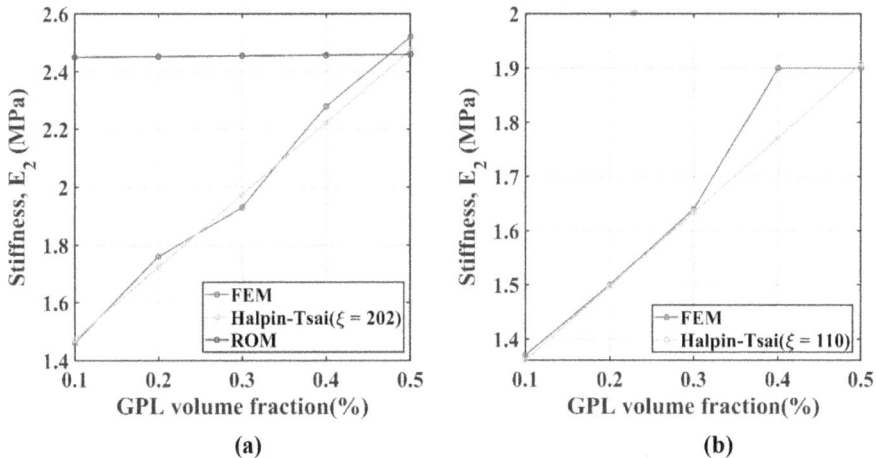

FIGURE 9.12 Variation of transverse stiffness with GPL volume fraction: (a) 2D aligned and (b) 3D random distribution.

process is iterative, FEM will not give the same results each time, rather it will always lie within ranges. In Figure 9.13a and b, the effect of AR of GPL on the stiffness is investigated for different volume fractions of GPL. Three ARs of 0.003, 0.004, and 0.005 are considered here. It is observed that the magnitude of stiffness reduces as the AR of inclusions increases which are found to be in coherence with the existing literature (Rouway et al. 2021; Elmarakbi, Azoti, and Serry 2017).

Figure 9.14 investigates the effect of microstructure, i.e., distribution of GPLs on the stiffness. From this, it can be concluded that the aligned RVEs are always stiffer

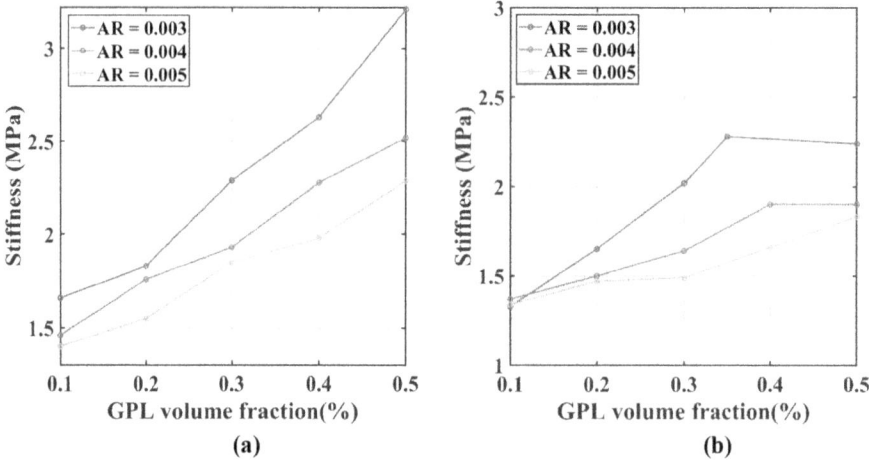

FIGURE 9.13   Variation of transverse stiffness with GPL aspect ratio: (a) 2D aligned and (b) 3D random distributions.

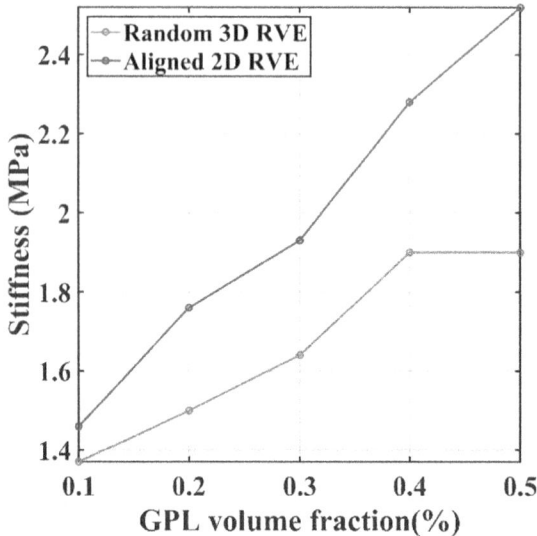

FIGURE 9.14   Effect of GPL distribution in RVE on transverse stiffness.

than random RVEs, the same observation was reported in the existing literature (López Jiménez and Pellegrino 2012).

From Figures 9.12–9.14, we can observe that the GPL/NR nanocomposite stiffens as the volume fraction of GPL increases. The same trend is also observed in the physical experiments on GE/NR nanocomposites (Xing et al. 2014; Mahmoud, Al-Ghamdi, and Al-Solamy 2012). The stiffness of rubber material is also increased due to the incorporation of GPLs into it.

Figures 9.15 and 9.16a–f depict the distribution of spatial displacement under six different loading patterns mentioned in Section 9.2.3 for 0.5% volume fraction of GPL in the aligned and random distributions, respectively. Figure 9.15g shows the displacements of internal graphene nanoplatelets when it is subjected to uniaxial loading in the X-axis. Almost all load studies show that displacements in the case of random distribution are greater than those of aligned distribution. This is owing to the fact that the latter's stiffness is greater (explained in Figure 9.14).

9.4 CONCLUDING REMARKS

The present study aims to use computational (FEM) modeling approaches to analyze the behavior of graphene nanoplatelets reinforced natural rubber (GPL/NR) nanocomposite in order to reduce physical experimental expenditures. For this, an effective FE software is used to perform the micromechanical analysis for two alternative distributions of GPLs within the rubber matrix, and the results obtained from FE simulation are validated using two analytical models, namely, ROM and HT. By means of cubic RVE with periodic boundary conditions, a 3D FE model is developed to examine the micromechanical properties of the material. The following main outcomes can be drawn from the present study:

- At a small strain zone, hyperelastic non-linear rubber (Ogden type) behaves like linear elastic material.
- The stiffness (strength) of the rubber can be increased by incorporating a small amount of GPLs.
- The stiffness and the strength of the composite material are observed to be increased with respect to the volume fraction of GPLs, which is also found in coherence with the available experimental observations.
- RVEs with aligned GPLs are significantly stiffer than RVEs with randomly aligned GPLs.
- The mechanical performance of GPL/NR nanocomposites is significantly influenced by shape parameters such as the nanoplatelet AR. It is worth noting that as the AR rises, the stiffness of the material decreases.

With the recent advances in nanoscale manufacturing and experimental capabilities, this chapter will offer essential physical insights into the mechanical/elastic behavior of nanocomposites using graphene-like multifunctional materials. Such nanocomposites can be utilized in different micro- and nano-electro-mechanical systems (M/NEMS) applications including sensors, actuators, and energy storage (energy harvesters).

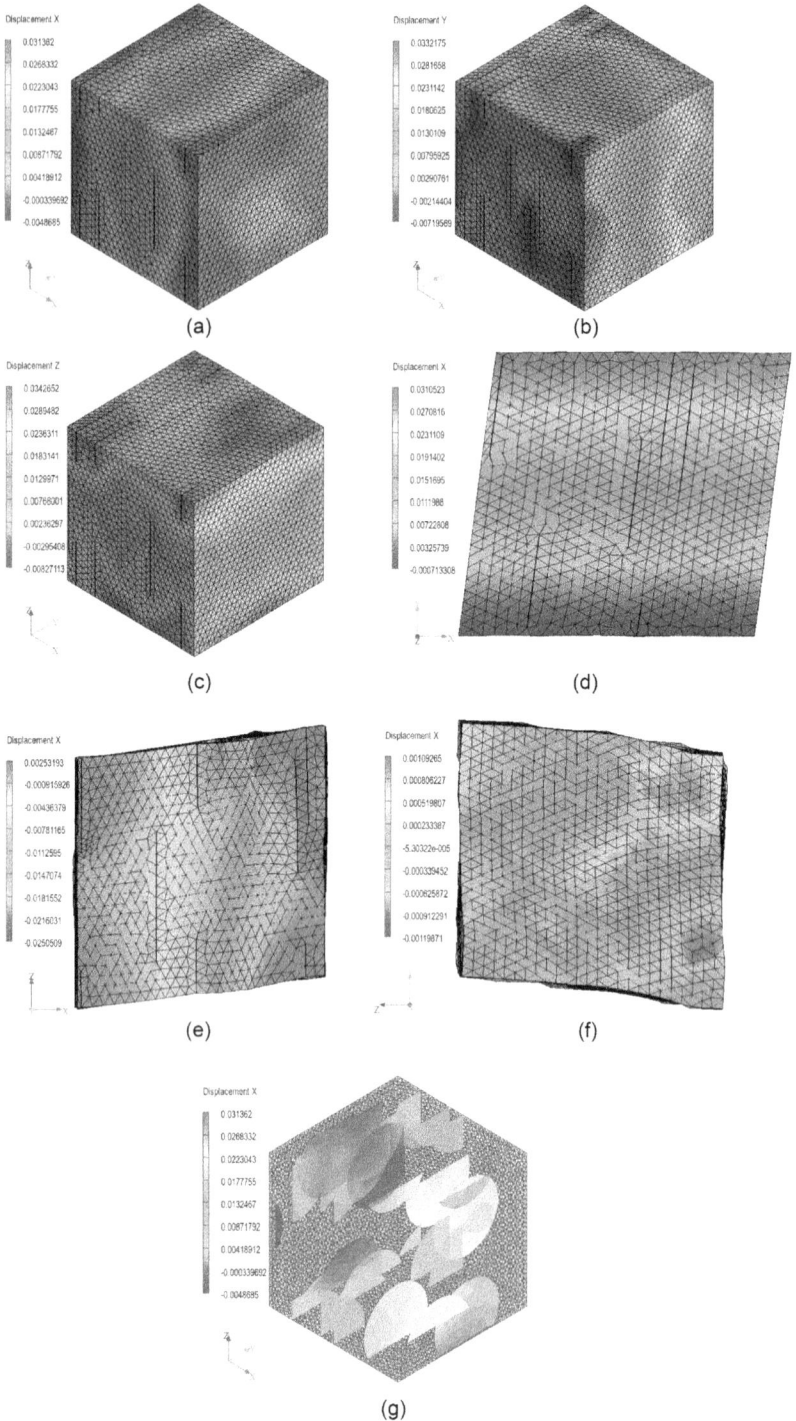

FIGURE 9.15 (a–f) Distribution of displacement in RVE with aligned GPLs under six load cases and (g) displacement of internal GPLs under uniaxial (X-axis) loading.

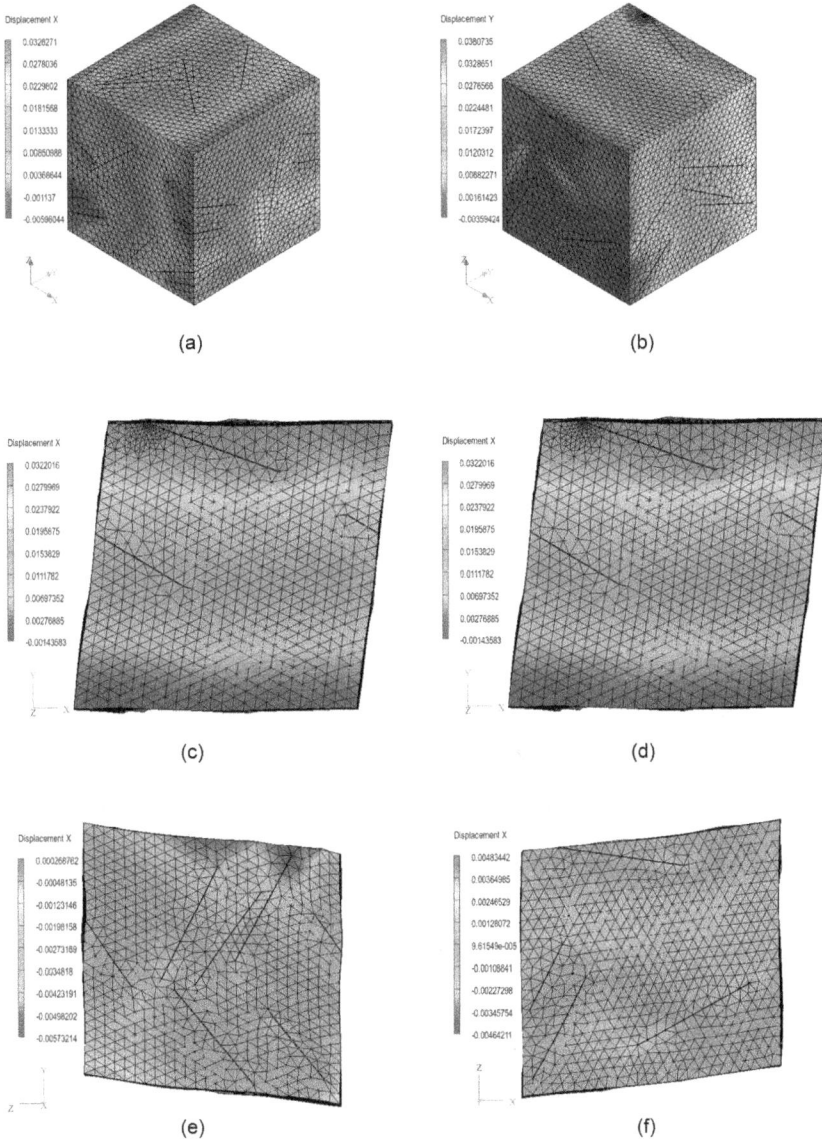

FIGURE 9.16 (a–f) Distribution of displacement in RVE with random GPLs under six load cases.

REFERENCES

Ahmed, K., Sirajuddin Nizami, S., Zahid Raza, N., Mahmood, K., 2013. Uticaj mikro mermernog mulja na fizičke osobine prirodnih kompozita gume. *Chem. Ind. Chem. Eng. Q.* 19, 281–293. doi:10.2298/CICEQ111225062A.

Alfonso, I., Figueroa, I.A., Rodriguez-Iglesias, V., Patiño-Carachure, C., Medina-Flores, A., Bejar, L., Pérez, L., 2016. Estimation of elastic moduli of particulate-reinforced composites using finite element and modified Halpin–Tsai models. *J. Brazilian Soc. Mech. Sci. Eng.* 38, 1317–1324. doi:10.1007/s40430-015-0429-y

Ali, A., Hosseini, M., Sahari, B.B., 2010. A review and comparison on some rubber elasticity models. *J. Sci. Ind. Res.* (India).

Arefi, M., Mohammad-Rezaei Bidgoli, E., Dimitri, R., Tornabene, F., 2018. Free vibrations of functionally graded polymer composite nanoplates reinforced with graphene nanoplatelets. *Aerosp. Sci. Technol.* 81, 108–117. doi:10.1016/j.ast.2018.07.036.

Bergström, J., 2015. Elasticity/hyperelasticity. *Mech. Solid Polym.* 209–307.

Elmarakbi, A., Azoti, W., Serry, M., 2017. Multiscale modelling of hybrid glass fibres reinforced graphene platelets polyamide PA6 matrix composites for crashworthiness applications. *Appl. Mater. Today* 6, 1–8. doi:10.1016/j.apmt.2016.11.003.

Feng, C., Wang, Y., Kitipornchai, S., Yang, J., 2017. Effects of reorientation of graphene platelets (GPLs) on young's modulus of polymer nanocomposites under uni-axial stretching. *Polymers (Basel).* 9. doi:10.3390/polym9100532.

Ji, X., Xu, Y., Zhang, W., Cui, L., Liu, J., 2016. Review of functionalization, structure and properties of graphene/polymer composite fibers. *Compos. Part A Appl. Sci. Manuf.* doi:10.1016/j.compositesa.2016.04.011.

Khisaeva, Z.F., Ostoja-Starzewski, M., 2006. On the size of RVE in finite elasticity of random composites. *J. Elast.* 85. doi:10.1007/s10659-006-9076-y.

Liu, H., Brinson, L.C., 2008. Reinforcing efficiency of nanoparticles: A simple comparison for polymer nanocomposites. *Compos. Sci. Technol.* 68, 1502–1512.

López Jiménez, F., Pellegrino, S., 2012. Constitutive modeling of fiber composites with a soft hyperelastic matrix. *Int. J. Solids Struct.* 49, 635–647.

Mahmoud, W.E., Al-Ghamdi, A.A., Al-Solamy, F.R., 2012. Evaluation and modeling of the mechanical properties of graphite nanoplatelets based rubber nanocomposites for pressure sensing applications. *Polym. Adv. Technol.* 23, 161–165. doi:org/10.1002/pat.1840.

Naskar, S., Mukhopadhyay, T., Sriramula, S., 2018. Probabilistic micromechanical spatial variability quantification in laminated composites. *Compos. Part B Eng.* 151, 291–325. doi:10.1016/j.compositesb.2018.06.002.

Naskar, S., Mukhopadhyay, T., Sriramula, S., Adhikari, S., 2017. Stochastic natural frequency analysis of damaged thin-walled laminated composite beams with uncertainty in micromechanical properties. *Compos. Struct.* 160, 312–334.

Ogden, R.W., 1972. Large deformation isotropic elasticity – On the correlation of theory and experiment for incompressible rubberlike solids. *Proc. R. Soc. London. A. Math. Phys. Sci.* 326, 565–584. doi:org/10.1098/rspa.1972.0026.

Ogden, R.W., Saccomandi, G., Sgura, I., 2004. Fitting hyperelastic models to experimental data. *Comput. Mech.* 34, 484–502. doi:10.1007/s00466-004-0593-y.

Reddy, J.N., 2003. Mechanics of laminated composite plates and shells, mechanics of laminated composite plates and shells. doi:10.1201/b12409.

Redzematovic, M., Kirane, K., 2021. Homogenization of the Mooney-Rivlin coefficients of graphene-based soft sandwich nanocomposites. *Mech. Soft Mater.* 3, 1–16.

Rintoul, M.D., Torquato, S., 1997. Reconstruction of the structure of dispersions. *J. Colloid Interface Sci.* 186, 467–476. doi:10.1006/jcis.1996.4675.

Rouway, M., Nachtane, M., Tarfaoui, M., Chakhchaoui, N., Omari, L.E.H., Fraija, F., Cherkaoui, O., 2021. Mechanical properties of a biocomposite based on carbon nanotube and graphene nanoplatelet reinforced polymers: Analytical and numerical study. *J. Compos. Sci.* 5, 234.

Shingare, K.B., Gupta, M., Kundalwal, S.I., 2020. Evaluation of effective properties for smart graphene reinforced nanocomposite materials, in: *Materials Today: Proceedings.* pp. 523–527.

Shingare, K.B., Kundalwal, S.I., 2020. Flexoelectric and surface effects on the electromechanical behavior of graphene-based nanobeams. *Appl. Math. Model.* 81, 70–91.

Shingare, K.B., Naskar, S., 2021. Probing the prediction of effective properties for composite materials. *Eur. J. Mech. A/Solids* 87. doi:10.1016/j.euromechsol.2021.104228.

Wang, F., Drzal, L.T., Qin, Y., Huang, Z., 2015. Mechanical properties and thermal conductivity of graphene nanoplatelet/epoxy composites. *J. Mater. Sci.* 50, 1082–1093.

Wang, J., Zhang, K., Bu, Q., Lavorgna, M., Xia, H., 2017. Graphene-rubber nanocomposites: Preparation, structure, and properties, in: *Carbon-Related Materials in Recognition of Nobel Lectures by Prof. Akira Suzuki in ICCE*, pp. 175–209.

Wang, Y., Huang, Z.M., 2018. Analytical micromechanics models for elastoplastic behavior of long fibrous composites: A critical review and comparative study. *Materials (Basel)* 11, 1919.

Xing, W., Wu, J., Huang, G., Li, H., Tang, M., Fu, X., 2014. Enhanced mechanical properties of graphene/natural rubber nanocomposites at low content. *Polym. Int.* 63, 1674–1681.

Young, R.J., Kinloch, I.A., Gong, L., Novoselov, K.S., 2012. The mechanics of graphene nanocomposites: A review. *Compos. Sci. Technol.* 72, 1459–1476.

Zhan, Y., Wu, J., Xia, H., Yan, N., Fei, G., Yuan, G., 2011. Dispersion and exfoliation of graphene in rubber by an ultrasonically- assisted latex mixing and in situ reduction process. Macromol. *Mater. Eng.* 296, 590–602. doi:10.1002/mame.201000358.

Zhang, B., Yu, X., Gu, B., 2017. Micromechanical modeling of large deformation in sepiolite reinforced rubber sealing composites under transverse tension. *Polym. Compos.* 38, 381–388.

Zhu, J., Lim, J., Lee, C.H., Joh, H.I., Kim, H.C., Park, B., You, N.H., Lee, S., 2014. Multifunctional polyimide/graphene oxide composites via in situ polymerization. *J. Appl. Polym. Sci.* 131. doi:10.1002/app.40177.

10 Simulation of Graphene Elastomer Composites

Sumit Sharma and Pramod Rakt Patel

Dr. BR Ambedkar National Institute of Technology

CONTENTS

10.1 INTRODUCTION

Natural rubber (NR) is a colloidal suspension of rubber latex particles, non-rubber constituents, and water that is extracted in the form of a latex. NR is extensively used in industrial applications. The properties of NR can be improved by adding various fillers, such as carbon black, carbon nanotubes (CNTs), and different types of nanoparticles. The quantum of improvement in the properties is highly dependent on the shape, size, aspect ratio, dispersion, and functionalization of the fillers. A lot of interest has been generated in using the nano-sized reinforcements that have the larger specific surface area and thus provide vastly improved properties to the resulting composites. Single-layer graphene (Gr) has revolutionized the rubber industry since its discovery by Novoselov et al. (2004). Gr has a large surface area due to its high aspect ratio. It is thermally stable, has high conductivity and in-plane stiffness, and can be modified by functionalization. All these qualities make Gr an ideal candidate for use in polymeric systems in the glassy state.

Though much experimental work has been performed related to the mechanical properties of NR-Gr composites, very few studies have been conducted at the atomic

DOI: 10.1201/9781003200444-10

level that can explain the mechanism of nano-reinforcements. Molecular dynamics (MD) simulation is one such tool that can be used effectively for predicting the properties of nanocomposites at the atomic level. Through MD, information is available on the interactions taking place between the Gr and the NR at the molecular level. MD uses Newton's equations of motion for predicting the interactions between various components of the nanocomposite. In this chapter, we have shown the procedure for predicting the mechanical properties of Gr-reinforced NR composites using MD. Young's, bulk, and shear modulus, the glass transition temperature, and the interfacial shear stress of Gr-NR composites have been predicted at the atomic level using MD.

10.1.1 WHAT IS MD?

MD is a technique used for solving Newton's equations of motion. Through MD, the displacement and interaction of all the atoms present in the system are described with respect to time to predict the new configuration of the system. A set of equations and force constants are used to define the molecular interactions between the constituents of the system. This collection is termed as "force-field". The force acting on each atom is calculated as the negative gradient of the potential energy which is a sum of bonded and non-bonded energies as shown in the following equation:

$$E_{\text{total}} = E_{\text{bonded}} + E_{\text{non-bonded}} \tag{10.1}$$

where the components of the bonded and non-bonded contributions are given by the following summations:

$$E_{\text{bonded}} = E_{\text{bond}} + E_{\text{angle}} + E_{\text{dihedral}} \tag{10.2}$$

$$E_{\text{non-bonded}} = E_{\text{electrostatic}} + E_{\text{vander Waals}} \tag{10.3}$$

where

$$E_b = \sum_b \left[k_2 (b - b_0)^2 + k_3 (b - b_0)^3 + k_4 (b - b_0)^4 \right] \tag{10.4}$$

$$E_\theta = \sum_\theta \left[k_2 (\theta - \theta_0)^2 + k_3 (\theta - \theta_0)^3 + k_4 (\theta - \theta_0)^4 \right] \tag{10.5}$$

$$E_\varnothing = \sum_\varnothing \left[k_1 (1 - \cos\varnothing) + k_2 (1 - \cos 2\varnothing) + k_3 (1 - \cos 3\varnothing) \right] \tag{10.6}$$

$$E_\chi = \sum_\chi k_2 \chi^2 \tag{10.7}$$

$$E_{b,b'} = \sum k (b - b_0)(b' - b_0') \tag{10.8}$$

$$E_{b,\theta} = \sum_{b,\theta} k(b - b_0)(\theta - \theta_0) \qquad (10.9)$$

$$E_{b,\varnothing} = \sum_{b,\varnothing} (b - b_0)[k_1\cos\varnothing + k_2\cos2\varnothing + k_3\cos3\varnothing] \qquad (10.10)$$

$$E_{\theta,\varnothing} = \sum_{\theta,\varnothing} (\theta - \theta_0)[k_1\cos\varnothing + k_2\cos2\varnothing + k_3\cos3\varnothing] \qquad (10.11)$$

$$E_{\theta,\theta'} = \sum_{\theta,\theta'} k(\theta - \theta_0)(\theta' - \theta_0') \qquad (10.12)$$

$$E_{\theta,\theta',\varphi} = \sum_{\theta,\theta',\varphi} k(\theta - \theta_0)(\theta' - \theta_0')\cos\varphi \qquad (10.13)$$

$$E_q = \sum_{ij} \frac{q_i q_j}{r_{ij}} \qquad (10.14)$$

$$E_{vdW} = \sum_{ij} \epsilon_{ij}\left[2\left(\frac{r_{ij}^0}{r_{ij}}\right)^9 - 3\left(\frac{r_{ij}^0}{r_{ij}}\right)^6\right] \qquad (10.15)$$

where k, k_1, k_2, k_3, and k_4 are force constants determined experimentally; b and θ are bond length and bond angle after stretching and bending respectively; b_0 and θ_0 are equilibrium bond length and equilibrium bond angle respectively; φ is the bond torsion angle; χ is the out-of-plane inversion angle; $E_{b, b'}$, $E_{\theta, \theta'}$, $E_{b, \theta}$, $E_{b, \varphi}$, $E_{\theta, \varphi}$, and $E_{\theta, \theta', \varphi}$ are cross-terms representing the energy due to interaction between bond stretch-bond stretch, bond bend-bond bend, bond stretch-bond bend, bond stretch-bond torsion, bond bend-bond torsion and bond bend-bond bend-bond torsion respectively; $\varepsilon_{i, j}$ is the well depth; r_{ij}^0 is the distance at which the interaction energy between the two atoms is zero; r_{ij} is the separation between the atoms and molecules; q_i and q_j are atomic charges on the atoms/molecules; and ε_0 is the permittivity of free space.

10.1.2 POTENTIALS IN MD

The interatomic interactions are described using the interatomic potentials. These are further used to develop the theoretical models of the given system which are then used to predict the characteristics of the system. The potentials commonly used in MD are the following:

   i. Tersoff potential
  ii. Brenner potential
 iii. Morse potential
 iv. Lennard-Jones potential

TABLE 10.1

Original Parameters for Tersoff EIP

(Tersoff 1988) for Carbon-Based Systems

$A = 1393.6 \, \text{eV}$	$B = 346.74 \, \text{eV}$
$\lambda_1 = 3.4879 \, \text{Å}^{-1}$	$\lambda_2 = 2.2119 \, \text{Å}^{-1}$
$\lambda_3 = 0.0000 \, \text{Å}^{-1}$	$n = 0.72751$
$c = 38049.0$	$\beta = 1.5724 \times 10^{-7}$
$d = 4.3484$	$h = -0.57058$
$R = 1.95 \, \text{Å}$	$D = 0.15 \, \text{Å}$

Tersoff potential: The analytical form for the pair-potential, V_{ij}, of the Tersoff model (Tersoff 1988) is given by the following functions with the corresponding parameters listed in Table 10.1.

$$V_{ij} = f_{ij}^C \left(a_{ij} f_{ij}^R - b_{ij} f_{ij}^A \right) \tag{10.16}$$

$$f_{ij}^R = A e^{-\lambda_1 r_{ij}} \tag{10.17}$$

$$f_{ij}^A = B e^{-\lambda_2 r_{ij}} \tag{10.18}$$

where r_{ij} is the distance between atoms i and j, f_{ij}^A and f_{ij}^R are competing attractive and repulsive pairwise terms, f_{ij}^C is the cutoff term which ensures only nearest-neighbor interactions, a_{ij} is a range-limiting term on the repulsive potential that is typically set equal to one.

The bond angle term, b_{ij}, depends on the local coordination of atoms around atom i and the angle between atoms i, j, and k:

$$b_{ij} = \left(1 + \beta^n \xi_{ij}^n \right)^{\frac{-1}{2n}} \tag{10.19}$$

$$\xi_{ij} = \sum_{k \neq i,j} f_{ik}^C g_{ijk} e^{\lambda_3^3 (r_{ij} - r_{ik})^3} \tag{10.20}$$

$$g_{ijk} = 1 + \frac{c^2}{d^2} - \frac{c^2}{d^2 + \left(h - \cos\left[\theta_{ijk} \right] \right)^2} \tag{10.21}$$

where θ_{ijk} is the angle between atoms i, j, and k.

Brenner potential: The Brenner potential (Brenner 1990) for solid-state carbon structures is given by the following functions with the corresponding parameters listed in Table 10.2.

$$V_{ij} = f_{ij}^C \left(f_{ij}^R - \overline{b_{ij}} f_{ij}^A \right) \tag{10.22}$$

TABLE 10.2

Original Parameters for Brenner EIP (Brenner 1990) for Solid-State Carbon Structures

$A = 10{,}953.544162170\,\text{eV}$	$B_1 = 12{,}388.79197798\,\text{eV}$
$B_2 = 17.5674064509\,\text{eV}$	$B_3 = 30.71493208065\,\text{eV}$
$\alpha = 4.746539060\,\text{Å}^{-1}$	$\lambda_1 = 4.7204523127\,\text{Å}^{-1}$
$\lambda_2 = 1.4332132499\,\text{Å}^{-1}$	$\lambda_3 = 1.3826912506\,\text{Å}^{-1}$
$Q = 0.3134602960833\,\text{Å}$	$R = 2\,\text{Å}$
$D = 1.7\,\text{Å}$	$T_0 = -0.00809675$
$\beta_0 = 0.7073$	$\beta_1 = 5.6774$
$\beta_2 = 24.0970$	$\beta_3 = 57.5918$
$\beta_4 = 71.8829$	$\beta_5 = 36.2789$

$$f_{ij}^R = \left(1 + \frac{Q}{r_{ij}}\right) A e^{-\alpha r_{ij}} \tag{10.23}$$

$$f_{ij}^A = \sum_{n=1}^{3} B_n e^{-\lambda_n r_{ij}} \tag{10.24}$$

where many of the terms are similar to the Tersoff model described above and the bond angle term, $\bar{b}ij$, is given by the following equations:

$$\bar{b}ij = \frac{1}{2}\left(b_{ij}^{\sigma-\pi} + b_{ji}^{\sigma-\pi}\right) + \Pi_{ij}^{RC} + b_{ij}^{DH} \tag{10.25}$$

$$b_{ij}^{\sigma-\pi} = \left(1 + \sum_{k \neq i,j} f_{ik}^C g_{ijk}\right)^{-1/2} \tag{10.26}$$

$$g_{ijk} = \sum_{i=0}^{5} \beta_i \cos^i\left[\theta_{ijk}\right] \tag{10.27}$$

Here, $b_{ij}^{\sigma-\pi}$ depends on the local coordination of atoms around atom i and the angle between atoms i, j, and k, θ_{ijk}. The coefficients, β_i, in the bond-bending spline function, g_{ijk}, fit experimental data for graphite and diamond and are also listed in Table 10.2. The term, Π_{ij}^{RC}, accounts for various radical energetics, such as vacancies, which are not considered here; thus, this term is taken to be zero. The term b_{ij}^{DH} is a dihedral bending function that depends on the local conjugation and is zero for the diamond but important for describing Gr and SWCNTs. This dihedral function involves the third nearest-neighbor atoms and is given by the following equation:

$$b_{ij}^{DH} = \frac{T_0}{2} \sum_{k,l \neq i,j} f_{ik}^C f_{jl}^C \left(1 - \cos^2\left[\Theta_{ijkl}\right]\right) \tag{10.28}$$

where T_0 is a parameter, f_{ij}^C is the cutoff function, and Θ_{ijkl} is the dihedral angle of four atoms identified by the indices, i, j, k, and l, and is given by the following equations:

$$\cos\left[\Theta_{ijkl}\right] = \bar{n}\,jik \cdot \overrightarrow{n}ijl \tag{10.29}$$

$$\bar{n}\,jik = \frac{\vec{r}ji \times \vec{r}ik}{\left|\vec{r}ji\right|\left|\vec{r}ik\right| \sin\left[\Theta_{ijk}\right]} \tag{10.30}$$

where $\bar{n}\,jik$ and $\overrightarrow{n}ijl$ are unit vectors normal to the triangles formed by the atoms given by the subscripts, $\vec{r}ij$ is the vector from atom i to atom j, and θ_{ijk} is the angle between atoms i, j, and k. In flat Gr, the dihedral angle, Θ_{ijkl}, is either 0 or π and the dihedral term is subsequently zero. The bending of the Gr layer leads to a contribution from this term.

Morse potential: The Morse potential (Morse 1929), named after physicist Philip M. Morse, is a convenient model for the potential energy of a diatomic molecule. It is a better approximation for the vibrational structure of the molecule than the quantum harmonic oscillator because it explicitly includes the effects of bond breaking, such as the existence of unbound states. It also accounts for the anharmonicity of real bonds. The Morse potential can also be used to model other interactions such as the interaction between an atom and a surface. Unlike the energy levels of the harmonic oscillator potential, which are evenly spaced by $\hbar \times \omega$ ($\hbar = h/2\pi$), where "h" is Planck's constant, the Morse potential level spacing decreases as the energy approaches the dissociation energy. The dissociation energy D_e is larger than the true energy required for dissociation D_0 due to the zero-point energy of the lowest ($v=0$) vibrational level. The Morse potential energy function is of the form:

$$V(r) = D_e\left(1 - e^{-a(r-r_e)}\right)^2 \tag{10.31}$$

Here, r is the distance between the atoms, r_e is the equilibrium bond distance, D_e is the well-depth (defined relative to the dissociated atoms), and a controls the "width" of the potential (the smaller a, the larger the well). The dissociation energy of the bond can be calculated by subtracting the zero-point energy E (0) from the depth of the well. The force constant of the bond can be found by Taylor expansion of $V(r)$ around $r=r_e$ to the second derivative of the potential energy function, from which it can be shown that the parameter, a, is given by the following equation:

$$a = \left(k_e / 2D_e\right)^{1/2} \tag{10.32}$$

where k_e is the force constant at the minimum of the well.

Lennard-Jones potential: The Lennard-Jones potential (Jones 1924) U_{LJ} is expressed as follows:

$$U_{LJ} = 4\varepsilon\left[\left(\sigma/r\right)^{12} - \left(\sigma/r\right)^6\right] \tag{10.33}$$

This potential is usually used as a model potential for rare gases such as Ar molecules. Here, ε is the well-depth, σ is the quantity corresponding to the particle diameter, and r is the separation between particles/molecules. The characteristics of the potential indicate that the interaction energy after a distance of approximately $r = 3\sigma$ can be assumed to be negligible. Hence, particle interaction energies or forces do not need to be calculated in the range of $r > 3\sigma$ in actual simulations. The distance for cutting off the calculation of energies or forces is known as the cutoff distance or cutoff radius, denoted by r_{coff}.

10.1.3 ENSEMBLES IN MD

A real system usually consists of thousands of atoms and solving the equations of motion for each atom will be a herculean task. The ensemble is a combination of microscopic states of a system, similar to the thermodynamic state. Classical thermo-dynamics is essentially particle-free. The only thing that matters to a thermodynami-cist is bulk properties such as the number of particles N, the temperature T, and the volume of the container V. Rather than worrying about the time development of the particles, we can make a very large number of copies of the system. We then calculate average values over this large number of replications and, according to the ergodic theorem, the average value we calculate is the same as the time average we would calculate by studying the time evolution of the original system. The two are the same. All the cells in the ensemble are not exact replicas at the molecular level. We just ensure that each cell has a certain number of thermodynamic properties that are the same. There is no mention of molecular properties at this stage. An ensemble of cells all with the same values of N, V, and T is said to form a canonical ensemble. There are three other important ensembles in the theory of statistical thermodynamics, and they are classified according to what is kept constant in each cell. Apart from the canoni-cal ensemble, where N, V, and T are kept constant, we have the following ensembles:

i. The micro-canonical ensemble where N, the total energy E, and V are kept constant in each cell. This is a very simple ensemble because energy cannot flow from one cell to another.
ii. In an isothermal-isobaric ensemble, N, T, and the pressure P are kept constant.
iii. Finally, we have the grand canonical ensemble, where V, T, and the chemical potential are kept constant. The grand canonical ensemble is a fascinating one because the number of particles is allowed to fluctuate.

10.1.4 THERMOSTATS

An equilibration procedure may be necessary to obtain the desired system tempera-ture T. This is achieved in MD simulations using any of the following thermostats:

Andersen's method: According to Andersen (1980), the temperatures calculated from the translational and angular velocities of particles are denoted by $T_{\text{cal}}^{(t)}$ and $T_{\text{cal}}^{(r)}$ respectively and written as follows:

$$T_{\text{cal}}^{(t)} = \frac{1}{3N} \sum_{i=1}^{N} \frac{mv_i^2}{k} \, , \, T_{\text{cal}}^{(r)} = \frac{1}{3N} \sum_{i=1}^{N} \frac{I\omega_i^2}{k} \tag{10.34}$$

where N is the total number of particles, assumed to be $N \gg 1$. $T_{cal}^{(t)}$ and $T_{cal}^{(r)}$ calculated from v_i and ω_i ($i = 1, 2,..., N$) are generally not equal to the desired temperature T. This equilibration procedure adjusts temperatures calculated from the translational and angular velocities of particles to T during the simulation by using the method of scaling the translational and angular velocities of each particle. If $T_{cal}^{(t)ave}$ and $T_{cal}^{(r)ave}$ denote the averaged values of $T_{cal}^{(t)}$ and $T_{cal}^{(r)}$ taken, for example, over 50 time-steps, then the scaling factors $c_0^{(t)}$ and $c_0^{(r)}$ are determined as follows:

$$c_0^{(t)} = \sqrt{\frac{T}{T_{cal}^{(t)ave}}} , \; c_0^{(r)} = \sqrt{\frac{T}{T_{cal}^{(r)ave}}} \tag{10.35}$$

With the scaling factors determined, the translational and angular velocities of all particles in a system are scaled as follows:

$$v_i' = c_0^{(t)} v_i, \; \omega_i' = c_0^{(r)} \omega_i \left(i = 1, 2, \ldots, N\right) \tag{10.36}$$

This treatment yields the desired system temperature T. In this example, the scaling procedure would be conducted at every 50 time-steps, but in practice, an appropriate time interval must be adopted for each simulation case. The above-mentioned equilibration procedure is repeated to give rise to the desired system temperature with sufficient accuracy.

Berendsen thermostat: The Berendsen thermostat (Berendsen et al. 1984) is an algorithm to re-scale the velocities of particles in MD simulations to control the simulation temperature. In this scheme, the system is weakly coupled to a heat bath with some temperature. The thermostat suppresses fluctuations of the kinetic energy of the system and therefore cannot produce trajectories consistent with the canonical ensemble. The temperature of the system is corrected such that the deviation exponentially decays with some time constant τ.

To maintain the temperature, the system is coupled to an external heat bath with fixed temperature T_0. The velocities are scaled at each step, such that the rate of change of temperature is proportional to the difference in temperature:

$$\frac{dT(t)}{dt} = \frac{1}{\tau}\left(T_0 - T(t)\right) \tag{10.37}$$

where τ is the coupling parameter that determines how tightly the bath and the system are coupled together. This method gives an exponential decay of the system toward the desired temperature. The change in temperature between successive time-steps is

$$\Delta T = \frac{\delta t}{\tau}\left(T_0 - T(t)\right) \tag{10.38}$$

Thus, the scaling factor for the velocities is

$$\lambda^2 = 1 + \frac{\delta t}{\tau}\left(\frac{T_0}{T\left(t - \frac{\delta t}{2}\right)} - 1\right) \tag{10.39}$$

The term $T\left(t - \dfrac{\delta t}{2}\right)$ is because the leap-frog algorithm is used for the time integration.

In practice, τ is used as an empirical parameter to adjust the strength of the coupling. Its value has to be chosen with care. In the limit $\tau \to \infty$, the Berendsen thermostat is inactive and the run is sampling a micro-canonical ensemble. The temperature fluctuations will grow until they reach the appropriate value of a micro-canonical ensemble. However, they will never reach the appropriate value for a canonical ensemble. On the other hand, too small values of τ will cause unrealistically low-temperature fluctuations. If τ is chosen the same as the time-step δt, the Berendsen thermostat is nothing else than the simple velocity scaling. Values of $\tau \approx 0.1\mathrm{ps}$ are typically used in MD simulations of condensed-phase systems. The ensemble generated when using the Berendsen thermostat is not a canonical ensemble.

Nosé–Hoover thermostat: The Berendsen thermostat is extremely efficient for relaxing a system to the target temperature, but once our system has reached equilibrium, it might be more important to probe a correct canonical ensemble. The extended system method was originally introduced by Nosé (1984) and subsequently developed by Hoover (1985). The idea of the method proposed by Nosé was to reduce the effect of an external system, acting as a heat reservoir, to an additional degree of freedom. This heat reservoir controls the temperature of the given system, i.e., the temperature fluctuates around the target value. The idea is to consider the heat bath as an integral part of the system by the addition of an artificial variable, $\tilde{s}$, associated with a "mass" $Q > 0$ as well as a velocity $\dot{\tilde{s}}$.

The magnitude of Q determines the coupling between the reservoir and the real system and so influences the temperature fluctuations. The artificial variable $\tilde{s}$ plays the role of a time-scaling parameter. More precisely, the timescale in the extended system is stretched by the factor, $\tilde{s}$:

$$d\tilde{t} = \tilde{s}\,dt \tag{10.40}$$

The atomic coordinates are identical in both systems. This leads to

$$\tilde{r} = r,\ \dot{\tilde{r}} = \tilde{s}^{-1}\dot{r},\ \tilde{s} = s \text{ and } \dot{\tilde{s}} = \tilde{s}^{-1}\dot{s} \tag{10.41}$$

The Lagrangian for the extended system is chosen to be

$$\mathcal{L} = \sum_i \frac{m_i}{2}\tilde{s}^2\,\dot{\tilde{r}}_i^{\,2} - U(\tilde{r}) + \frac{1}{2}Q\dot{\tilde{s}}^2 - gk_bT_0\ln\tilde{s} \tag{10.42}$$

The first two terms of the Lagrangian represent the kinetic energy minus the potential energy of the real system. The additional terms are the kinetic energy of $\tilde{s}$ and the potential, which is chosen to ensure that the algorithm produces a canonical ensemble where $g = N_{df}$ in real-time sampling (Nosé–Hoover formalism) and $g = N_{df} + 1$ for virtual-time sampling (Nosé formalism). This leads to the Nosé equations of motion:

$$\ddot{\tilde{r}}_i = \frac{\tilde{F}_i}{m_i\tilde{s}^2} - \frac{2\dot{\tilde{s}}\dot{\tilde{r}}_i}{\tilde{s}} \tag{10.43}$$

$$\ddot{\tilde{s}} = \frac{1}{Q\tilde{s}}\left(\sum_i m_i \tilde{s}^2 \dot{\tilde{r}}_i^2 - gk_bT_0\right)$$ (10.44)

These equations sample a micro-canonical ensemble in the extended system $(\tilde{r}, \tilde{p}, \tilde{t})$. However, the energy of the real system is not constant. Accompanying the fluctuations of $\tilde{s}$, heat transfers occur between the system and a heat bath, which regulates the system temperature. It can be shown that the equations of motion sample a canonical ensemble in the real system. The Nosé equations of motion are smooth, deterministic, and time-reversible. However, because the time-evolution of the variable $\tilde{s}$ is described by a second-order equation, heat may flow in and out of the system in an oscillatory fashion, leading to nearly periodic temperature fluctuations. The stretched timescale of the Nosé equations is not very intuitive and the sampling of a trajectory at uneven time intervals is rather impractical for the investigation of dynamical properties of a system. However, as shown by Nosé and Hoover, the Nosé equations of motion can be reformulated in terms of real system variables. The transformation is achieved through the following equation:

$$s = \tilde{s}, \ \dot{s} = \tilde{s}\dot{\tilde{s}}, \ \ddot{s} = \tilde{s}^2\ddot{\tilde{s}} + \tilde{s}\dot{\tilde{s}}^2,$$

$$r = \tilde{r}, \ \dot{r} = \tilde{s}\dot{\tilde{r}}, \ \ddot{r} = \tilde{s}^2\ddot{\tilde{r}} + \tilde{s}\dot{\tilde{r}}^2$$ (10.45)

and with substituting

$$\gamma = \frac{\dot{s}}{s}$$ (10.46)

the Lagrangian equations of motion can be written as follows:

$$\ddot{r}_i = \frac{F_i}{m_i} - \gamma r_i \ \ddot{r}_i = \frac{F_i}{m_i} - \gamma r_i$$ (10.47)

$$\dot{\gamma} = \frac{-k_B N_{df}}{Q}T(t)\left(\frac{g}{N_{df}}\frac{T_0}{T(t)} - 1\right)$$ (10.48)

In both algorithms, some care must be taken in the choice of the fictitious mass Q and extended system energy E_e. On one hand, too large values of Q (loose coupling) may cause poor temperature control (Nosé–Hoover thermostat with $Q \to \infty$ is MD which generates a micro-canonical ensemble). Although any finite (positive) mass is sufficient to guarantee in principle the generation of a canonical ensemble, if Q is too large, the canonical distribution will only be obtained after very long simulation times. On the other hand, too small values (tight coupling) may cause high-frequency temperature oscillations. The variable $\tilde{s}$ may oscillate at a very high frequency. It will tend to be off-resonance with the characteristic frequencies of the real system and effectively decouple from the physical degrees of freedom (slow exchange of kinetic energy). As a more intuitive choice for the coupling strength, the Nosé equations of motion can be expressed as follows:

$$\dot{\gamma} = \frac{-1}{\tau_{NH}} \left(\frac{g}{N_{df}} \frac{T_0}{T(t)} - 1 \right) \tag{10.49}$$

with the effective relaxation time

$$\tau_{NH}^2 = \frac{Q}{N_{df} k_B T_0} \tag{10.50}$$

The relaxation time can be estimated when calculating the frequency of the oscillations for small deviations $\delta\tilde{s}$ from the average $\tilde{s}$.

10.2 MOLECULAR DYNAMICS METHODOLOGY

10.2.1 MODELING OF MATERIALS

The present MD study has been carried out to predict the elastic moduli, tensile behavior, glass transition temperature (T_g), and interfacial shear strength (IFSS) of the Graphene(Gr) reinforced NR composite. The "Materials Studio 2017" software package (Biovia 2017) was used for modeling materials and MD simulations. First of all, a monomer of NR (cis-1,4 isoprene) was modeled as shown in Figure 10.1a using a sketch toolbar to create matrix materials. Then, head & tail atoms were marked to create the NR homopolymer chain with a polymerization index of 10 as depicted in Figure 10.1b. A single NR chain was created with 132 atoms.

Thereafter, a crystal of $40\times40\times40\,\text{Å}^3$ was created in which a Gr sheet of size $40\times40\,\text{Å}^2$ was placed centrally in the XY plane as shown in Figure 10.2a. The armchair and zig-zag edge of the Gr sheet were aligned along the X- and Y-axis respectively. The Gr sheet was formed with 680 atoms of carbon. Then, a Connolly surface was created at a distance of $1\,\text{Å}$, parallel to the plane of Gr on both sides

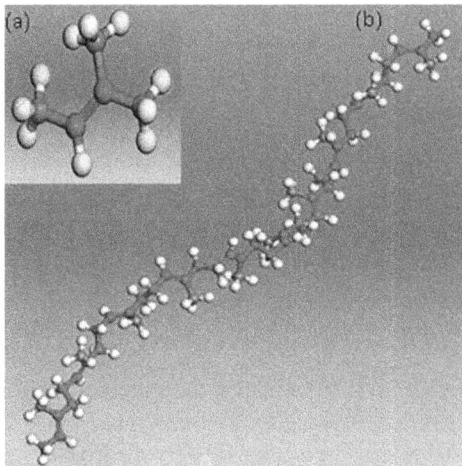

FIGURE 10.1   (a) Monomer of NR (cis-1,4 isoprene) and (b) NR homopolymer.

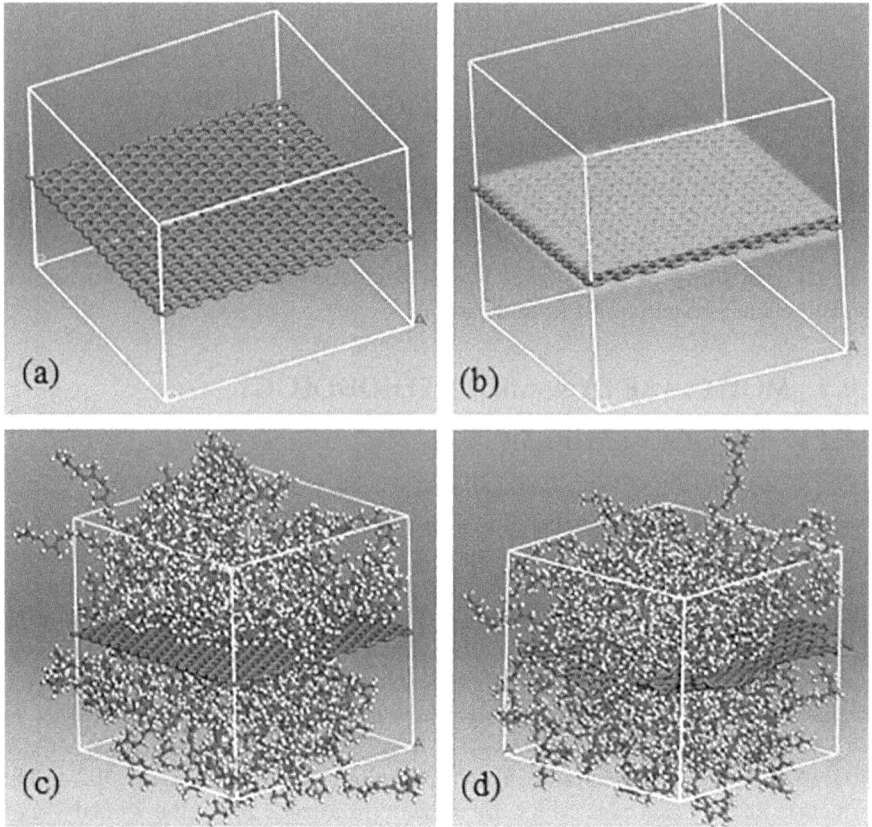

FIGURE 10.2 (a) Gr placed in the XY plane in the center of $40 \times 40 \times 40\,\text{Å}^3$ crystal, (b) iso-surface created on both sides of Gr sheet, (c) Gr-reinforced NR composite after packing, and (d) optimized geometry of composite.

of the Gr sheet to segregate different volumes in crystal, and then isosurface was created on it to avoid the packing of NR inside the Gr hexagonal rings as shown in Figure 10.2b. The estimation of energy was made with the COMPASS forcefield (Sun 1998). The 48 NR chain was packed inside this crystal with a density of 0.93 g cc^{-1} using an amorphous cell module. The packed crystal was composed of 7,016 atoms of Gr and NR which is shown in Figure 10.2c. The packed crystal was optimized for 5,000 iterations using the conjugate gradient method (Fletcher et al. 1964) with a convergence tolerance of 0.0001 kcal mol^{-1} for energy, 0.005 kcal $mol^{-1}\text{Å}^{-1}$ for force, and $5 \times 10^{-5}\text{Å}$ for displacement. The optimized geometry of the Gr-NR composite is shown in Figure 10.2d. Further, the system was equilibrated for 60 ps at room temperature and pressure, using an NPT ensemble with a time-step of 1 fs. The "Berendsen" thermostat (Berendsen et al. 1984) and barostat were used for temperature and pressure equilibration. After dynamics, a final density of 0.976 g cc^{-1} was achieved and shown in Figure 10.3.

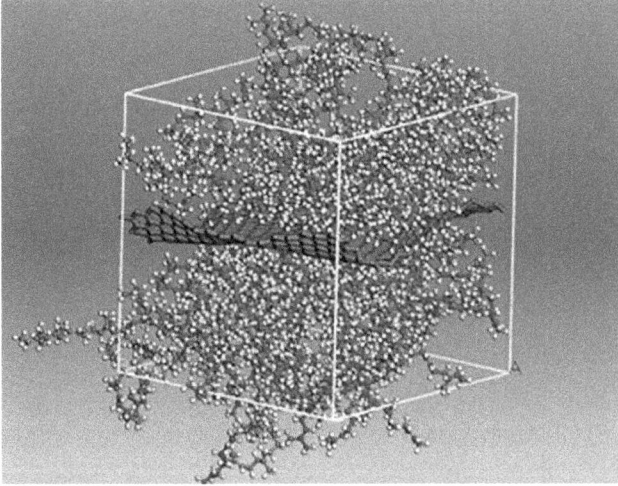

FIGURE 10.3 Gr-NR crystal after equilibration.

TABLE 10.3
Mechanical Properties of Gr-Reinforced NR Composite

Young's Modulus (GPa)	Bulk Modulus (GPa)	Shear Modulus (GPa)
$X = 4.7330$	Reuss $= 2.1593$	Reuss $= 0.8617$
$Y = 3.8108$	Voigt $= 3.4297$	Voigt $= 1.1859$
$Z = 1.6967$	Hill $= 2.7945$	Hill $= 1.0238$

10.2.2 Modeling for Elastic Moduli, Tensile Behavior, and T_g

Thereafter, the equilibrated system was strained in four steps with a maximum amplitude of 0.003 and the geometry was optimized after each strain to evaluate the elastic moduli. The evaluated elastic moduli are listed in Table 10.3. Further, the study was extended to predict the behavior of the material under tensile loading. For this purpose, a Perl script provided by Materials Studio was used to stretch the system under a sequential loading and corresponding strains were recorded. To obtain T_g, the trajectory obtained after the dynamics of the system was equilibrated at NPT ensemble many times at different temperatures with a step of 10 K ranging from 150 to 400 K. The obtained density from different frames was averaged over each temperature and a variation in density was recorded with respect to temperature. The sudden change in slope of the density-temperature curve was marked as T_g.

10.2.3 Modeling for Pull-Out of Graphene(Gr) from NR

First of all, the equilibrated crystal was made nonperiodic in the pull-out direction by adding a vacuum slab in the X-direction. The right end (pulling end highlighted

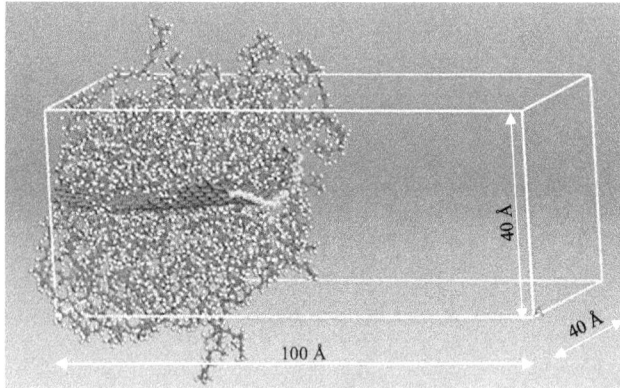

FIGURE 10.4 Pull-out model of Gr from NR with vacuum slab of 60 Å when the Gr was fully embedded in NR.

in yellow) of the Gr sheet as shown in Figure 10.4 was constrained in all the directions to avoid the bending of the Gr sheet during geometry optimization. Thereafter, the geometry was optimized again using the previous convergence tolerance. Further, the system was equilibrated again using an NVT ensemble for 20 ps with the previously used parameters. After dynamics, the non-bonded energy of the system was recorded. In the next step, the Gr sheet was displaced by 5 Å in the X-direction from the optimized geometry in the previous step and the dynamics were performed again for 20 ps. This incremental displacement ($\Delta x = 5$ Å) and dynamics sequence were followed repeatedly until the Gr sheet was completely pulled out from NR as shown in Figure 10.5. The total change in non-bonded energy was used to evaluate the pull-out force and IFSS.

10.3   RESULTS AND DISCUSSION

In this section, the results obtained by using the MD simulation have been discussed in detail. The properties obtained are the mechanical properties such as Young's moduli, bulk moduli, and shear moduli, the glass transition temperature, and the interfacial shear stress of Gr-NR composites.

10.3.1 MECHANICAL PROPERTIES

Table 10.3 shows the mechanical properties of Gr-reinforced NR composites. Young's modulus of the Gr-NR composite in the x-y plane for a cell density of 0.976 g cc^{-1} was found to be higher in comparison to Young's modulus in the transverse direction (Z-direction). The reinforcement of Gr improved the elastic moduli of NR in the plane of Gr from 2.93 GPa (Wu et al. 2021) to 4.73 and 3.26 in the X- and Y-directions by 61% and 11% respectively. The Gr was reinforced in NR such that its zig-zag and armchair edges were aligned along the X- and Y-axis respectively and the chirality of

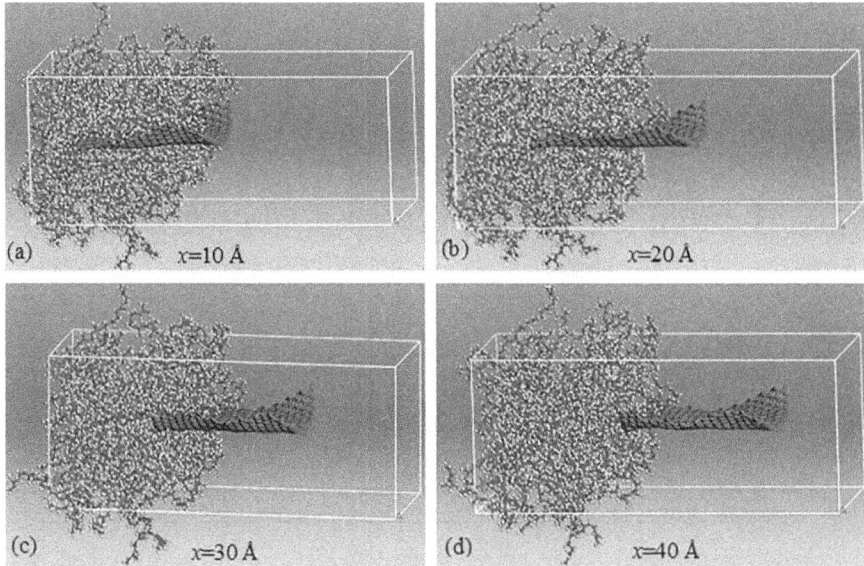

FIGURE 10.5   Pulled-out Gr sheet from NR at different displacement positions, x.

the Gr sheet affects the stiffness of the composite. Hence, there was a difference in Young's modulus in the X- and Y-directions. While in the Z-direction, the modulus was decreased to 1.70 GPa by 42%, due to the discontinuity caused by the Gr sheet in the Z-direction. Young's modulus of the composite in the plane of the Gr sheet (XY) was found to be approximately 65% higher in comparison to the Z-direction. The average bulk modulus of the Gr-NR composite was enhanced to 2.8 GPa from 1.84 GPa (Wu et al. 2021) and 2.25 GPa (Diani et al. 2008) of NR by 52%.

The average shear modulus of 1.02 GPa was achieved by Gr-NR composite with an improvement of 44%, which was much higher in comparison to the shear modulus of NR's, i.e., 0.71 GPa (Wu et al. 2021). Hence, the reinforcement of Gr in NR enhanced Young's modulus, Bulk Modulus, and Shear modulus except for Young's modulus in the direction perpendicular to the plane of Gr. Further, the tensile behavior of the Gr-NR composite in the X-direction is shown in Figure 10.6. The stress-strain characteristics were found to be linear up to 10% strain, and after that, it shows the non-linear response with an increase in tensile stress. The ultimate tensile strength (UTS) of 0.295 GPa was achieved at a strain of 24% with reinforcement of Gr in NR. After achieving the UTS, the material keeps on straining because there is no bond breakage considered in COMPASS forcefield and atoms used to interact even after failure. The atomic arrangement of the Gr-NR composite at a strain of 24% is shown in Figure 10.7, which reveals that the weak interfacial bonding between Gr and NR was the main factor of failure. The ultimate tensile strength of 0.295 GPa is obtained without considering the bond breakage using the COMPASS forcefield. The tensile strength value is highly dependent on the type of force field used. MD gives us the

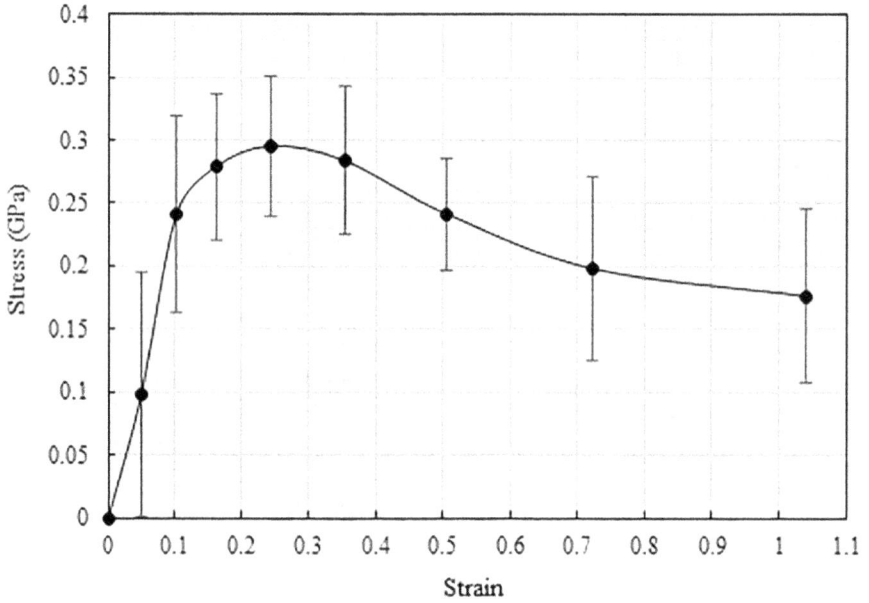

FIGURE 10.6   Stress-strain behavior of Gr-reinforced NR nanocomposites.

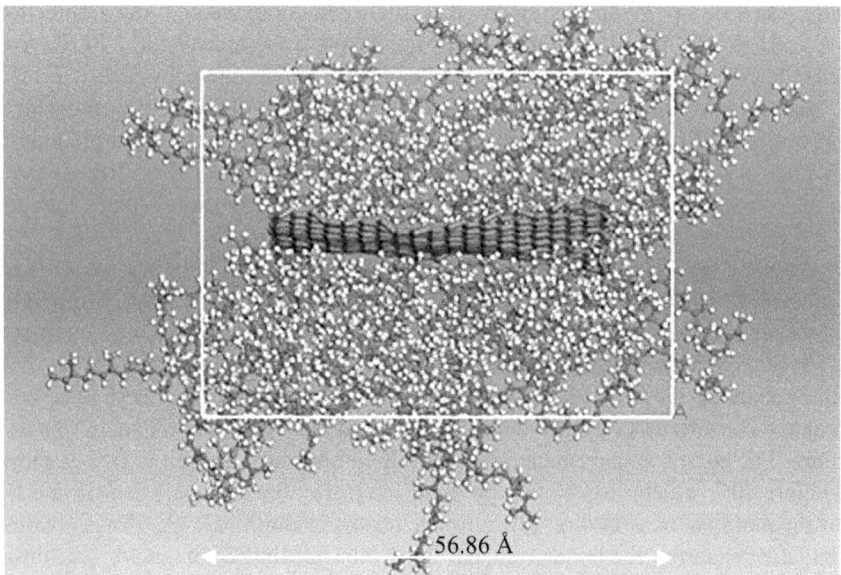

FIGURE 10.7   The Gr-NR nanocomposite under ultimate tensile stress (UTS) at strain = 0.24.

trend of variation but the individual values are higher than the experimental values because of various assumptions as follows:

i. The material is assumed to be free of any defects (in MD) whereas, in experiments, the material generally has a lot of defects.
ii. The properties at the nanoscale are considerably different from experimental values. We cannot compare the two. However, we can get a trend of variation of the properties. The trend variation will be the same for both MD as well as experiments.
iii. The authors cannot find any paper where the properties of Gr-reinforced rubber composites have been predicted using MD. This study will be helpful for researchers working in the molecular modeling of these composites.

10.3.2 GLASS TRANSITION TEMPERATURE

The variation in density of Gr-NR composite with temperature is shown in Figure 10.8. The density was found to be decreasing with an increase in temperature. The composite was heated from 150 to 400 K which results in a drop of density from 1.05 to 0.93 g cc^{-1}. During heating, it was observed that the density gradient with respect to temperature up to 280 K is approximately the same but if the temperature was further increased there was an increase in the density gradient, which revealed that the gelation process (rubbery state) has been started at 280 K. So, T_g was estimated by the intersection of two different fitting lines, i.e., 280 K,

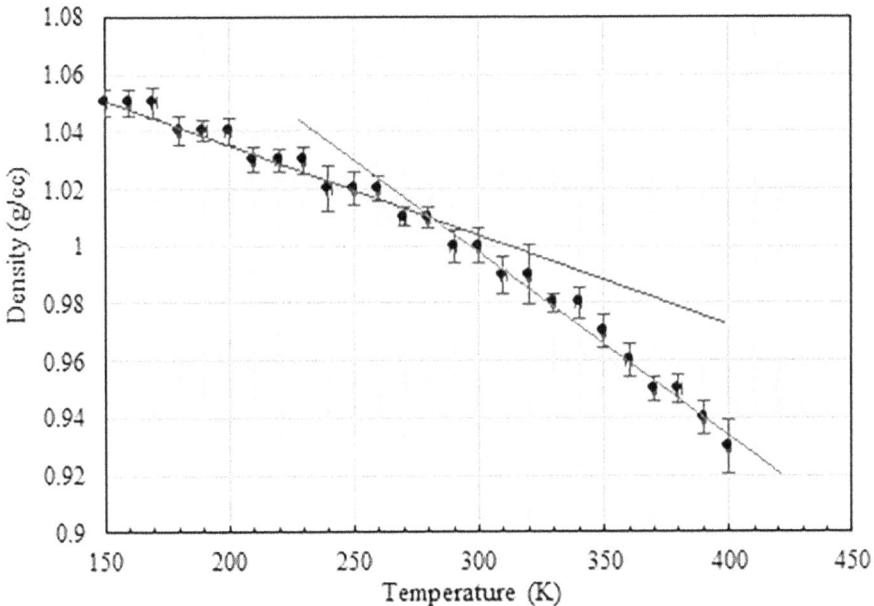

FIGURE 10.8 Variation of density with the temperature of Gr-NR nanocomposite.

as shown in Figure 10.8. The glass transition temperature T_g of NR was improved by 40% from 200 K (Loadman 1985) to 280 K with Gr reinforcement. The threshold density was found to be 1.01 gm cc^{-1} which was much higher than the density of NR at room temperature. The threshold density was found to be closer with Yolong et al. (2019). The NR chains nearer to Gr were glassy and ordered but rubbery when it was far from the Gr which may induce the local glass transition at a temperature above T_g.

10.3.3 INTERFACIAL PROPERTIES

The interfacial properties have been evaluated with a pull-out of the Gr sheet from the NR matrix and shown in Figure 10.5. Since there was no interfacial bonding at the interface of Gr and NR except van der Waals interaction, i.e., a type of non-bonded energy. Hence, the change in non-bonding energy was the main considerable factor to predict the interfacial bonding of Gr and NR. Figure 10.9 illustrates the variation in non-bonded energy and potential energy with Gr's pull-out displacement. The energy of the system was found to be increasing with an increase in pull-out displacement of Gr sheet from NR which revealed less stability of the system. The non-bonded energy increases linearly from fully embedded Gr in NR to 30 Å pull-out of Gr from NR, and after that, there was a slight increase. When the Gr sheet was pulled out from NR, there were formation and breaking of van der Waals interaction that took place due to which the change in non-bonded energy is higher at initial stages rather than beyond 30 Å pull-out. The cutoff distance for van der Waals interaction was set to be 15.5 Å, which deduce the interaction of NR and Gr significantly if the

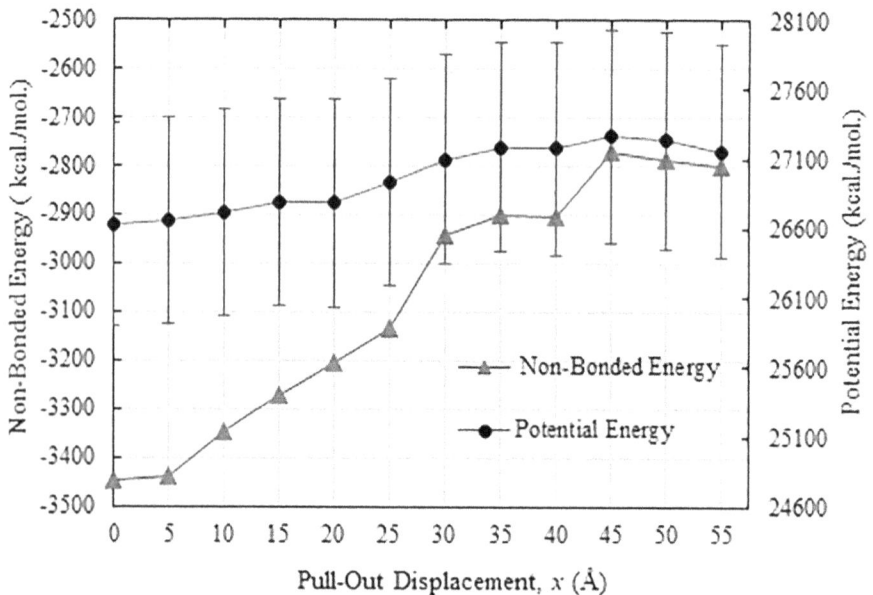

FIGURE 10.9 Variation of the non-bonded and potential energy of Gr-NR with pull-out displacement.

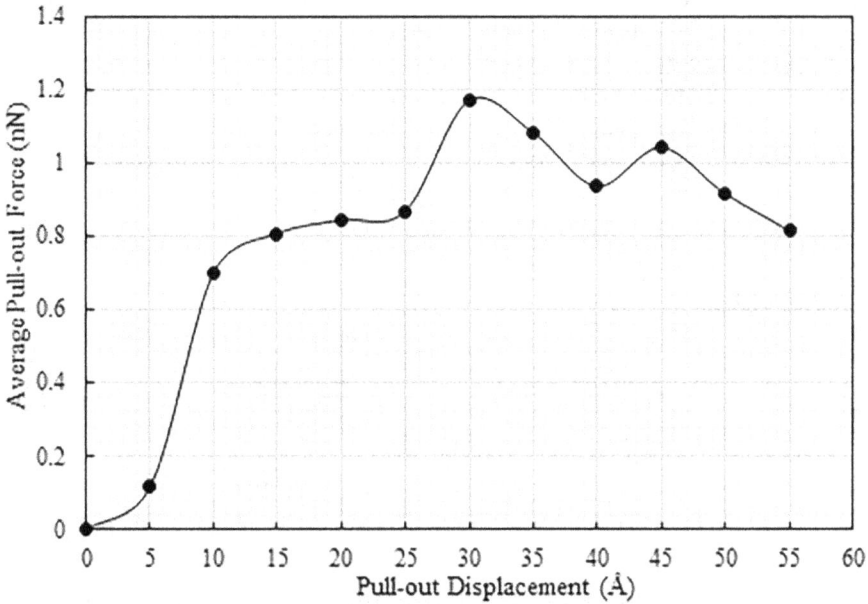

FIGURE 10.10   Variation of the average pull-out force of Gr from NR with pull-out displacement.

Gr was pulled out more than $30\,\text{Å}$. After the complete pull-out of the Gr sheet, there was a slight variation in non-bonded energy due to nominal interaction.

A similar trend was observed for change in potential energy. Figure 10.10 revealed the average pull-out force required to pull the Gr from NR to increase up to a pull-out displacement of $30\,\text{Å}$ and to decrease after that. Hence, there was a continuous increase in the pull-out effort up to $30\,\text{Å}$ pull-out of Gr, and after that, the effort kept on decreasing. Further, IFSS was evaluated by considering the change in non-bonded energy ($506\,\text{kcal mol}^{-1}$) which was caused by work done in pulling the Gr from NR by using the following relation:

$$\text{IFSS (Pa)} = \frac{\text{Change in non bonded energy to completely pullout Gr from NR (Joule)}}{\text{Length of Gr(m)} \times 2 \times \text{Area of Gr sheet} (\text{m}^2)}$$

$$(10.51)$$

The estimated IFSS was found to be $28.8\,\text{MPa}$.

10.4   CONCLUSIONS

In this study, MD simulations were carried out to estimate the elastic moduli, tensile behavior, glass transition temperature, and IFSS of Gr-reinforced NR composites. The study revealed that Gr is an effective reinforcement to improve all the above-mentioned properties except the transverse modulus in a direction perpendicular to the plane of the Gr sheet. Young's modulus of NR was enhanced by 61% in the X-direction and 11% in the Y-direction, while the shear modulus and bulk modulus by 44% and 52% respectively due to Gr reinforcement. The Gr-reinforced NR achieved a UTS of 0.295 GPa. The Gr-NR composite exhibits a T_g of 280 K, which was 40% higher than that of NR and achieved a threshold density of 1.01 gm cc^{-1}. Finally, the interfacial shear strength was predicted using pull-out of Gr from NR, i.e., 28.8 MPa. Thus, this study revealed that the Gr is a promising reinforcement to improve the mechanical properties of NR significantly.

REFERENCES

Andersen, H.C., 1980. Molecular dynamics simulations at constant pressure and/or temperature. *J. Chem. Phys.* 72(4), 2384–2393. doi:10.1063/1.439486.

Berendsen, H.J., Postma, J.V., Van Gunsteren, W.F., DiNola, A.R.H.J., Haak, J.R., 1984. Molecular dynamics with coupling to an external bath. *J. Chem. Phys.* 81(8), 3684–3690. doi:10.1063/1.448118

Biovia, D. S. 2017. *Materials Studio*. R2 Dassault Systèmes BIOVIA, San Diego.

Brenner, D.W., 1990. Empirical potential for hydrocarbons for use in simulating the chemical vapor deposition of diamond films. *Phys. Rev. B* 42(15), 9458. doi:10.1103/PhysRevB.42.9458.

Diani, J., Fayolle, B., Gilormini, P., 2008. Study on the temperature dependence of the bulk modulus of polyisoprene by molecular dynamics simulations. *Mol. Simul.* 34(10–15), 1143–1148. doi:10.1080/08927020801993388.

Fletcher, R., Reeves, C.M., 1964. Function minimization by conjugate gradients. *Comput. J.* 7(2), 149–154. doi:10.1093/comjnl/7.2.149.

Hoover, W.G., 1985. Canonical dynamics: Equilibrium phase-space distributions. *Phys. Rev. A* 31(3), 1695. doi:10.1103/PhysRevA.31.1695.

Jones, J.E., 1924. On the determination of molecular fields.—II. From the equation of state of a gas. *Proc. R. Soc. Lond. A Math Phys. Sci.* 106(738), 463–477. doi:10.1098/rspa.1924.0082.

Loadman, M.J.R., 1985. The glass transition temperature of natural rubber. *J. Therm. Anal. Calorim.* 30(4), 929–941. doi:10.1007/bf01913321.

Morse, P.M., 1929. Diatomic molecules according to the wave mechanics. II. Vibrational levels. *Phys. Rev.* 34(1), 57. doi:10.1103/PhysRev.34.57.

Nosé, S., 1984. A unified formulation of the constant temperature molecular dynamics methods. *J. Chem. Phys.* 81(1), 511–519. doi:10.1063/1.447334.

Novoselov, K.S., Geim, A.K., Morozov, S.V., Jiang, D.E., Zhang, Y., Dubonos, S.V., Grigorieva, I.V., Firsov, A.A., 2004. Electric field effect in atomically thin carbon films. *Science* 306(5696), 666–669. doi:10.1126/science.1102896.

Sun, H., 1998. COMPASS: An ab initio force-field optimized for condensed-phase applications overview with details on alkane and benzene compounds. *J. Phys. Chem. B* 102(38), 7338–7364. doi:10.1021/jp980939v.

Tersoff, J., 1988. New empirical approach for the structure and energy of covalent systems. *Phys. Rev. B* 37(12), 6991. doi:10.1103/PhysRevB.37.6991.

Wu, J., Teng, F., Su, B., Wang, Y., 2021. Molecular dynamics study on tribological properties of EUG/NR composites. *Comput. Mater. Sci.* 199, 110732. doi:10.1016/j.commatsci.2021.110732.

Yolong, T., Sutthibutpong, T., 2019. Local glass transition of polyisoprene induced by graphene planes observed in silico through atomistic molecular dynamics simulations. *J. Phys. Conf. Ser.* 1380(1), 012081. doi:10.1088/1742-6596/1380/1/012081.

11 Graphene-Elastomer Composites for Barrier Applications

Aswathy T. R., Asit Baran Bhattacharya, and Kinsuk Naskar
IIT Kharagpur

CONTENTS

11.1 INTRODUCTION

Since the first discovery of graphene by Geim and his co-workers in 2004 (Novoselov et al., 2014), the graphene got tremendous attention in various fields such as physics and material science. Graphene is an allotrope of carbon, where sp^2 hybridized carbon atoms are arranged in one atom thickness planar sheets in a honeycomb-like

DOI: 10.1201/9781003200444-11

lattice. Because of the high aspect ratio structure of graphene, it possesses extraordinary characteristics such as high mechanical properties, thermal conductivity, and very high air impermeability (Kaneko et al., 2017). Because of the presence of closely spaced carbon atoms in graphene, it is the most air-impermeable material discoursed so far (Sun et al., 2020). A study by Geim et al. shows that the membranes containing graphene are impermeable to all gases and liquids (Schedin et al., 2007). If these properties can successfully couple with elastomers, the resulting elastomer-graphene composites will have high air impermeability and better mechanical properties. Gusev et al. (2001) studied the effect of various fillers having different shapes on the permeability properties of polymer composites and confirmed that, as the aspect ratio and the volume fraction of the filler increase, the air impermeability of the nanocomposites also increases. Graphene, has comparatively higher aspect ratio compared to other flake-like fillers. When comparing with nanoclay (NC), the aspect ratios of graphene and NC are 130 and 108 respectively (Wu et al., 2013). The higher aspect ratio of graphene can induce higher level of air impermeability in various rubber matrices.

Modification of rubber using NC is usually applied for the fabrication of highly air-impermeable rubber composites for various applications. After the establishment of graphene, from 2013 onwards, researches on graphene-modified rubber nanocomposite became one of the significant topics. The combined properties of conductivity of carbonaceous materials with the structural characteristics of clay sheets make graphene capable of providing multifunctional rubber nanocomposites. In 2015, with collaboration of Chengdu Trustwell company with Sichuan University, fabricated graphene-based NR composites were with high electrical conductivity and barrier performance (Wang et al., 2018). In addition to that, a study by Compton et al. (2010) shows that the barrier properties of graphene nanoplatelets are 25–100 times superior than the barrier properties of NC. Even though, for the preparation of graphene/rubber nanocomposites, few problems need to be addressed, the main issue concerning the development of graphene nanocomposites is in obtaining a good dispersion of graphene layers in the rubber matrix. The strong π–π (pi–pi) interactions between the stacked graphene monolayers and the high viscosity of the molten rubber matrix make the dispersion of graphene still challenging (Mondal et al., 2012). Due to the difficulties in developing an even dispersion graphene in rubber matrix, there were no studies reported till 2011 regarding the development and characterization of graphene/rubber nanocomposites. The aggregation of graphene in the elastomer can adversely affect the properties of the composites; even it can lead to the deterioration of barrier properties.

The mixing of graphene with rubber matrix can result in two types of morphologies: Intercalated, and exfoliated (Figure 11.1). The formation of either intercalated or exfoliated morphology of graphene in the elastomer matrix can lead to an increment in the physical as well as barrier properties. The high aspect ratio fillers interact with polymer chains, thus hindering the relaxation of polymer chains, and the presence of such high aspect ratio fillers on the direction perpendicular to the direction of motion of gas molecules will make the gas particles to travel along pathway. This effects will in turn reduce the air permeability of matrix and increase the physical properties of the composites.

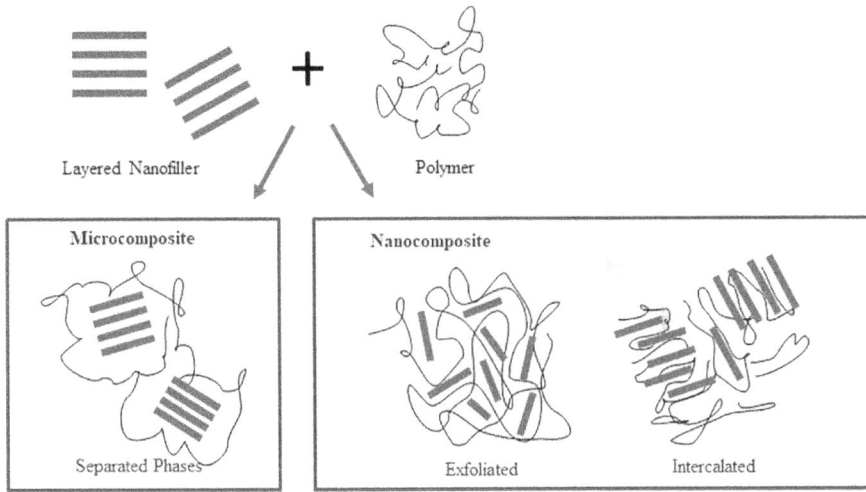

FIGURE 11.1 The schematic representation of various morphologies developed in rubber nanocomposites.

Graphene is currently produced by a variety of approaches, such as chemical vapour deposition (CVD), epitaxial growth, micromechanical exfoliation, liquid exfoliation, and oxidation-reduction as discussed in the earlier chapters in this book. From these methods, liquid exfoliation and oxidation-reduction methods are widely used for the industrial production of graphene. However, there is an emerging tendency in using graphene and its various derivatives for the fabrication of rubber nanocomposites for various barrier applications. One of the main derivatives used for the development of rubber nanocomposites is graphene oxide (GO) and it is produced by exfoliating and reducing the graphite through various methods.

11.2 THE EFFECT OF GRAPHENE ON THE AIR PERMEABILITY OF THE RUBBER COMPOSITES

The process of passage of gas molecules through rubber composites occurs mainly in two stages: in the first stage, the gas molecules will get adhered to the holes present in the surfaces of the rubber composites, and in the second stage, the gas molecules will diffuse through the rubber matrix by jumping into the neighbouring holes. When it comes to the rubber nanocomposites containing high aspect ratio fillers like graphene, the passage of gas molecules through it becomes more complex. The usage of high aspect ratio filler at its completely dispersed state can increase the air impermeability by 50 times of the initial value. The layered fillers in its dispersed state will act as obstacles for the diffusing gas molecules in the rubber matrix, thus creating a longer pathway than the usual, known as tortuosity effect (Wolf et al., 2018). The schematic representation for the passage of gas molecules in the presence of various fillers is represented in Figure 11.2.

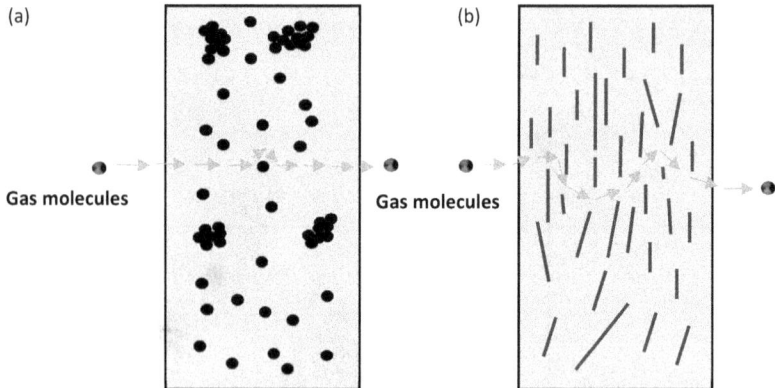

FIGURE 11.2   The schematic representation of passage of gas molecules different fillers.

The main applications of air-impermeable rubber nanocomposites lie in the production of tyre inner-liners and pharmaceutical stoppers. For the production of impermeable rubber nanocomposites, the selection of the elastomer for barrier applications typically considers the permeation, physical, economic, and recycling properties. Halogenated butyl rubber, especially brominated butyl rubber (BIIR), which is having excellent air retention properties is extensively used in tyre inner-liner applications. In addition to IIR/(X) IIR, there are other elastomers which are having high air impermeability properties, for example, epoxidized natural rubber (ENR), epichlorohydrin rubber (ECH), and hydrogenated acrylonitrile butadiene rubber (HNBR). For dynamic applications such as tyre inner-liners, the development of a highly air-impermeable rubber nanocomposite with adequate physical properties is very important. So, the rubbers such as ENR and ECH are not used in tyre industries due to the absence of required physical characteristics.

It is well-known that the fillers with high aspect ratio can improve the air impermeability properties with the elastomer by creating a tortuous pathway. By using butyl rubber as the matrix and graphene as the filler, the air impermeability properties of the butyl rubber can be improved 50 times better than the usual butyl-carbon black (CB) composites. However, the agglomeration of graphene in the graphene/rubber nanocomposites still remains as the main issue for industrial applications. To overcome this issue, several works have been carried out including the modification of graphene and the experimentation of different compounding methods. One of the promising ways to enhance the dispersion of graphene in the rubber matrix is by increasing the chemical/physical interactions between the rubber and the graphene. The grafting of graphene or rubber using various chemical entities makes the dispersion easy. Even though the production time and cost increases extremely, in the last few years, making blends of bromobutyl rubber with various rubbers and usage of hybrid nanofillers came as most economical and viable way for preparing a fine BIIR/graphene-rubber nanocomposite. The important discoveries in terms of getting a better dispersion of graphene in the BIIR rubber matrix for improving the air impermeability lead to revolution of traditional rubber nanocomposites.

This chapter will focus on the state of the art on graphene/bromobutyl rubber (BIIR) nanocomposites with specific focus on the preparation approaches, structure, and properties, mainly for inner-liner applications.

11.3 VARIOUS RUBBER MATERIALS USED FOR BARRIER APPLICATIONS

11.3.1 BUTYL RUBBER

The discussion about gas permeability in elastomers will eventually lead to butyl rubber (Figure 11.3) because this is the only polymer replaced with natural rubber in tyre making for air retention properties of pneumatic tyres. In 1930, Mr. I.G. Farben, a prominent scientist in Germany, synthesized high molecular weight Polyisobutylene at a very low temperature by homopolymerization of isobutylene. The polymer showed very good ageing and low gas permeability properties. Higher air impermeability of butyl rubber is mainly due to the presence of isobutylene group present on the backbone of the polymer. Polyisobutylene, however, was not found curable with sulphur /accelerators system and hence was not suitable for tyre industries. To make polyisobutene rubber commercially applicable, the isobutene monomer is copolymerized with 1%–2% of isoprene monomer and the resultant polymer is known as butyl rubber. Isoprene is copolymerized mainly by a trans-1,4 addition and introduces unsaturation in the polymer, that is, one double bond for each isoprene unit (Long et al., 2001).

11.3.1.1 Halogenated Butyl Rubber

With halogenations of butyl, both chlorobutyl and bromobutyl are formed. Chlorine (Cl_2) or Bromine (Br_2) gas is passed through the butyl polymer solution for the production of chlorobutyl rubber (CIIR) and bromobutyl rubber (BIIR) (Figure 11.4). Prior to halogenation, the butyl must be dissolved in a suitable solvent (hexane, pentane, etc.) and all unreacted monomers need to be removed. Several different processes are currently used these days to prepare butyl solution for halogenation. Either reactor effluent polymer, in-process rubber crumb, or butyl product bales may be dissolved in solvent in preparation for halogenation. Both Bromine and chlorine vapour should be added to the butyl solution in a very controlled condition in a highly agitated

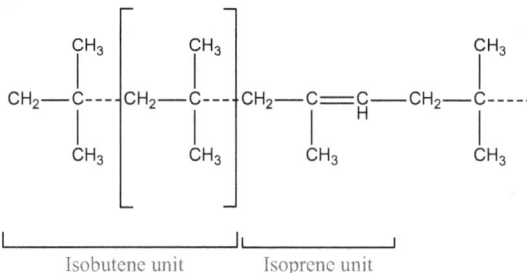

FIGURE 11.3 The structure of butyl rubber (IIR).

(a)

structure 1   structure 2

(b)

structure 1   structure 2

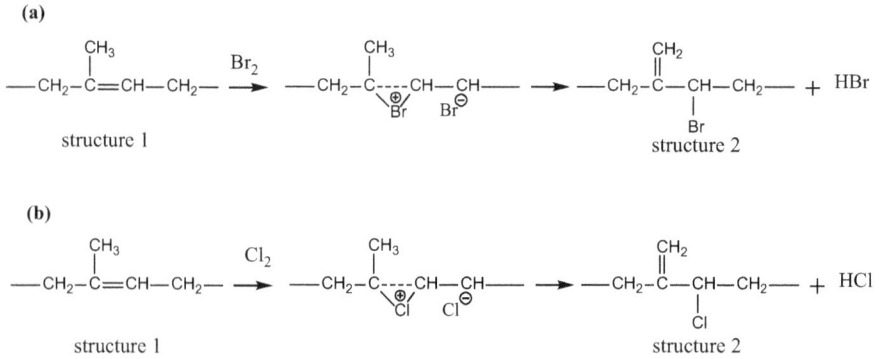

FIGURE 11.4   Bromination (a) and Chlorination (b) chemistry of butyl rubber.

TABLE 11.1
Various Characteristics of BIIR and CIIR

Characteristics	BIIR	CIIR
Vulcanization systems	Sulphur alone, ZnO, conventional EV/semi-EV system, Phenol formaldehyde resin, Peroxide	ZnO, conventional EV/semi-EV system, Phenol formaldehyde resin, hexamethylenediamine carbamate
Cure rate	Fast cure rate	Slower compared to BIIR
Co-Curing	Good	Poor
Adhesion	Better adhesion to metals and other diene rubbers	Poor adhesion

reaction vessels. The halogenation of the IIR allows co-vulcanization and enhances the compatibility with rubber having larger amount of diene content (Threadingham et al., 2011).

Compared to CIIR, BIIR is most commonly used for the production of rubber composites for barrier application. BIIR has advantages over the CIIR due to its greater cure versatility and faster cure rate, and BIIR will offer a better adhesion towards unsaturated rubbers such as natural rubber (NR), styrene-butadiene rubber (SBR), chloroprene rubber (CR), and acrylonitrile butadiene rubber (NBR). Table 11.1 describes the differences between CIIR and BIIR.

11.3.2   EPOXIDIZED NATURAL RUBBER (ENR)

When it comes to mechanical properties, NR always stand superior to other rubbers. Even though the NR has some drawbacks in terms of its oil resistance and permeability properties, chemical modifications of NR are always chosen for overcoming these downsides in which the epoxidation of NR gives the best results compared to the other ones. The ENR is prepared from NR latex using hydrogen peroxide/formic acid in an in situ technique (Figure 11.5). The introduction of epoxidation into the

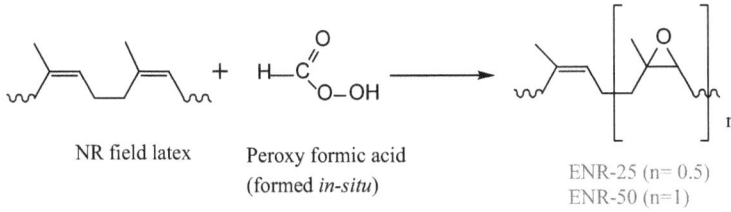

FIGURE 11.5   Preparation of epoxidized natural rubber from NR latex.

NR backbone will enhance the properties of NR such as rolling resistance, wet grip, damping, adhesion, and oil resistance.

As the epoxidation increases, the air impermeability of the NR also increases with it. For 70% epoxidation, the air impermeability of ENR is nearly closer to IIR (Figure 11.6). However, for 70% epoxidized ENR, the glass transition temperature is too high, which leads to processing difficulties of the rubber. So, ENR 25 and ENR 50 are the commonly used grades for industrial applications. In the past few years, the research articles published with the usage of ENR for various barrier applications. Wei et al. (2019) studied the effect of ENR 50 as compatibilizer in Epoxy/vulcanized natural rubber/GNP systems and the results show an improved dispersion of GNP with the increase in ENR-50 content. The enhanced dispersion of GNP in the matrix results in an improved mechanical properties of the end product.

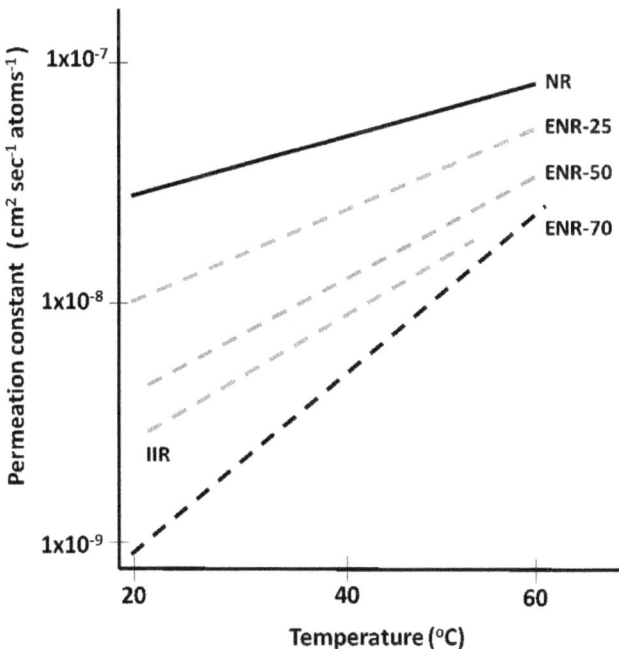

FIGURE 11.6   Comparative air permeability of ENR with IIR.

$$\left(\!\!-\overset{\displaystyle CH}{\underset{\displaystyle CH_2Cl}{|}}\!\!-CH_2\!-\!O\!-\!\right)_{\!n}$$

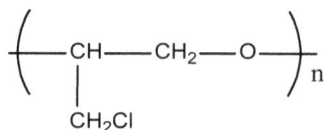

FIGURE 11.7   The structure of Polyepichlorohydrin rubber (ECH).

11.3.3   POLYEPICHLOROHYDRIN RUBBER (ECH)

High molecular weight polymers of epichlorohydrin (Figure 11.7) were prepared by E.J Vandenberg in 1957 at Hercules Powder Co. using aluminum alkyl catalyst systems. Due to the presence of highly polar chloromethyl groups in the backbone of the epichlorohydrin rubber (ECH), it has a very high impermeability towards various gases. The gas impermeability of the ECH is comparable with IIR. So, ECH became one of the potential rubbers which can replace BIIR and CIIR in various applications.

The reason for the high air impermeability properties can be explained with the cohesive energy density and the crystallinity of ECH. The crystalline domains present inside the rubber matrix act as huge cross-links inside the rubber matrix which in turn act as obstacles for the passage of gas molecules through the matrix. ECH has good resistance to swelling in oils, good heat resistance, and excellent temperature flexibility. It is used in the automotive industry for making seals, diaphragms, and hoses. ECH is especially used in applications where the membranes come in contact with aromatic hydrocarbons. Lu et al. (2007) studied the microstructure and properties of highly filled ECH/ NC composites and reported that the composites developed are suitable as sealant material in aerospace applications.

11.4   PREPARATION OF GRAPHENE-ELASTOMER COMPOSITES THROUGH DIFFERENT METHODS

Since graphene have the properties required for an ideal nanofiller for the rubber matrix, several mixing methods are developed in order to obtain an efficient rubber/ graphene nanocomposite as shown in Table 11.2 in which in situ polymerization, solution mixing, latex mixing, and melt mixing became more common. Although in situ polymerization, solution mixing, and latex mixing give a thorough dispersion of graphene in the rubber matrix, there are still some major limitations including solvent handling and evaporation, which is environmentally unsafe. In the case of in situ polymerization technique, there is a chance for graphene to act negatively during the polymerization process which in turn reduces the output of the polymerization process. In latex mixing method, graphene should be added into the latex either directly or in the form of slurry followed by stirring and drying. Kim et al. (2010) studied the properties of multilayered graphene (MLG)/SBR composites, prepared with latex mixing method and the composites prepared show significant improvement in thermal stability and electric conductivity properties. The main drawbacks of latex mixing are that it is time consuming and it requires large amount of water for the process. For solvent mixing method, the rubber needs to be dissolved in a suitable solvent before the addition of graphene. After mixing the graphene into the solvent,

TABLE 11.2

The Advantages and Disadvantages of Various Mixing Methods Used for the Preparation of Graphene-Rubber Nanocomposites

Mixing Method	Advantages	Disadvantages
In situ polymerization	• Better dispersion	• Side effects, such as graphene to act negatively during the polymerization process
Solution mixing	• Uniform dispersion of graphene can be achieved up to some degree	• The dispersion of graphene down to single or few layers in organic solvent is difficult to realize • Organic solvents need to recycled • Environmentally unsafe
Latex mixing	• Well dispersion • Low cost • Environmental friendly	• Too much water is required • Energy consumption is high
Melt mixing	• Solvent free • Low cost • Simple technical process	• Difficulty in obtaining a uniform dispersion

the solvent should be removed by evaporation method. Bai et al. prepared the dispersion of GO in dimethyl formamide (DMF) using ultrasonically assisted solution mixing. Then, the graphene dispersion as received was added to solution of nitrile-butadiene rubber (NBR) in tetrahydrofuran and then the mixture was subjected to ultrasonic dispersion, drying, and two-roll milling (Bai et al., 2011) (Figure 11.8). The removal of the solvent after mixing makes the solvent mixing method unsuitable for commercial production methods.

Given this, melt mixing becomes the most economical method for the production of graphene-rubber composites in bulk. In melt compounding, the rubber and all the other additives were added into an internal mixer and mixing was done by applying mechanical shearing. Song et al. prepared the graphene/silicone rubber composites by mechanical mixing using the Flacktek Speed Mixer and found that the mechanical properties and thermal stability of silicone rubber composites were enhanced even though the self-aggregation of graphene nanoplatelets due to the π–π interactions is the main concern. However, the melt compounding has its practical significance because it requires only the conventional rubber compounding machine. In order to achieve a fine dispersion of graphene in rubber through melt mixing method, modifications should be done in both raw materials and the mixing procedure. In the last decade, the researchers developed several methods including modification of graphene and compounding procedures. For improving the barrier properties of in IIR/(X)IIR rubber matrix, the graphene needed to be dispersed evenly in the matrix. This chapter mainly discusses the recent developments in IIR/(X)IIR formulation, in order to obtain a fine dispersion of GNP through melt mixing method with unique reference to barrier applications.

TEM image of the
rubber-GNP Nanocomposite

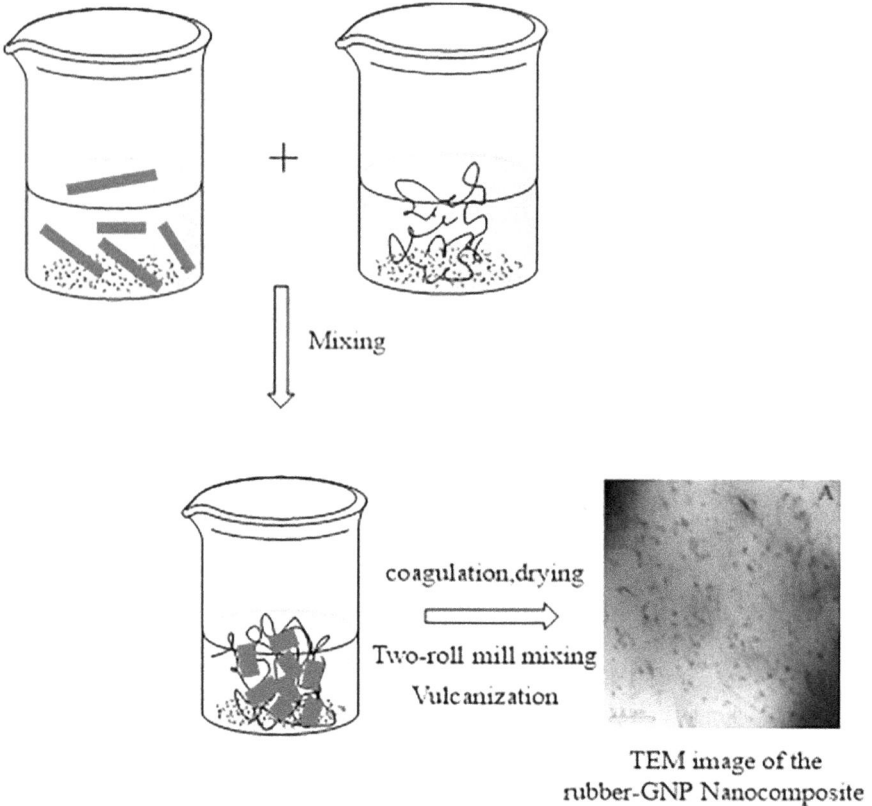

FIGURE 11.8   Schematic diagram for the preparation of GNP-rubber nanocomposites using solution mixing method.

11.5   GRAPHENE IN BUTYL RUBBER (IIR)-HALOGENATED BUTYL RUBBER (X)IIR AND THEIR BLENDS

11.5.1   IIR/Graphene-Rubber Nanocomposites

IIR is useful for various applications such as tyre inner liners, sporting goods, and pharmaceutical closures due to its outstanding air impermeability properties. Even though the reinforcement of IIR for further improving the mechanical and barrier properties remains as a main challenge, the addition of plate-like fillers into the IIR matrix in a well-dispersed form can increase the torturous path of gas molecules and also can enhance the physical properties. Sadasivuni et al. showed graphene-filled butyl rubber has a lower oxygen transmission rate (OTR) value than clay-filled rubber composites. The aspect ratios of graphene and clay determined from experiments are 130 and 108 respectively (Feller et al., 2015). The higher aspect ratio of graphene can induce lower air permeability values for IIR composites. Moreover, the larger specific area and the uniform dispersion of graphene also contribute to the lower permeability values of graphene/IIR composites compared to that of clay/IIR composites. The main concern regarding the development of IIR/graphene-rubber nanocomposite is the formation of graphene agglomerates in the rubber matrix. So, for achieving a fine dispersion of graphene in IIR matrix, graphene must be properly modified for giving a good interaction with the elastomers. The commonly used modifiers for graphene are amine, isocyanate, and ionic liquids. Besides, in some cases, elastomer itself is also used to modify the graphene.

Sadasivuni et al. (2013) prepared maleic anhydride grafted butyl rubber (MA-g-IIR) and produced graphene/MA-g-IIR nanocomposites by solution mixing method and reported improved mechanical and barrier properties. Lian et al. prepared modified graphene (MG)/butyl rubber (IIR) composites by solution mixing method. MG was prepared using Cetyltrimethylammonium bromide solution and the addition of 10 wt.% of MG exhibits significant increase in mechanical properties, for example, the modulus of GE/IIR increases 16-folds compared with pristine IIR (Lian et al., 2011). Zheng et al. prepared graphene/ IIR nanocomposites using latex stage mixing method and the morphological analysis confirms an even dispersion of graphene in IIR matrix with an improvement in properties such as tear (44%), tensile strength (102%), and decrease in diffusion coefficient up to 46%. Kim et al. (2016) prepared modified graphene-reinforced IIR nanocomposites using shear mixing method. The graphene was modified using octadecylamine and the nanocomposites were prepared by employing Haake torque rheometer. The properties of MG-reinforced composites were compared with pure graphene-reinforced composites and the results are showing an improvement in both physical and morphological properties for MG/IIR composites over G/IIR composites. Chinnasamy et al. (2020) studied the properties of graphene-reinforced IIR matrix prepared by various mixing procedures. The composites were prepared using direct addition approach (DAAM), single-step method (SSM), and two-step method (TSM). The study also considered the effect of compatibilizer on the dispersion of graphene in the matrix. The compatibilizer used for improving the dispersion of graphene is CIIR and it was used in both SSM and TSM procedures. The properties of the IIR increase with the addition of graphene and the properties show a maximum value for 1.6% of graphene in the matrix. Further, the addition of graphene deteriorates the properties due to the agglomeration. Comparing the various mixing procedures, TSM mixing procedure with the use of compatibilizer gives efficient dispersion of graphene in the rubber matrix.

11.5.1.1 CIIR/Graphene-Rubber Nanocomposites

The properties of chlorobutyl rubber (CIIR) are very similar to the properties of butyl rubber. The most important consumption of CIIR is for the manufacture of tyre inner tubes and inner-liner. In addition to that, with non-toxic vulcanization, the CIIR is also used in medical and food applications. Studies are showing that the CIIR rubber reinforced with sheet-like nanofillers has the potential to further improve the permeability properties.

Daniele et al. studied the properties of CIIR/CB/GNP nanocomposites by replacing a part of CB with GNP. The physical properties of the CIIR rubber increased with the replacement of CB with graphene and the increment was prominent in terms of Young's modulus. The morphological analysis also shows an even dispersion of graphene in the CIIR matrix (Frasca et al., 2016). Sreehari et al. (2021) evaluated the solvent transportation properties of CIIR/GO nanocomposites and reported a restriction of passage of solvent molecules through the CIIR matrix in the presence of GO platelets. Jahankhanemlou et al. investigate the effect of graphene and NC on permeability properties of CIIR. The studies were carried out with CIIR/NR blends in 80:20 ratios and the results confirmed that when comparing both NC and graphite, the graphene-reinforced CIIR/NR nanocomposite shows better mechanical and impermeability properties (Hermenegildo et al., 2014). Jiang et al. (2017) studied the viscoelastic properties of CIIR matrix on addition with graphene and confirmed the physical interactions between graphene and polymer chain using FT-IR and bound rubber content analysis. Keloth Paduvilan et al. (2021) analyze the synergistic effect of graphene and NC in CIIR/NR blends. The synergism of graphene and NC increases the air impermeability better than the single loading of graphene and NC. Due to the drawbacks of CIIR over BIIR, the researches on CIIR-based composites are limited.

11.5.1.2 BIIR/Graphene Nanocomposites

BIIR is having faster cure rate and better adhesion properties compared to both IIR and CIIR. The better cure properties of BIIR make it more useful for various applications including tyre inner liners, sealants, gaskets, and sporting goods. However, major challenge has always been in reinforcing the BIIR using various fillers for improving the mechanical and barrier properties. For the current requirements, the need of BIIR composite with higher mechanical and barrier properties is essential, especially for tyre applications. Nanofillers with platelet type structure can increase the physical as well as mechanical properties of the BIIR matrix. There were a lot of studies reported by researchers where NC is used to improve the barrier properties of the BIIR composites even though Zhan et al. reported an improved water vapor barrier properties for NR/graphene nanocomposites compared with NR/ NC composites (Park et al., 2020). So, the introduction of graphene into the BIIR matrix can enhance both physical and barrier properties of the BIIR composites, provided it is properly dispersed in the rubber matrix.

The researchers all over the globe are trying out various strategies for the development of graphene/BIIR composites for further improving the air barrier properties of the BIIR matrix for recent applications in which functionalizing graphene using grafting small molecules gives more attraction due to the proper dispersion and better barrier properties obtained for the BIIR matrix. Kotal et al. (2016) studied the efficiency of BIIR-grafted graphene on the mechanical and permeability properties of BIIR composites (Figure 11.9). The grafting of BIIR on graphene was carried out with the aid of p-phenylenediamine (PPD). The modification of graphene followed by the solution blending of the modified graphene with BIIR will result in enhancement of the properties, such as tensile strength (200%), storage modulus (189%), permittivity (460%), and drastic improvement in gas impermeability (44%). Yang et al. prepared the BIIR/GO modified with octadecylamine composites by solution mixing method and the well-dispersed morphology of modified graphene oxide (MGO) in BIIR matrix was reported using TEM. The gas barrier properties of the MGO/ BIIR nanocomposites show 21.2% improvement in air impermeability compared

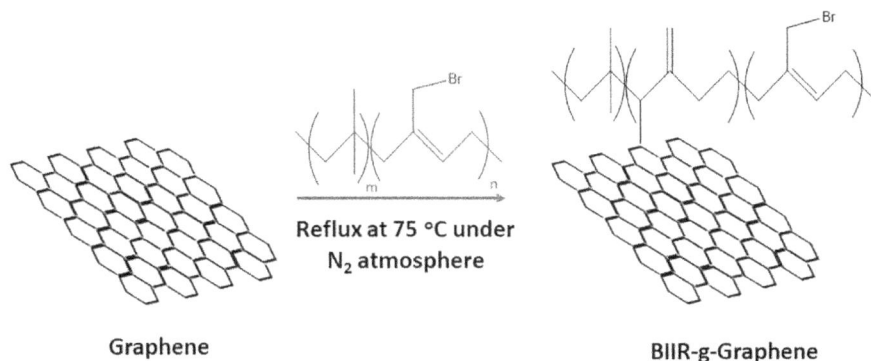

FIGURE 11.9 Schematic diagram for the tethering of BIIR on graphene (BIIR-g-graphene) (Kotal et al., 2016).

to regular BIIR composites (Yang et al., 2017). He et al. studied the properties of BIIR/graphene-rubber nanocomposites prepared by latex stage mixing method. The morphology of nanocomposites was measured using TEM and a segregated structure of graphene is obtained in the observed area. The formation of such structures improved the oxygen and carbon dioxide impermeability (Fangfang et al., 2017).

Ionic liquid (IL)-modified graphene-BIIR nanocomposites were prepared by Das et al. and the barrier properties of prepared the BIIR/IL-GO nanocomposites were studied. The composites were prepared by melt mixing method and the exfoliated structure of IL-graphene in the BIIR matrix is confirmed. The presence of exfoliated GNP in the BIIR matrix also leads to an improvement in the air impermeability of the BIIR matrix (Das et al., 2016).

However, in the above-mentioned studies, the BIIR-graphene nanocomposites were prepared by mainly using either solution mixing method or latex stage mixing method. As already mentioned in Section 11.4, the solution mixing process is environmentally unsafe and often leads to various health hazards and the latex stage mixing is more time consuming. So, the preparation of graphene-rubber nanocomposites using solution and latex stage mixing is not practically viable for industrial applications. The normal melt mixing of BIIR with graphene always ends up with the reaggregation of the graphene in the rubber matrix. On the contrary, the reaggregation of the graphene in the BIIR could also lead to the deterioration of the barrier properties. The objective of developing a highly air-impermeable BIIR/graphene-rubber nanocomposite using a method which will reduce the cost of the production and not adversely affect the environment is one of the main concerns. The modification of the BIIR matrix is the next best possible way to achieve a fine dispersion of graphene. BIIR itself is an expensive material. So, the modification of BIIR using chemicals will increase the cost of the production which will in turn increase the cost of the end product. Though, the blending of the BIIR rubber with other rubbers, which is having high air impermeability and the chemical entities which can interact with graphene monolayers is the best way to enhance the dispersion of graphene in BIIR matrix. Following this, we have successfully developed graphene/BIIR rubber nanocomposites based on various BIIR blends and studied the properties.

11.5.2 BROMOBUTYL RUBBER/EPOXIDIZED NATURAL RUBBER/GRAPHENE-RUBBER NANOCOMPOSITES

Bromobutyl rubber has a number of advantages over other rubbers. For example, its low gas permeability, good thermal and oxidative stability as well as excellent moisture barrier. Even though, due to the absence of covalent network structure in BIIR it generally possesses lower mechanical properties, in particular, pure tensile and creep resistance properties, the blending of BIIR with other rubbers, reinforcement of BIIR with various fillers, or a combination of both can enhance the properties required for the end applications. Since the introduction of GNP into rubber, the researchers are trying to find out a method to make fine dispersion of graphene in the BIIR matrix for obtaining improved physical and barrier properties. In the former section, we have been discussed about a few studies to develop a BIIR/graphene nanocomposite with various methods for barrier applications. However, the agglomeration of GNP in the BIIR matrix developed using melt mixing method remains as the main problem for industrial usage. So, blending of BIIR with other rubbers which are having good barrier as well as better interaction with GNP can enhance the dispersion of GNP in the BIIR matrix without compromising the barrier properties.

In recent years, the practice of using ENR as a compatibilizer for improving the dispersion of graphene is studied by various researches. Ismail et al. studied the effect of ENR 50 as reinforcement modifier in NR/GO/CB nanocomposites. The results suggest the enhancement in the reinforcing efficiency of GO in the presence of ENR. The presence of polar functional epoxide group significantly enhances the dispersion of GO in NR (Ismail et al., 2017). Esmizadeh et al. (2019) investigated the effect of ENR 50 on morphological properties of NR/Poly lactic acid (PLA)/ GNP nanocomposites and confirmed the effectiveness of ENR 50 for improving the dispersion of GNP in NR/PLA blends. Wei et al. (2019) studied the effect of ENR as compatibilizer in epoxy/vulcanized natural rubber blend and the results are showing an increase in the d-spacing between the GNPs with an increase in the addition of ENR 50.

The ENR is used in various applications since the commercial production of ENR began by Guthrie in 1987. The epoxidization is carried out to overcome the drawbacks of NR, especially the poor oil resistance. In addition to that, the epoxidization also enhances the air impermeability properties of the pure NR. Besides the epoxidization process, the ENR is also capable of crystallizing like natural rubber and thus giving good mechanical properties. A study by Yaragalla et al. shows dispersion behaviour of GNP in ENR having different amount of epoxidization and they reported the possible chemical interactions between the GNP and the ENR in the work (Figure 11.10). The addition of 2% GNP increased the air impermeability by 59% in ENR 25 and 62% in ENR 50. The presence of epoxidized units in the ENR chain can in fact enhance the dispersion of GNP in the matrix (Yaragalla et al., 2015).

Followed by this, a number of studies were reported where the ENR is used as a compatibilizer in BIIR-graphene nanocomposites for improving the dispersion of GNP in the BIIR matrix. Kumar et al. examined the properties of BIIR/ENR 50 blend at a ratio of 75:25. The physical and morphological properties of BIIR/ENR

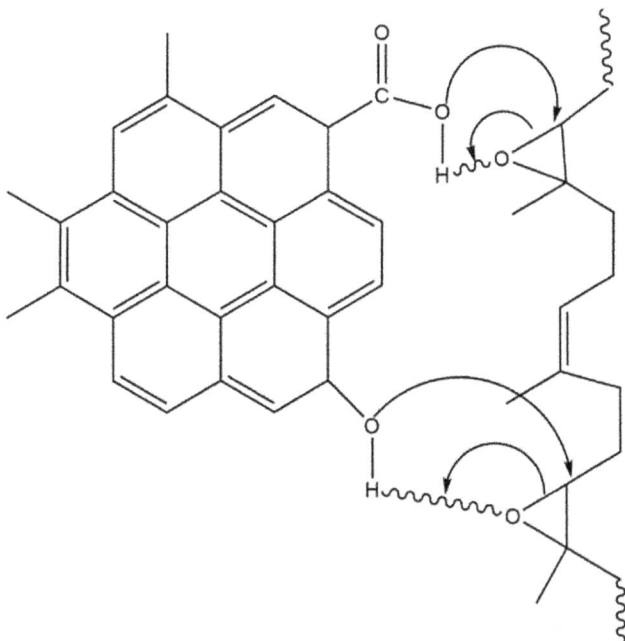

FIGURE 11.10 Schematic representation of the hydrogen bonding interactions between ENR and GNP.

50 blend were compared with BIIR composites and an increment in the mechanical properties was reported with the reinforcement of CB and NC (Sankaran et al., 2016). The blending of ENR into the BIIR can increase the dispersion GNP without compromising the mechanical and permeability properties of raw BIIR. The next main challenge is the selection of the suitable ENR for obtaining an even dispersion of GNP. Even though, the report by Yaragalla et al. shows an independency of GNP dispersion on the amount of epoxidization units, few recent reports show results in an opposite way. ENR is available with an epoxidization level varying from 25% to 70%. The 70% epoxidized ENR is having an air impermeability even higher than the IIR. However, the high Tg value of ENR 70 makes it not applicable for commercial usages. ENR-25 and ENR-50 with glass transitions of −23°C and −48°C are mainly chosen for most of the common applications. The selection of ENR for blending with BIIR for the preparation of BIIR-GNP nanocomposites highly depends upon the level of epoxidization. The level of epoxidization does have an effect on the dispersion of GNP in the BIIR matrix. The FT-IR spectra for the GNP-based BIIR-ENR 25 and BIIR-ENR 50 composites are given in Figure 11.11 and it confirms the interactions between epoxy units and GNP in BIIR-ENR 50 based composites. The broadening of the peak corresponding to oxirane ring ($1,250\,\mathrm{cm^{-1}}$) without changing the peak position indicates the formation of excessive hydrogen bonding. However, such interactions are absent in the case of BIIR-ENR 25 based graphene nanocomposites (Aswathy et al., 2020).

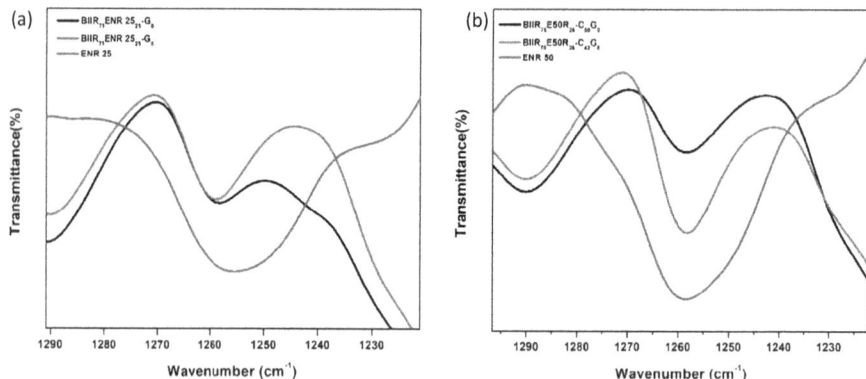

(a) [legend: BIIR$_n$ENR 25$_{21}$-G$_3$ / BIIR$_n$ENR 25$_{21}$-G$_3$ / ENR 25]

(b) [legend: BIIR$_n$E50R$_{26}$-C$_{20}$G$_3$ / BIIR$_n$E50R$_{26}$-C$_{41}$G$_3$ / ENR 50]

FIGURE 11.11   The FT-IR spectra for showing the dependency of epoxidization level on dispersion GNP in BIIR-ENR 25 and BIIR-ENR 50 composites.

For the addition of same quantity of GNP into BIIR-ENR 25 and BIIR-ENR 50 matrix, the BIIR-ENR 50 based composites show an increment of 30% air imperme-ability. However, the BIIR-ENR 25 based GNP composites show a decrement in air impermeability with respect to the reference composites. This again confirms the adverse effect of GNP on the properties of the rubber matrix due to agglomeration. So, this study confirms the effect of epoxidization on the dispersion of graphene nanoplatelets in various rubber matrices.

11.5.3   BROMOBUTYL RUBBER/POLYEPICHLOROHYDRIN RUBBER/ GRAPHENE-RUBBER NANOCOMPOSITES

In 1975, E. J. Vandenberg developed high molecular weight polymers of epichlorohy-drin and it is widely used for various applications because of its excellent ozone, oil and petrol resistance. Besides, the epichlorohydrin rubber also provides a very low gas permeability together with a good heat resistance. The usage of ECH is limited in commercial applications because of its high brittle point of the polymer around −17°C. The brittle nature of ECH rubber limits the usage of rubber in applications such as inner liners and sealing applications even though, in recent years, few studies are showing the advantages of blending ECH with BIIR to obtain the properties such as heat and oil resistance required for the end applications.

The study of blending ECH with BIIR for enhancing the dispersion of GNP was carried out by Kumar et al. (2015), and the results are showing a noble improvement of air impermeability by blending and reinforcing the BIIR with ECH and GNP, respectively. As mentioned in the case of ENR rubbers, the presence of polar groups in the side chains of ECH rubber can increase the dispersion of GNP on blending with BIIR. Due to the high air impermeability of ECH, the blending ECH with BIIR will not affect the air impermeability of the clean rubber matrix. In the study by Kumar et al., BIIR was blended with ECH in 90:10 ratios and the phr of GNP varied from 0 to 7. The variation of air impermeability with the increase in the amount of GNP added to BIIR–ECH is given in Figure 11.12. Thus, the blends of BIIR with

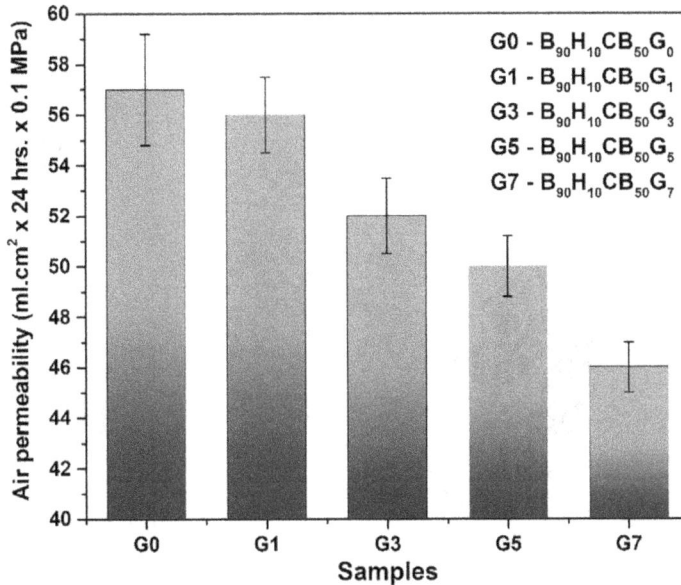

FIGURE 11.12   Air permeation characteristics of the BIIR-ECH samples with the addition of GNP (Kumar et al., 2015).

other elastomers such as BIIR/ENR and BIIR/ECH can enhance the dispersion of GNP without affecting the physical and barrier properties of the BIIR when it is used in its pure state. Hence, the composites developed with such blends can be used to replace the usual BIIR-CB composites for applications like tyre inner-liners for improved properties and enhanced life span.

11.5.4   Synergism of Various Nanofillers for Improving the Dispersion of Graphene in BIIR

Recently, the introduction of different types of nanofillers into the BIIR or BIIR blends along with graphene also achieved an even attention due to the extraordinary properties reported by the resultant nanocomposites. This area of research begins with the introduction of graphene with CB to various rubber matrix and the combination of GNP and CB develops different types of microstructures in the rubber, which can also increase the physical and barrier properties of the nanocomposites. The microunits developed can act as obstacles during the passage of gas molecules through the rubber matrix and improve the impermeability properties. The formation of such microstructures is more useful compared to the tortuous pathway created by the graphene nanoplatelets alone. Compared to the pathway created by GNP, the one created by the microunits is found to be longer and more efficient. This type of microstructures was reported in rubber NC composites, where there is a good interaction between the rubber chains and NC (Aswathy et al., 2021). The presence of such units in BIIR-ENR 50-CB-GNP nanocomposites confirms a better interaction

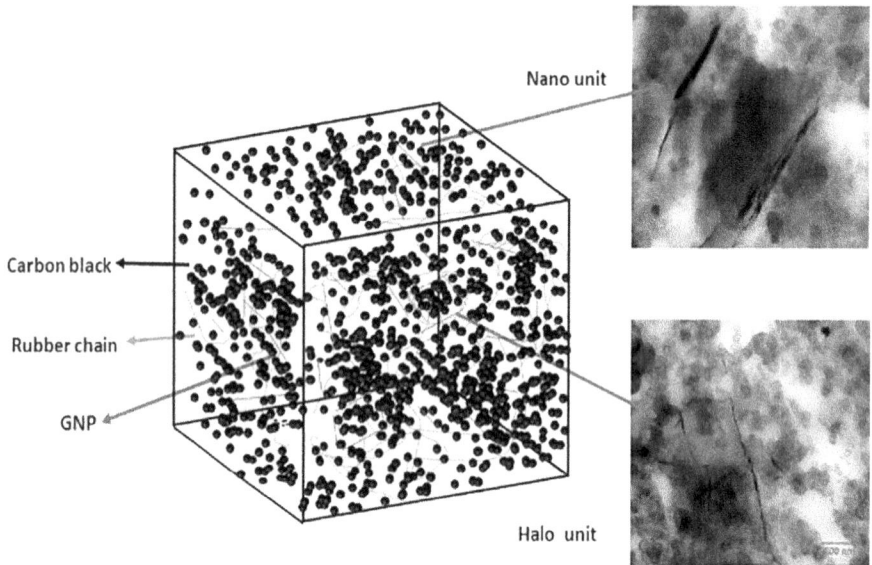

FIGURE 11.13   The schematic representation for the formation halo and nano units in BIIR-ENR 50 –GNP nanocomposites.

of GNP with rubber matrix. The transmission electron microscopy (TEM) analysis of the BIIR-ENR 50-GNP nanocomposites (Figure 11.13) confirms the presence of such micro units in the BIIR-ENR 50 matrix. The air impermeability properties of the rubber nanocomposites with BIIR- ENR 50-CB-GNP-based nanocomposites show an improvement over BIIR-CB and BIIR-GNP-CB composites (Aswathy et al., 2021).

In addition to CB-GNP synergy, lately, the researchers are trying to study the synergistic effect of various nanofillers and receiving greater attention due to their attractive properties. Synergetic effect of graphene with CNTs is reported by various researchers for different applications and the studies show that the introduction of CNT inhibits the aggregation of graphene in the rubber matrix (Nguyen et al., 2016; Li et al., 2015). The synergistic effect of graphene with NC, GO, and nanosilica (NS) was studied in site of permeability properties, and on addition of hybrid nanofillers into the BIIR-ENR 50 matrix, a 15%–25% improvement is obtained in reference with the BIIR-ENR 50-CB-GNP nanocomposite (Aswathy et al., 2021) (Figure 11.14). The addition of nanofillers such as NC and GO will reduce the pi–pi interactions between the monolayers of GNP and will increase the dispersion of nanofillers in the BIIR-ENR 50 matrix.

Hence, the blending of BIIR with other elastomers which is having good impermeability properties along with the reinforcement of the resultant blends with graphene or hybrid nanofiller combinations can enhance the properties of the BIIR matrix for meeting the present requirements.

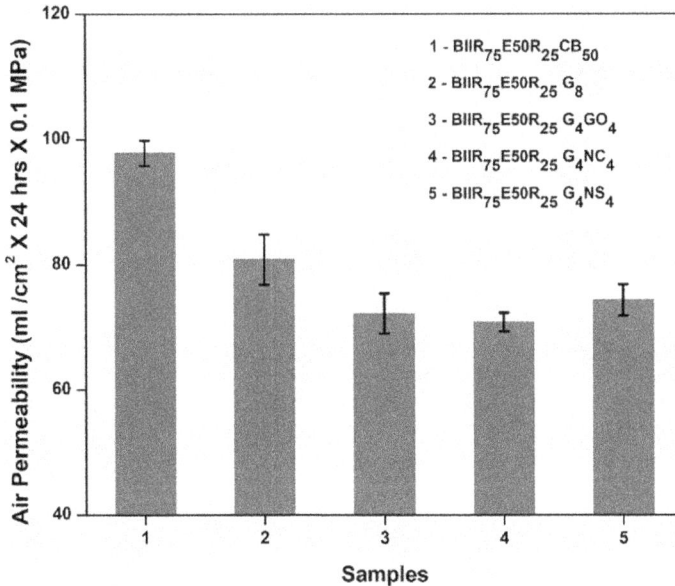

FIGURE 11.14 Air permeability results of various rubber nanocomposites. 1 – BIIR-ENR-CB, 2 – BIIR-ENR 50-GNP,3 – BIIR-ENR 50-GNP-GO,4 – BIIR-ENR 50-GNP-NC, and 5 – BIIR-ENR 50-GNP-NS.

11.6 SUMMARY

Elastomers have been used for variety of applications including automobile, aerospace, and electrical products. For some applications, many rubber products require extremely high standards in terms of gas barrier properties. There are mainly two methods to develop highly air-impermeable rubber material, one consists of special rubbers or blends of special rubbers which have high air impermeability properties. Example for such rubbers is Butyl rubber, halogenated butyl rubber, chloroprene rubber, and functionalized NR. The other one is the nanofiller-modified rubber or rubber-based composites. Nanofillers such as graphene have the structural properties such as high aspect ratio and high surface area which leads to an improved air permeability property for the resultant rubber composite. So, the modification of special rubbers such as butyl rubber and the blends of such special rubber with GNP can lead to a product which has extreme gas barrier properties. When these rubbers were reinforced with GNP, the gas barrier properties obtained were better than the same blends reinforced with NC. Yet, the main problem regarding the industrialization of BIIR-GNP composites is due to the high cost of the GNP which will in turn increase the cost of the end product.

However, the in-depth studies on preparation of graphene, which have been carried out by the researchers, can solve these problems eventually in the future.

REFERENCES

Aswathy, Thuruthil Raju, Asit Baran Bhattacharya, Biswaranjan Dash, Pranab Dey, Sujith Nair, and Kinsuk Naskar. 2021a. "Assessment of various nano-clays in bromobutyl rubber hydrogenated acrylonitrile butadiene rubber blend for improved gas barrier applications." *Journal of Applied Polymer Science* 138, 13: 50086.

Aswathy, Thuruthil Raju, Biswaranjan Dash, Pranab Dey, Sujith Nair, and Kinsuk Naskar. 2020. "Evaluation of air permeability characteristics on the hybridization of carbon black with graphene nanoplatelets in bromobutyl rubber/epoxidized natural rubber composites for inner-liner applications." *Polymers for Advanced Technologies* 31, 10: 2390–2402.

Aswathy, Thuruthil Raju, Biswaranjan Dash, Pranab Dey, Sujith Nair, and Kinsuk Naskar. 2021b. "Synergistic effect of graphene with graphene oxide, nanoclay, and nanosilica in enhancing the mechanical and barrier properties of bromobutyl rubber/epoxidized natural rubber composites." *Journal of Applied Polymer Science* 31: 50746.

Bai, Xin, Chaoying Wan, Yong Zhang, and Yinghao Zhai. 2011. "Reinforcement of hydrogenated carboxylated nitrile–butadiene rubber with exfoliated graphene oxide." *Carbon* 49, 5: 1608–1613.

Chinnasamy, Sathishranganathan, Rajasekar Rathanasamy, Harikrishna Kumar Mohan Kumar, Prakash Maran Jeganathan, Sathish Kumar Palaniappan, and Samir Kumar Pal. 2020. "Reactive compatibilization effect of graphene oxide reinforced butyl rubber nanocomposites." *Polímeros: Ciência e Tecnologia,* 30, 3: e2020032.

Compton, Owen C., Soyoung Kim, Cynthia Pierre, John M. Torkelson, and SonBinh T. Nguyen. 2010. "Crumpled graphene nanosheets as highly effective barrier property enhancers." *Advanced Materials* 22, 42: 4759–4763.

Das, A., A. Leuteritz, P. Kavimani Nagar, B. Adhikari, K. W. Stöckelhuber, R. Jurk, and G. Heinrich. 2016. "Improved gas barrier properties of composites based on ionic liquid integrated graphene nanoplatelets and bromobutyl rubber." *International Polymer Science and Technology* 43, 6: 1–8.

Esmizadeh, Elnaz, Taha Sadeghi, Ali Vahidifar, Ghasem Naderi, Mir Hamid Reza Ghoreishy, and Seyed Mohammad Reza Paran.2019. "Nano graphene-reinforced bio-nanocomposites based on NR/PLA: The morphological, thermal and rheological perspective." *Journal of Polymers and the Environment* 27, 7: 1529–1541.

Feller, J. F., K. K. Sadasivuni, M. Castro, H. Bellegou, I. Pillin, S. Thomas, and Y. Grohens. 2015. "Gas barrier efficiency of clay-and graphene-poly (isobutylene-co-isoprene) nanocomposite membranes evidenced by a quantum resistive vapor sensor cell." *Nanocomposites* 1, 2: 96–105.

Frasca, Daniele, Dietmar Schulze, Volker Wachtendorf, Bernd Krafft, Thomas Rybak, and Bernhard Schartel. 2016 "Multilayer graphene/carbon black/chlorine isobutyl isoprene rubber nanocomposites." *Polymers* 8, 3: 95.

Gusev, Andrei A., and Hans Rudolf Lusti. 2001. "Rational design of nanocomposites for barrier applications." *Advanced Materials* 13, 21: 1641–1643.

He, Fangfang, Giuseppe Mensitieri, Marino Lavorgna, Martina Salzano de Luna, Giovanni Filippone, Hesheng Xia, Rosario Esposito, and Giuseppe Scherillo. 2017. "Tailoring gas permeation and dielectric properties of bromobutyl rubber–Graphene oxide nanocomposites by inducing an ordered nanofiller microstructure." *Composites Part B: Engineering* 116: 361–368.

Hermenegildo, Gislaine. 2014. "Desenvolvimento de nanocompósitos de borracha clorobutílica/borracha natural com montmorilonita para aplicação em revestimentos internos de pneus sem câmara de ar."

Ismail, Nik Intan Nik, Siti Salina Sarkawi, Fatima Rubaizah, and Azira Abdul Aziz. 2017. "Ekoprenatm as a reinforcement modifier for graphene oxide/carbon black filled natural rubber." *International Proceedings of IRC* 2017: 704–713.

Jiang, Ping, Chunhua Yang, Xianru He, Alisson M. Rodrigues, and Rui Zhang. 2017. "Viscoelastic changes in chlorinated butyl rubber modified with graphene oxide." *Iranian Polymer Journal* 26, 11: 861–870.

Kaneko, Satoru, Paolo Mele, Tamio Endo, Tetsuo Tsuchiya, Katsuhisa Tanaka, Masahiro Yoshimura, and David Hui, 2017. eds. *Carbon-Related Materials in Recognition of Nobel Lectures by Prof. Akira Suzuki in ICCE*. Springer, Switzerland.

Keloth Paduvilan, Jibin, Prajitha Velayudhan, Ashin Amanulla, Hanna Joseph Maria, Allisson Saiter-Fourcin, and Sabu Thomas. 2021. "Assessment of graphene oxide and nanoclay based hybrid filler in chlorobutyl-natural rubber blend for advanced gas barrier applications." *Nanomaterials* 11, 5: 1098.

Kim, Jin Sil, Sanghyun Hong, Dong Wha Park, and Sang Eun Shim. 2010. "Water-borne graphene-derived conductive SBR prepared by latex heterocoagulation." *Macromolecular Research* 18, 6: 558–565.

Kim, Young-Min, Saravanan Nagappan, Jae-Hoon Jeong, Eun-Ji Park, Kyung-Man Choi, Ji-Eun Lee, and Ildoo Chung. 2016. "Properties of hydrophobically-modified graphene oxide (HG)/butyl rubber (IIR) nanocomposites prepared by shear mixing process." *Composite Interfaces* 23, 8: 819–829.

Kotal, Moumita, Shib Shankar Banerjee, and Anil K. Bhowmick. 2016. "Functionalized graphene with polymer as unique strategy in tailoring the properties of bromobutyl rubber nanocomposites." *Polymer* 82: 121–132.

Kumar, S., S. Chattopadhyay, A. Sreejesh, Sujith Nair, G. Unnikrishnan, and G. B. Nando. 2015. "Analysis of air permeability and WVTR characteristics of highly impermeable novel rubber nanocomposite." *Materials Research Express* 2, 2: 025001.

Li, Xueqin, Lu Ma, Haiyang Zhang, Shaofei Wang, Zhongyi Jiang, Ruili Guo, Hong Wu, XingZhong Cao, Jing Yang, and Baoyi Wang. 2015. "Synergistic effect of combining carbon nanotubes and graphene oxide in mixed matrix membranes for efficient CO_2 separation." *Journal of Membrane Science* 479: 1–10.

Lian, Huiqin, Shuxin Li, Kelong Liu, Liangrui Xu, Kuisheng Wang, and Wenli Guo. 2011. "Study on modified graphene/butyl rubber nanocomposites. I. Preparation and characterization." *Polymer Engineering & Science* 51, 11: 2254–2260.

Long, Joan C. 2001. "The history of rubber—A survey of sources about the history of rubber." *Rubber Chemistry and Technology* 74, 3: 493–508.

Lu, Yong-Lai, Zhao Li, Zhong-Zhen Yu, Ming Tian, Li-Qun Zhang, and Yiu-Wing Mai. 2007. "Microstructure and properties of highly filled rubber/clay nanocomposites prepared by melt blending." *Composites Science and Technology* 67, 14: 2903–2913.

Mondal, Titash, Anil K. Bhowmick, and Ramanan Krishnamoorti. 2012. "Chlorophenyl pendant decorated graphene sheet as a potential antimicrobial agent: Synthesis and characterization." *Journal of Materials Chemistry* 22, 42: 22481–22487.

Nguyen, Duc Dung, Thi Thanh Cao, Phuoc Huu Le, and Ngoc Minh Phan. 2016. "Recent trends in preparation and application of carbon nanotube–graphene hybrid thin films." *Advances in Natural Sciences: Nanoscience and Nanotechnology* 7, 3: 033002.

Novoselov, Kostya S., Andre K. Geim, Sergei V. Morozov, De-eng Jiang, Yanshui Zhang, Sergey V. Dubonos, Irina V. Grigorieva, and Alexandr A. Firsov. 2004. "Electric field effect in atomically thin carbon films." *Science* 306, 5696: 666–669.

Park, Sang-Yu, Ji-Young Hwang, Young Su Park, and Seung Beom Kang. 2020. "A review of graphene nanoplatelets in nanocomposites: Dispersion." *Composites Research* 33, 6: 321–328.

Sadasivuni, Kishor Kumar, Allisson Saiter, Nicolas Gautier, Sabu Thomas, and Yves Grohens. 2013. "Effect of molecular interactions on the performance of poly (isobutylene-co-isoprene)/graphene and clay nanocomposites." *Colloid and Polymer Science* 291, 7: 1729–1740.

Sankaran, Kumar, Partheban Manoharan, Santanu Chattopadhyay, Sujith Nair, Unnikrishnan Govindan, Sreejesh Arayambath, and Golok B. Nando. 2016. "Effect of hybridization of organoclay with carbon black on the transport, mechanical, and adhesion properties of nanocomposites based on bromobutyl/epoxidized natural rubber blends." *RSC Advances* 6, 40: 33723–33732.

Schedin, Fredrik, Andrei Konstantinovich Geim, Sergei Vladimirovich Morozov, E. W. Hill, Peter Blake, M. I. Katsnelson, and Kostya Sergeevich Novoselov. 2007. "Detection of individual gas molecules adsorbed on graphene." *Nature Materials* 6, 9: 652–655.

Sreehari, H., Asok Aparna, Jitha S. Jayan, A. S. Sethulekshmi, Venu Gopika, K. P. Anjali, N. Parvathy, and Appukuttan Saritha. 2021. "Evaluation of solvent transport and cure characteristics of chlorobutyl rubber graphene oxide nanocomposites." *Materials Today: Proceedings* 49, 5: 1431–1435.

Sun, P. Z., Qian Yang, W. J. Kuang, Y. V. Stebunov, W. Q. Xiong, Jin Yu, Rahul Raveendran Nair et al. 2020. "Limits on gas impermeability of graphene." *Nature* 579, 7798: 229–232.

Threadingham, Desmond, Werner Obrecht, Jean-Pierre Lambert, Michael Happ, Christiane Oppenheimer-Stix, John Dunn, Ralf Krüger et al. 2000. "Rubber, 3. Synthetic." in *Ullmann's Encyclopedia of Industrial Chemistry*.

Threadingham, Desmond, Obrecht, Werner, Wieder, Wolfgang, Wachholz, Gerhard and Engehausen Rüdiger. 2011. "Rubber, 3. Synthetic Rubbers, Introduction and Overview." *Ullmann's Encyclopedia of Industrial Chemistry* 31, 597-622.

Wang, Jian, Kaiye Zhang, Zhengang Cheng, Marino Lavorgna, and Hesheng Xia. 2018. "Graphene/carbon black/natural rubber composites prepared by a wet compounding and latex mixing process." *Plastics, Rubber and Composites* 47, 9: 398–412.

Wei, Kam Ka, Teh Pei Leng, Yeoh Cheow Keat, Hakimah Osman, and Lim Bee Ying. 2019. "Enhancing compatibility in epoxy/vulcanized natural rubber (VNR)/Graphene nanoplatelets (GNP) system using epoxidized natural rubber (ENR-50)." *Composites Part B: Engineering* 174: 107058.

Wolf, Caroline, Helene Angellier-Coussy, Nathalie Gontard, F. Doghieri, and Valérie Guillard. 2018. "How the shape of fillers affects the barrier properties of polymer/non-porous particles nanocomposites: A review." *Journal of Membrane Science* 556: 393–418.

Wu, Siwu, Zhenghai Tang, Baochun Guo, Liqun Zhang, and Demin Jia. 2013. "Effects of interfacial interaction on chain dynamics of rubber/graphene oxide hybrids: a dielectric relaxation spectroscopy study." *RSC Advances* 3, 34: 14549–14559.

Yang, Xinya, Yong Zhang, Yan Xu, Steven Gao, and Sharon Guo. 2017. "Effect of octadecylamine modified graphene on thermal stability, mechanical properties and gas barrier properties of brominated butyl rubber." *Macromolecular Research* 25, 3: 270–275.

Yaragalla, Srinivasarao, C. Sarath Chandran, Nandakumar Kalarikkal, R. H. Y. Subban, Chin Han Chan, and Sabu Thomas. 2015. "Effect of reinforcement on the barrier and dielectric properties of epoxidized natural rubber–graphene nanocomposites." *Polymer Engineering & Science* 55, 11: 2439–2447.

12 Graphene-Thermoplastic Polyurethane Elastomer Composites

Fundamentals and Applications

Madhab Bera and Pradip K. Maji
Indian Institute of Technology Roorkee

CONTENTS

DOI: 10.1201/9781003200444-12

12.1 INTRODUCTION

Thermoplastic Elastomers (TPEs) are a particular category of copolymers with thermoplastic material properties and an elastomer. There are several types of thermoplastic elastomers, one of the most important members of this family is thermoplastic polyurethane (TPU). Like all thermoplastic elastomers, TPU also possesses a hard segment (HS) and soft segment in its structure; the HS gives strength and rigidity to the material, whereas the soft segment provides flexibility to it (Hepburn, 2012). Extensive use and acceptance of this material are due to its ease of synthesis, wide range of properties, ease of processing, biocompatibility, and most importantly, its properties can be tailor-made as per the requirement. Various nanomaterials have been incorporated into the TPU matrix (Bera et al., 2020; Bera & Maji, 2017a; Farzaneh et al., 2021; Maji et al., 2009; Maji & Bhowmick, 2013) to mitigate the growing demand for new technologies like coatings, adhesives, fibers, foams, rubbers, thermoplastic elastomers, and composites. Among different available nanomaterials, graphene is one of the important nanomaterials in TPU-nanocomposites. Graphene or graphene-based materials have a unique combination of properties that can be utilized for making their polymer nanocomposites. High-performance and multifunctional composite material can be synthesized by incorporating a very small amount of this filler into the polymer matrix. However, due to the hydrophobic nature and very high surface energy of pristine graphene, it is incompatible with almost all available organic polymers, including TPU, and can't be dispersed uniformly. Suspended graphene will stick together due to high van der Waals attractive force (5.9 kJ mol^{-1}). So, uniform dispersion of graphene sheets within the polymer matrix is a big challenge to researchers (Zacharia et al., 2004). Chemical modification or functionalization is the only way to make stable suspension and uniform dispersion of graphene. Graphene oxide (GO), reduced graphene oxide (RGO), and graphite nanoplatelets (GNPs) are the three most important graphene-based materials which recently come into the limelight because they are relatively easy to synthesize and modify compared to other carbon-based nanofillers like carbon nanotubes (CNTs) and carbon nanofibers. Furthermore, GO and RGO have better compatibility with most of the organic polymers, including TPU, compared to pristine graphene. The GO and RGO have relatively higher thicknesses compared to pristine graphene. The thickness of pristine graphene is around 0.34 nm (Novoselov et al., 2004), whereas the same for GO and RGO is 1.0 ± 0.2 nm (Bera et al., 2018) and 0.55 ± 0.2 nm (Stankovich et al., 2007), respectively, depending on the degree of oxidation/reduction. Graphene/graphene-based materials have a unique combination of properties that are being utilized in their polymer nanocomposites to fulfill the growing demand for hi-tech materials for a diverse range of applications. Morphology and structure of graphene-based materials play a crucial role in the properties of graphene/TPU nanocomposites (Bera et al., 2020; Bera & Maji, 2017a; Strankowski, 2018; Zahid et al., 2020a). Stankowski (2018) proved that morphologies, thermal and mechanical properties of graphene/TPU nanocomposite are dependent on the percentage of a HS in TPU as well as the type of graphene-based nanofiller and its concentration. Both RGO and graphene nanoplatelets (GNPs) influence the thermal and mechanical properties of TPU but to a different extent. At 1% loading of reinforcement, GNPs provide better thermal stability to the polyurethane having a 40% HS than the polyurethane with a 20% HS. Whereas for RGO, the effect is just the opposite. Further, both the nanofillers showed better interaction with 40% HS TPU, causing better improvement in mechanical properties. The effect was more prominent with GNP. Bera et al. (Bera et al., 2020; Bera & Maji, 2017a) also proved that, for graphene/TPU nanocomposites, thermal, mechanical, and other properties depend on the structure of graphene-based material and their loading. According to their observation, at ultra-low loading (0.1 wt.%), GO improves mechanical properties and thermal stability better than RGO loading. The same can be further improved by modifying the GO with paraphenylenediamine (PPD). For example, the maximum increase in thermal stability of TPU/GO nanocomposite with 0.1% loading of filler was 7.6°C, whereas the same with TPU/RGO at similar loading was 5.6°C and for TPU/GO-PPD was 13°C. Graphene/TPU nanocomposites are being used in high moisture, vapor and gas barrier applications (Adak et al., 2019; Maldonado-Magnere et al., 2021; Yousefi et al., 2013; Zahid et al., 2020a), electromagnetic interface shielding (EMI) applications (Zahid et al., 2020b; Jiang et al., 2019, 2020b; Carbone et al., 2021), thermal management (Guo et al., 2021; Li et al., 2017; Zhang et al., 2020a), flame and fire retardant composites (Cao et al., 2021; Sabet et al., 2021; Wang et al., 2019) (Cao et al., 2019; Du et al., 2019, 2020; Liu et al., 2020), smart textiles and wearable electronics (Hu et al., 2015; Li et al., 2019; Sadanandan et al., 2021; Sáenz-Pérez et al., 2018; Salavagione et al., 2018; Soong & Chiu, 2021; Thakur, 2017), shape memory application (Du et al., 2018; Kumar Patel & Purohit, 2019; Sofla et al., 2019; Panahi-Sarmad et al., 2019; Zhou et al., 2018), self-healing applications (Du et al., 2020; Huang et al., 2013; Lin et al., 2019; Luan et al., 2018; Wan & Chen, 2018; Zhou et al., 2021), corrosion and abrasion-resistant coating (Chen et al., 2019; Fechine et al., 2020; Li et al., 2016; Naderizadeh et al., 2018; Zhang et al., 2020b), antibacterial coating (An et al., 2013; Borges et al., 2020; Perreault et al., 2015), solar water desalination (Awad et al., 2018; Wang et al., 2017), water purification (Sundaran et al., 2020), oil spill cleaning (Guan et al., 2019; Kong et al., 2017; Li et al., 2013; Liu et al., 2013b; Oribayo et al., 2017; Ye et al., 2021; Zhang et al., 2018; Zhang et al., 2017), biomedical applications (Bahrami et al., 2019; Ghorbani et al., 2019; Jing et al., 2015; Kaur et al., 2015), and sensors (Feng et al., 2020; Khalifa et al., 2020; Jia et al., 2020; Liu et al., 2016; Xie et al., 2020; Mei et al., 2021; Jasmi et al., 2018a).

Looking at the increasing research trend in graphene/TPU nanocomposites, this chapter will focus mainly on fundamentals of graphene/TPU nanocomposites, such as understanding TPU, graphene, graphene-based materials (GO, RGO, and GNPs). The synthesis methodologies of graphene/TPU nanocomposites, the microstructure of nanocomposites, properties of nanocomposites, and most importantly, application opportunities of graphene/TPU nanocomposites are also discussed in this chapter.

12.2 GRAPHENE AND GRAPHENE-BASED MATERIALS

Graphene is a carbon-based two-dimensional (2D) nanomaterial having a thickness of one atom (0.34 nm) (Novoselov et al., 2004). It was discovered in 2004 by two noble laureates, K.S. Novoselov and A.K. Geim from Manchester University (Novoselov et al., 2004). It is a single-layer material with a hexagonal closed-packed structure. In graphene, every carbon atom is sp^2 hybridized and linked together to form a honeycomb-like structure. The presence of long-range π conjugation in its structure provides outstanding physical and chemical properties. Graphene is about 100 times stronger than steel with Young's modulus of 1 TPa and tensile strength of 130 GPa (Lee et al., 2008). It has a very high surface area (2,630 $m^2 g^{-1}$) (Stoller et al., 2008; Zhu et al., 2010). This value is quite higher than that of CNTs and carbon nanofibers. Graphene's thermal conductivity could be as high as 5,000 W m^{-1} K^{-1} (Balandin et al., 2008; Saito et al., 2007). Being a zero-overlap semi-metal, it has very high electrical conductivity, which is greater than 200,000 $cm^2 V^{-1} S^{-1}$ in the suspended form (Morozov et al., 2008). It is optically transparent and has antibacterial properties (Perreault et al., 2015). Graphene is synthesized by separating individual layers of the multilayered carbon material, graphite, by a variety of methods. In the ideal case, graphene possesses only one layer, but in some cases, the number of layers may be more than one, and according to that, it is called bilayer (2 layers), few layers (2–5 layers), and multilayers (5–10 layers) graphene (Bianco et al., 2013). Graphene is the basic structural unit for other carbon-based nanomaterials. For example, several layers of graphene can be stacked together to form 3D graphite; it can be rolled to 1D CNTs and it can be wrapped to make 0D Fullerene (Figure 12.1) (Liu et al., 2013a). Graphene possesses some defects in its structure, which are generated during its synthesis in a variety of methods (Banhart et al., 2010; Hashimoto et al., 2004). All these defects influence the properties of graphene and thus affect various important properties of its polymer nanocomposites (Hashimoto et al., 2004).

12.2.1 GRAPHENE OXIDE (GO)

Graphene oxide (GO) is one of the most important graphene-based materials. It is the single layer of graphite oxide (GtO) prepared by mechanical exfoliation and delamination of GtO. Among the possible synthesis routes, the most popular one is the Hummers method (Hummers & Offeman, 2002). With time, the original Hummers method has been modified for more simplicity and productivity. Graphene oxide synthesized by the modified Hummers method (Chen et al., 2013) has 1.1 ± 0.2 nm of flake thickness. The value is slightly greater than the single graphene sheet (0.34 nm). This is due to the existence of various oxygenated functional

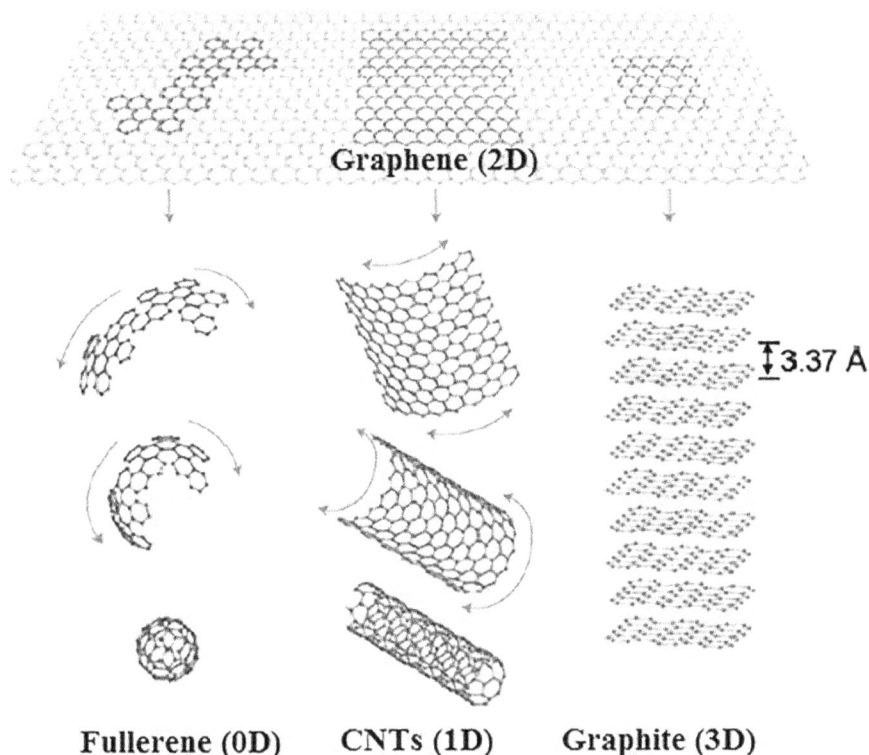

FIGURE 12.1 Schematic representation of carbon allotropes based on 2D graphene. (Reproduced with permission from Liu et al. (2013a). Copyright 2013, Elsevier.)

groups (e.g. carboxyl, hydroxyl, and epoxide) within the GO structure (Bera et al., 2018). Carboxyl, hydroxyl, and epoxide groups are localized at different positions in GO sheets. Carboxyl groups are generally positioned at the edges, while –OH and epoxide groups reside on the basal plane. As a result, the basal plane of GO is less hydrophilic compared to its edge (Figure 12.2). This provides an amphiphilic character to the 2D GO. Functionalization helps to improve the interaction between graphene and polymer matrix via van der Waals interactions and/or covalent bonding (Anagnostopoulos et al., 2015). Thus, easy dispersion of GO in water, organic solvent, or polymer matrixes is possible. GO dispersibility also depends on the C:O ratio or extent of oxygenated surface groups and the solvent.

12.2.2 Reduced Graphene Oxide (RGO)

RGO is synthesized by reduction of GO in a variety of ways: thermal (Huh, 2011; Jahanshahi et al., 2013); hydrothermal (Bao et al., 2012); chemical (Chua & Pumera, 2015; Stankovich et al., 2007); electrochemical (Zhou et al., 2009); and bio-reduction (Jana et al., 2014; Kuila et al., 2012; Wang et al., 2011b) methods. The transformation from GO to RGO is confirmed primarily by a visual color change from brown (GO)

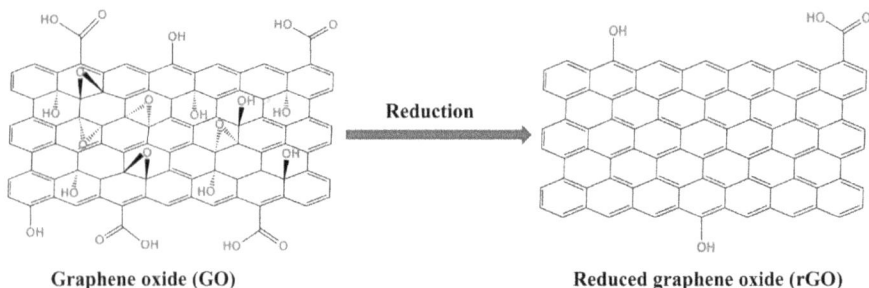

FIGURE 12.2   Chemical structure of GO and RGO.

to black (RGO). It is further confirmed by elemental analysis, where an increased C:O ratio ~6 indicates the formation of RGO. The thermal reduction has drawn tremendous attention to the researchers to produce graphene-like material, RGO. There are various reasons behind the wide acceptance of the process. It is less hazardous, less time-consuming, economical, and environmentally friendly compared to the chemical reduction process (Huh, 2011). Although RGO looks analogous to pristine graphene, they are not identical. From a very minute observation, it has been noticed that the π network structure is restored by reducing GO, but the complete restoration of the π network is not possible. Few oxygen-containing functional groups remain within the structure, differentiating RGO from pristine Graphene (Dreyer et al., 2009). RGO contains a significantly less number of –OH, –COOH, C=O, and –C–O–C– groups (Figure 12.2) and restores the π-network. RGO sheets are 0.34–0.70 nm thick, depending on the degree of reduction.

12.2.3   GRAPHITE NANOPLATELETS (GNPs)

Graphite nanoplatelets (GNPs) are one of the key members of the graphene-based materials family, drawing considerable interest to the scientists as a cost-effective alternative of the well-known nanomaterial, carbon nanotubes (Raza et al., 2011). Various methods are available for the synthesis of GNPs. It is synthesized by chemical vapor deposition (CVD), ball milling, arc discharge, and other methods. But the easiest, cost-effective. and popular method is direct exfoliation of graphite flakes or powder. Geng et al. (2009) introduced the synthesis procedure of GNPs by exfoliation of natural graphite flakes in the presence of formic acid. Thu et al. (2013) reported the synthesis of GNPs by the electrochemical method where the material was obtained by filtering a suspension of GNPs produced by electrochemical exfoliation of carbon rods. The structure of GNPs consists of several layers of atomically thick graphene sheets (usually greater than 10 layers) (Bianco et al., 2013) stacking together. Depending upon the number of layers, its thickness varies from 3.5 to 100 nm (Raza et al., 2011). Individual layers are connected by a weak van der Waal's force and thus maintain a constant interlayer distance of 0.34 nm (Raza et al., 2011). GNPs are also called graphite nanoflakes (GNFs), graphite nanosheets (GNs), or simply expanded graphite (EG) (Bianco et al., 2013). Structurally, graphite and GNPs are analogous, only the number of layers varies. GNPs have excellent mechanical, thermal, and electrical properties (Geng et al., 2009).

12.3 THERMOPLASTIC POLYURETHANE ELASTOMER (TPE)

Polyurethane is an extremely versatile multi-block synthetic polymer having great industrial importance (Oertel & Abele, 1994). German scientist Prof. Otto Bayer discovered polyurethane in 1937. It is a special thermoplastic elastomer with urethane linkage (–NH–C–O–O–) in the main chain (Figure 12.3). The repeating urethane linkage is the basis for the generic name polyurethane (Ionescu 2005). However, typical polyurethane contains urethane linkage and ester, ether, urea, amide, and biuret linkages (Hepburn, 2012).

Although polyurethane chemistry is complex, the basics are easy to understand. It is prepared by step-growth polymerization of a diisocyanate with a polyol and a chain extender. Several other by-products are also generated during the synthesis of polyurethane (Figure 12.4), depending on the reaction condition, duration of reaction, and the molar ratio of the reactants. Polyurethane is a block copolymer having hard and soft segments. The HS consists of diisocyanates and chain extenders, whereas the soft segment consists of polyol only. Soft segment influences the elastic nature and low-temperature properties of PU, whereas HS acts as a filler and is very important for mechanical properties (Ionescu, 2005; Prisacariu, 2011).

12.4 SYNTHESIS METHODOLOGIES OF GRAPHENE/TPU NANOCOMPOSITE

Various synthesis methods are being used for the dispersion of nanofiller into the polymer matrix and for the preparation of graphene/TPU nanocomposites. The properties of a polymer nanocomposite depend on the processing method used. So, the synthesis process plays a crucial role in the properties of the nanocomposite. The following three synthetic routes are mostly used to prepare graphene/TPU nanocomposites.

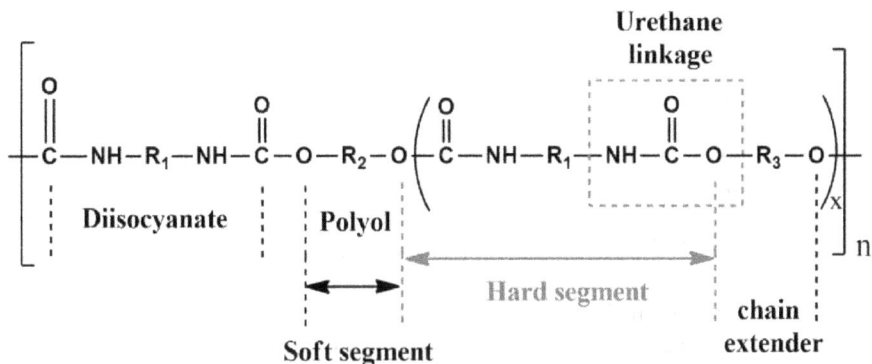

FIGURE 12.3 Schematic representation of the repeating unit of TPU.

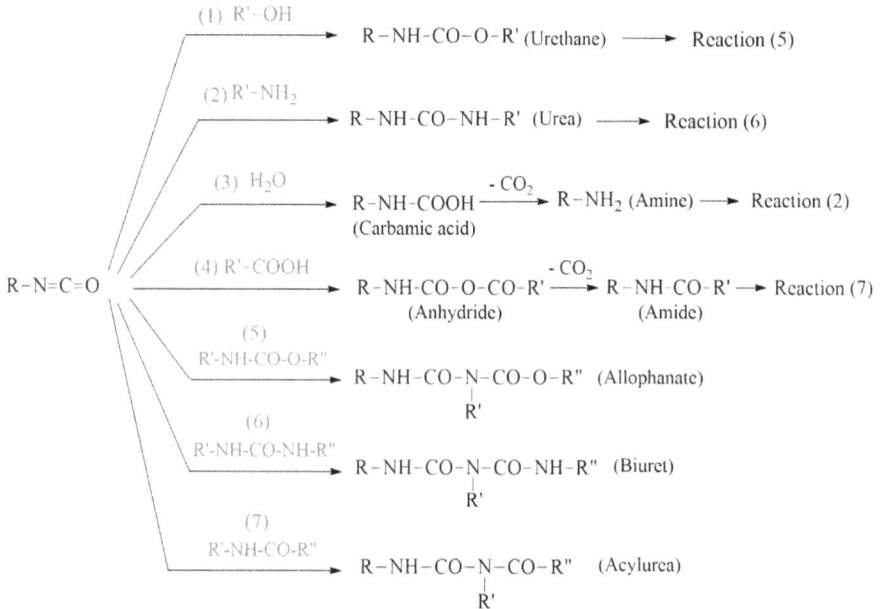

FIGURE 12.4 Basic chemical reactions of isocyanate with different chemical compounds.

12.4.1 IN SITU POLYMERIZATION

In situ polymerization means "in the polymerization mixture." In this method, graphene or its derivatives are first dispersed in a suitable monomer, and then, the resulting mixture is polymerized by a standard polymerization technique using an initiator. There are various advantages of in situ polymerization method over solution method or melt mixing method: (i) possibility of grafting the polymer onto the nanoparticle surface, (ii) since no solvent is used in this process, it is comparatively less hazardous process compared to solution method, and (iii) very good dispersion of the filler within liquid monomer. Because of these advantages, TPU and other polymer nanocomposites are also prepared by the in situ polymerization method (Wang et al., 2011a).

12.4.2 SOLUTION MIXING

In this method, a solvent is used, in which the polymers are soluble, and the graphene/graphene-based materials are dispersible. Graphene or its derivatives are first dispersed in a suitable solvent, such as water, tetrahydrofuran (THF), acetone, dimethylformamide (DMF), toluene, and chloroform, based on the polarity of the material. Then, the colloidal suspension of graphene-based materials is mixed with the polymer dissolved in the same solvent. The mixing is carried out by a simple stirring or shear mixing technique. During this process, the polymer chain intercalates and displaces the solvent molecules from the interlayer spaces and is adsorbed on the delaminated sheets. A non-solvent is added to precipitate the nanocomposite.

It is then separated, dried to remove the solvent, and further processed for testing and applications. Alternatively, the polymer/filler suspension can be directly cast into a mold followed by solvent removal. The solution method is widely used for the preparation of nanocomposites because of the simplicity of the process (Bera et al., 2020; Bera & Maji, 2017a; Maldonado-Magnere et al., 2021; Yousefi et al., 2013). Further, the instruments required for this process are less expensive compared to other techniques. It reduces the tendency of nanoparticle agglomeration. In general, the solution method is a very easy and efficient technique for the synthesis of polymer nanocomposites. The only disadvantage of this method is the solvent hazards and environmental pollution. This process requires a huge amount of solvent and its removal through evaporation is also an energy-consuming process. Hence, solution mixing is not a commercially viable option.

12.4.3 Melt Mixing

In the melt mixing technique, no solvent is used. GO, RGO, and GNPs are directly mixed with the TPU in the molten condition. Composites are made by high shear mixing of polymer and filler using injection or extrusion techniques. From the economic point of view, melt mixing is more advantageous than the solution method since it does not require any solvent. On the other hand, it is also a commercially viable route from the industrial angle. But several studies suggest that melt mixing cannot provide the same level of dispersion of fillers as a solution method or in situ polymerization method can do (Kim et al., 2010a). The layered nanofillers such as layered silicate, montmorillonite, and graphene are usually exfoliated before mixing with polymer melt to get good intercalation with the polymer chain. Thus, good properties of the finished composite are achieved (Adak et al., 2019).

12.4.4 Other Methods

Among other less commonly used methods, the most noticeable ones are the sol-gel method (Wang et al., 2012) and Mussel-Inspired Chemistry (Liu et al., 2021).

12.5 MICROSTRUCTURE OF NANOCOMPOSITES

Since the property improvement is directly related to the microstructure of the nanocomposite, understanding the microstructure of the nanocomposite is very crucial. TEM and wide-angle X-ray scattering (WAXS) are the two most commonly used techniques to study the dispersion of the graphene-based filler in polymer nanocomposites. Although the TEM image of a microtomed thin section of polymer nanocomposite provides clear information about the dispersion of graphene-based materials, single-layer graphene is hard to visualize within the polymer matrix (Zhang et al., 2010). On the other hand, WAXS is a relatively faster technique and provides dispersion information over a large volume of the composite. But sometimes, morphological information of the dispersed filler may be missed out due to a very small size and poor scattering intensity. Filler aggregation can be determined by small-angle X-ray scattering (SAXS) and ultra SAXS (USAXS) techniques. It has been

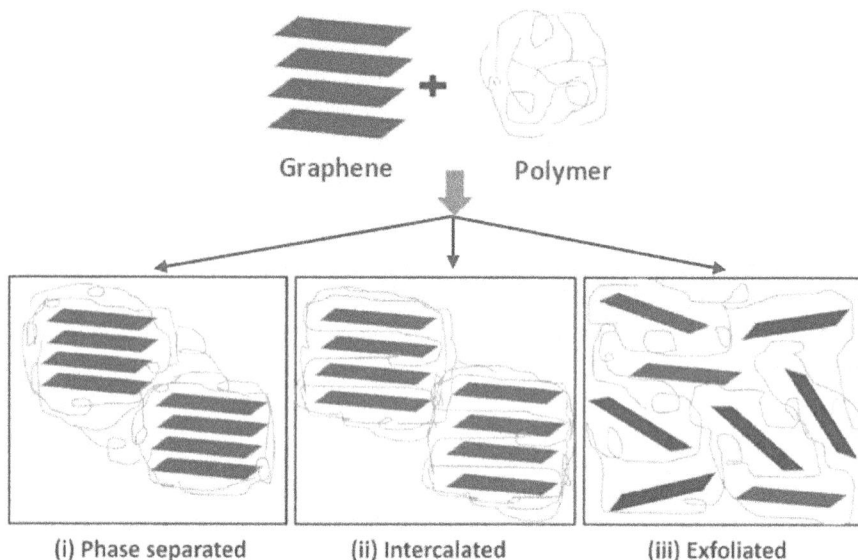

FIGURE 12.5 Different states of dispersion of graphene in graphene/TPU nanocomposites: (a) Phase separated, (b) intercalated, and (c) exfoliated.

observed that GNPs, GO, and their various derivatives have a silicate-like layered structure. Both nano clay and graphene exhibit similar microstructure in their polymer nanocomposites.

Various studies proved the existence of three different states of dispersion of the layered filler (Figure 12.5) (phase-separated, intercalated, and exfoliated) in the polymer matrix (Alexandre & Dubois, 2000; Bera & Maji, 2017b) depending on the processing technique and the polymer-filler interaction parameter. Since the GO- and GNPs-based composites have the same morphology, one can extend these terminologies for graphene-based polymer nanocomposite systems. Increased interlayer spacing is observed in the intercalated structure compared to the phase-separated one. In exfoliated structures, graphene layers have maximum interaction with the polymer matrix giving superior properties to the composites.

12.6 PROPERTIES OF GRAPHENE/TPU NANOCOMPOSITES

12.6.1 MECHANICAL PROPERTIES

Mechanical properties of graphene/TPU nanocomposites depend on the size and the aspect ratio of graphene sheet, its dispersion in TPU, and interaction between TPU and graphene. According to Terrones et al. (2011), three different polymer-filler interactions exist in a polymer nanocomposite. They are covalent interaction, non-covalent interaction (π–π interactions, electrostatic interactions), and polymer blending. Incorporating a minimal amount of graphene for a well-dispersed graphene/TPU nanocomposite leads to many-fold improvements in tensile strength, modulus, and dynamic mechanical properties (Bera & Maji, 2017b). Zhang and Gu (2013) reported that the mechanical properties of graphene are dependent on the number of layers in graphene. Multilayer graphene (MLG) having a large number of layers provides superior modulus to graphene and its polymer nanocomposites. The aspect ratio of graphene-based nanomaterials is also very important for tailoring the mechanical properties. With an increase in size and aspect ratio of graphene sheet, mechanical properties increase (Vallés et al., 2014). Strankowski (2018) proved that the mechanical properties of graphene/TPU nanocomposite depend on the percentage of the HS in TPU and the type of graphene-based nanofiller and its concentration. At 1% loading, both RGO and GNPs showed better interaction with 40% HS TPU compared to 20% HS TPU, causing better improvement in mechanical properties. But the effect was more prominent with GNP.

Mechanical properties are divided into two categories such as tensile property and dynamic mechanical property. The latter is usually used for rubbers and elastomers.

12.6.1.1   Tensile Properties

Tensile properties of graphene/TPU nanocomposites have been reported in many works of literature (Bera & Maji, 2017a; Ma et al., 2013). In most cases, GO and RGO are used as reinforcement instead of graphene because of the ease of synthesis and their effect on tensile strength, Young's modulus and elongation at break have been investigated. An enthralling report by Bera and Maji (2017a) shows that with the addition of just 0.1 wt.% GO in TPU, tensile strength and elongation at break increased by 280% and 71%, respectively. These could be due to ease of load transfer between matrix and filler. Tensile properties of graphene/TPU nanocomposite depend on the combination of polymer and filler or structure/morphology of the graphene-based materials used. For example, the tensile strength of TPU/GO nanocomposite is higher than the TPU/RGO nanocomposites (Bera & Maji, 2017a) at a similar loading. Further, at 0.1% loading, the tensile properties of TPU/GO-PPD are higher than TPU/GO. Tensile strength and Young's modulus of TPU/GO-PPD-0.1 were 41.7 MPa and 5.4 MPa, respectively, whereas the same for TPU/GO-0.1 were 31.7 and 3.5 MPa, respectively. This indicates a positive effect of the PPD functionalization of GO (Bera et al., 2020). Modification of GO with polyisocyanate (PI) enhances the modulus and tensile strength of acrylic polyol-based polyurethane compared to unmodified GO/polyurethane nanocomposites (Sadasivuni et al., 2014). Grafting of PI on the GO surface and the chemical interaction between GO-PI and PU are responsible for this kind of improvement. Tensile properties also improved notably with the addition of hyperbranched polymer functionalized graphene to the polyurethane matrix (Wu et al., 2012). Adak et al. (2019) also investigated that the tensile property of TPU increases with increasing graphene loading.

12.6.1.2   Dynamic Mechanical Property

The dynamic mechanical study can determine the influence of graphene-based materials on molecular structure and viscoelastic characteristics of graphene/TPU nanocomposites. In TPU, the HS is responsible for storage modulus (E') and the soft segment is responsible for loss modulus (E''). Storage modulus (E') and glass transition temperature (T_g) are obtained from the dynamic mechanical analysis.

The parameter E' indicates about the mobility of polymer chains, while E'' gives information about the reinforcing efficiency of the graphene-based fillers. Liao et al. (2013) reported that elastic modulus (E') of TPU increases with the loading of ARG, and a huge increment (561%) is obtained with 3% loading. Sadasivuni et al. (2014) investigated the effect of GO and 4,4′-MDI modified GO (mGO) in the dynamic mechanical property of polyurethane. Storage modulus (E') increases with increasing GO/mGO content, and the effect is more prominent for mGO. Restriction in the mobility of polymer chains and reinforcement exerted by GO and mGO are mainly responsible for such improvement.

12.6.2 Thermal Properties

The thermal properties of graphene/TPU polymer nanocomposite primarily depend on the nature of nanomaterial and interaction with TPU. Based on the application, thermal properties are divided into two categories: thermal stability and thermal conductivity.

12.6.2.1 Thermal Stability

The performance of a material can be determined from the thermal stability study. The thermal stability of graphene/TPU nanocomposites with different graphene-based materials has been investigated here. Bera et al. (Bera et al., 2020; Bera & Maji, 2017a) reported that the thermal stability of TPU depends on the structure of graphene-based materials and their loading. At just 0.2% GO loading, the thermal stability of TPU increases by 12°C, and the same for RGO was only 6°C (Bera & Maji, 2017a). In another study, Bera et al. (2020) also investigated that the thermal stability of PPD functionalized GO (GO-PPD)/TPU nanocomposite increased by 13°C at just 0.1% loading of GO-PPD and the value is almost double compared to the TPU/GO nanocomposite at similar loading. They also concluded that amine functionalization of GO positively affects the thermal stability of TPU. Adak et al. (2019) reported that the initial degradation temperature (IDT) for graphene/TPU nanocomposite increased by 12°C with the incorporation of 3 wt.% graphene. This indicates that the thermal stability of TPU increases with graphene loading. Different kinds of physicochemical interaction between polymer and nanofillers are responsible for improved thermal stability. The high barrier property of graphene-based material also plays a crucial role in it. Graphene-based materials limit the passage of oxygen through the nanocomposites, which are very essential for the combustion of the material. It also helps to retain the volatile gases generated during combustion. Thus, the thermal stability of nanocomposites is improved (Bera & Maji, 2017a). Zahid et al. (2020a) investigated the effect of MLG and FLG in the thermal properties of TPU and found that the thermal stability of TPU/MLG is more than TPU/FLG. So, the number of layers plays an important role in the overall thermal stability of TPU nanocomposites.

12.6.2.2 Thermal Conductivity

The graphene has a very high thermal conductivity (5,000 $Wm^{-1}s^{-1}$) (Balandin et al., 2008; Saito et al., 2007). So, one of the major focuses of graphene-based polymer nanocomposites is to fabricate a material having a high thermal conductivity that can quickly dissipate the thermal energy generated from the electronic devices. Thermal conductivity in graphene-based polymer nanocomposite is a very complicated process, governed by various parameters like size and defect in graphene, orientation and loading of graphene, and nature of the interface between graphene and polymer. When graphene is added to the polymer matrix, it creates a large number of interfaces that cause scattering of phonon, making the heat transfer difficult. In the pristine polymer, this kind of phonon scattering does not occur, and hence, the heat transfer becomes smooth. In polymer nanocomposite, when the filler loading is below the percolation threshold limit, fillers are not well connected to form a conducting pathway (Burger et al., 2016). In that case, interfacial thermal resistance becomes high, making the material thermally less conducting. So, a major part of research in thermal management is focused on minimizing the interfacial thermal resistance. This can be done by adding filler at the threshold limit and by modifying the filler surface. The graphene aligned in a particular direction provides superior thermal conductivity to the polymer compared to randomly oriented graphene (Li et al., 2017). Sometimes, the presence of defects in the graphene structure alters the thermal conductivity of the polymer nanocomposites. Xin et al. (2014) observed that thermally annealed (annealing temperature = 2,200°C) defect-free graphene-filled polymer nanocomposite shows the highest thermal conductivity (~3.55 W m^{-1} K^{-1}). Thermal annealing heals the defects and reduces the number of oxygen functionalities. Thus, it decreases the phonon scattering and increases the thermal conductivity. Zhang et al. (2020a) reported that with the incorporation of just 0.8 wt.% RGO, the thermal conductivity of RGO-coated PU foam (RGOF)/segmented polyurethane (SPU) nanocomposite increased by 63% compared to SPU, making it suitable for thermal management application. Various other researchers have reported a positive effect of graphene in the thermal conductivities of graphene/TPU nanocomposite (Awad et al., 2018; Guo et al., 2021; Wang et al., 2017).

12.6.3 ELECTRICAL PROPERTIES

The electrical property of TPU/graphene nanocomposite material is very important for energy storage devices and dielectrics. It is also divided into two categories: electrical conductivity and dielectric property.

12.6.3.1 Electrical Conductivity

Single-layer graphene has exceptionally high electrical conductivity (Morozov et al., 2008). In graphene-based polymer nanocomposite, it is not only the graphene-based materials that alter the electrical properties of the composite but also several other factors like concentration of filler, the number of layers in graphene, the aspect ratio of graphene sheets, processing technique, wrinkles and folds in sheets, presence of functional groups, and distribution of filler in the matrix play a crucial role (Al-Hartomy et al., 2012; Sharma et al., 2010). Galindo et al. (2014) reported that the number of layers in graphene plays an important role in the electrical conductivity of TPU/graphene nanocomposites. At a particular loading (0.25 wt.%), the electrical conductivity of single-layer graphene-filled TPU nanocomposite is 10,000 times greater than few-layer graphene (FLG)/TPU nanocomposites. Further, the electrical conductivity depends on the composite fabrication technique also. Composite prepared by solution casting technique shows ten times more electrical conductivity compared to melt-processed one. There are several other reports available where researchers have proved that incorporating graphene or graphene-based materials improves the electrical conductivity of TPU (Yousefi et al., 2012; Yuehui Wang et al., 2020; Fu et al., 2021; Jiang et al., 2020b). Fu et al. (2021) reported that the electrical conductivity of melamine modified graphene/TPU nanocomposite foam reached as high as 45.2 S m^{-1} at 2.01 vol.% loading of graphene. Jiang et al. (2020b) explored that the electrical conductivity of graphene/TPU nanocomposite foam can reach up to 50 S m^{-1} at just 2 wt.% loading of graphene. Graphene/TPU nanocomposite can be used in the flexible conducting film (Wang et al., 2020).

12.6.3.2 Dielectric Properties

Dielectric properties are specifically important for those materials which are intended to be used in electrical applications. Electrons are strongly bonded to the structure through a covalent bond for insulator and dielectric materials. Polymer nanocomposites usually have a higher dielectric constant compared to pristine polymer and undergo different types of polarization. Sadasivuni et al. (2014) investigated the influence of graphene oxide (GO) and diisocyanate modified graphene oxide (mGO) on the dielectric properties of polyurethane nanocomposites. The dielectric constant of PU nanocomposites at room temperature was found to increase over the frequency range of 10^{-2} to 10^6 Hz but the dielectric loss was decreased which claims their applications in electronic devices (Figure 12.6). Dielectric property can be observed even at ultralow loading also (Wang et al., 2013). Yousefi et al. (2014) proved that the dielectric properties of polymer nanocomposites are dependent on the alignment and aspect ratio of the filler used. It can be concluded that the dielectric property of a material is interrelated to its EMI shielding property and depends on the size of the nanomaterial and the frequency range used.

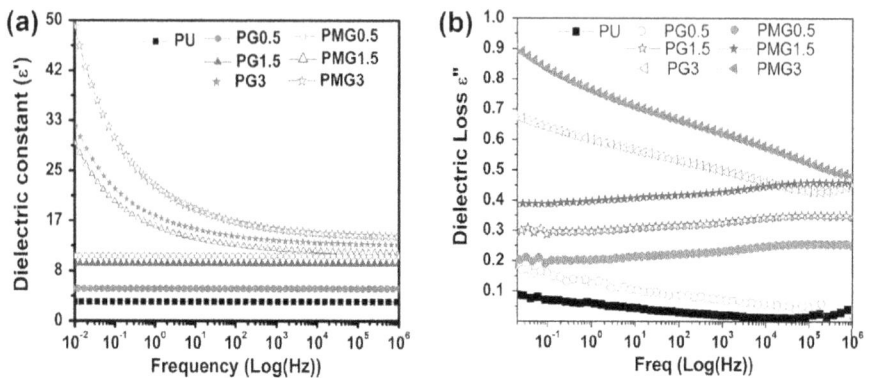

FIGURE 12.6 Change of (a) dielectric constant and (b) dielectric loss with frequency for PU nanocomposites at 25°C. (Reproduced with permission from Sadasivuni et al. (2014). Copyright 2014, Elsevier.)

12.6.4 EMI SHIELDING PROPERTY

EMI shielding implies the blocking of electromagnetic radiation. The terminology is very familiar in the electronics, instrumentation, and telecommunication sector (Yousefi et al., 2014). The "shield" typically indicates a protective covering of an electronic instrument or a part of it from the interferences arising from other appliances so that no electromagnetic wave can come out from them. Mutual interference among the appliances frequently causes interruption or sometimes complete failure of the device. It is a kind of pollution, which decreases the performance of a device simultaneously and causes an adverse effect on human health (Choudhary et al., 2012). EMI shielding is particularly important to protect electronic appliances and to avoid spying on any form of electromagnetic radiation. For the last few decades, metals or metal-coated materials were exclusively used as EMI shielding material, but the basic problem was that they were heavy in weight. This limits their use, particularly for mobile devices like smartphones and laptops. Graphene/TPU nanocomposite is an effective candidate for EMI shielding because carbon materials have high reflectivity (Chung, 2012). The performance of an EMI shield is measured by its shielding efficiency (SE) value and expressed in decibels (dB). Zahid et al. (2020b) reported the EMI shielding behavior of TPU/RGO nanocomposite. EMI shielding effectiveness of TPU/RGO nanocomposite was assessed in 700–2,500 nm range of near-infrared (NIR) wavelength and 11–20 GHz microwave range. For TPU/RGO nanocomposite, the maximum shielding effectiveness (53 dB) was observed in the 12–14 GHz frequency range and at 2.5% loading of RGO. Transmission in NIR region is only 0.5% making it a very good EMI shielding material. Various researchers have investigated the EMI shielding behavior of TPU/graphene nanocomposite foam (Jiang et al., 2019, 2020b; Carbone et al., 2021). Jiang et al. (2019) reported EMI shielding behavior of flexible TPU/RGO nanocomposite foam. With the incorporation of 3.18 vol.% RGO, the shielding effectiveness of the composite reached 21.8 dB. In another work, Jiang et al. (2020) demonstrated the shielding effectiveness of TPU/graphene composite foam. At 2% graphene loading, EMI shielding of graphene/TPU nanocomposite reached 34.3 dB. Fu et al. (2021) reported a similar observation for TPU/graphene three-dimensional nanocomposite foam. So, we can conclude that graphene provides better EMI shielding property to TPU than RGO.

12.6.5 BARRIER PROPERTY

The barrier property of TPU/graphene nanocomposites is important for their use in food packaging, paint, chemical packaging, medical use, and electronic applications (Compton et al., 2010; Maji et al., 2010). Barrier property includes the gas and moisture vapor barrier. Much work has been done to improve the barrier property of TPU (Adak et al., 2019; Maldonado-Magnere et al., 2021; Yousefi et al., 2013; Zahid et al., 2020a). The main aim of these works was to lengthen the diffusion path of the diffusing gas molecules by creating a "tortuous path" (Figure 12.7). The barrier property of graphene/TPU nanocomposites depends on several parameters like volume fraction graphene, morphology, aspect ratio, and dispersion and orientation of filler (Cao et al., 2019; Du et al., 2019, 2020; Liu et al., 2020). Perpendicular

(a)   Gas molecules   (b)   Gas molecules

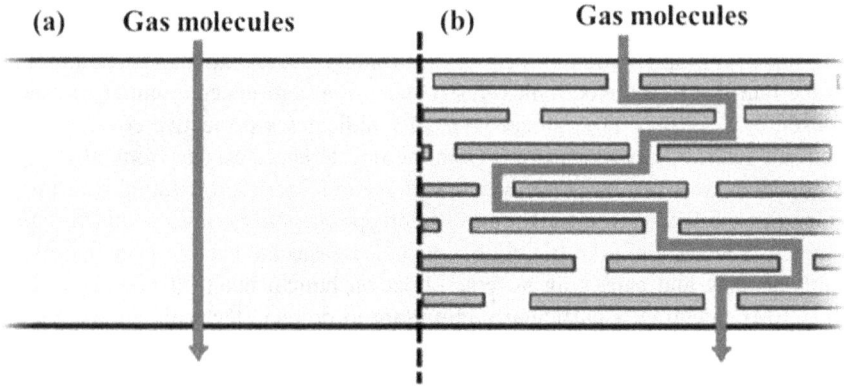

FIGURE 12.7 Schematic illustration showing the difference in the diffusion of gas molecules through (a) pristine polymer film and (b) polymer nanocomposites. (Reproduced with permission from Duncan (2011). Copyright 2011, Elsevier.)

orientation of the impermeable graphene nanoplatelets with respect to the direction of diffusing molecules compels them to follow a tortuous pathway, making the diffusion very difficult.

Due to the high aspect ratio (α), GNPs, GO, and RGO are more efficient in enhancing the moisture-vapor and gas barrier property of polymer nanocomposites than inorganic fillers (Yoo et al., 2014). Zahid et al. (2020a) investigated the morphology of a FLG and MLG in the oxygen gas permeability of TPU. According to his observation, at 4% loading, MLG with wide size distribution (0.1–25 μm) and low bulk density (0.012 g cc^{-1}) exhibits an 81% reduction in oxygen gas permeability compared to the pristine TPU. However, FLG and MLG having narrow lateral size distribution (0.1–14.5 μm) and higher bulk density (0.04–0.06 g cc^{-1}) showed a 50% reduction in oxygen gas permeability at similar loading. So, bulk density, lateral size, and distribution play a vital role in the gas permeability of TPU. The oxygen gas barrier property of TPU was also investigated by Maldonado-Magnere et al. (2021) for thermally reduced graphene oxide (TR-RGO)/TPU nanocomposite. With the incorporation of 3% TR-RGO, oxygen gas permeability was reduced by 46.1%. They also concluded that the effect of TR-RGO in the reduction of gas permeability depends on the reduction temperature and TR-RGO loading. Adak et al. (2019) reported the helium gas barrier property of TPU. With an increase in loading of hydroxyl functionalized graphene, the helium gas barrier property of TPU increased, and at 3% loading, 30% reduction in helium gas permeability was observed. Moisture permeability or water vapor transmission (WVT) of TPU was investigated by Yousefi et al. (2013). A 76% reduction in moisture permeability was reported by incorporating 3% RGO in TPU. This is due to the large size and impermeable RGO sheets in the RGO/TPU nanocomposite.

12.6.6 Flame and Fire Retardant Property

Polyurethane is inherently flammable, being an organic polymer. Scientists are working hard to reduce the flammability of organic polymers for their use in electrical applications. Flame retardant property can be determined by using cone calorimeter experiments where limiting oxygen index (LOI) and peak heat release rate (HRR) are the two most important parameters. A material is considered flammable when its LOI $\leq 26\%$ (Van Krevelen, 1975). An international standard UL-94 expresses the flammability rating of the material. Hu and Zhang (2014) reported that the addition of 1% RGO in waterborne polyurethane leads to a decrease in the total smoke released by 25% and smoke factor by 38%. This is due to the high barrier property of RGO, which decreases the passage of oxygen and volatile products of burning. Sabet et al. (2021) investigated the effect of GO on the flame retardancy of polyurethane. According to their observation, LOI increases with an increase in GO loading, and at 3 wt.% loading of GO, TPU becomes flame retardant with an LOI value of 26.8% and V-2 rating as per UL-94 standard method. Pristine TPU was inherently flammable with an LOI value of 22%. Peak HRR of TPU also decreases from 1,370.3 to 797.8 kW m^{-2} with the incorporation of 3 wt.% GO. To improve the flame retardancy of TPU further, some researchers have used functionalized graphene (Cao et al., 2019; Du et al., 2019) and some researchers have used hybrid materials (Cao et al., 2019; Du et al., 2019; Liu et al., 2020; Wang et al., 2019). Wang et al. (2019) reported the effect of cerium oxide (CeO_2)/RGO hybrid material in the fire safety of TPU. At 2% loading of CeO_2/RGO, peak HRR decreased by 41%, smoke production rate (SPR) decreased by 50%, and CO and CO_2 yield decreased by 42% each compared to neat TPU (Figure 12.8).

This indicates that the toxicity of TPU reduces significantly after incorporating the CeO_2/RGO hybrid material. Du et al. (2019) investigated the effect of urethane-silica functionalized graphene oxide (FGO) in the fire safety property of waterborne polyurethane (WPU). With the incorporation of 3% FGO, the LOI value of WPU increased from 17.8% to 23.5%, with a UL-94 rating V2. This clearly indicates fire safety of WPU improved a lot with the incorporation of FGO. Sometimes, hetero-atom-doped GO can improve the flame retardancy of TPU to a great extent. Cao et al. (2021) reported the effect of nitrogen and phosphorus-doped GO (GO-DOPO) in the flame retardance property of TPU. Their observation showed that peak HRR and peak SPR decreased by 35.8% and 50% after adding 2 wt.% GO-DOPO. Also, the CO production rate decreased by 57.1% and the CO_2 production rate decreased by 35.9%, making it a fire-safe material. In some cases, it is very difficult to prepare the sample required for cone calorimetric experiment, especially for the nanomaterials and some other expensive materials, since the amount of sample required for the test is large enough. The scientists have developed an advanced testing method called the micro-cone calorimeter test (pyrolysis-combustion flow calorimeter), which required only a few milligrams of the sample (Cao et al., 2021; Sabet et al., 2021; Wang et al., 2019).

TGA also indicates the flame retardance property of the material. In 1975, D.W. van Krevelen (1975) developed a significant correlation between the char residue (CR) and the LOI to determine the flammability of the polymer. The linear relationship between CR and LOI is shown in the following equation:

$$LOI \times 100 = 17.5 + 0.4\,CR \tag{12.1}$$

where CR is the char residue from TGA in wt.% at 850°C temperature in the inert atmosphere.

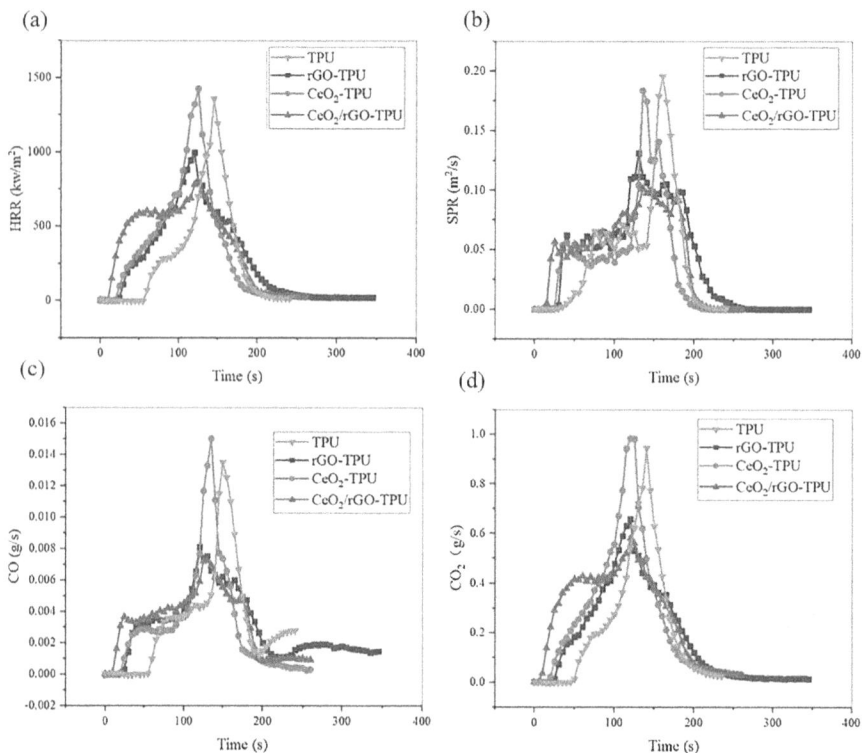

FIGURE 12.8 (a) HRR, (b) SPR, (c) CO, and (d) CO_2 curves of TPU and TPU nanocomposites. (Reproduced with permission from Wang et al. (2019). Copyright 2019, Elsevier.)

12.6.7 SHAPE MEMORY PROPERTY

For graphene/TPU nanocomposite, the shape-memory property is one of the most important ones for their applications in changeable atmospheres like variation in stress, light intensity, temperature, solutions, magnetic or electric field. This type of material can be used in the biomedical field, robotics, orthopedic surgery, automotive parts, etc. The shape memory property of graphene/TPU nanocomposite has been investigated by incorporating graphene or functionalized graphene material. Kumar Patel and Purohit (2019) reported the microwave irradiation-induced shape memory behavior of TPU/GNPs composites. Shape recovery improves under microwave irradiation compared to normal light and TPU/GNPs recover fully within 30 seconds in the microwave but pristine TPU cannot recover. Shape recovery increases with an increase in GNPs loading and frequency of microwave irradiation. Similarly, shape memory behavior of TPU/functionalized GO under near-infrared light was investigated by Du et al. (2018), and the electroactive shape memory behavior of TPU/graphene nanocomposites was reported by Sofla et al. (2019). The shape memory property of TPU was also investigated by incorporating hybrid material. Panahi-Sarmad et al. (2019) investigated the heat-assisted shape memory property of TPU by using a 1:1 mixture of GO and RGO. With 5% loading of hybrid filler, shape recovery increased by 96.7% and shape fixity improved by 99.1% compared to neat TPU. Zhou et al. (2018) reported the effect of GO/montmorillonite (MMT) hybrid material in the shape memory behavior of TPU. According to their observation at the 1:1 ratio of GO and MMT, shape fixing ratio (Rf) increases significantly at 0.3% loading. The effect is more prominent than either GO or MMT-filled composite.

12.7 POTENTIAL APPLICATIONS OF GRAPHENE/ TPU NANOCOMPOSITES

12.7.1 SOLAR WATER DESALINATION

Solar energy is the most abundant, renewable, and green energy source. It is being utilized in many of our daily activities. Looking into the increasing threat of environmental pollution, scientists all over the world are trying hard to convert and utilize solar energy as much as possible. Solar steam generation and solar water desalination are the two important technologies that have been developed as a result of hard work. Traditionally, copper or aluminum was used as an absorbent of solar energy. Non-availability and high cost limit the use of these metallic materials. Wang et al. (2017) developed nanocomposite foam based on polyurethane/rGO, which can be used effectively for solar steam generation. Solar steam generation is dependent on three factors: absorbent, light density, and illumination time. According to Wang et al. (2017), when $1\,kW\,m^{-2}$ light source is exposed on rGO/PU for 30 minutes, $0.568\,g$ water evaporates, the amount is 1.16 times more than GO/PU and 3.42 times more than pure water. So, from this experiment, it can be concluded that both rGO/PU and GO/PU can act as a very good absorbent for solar steam generation. But the efficiency is higher for a rGO/PU system. This could be due to the stronger light absorption capacity of rGO/PU. Solar steam generation ability of rGO/PU and GO/PU is a function of illumination time and light density. So, RGO/PU nanocomposite foam exhibits a solar photothermal efficiency of 81% at a light density of $10\,kW\,m^{-2}$. A cost-effective, light-weight, flexible, and highly efficient plasmonic graphene polyurethane (PGPU) nanocomposite material was developed by Awad et al. (2018).

The PGPU nanocomposite containing a small amount of metallic nanoparticles acts as a very good absorbent of solar energy (Figure 12.9). With the incorporation of $0.2\,wt.\%$ Au and Ag nanoparticles, PGPU nanocomposite exhibits a very high water evaporation rate of $11.34\,kg\,m^{-2}h^{-1}$ and high solar thermal efficiency up to 96.5% at 8 sun illumination. The nanocomposites foam can be used up to 10 cycles without any decline in its performance. So, the material can be useful for water desalination, especially in remote areas.

12.7.2 WATER PURIFICATION

Pure water is one of the most essential things for any life in this world. Unfortunately, the water is being polluted day by day due to increasing population and industrialization. Various types of filtration membranes have been used to remove the impurities and make them useable. Recently, an electrospun nanofibrous membrane made of

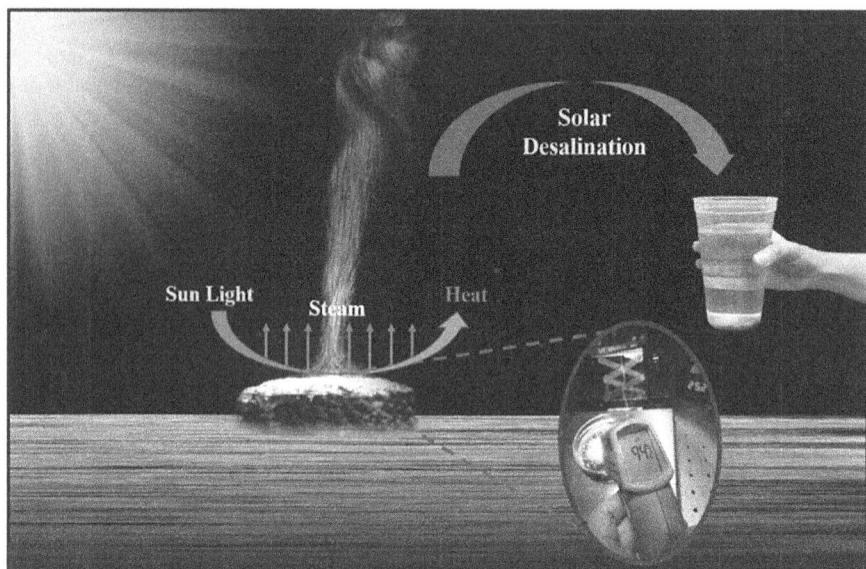

FIGURE 12.9 Schematic representation of solar water desalination process. (Reproduced with permission from Awad et al. (2018). Copyright 2018, ACS.)

polymeric material has drawn significant attention for wastewater treatment due to its light-weight, tailor-made pore structure, high surface area, durability, and antifouling property. Sundaran et al. (2020) developed PU/rGO-TiO$_2$ nanofibrous membranes with varying wt.% of rGO-TiO$_2$ and found that at 10 wt.% loading of rGO-TiO$_2$, nanocomposite membrane shows water flux of $12,810 \pm 49.65$ Lm^{-2}h^{-1} and oil rejection becomes 85.50%. The fouling behavior of a membrane is determined by the flux recovery ratio (FRR). As the FRR increases, the fouling resistance of the membrane increases and makes it appropriate for filtration. The FRR of PU/10rGO-TiO$_2$ membrane becomes 88.10%, making it a perfect filtering membrane. The nanofibrous membrane also exhibits good photocatalytic performance against methylene blue dye when visible light is irradiated.

12.7.3 SMART TEXTILES AND WEARABLE ELECTRONICS

Cotton fibers are commonly used as a fabric material in textiles and clothing. With the rapid development of technology, we are now in the era of robotics and wearable electronics technology (Li et al., 2019; Cataldi et al., 2017). Recently, flexible fibrous textile materials or smart textiles and wearable electronics have drawn great attention due to their various advantageous properties (Li et al., 2019; Sadanandan et al., 2021; Salavagione et al., 2018; Soong & Chiu, 2021). Li et al. (2019) developed a wearable pressure sensor based on graphene and cotton fiber to monitor human physiological signals. Cataldi et al. (2017) developed flexible, breathable, and electrically conductive material based on cotton, TPU, and graphene. The cotton fabric was impregnated with graphene and TPU. The resultant material exhibits high electrical conductivity ($\sim 10\ \Omega\ \text{sq}^{-2}$), breathability, and self-healing property which are required for wearable electronics.

Smart textiles are similar to normal textile material but with added features like energy storage and conversion, sensor, color change, UV light blocking (Hu et al., 2015), stretchability, and washability (Thakur, 2017). TPU is being used in smart clothing due to its certain advantageous properties like elasticity, flexibility to use with a number of fabrics, shape memory, and self-healing properties (Sáenz-Pérez et al., 2018). Shape memory polyurethane-coated fabrics are waterproof and can be used for a number of advanced applications like thermal protective clothing, smart breathable garments, sportswear, leisurewear, and gloves (Sáenz-Pérez et al., 2018; Thakur, 2017). The addition of graphene or graphene-based materials into the cotton/TPU system improves the properties further and introduces multifunctional properties that make TPU/graphene nanocomposite an ideal choice for smart textile (Hu et al., 2015; Salavagione et al., 2018; Soong & Chiu, 2021). Hu et al. (2015) developed a multifunction fabric material by coating cotton fabric with graphene/TPU composite. The newly developed fabric material can exhibit protection against UV light, remarkable electrical conductivity, and the ability to absorb far-infrared light. In another report, Soong and Chiu (2021) developed a multifunctional nanocomposite material based on TPU, multilayered graphene (GNP), and boron nitride (BN). With the addition of 20% GNP and 20% BN in TPU, the thermal conductivity of the composite increases by 2,844% compared to neat TPU and it remained almost the same even after 10 cycles of home laundering. The water contact angle of the nanocomposite increases to 89.2°, making it highly hydrophobic. Also, the thermal stability and cooling behavior of BN-GNP/TPU nanocomposite increases compared to neat TPU. All these factors made BN-GNP/TPU nanocomposite an ideal choice for breathable smart textile material.

12.7.4 OIL SPILL CLEANING

Pure water is a basic need for all living organisms like animals and creatures. But, due to rapid industrialization, the river, pond, and seawater become contaminated with various organic substances, including chemicals, organic solvents, and oils. This is a serious threat to the lives of living organisms and our ecological system. Cleaning of the oil spill from the water bodies is one of the major challenges. Various traditional methods have been applied, but they become ineffective due to low efficiency and high cost (Zhang et al., 2017). In the quest for a low cost, high efficiency, and rapid oil removing material, various types of 3D porous polymeric materials were used (Guan et al., 2019), oleophilic polyurethane foam is one of them. It is prepared by graft copolymerization of PU with an oleophilic monomer like Lauryl methacrylate (LMA), Butyl methacrylate (BMA) (Li et al., 2013). The grafting reaction is carried out in the presence of a cross-linker (divinylbenzene) and an initiator (benzoyl peroxide (BPO)). But it was not easy as the material was not so effective. Oribayo et al. (2017) developed a superhydrophobic and superoleophilic foam-like material using lignin-based polyurethane (LPU) and functionalized graphene material. Polyurethane foam surface was grafted with polydopamine reduced graphene oxide (rGO) and octadecyl amine (ODA), represented as LPU-rGO-ODA. The resultant composite foam surface exhibits a water contact angle of 152°. LPU-rGO-ODA foam can absorb oil 26–68 times of its own weight and the sorption capacity remains unaltered even after 20 cycles of sorption–squeezing.

In another study, Zhang et al. (2017) reported that a graphene-coated polyurethane sponge could be an effective material for cleaning oil spillage. In their experiment, they used cellulose nanowhiskers (CNWs) to assist the dispersion stability of graphene in the aqueous medium. 1:20 ratios of graphene and CNWs aqueous dispersion were prepared by ultra-sonication, and PU sponge was dip-coated, dried, and again dip-coated in graphene dispersion. The resultant superhydrophobic GN@PU sponge can absorb 31 gram of oil per gram of sponge and be reused up to 100 cycles. In addition to oil, it can also remove chemicals such as acetone, ethanol, and toluene,. A very high oil absorbent material was developed by Liu et al. (2013b).

As per their investigation, the reduced graphene oxide coated PU (rGPU) sponge exhibited an absorption capacity of more than $80\,g\,g^{-1}$ for all the tested liquids and became $160\,g\,g^{-1}$ for chloroform (Figure 12.10). Continuous removal of oil from water has been studied and reported by other researchers also by using graphene-coated porous and flexible polyurethane sponges (Kong et al., 2017; Zhang et al., 2018). In 2021, Ye et al. (2021) prepared a 1.5% rGO-coated superhydrophobic polyurethane monolith which can be used at least 40 times. The sponge materials can remove the oil, which is 15,000 times its weight.

12.7.5 Self-Healing Coating

Self-healing materials can heal by themselves, i.e., the material can repair the damages automatically and regain its original structure and function. It is one of the most important properties of a material, extending its application opportunities in

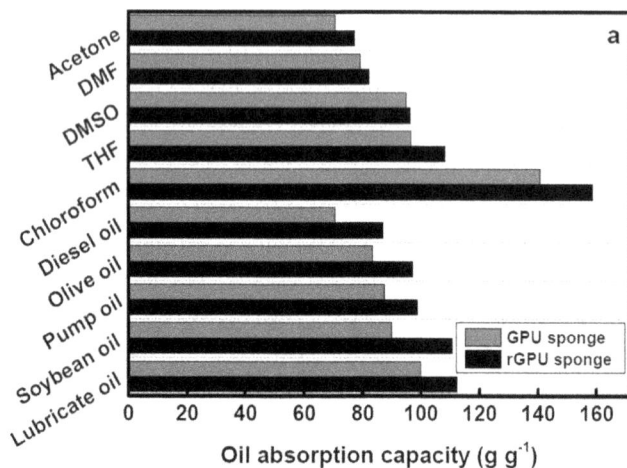

FIGURE 12.10 Absorption capacity of GO-coated PU (GPU) and RGO-coated PU (RGPU) sponges for various organic liquids. (Reproduced with permission from Liu et al. (2013b). Copyright 2013, ACS.)

diversified fields like self-healing coating, electronic screen, wearable electronics, and electronic sensors (Jasmi et al., 2018; Sadanandan et al., 2021). Graphene/polyurethane nanocomposites possess this property and various researchers have proved it. Lin et al. (2019) informed that the self-healing property of graphene/polyurethane nanocomposite is dependent on the size of 2D nanomaterials used. As per their investigation, GO sheet with a smaller size gives better self-healing performance than a larger one at the same loading. Huang et al. (2013) investigated the self-healing property of FLG/TPU nanocomposites by three different methods such as electricity, IR light, and electromagnetic wave. The maximum healing efficiency of FLG/TPU nanocomposite is around 99% in all cases and zero in the case of TPU without any FLG. Moreover, healing efficiency is dependent on the condition used. In case of IR light, the healing efficiencies of FLG/TPU nanocomposites are 99% irrespective of filler loading. This could be due to the very good IR absorbing capacity of graphene. Healing time, in this case, is a few minutes depending on the loading of FLG. Healing time is minimum at 5% loading of FLG. In the case of electrical healing, FLG/TPU nanocomposites can heal 98% or more when FLG is 5% or more. Healing time, in this case, is 3 minutes. Nanocomposites with 4% or lower FLG loading cannot be healed. Healing time decreases with an increase in applied voltage. Graphene can absorb electromagnetic waves also. Healing efficiencies in this case for all FLG/TPU nanocomposites are 98% and higher, and the healing time depends on the FLG loading. Healing time decreases linearly with increasing FLG loading. Zhou et al. (2021) also reported IR light-responsive self-healing behavior of TPU/GO nanocomposites film prepared by solvent casting method. According to their observation, the scratches on the nanocomposite film healed completely in 10 minutes and the same has been proved by optical microscope and SEM images. The effect of temperature on the healing efficiency of waterborne polyurethane was reported by Wan and Chen (2018). Nanocomposite with 0.5% GO showed maximum self-healing properties and increases with increasing temperature. The synergistic effect of graphene-CNT in the microwave-assisted self-healing behavior of TPU was investigated by Luan et al. (2018). The healing efficiency is higher than graphene/TPU or CNT/TPU. This is due to the high microwave absorption capacity by both graphene and CNT.

12.7.6 CORROSION- AND ABRASION-RESISTANT COATING

A number of factors determine the performance of coating material; abrasion resistance and corrosion resistance are the two most important ones. Zhang et al. (2020b) reported corrosion-resistant coating based on WPU and triethylene tetramine-polyethylene glycol diglycidyl ether (TETA-DGPEG) functionalized GO (TDPG). The reasons behind the selection of WPU were stronger adhesion with the matrix, no emission of toxic materials, and good abrasion resistance. The TDPG facilitates corrosion resistance property and the resultant composite coating exhibits very good adhesion strength. Coating adhesion strength is measured by a pull-off adhesion strength test before and after immersing in 3.5% NaCl solution for 30 days. It was observed that adhesion strength increases with the increase in TDPG loading and becomes the maximum at 0.1% loading. Further, salt spray test, electrochemical impedance spectroscopy (EIS), and Tafel curve proved that the corrosion resistance of WPU/TDPG is much better than WPU. The corrosion resistance property of the graphene/polyurethane system was also investigated by other researchers (Chen et al., 2019; Fechine et al., 2020; Li et al., 2016; Naderizadeh et al., 2018). Chen et al. (2019) reported the anticorrosion behavior of polyvinyl sesquisiloxane (PVSQ)-functionalized GO (PVSQ-GO) and WPU. According to their observations, 0.5% PVSQ-GO-incorporated WPU coating showed superior corrosion resistance.

Abrasion resistance is another important property of coating material. Fechine et al. (2020) reported ductile and abrasion resistance TPU/GO nanocomposite having improved mechanical properties. Abrasion resistance was determined by DIN abrasion test and found to improve by 45%. Naderizadeh et al. (2018) investigated the robustness of a coating by wear abrasion resistance test. Hydrophobic fumed silica and graphene nanoplatelets were incorporated into the TPU matrix by solution mixing and then nanocomposite film was prepared by spray coating on aluminum surface. The resultant nanocomposite film exhibits superior wear resistance under a constant load of 20 kPa compared to the film without graphene nanoplatelets' thermal interface.

12.7.7   ANTIBACTERIAL COATING

The antibacterial property of graphene-based materials (GO, rGO) was first proved by Liu et al. (2011) in 2011. An aqueous dispersion of GO or rGO was used to measure their antibacterial activity toward *Escherichia. coli* bacteria by the Colony counting method and proved that GO has better antibacterial activity than rGO. Also, the antibacterial activity is concentration- and time-dependent.

In 2015, Perreault et al. (2015) proved that the antibacterial property is size-dependent; smaller nanomaterials exhibit better antibacterial properties than larger size materials. Antibacterial property is an essential criterion for packaging materials, especially for the food and medical industries and for coating materials. An et al. (2013) reported the excellent antibacterial activity of PLA/PU/GO ternary composites against gram-positive *Staphylococcus aureus* and the gram-negative *E. coli* bacteria. With the incorporation of 5% GO on PLA/PU blend, the growth of *E. coli* and *S. aureus* was reduced up to 100% within 24 hours. Again, incorporation of 3% GO reduces the growth of gram-positive bacteria up to 99% and gram-negative bacteria up to 100% at 24 hours. These types of nanocomposite materials have application opportunities in different areas like packaging and biomedical. Borges et al. (2020) investigated the effect of graphene nanoplatelets (GNPs) and their oxidized form in the antibacterial property of TPU. Testing samples were prepared by two different techniques: melt blending of TPU and GNP and dip coating of TPU in GNP dispersion. As per their investigation, dip coating provides better antibacterial properties to TPU compared to the melt blending method because most of the GNPs are exposed on the PU surface. Further, the oxidized form of GNP having smaller sizes provides better antibacterial properties toward *Staphylococcus epidermidis* than non-oxidized GNP having a similar size. Oxidized and smaller size GNP exhibits a 70% reduction in bacterial adhesion and 70% bacterial death to TPU. This material can be used in catheters to prevent infections.

12.7.8 BIOMEDICAL APPLICATIONS

With the advancement of technologies, polymeric materials are being used in bio-medical applications, mostly tissue engineering. Among different types of poly-mers used in the biomedical field, TPU is one of the most important members. This is because of their excellent biocompatibility and mechanical properties. With the discovery of graphene, scientists/researchers are also trying to utilize the multifunctional nanomaterial in biomedical applications by making polymer nano-composites. Graphene/TPU nanocomposite is the most effective one. Ghorbani et al. (2019) reported polydopamine (PDA)-coated electrospun polyurethane/GO scaf-fold for bone tissue engineering application. PDA coating was performed by dip-ping the polyurethane/GO composite scaffold in dopamine hydrochloride solution under alkaline conditions. TPU-GO-PDA nanocomposite showed enhanced water absorption capacity and wettability compared to the control or TPU/GO. Moreover, cell attachment and proliferation were increased greatly in TPU-GO-PDA system compared to the TPU-GO system. In another study, Bahrami et al. (2019) reported graphene/polyurethane nanocomposites prepared using conventional solvent casting and electro-spinning techniques. The mechanical properties of the composites pre-pared by the electro-spinning technique are much better than those prepared by the conventional solvent casting method. The electroconductive graphene/TPU nano-composite scaffold is non-toxic and can assist cell adhesion and proliferation. This composite material might be suitable for tissue engineering applications to replace and regenerate damaged tissue. Electrospinning TPU/GO nanocomposites' scaffolds can be used for small diameter vascular tissue engineering applications (Jing et al., 2015). At 0.5% loading of GO, cell viability and mechanical properties improved, and the burst pressure of TPU/GO-0.5 tubular scaffold exceeded human blood vessels' requirements. So, the electrospun TPU/GO nanocomposite scaffold can be useful in small-diameter vascular graft tissue engineering applications. Kaur et al. (2015) also investigated the biomedical applications of TPU/graphene nanocomposites. The research in this field is going on to uncover many more possibilities.

12.7.9 SENSOR APPLICATION

In the technological era, sensors have played a significant role in making our lifestyle more easy and comfortable. Among the synthetic materials, graphene/TPU nano-composite can be used as sensor material because of their high mechanical strength, shape-memory property, and high flexibility. Graphene/TPU nanocomposites are used for sensing temperature (Jasmi et al., 2018), pressure (Feng et al., 2020), strain (Khalifa et al., 2020; Mei et al., 2021; Xie et al., 2020), human motion detection (Jia et al., 2020), Organic Vapor (Liu et al., 2016), etc. Graphene/TPU nanocomposite was used as a coating material for fiber Bragg grating (FBG). Very good linearity was observed in the composite-coated FBD with 6 pm/°C sensitivity, making it suit-able for the temperature sensor. Feng et al. (2020) developed porous polyurethane/graphene composite foam, which can exhibit very good pressure sensing behavior within a wide range of up to 500 kPa and with a sensitivity as high as 7.62 kPa^{-1}. It can also detect human motion such as detecting the movement/bending of a finger,

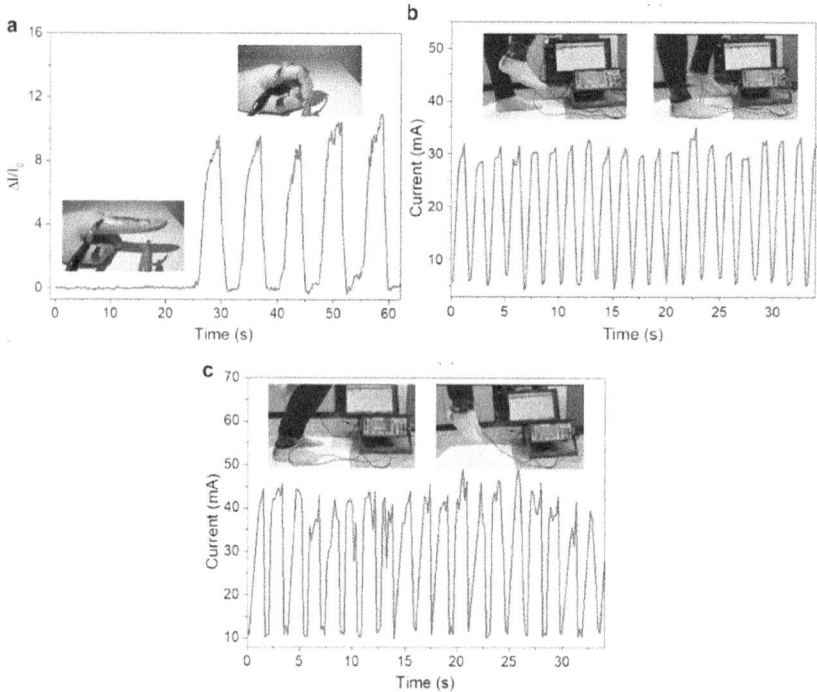

FIGURE 12.11 Graphene/TPU nanocomposite foam (with 30 wt.% graphene) pressure sensor for human motion detection. Change of current signal with (a) bending of a finger, (b) walking and (c) jumping. (Reproduced with permission from Feng et al. (2020). Copyright 2020, Elsevier.)

walking, jumping (Figure 12.11). The pressure sensing behavior of graphene/TPU nanocomposites was also investigated by Tung et al. (2016). Piezoresistive strain sensing behavior of graphene/TPU composite foam was reported by Mei et al. (2021) and Khalifa et al. (2020). Strain sensing behavior was evaluated by tapping and bending mode and found excellent sensitivity. Sensitivity remained unchanged even after 10,000 cycles under constant load. Jia et al. (2020) developed a multifunctional composite material based on PDA, RGO, and electrospun TPU. The developed composite material has very good sensitivity even after 9,000 cycles at 50% strain. This type of material is found excellent in sensing the movement of human body parts such as finger, leg, and elbow, when the sensor is attached with that body part. TPU/graphene nanocomposite can also be used for sensing organic vapor. Liu et al. (2016) developed a nanocomposite sensor that can sense acetone, CCl_4 cyclohexane, and ethyl acetate vapor.

12.7.10 OTHER APPLICATIONS

Graphene/TPU nanocomposites can also be used in various other applications. Because of high thermal conductivity, it can be used for thermal management and thermal interface material (TIM) (Carbone et al., 2021; Jiang et al., 2020; Li et al., 2017). The high electrical conductivity of TPU/graphene nanocomposite can be exploited in supercapacitor electrodes (Tai et al., 2012).

12.8 CONCLUSIONS

This chapter describes the fundamentals of graphene/TPU nanocomposites, such as understanding TPU, graphene, graphene-based materials (GO, RGO, and GNPs). The synthesis methodology of graphene/TPU nanocomposites, the microstructure of nanocomposites, the essential properties of nanocomposites, and finally, the important applications of the nanocomposites have been discussed. The effect of the structure of the graphene-based nanomaterials on the properties of their nanocomposites has also been discussed and found that the structure of the nanomaterials has a great influence in the properties and applications. Polymer nanocomposites based on TPU and graphene represent one of the most technologically promising developments of the 21st century. In the technologically advancing era, the application of this kind of composite material is ubiquitous. Whether it is decorative or industrial paints, biomedical field, solar water desalination, water purification, oil spill cleaning, smart textiles and wearable electronics, sensors, thermal management, or electrical applications, TPU/graphene nanocomposite has left its impression everywhere.

REFERENCES

Adak, B., Joshi, M., & Butola, B. S. (2019). Polyurethane/functionalized-graphene nanocomposite films with enhanced weather resistance and gas barrier properties. *Composites Part B: Engineering, 176*, 107303. doi:10.1016/J.COMPOSITESB.2019.107303.

Alexandre, M., & Dubois, P. (2000). Polymer-layered silicate nanocomposites: Preparation, properties and uses of a new class of materials. *Materials Science and Engineering: R: Reports, 28*(1–2), 1–63. doi:10.1016/S0927-796X(00)00012-7.

Al-Hartomy, O. A., Al-Ghamdi, A., Dishovsky, N., Shtarkova, R., Iliev, V., Mutlay, I., & El-Tantawy, F. (2012). Dielectric and microwave properties of natural rubber based nanocomposites containing graphene. *Materials Sciences and Applications, 3*, 453–459. doi:10.4236/msa.2012.37064.

An, X., Ma, H., Liu, B., & Wang, J. (2013). Graphene oxide reinforced polylactic acid/polyurethane antibacterial composites. *Journal of Nanomaterials, 2013*. doi:10.1155/2013/373414.

Anagnostopoulos, G., Androulidakis, C., Koukaras, E. N., Tsoukleri, G., Polyzos, I., Parthenios, J., Papagelis, K., & Galiotis, C. (2015). Stress transfer mechanisms at the submicron level for graphene/polymer systems. *ACS Applied Materials and Interfaces, 7*(7), 4216–4223. doi:10.1021/AM508482N/SUPPL_FILE/AM508482N_SI_001.PDF.

Awad, F. S., Kiriarachchi, H. D., Abouzeid, K. M., Özgür, Ü., & El-Shall, M. S. (2018). plasmonic graphene polyurethane nanocomposites for efficient solar water desalination. *ACS Applied Energy Materials, 1*(3), 976–985. doi:10.1021/ACSAEM.8B00109/SUPPL_FILE/AE8B00109_SI_002.AVI.

Bahrami, S., Solouk, A., Mirzadeh, H., & Seifalian, A. M. (2019). Electroconductive polyurethane/graphene nanocomposite for biomedical applications. *Composites Part B: Engineering, 168*, 421–431. doi:10.1016/J.COMPOSITESB.2019.03.044.

Balandin, A. A., Ghosh, S., Bao, W., Calizo, I., Teweldebrhan, D., Miao, F., & Lau, C. N. (2008). Superior thermal conductivity of single-layer graphene. *Nano Letters, 8*(3), 902–907. doi:10.1021/NL0731872.

Banhart, F., Kotakoski, J., & Krasheninnikov, A. v. (2010). Structural defects in graphene. *ACS Nano*, *5*(1), 26–41. doi:10.1021/NN102598M.

Bao, C., Song, L., Xing, W., Yuan, B., Wilkie, C. A., Huang, J., Guo, Y., & Hu, Y. (2012). Preparation of graphene by pressurized oxidation and multiplex reduction and its polymer nanocomposites by masterbatch-based melt blending. *Journal of Materials Chemistry*, *22*(13), 6088–6096. doi:10.1039/C2JM16203B.

Bera, M., Chandravati, Gupta, P., & Maji, P. K. (2018). Facile one-pot synthesis of graphene oxide by sonication assisted mechanochemical approach and its surface chemistry. *Journal of Nanoscience and Nanotechnology*, *18*(2), 902–912. doi:10.1166/JNN.2018.14306.

Bera, M., & Maji, P. K. (2017a). Effect of structural disparity of graphene-based materials on thermo-mechanical and surface properties of thermoplastic polyurethane nanocomposites. *Polymer*, *119*, 118–133. doi:10.1016/J.POLYMER.2017.05.019.

Bera, M., & Maji, P. K. (2017b). Graphene-based polymer nanocomposites: Materials for future revolution. *MOJ Polymer Science, Volume 1*(3). doi:10.15406/MOJPS.2017.01.00013.

Bera, M., Prabhakar, A., & Maji, P. K. (2020). Nanotailoring of thermoplastic polyurethane by amine functionalized graphene oxide: Effect of different amine modifier on final properties. *Composites Part B: Engineering*, *195*, 108075. doi:10.1016/J.COMPOSITESB.2020.108075.

Bianco, A., Cheng, H. M., Enoki, T., Gogotsi, Y., Hurt, R. H., Koratkar, N., Kyotani, T., Monthioux, M., Park, C. R., Tascon, J. M. D., & Zhang, J. (2013). All in the graphene family – A recommended nomenclature for two-dimensional carbon materials. *Carbon*, *65*, 1–6. doi:10.1016/J.CARBON.2013.08.038.

Borges, I., Henriques, P. C., Gomes, R. N., Pinto, A. M., Pestana, M., Magalhães, F. D., & Gonçalves, I. C. (2020). Exposure of smaller and oxidized graphene on polyurethane surface improves its antimicrobial performance. *Nanomaterials*, *10*(2), 349. doi:10.3390/NANO10020349.

Burger, N., Laachachi, A., Ferriol, M., Lutz, M., Toniazzo, V., & Ruch, D. (2016). Review of thermal conductivity in composites: Mechanisms, parameters and theory. *Progress in Polymer Science*, *61*, 1–28. doi:10.1016/J.PROGPOLYMSCI.2016.05.001.

Cao, X., Zhao, W., Huang, J., He, Y., Liang, X., Su, Y., Wu, W., & Li, R. K. Y. (2021). Interface engineering of graphene oxide containing phosphorus/nitrogen towards fire safety enhancement for thermoplastic polyurethane. *Composites Communications*, *27*. doi:10.1016/J.COCO.2021.100821.

Cao, Z. J., Liao, W., Wang, S. X., Zhao, H. B., & Wang, Y. Z. (2019). Polyurethane foams with functionalized graphene towards high fire-resistance, low smoke release, superior thermal insulation. *Chemical Engineering Journal*, *361*, 1245–1254. doi:10.1016/J.CEJ.2018.12.176.

Carbone, M. G. P., Beaugendre, M., Koral, C., Manikas, A. C., Koutroumanis, N., Papari, G. P., Andreone, A., di Maio, E., & Galiotis, C. (2021). Thermoplastic polyurethane-graphene nanoplatelets microcellular foams for electromagnetic interference shielding. *Graphene Technology*, *5*(3–4), 33–39. doi:10.1007/s41127-020-00034-0.

Cataldi, P., Ceseracciu, L., Athanassiou, A., & Bayer, I. S. (2017). Healable cotton-graphene nanocomposite conductor for wearable electronics. *ACS Applied Materials and Interfaces*, *9*(16), 13825–13830. doi:10.1021/ACSAMI.7B02326/SUPPL_FILE/AM7B02326_SI_003.AVI.

Chen, C., Wei, S., Xiang, B., Wang, B., Wang, Y., Liang, Y., & Yuan, Y. (2019). Synthesis of silane functionalized graphene oxide and its application in anti-corrosion waterborne polyurethane composite coatings. *Coatings*, *9*(9), 587. doi:10.3390/COATINGS9090587.

Chen, J., Yao, B., Li, C., & Shi, G. (2013). An improved Hummers method for eco-friendly synthesis of graphene oxide. *Carbon*, *64*, 225–229. doi:10.1016/J.CARBON.2013.07.055.

Choudhary, V., Saini, P., Jaroszewski, M., Ziaja, J., & Dhawan, S. K. (2012). Polymer based Nanocomposites for Electromagnetic Interference (EMI) shielding. *EMI Shielding Theory and Development of New Materials; Research Signpost*. Kerala, India, 67–100.

Chua, C. K., & Pumera, M. (2015). The reduction of graphene oxide with hydrazine: Elucidating its reductive capability based on a reaction-model approach. *Chemical Communications*, *52*(1), 72–75. doi:10.1039/C5CC08170J.

Chung, D. D. L. (2012). Carbon materials for structural self-sensing, electromagnetic shielding and thermal interfacing. *Carbon*, *50*(9), 3342–3353. doi:10.1016/J. CARBON.2012.01.031.

Compton, O. C., Kim, S., Pierre, C., Torkelson, J. M., & Nguyen, S. T. (2010). Crumpled graphene nanosheets as highly effective barrier property enhancers. *Advanced Materials*, *22*(42), 4759–4763.

Dreyer, D. R., Park, S., Bielawski, C. W., & Ruoff, R. S. (2009). The chemistry of graphene oxide. *Chemical Society Reviews*, *39*(1), 228–240. doi:10.1039/B917103G.

Du, W., Jin, Y., Lai, S., Shi, L., Fan, W., & Pan, J. (2018). Near-infrared light triggered shape memory and self-healable polyurethane/functionalized graphene oxide composites containing diselenide bonds. *Polymer*, *158*, 120–129. doi:10.1016/J.POLYMER.2018.10.059.

Du, W., Jin, Y., Lai, S., Shi, L., Shen, Y., & Pan, J. (2019). Urethane-silica functionalized graphene oxide for enhancing mechanical property and fire safety of waterborne polyurethane composites. *Applied Surface Science*, *492*, 298–308. doi:10.1016/J. APSUSC.2019.06.227.

Du, W., Jin, Y., Lai, S., Shi, L., Shen, Y., & Yang, H. (2020). Multifunctional light-responsive graphene-based polyurethane composites with shape memory, self-healing, and flame retardancy properties. *Composites Part A: Applied Science and Manufacturing*, *128*, 105686. doi:10.1016/J.COMPOSITESA.2019.105686.

Duncan, T. v. (2011). Applications of nanotechnology in food packaging and food safety: Barrier materials, antimicrobials and sensors. *Journal of Colloid and Interface Science*, *363*(1), 1–24. doi:10.1016/J.JCIS.2011.07.017.

Farzaneh, A., Rostami, A., & Nazockdast, H. (2021). Thermoplastic polyurethane/multi-walled carbon nanotubes nanocomposites: Effect of nanoparticle content, shear, and thermal processing. *Polymer Composites*, *42*(9), 4804–4813. doi:10.1002/PC.26190.

Fechine, G. J. M., Maia, J. M., Danda, C., Amurin, L. G., Muñoz, P. A. R., Nagaoka, D. A., Schneider, T., Troxell, B., Khani, S., Domingues, S. H., & Andrade, R. J. E. (2020). Integrated computational and experimental design of ductile, abrasion-resistant thermoplastic polyurethane/graphene oxide nanocomposites. *ACS Applied Nano Materials*, *3*(10), 9694–9705. doi:10.1021/ACSANM.0C01740/SUPPL_FILE/AN0C01740_SI_001.PDF.

Feng, C., Yi, Z., Jin, X., Seraji, S. M., Dong, Y., Kong, L., & Salim, N. (2020). Solvent crystallization-induced porous polyurethane/graphene composite foams for pressure sensing. *Composites Part B: Engineering*, *194*, 108065. doi:10.1016/j.compositesb.2020.108065.

Fu, B., Ren, P., Guo, Z., Du, Y., Jin, Y., Sun, Z., Dai, Z., & Ren, F. (2021). Construction of three-dimensional interconnected graphene nanosheet network in thermoplastic polyurethane with highly efficient electromagnetic interference shielding. *Composites Part B: Engineering*, *215*, 108813. doi:10.1016/J.COMPOSITESB.2021.108813.

Galindo, B., Gil Alcolea, S., Gómez, J., Navas, A., Murguialday, A. O., Fernandez, M. P., & Puelles, R. C. (2014). Effect of the number of layers of graphene on the electrical properties of TPU polymers. doi:10.1088/1757-899X/64/1/012008.

Geng, Y., Wang, S. J., & Kim, J. K. (2009). Preparation of graphite nanoplatelets and graphene sheets. *Journal of Colloid and Interface Science*, *336*(2), 592–598. doi:10.1016/J. JCIS.2009.04.005.

Ghorbani, F., Zamanian, A., & Aidun, A. (2019). Bioinspired polydopamine coating-assisted electrospun polyurethane-graphene oxide nanofibers for bone tissue engineering application. *Journal of Applied Polymer Science*, *136*(24), 47656. doi:10.1002/APP.47656.

Guan, Y., Cheng, F., & Pan, Z. (2019). Superwetting polymeric three dimensional (3D) porous materials for oil/water separation: A review. *Polymers*, 11(5), 806. doi:10.3390/ POLYM11050806.

Guo, H., Zhao, H., Niu, H., Ren, Y., Fang, H., Fang, X., Lv, R., Maqbool, M., & Bai, S. (2021). Highly thermally conductive 3D printed graphene filled polymer composites for scalable thermal management applications. *ACS Nano*, 15(4), 6917–6928. doi:10.1021/ ACSNANO.0C10768.

Hashimoto, A., Suenaga, K., Gloter, A., Urita, K., & Iijima, S. (2004). Direct evidence for atomic defects in graphene layers. *Nature*, 430(7002), 870–873. doi:10.1038/ nature02817.

Hepburn. (2012). *Polyurethane Elastomers - Google Books*. https://www.google.co.in/books/ edition/Polyurethane_Elastomers/7WjuCAAAQBAJ?hl=en&gbpv=1&dq=1.+Hepbur n,+C.+2012.+Polyurethane+elastomers.+Elsevier+Science+Publishers+Ltd.+Englan d&pg=PA1&printsec=frontcover.

Hu, J., & Zhang, F. (2014). Self-assembled fabrication and flame-retardant properties of reduced graphene oxide/waterborne polyurethane nanocomposites. doi:10.1007/ s10973-014-4078-7.

Hu, X., Tian, M., Qu, L., Zhu, S., & Han, G. (2015). Multifunctional cotton fabrics with graphene/polyurethane coatings with far-infrared emission, electrical conductivity, and ultraviolet-blocking properties. *Carbon*, 95, 625–633. doi:10.1016/J. CARBON.2015.08.099.

Huang, L., Yi, N., Wu, Y., Zhang, Y., Zhang, Q., Huang, Y., Ma, Y., Chen, Y., Huang, L., Yi, N., Wu, Y., Zhang, Y., Zhang, Q., Huang, Y., Ma, Y., & Chen, Y. (2013). Multichannel and repeatable self-healing of mechanical enhanced graphene-thermoplastic polyurethane composites. *Advanced Materials*, 25(15), 2224–2228. doi:10.1002/ ADMA.201204768.

Huh, S. H. (2011). Thermal reduction of graphene oxide. *Physics and Applications of Graphene - Experiments*. doi:10.5772/14156.

Hummers, W. S., & Offeman, R. E. (2002). Preparation of graphitic oxide. *Journal of the American Chemical Society*, 80(6), 1339. doi:10.1021/JA01539A017.

Ionescu, M. (2005). Chemistry and technology of polyols for polyurethanes. In *Rapra Technology* (Vol. 56). Rapra Technology. doi:10.1002/pi.2159.

Jahanshahi, M., Jabari, R., Rashidi, A. M., & Ghoreyshi, A. A. (2013). Synthesis and characterization of thermally-reduced graphene. doi:10.5829/idosi.ijee.2013.04.01.09.

Jana, M., Saha, S., Khanra, P., Murmu, N. C., Srivastava, S. K., Kuila, T., & Lee, J. H. (2014). Bio-reduction of graphene oxide using drained water from soaked mung beans (Phaseolus aureus L.) and its application as energy storage electrode material. *Materials Science and Engineering: B*, 186(1), 33–40. doi:10.1016/J.MSEB.2014.03.004.

Jasmi, F., Azeman, N. H., Bakar, A. A. A., Zan, M. S. D., Haji Badri, K., & Su'ait, M. S. (2018). Ionic conductive polyurethane-graphene nanocomposite for performance enhancement of optical fiber Bragg grating temperature sensor. *IEEE Access*, 6, 47355–47363. doi:10.1109/ACCESS.2018.2867220.

Jia, Y., Yue, X., Wang, Y., Yan, C., Zheng, G., Dai, K., Liu, C., & Shen, C. (2020). Multifunctional stretchable strain sensor based on polydopamine/reduced graphene oxide/electrospun thermoplastic polyurethane fibrous mats for human motion detection and environment monitoring. *Composites Part B: Engineering*, 183, 107696. doi:10.1016/J.COMPOSITESB.2019.107696.

Jiang, Q., Liao, X., Li, J., Chen, J., Wang, G., Yi, J., Yang, Q., & Li, G. (2019). Flexible thermoplastic polyurethane/reduced graphene oxide composite foams for electromagnetic interference shielding with high absorption characteristic. *Composites Part A: Applied Science and Manufacturing*, 123, 310–319. doi:10.1016/J.COMPOSITESA. 2019.05.017.

Jiang, Q., Liao, X., Yang, J., Wang, G., Chen, J., Tian, C., & Li, G. (2020). A two-step process for the preparation of thermoplastic polyurethane/graphene aerogel composite foams with multi-stage networks for electromagnetic shielding. *Composites Communications*, *21*, 100416. doi:10.1016/J.COCO.2020.100416.

Jing, X., Mi, H. Y., Salick, M. R., Cordie, T. M., Peng, X. F., & Turng, L. S. (2015). Electrospinning thermoplastic polyurethane/graphene oxide scaffolds for small diameter vascular graft applications. *Materials Science and Engineering: C*, *49*, 40–50. doi:10.1016/J.MSEC.2014.12.060.

Kaur, G., Adhikari, R., Cass, P., Bown, M., Evans, M. D. M., Vashi, A. v., & Gunatillake, P. (2015). Graphene/polyurethane composites: Fabrication and evaluation of electrical conductivity, mechanical properties and cell viability. *RSC Advances*, *5*(120), 98762–98772. doi:10.1039/C5RA20214K.

Khalifa, M., Ekbote, G. S., Anandhan, S., Wuzella, G., Lammer, H., & Mahendran, A. R. (2020). Physicochemical characteristics of bio-based thermoplastic polyurethane/ graphene nanocomposite for piezoresistive strain sensor. *Journal of Applied Polymer Science*, *137*(44), 49364. doi:10.1002/APP.49364.

Kong, Z., Wang, J., Lu, X. et al. (2017) *In situ* fastening graphene sheets into a polyurethane sponge for the highly efficient continuous cleanup of oil spills. *Nano Res. 10*, 1756–1766. doi:10.1007/s12274-017-1484-8.

Kuila, T., Bose, S., Khanra, P., Mishra, A. K., Kim, N. H., & Lee, J. H. (2012). A green approach for the reduction of graphene oxide by wild carrot root. *Carbon*, *50*(3), 914–921. doi:10.1016/J.CARBON.2011.09.053.

Kumar Patel, K., & Purohit, R. (2019). Improved shape memory and mechanical properties of microwave-induced thermoplastic polyurethane/graphene nanoplatelets composites. *Sensors and Actuators, A: Physical*, *285*, 17–24. doi:10.1016/J.SNA.2018.10.049.

Lee, C., Wei, X., Kysar, J. W., & Hone, J. (2008). Measurement of the elastic properties and intrinsic strength of monolayer graphene. *Science*, *321*(5887), 385–388. doi:10.1126/SCIENCE.1157996/SUPPL_FILE/LEE-SOM.PDF.

Li, J., Cui, J., Yang, J., Li, Y., Qiu, H., & Yang, J. (2016). Reinforcement of graphene and its derivatives on the anticorrosive properties of waterborne polyurethane coatings. *Composites Science and Technology*, *129*, 30–37. doi:10.1016/J.COMPSCITECH.2016.04.017.

Li, H., Liu, L., & Yang, F. (2013). Oleophilic polyurethane foams for oil spill cleanup. *Procedia Environmental Sciences*, *18*, 528–533. doi:10.1016/j.proenv.2013.04.071.

Li, A., Zhang, C., & Zhang, Y. F. (2017). Thermal conductivity of graphene-polymer composites: Mechanisms, properties, and applications. *Polymers*, *9*(9), 437. doi:10.3390/POLYM9090437.

Li, P., Zhao, L., Jiang, Z., Yu, M., Li, Z., Zhou, X., & Zhao, Y. (2019). A wearable and sensitive graphene-cotton based pressure sensor for human physiological signals monitoring. *Scientific Reports*, *9*(1). doi:10.1038/S41598-019-50997-1.

Liao, K. H., Park, Y. T., Abdala, A., & Macosko, C. (2013). Aqueous reduced graphene/thermoplastic polyurethane nanocomposites. *Polymer*, *54*(17), 4555–4559. doi:10.1016/J.POLYMER.2013.06.032.

Lin, C., Sheng, D., Liu, X., Xu, S., Ji, F., Dong, L., Zhou, Y., & Yang, Y. (2019). Effect of different sizes of graphene on Diels-Alder self-healing polyurethane. *Polymer*, *182*, 121822. doi:10.1016/J.POLYMER.2019.121822.

Liu, J., Cui, L., & Losic, D. (2013a). Graphene and graphene oxide as new nanocarriers for drug delivery applications. *Acta Biomaterialia*, *9*(12), 9243–9257. doi:10.1016/J.ACTBIO.2013.08.016.

Liu, H., Huang, W., Yang, X., Dai, K., Zheng, G., Liu, C., Shen, C., Yan, X., Guo, J., & Guo, Z. (2016). Organic vapor sensing behaviors of conductive thermoplastic polyurethane–graphene nanocomposites. *Journal of Materials Chemistry C*, *4*(20), 4459–4469. doi:10.1039/C6TC00987E.

Liu, Y., Ma, J., Wu, T., Wang, X., Huang, G., Liu, Y., Qiu, H., Li, Y., Wang, W., & Gao, J. (2013b). Cost-effective reduced graphene oxide-coated polyurethane sponge as a highly efficient and reusable oil-absorbent. *ACS Applied Materials and Interfaces, 5*(20), 10018–10026. doi:10.1021/AM4024252/SUPPL_FILE/AM4024252_SI_001.PDF.

Liu, C., Wu, W., Shi, Y., Yang, F., Liu, M., Chen, Z., Yu, B., & Feng, Y. (2020). Creating MXene/reduced graphene oxide hybrid towards highly fire safe thermoplastic polyurethane nanocomposites. *Composites Part B: Engineering, 203*, 108486. doi:10.1016/J.COMPOSITESB.2020.108486.

Liu, S., Zeng, T. H., Hofmann, M., Burcombe, E., Wei, J., Jiang, R., Kong, J., & Chen, Y. (2011). Antibacterial activity of graphite, graphite oxide, graphene oxide, and reduced graphene oxide: Membrane and oxidative stress. *ACS Nano, 5*(9), 6971–6980. doi:10.1021/NN202451X/SUPPL_FILE/NN202451X_SI_001.PDF.

Liu, Y., Zheng, J., Zhang, X., Li, K., Zhang, Y., Du, Y., Yu, G., & Jia, Y. (2021). Mussel-inspired waterproof and self-healing polyurethane with enhanced mechanical properties. *European Polymer Journal, 159*, 110751. doi:10.1016/J.EURPOLYMJ.2021.110751.

Luan, Y., Gao, F., Li, Y., Yang, J., Hu, Y., Guo, Z., Wang, Z., & Zhou, A. (2018). Healing mechanisms induced by synergy of graphene-CNTs and microwave focusing effect for the thermoplastic polyurethane composites. *Composites Part A: Applied Science and Manufacturing, 106*, 34–41. doi:10.1016/J.COMPOSITESA.2017.12.009.

Ma, W., Wu, L., Zhang, D., & Wang, S. (2013). Preparation and properties of 3-aminopropyltriethoxysilane functionalized graphene/polyurethane nanocomposite coatings. *Colloid and Polymer Science, 291*(12), 2765–2773. doi:10.1007/S00396-013-3014-X/FIGURES/10.

Maji, P. K., & Bhowmick, A. K. (2013). Structure–property correlation of polyurethane nanocomposites: Influence of loading and nature of nanosilica and microstructure of hyperbranched polyol. *Journal of Applied Polymer Science, 127*(6), 4492–4504. doi:10.1002/APP.38063.

Maji, P. K., Das, N. K., & Bhowmick, A. K. (2010). Preparation and properties of polyurethane nanocomposites of novel architecture as advanced barrier materials. *Polymer, 51*(5), 1100–1110. doi:10.1016/J.POLYMER.2009.12.040.

Maji, P. K., Guchhait, P. K., & Bhowmick, A. K. (2009). Effect of nanoclays on physico-mechanical properties and adhesion of polyester-based polyurethane nanocomposites: Structure-property correlations. *Journal of Materials Science, 44*(21), 5861–5871. doi:10.1007/S10853-009-3827-7.

Maldonado-Magnere, S., Yazdani-Pedram, M., Aguilar-Bolados, H., & Quijada, R. (2021). Thermally reduced graphene oxide/thermoplastic polyurethane nanocomposites: Mechanical and barrier properties. *Polymers, 13*(1), 1–10. doi:10.3390/POLYM13010085.

Mei, S., Zhang, X., Ding, B., Wang, J., Yang, P., She, H., Cui, Z., Liu, M., Pang, X., & Fu, P. (2021). 3D-Printed thermoplastic polyurethane/graphene composite with porous segregated structure: Toward ultralow percolation threshold and great strain sensitivity. *Journal of Applied Polymer Science, 138*(14), 50168. doi:10.1002/APP.50168.

Morozov, S. v, Novoselov, K. S., Katsnelson, M. I., Schedin, F., Elias, D. C., Jaszczak, J. A., & Geim, A. K. (2008). Giant intrinsic carrier mobilities in graphene and its bilayer. doi:10.1103/PhysRevLett.100.016602.

Naderizadeh, S., Athanassiou, A., & Bayer, I. S. (2018). Interfacing superhydrophobic silica nanoparticle films with graphene and thermoplastic polyurethane for wear/abrasion resistance. *Journal of Colloid and Interface Science, 519*, 285–295. doi:10.1016/J.JCIS.2018.02.065.

Novoselov, K. S., Geim, A. K., Morozov, S. v., Jiang, D., Zhang, Y., Dubonos, S. v., Grigorieva, I. v., & Firsov, A. A. (2004). Electric field effect in atomically thin carbon films. *Science (New York, N.Y.), 306*(5696), 666–669. doi:10.1126/SCIENCE.1102896.

Oertel, G., & Abele, L. (1994). *Polyurethane Handbook: Chemistry, Raw Materials, Processing, Application, Properties*. 688. Munich ; New York : Hanser ; Cincinnati : Hanser/Gardner [distributor], ©1994.

Oribayo, O., Feng, X., Rempel, G. L., & Pan, Q. (2017). Synthesis of lignin-based polyurethane/graphene oxide foam and its application as an absorbent for oil spill clean-ups and recovery. *Chemical Engineering Journal, 323*, 191–202. doi:10.1016/j. cej.2017.04.054.

Panahi-Sarmad, M., Goodarzi, V., Amirkiai, A., Noroozi, M., Abrisham, M., Dehghan, P., Shakeri, Y., Karimpour-Motlagh, N., Poudineh Hajipoor, F., Ali Khonakdar, H., & Asefnejad, A. (2019). Programing polyurethane with systematic presence of graphene-oxide (GO) and reduced graphene-oxide (rGO) platelets for adjusting of heat-actuated shape memory properties. *European Polymer Journal, 118*, 619–632. doi:10.1016/J. EURPOLYMJ.2019.06.034.

Perreault, F., de Faria, A. F., Nejati, S., & Elimelech, M. (2015). Antimicrobial properties of graphene oxide nanosheets: Why size matters. *ACS Nano, 9*(7), 7226–7236. doi:10.1021/ ACSNANO.5B02067/SUPPL_FILE/NN5B02067_SI_001.PDF.

Prisacariu, C. (2011). Polyurethane elastomers. doi:10.1007/978–3–7091–0514–6.

Raza, M. A., Westwood, A., Brown, A., Hondow, N., & Stirling, C. (2011). Characterisation of graphite nanoplatelets and the physical properties of graphite nanoplatelet/silicone composites for thermal interface applications. *Carbon, 49*(13), 4269–4279. doi:10.1016/J. CARBON.2011.06.002.

Sabet, M., Soleimani, H., Mohammadian, E., & Hosseini, S. (2021). The effect of graphene oxide on the mechanical, thermal characteristics and flame retardancy of polyurethane. *Plastics, Rubber and Composites, 50*(2), 61–70. doi:10.1080/14658011.2020.1833557.

Sadanandan, K. S., Bacon, A., Shin, D. W., Alkhalifa, S. F. R., Russo, S., Craciun, M. F., & Neves, A. I. S. (2021). Graphene coated fabrics by ultrasonic spray coating for wearable electronics and smart textiles. *Journal of Physics: Materials, 4*(1). doi:10.1088/2515-7639/ABC632.

Sadasivuni, K. K., Ponnamma, D., Kumar, B., Strankowski, M., Cardinaels, R., Moldenaers, P., Thomas, S., & Grohens, Y. (2014). Dielectric properties of modified graphene oxide filled polyurethane nanocomposites and its correlation with rheology. *Composites Science and Technology, 104*, 18–25. doi:10.1016/J.COMPSCITECH.2014.08.025.

Sáenz-Pérez, M., Bashir, T., Laza, J. M., García-Barrasa, J., Vilas, J. L., Skrifvars, M., & León, L. M. (2018). Novel shape-memory polyurethane fibers for textile applications. *Textile Research Journal, 89*(6), 1027–1037. doi:10.1177/0040517518760756.

Saito, K., Nakamura, J., & Natori, A. (2007). Ballistic thermal conductance of a graphene sheet. doi:10.1103/PhysRevB.76.115409.

Salavagione, H. J., Gómez-Fatou, M. A., Shuttleworth, P. S., & Ellis, G. J. (2018). New perspectives on graphene/polymer fibers and fabrics for smart textiles: The relevance of the polymer/graphene interphase. *Frontiers in Materials, 5*. doi:10.3389/ FMATS.2018.00018.

Sharma, R., Baik, J. H., Perera, C. J., & Strano, M. S. (2010). Anomalously large reactivity of single graphene layers and edges toward electron transfer chemistries. *Nano Letters, 10*(2), 398–405. doi:10.1021/NL902741X/SUPPL_FILE/NL902741X_SI_001.PDF.

Sofla, R.L.M, Rezaei, M., Babaie, A., & Nasiri, M. (2019). Preparation of electroactive shape memory polyurethane/graphene nanocomposites and investigation of relationship between rheology, morphology and electrical properties. *Composites Part B: Engineering, 175*, 107090. doi:10.1016/J.COMPOSITESB.2019.107090.

Soong, Y. C., & Chiu, C. W. (2021). Multilayered graphene/boron nitride/thermoplastic polyurethane composite films with high thermal conductivity, stretchability, and washability for adjustable-cooling smart clothes. *Journal of Colloid and Interface Science, 599*, 611–619. doi:10.1016/J.JCIS.2021.04.123.

Stankovich, S., Dikin, D. A., Piner, R. D., Kohlhaas, K. A., Kleinhammes, A., Jia, Y., Wu, Y., Nguyen, S. B. T., & Ruoff, R. S. (2007). Synthesis of graphene-based nanosheets via chemical reduction of exfoliated graphite oxide. *Carbon, 45*(7), 1558–1565. doi:10.1016/J.CARBON.2007.02.034.

Stoller, M. D., Park, S., Yanwu, Z., An, J., & Ruoff, R. S. (2008). Graphene-based ultracapacitors. *Nano Letters, 8*(10), 3498–3502. doi:10.1021/NL802558Y.

Strankowski, M. (2018). Effect of variation of hard segment content and graphene-based nanofiller concentration on morphological, thermal, and mechanical properties of polyurethane nanocomposites. *International Journal of Polymer Science, 2018*. doi:10.1155/2018/1090753.

Sundaran, S. P., Reshmi, C. R., Sagitha, P., & Sujith, A. (2020). Polyurethane nanofibrous membranes decorated with reduced graphene oxide–TiO$_2$ for photocatalytic templates in water purification. *Journal of Materials Science, 55*(14), 5892–5907. doi:10.1007/S10853-020-04414-Y/TABLES/7.

Tai, Z., Yan, X., & Xue, Q. (2012). Shape-alterable and -recoverable graphene/polyurethane bi-layered composite film for supercapacitor electrode. *Journal of Power Sources, 213*, 350–357. doi:10.1016/J.JPOWSOUR.2012.03.086.

Terrones, M., Martín, O., González, M., Pozuelo, J., Serrano, B., Cabanelas, J. C., Vega-Díaz, S. M., Baselga, J., Martín, O., González, M., Pozuelo, J., Serrano, B., Cabanelas, J. C., Baselga, J., Terrones, M., & Vega-Díaz, S. M. (2011). Interphases in graphene polymer-based nanocomposites: Achievements and challenges. *Advanced Materials, 23*(44), 5302–5310. doi:10.1002/ADMA.201102036.

Thakur, S. (2017). Shape memory polymers for smart textile applications. *Textiles for Advanced Applications.* doi:10.5772/INTECHOPEN.69742.

Thu, T. v., Tanizawa, Y., Phuc, N. H. H., Ko, P. J., & Sandhu, A. (2013). Synthesis and characterization of graphite nanoplatelets. *Journal of Physics: Conference Series, 433*(1), 012003. doi:10.1088/1742-6596/433/1/012003.

Tung, T. T., Robert, C., Castro, M., Feller, J. F., Kim, T. Y., & Suh, K. S. (2016). Enhancing the sensitivity of graphene/polyurethane nanocomposite flexible piezo-resistive pressure sensors with magnetite nano-spacers. *Carbon, 108*, 450–460. doi:10.1016/J.CARBON.2016.07.018.

Vallés, C., Abdelkader, A. M., Young, R. J., & Kinloch, I. A. (2014). Few layer graphene–polypropylene nanocomposites: The role of flake diameter. *Faraday Discussions, 173*(0), 379–390. doi:10.1039/C4FD00112E.

Van Krevelen, D. W. (1975). Some basic aspects of flame resistance of polymeric materials. *Polymer, 16*(8), 615–620. doi:10.1016/0032-3861(75)90157-3.

Wan, T., & Chen, D. (2018). Mechanical enhancement of self-healing waterborne polyurethane by graphene oxide. *Progress in Organic Coatings, 121*, 73–79. doi:10.1016/J.PORGCOAT.2018.04.016.

Wang, G., Fu, Y., Guo, A., Mei, T., Wang, J., Li, J., & Wang, X. (2017). Reduced graphene oxide-polyurethane nanocomposite foam as a reusable photoreceiver for efficient solar steam generation. *Chemistry of Materials, 29*(13), 5629–5635. doi:10.1021/ACS.CHEMMATER.7B01280/SUPPL_FILE/CM7B01280_SI_001.PDF.

Wang, S., Gao, R., & Zhou, K. (2019). The influence of cerium dioxide functionalized reduced graphene oxide on reducing fire hazards of thermoplastic polyurethane nanocomposites. *Journal of Colloid and Interface Science, 536*, 127–134. doi:10.1016/J.JCIS.2018.10.052.

Wang, X., Hu, Y., Song, L., Yang, H., Xing, W., & Lu, H. (2011a). In situ polymerization of graphene nanosheets and polyurethane with enhanced mechanical and thermal properties. *Journal of Materials Chemistry, 21*(12), 4222–4227. doi:10.1039/C0JM03710A.

Wang, Y., Shi, Z. X., & Yin, J. (2011b). Facile synthesis of soluble graphene via a green reduction of graphene oxide in tea solution and its biocomposites. *ACS Applied Materials and Interfaces*, *3*(4), 1127–1133. doi:10.1021/AM1012613/SUPPL_FILE/AM1012613_SI_001.PDF.

Wang, X., Xing, W., Song, L., Yang, H., Hu, Y., & Yeoh, G. H. (2012). Fabrication and characterization of graphene-reinforced waterborne polyurethane nanocomposite coatings by the sol–gel method. *Surface and Coatings Technology*, *206*(23), 4778–4784. doi:10.1016/J.SURFCOAT.2012.03.077.

Wang, D., Zhang, X., Zha, J. W., Zhao, J., Dang, Z. M., & Hu, G. H. (2013). Dielectric properties of reduced graphene oxide/polypropylene composites with ultralow percolation threshold. *Polymer*, *54*(7), 1916–1922. doi:10.1016/J.POLYMER.2013.02.012.

Wang, Y., Zhou, Z., Zhang, J., Tang, J., Wu, P., Wang, K., & Zhao, Y. (2020). Properties of graphene-thermoplastic polyurethane flexible conductive film. *Coatings*, *10*(4), 400. doi:10.3390/COATINGS10040400.

Wu, C., Huang, X., Wang, G., Wu, X., Yang, K., Li, S., & Jiang, P. (2012). Hyperbranched-polymer functionalization of graphene sheets for enhanced mechanical and dielectric properties of polyurethane composites. *Journal of Materials Chemistry*, *22*(14), 7010–7019. doi:10.1039/C2JM16901K.

Xie, X., Huang, H., Zhu, J., Yu, J., Wang, Y., & Hu, Z. (2020). A spirally layered carbon nanotube-graphene/polyurethane composite yarn for highly sensitive and stretchable strain sensor. *Composites Part A: Applied Science and Manufacturing*, *135*, 105932. doi:10.1016/J.COMPOSITESA.2020.105932.

Xin, G., Sun, H., Scott, S. M., Yao, T., Lu, F., Shao, D., Hu, T., Wang, G., Ran, G., & Lian, J. (2014). Advanced phase change composite by thermally annealed defect-free graphene for thermal energy storage. *ACS Applied Materials and Interfaces*, *6*(17), 15262–15271. doi:10.1021/AM503619A/SUPPL_FILE/AM503619A_SI_001.PDF.

Ye, S., Wang, B., Pu, Z., Liu, T., Feng, Y., Han, W., Liu, C., & Shen, C. (2021). Flexible and robust porous thermoplastic polyurethane/reduced graphene oxide monolith with special wettability for continuous oil/water separation in harsh environment. *Separation and Purification Technology*, *266*, 118553. doi:10.1016/J.SEPPUR.2021.118553.

Yoo, B. M., Shin, H. J., Yoon, H. W., & Park, H. B. (2014). Graphene and graphene oxide and their uses in barrier polymers. *Journal of Applied Polymer Science*, *131*(1). doi:10.1002/APP.39628.

Yousefi, N., Gudarzi, M. M., Zheng, Q., Aboutalebi, S. H., Sharif, F., & Kim, J. K. (2012). Self-alignment and high electrical conductivity of ultralarge graphene oxide–polyurethane nanocomposites. *Journal of Materials Chemistry*, *22*(25), 12709–12717. doi:10.1039/C2JM30590A.

Yousefi, N., Gudarzi, M. M., Zheng, Q., Lin, X., Shen, X., Jia, J., Sharif, F., & Kim, J. K. (2013). Highly aligned, ultralarge-size reduced graphene oxide/polyurethane nanocomposites: Mechanical properties and moisture permeability. *Composites Part A: Applied Science and Manufacturing*, *49*, 42–50. doi:10.1016/J.COMPOSITESA.2013.02.005.

Yousefi, N., Sun, X., Lin, X., Shen, X., Jia, J., Zhang, B., Tang, B., Chan, M., Kim, J.-K., Yousefi, N., Sun, X. Y., Lin, X. Y., Shen, X., Jia, J. J., Zhang, B., Kim, J. K., Tang, B. Z., & Chan, M. (2014). Highly aligned graphene/polymer nanocomposites with excellent dielectric properties for high-performance electromagnetic interference shielding. *Advanced Materials*, *26*(31), 5480–5487. doi:10.1002/ADMA.201305293.

Zacharia, R., Ulbricht, H., & Hertel, T. (2004). Interlayer cohesive energy of graphite from thermal desorption of polyaromatic hydrocarbons. *Undefined*, *69*(15). doi:10.1103/PHYSREVB.69.155406.

Zahid, M., del Río Castillo, A. E., Thorat, S. B., Panda, J. K., Bonaccorso, F., & Athanassiou, A. (2020a). Graphene morphology effect on the gas barrier, mechanical and thermal properties of thermoplastic polyurethane. *Composites Science and Technology, 200.* doi:10.1016/J.COMPSCITECH.2020.108461.

Zahid, M., Nawab, Y., Gulzar, N., Rehan, Z. A., Shakir, M. F., Afzal, A., Abdul Rashid, I., & Tariq, A. (2020b). Fabrication of reduced graphene oxide (RGO) and nanocomposite with thermoplastic polyurethane (TPU) for EMI shielding application. *Journal of Materials Science: Materials in Electronics, 31*(2), 967–974. doi:10.1007/S10854-019-02607-Z.

Zhang, Y. Y., & Gu, Y. T. (2013). Mechanical properties of graphene: Effects of layer number, temperature and isotope. *Computational Materials Science, 71*, 197–200. doi:10.1016/J.COMMATSCI.2013.01.032.

Zhang, X., Liu, D., Ma, Y., Nie, J., & Sui, G. (2017). Super-hydrophobic graphene coated polyurethane (GN@PU) sponge with great oil-water separation performance. *Applied Surface Science, 422*, 116–124. doi:10.1016/J.APSUSC.2017.06.009.

Zhang, F., Liu, W., Liang, L., Wang, S., Shi, H., Xie, Y., Yang, M., & Pi, K. (2020b). The effect of functional graphene oxide nanoparticles on corrosion resistance of waterborne polyurethane. *Colloids and Surfaces A: Physicochemical and Engineering Aspects, 591.* doi:10.1016/j.colsurfa.2020.124565.

Zhang, C., Shi, Z., Li, A., & Zhang, Y. F. (2020a). RGO-coated polyurethane foam/segmented polyurethane composites as solid–solid phase change thermal interface material. *Polymers, 12*(12), 3004. doi:10.3390/POLYM12123004.

Zhang, T., Xiao, C., Zhao, J., Chen, K., Hao, J., & Ji, D. (2018). Continuous separation of oil from water surface by a novel tubular unit based on graphene coated polyurethane sponge. *Polymers for Advanced Technologies, 29*(8), 2317–2326. doi:10.1002/PAT.4343.

Zhang, H. B., Zheng, W. G., Yan, Q., Yang, Y., Wang, J. W., Lu, Z. H., Ji, G. Y., & Yu, Z. Z. (2010). Electrically conductive polyethylene terephthalate/graphene nanocomposites prepared by melt compounding. *Polymer, 51*(5), 1191–1196. doi:10.1016/J.POLYMER.2010.01.027.

Zhou, M., Wang, Y., Zhai, Y., Zhai, J., Ren, W., Wang, F., & Dong, S. (2009). Controlled synthesis of large-area and patterned electrochemically reduced graphene oxide films. *Chemistry (Weinheim an Der Bergstrasse, Germany), 15*(25), 6116–6120. doi:10.1002/CHEM.200900596.

Zhou, X., Hu, B., Xiao, W. Q., Yan, L., Wang, Z. J., Zhang, J. J., Lin, H. L., Bian, J., & Lu, Y. (2018). Morphology and properties of shape memory thermoplastic polyurethane composites incorporating graphene-montmorillonite hybrids. *Journal of Applied Polymer Science, 135*(15), 46149. doi:10.1002/APP.46149.

Zhou, Z. M., Wang, K., & Wang, Y. H. (2021). High performance of thermoplastic polyurethane-graphene oxide self-healing composite film. *Coatings, 11*(2), 128. doi:10.3390/COATINGS11020128.

Zhu, Y., Murali, S., Cai, W., Li, X., Suk, J. W., Potts, J. R., & Ruoff, R. S. (2010). Graphene and graphene oxide: Synthesis, properties, and applications. *Advanced Materials, 22*(35), 3906–3924. doi:10.1002/ADMA.201001068.

13 Role of Graphene in Tire Tread Wear Improvement

Jagannath Chanda, Prasenjit Ghosh, Saikat Das Gupta, Rabindra Mukhopadhyay
Hari Shankar Singhania Elastomer and
Tire Research Institute (HASETRI)

CONTENTS

DOI: 10.1201/9781003200444-13

13.1 INTRODUCTION

Growing demand for performance excellence of automotive tire requires newer and advanced technology to meet customer expectations, legislation and regulation of government authority. Reduced rolling resistance along with better traction and millage is the key criteria of high performance tire. To optimize these properties, tire industries require advanced nanocomposites with new generation fillers (NGFs) which can exceptionally improve polymer-filler interaction by exposing their higher surface area and lower particle size.

13.1.1 ROLE OF FILLER IN RUBBER COMPOUNDS

Filler is the second most important material in the tire industry after rubber (Das Gupta et al., 2019, Waddell and Evans, 1996, Wolff, 1996) to optimize price, processing and performance. Fillers were generally added to the rubber in order to make the articles cheaper. Further, the addition of filler (~50–60 per hundred g of rubber) improves the performance properties of rubber vulcanizates such as stiffness, toughness, tensile properties and durability. Cost reduction of the rubber compound with the introduction of non-reinforcing filler can result in sacrifice of required performance properties. Therefore, balance between cost and properties is very much essential for efficient compounding. Fillers can be typically classified by four categories as diluents or degrading fillers (non-reinforcing), extending or semi-reinforcing fillers and reinforcing fillers. Reinforcement can be defined as the improvement of the mechanical properties of compounds, particularly their strength, hardness, stiffness, abrasion and tear resistance, which consequently improves the service life of rubber products (Hilonga et al., 2012, Donnet, 1998). The extent of reinforcement depends on the hydrodynamic volume of filler particles, filler-filler interaction (Payne effect) and polymer-filler interaction. Above all, the degree of dispersion of filler particles by efficient mixing is also an important criterion to achieve better reinforcing effect. Four different steps are followed during mixing of filler particles into the rubber matrix such as incorporation, plasticization, dispersion and distribution. This process depends on filler particle size, surface area, structure and surface activity to control aggregate size distribution throughout the rubber matrix. Figure 13.1 represents the schematic diagram of filler particle, aggregate and agglomerate as captured by morphological analysis.

In rubber industries, especially in tire industry, carbon black (CB) is widely used filler. However, CB possesses some limitations with respect to high heat generation due to its mechanical bonding with rubber which induce more hysteresis. Silica with silane coupling agent exploited chemical bonding between rubber and silica and addressed the hysteresis issue to a great extent. It is worth mentioning here that dispersion and processibility of silica-based rubber composites is a challenging task, especially when it is used in large quantity. This limits the usage of silica up to a certain amount due to the requirement of modified mixing and process technology. Compared to CB, silica particles are less compatible with hydrocarbon rubbers and tend to form filler-filler networks via hydrogen bonding of the silanol groups present on the particle surfaces (Hilonga et al., 2012, Wang, 1998, Moaddab et al., 2015, Kralevich et al., 1998). The development of silica treads in passenger tires inspired extensive research on

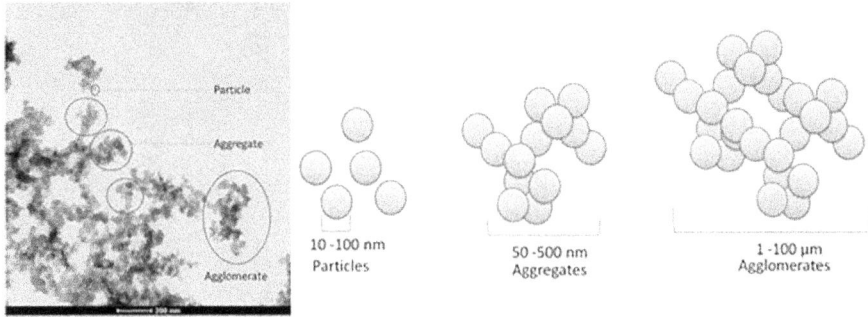

FIGURE 13.1 Schematic diagram of filler particle, aggregate and agglomerates.

non-black fillers other than silica. Commercial clay has long been used as inexpensive, non-reinforcing filler, but a breakthrough has happened when Toyota Co. invented a layered silicate nanocomposite for automotive application (Laskowska et al., 2010, Sokolowska et al., 2010). Substantial research was carried out on the use of layered mineral fillers as reinforcing fillers in many other polymer matrices. Not only CB and silica but also layered silicates, layered double hydroxides, calcium carbonates, calcium silicates, zeolites, alumina, starch and many others have been studied extensively as potential reinforcing additives (Sadasivuni et al., 2014, Bokobza et al., 2007, Mazumder et al., 2021, Sushmitha et al., 2021, Basu et al., 2014, Rooj et al., 2013). In addition to classifications based on their reinforcing effect in the elastomer matrix, fillers can also be distinguished based on their structure and different dimensions:

Zero-dimensional (0D): low aspect ratio, isotropic spheres, cubes and polyhedrons, e.g., CB, silica, calcium carbonate and fullerene.

One-dimensional (1D): filler particles in which one dimension is considerably longer than the others (rods, wires and tubes), e.g., multiwalled carbon nanotubes, carbon nanowires and sepiolite.

Two-dimensional (2D): filler particles in which two dimensions are of considerable length (disks, prisms and plates), e.g., layered silicate, graphene and graphite.

Three-dimensional (3D): filler particles in which three dimensions are of considerable length, e.g., boehmite and diamond.

The reinforcing effect and mechanical properties of elastomeric composites can vary dramatically depending on their shape, size and dimensionality of the filler (1D, 2D and 3D) (Hohenberger et al., 2001, Moczo et al., 2008, Fröhlich et al., 2005) as shown in Figure 13.2.

13.1.2 Development of Advanced Composites with New Generation Filler

Future development of tire industries requires robust and sustainable technology to address magic triangle as shown in Figure 13.3 without affecting the cost and quality by incorporation of NGF.

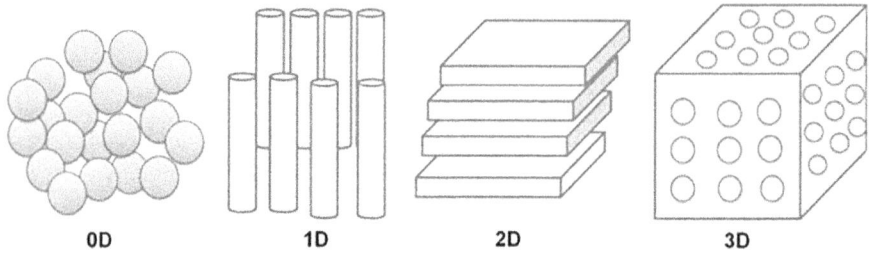

FIGURE 13.2   Schematic diagram of 0D, 1D, 2D and 3D fillers.

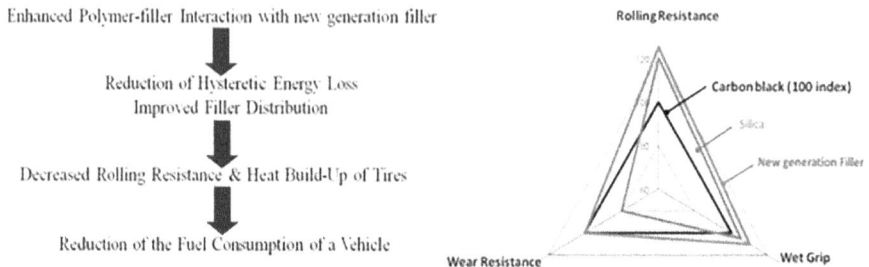

FIGURE 13.3   Schematic mechanism of reduced fuel efficiency with new generation filler.

NGF is capable of improvement in polymer-filler interaction by exposing their higher surface area and structure leading to decrease in hysteresis loss of rubber composites and thereby reduction of tire rolling resistance (RR) as mentioned in Figure 13.3. Optimization of rolling resistance, wet grip and wear can also be achieved by the incorporation of NGF with appropriate size and surface activity along with proper mixing technology. Additional advantage of developing the advanced composites with NGF increases wet traction with lowering the RR value as presented in Figure 13.4.

In recent years, nanotechnology is in the forefront and widely used to optimize the performance properties of tire by developing hybrid rubber composites with combination of CB and silica. Among several new materials being tested, the trend is already toward using Functionalized Solution SBR modified with different coupling agents, Rubber containing functional groups over the entire length of the backbone, Latex – CB/Silica Master Batch, Magneto-Rheological Materials, etc. Based on the principle of bionics, natural materials such as Natural Silica, Nano clay, Nano Calcium Carbonate, Montmorillonite (MMT) Clays, Carbon Nanofibers (CNFs), Polyhedral Oligomeric Silsesquioxanes (POSS), Carbon Nanotubes (SWNT and MWNT), and graphene are being tested. In this direction, two-dimensional (2D) graphite or graphene oxide or reduced graphene oxide is a very useful filler to achieve all the performance parameters. Graphene, being the youngest member of the carbon allotrope family, is often referred to as the mother of all carbon nanomaterials (Geim et al., 2010, Papageorgiou et al., 2015, Mondal et al., 2016). Zero-dimensional fullerenes, one-dimensional carbon nanotubes and three-dimensional graphite can be easily generated through the wrapping, rolling and stacking of the graphene sheet respectively (Papageorgiou et al., 2015,

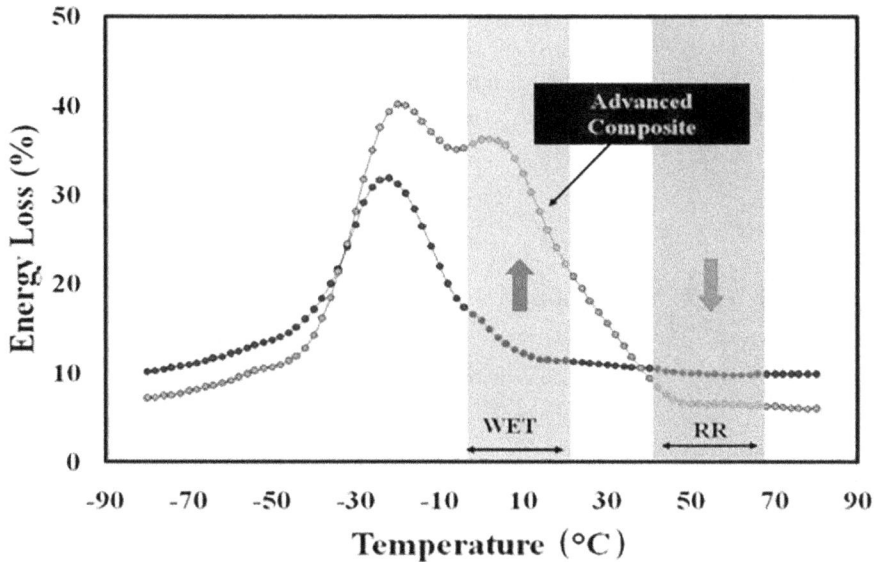

FIGURE 13.4 Graphical representation of energy loss versus temperature for conventional and advanced composites.

Kuilla et al., 2010). Graphene got much attention due to its exceptionally high surface area (500–750 m^2g^{-1}), very high elastic modulus (up to 1 TPa), high thermal conductivity (up to 5,000 W mK^{-1}), very high electron mobility and strong barrier properties (Novoselov et al., 2005, Balandin et al., 2008, Lee et al., 2008). Thus, it can be considered as ideal multifunctional nanofiller to improve physico-mechanical properties of elastomeric hybrid composites. Many researchers have explored that the extent of property enhancement depends on size, shape, number of layers and functionality of graphene. It also depends on formulation and method of composite preparation. There are several techniques to develop graphene-based elastomeric composites with thermo-mechanical exfoliation that includes mechanical agitation and chemical reaction during composite preparation (Zhang et al., 2019, Potts et al., 2012, Maiti et al., 2008, Xing et al., 2014). This chapter summarizes our recent work to develop hybrid elastomeric nanocomposites with various grades of graphene by varying their morphology and particle size. It also includes the comparison of regularly used fillers in tire industries such as CB or silica where dispersion, physical properties and morphological analysis of the said composites have been presented. This fundamental research gives us an idea to design our composites with absolute method of preparation and optimization of compounding recipes.

13.2 MATERIALS AND EXPERIMENTS

13.2.1 PREPARATION OF GRAPHENE NANOCOMPOSITES

In this study, three different grades of graphene oxide (GO) or graphite such as G1, G2 and G3 have been used to prepare the composite by varying their particle size and shape. As per the supplier data, four to six layers are present in each grade. As whole study has been performed in passenger car radial (PCR) compound, the reference compound consists of 50 phr of silica, whereas three experimental compounds EG-1, EG-2 and EG-3 contain 47 phr silica and 3 phr graphene. Formulation and mixing sequence are presented in Tables 13.1 and 13.2 respectively.

TABLE 13.1
Formulation of Graphene-Based Elastomeric Nanocomposites

Ingredients[a] (phr)	Ref	EG-1	EG-2	EG-3
NR	25	25	25	25
SSBR	75	75	75	75
Silica	50	47	47	47
Graphene-1	0	3	0	0
Graphene-2	0	0	3	0
Graphene-3	0	0	0	3
Silane coupling agent	6	6	6	6

[a] All other ingredients are keeping same like reference compound.

TABLE 13.2
Mixing Sequence of Graphene-Based Elastomeric Nanocomposite

Time (s)	Ref	EG-1, EG-2 and EG-3
	Master-I	
0	NR + SBR	NR + SBR + graphene
30	1/3 Silica + silane + other chemicals	
120	2/3 Silica + silane + other chemicals	1/3 Silica + silane + other chemicals
240	Silanization	2/3 Silica + silane + other chemicals
360	Dump	Silanization
480		Dump
	Repass	
0	Master-I compound	
90	Ram sweep	
180	Dump	
	Final	
0	Repass compound + curatives	
90	Ram sweep	
180	Dump	

13.2.2 Method of Preparation and Mixing Sequence

As tire industries are very much convenient to prepare the composite in solid phase mixing, all the compounds are prepared by following conventional dry mixing process in 1.6 L laboratory internal mixture followed by two roll mill to make uniform rubber sheet. Three stage mixing sequence has been followed as per the details given in Table 13.2. The rotor RPM (rotation per minute) was 50 and the TCU temperature was 90°C for master and repass, whereas for final batch mixing rotor RPM was 30 and the TCU temperature was 70°C. In the case of EG-1, EG-2 and EG-3 compounds, 120 seconds of dry mixing of rubber with the graphene was allowed for better polymer filler interaction as we did not adapt any pretreatment of nanofiller for the exfoliation of graphitic layers. For all the compounds, 2 minutes silanization time was given for better polymer-filler interaction.

13.2.3 Characterization of Fillers and Rubber Composites

13.2.3.1 Fourier Transform Infrared (FTIR) Analysis Spectroscopic

Fourier Transform Infrared (FTIR) Spectroscopy was performed on the raw graphene sample by using FTIR spectrometer of model SPECTRUML 125150D from M/s Perkin Elmer, Massachusetts, United States. Sample preparation was done by using KBr powder to make the pellet for FTIR characterization. The study was carried out by Attenuated Total Reflectance technique with Germanium and specific peaks were investigated as per the requirement.

13.2.3.2 X-Ray Diffraction (XRD) Analysis

X-ray analysis of graphene and rubber composites was conducted by using D8 advance (Bruker AXS, Germany) with Cu-Kα irradiation at a generator voltage of 40 kV. The crystallographic spacing (d-spacing) between the particles was calculated by using Bragg's law as given in the following equation:

$$n\lambda = 2d\sin\theta \tag{13.1}$$

where λ is the wavelength of the radiation; d is the inter-planar spacing involved; θ is the angle between the incident (or diffracted) ray and the relevant crystal planes and n is the Integer (Order of diffraction). Intercalation and exfoliation of graphene into the rubber matrix was also confirmed by this analysis where interlayer distance among the graphite layers was measured. Samples were scanned in the 2θ range of 3°–60° by step mode with a scan rate of 3°C min^{-1}.

13.2.3.3 Transmission Electron Microscopy (TEM) Analysis

Microstructure of the graphene material can be studied using electron microscopic techniques. High-resolution TEM (HRTEM) analysis is commonly used to understand their size, shape, structure and geometry. TEM imaging is captured when the sample is illuminated by an electron beam over a wide area, and electron optical lenses project a magnified image on the screen or CCD camera. A slightly defocused image produces a contrast from a phase object differentiating the sample boundaries.

TEM has a sole ability to reveal microscopic structure of materials, potentially down to atomic level (Jiang and Spence, 2009, Buseck et al., 1989). Structure of graphene and its defects are of exceptional interest to scientific community (Geim et al., 2010, Meyer et al., 2007) and applications toward NGF in tire industry. In addition, microscopic studies cater more generalized insights into the nature of dispersion of particulates, their size and shape and tendency of aggregation. Therefore, TEM can be considered as a versatile tool to structure and study the monolayers of graphene with highest resolution (Meyer et al., 2007, Peng et al., 2012). Samples for TEM studies were prepared by dispersing in acetone followed by ultrasonication as per ASTM method D3849-14a. Top layer of the dispersed solution was drop casted on a Carbon support Cu TEM grid, dried and mounted on a single tilt holder. TEM analysis for this work has been performed at 200 kV on a Talos-200S model instrument from Thermoscientific, USA. The dispersion and distribution of filler into the rubber matrix was confirmed by HRTEM analysis. For this study, thin section of rubber composites (less than 100 nm) was prepared by sectioning at −100°C using a cryo-ultramicrotome (Leica EM UC7, Germany) with a glass knife.

13.2.3.4   Field Emission Scanning Electron Microscopy (FESEM) Analysis

Field Emission Scanning Electron Microscopy (FESEM) analysis for this work has been carried out at 20 kV on Apreo S model instrument from Thermoscientific, USA. It was used to capture the micrographs of surface topography of raw filler and dispersion and distribution of these fillers into the rubber matrix. Samples were coated with Platinum by using a sputter coater Q150T ES (Quorum Technologies, UK) to avoid the charging effect.

13.2.3.5   Atomic Force Microscopy (AFM) Analysis

Atomic Force Microscopic images were obtained in tapping mode by using NX10, Park Systems from South Korea. The resonance frequency of the tip was 300 kHz along with 26 N m^{-1} force constant. Silicon nitride tip having 10 nm radius was used. Surface analysis was carried out from 20 μm × 20 μm images with the help of Smart Scan software. The topographic images were captured in tapping mode and the modulus mapping throughout the sample surface was performed by pin-point analysis.

13.2.3.6   Measurement of Cure Characteristics

The optimum cure time was measured using a moving die rheometer (MDR 2000, M/S Alpha Technologies, USA) at 160°C for 30 minutes according to ASTM D5289. The optimum cure time corresponds to the time required to achieve 90% (t_{90}) of the cure calculated from the formula:

$$\text{Optimum cure time,}(t_{90}) = \left[0.9(M_H - M_L) + M_L\right] \qquad (13.2)$$

$$\text{Cure Rate Index} = \frac{100}{\left(t_{c90} - t_{s2}\right)} \qquad (13.3)$$

where MH and ML are the maximum and the minimum torque. The Payne effect was studied at 50°C for 0.1%–100% strain sweep according to ASTM D604 in RPA

2000 by M/s Alpha Technologies, USA. Mooney viscosity (ML1+4 at 100°C) was tested by Premier MV by M/s Alpha Technologies according to ASTMD1646.

13.2.3.7 Measurement of Physical Properties

Physical properties of rubber composites were measured according to ASTM D 412 at 500 mm min^{-1} test speed using ASTM die C type dumbbell specimens. Tensile slabs were molded at 160°C at $2*t_{c90}$ min.

13.2.3.8 Dynamic Mechanical Properties

Dynamic mechanical analysis of rubber composites was carried out by DMA VA 1000 (M/s Metravib, France). The test was conducted for temperature sweep from 0°C to 70°C in tension compression mode. Ten Hz frequency and 1% strain were maintained throughout the test to determine the values of storage modulus (E^2), loss modulus (E^2) and tan δ at different temperatures.

13.2.3.9 Measurement of Wear Resistance by Laboratory Abrasion Tester-100 (LAT-100)

Abrasion resistance of the developed composites was measured by using LAT100, from M/s VMI, Netherlands (Figure 13.5). A small solid rubber wheel was prepared by compression molding and placed on a rotating disk, where the speed, slip angle, surface of the disk, load, and temperature could be varied and the wear resistance could be measured. Tests were performed at room temperature on a Corundum 60 type disk and a powder of 2:1 Aluminum Oxide and Magnesium Oxide was used to avoid the sticking of rubber on the corundum disk. Severity of test conditions was also classified as mentioned in Table 13.3. The rating of the compounds was done with respect to the reference compound. The higher the rating over 100, the better is the compound.

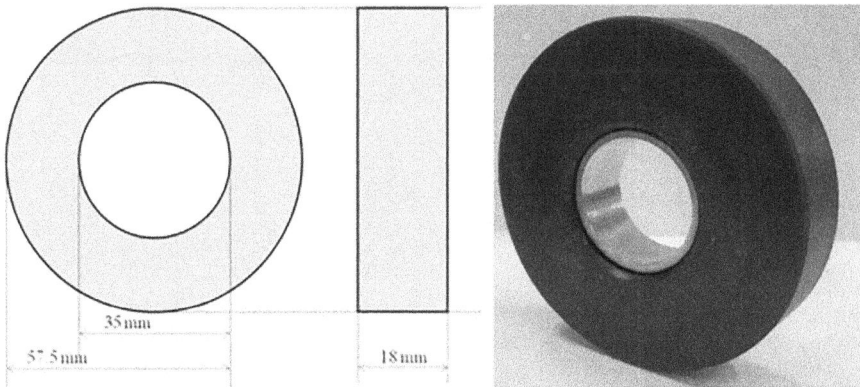

FIGURE 13.5 CAD image and molded wheel for LAT100 testing.

TABLE 13.3
Classification of Severity for LAT 100 Abrasion Tests

	Low Severity	Medium Severity	High Severity
Load, N	20	40	70
Slip angle, °	2	6	12
Speed, kmph	40	20	20

13.3 RESULTS AND DISCUSSION

13.3.1 MEASUREMENT OF SURFACE FUNCTIONALITY THROUGH FTIR

Figure 13.6 depicts the room temperature FTIR spectra of silica and different grades of graphene samples, viz. G1, G2 and G3, ranging from 4,000 to 500 cm^{-1}. For the silica sample, the band at 1,090 cm^{-1} is due to the stretching vibration of O–Si–O bond. For the graphene samples, the band around 3,450 cm^{-1} is attributed to the stretching vibrations of –O–H group present in the samples. The alkyl chain features are observed at wave numbers 2,950–2,800 cm^{-1} (Bera et al., 2018). The band near 2,930 cm^{-1} can be ascribed as the symmetric stretching vibration of =C–H bond; whereas the peak at 2,815 cm^{-1} can be assigned as the stretching vibration peak of –C–H alkyl chains. The peak at 1,640 cm^{-1} is due to the stretching frequency of –C=C– bond. The intensity of the same peak is higher for G1 than the others. This may be due to the difference in dipole moment of –C=C– bond between the samples. G1 contains more number of functional groups (–O–H) which pulls the electron crowd toward itself more resulting in slight dipole moment in the –C=C– bond. There is no shifting or any extra peak in the spectra of different samples.

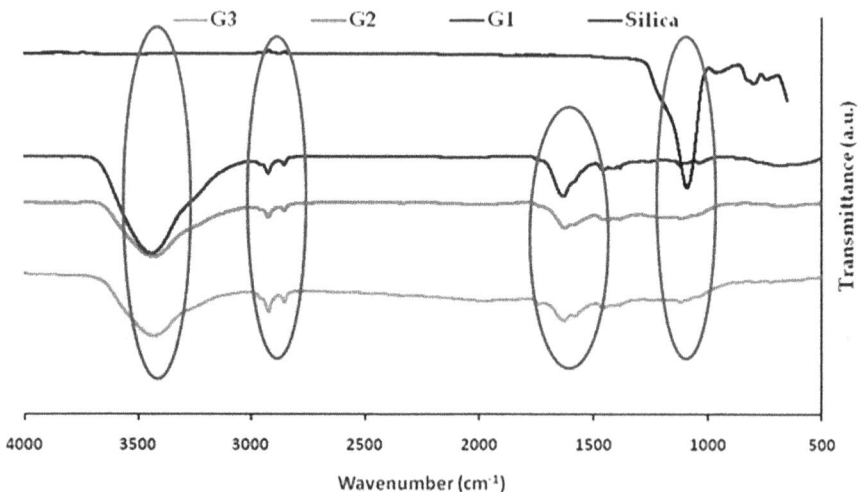

FIGURE 13.6 FTIR spectra of silica and three different grades of graphenes G1, G2 and G3.

13.3.2 Crystallographic Studies of Graphene and Rubber Nanocomposites

The XRD pattern of the graphene samples in Figure 13.7 exhibits that there is a sharp peak near $2\theta \sim 26°$ due to the diffraction of the (002) plane which is the characteristics diffraction peak of the carbonaceous materials specially graphite like materials and a less intense broad peak at $2\theta \sim 43°$. The first peak is referred to as Π-band and appears due to the presence of aromatic ring structure, whereas the second peak reveals the presence of sp^2 hybridized hexagonal graphite-like carbon lattice (Siburian et al., 2018, Mohan et al., 2013). The d-spacing and crystallinity of the samples are summarized in Table 13.4. The d-spacing of all the samples is quite similar, whereas the crystallinity of G2 is bit higher than that of the others. Silica being a highly amorphous material, no intense peak could be observed in the spectra of the silica. Only a background hump at $2\theta \sim 22°$ is observed which proves the amorphous nature of silica (Rozainee et al., 2008).

Figure 13.8 illustrates the XRD pattern of reference as well as graphene-based compounds in a 2θ range of $3°–60°$. In the compounds, small peaks at $2\theta \sim 26°$ confirm the presence of graphene in these composites. The intensity of the peak

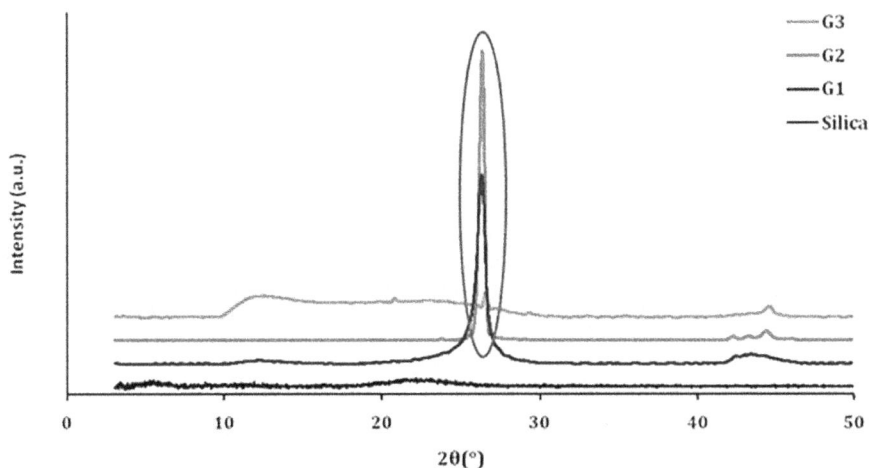

FIGURE 13.7   XRD pattern of silica and three different grades of graphenes G1, G2 and G3.

TABLE 13.4

d-Spacing and Crystallinity of Silica and Three Grades of Graphenes G1, G2 and G3

Sample	2θ (°)	d-Spacing (Å)	Plane	Crystallinity (%)
Silica	22.21	4.00	(101)	15
G1	26.38	3.38	(002)	58
G2	26.39	3.37	(002)	69
G3	26.59	3.35	(002)	48

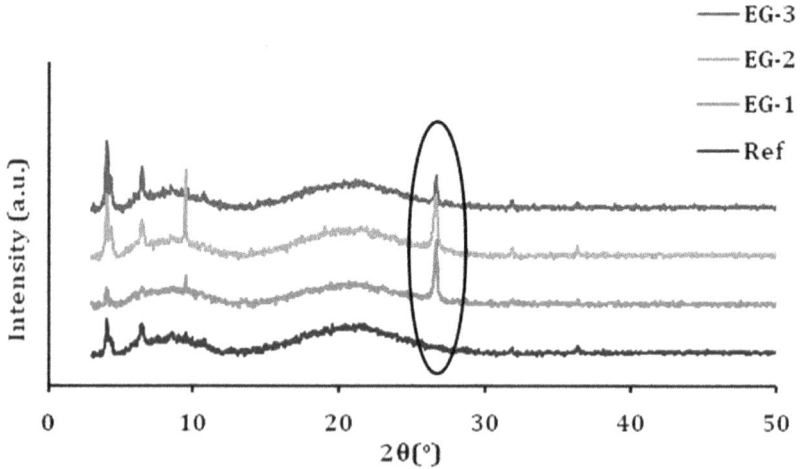

FIGURE 13.8 XRD pattern of reference and graphene-based nanocomposites EG-1, EG-2 and EG-3.

decreases significantly which can be attributed to the exfoliation of the graphene layers in the composites. Though the peak does not shift a lot as compared to the raw graphene, intercalation and exfoliation of the graphene layers can be responsible for this kind of phenomena.

13.3.3 TEM ANALYSIS OF GRAPHENE AND RUBBER COMPOSITES

TEM analysis elucidates the surface morphology of silica and three different graphenes such as G1, G2 and G3. Images are self-explanatory as one can view a replica image of how the particulates aggregate or orient at micro and nano scales. TEM micrographs of raw material samples clearly show the difference in morphology of Silica, G1, G2 and G3 where they are exhibiting spherical particulate, spherical aggregate, flake-wrinkled, thin continuous sheet and plate-like topography respectively as seen from Figure 13.9. TEM micrographs of CB and Si are exhibiting the spherical particulate morphology with 30–40 and 10–20 nm diameters respectively. TEM micrographs of Sample G1 infer monolayer wrinkled graphene flakes of approximately 300–500 nm in width. Micrographs of Sample G2 infer spread-out few layer graphene sheets of approximately 500–800 nm in width. Micrographs of Sample G3 infer graphene plates that are comparatively showcasing poor electron transparency indicating greater thickness of sheets and approximately 400–600 nm in width.

TEM micrographs of ultrathin sections from compounds EG1, EG2 and EG3 exhibit distinguishable topography of graphene retaining its original topographical behavior even after the process of compounding. In Figure 13.10, a reference TEM micrograph without graphene is shown for comparison purpose. From this figure, it can be observed that samples G2 and G3 retain its property of spread-out morphology providing greater surface area for reinforcing. Sample G1 provides a uniform dispersion of graphene in the compound matrix.

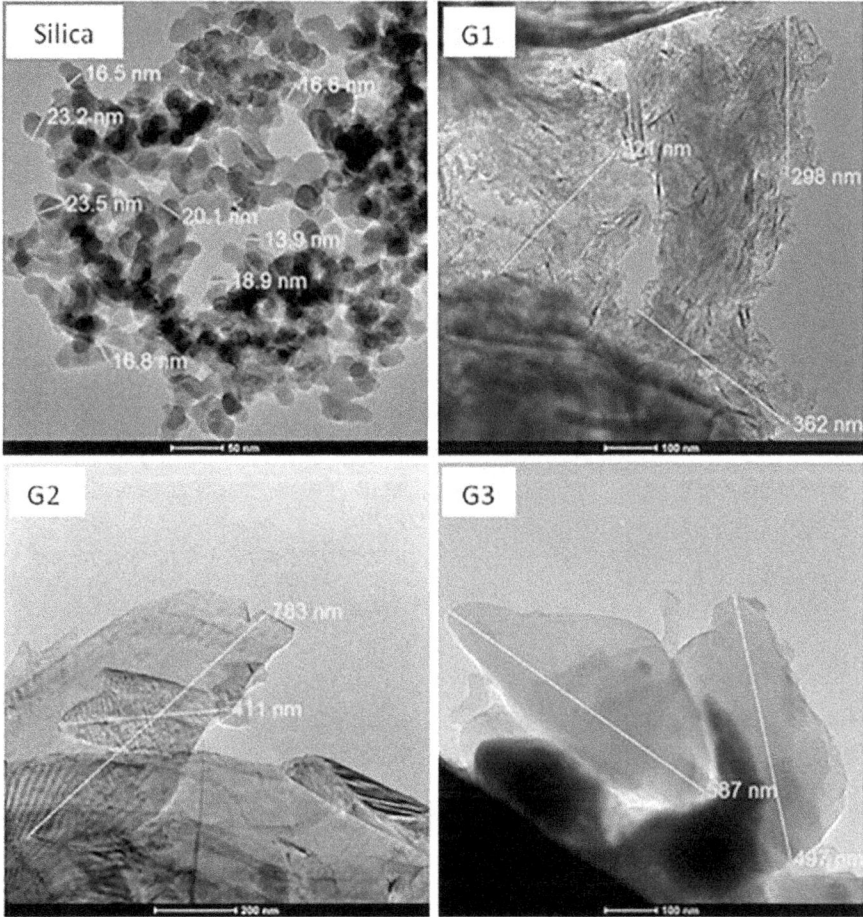

FIGURE 13.9 Representative HRTEM images of silica and three different grades of graphene G1, G2 and G3.

13.3.4   FESEM ANALYSIS OF GRAPHENE AND RUBBER COMPOSITES

FESEM micrographs provide the details of surface topography information of raw fillers and their aggregation in rubber composites including the shape and size of particulates. Results obtained from FESEM have complemented the TEM analysis and provides a 3D image of G1, G2 and G3 with flake-, sheet- and plate-like morphologies as presented in Figure 13.11.

FESEM images of reference and graphene nanocomposites are presented in Figure 13.12 where appreciable difference in dispersion was observed in experimental compounds, EG-1, EG-2 and EG-3. Composite having G1 graphene showed better dispersion in comparison to the G2- and G3-based compounds. As reference compound does not have graphene in the formulation, it appears as smoother surface

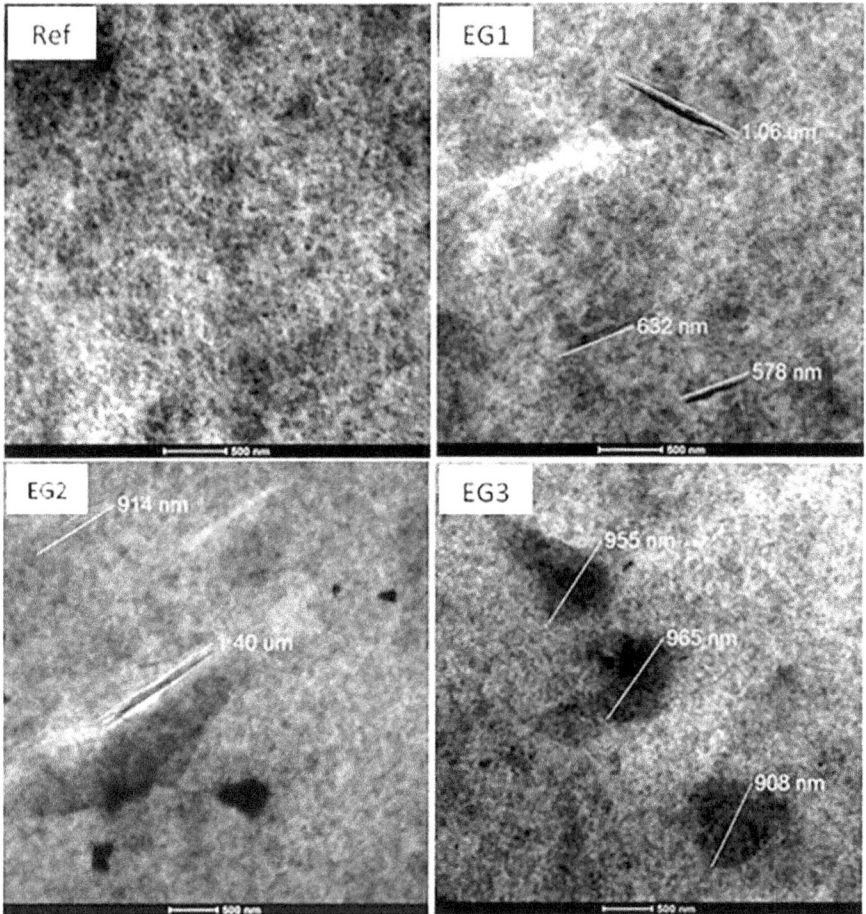

FIGURE 13.10 TEM micrographs of reference and experimental compounds EG-1, EG-2 and EG-3.

among others. While comparing EG-1, EG-2 and EG-3, EG-3 is displaying bigger aggregates as compared to EG-1 and EG-2 as G3 graphene exhibited larger particle size and higher number of layers determined from the earlier section.

13.3.5   AFM ANALYSIS OF RUBBER COMPOSITES

AFM has also been used to study the dispersion of filler in rubber matrix. From surface morphology of the height images, the filler aggregates and its dispersion (Sadhu et al., 2005) can easily be observed in Figure 13.13. In silica compound, individual aggregates are clearly visible and the fillers seem to be well dispersed in the matrix. The morphologies are observed with particle diameters ranging from 2 to 4 μm. A good uniform and well dispersion of graphene in compound is observed in the case of G1 followed by G2 and G3. In the G1 compound, as the graphene flakes are too small,

FIGURE 13.11 Representative FESEM images of silica and three different grades of graphene G1, G2 and G3.

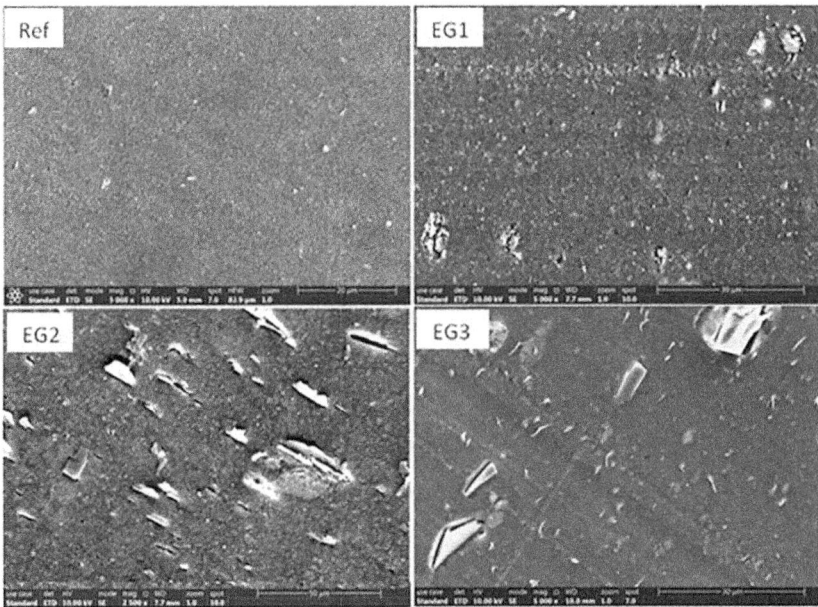

FIGURE 13.12 FESEM micrographs of reference and experimental compounds EG-1, EG-2 and EG-3.

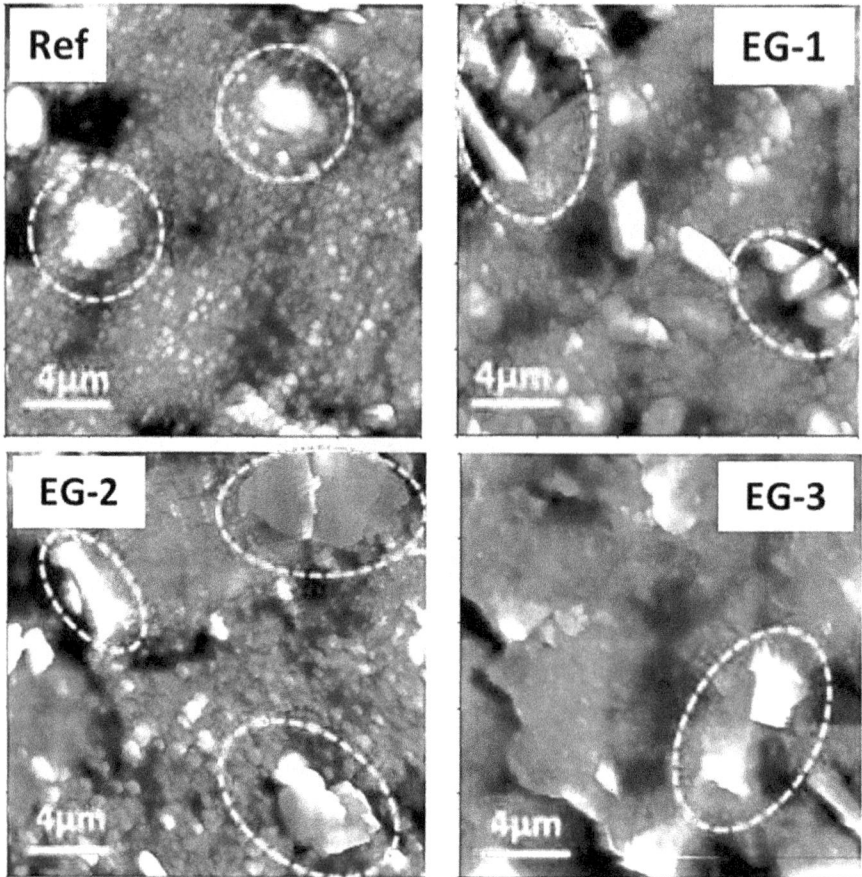

FIGURE 13.13   AFM images of reference and experimental compounds EG-1, EG-2 and EG-3.

they are rolled up and have formed small tube-like structures; whereas, in the G2 compound, the graphene flakes have retained their structure as thin sheets. Graphene flakes of the G3 compound were of bigger size hence similar plate-like structures are visible in the compound. This might adversely affect the mechanical properties of the compound.

13.3.6   Effect of Graphene on Processing Parameters

Table 13.5 summarized the processing parameter such as Mooney viscosity, scorch safety, cure characteristics and Payne effect of silica-filled reference compound and graphene-based nanocomposites EG-1, EG-2 and EG-3.

From Table 13.5, it can be concluded that graphene-based nanocomposite showed slightly faster curing rate as compared to the reference one as t_{c90} of experimental compound is always in the lower side. While comparing the values among experimental compounds, EG-1 exhibits lower scorch safety and lower t_{c90} value among

TABLE 13.5

Processing Parameter of Silica and Graphene-Based Rubber Composites

Processing Parameter	Ref	EG-1	EG-2	EG-3
Minimum torque M_L (lb-in)	1.61	1.8	1.66	1.57
Maximum torque M_H (lb-in)	14.53	14.55	13.92	14.37
ΔTorque (lb-in)	12.92	12.75	12.26	12.8
t_{s2} [min]	3.06	2.33	2.78	3.01
t_{c90} [min]	15.4	13.31	14.28	13.59
Cure Rate Index [min^{-1}]	8.10	9.10	8.70	9.44
Mooney Viscosity [MU]	63	68	65	62
Mooney Scorch at 135°C [min]	11	7	8	10
$\Delta G'$ [0.1%–100% strain] [MPa]	0.65	0.57	0.56	0.64

others. This might be due to the presence of –O–H functional group on graphene surface as confirmed by the FTIR studies in the earlier section. Other properties like M_H, M_L, ΔTorque and Mooney viscosity are comparable with respect to the reference compound. Payne green effect of graphene composites EG-1 and EG-2 is quite low compared to the reference compound. Graphical representation of $\Delta G'$ value in the specified range of 0.1%–20%, 0.1%–50% and 0.1%–100% is displayed in Figure 13.14. The shear modulus of the filler-reinforced rubber is sensitive to strain. With increase in the strain percentage, it decreases rapidly which is due to the collapsing of filler-filler network in rubber compounds (Payne, 1962). As compared to the reference sample, the experimental compounds exhibited lower Payne effect due to improved rubber- filler interaction with the aid of GO nanosheets which is reflected in the $\Delta G'$ values. EG-1 compound shows lower $\Delta G'$ value due to strong interfacial interaction of graphene layers with SBR which plays the key role to reduce the same (Yang et al., 2015).

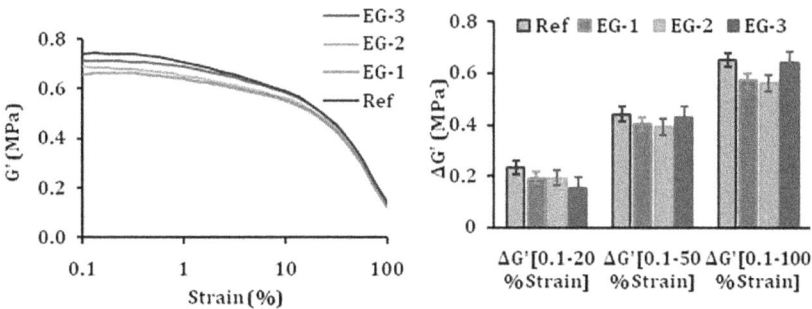

FIGURE 13.14 Plot of G' (MPa) as a function of strain (0.1%–100%), and $\Delta G'$ (MPa) values at specified range of 0.1%–20%, 0.1%–50% and 0.1%–100% for Ref and EG-1, EG-2 and EG-3 compounds.

13.3.7 EFFECT OF GRAPHENE ON PHYSICO-MECHANICAL PROPERTIES

Tensile properties of the reference and graphene nanocomposites are presented in Table 13.6.

The mechanical properties of the compounds are summarized in Table 13.6. It elucidates that there is a marginal increase in tensile strength in the graphene-based compounds, whereas the modulus and elongation at break improve significantly as compared to that of the reference one. An optimum static mechanical property was observed in the case of EG-1 compound. This improvement in reinforcing efficiency by graphene sheets is attributed to mainly two factors. First, the large specific area of graphene oxide provides higher contact area. Second, a glassy layer in the vicinity of the graphene oxide sheets can be formed as a result of the strong ionic interactions between the functional groups present on the graphene oxide and the functionalized S-SBR. Under external stresses, the rubber chains in glassy layer may slip along the graphene sheets to form stretched and parallel-arraying straight chains. As a result of that orientation of the chains, the stresses can be uniformly shared to avoid stress concentrations. Also, the glassy layer acts as a mediator to facilitate the transfer of stress from the SBR matrix to the graphene oxide sheets (Mao et al., 2013).

DMA results in Figure 13.15 show that there is a prominent increment in storage moduli in the graphene-based compounds which might be due to the well-dispersed graphene that can act as the physical cross-linker for the elastomer (Zhang et al., 2017). Also, there is a decrease in the tan δ peak which signifies low heat build-up and lower damping capability of the composites which is beneficial for achieving low rolling resistance. The foremost dynamic mechanical properties were observed in the case of EG-1 compound. It shows a beneficial effect in the case of both grip and rolling resistance properties as compared to the reference as well as other graphene-based composites.

13.3.8 EFFECT OF GRAPHENE ON WEAR RESISTANCE

Wear resistance of reference and graphene-based nanocomposites is presented in Figure 13.16, where abrasion loss of specific compound was measured at three different conditions as detailed below.

TABLE 13.6

Tensile Properties of Reference and Graphene-Based Nanocomposites EG-1, EG-2 and EG-3

Test Parameter	Ref	EG-1	EG-2	EG-3
Hardness	60	63	59	62
Modulus at 100% elongation [MPa]	2.1	2.8	2.6	2.4
Modulus at 200% elongation [MPa]	5.6	7.1	6.6	6.3
Modulus at 300% elongation [MPa]	11.8	13.2	12.8	12.1
Tensile strength, TS [MPa]	18.7	21.6	20.8	19.2
Elongation at break, EB [%]	389	438	426	403

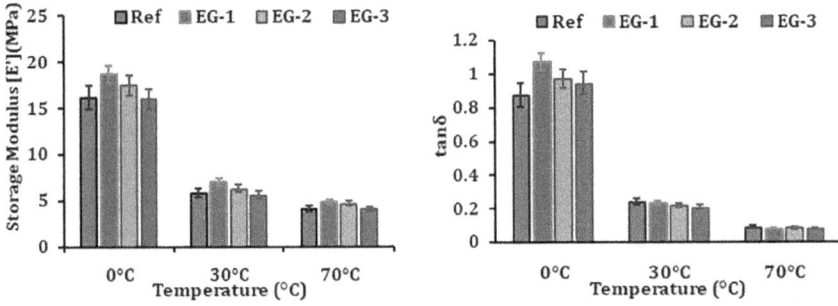

FIGURE 13.15 Plot of storage modulus and tan δ values at 0°C, 30°C and 70°C of reference and experimental compounds EG-1, EG-2 and EG-3.

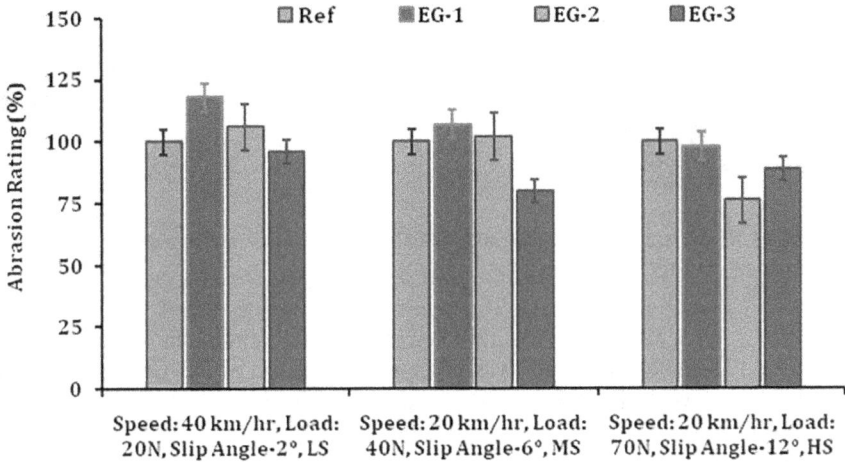

FIGURE 13.16 Abrasion rating of reference and experimental compounds EG-1, EG-2 and EG-3 at low severity, medium severity and high severity.

The wear resistance of the graphene-based compounds was determined at three different slip angles, loads and speeds to understand the wear characteristics of composites in service condition. Laboratory prediction showed that the addition of graphene improved the abrasion resistance of composites compared to the reference one as seen from Figure 13.16. Abrasion rating at both low severity as well as medium severity is higher in the case of EG-1 and EG-2 compounds, whereas that at high severity is almost comparable with that of the reference. This signifies less amount of wear loss in the compounds. Surface appearance of abraded wheel was confirmed by the optical and scanning electron microscope.

Figure 13.17 represents the optical microscope images of the wheels tested for abrasion resistance property by LAT 100. A large amount of debris was found on the surface of the reference and EG-3 compounds as compared to others. Also the wear pattern is quite deep which indicates the poorer abrasion resistance property

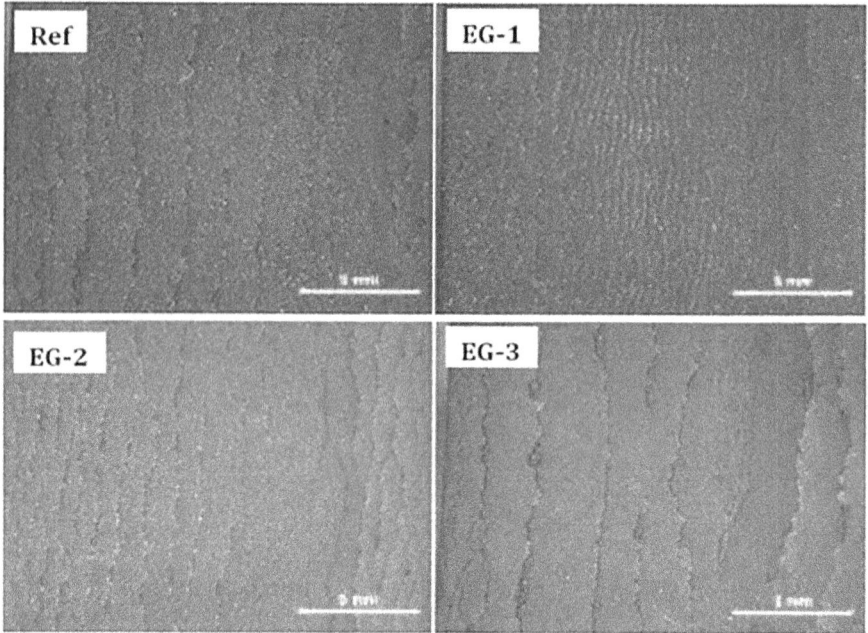

FIGURE 13.17 Optical microscope images of abraded surface for reference and experimental compounds EG-1, EG-2 and EG-3.

(Mao et al., 2013, Sridharan et al., 2019). On the other hand, a narrower wear pattern as well as smoother surface was observed in the case of EG-1 and EG-2 compounds which reveals an improved wear resistance property. SEM was also engaged to examine the abraded surface of the tested samples. The images show that the EG-1 composite exhibits a series of narrower spaced ridges perpendicular to the sliding direction in its worn surface as compared to that of the reference one as well as the other graphene-based composites, viz. EG-2 and EG-3, as presented in Figure 13.18. The ridge heights and widths are also found to be quite small. The layered structure of graphene plays the key role to improve the wear resistance property (Xing et al., 2014).

13.4 CONCLUSIONS

Tire technology will play a vital role in meeting green mobility norms, and research into various materials is elemental in tire transformation. With the availability of alternative natural sources and cutting-edge technology, the tire of today can be completely redesigned to score on several parameters of cost, durability, performance and sustainability. Sustainable growth in tire industries requires robust and advanced technology to develop new generation compound where the balance between traction (Safety), rolling resistance (Fuel efficiency) and wear (Millage) has to be made without sacrificing cost and quality. In this chapter, we have summarized potentiality of three different grades of graphene to be used in tire industries to improve wear resistance of rubber composites. Morphology, functionality and distance between

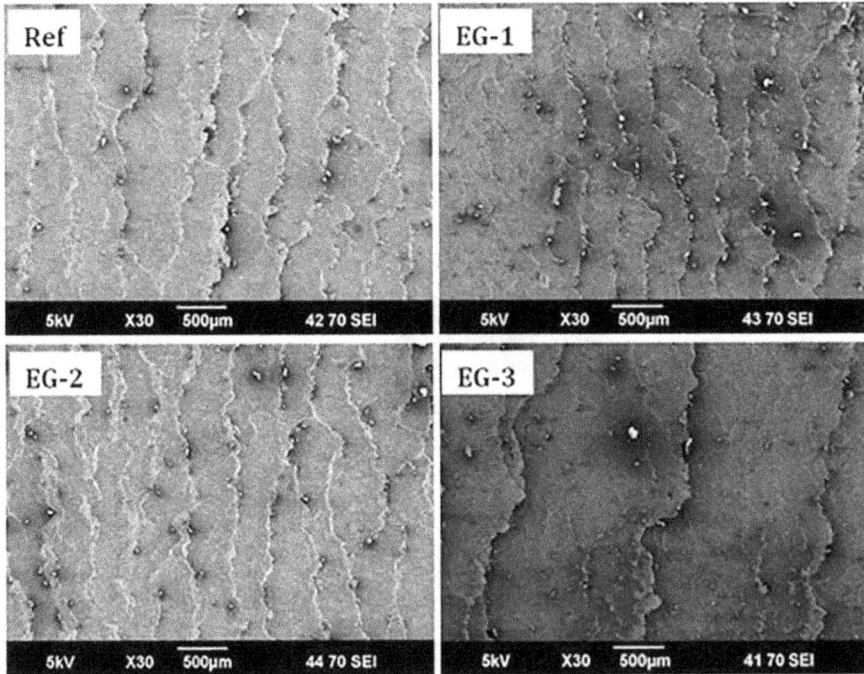

FIGURE 13.18 SEM images of abraded surface of reference and experimental compounds EG-1, EG-2 and EG-3.

layers are identified by several characterization techniques such as SEM, TEM, AFM, XRD and FTIR. Method of preparation of graphene nanocomposites also revealed in this chapter which is industrially viable and cost effective proposition for future development. Processing parameters and physico-mechanical behavior of the composites also determined and compared with the reference compound exhibiting no additional filler. Distribution and dispersion of graphene into the rubber matrix is confirmed by microscopic analysis and concludes that EG-1 compound showed best dispersion among others. Same has also reflected in dynamic mechanical properties and wear resistance where composites with G1 graphene displayed highest dynamic modulus and lower tan δ compared to other composites. Thus, selecting a suitable grade of graphene with best performance properties has been described here. This will be beneficial for the researchers to develop advanced rubber composites with NGFs such as graphite, graphene oxide and reduced graphene oxide or graphene.

ACKNOWLEDGMENTS

The authors are very grateful to Hari Shankar Singhania Elastomer and Tyre Research Institute (HASETRI) for the great support and giving permission to publish this research work. They would also like to thank Mr. Arun D, Mr. Tuhin Dolui and Ms. Sushmitha H for their assistance in conducting the experiments and data collection.

REFERENCES

Balandin, Alexander A., Suchismita Ghosh, Wenzhong Bao, Irene Calizo, Desalegne Teweldebrhan, Feng Miao, and Chun Ning Lau. 2008. "Superior thermal conductivity of single-layer graphene." *Nano Letters* 8(3): 902–907.

Basu, Debdipta, Amit Das, Klaus Werner Stöckelhuber, Udo Wagenknecht, and Gert Heinrich. 2014. "Advances in layered double hydroxide (LDH)-based elastomer composites." *Progress in Polymer Science* 39(3): 594–626.

Bera, Madhab, Pragya Gupta, and Pradip K. Maji. 2018. "Facile one-pot synthesis of graphene oxide by sonication assisted mechanochemical approach and its surface chemistry." *Journal of Nanoscience and Nanotechnology* 18(2): 902–912.

Bokobza, Liliane. 2007. "Multiwall carbon nanotube elastomeric composites: A review." *Polymer* 48 (17): 4907–4920.

Buseck, Peter, John Cowley, and LeRoy Eyring, eds. 1989. "High-resolution transmission electron microscopy: and associated techniques." *Oxford University Press.*

Das Gupta, Saikat, Rabindra Mukhopadhyay, Krishna C. Baranwal, and Anil K. Bhowmick. 2019. *Reverse Engineering of Rubber Products: Concepts, Tools, and Techniques.* CRC Press Taylor & Francis Group, USA.

Donnet, Jean-Baptiste. 1998. "Black and white fillers and tire compound." *Rubber Chemistry and Technology* 71 (3): 323–341.

Fröhlich, J., W. Niedermeier, and H-D. Luginsland. 2005. "The effect of filler–filler and filler–elastomer interaction on rubber reinforcement." *Composites Part A: Applied Science and Manufacturing* 36 (4): 449–460.

Geim, Andre K., and Konstantin S. Novoselov. 2010. "The rise of graphene." *Nanoscience and Technology: A Collection of Reviews from Nature Journals*: 11–19. doi:10.1142/9789814287005_0002.

Hilonga, Askwar, Jong-Kil Kim, Pradip B. Sarawade, Dang Viet Quang, Godlisten N. Shao, Gideon Elineema, and Hee Taik Kim. 2012. "Synthesis of mesoporous silica with superior properties suitable for green tire." *Journal of Industrial and Engineering Chemistry* 18(5): 1841–1844.

Hohenberger, Walter. 2001. "Fillers and reinforcements/coupling agents." *Plastics Additives Handbook*, 5th Edition, (ed. H. Zweifel), Hanser Publishers, Munich: 901–948. Hanser Publishers, Munich, Germany.

Jiang, Nan, and John C.H. Spence. 2009. "Radiation damage in zircon by high-energy electron beams." *Journal of Applied Physics* 105(12): 123517–123524.

Kralevich, Mark L., and Jack L. Koenig. 1998. "FTIR analysis of silica-filled natural rubber." *Rubber Chemistry and Technology* 71 (2): 300–309.

Kuilla, Tapas, Sambhu Bhadra, Dahu Yao, Nam Hoon Kim, Saswata Bose, and Joong Hee Lee. 2010. "Recent advances in graphene based polymer composites." *Progress in Polymer Science* 35(11): 1350–1375.

Laskowska, Anna, Magdalena Lipinska, Marian Zaborski, and Jolanta Sokolowska. 2010. "Effect of ionic liquids on elastomeric composites filled with a layered filler." *Przemysl Chemiczny* 89 (11): 1459–1463.

Laskowska, Anna, and Marian Zaborski. 2010. "Use of a mineral layered silicate in elastomeric composites." *Przemysl Chemiczny* 89(4): 458–461.

Lee, Changgu, Xiaoding Wei, Jeffrey W. Kysar, and James Hone. 2008. "Measurement of the elastic properties and intrinsic strength of monolayer graphene." *Science* 321(5887): 385–388.

Maiti, Madhuchhanda, Mithun Bhattacharya, and Anil K. Bhowmick. 2008. "Elastomer nanocomposites." *Rubber Chemistry and Technology* 81(3): 384–469.

Mao, Yingyan, Shipeng Wen, Yulong Chen, Fazhong Zhang, Pierre Panine, Tung W. Chan, Liqun Zhang, Yongri Liang, and Li Liu. 2013. "High performance graphene oxide based rubber composites." *Scientific Reports* 3(1): 1–7.

Mazumder, Amrita, Jagannath Chanda, Sanjay Bhattacharyya, Saikat Dasgupta, Rabindra Mukhopadhyay, and Anil K. Bhowmick. 2021. "Improved tire tread compounds using functionalized styrene butadiene rubber-silica filler/hybrid filler systems." *Journal of Applied Polymer Science* 138: 51236–51255.

Meyer, Jannik C., Andre K. Geim, Mikhail I. Katsnelson, Konstantin S. Novoselov, Tim J. Booth, and Siegmar Roth. 2007. "The structure of suspended graphene sheets." *Nature* 446 (7131): 60–63.

Moaddab, Ahmad, Mohammadreza Kalaee, Saeedeh Mazinani, Ali Aghajani, and Mohammad M. Rajab. 2015. "Cure kinetics and final performance of styrene butadiene styrene block copolymer/silica nanocomposites." *Rubber Chemistry and Technology* 88(1): 53–64.

Moczo, Janos, and Bela Pukanszky. 2008. "Polymer micro and nanocomposites: structure, interactions, properties." *Journal of Industrial and Engineering Chemistry* 14(5): 535–563.

Mohan, Anu N., B. Manoj, Jerin John, and A. V. Ramya. 2013. "Structural characterization of paraffin wax soot and carbon black by XRD." *Asian Journal of Chemistry* 25: 76–78.

Mondal, Titash, Anil K. Bhowmick, Ranjan Ghosal, and Rabindra Mukhopadhyay. 2016. "Graphene-based elastomer nanocomposites: Functionalization techniques, morphology, and physical properties." *Designing of Elastomer Nanocomposites: From Theory to Applications*: 267–318.

Novoselov, Kostya S., Andre K. Geim, Sergei Vladimirovich Morozov, Dingde Jiang, Michail I. Katsnelson, IVa Grigorieva, SVb Dubonos, and AA Firsov. 2005. "Two-dimensional gas of massless Dirac fermions in graphene." *Nature* 438(7065): 197–200.

Papageorgiou, Dimitrios G., Ian A. Kinloch, and Robert J. Young. 2015. "Graphene/elastomer nanocomposites." *Carbon* 95: 460–484.

Payne, Arthur R. 1962. "The dynamic properties of carbon black-loaded natural rubber vulcanizates. Part I." *Journal of Applied Polymer Science* 6(19): 57–63.

Peng, Zhenmeng, Ferenc Somodi, Stig Helveg, Christian Kisielowski, Petra Specht, and Alexis T. Bell. 2012. "High-resolution in situ and ex situ TEM studies on graphene formation and growth on Pt nanoparticles." *Journal of Catalysis* 286: 22–29.

Potts, Jeffrey R., Om Shankar, Ling Du, and Rodney S. Ruoff. 2012. "Processing–morphology–property relationships and composite theory analysis of reduced graphene oxide/natural rubber nanocomposites." *Macromolecules* 45(15): 6045–6055.

Rooj, Sandip, Amit Das, Klaus Werner Stöckelhuber, De-Yi Wang, Vassilios Galiatsatos, and Gert Heinrich. 2013. "Understanding the reinforcing behavior of expanded clay particles in natural rubber compounds." *Soft Matter* 9(14): 3798–3808.

Rozainee, M., S. P. Ngo, Arshad A. Salema, and K. G. Tan. 2008. "Fluidized bed combustion of rice husk to produce amorphous siliceous ash." *Energy for Sustainable Development* 12(1): 33–42.

Sadasivuni, Kishor Kumar, Deepalekshmi Ponnamma, Sabu Thomas, and Yves Grohens. 2014. "Evolution from graphite to graphene elastomer composites." *Progress in Polymer Science* 39(4): 749–780.

Sadhu, S., and A. K. Bhowmick. 2005. "Morphology study of rubber based nanocomposites by transmission electron microscopy and atomic force microscopy." *Journal of Materials Science* 40(7): 1633–1642.

Siburian, R., H. Sihotang, S. Lumban Raja, M. Supeno, and C. Simanjuntak. 2018. "New route to synthesize of graphene nano sheets." *Oriental Journal of Chemistry* 34(1): 182–187.

Sridharan, Harini, Abhilash Guha, Sanjay Bhattacharyya, Anil K. Bhowmick, and R. Mukhopadhyay. 2019. "Effect of silica loading and coupling agent on wear and fatigue properties of a tread compound." *Rubber Chemistry and Technology* 92(2): 326–349.

Sushmitha, H., Jagannath Chanda, Amitabha Saha, Prasenjit Ghosh, and Rabindra Mukhopadhyay. "Influence of organoclay dispersion on air retention and fatigue resistance of tyre inner liner compound." *Journal of Applied Polymer Science* 138 (20): 50419–50430.

Waddell, Walter H., and Larry R. Evans. 1996. "Use of nonblack fillers in tire compounds." *Rubber Chemistry and Technology* 69(3): 377–423.

Wang, Meng-Jiao. 1998. "Effect of polymer-filler and filler-filler interactions on dynamic properties of filled vulcanizates." *Rubber Chemistry and Technology* 71(3): 520–589.

Wolff, Siegfried. 1996. "Chemical aspects of rubber reinforcement by fillers." *Rubber Chemistry and Technology* 69(3): 325–346.

Xing, Wang, Maozhu Tang, Jinrong Wu, Guangsu Huang, Hui Li, Zhouyue Lei, Xuan Fu, and Hengyi Li. 2014. "Multifunctional properties of graphene/rubber nanocomposites fabricated by a modified latex compounding method." *Composites Science and Technology* 99: 67–74.

Yang, Ganwei, Zhenfei Liao, Zhijun Yang, Zhenghai Tang, and Baochun Guo. 2015. "Effects of substitution for carbon black with graphene oxide or graphene on the morphology and performance of natural rubber/carbon black composites." *Journal of Applied Polymer Science* 132(15): 41832–41841

Zhang, Hao, Wang Xing, Hengyi Li, Zhengtian Xie, Guangsu Huang, and Jinrong Wu. 2019. "Fundamental researches on graphene/rubber nanocomposites." *Advanced Industrial and Engineering Polymer Research* 2(1): 32–41.

Zhang, Chunmei, Tianliang Zhai, Yi Dan, and Lih-Sheng Turng. 2017. "Reinforced natural rubber nanocomposites using graphene oxide as a reinforcing agent and their in situ reduction into highly conductive materials." *Polymer Composites* 38: 199–207.

14 Graphene-Based Elastomer Nanocomposites
A Fascinating Material for Flexible Sensors in Health Monitoring

Mohammed Khalifa
Kompetenzzentrum Holz GmbH, Wood K plus

Selvakumar Murugesan and S. Anandhan
National Institute of Technology Karnataka

CONTENTS

14.1 INTRODUCTION

Elastomers are polymers that exhibit a rubber-like elasticity and are known for its viscoelastic behavior, low elastic modulus and high stretchable characteristics. Elastomer offers good heat resistance along with its excellent deformability and flexibility in ambient conditions. Elastomers are commonly used in automobile, industrial, packaging and medical applications. This interesting and versatile material has inevitably been elaborated in its area in the form of blends and nanocomposites (Maiti, Bhattacharya, & Bhowmick, 2008; Nardo, 1987; Sedlačík, 2021). Recently, flexible and stretchable smart sensors have attracted great interest due to their versatility

DOI: 10.1201/9781003200444-14

and adaptability. Smart sensors deliver user feedback when responding to external stimuli, such as pressure, strain, force and temperature, and have seen a tremendous potential in wearable electronics devices for continuous real-time health monitoring. In wearable electronic devices, it is vital for the sensors to have good compatibility with the human skin with lightweight and flexible characteristics.

Elastomers with conductive nanofillers with the ability of electron transport and elasticity play a vital role in wearable sensors. There are various reports wherein elastomer-based sensor materials have been used in various fields, including human locomotion monitoring, energy harvesting, electronic skins and strain sensors. Elastomers-based sensors offer excellent sensitivity and stretchability and have the ability to respond to small changes. Generally, flexible sensors require tortuously intricate nanostructured schemes, which are rather complex and expensive. Therefore, designing highly sensitive and low-cost flexible sensors remains a great challenge.

Flexible and stretchable sensor technology is quite possibly the area in which nanotechnology has a significant impact. Development of sensor materials enables nanostructured materials that offer significant advantages over traditional sensor materials. The benefits may be in terms of higher sensitivity, selectivity and lower power consumption and stability. Such materials with flexibility and stretchability are ideal for wearable technologies. Also, nanostructured materials enable to tailor sensor properties dependent on the application and could form new devices. The market for nanostructured-enabled sensors is constantly growing and the development in the field of sensor technology enables flexible and stretchable sensors for wearable applications. Till date, many stretchable strain sensors have been developed that incorporate functional nanoparticles with good electrical properties that serve as an active sensing material (Ni & Zhang, 2017; Noh, 2016).

Elastomer-based nanocomposites have attracted both academics and industries. The introduction of nanoparticles as fillers offers ample opportunity to tune the properties of the elastomer due to their excellent surface area and strong filler-matrix interaction. Various nanofillers, including silicates, carbon nanotube (CNT), fullerenes, silica, metal oxides, metal nanoparticles, and biomaterials, have been incorporated to modulate the properties of elastomers. Conductive elastomer nanocomposite usually does not undergo large deformations without special geometry, designs and processing method. Ionic or electronics hydrogels based on hybrids offer large elongation of the elastomer. However, they significantly lack necessary electrical conductivity and strain sensitivity. Comparatively, elastomers filled with nanoparticles prepared in the form of nanocomposite via melt blending, solution cast, coating, nanomaterial decoration have been extensively reported. Although nanomaterial decorated elastomers possess high sensitivity as strain sensors, they significantly lack stretchability (Lee et al., 2014; Liu et al., 2018; Shyu et al., 2015; Wu et al., 2019). In particular, stretchable strain sensors comprising elastomer-based nanocomposites have drawn tremendous attention due to their ease of synthesis and wearability. On the other hand, soft strain sensors have shown tremendous potential in highly durable robotic skins to accommodate multiscale and dynamic deformations. There are several sensors available in the market, which utilize nanotechnology. However, future nanotechnology-enabled sensors, which possess smarter, inexpensive, selective and sensitive sensors along with wearable characteristics, will be of prime focus in sensor industries. There are various nanostructured materials, such as CNT, graphene, carbon black, silver nanowires (Ag NWs) and copper nanowires (Cu NWs), that have been incorporated into elastomer as an active sensing material. Among various elastomer-based nanocomposites, nanocomposites embedded with graphene have shown significant advances in sensor technology (Avilés, Oliva-Avilés, & Cen-Puc, 2018; Costa et al., 2014; He et al., 2019; Liu, Han, Du, Liu, & Liu, 2018; Tallman, 2018; Yamada et al., 2011).

Graphene and its derivatives are phenomenal materials in the field of science and technology. It has received significant attention in the area of sensor technology, bio-medical and wearable devices. Graphene exhibits unique physical and chemical properties, including mechanical stiffness, elasticity and modulus with excellent electrical and thermal conductivity. Therefore, it is considered a potential material in transparent conductive films, chemical sensors, thin-film transistors, quantum dots and coatings applications. Furthermore, the combination of optical transparency, electrical and thermal conductivity, and excellent mechanical properties makes graphene and its derivatives a perfect selection of flexible electronics. By comparison with CNT, graphene has low coefficient of friction, which facilitates high mechanical integrity. On the other hand, graphene-filled elastomer nanocomposites show percolation threshold at a very low loading compare with CNT (Filleter et al., 2009; Sadasivuni et al., 2015).

Graphene and its derivative is an ideal filler for the fabrication of electrically conductive elastomer composite due to its excellent inherent electrical conductivity and high surface area. There are several factors, such as lateral size, number of layers, defects, purity and dispersion, that affect the electrical properties of the elastomer nanocomposite (Cataldi et al., 2017). Similar to all other physico-chemical characteristics of elastomer/graphene nanocomposites, the degree of dispersion of graphene is vital in achieving high electrical properties of nanocomposite. An ideal graphene dispersion encompasses the formation of conductive network in which the graphene nanoparticles are close by and the conduction takes place by tunneling effect through the elastomer layers. Contrary to other properties such as tensile strength, the electrical conductivity of elastomer/graphene nanocomposites is augmented by partial segregation and is less dependent on the homogenous dispersion. Using percolation threshold theory, the dispersion characteristics of graphene can be predicted. On the other hand, to improve the surface area and interaction, graphene is often modified/functionalized, which helps in the enhancement of electrical conductivity of elastomer/graphene nanocomposite (Bauhofer & Kovacs, 2009; Payandehpeyman et al., 2020). It is important to note that graphene-based elastomer nanocomposites offer high electrical conductivity at lower loading than other fillers such as carbon black, CNT, Polyaniline (PANI) and graphite. During the crosslinking reaction of the elastomer especially in solvent evaporation process, the elastomer gradually shrinks and brings the graphene sheets to more tightly overlapped and stacked. Hence, processing method of the nanocomposite is vital in defining the final electrical conductivity. On the other hand, with increase in the graphene loading, more number of conductive networks are formed, which increase the electrical conductivity of the elastomer/graphene nanocomposite. However, overloading of graphene tends to form large agglomerations of graphene within the matrix, which results in poor dispersion. To achieve high degree of dispersion, various methods including physical dispersion, covalent and non-covalent bonding have been used. However, at higher loading of graphene, the formation of agglomerates or lumps within the elastomer matrix is inevitable (Bauhofer & Kovacs, 2009; Dhakal et al., 2019; Tung et al., 2017).

14.2 STRAIN SENSORS BASED ON GRAPHENE/ELASTOMER NANOCOMPOSITE

Strain sensors mounted on different parts have been widely used in aerospace, health monitoring, construction, structural health monitoring, etc. Recently, flexible strain sensors capable of adapting to complex shapes or surfaces that undergo large deformations are seen to be most effective for continuous strain monitoring systems. Flexible strain sensors should possess high sensitivity, stretchability, stability, linearity and good reproducibility. Among various features, sensitivity and ultra-stretchability of the sensor are considered essential factors that define the performance of the wearable strain sensors. Also, it is essential to maintain good electrical conductivity of the sensor, which helps in achieving a wide range of strain sensing. Graphene is most commonly used in piezoelectric strain sensors due to its excellent sensitivity and mechanical properties and tunable band structure. A flexible strain sensor with high sensitivity based on the nanocomposite has been developed to detect strain. For stretchable strain sensor, a spin coating method was adopted to prepare conductive (polydimethylsiloxane) PDMS/graphene sandwich structure. At a graphene loading of 25 wt.%, the electrical resistance decreased to 9.4 Ωm^{-1} due to the formation of a three-dimensional conductive network, which facilitated the effective transport of the electrons. As a result, the sensor was sensitive to the repeated bending and stretching cycles and showed a good response in the strain range of 5–9% (Zhang et al., 2019). Graphene aerogel/PDMS nanocomposite prepared by vacuum-assisted infiltration method was adopted to fabricate the strain sensors with adjustable sensitivity characteristics. The freeze-drying process altered the graphene aerogel microstructure, cell density and surface area. As a result, the electrical conductivity of graphene aerogel/PDMS nanocomposite increased with decreasing the temperature (up to −196°C). Further, the sensitivity of the strain sensor was modulated with the aid of the freeze-drying process and varying the concentration of graphene aerogel (Figure 14.1a).

When the sensor underwent repeated cyclic loads, the electrical resistance of the sensor varied linearly. The graphene aerogel/PDMS nanocomposite containing graphene treated at a freezing temperature of −50°C exhibited a large electrical resistance change (310%) upon the application of cyclic loading (Figure 14.1b). Three mechanisms generally govern the response of the graphene aerogel/PDMS nanocomposite-based strain sensor: First, formation and disruption of the conductive path upon the application of external load. Second, the tunneling resistance change due to the alteration in the distance between the neighboring graphene particles. Third, the intrinsic piezoresistive characteristics of the graphene aerogels. Due to the rigid and stable crystal structure of graphene, the piezoresistance effect is not significantly high. The use of PDMS as a matrix material facilitated excellent load transfer that

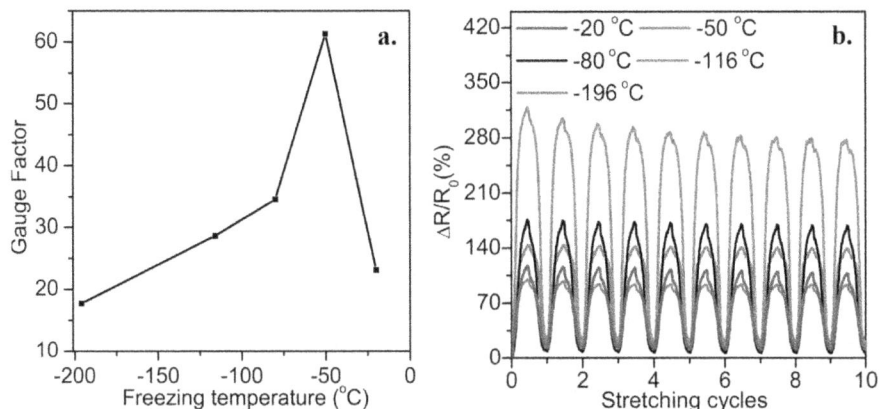

FIGURE 14.1 (a) Influence of freezing temperature on the gauge factor (sensitivity) of graphene aerogel/PDMS nanocomposite. (b) Response of the strain sensor upon repeated cyclic loading (Wu et al., 2016). (Reproduced with permissions ©2016, American Chemical Society.)

may have caused strain in the graphene. Moreover, it is widely accepted that the piezoresistive behavior of graphene is a dominating factor in defining the sensitivity of the strain sensor, but other factors also play a vital role in providing the necessary linearity and stability to the strain sensor (Wu et al., 2016).

The microstructure and the cell wall size of graphene also influence the electrical resistance change. The conductive path networks within the PDMS matrix rely on the cell size, dispersion and number of crosslinking sites. The crosslinking sites per unit volume increase with smaller cell size, which helps in forming a large number of conductive paths. When the cell thickness is high, the sensor response is weak upon applying strain due to the less change in the overlap area. Therefore, controlling graphene cell wall size helps to exhibit high sensitivity of graphene aerogel/PDMS nanocomposite-based strain sensor.

Thermoplastic polyurethane (TPU) is a well-known elastomer exhibiting excellent stretchability, strength and thermal stability, which makes it suitable for the construction of smart sensors. In particular, TPU/conductive filler nanocomposite in porous structure, fibers or foams are particularly interested in achieving highly stretchable sensors. 3D porous structure comprising TPU-based conductive nanocomposite offers lightweight, high flexibility and piezoresistive sensor performance (Lü et al., 2021). There are various approaches such as thermally induced phase separation (Jing et al., 2014), solution cast (Wu & Chen, 2017), salt leaching (Pang et al., 2013), vapor-induced phase separation (Guo et al., 2020) and water-induced phase separation method (Sang et al., 2021) that have been adopted to fabricate TPU-based conductive nanocomposite. A strain sensor based on graphene/TPU nanocomposite was fabricated by co-coagulation and compression molding technique. At a graphene loading of 0.1 wt.%, the nanocomposite showed typical percolation behavior and displayed eightfold increase in the volume electrical conductivity. The strong hydrogen interaction between the TPU and the graphene and the uniform dispersion of graphene led to the formation of a conductive path. The interconnected graphene sheets within the TPU matrix are disturbed when it is stretched/compressed. As a result, the electrical conductivity was altered. Greater electrical resistance change is achieved at a higher strain rate, which leads to greater variation in the sensor response (Figure 14.2).

A highly sensitive strain sensor based on styrene—ethylene—butylene—styrene (SEBS) with graphene oxide (GO), reduced graphene oxide (rGO) and graphene nanoplatelets was fabricated for finger moment monitoring. The nanocomposite films were prepared via the doctor blade technique. The combination of good mechanical and electrical properties of elastomer/graphene nanocomposite is vital in defining the performance of the sensor. The percolation threshold of GO SEBS/ and rGO/SEBS was seen around 2 wt.% loadings, while for graphene nanoplatelets/SEBS nanocomposite, it was around 6 wt.%. The intrinsic electrical conductivity, SEBS and filler interaction and the aspect ratio of fillers strongly influence the electrical properties of the nanocomposite films. The size of the conductive graphene is vital in defining the electrical properties of the nanocomposite. Elastomers generally show non-linear hyper-elastic characteristics and high deformation levels. SEBS shows a high stretchability of 800% (strain), while it decreases slightly (~100%) upon adding nanofillers. Generally, the reinforcement effect of graphene may increase the mechanical properties of the nanocomposite. Higher graphene content lowers the stretchability of the SEBS. However, the lower content of graphene may affect the electrical properties of the nanocomposite. Also, graphene at higher loading tends to agglomerate in the matrix, which may act as rupture points due to the formation of defects at the graphene-SEBS interface.

Augmented mechanical and electrical properties of graphene/elastomer nanocomposite significantly help implement the sensor for wearable application. One such suitability of sensors has been demonstrated by attaching the rGO/SEBS-based strain sensor to detect the movement of human fingers (Figure 14.3a). Strain sensors in the form of strips (12 cm×0.8 cm) were attached to each finger with adhesive tapes. Figure 14.3b shows the response of the graphene/SEBS nanocomposite-based strain sensors attached to the gloves to record the movement of the fingers. Sensors showed a quick response when stretched, while the recovery was slower due to its viscoelastic

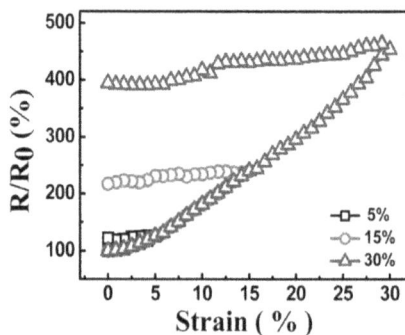

FIGURE 14.2 Response of graphene/TPU-based sensor at different rates (Liu et al., 2015). (Reproduced with permissions ©2015 Royal Society of Chemistry.)

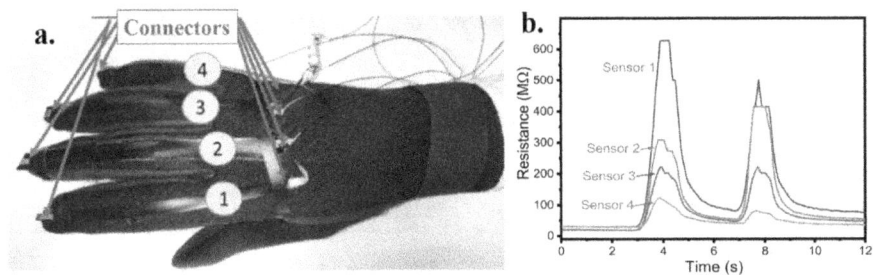

FIGURE 14.3 (a) Hand glove attached with rGO/SEBS sensors placed over fingers. (b) Electric resistance change of the sensor upon hand movement (Costa et al., 2019). (Reproduced with permissions ©2019 American Chemical Society.)

behavior. In addition, the sensor responded distinctly for each finger, which could be attributed to the difference in the strain input due to diverse hand anatomy.

The exploration of materials and suitable synthesis processes for the procurement of flexible and stretchable conductors has been driven by technological needs. The steep increase in flexible devices, wearable sensors, epidermal electronics, biointerfaces, soft robots, prosthetics and energy harvesting nanogenerators has boosted the research in developing new materials and methods. Most often, wearable sensors show a decline in electrical conductivity, stability and sensitivity due to various factors, including fatigue cracks and damages to the sensor. Hence, self-healing materials extended their opportunities in the area of wearable sensors. Basically, self-healing of polymeric materials relies on two approaches: First, the extrinsic healing process by the bursting of capsules/vascular networks embedded within the matrix. Second, the intrinsic self-healing characteristics due to reversible bonds, ionic bonds, hydrogen bonds, etc. (Li et al., 2016; Ma et al., 2021; Tee et al., 2012).

Zhao et al. (2019) fabricated a pressure sensor based on a polysiloxane elastomer network crosslinked by dynamic Diels-Alder bonds followed by the incorporation of graphene. The addition of PDMS tuned the flexibility, mechanical performance and stretchability of the nanocomposite. As a result, the polysiloxane nanocomposite exhibited good self-healing efficiency, mechanical performance and sensing behavior. The prepared nanocomposite-based pressure sensor was tested under different modes through human locomotion and exhibited a high sensitivity of 0.765 kPa^{-1} and a gauge factor of 4.9. In order to achieve high sensitivity, good electrical conductivity and durability of self-healing characteristics of the sensor, it is vital to consider selecting appropriate conductive fillers. It is well-known that silver (Ag) nanoparticles have been widely adopted as excellent conductive fillers for sensor application. However, the weak interaction of Ag with elastomer and poor stability create issues for wearable applications. To resolve, elastomer/graphene/Ag hybrid structures have been proposed by various researchers. Graphene not only improves the electrical conductive path but also improves the adhesion with the elastomer chains due to abundant functional groups on its surface (Lai et al., 2016; Luan et al., 2012). In this regard, a two-layered flexible self-healing strain sensor was fabricated using poly(ε-caprolactone)/GO/Ag hybrid thin film obtained by pumping filtration method and followed by the PDMS encapsulation. GO facilitates in bridging the conductive

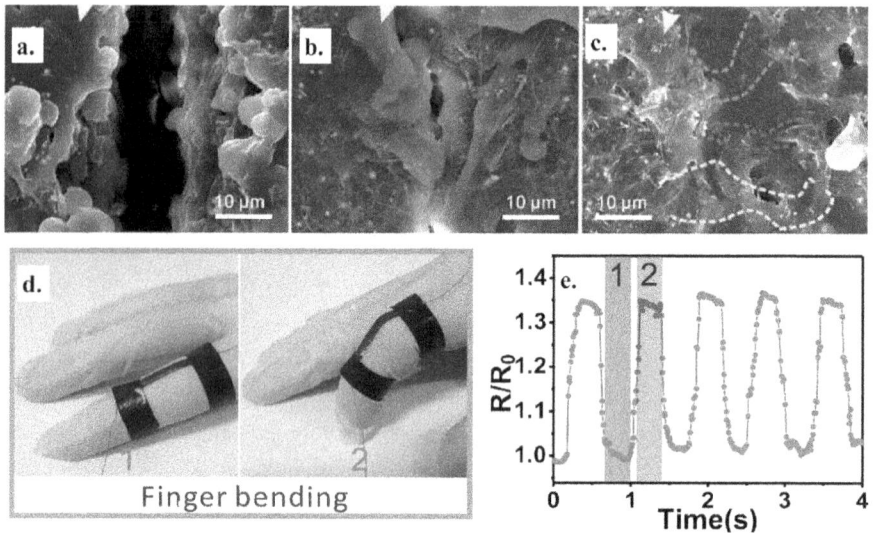

FIGURE 14.4 SEM images of poly(ε-caprolactone)/GO/Ag hybrid thin film encapsulated with PDMS-based strain sensor demonstrating the self-healing process at 80°C: (a) 0 minute; (b) 1 minute and (c) 3 minutes. (d) Digital image of the strain sensor attached to finger. (e) Response of the strain sensor upon bending and straightening (Liu et al., 2017). (Reproduced with permissions © copyright 2017 Elsevier.)

networks created by Ag nanowires. The PDMS was cured at 50°C for 24 hours to obtain a two-layered structure. The PDMS was fully penetrating the microholes of the poly(ε-caprolactone), which makes a good interface between the layers. The strain sensor showed excellent self-healing properties that resulted in augmenting the durability of the sensor. The self-healing process was demonstrated by making an artificial scratch on the strain sensor surface. The strain sensor was self-healed completely within 3 minutes when conditioned at 80°C (Figure 14.4a–c). Interestingly, the conductive network was rebuilt after the healing process and the sensor morphology was seemed to be more compact. As a result, the electrical resistance decreased by 25%, while the sensitivity decreased slightly. After the repeated healing process, the sensitivity of the strain sensor decreased by 35% due to the disruption of the Ag nanowires network. On the other hand, the bending sensitivity and gauge factor increased after the repeated healing cycles. The human health monitoring of the strain sensor was demonstrated by attaching the sensors to the fingers, elbow and neck. The resistance increased as the finger was bent and recovered when straightened (Figure 14.4d). The sensor showed a repeatable response when the finger was bent and straightened repeatedly (Figure 14.4e). Overall, the sensor showed excellent repeatability and stability (>2,400 cycles) under constant strain. High self-healing efficiency of poly(ε-caprolactone)/GO/Ag hybrid thin-film encapsulated with PDMS with excellent strain sensing characteristics could be promising in wearable devices to detect human locomotion.

Recently, two different types of conductive fillers have been used to obtain high electrical properties at lower loading to retain the mechanical integrity of the elastomer matrix. A combination of 1D and 2D nanostructures such as CNT/graphene and Ag/GO offers excellent electrical conductivity. Since one of the fillers bridge the gap between the other conductive particles, it helps forming long and continuous conductive paths. Additionally, incorporating two fillers usually helps in achieving good dispersion within the matrix (Ji et al., 2016; Luan et al., 2012; Oh et al., 2016; Verma et al., 2017). Such systems have also been adopted in the spandex matrix to achieve highly sensitive strain sensors for human health monitoring (Vo et al., 2020). Spandex is a urethane-based elastomer comprising soft and hard segments and offers excellent stretchability up to 700% and superior elastic recovery. GO/Ag nanowires/spandex nanocomposite prepared via solution cast process to develop a highly stretchable strain sensor. The hybrid nanocomposite was cut into rectangular strips and copper electrodes were attached at the two ends of the composite strip. The addition of Ag nanowires altered the electrical and mechanical integrity of the GO/spandex nanocomposite. It is important to note that the electrical and mechanical properties of the nanocomposite were sensitive to the Ag nanowire loading. The nanocomposite showed excellent electrical and mechanical properties when the ratio of Ag nanowires to GO was maintained at 1:4. The strain sensor was evaluated when repeatedly stretched and released at a strain ranging from 10% to 140%. The response of the sensor increased upon increasing the strain. The strain sensor based on GO/spandex (without Ag) struggled to exhibit stable response when stretched above 70%, while upon the addition of Ag nanowires to the nanocomposite, the sensor exhibited a stable response up to 140% (Figure 14.5a). The stable response could be attributed to the formation of long conductive paths within the matrix. However, the long conductive path could be detrimental in achieving high sensitivity. The sensitivity (gauge factor) of the GO/spandex sensor was increased up to 963 when stretched up to 60%, while the GO/Ag nanowires/spandex hybrid nanocomposite sensor exhibited a gauge factor of 150.3 at 60% stretching. However, the GO/Ag nanowires/spandex hybrid nanocomposite sensor offers a stable response even at large strain values up to 140%. Figure 14.5b shows the ability of the sensor to detect the heartbeats by attaching it onto the wrist of human hand. An electrical

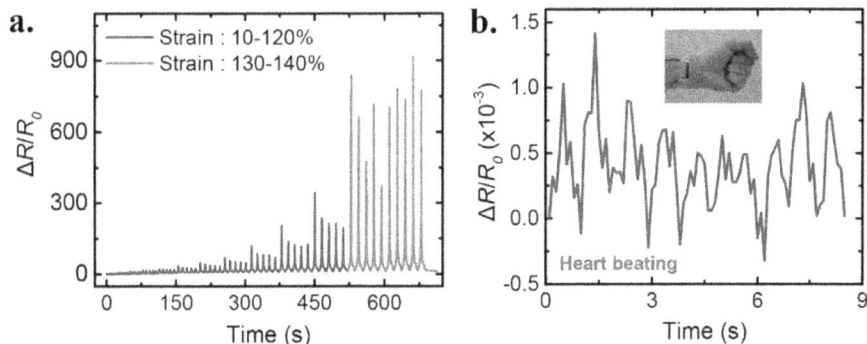

FIGURE 14.5 GO/Ag nanowires/spandex hybrid nanocomposite sensor. (a) Response of the sensor at different strains (10%–140%). (b) Sensor attached to the wrist to monitor heartbeat (Vo et al., 2020). (Reproduced with permission ©2020 MDPI.)

resistance change waveform pattern provides information of ten pulses in 8 seconds, presenting a heartbeat rate of ~71, signifying the sensitivity of the sensor to subtle movements. Also, the sensor was able to exhibit stable and distinct signals upon bending the wrist and knee.

14.3 HUMIDITY AND GLUCOSE DETECTION SENSORS

Water plays an important role in human health and is an indispensable substance to life. The water molecules are one of the main sources in the human metabolism process. Hence, obtaining the physiological data through real-time monitoring of the water content on human skin or exhaled air provides important information. Wearable humidity sensors should possess skin-friendly characteristics, high flexibility, low humidity hysteresis, high stability, linear response, wide sensing range, excellent sensitivity and selectivity. Sensors for respiratory and skin dryness monitoring are vital in providing information for patients and athletes. Flowmeter and spirometer are the most commonly used techniques for the monitoring of respiration. However, these techniques are not mobile and do not provide real-time data. In addition, most of the sensors do not distinguish the inhalation and exhalation data, which may be tricky for the diagnosis. In the last decade, various wearable humidity sensors have been developed based on transducing methods of resistance, capacitance, impedance and optical fibers. Among them, capacitive and resistance-based humidity are widely exploited due to their easy readout. Various advanced nanomaterials have been adopted in fabricating humidity sensors such as metal oxides, electrolytes, polymers, dielectric materials and carbon materials. It is noteworthy that the 2D materials, including graphene, graphene oxide, tungsten diselenide, and molybdenum diselenide and graphitic-carbon nitride, have received great interest due to their excellent sensor characteristics. On the other hand, to achieve high flexibility, elastomers have been broadly utilized, including PDMS, TPU and PEDOT: PSS (Gong et al., 2021; Kantoch et al., 2014; Khalifa et al., 2020; Zhou et al., 2017).

Graphene and its derivatives are seen to be an excellent material for humidity sensing. Due to oxidization, GO contains a large number of oxygenated groups, which can readily interact with the water molecules. GO agglomerates into a molecular water film at high humidity conditions due to the interaction with water molecules, making it an ideal material for humidity sensing. Due to the good hydrophilicity of GO, oxygen-containing groups form a hydrogen bond with the water molecules. As a result, the electrical properties of GO are altered. In low humidity conditions, the physical adsorption of water molecules takes place on the GO surface and the charge transfer takes place through the hopping of protons. With the increase in the humidity levels, layers of water molecules are stacked on the GO surface. Water molecules form the second layer by physical adsorption through a single hydrogen bond on the hydroxyl and becomes mobile. At high humidity conditions, the water molecules ionize due to the electrostatic field to form a large number of charge carriers. Depending on the semiconductor type (p or n-type), the electrical properties of GO change (Borini et al., 2013; Liang et al., 2020). A wearable, transparent and stretchable humidity sensor based on rGO/TPU composite was fabricated via the liquid phase blending method. The PDMS substrate with source-drain electrodes is treated with oxygen plasma and coated with GO/TPU thin film using spin coating process. The fabricated device showed transparency of 78% with good mechanical integrity and can be wrapped around a human finger for continuous humidity monitoring of skin and surroundings. The sensor showed good response and repeatability with a sensitivity of 4.27. Also, the sensor showed excellent wearability due to its excellent flexibility, lightweight and stretchable characteristics (Trung et al., 2017).

A multifunctional and stretchable sensor based on polydopamine (PDA)/rGO/ electrospun TPU fibrous mats was developed for human motion and environment monitoring. Electrospun TPU fibrous mat was immersed in rGO/sodium dodecyl sulfate (SDS) solution for 1 hour. Once a sufficient amount of rGO was coated onto the surface of TPU, the rGO-coated fibrous mat was immersed in dopamine solution. The dopamine polymerized and the final product was obtained in the form of hybrid film. The TPU fibers showed a wrinkled structure upon coating with PDA and rGO (Figure 14.6a). The morphology showed different pore sizes and interconnected structures, which are typically observed in electrospun fibers that facilitate good flexibility and sensing characteristics. As predicted, the rGO/PDA/TPU-based sensor showed excellent stretchability (550%). The addition of PDA to rGO/TPU helped the sensor to exhibit excellent response under different stretching conditions with an increase of gauge factor by ~100%. The presence of PDA partially blocked the movement of the rGO, which increased the electrical resistance of the sensor when stretched. With the increase in the strain % from 100% to 140%, the gauge factor increased to 1,314.8 (Figure 14.6b). At lower stretching conditions, the fibers tend to orient along the direction of stretching, while at higher stretching conditions (>100%), the fibers elongate and become thin and the distance between the adjacent rGO nanosheet increases. As a result, a large change in the resistance was observed. The rGO/PDA/TPU-based sensor showed excellent wearable characteristics. The sensor was attached to various body parts such as fingers, wrist, elbow and leg. The sensor showed distinct signals at different human movements. On the other hand, the sensor was also seen to be sensitive to different humidity conditions. In Figure 14.6c, the response of the sensor at different humidity conditions is shown. The resistance of the sensor increased upon increasing the humidity, which is likely due to the interaction between the functional groups of rGO and water molecules. The results demonstrate that the rGO/PDA/TPU-based multifunctional sensor can be used for wearable electronics for human health monitoring (Jia et al., 2020).

A graphene/PDMS composite-based wireless sensor was fabricated via a molding process for respiration monitoring based on respiratory induction plethysmography (RIP). The sensor showed high sensitivity (up to 150 cm^{-1}) and durability when repeatedly stretched for 3,600 cycles. The respiratory sensor was compared with the polysomnography signals and the data were acquired by attaching a sensor near the abdomen region of a male volunteer. As a result, the sensor acquired respiratory signals with high precision and in line with the commercial respiratory monitoring system. Given the flexibility, facile synthesis and wireless characteristics along with highly accurate data procurement, graphene/PDMS composite could be a great potential in health monitoring applications (Chen et al., 2019).

Continuous monitoring of glucose is vital for diabetic patients. Generally, implantable enzyme electrodes are often used for glucose monitoring. However, the

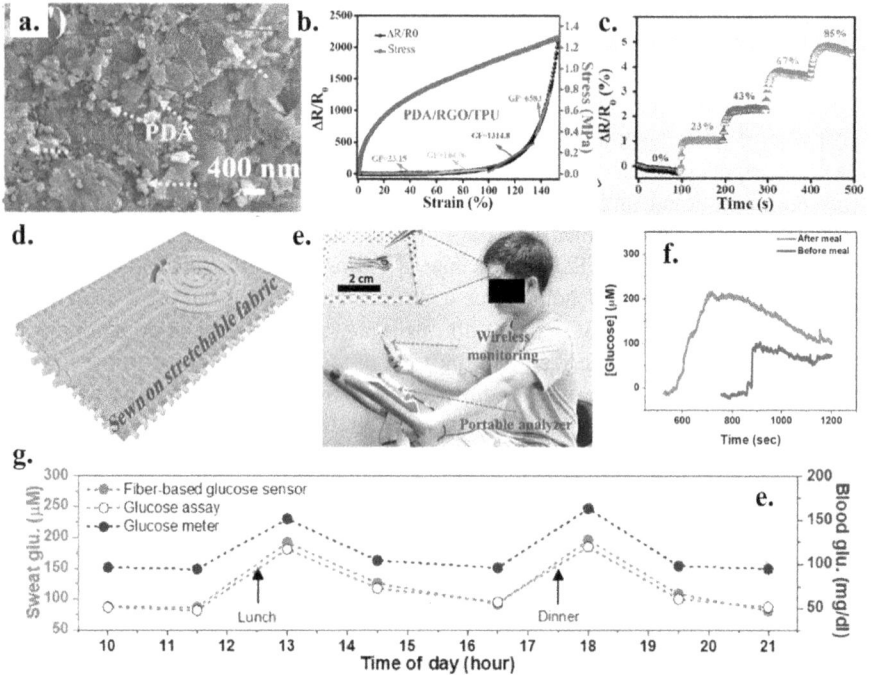

FIGURE 14.6 (a) SEM image of rGO/PDA/TPU multifunctional sensor. (b) The response of the rGO/PDA/TPU multifunctional sensor at different strain values. (c) Humidity sensing performance of rGO/PDA/TPU multifunctional sensor at different humidity concentrations (Jia et al., 2020). (Reproduced with permissions ©2020 Elsevier.) (d) Schematic illustration of free standing rGO/TPU/Au fibers embedded with PDMS substrate. (e) rGO/TPU/Au fibers-based sensor attached to the forehead of the volunteer for glucose monitoring. (f) Monitoring of glucose levels before and after meals. (g) Glucose levels measured in 1 day using rGO/TPU/Au fibers-based sensor (Toi et al., 2019). (Reproduced with permissions ©2019 American Chemical Society.)

lifespan of such monitoring is short, which may be fatal and lead to foreign body reactions. Recently, portable devices are being used by invasive sampling methods. For non-invasive and continuous glucose monitoring, various technologies have been proposed, including contact lenses, watches, tattoos and patches, which collect the sample from tears, interstitial fluid or sweat. Thus, flexible, wearable patches have been seen with tremendous growth for continuous glucose monitoring. Metal oxides, alloys, nanomaterials and polymer have been adopted in fabricating the glucose sensor patches (Pu et al., 2016; Senior, 2014; Tierney et al., 2001; Zhao et al., 2019).

A highly flexible and stretchable patch consisting of rGO/TPU nanocomposite fibers was produced via the wet-spinning process. Further, the fibers were coated with gold (Au) via the thermal evaporation process and the glucose sensor was prepared on a PDMS substrate. The free-standing elastomeric behavior of rGO/TPU fibers with an excellent surface area has several benefits in developing a high-performance wearable glucose sensor patch. The patch was further integrated into fabric and interfaced with an electrochemical analyzer, which can be readily attached to various body parts to detect the glucose levels based on sweat. The Au particles on the surface of the fibers, which acts as an electrocatalytic material, speed up the electro-oxidation of glucose. On the other hand, the functional groups present on the surface of rGO/TPU fibers readily interact with glucose and increase the dehydrogenation process. As a result, the sensor showed excellent reproducibility and selectivity toward glucose. Also, the biocompatible and stretchable characteristics of the sensor provide an excellent platform for real-time monitoring of glucose levels. The sensor patch was attached to the forehead of the human volunteer for glucose monitoring in sweat (Figure 14.6d). To generate the sweat, the volunteer was cycled at different speeds and times. The generated sweat was absorbed by the fabric and interacted with the sensor. As a result, the glucose signal increased. The sensor response was also tested before and after the meal, and the sensitivity of the sensor was in-line with the commercially available glucose meter (Figure 14.6e and f). Also, the sensor was able to read out accurate glucose levels for 24 hours, which confirms its real-time glucose monitoring ability (Figure 14.6g).

14.4 TEMPERATURE SENSORS

Maintaining the body temperature is important for the effective functioning of the immune system and blood circulation for the human body. If the body temperature increases by 1%, then the immunity decreases by 30%. Therefore, it is vital to maintain the body temperature to prevent the risk of developing viral diseases. Also, if the body temperature drops below 34°C for a long time, cardiac arrest chances increase drastically. On the other hand, blood circulation is also sensitive to skin temperature. Hence, it is important to monitor the body temperature regularly. Traditional temperature sensors based on platinum and gold have been used commonly for temperature sensors. However, they do not have the necessary stretchable and flexible characteristics for wearable applications. Also, for special groups (newborns, kids), traditional methods lack feasibility. Therefore, skin-attachable stretchable temperature sensors have attracted significant attention, which offer comfort to the skin and provide a linear response, air permeability, sensitivity and selectivity. Stretchable temperature sensors can be easily attached to the body. Sensors in the form of patches that can be easily adhered to the skin or stitched to the fabric to breathe are of great interest. Wearable temperature sensors should be durable, lightweight besides an easy readout technique. To achieve lightweight, temperature sensors are prepared using thin films, coatings, fibers, yarns and fabrics. Wearable temperature sensors based on electrical resistance, dielectric properties, capacitance and impedance-based sensors are used. Among them, resistance-based temperature sensors are widely used due to stability, linearity and reliable output (Hudec et al., 2020; Park et al., 2020, 2021; Yu et al., 2020). Flexible temperature sensors are based on gold and PIN diode using silica nanoribbons on elastomers, nickel-polymer composite and lanthanum-doped aluminum oxide. Also, a bimodal sensor based on P(VDF-TrFE) and $BaTiO_3$ nanocomposite have been explored (Jeon et al., 2013; Tien et al., 2014). These temperature sensors are capable of providing accurate data, but lack thermal response and require complex electronic circuits to achieve precise detection. Besides, such sensors also failed to fulfill the requirements of biocompatibility due to the usage of metals. There are various reports wherein skin attachable temperature sensors were studied with different materials, including metal nanoparticles, graphene, CNT and polymers (PEDOT:PSS). Recently, carbon-based nanomaterials have been explored due to their excellent electron transport mechanism. Graphene and its derivatives have received considerable attention due to its thermal conductivity, electrical properties and thermal responsive behavior. In the case of graphene-based temperature sensor, the charge carriers and phonon play a vital role in the conducting channel medium. The electrical transport characteristics of graphene are an effect of the linear dispersion of Dirac electrons. The electrical conductivity of graphene is lower due to the Fermi energy lying at the charge neutrality point. Application of temperature gradient on graphene may either disturb electrons or holes, which in turn change the electrical properties. In rGO, the effect of its structure and electron scattering, surface defects and functional groups may also affect the electrical conductivity upon the application of temperature gradient.

Graphene nanowalls (GNWs)/PDMS-based wearable temperature sensor was developed using a plasma-enhanced chemical vapor deposition and polymer-assisted transfer method. The temperature coefficient of resistance (TCR) of the sensor was $0.214°C^{-1}$, which is three orders higher than the traditional wearable temperature sensor. The sensor not only offers a high thermal response but also shows excellent stretchability and sensitivity enough to be used for monitoring the temperature of a human body (Yang et al., 2015). Graphene nanoplatelets (GNPs)/PDMS nanocomposite film fabricated via molding process can be used as a multifunctional flexible sensor due to their excellent electrical and mechanical integrity. The electrical resistivity of the GNP/PDMS nanocomposite film was dependent on the thermal expansion of the PDMS and the alteration in the tunneling conductivity. The electrical resistance is sensitive in the wide temperature range from −25°C to 160°C. The response of the nanocomposite film was linear with a response of ~80%. The GNP/PDMS nanocomposite films can also be used as a strain sensor with a good sensitivity of 1.6 (Han et al., 2019).

Sensors based on fibers or yarns that can be integrated with textiles are attracted due to their feasibility of attaching it to clothes or gloves for continuous monitoring of physiological and biological signals from the human body. However, temperature sensors based on fibers are still limited. Temperature sensors in the form of fibers exhibiting high thermal responsivity and mechanical integrity could be unique and convenient for wearers. Several studies have fabricated stretchable temperature sensors by utilizing stretchable materials or the geometric engineering of flexible materials. However, these temperature sensors are affected by the strain and pressure signals, which give out unstable output and hinder the accuracy of the sensor (Hong et al., 2016; Wu et al., 2018; Yan et al., 2015). To overcome the issues, Trung et al. (2019) proposed a new approach to designing fiber-based stretchable temperature sensors that can reduce strain-induced interference by utilizing the synergism of geometric engineering free-standing stretchable fiber. Sensor comprising rGO integrated into TPU in the form of composite and encapsulated with PDMS. The GO/TPU composite was prepared via wet-spinning method and then the graphene/TPU fibers were reduced to rGO/TPU by using vitamin C. The fiber diameter was in the range of ~160–550 μm, which is feasible for knotting the fibers. To eliminate the strain/pressure interference, the fibers were placed in a serpentine form. The rGO/TPU nanocomposite-based stretchable sensor can attach to different parts of the body to monitor the temperature of the human skin. One such capability of the sensor was demonstrated by attaching it to the skin during cycling (Figure 14.7a and b). The sensor showed excellent response to skin temperature with a resolution of 0.1°C. Moreover, the sensor showed responsivity of 0.8%/°C with a response time of 7 seconds and 90% stretchability. The sensor readily responded when attached to different parts of the body including wrist, palm and forearm in the form of bandage. The skin temperature was also monitored by applying the antiphlamine lotion on the skin. The sensor was even sensitive to the small increase in the skin temperature (Figure 14.7c). The temperature sensing of rGO/TPU nanocomposite is attributed to

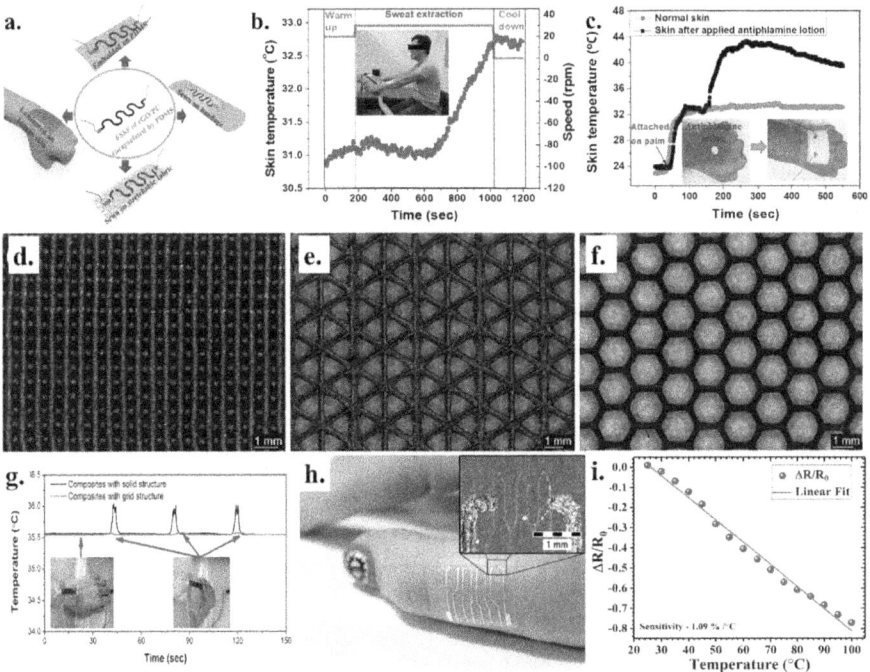

FIGURE 14.7 (a) Schematic depiction of wet-spun rGO/TPU-based temperature sensor encapsulated into PDMS and sewed on a bandage. (b) Temperature monitoring of skin during cycling using rGO/TPU-based temperature sensor. (c) Temperature monitoring of skin before and after the application of antiphlamine lotion (Trung et al., 2019). (Reproduced with permission ©2018 American Chemical Society.) Optical images of the 3D-printed graphene/PDMS composites with (d) grid structure; (e) triangular porous structure and (f) hexagonal porous structure. (g) Skin temperature monitoring of wrist using 3D-printed graphene/PDMS composite (Wang et al., 2019). (Reproduced with permission ©2018 American chemical society.). (h) Patterned graphene/PEDOT:PSS-based temperature sensor (Vuorinen et al., 2016). (Reproduced with permission ©2016 Springer Nature.). (i) The response (ΔR/R%) of GO/PEDOT:PSS-based temperature sensor as a function of temperature (Soni et al., 2020). (Reproduced with permission ©2019 IEEE.)

the comportment of rGO at different temperatures. With the increase in the temperature, the TPU expands leading to increase in the distance between the adjacent rGO nanosheets, which alters the electrical conductivity of the nanocomposite. However, hopping mechanism in rGO was a dominant factor in achieving high resolution of the sensor. The electrical conductivity of the sensor increased upon increasing the temperature, which could be ascribed to the charge transport within the rGO and at the junctions. On the other hand, the presence of oxygen-containing functional groups offers large number of activated trap sites, which contributes to the hopping phenomenon. It is noteworthy that the response of the rGO/TPU nanocomposite sensor is tunable by changing the density of functional groups.

In another study (Wang et al., 2019), 3D-printed graphene/PDMS nanocomposite with different structures (grid, triangular and hexagonal) and designs have been fabricated for a highly stretchable wearable temperature sensor with insensitive strain response (Figure 14.7d–f). All three-dimensional structures (3D-printed graphene/PDMS nanocomposite) showed excellent response and linearity as a function of temperature (25°C–75°C). A linear trend in the electrical resistance was observed upon increasing the temperature, indicating its positive temperature coefficient (PTC) behavior. The electrical resistance increased significantly upon increasing the temperature, suggesting the metallic nature of graphene. Such a phenomenon occurs due to augmented phonon/charge carrier scattering and compact charge carried path, resulting in decreased mobility. As a result, the response and recovery time was less than 4 and 5 seconds, respectively, comparable with the standard temperature sensors. Also, the resolution of the grid structure was 0.5°C. The temperature sensitivity of the sensor can further be achieved, but the sensor may suffer from selectivity due to external interference. In order to achieve a stable electrical resistance, long-range ordered porous graphene/PDMS structures are desirable. In this regard, the 3D printing technique is efficient in accurately controlling the architecture of the sensor. Further, the sensor was attached to the wrist to monitor the skin temperature (Figure 14.7g). The sensor was able to respond accurately to the skin temperature. The grid structure of the graphene/PDMS nanocomposite sensor did not have any effect in its response upon the application of strain, while on the other hand, solid structure based on graphene/PDMS nanocomposite was sensitive to strain. The high resolution, sensitivity, stretchability and selective temperature sensor could be a promising material for artificial skins, robot sensing and human health monitoring systems.

Transparent, flexible graphene/PEDOT: PSS-based temperature sensors attached to TPU bandages were fabricated through screen printing. Graphene/PEDOT:PSS was used as the active sensing material, which was screen printed directly onto the bandage (substrate). Screen-printed PEDOT: PSS-based temperature sensors are known to offer high TCR up to 0.61%. Printing directly onto the bandage provides an opportunity to fabricate bio-compatible and disposable devices and can be attached easily to the skin (Figure 14.7h). The sensor was heated from 35°C to 45°C using a Peltier element. The sensor electrical resistance decreased upon increasing the temperature. The response and recovery time of the sensor was 20 and 18 seconds, respectively. The temperature range chosen for the evaluation was not wide to be realized for human body temperature monitoring. In addition to this, the sensor showed issues related to selectivity, which hinders its applicability (Vuorinen et al., 2016).

Printed temperature sensors offer several advantages since they are facile, cost-effective, scalable and most importantly flexible. In addition, a skin conformable printed temperature sensor on a flexible substrate offers stretchable characteristics, which can be used for human health monitoring. GO/PEDOT:PSS composite printed temperature sensor was demonstrated as a highly temperature-sensitive material. The response of the sensor was linear, with a sensitivity of 1.09%/°C. Figure 14.7i shows the sensor response as a function of temperature, which shows linearity with a coefficient value of 0.988. On the other hand, the response and recovery time of sensor was 18 and 32 seconds, respectively.

Muscle spasm or swellings are the most common complications of stroke patients triggering functional damage and pain. Such patients require a rehabilitation process to counter the pain. Among various rehabilitation processes, thermotherapy is the most basic process, which provides relief to the patients. Thermotherapy pads are one of the most widely used tools in orthopedics and physiotherapist. Thermotherapy helps in providing relief against chronic inflammation, joint stiffness, blood circulation and pain caused by obesity. However, traditional thermotherapy pads have limitations that hinder their wearability.

Recently, nanomaterials such as metallic nanowires, graphene, CNT and hybrid composites have been reported as active materials for wearable heaters. Among them, Ag nanowires are well-known material for wearable heaters. Most of the wearable heaters work on the principle of resistive Joule heating. Thermotherapy pad involves heaters in the temperature range of 30°C–70°C for pain and stiffness relief. Currently, thermotherapy pads made of metal nanoparticles lack the necessary flexibility of wearable applications. On the other hand, the oxidization of the metal nanoparticles reacts with water and air. Also, the thermotherapy pads do not have a controller for maintaining the temperature. Therefore, alternative materials, which possess high conductivity, stretchability and chemical inertness, are necessary for thermotherapy pads. Recently, a number of carbon-based materials have been reported to develop stretchable thermal heaters that comprise CNT, hybrid fillers, graphene films, graphite patterned thins films, sacrificial nanofibers, etc. The wearable thermotherapy pads must have lightweight, skin-friendliness, soft and biocompatible characteristics. In this regard, elastomer-based conductive composites with excellent elasticity are considered the most suitable material for the thermotherapy pad. Elastomers such as PEDOT:PSS, PDMS and TPU are considered most suitable materials for thermotherapy pads (Jang et al., 2017; Kim et al., 2019; Li et al., 2020; Liu et al., 2021).

A highly stretchable nanocomposite made of GO/PEDOT/waterborne polyurethane (WPU) was developed via solution cast process (Zhou et al., 2017). Further, the GO was reduced via chemical reduction process to form rGO/PEDOT/WPU. rGO is considered as an exceptional thermal conductivity up to 5,000 W mK^{-1} that could improve the thermal response of the conductive elastomer composite. Two copper electrodes were attached on rGO/PEDOT/WPU nanocomposite film and the temperature was modulated using DC voltage supply. Upon reducing the GO, the electrical conductivity of the nanocomposite was 14 S cm^{-1}, while the elongation was achieved up to 357%. The performance of the nanocomposite-based electric heater generally relies on heating rate, temperature distribution, stability, durability and mechanical integrity. A constant DC voltage was applied at the ends of the nanocomposite heater to induce resistive joule heating. The temperature of the nanocomposite heater increased upon increasing the applied voltage. The uniform dispersion of rGO within the matrix defines the uniform heat distribution. The effect of incorporation of rGO on the temperature distribution of nanocomposite film was evaluated using thermal imaging. Figure 14.8a shows the temperature distribution along the length of the film. The nanocomposite film exhibited the highest temperature in the middle and gradually decreased at the end. Moreover, the sensor was attached to the human wrist to demonstrate its performance under mechanical disturbances (Figure 14.8b).

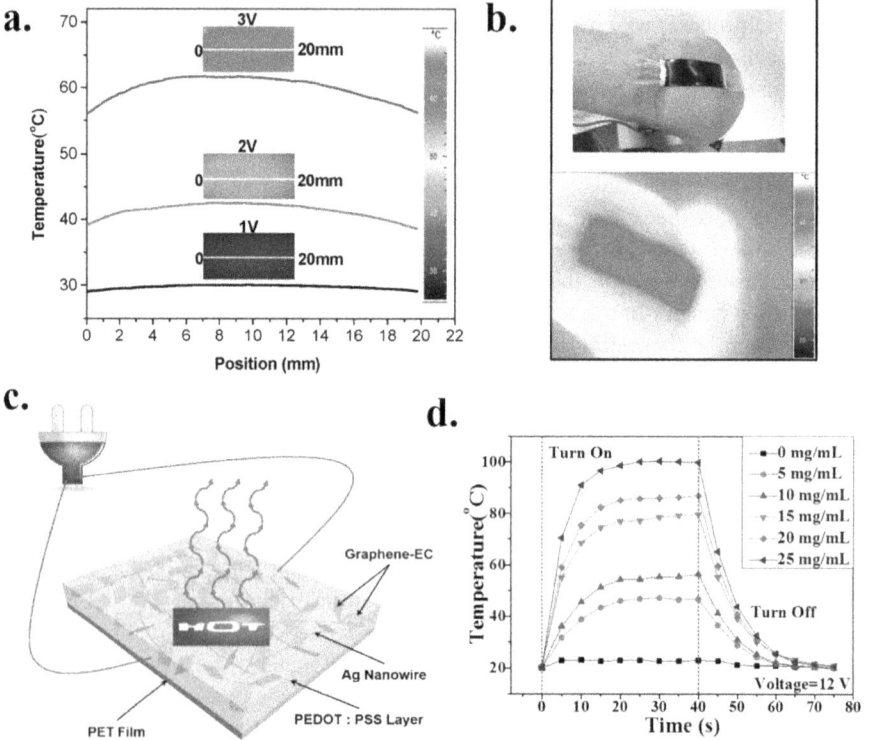

FIGURE 14.8 (a) Position-dependent temperature distribution profiles of rGO/PEDOT:PSS/ WPU nanocomposite-based thermotherapy pad at different applied voltages [The inset showing the IR images of nanocomposite pad at different applied voltages]. (b) Digital photograph of rGO/PEDOT:PSS/WPU nanocomposite-based thermotherapy pad attached to the human wrist in bending condition (Zhou et al., 2017). (Reproduced with permissions © 2017 Royal society of chemistry.). (c) Schematic illustration of the modified rGO/Ag NW/PEDOT:PSS nanocomposite-based heater. (d) Temperature-time dependent curves of modified rGO/Ag NW/PEDOT:PSS nanocomposite-based heater as a function of Ag nanowire loading (Cao et al., 2018). (Reproduced with permissions © 2017 American Chemical Society.)

The nanocomposite film exhibited uniform heat distribution and the same performance level at different humidity and strain conditions. The high performance of the nanocomposite film was attributed to the high electrical conductivity and mechanical properties, which could be a potential material as wearable thermotherapy pads.

In another study, to achieve a high degree of dispersion and control the size, rGO was modified with ethyl cellulose. Further, transparent, flexible with uniform dispersed nanocomposite film comprising modified rGO/Ag NW/PEDOT:PSS was prepared via spin coating process. The presence of modified rGO forms a strong hydrogen bonding interaction with PSS chains leading to a weakening of the Columbic interaction between PEDOT and PSS. Further, the interaction may help in the rearrangement of the PEDOT and facilitate more conductive paths for the charge carriers. The modified rGO/Ag NW/PEDOT:PSS nanocomposite film heater showed excellent stability, quick response and offered steady-state values under an applied voltage of 12 V. A schematic illustration of the modified rGO/Ag NW/PEDOT:PSS nanocomposite-based heater is shown in Figure 14.8c. The performance of the nanocomposite heater was sensitive to the Ag nanowire loading. The time required to reach the steady state was shortened and the steady-state temperature values increased from 46°C to 99.8°C upon increasing the Ag nanowire concentration from 5 to 25 mg mL^{-1} (Figure 14.8d). Also, the temperature increased proportionally to the voltage change. Such ability of modified rGO/Ag NW/PEDOT:PSS nanocomposite-based heater is a viable alternative to traditional heaters in various applications (Cao et al., 2018).

14.5 PIEZORESISTIVE SENSORS BASED ON GRAPHENE/ ELASTOMER NANOCOMPOSITE

The piezoresistive effect is among the most dominant sensing effects and widely employed in sensing applications besides thermoelectric and piezoelectric effects. The piezoresistive effect has been widely adopted in force sensors, strain sensors, pressure sensors, tactile sensors, etc. due to their low cost, small, low power consumption, simple readout and high-performance characteristics. In the last decade, the piezoresistive effect has been extensively explored with new methods and materials. To improve the piezoresistive effect, facile mechanisms, structures and phenomena have been explored. Recently, the synergism of physical effects and coupling method has been introduced to modulate the sensitivity of the piezoresistive effect. The basic working principle of the piezoresistive effect is based on the pressure/force applied to the piezoresistive material, which reflects as a change in the electrical resistance of the material. With nanotechnology and smart materials advancement, piezoresistive sensors made themselves a fascinating material for flexible, stretchable wearable electronics. Flexible piezoresistive sensors are basically comprised of two flexible electrodes along with micro- and nanoscale active structures. Owing to their excellent simplicity, facile readout, high sensitivity, linearity and flexibility, piezoresistive strain and pressure sensors are promising candidates for the future generation of smart wearable electronics. Elastomers filled with nanostructured conductive fillers have shown appreciable electrical, mechanical and sensing properties. A wide range of elastomers, including silicone, SBR, EPDM, TPU and PDMS, have been used to achieve high stretchability and flexibility. On the other hand, sensor materials such as CNT, graphene, carbon black, polyaniline, Ag nanoparticles and PVDF have been utilized. Graphene has been extensively used in piezoresistive material due to its low density, mechanical strength, electrical conductivity and flexibility (Dao et al., 2016; Dziuban et al., 1994; Gupta et al., 2021; Li et al., 2020; Xing et al., 2014). Flexible piezoresistive pressure sensor based on modified graphite/TPU composite developed via solution compounding method. The modified graphite/TPU composite exhibited a sensitivity of 0.275 kPa^{-1} and offered good stability and linearity. Given the flexibility and lightweight characteristics, modified graphite/TPU composite has great potential in wearable devices for pressure sensing, body motion detection, etc. (He et al., 2017). Graphene-wrapped PU sponge (rGO/PU) prepared by coating rGO onto the backbone of PU sponge via solution dipping process for the piezoresistive pressure sensor. To improve the sensitivity of the sensor, a dense and fractured microstructure was formed via hydrothermal process (140°C) and compressed (95%) for 2 hours. As a result, a three-dimensional cellulose-like network was formed. The sensor offered excellent sensitivity (0.26 kPa^{-1}) and durability (10,000 cycles) that makes it a potential candidate for artificial skins and wearable sensor applications (Yao et al., 2013). Guo et al. (2018) presented a simple, cost-effective fabrication of a wearable piezoresistive pressure sensor whose microstructure was bioinspired from crack-shaped mechano-sensory systems of spiders and beetles. rGO was developed on the surface of the TPU sponge by the supramolecular assembly. Further, polyaniline arrays were coated onto the surface of rGO/TPU via in situ chemical oxidative polymerization process to mimic the wing-locking sensing system of beetles. rGO/polyaniline/TPU sponges can be easily installed for the comprehensive and real-time health care monitoring of pressure (27 Pa to 25 kPa) and offer excellent response-recovery time. The sensor showed a great potential to detect the early stages of Parkinson's diseases. Further, such microstructures can also be assembled as artificial skins for mapping pressure distribution, location, etc.

Niu et al. (2018) reported that graphene nanoplatelets/PDMS nanocomposite was fabricated via two-step mechanical mixing for piezoresistive strain and pressure detection. The graphene nanoplatelets/PDMS nanocomposite-based sensor showed a positive piezoresistive characteristic with a strain sensitivity up to 140%. On the other hand, the piezoresistive pressure sensor showed negative piezoresistive characteristics with pressure sensitivity up to 260. The pressure sensitivity of the nanocomposite deteriorated when the sensor was loaded above 0.6 N, which could be attributed to the destruction of conductive networks. In addition, the sensor was attached to the human fingers to detect its bending motion. Upon bending, the sensor was able to respond distinctly and was seen to be sensitive to a subtle change in input.

Nanofiber aerogels possess high porosity with an excellent surface area and aspect ratio. This would facilitate hierarchical self-entanglement that forms a 3D nanofiber-based network and provide the foundation for the interconnection of conductive fillers under compression. With the above motivation, it is able to realize a highly sensitive piezoresistive pressure sensor. However, the poor mechanical properties could hinder its realization as pressure sensors. Incorporating the carbon nanostructures could enhance the mechanical properties of the fiber aerogel and facilitate 3D structure with mechanical integrity. Zhong et al. (2019) fabricated 3D rGO/polyolefin nanofiber composite aerogel by the chemical reducing process. Prior to the reduction,

GO/polyolefin nanofiber composite aerogel was obtained via freeze-drying. Upon the addition of rGO, the compressive strength of the composites improved. The composite exhibited excellent sensitivity of 223 kPa^{-1} with a wide range of working pressure and durability (Zhong et al., 2019).

Tan et al. (2020) reported graphene nanoribbons/electrospun TPU/TPU-boron nitride-based sandwiched composite structure as a highly stretchable and breathable sensor with an advance thermal management system. The structure showed excellent stretchability (>100%) with excellent electrical properties and a gauge factor of 35.7. With the formation of a conductive network, the sandwich structure can be readily used for human motion detection. The presence of boron nitride/TPU nanocomposite ensured rapid dissipation of thermal heat. Further, the sandwich structure-based sensor was tested for real-time monitoring of human motion. The sensor was attached to different parts of the body, including fingers, knee, elbow, etc. The sensor precisely detected the change in the motion in the form of electrical resistance change and showed excellent reproducibility in signal values. Further, the applicability of the sensor was demonstrated by attaching the sensor to the professional dragon boat paddler (Figure 14.9a). The paddler usually undergoes injuries related to sprain/strain in the upper body, shoulder joints and knee due to inappropriate paddling movement. Hence, the paddling movements were monitored by attaching the sensor at different points of the body. The sensor was intelligent enough to provide distinct signals for different movements made by the paddler, which helped accurately predict an paddler's improper paddling. Figure 14.9b and c shows the electrical resistance change in the sensor when the paddler makes the shoulder movement. Two peaks appear when the paddler stretches and rotates the shoulder joint. The response signals were also monitored when the paddler does a non-standard paddling. Interestingly, the sensor provided distinct signals when there was a change in the paddler's technique. The ability of the sensor to detect subtle changes in the paddler's technique could be of significant importance in preventing the paddler from injuries.

Khalifa et al. (2020) fabricated graphene/bio-based TPU nanocomposite thin films via solution cast process as a highly sensitive piezoresistive sensor for strain and pressure monitoring. Incorporation of graphene augmented the mechanical and electrical properties of the nanocomposite. The graphene/TPU nanocomposite thin film-based sensor was tested for its piezoresistive response by applying stress through finger pressure, bending stress and finger touch. The sensor was able to respond to the various modes of input pressure distinctly. The sensor was further attached to the fingers to monitor its motion (Figure 14.9d). Upon bending the finger at a particular angle, the sensor showed a discrete change in the electrical resistance. The sensor showed excellent sensitivity of 11 besides excellent reproducibility even after undergoing 10,000 cycles. Figure 14.9e shows the mechanism of pressure sensing of graphene/bio-based TPU nanocomposite-based sensor. When the pressure was applied to the sensor, the graphene particles came close to each other and the contact between the adjacent particles formed conductive networks. As a result, the electrical conductivity decreased. When the pressure was removed, the conductive network was disrupted and returned to its initial resistance value. The combination of bio-based matrix and graphene with improved electrical and mechanical properties could be a potential material in flexible electronics and human health monitoring.

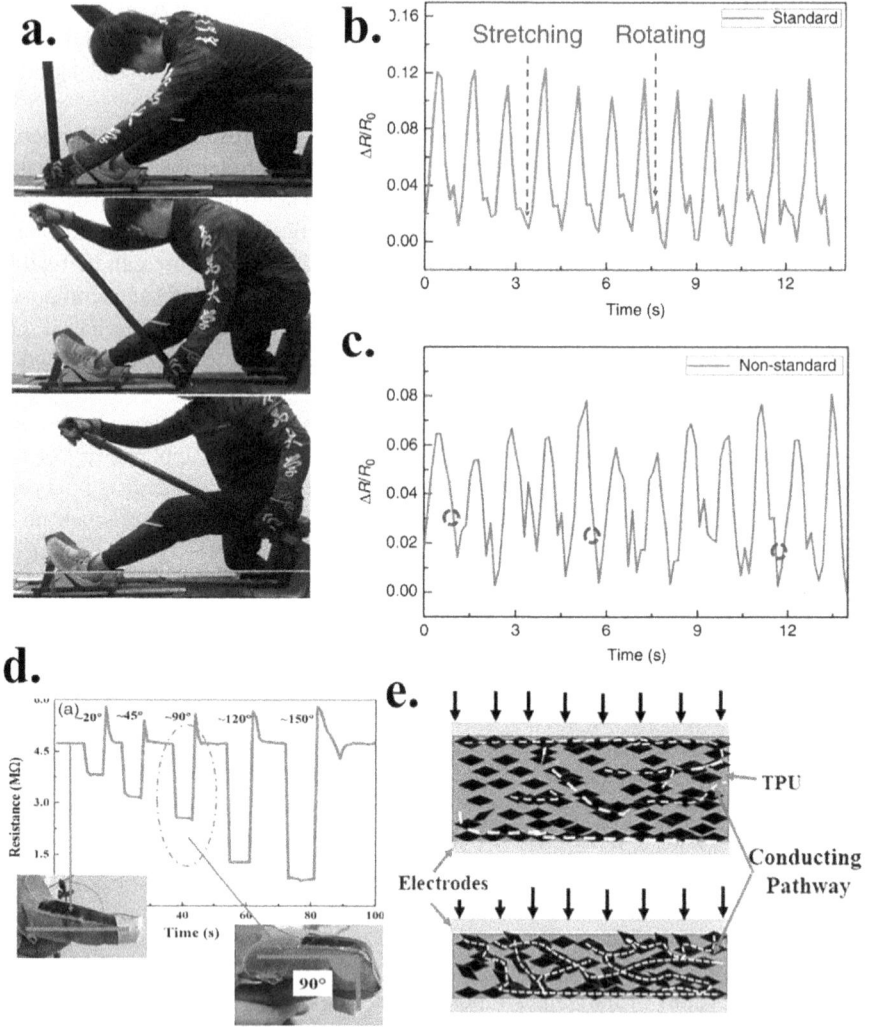

FIGURE 14.9 (a) Digital photograph depicting the movements of paddler made during dragon boat paddling. Monitoring of movements of the paddler using graphene nanoribbons/electrospun TPU/TPU-boron nitride based sandwiched composite structure based sensor attached to the paddler joints. (b) standard paddling and (c) non-standard paddling (Tan et al., 2020). (Reproduced with permissions © copyright 2020 Springer Nature). (d) Graphene/bio-based TPU nanocomposite-based sensor demonstrating the sensitivity upon bending at different angles attached to human finger. (e) Plausible tunneling mechanism responsible for the piezoresistive effect (Khalifa et al., 2020). (Reproduced with permissions © copyright 2020 Wiley publications.)

14.6 SUMMARY

To summarize, graphene/elastomer nanocomposites are a fascinating material for wearable sensors that have good compatibility, flexibility, stretchability, lightweight and desirable performance. The demand for the development of graphene/elastomer nanocomposite with low content of graphene is still a challenge. On the other hand, agglomeration of graphene is still an issue to be embarked on to achieving desirable sensor performance. Current studies have shown that tailoring the conductive network with special designs, geometries and special processing methods not only improve the necessary stretchability but also offer high sensitivity of the sensor. There are several reports that reported high performance strain and piezoresistive pressure sensors with excellent wearable characteristics. Further, self-healable graphene/elastomer composite based sensors provide necessary stability and durability for the strain sensors. On the one hand, temperature, glucose and humidity sensors are required to fill the gap to reach the desirable performance as wearable sensors. On the other hand, the major challenge with elastomer-graphene nanocomposite-based sensors is its selective response, which requires significant focus. Overall, graphene/elastomer nanocomposites demonstrate as a promising material for wearable application for human health monitoring system.

ACKNOWLEDGMENT

M.K. would like to thank Kompetenzzentrum Holz GmbH, Austria for their constant support during the preparation of this book chapter.

REFERENCES

Avilés, F., Oliva, A., Andreas, I., & Cen-Puc, M. (2018). Piezoresistivity, strain, and damage self-sensing of polymer composites filled with carbon nanostructures. *Advanced Engineering Materials*, 20(7), 1–23. doi:10.1002/adem.201701159.

Bauhofer, W., & Kovacs, J. (2009). A review and analysis of electrical percolation in carbon nanotube polymer composites. *Composites Science and Technology*, 69(10), 1486–1498. doi:10.1016/j.compscitech.2008.06.018.

Borini, S., White, R., Wei, D., Astley, M., Haque, S., Spigone, E., Nadine, H., Jani, K., Ryhänen, T. (2013). Ultrafast graphene oxide humidity sensors. *ACS Nano*, 7(12), 11166–11173. doi:10.1021/nn404889b.

Cao, M., Wang, M., Li, L., Qiu, H., & Yang, Z. (2018). Effect of graphene-EC on Ag NW-based transparent film heaters: Optimizing the stability and heat dispersion of films. *ACS Applied Materials and Interfaces*, 10(1), 1077–1083. doi:10.1021/acsami.7b14820.

Cataldi, P., Ceseracciu, L., Marras, S., Athanassiou, A., & Bayer, I. (2017). Electrical conductivity enhancement in thermoplastic polyurethane-graphene nanoplatelet composites by stretch-release cycles. *Applied Physics Letters*, 110(12). doi:10.1063/1.4978865.

Chen, H., Bao, S., Ma, J., Wang, P., Lu, H., Oetomo, B., & Chen, W. (2019). A wearable daily respiration monitoring system using pdms-graphene compound tensile sensor for adult. *Proceedings of the Annual International Conference of the IEEE Engineering in Medicine and Biology Society, EMBS*, 1269–1273. doi:10.1109/EMBC.2019.8857144.

Costa, P., Gonçalves, S., Mora, H., Carabineiro, S. A.C., Viana, J. C., & Lanceros-Mendez, S. (2019). Highly sensitive piezoresistive graphene-based stretchable composites for sensing applications. *ACS Applied Materials and Interfaces*, 11(49), 46286–46295. doi:10.1021/acsami.9b19294.

Costa, P., Silva, J., Ansón-Casaos, A., Martinez, T., Abad, J., Viana, J., & Lanceros-Mendez, S. (2014). Effect of carbon nanotube type and functionalization on the electrical, thermal, mechanical and electromechanical properties of carbon nanotube/styrene-butadiene-styrene composites for large strain sensor applications. *Composites Part B: Engineering*, *61*, 136–146. doi:10.1016/j.compositesb.2014.01.048.

Dao, V., Phan, P., Qamar, A., & Dinh, T. (2016). Piezoresistive effect of p-type single crystalline 3C-SiC on (111) plane. *RSC Advances*, *6*. doi:10.1039/c5ra28164d.

Dhakal, R., Lamichhane, P., Mishra, K., Nelson, L., & Vaidyanathan, K. (2019). Influence of graphene reinforcement in conductive polymer: Synthesis and characterization. *Polymers for Advanced Technologies*, *30*(9), 2172–2182. doi:10.1002/pat.4650.

Dziuban, J., Górecka-Drzazga, A., Lipowicz, U., Indyka, W., & Wąsowski, J. (1994). Self-compensating piezoresistive pressure sensor. *Sensors and Actuators: A. Physical*, *42*(1–3), 368–374. doi:10.1016/0924–4247(94)80014–6.

Filleter, T., McChesney, L., Bostwick, A., Rotenberg, E., Emtsev, V., Seyller, T., Horn, K., Bennewitz, R. (2009). Friction and dissipation in epitaxial graphene films. *Physical Review Letters*, *102*(8), 1–4. doi:10.1103/PhysRevLett.102.086102.

Gong, L., Wang, X., Zhang, D., Ma, X., & Yu, S. (2021). Flexible wearable humidity sensor based on cerium oxide/graphitic carbon nitride nanocomposite self-powered by motion-driven alternator and its application for human physiological detection. *Journal of Materials Chemistry A*, *9*(9), 5619–5629. doi:10.1039/d0ta11578a.

Guo, T., Li, C., Yang, J., Wang, P., Yue, J., Huang, X., Huang, X., Wang, J., Tang, Z. (2020). Holey, anti-impact and resilient thermoplastic urethane/carbon nanotubes fabricated by a low-cost "vapor induced phase separation" strategy for the detection of human motions. *Composites Part A: Applied Science and Manufacturing*, *136*(May), 105974–105983. doi:10.1016/j.compositesa.2020.105974.

Guo, Y., Guo, Z., Zhong, M., Wan, P., Zhang, W., & Zhang, L. (2018). A flexible wearable pressure sensor with bioinspired microcrack and interlocking for full-range human–machine interfacing. *Small*, *14*(44), 1–9. doi:10.1002/smll.201803018.

Gupta, L., Singh, G., & Pandey, V. (2021). A study of piezoresistive pressure sensor technology. *AIP Conference Proceedings*, *2327*(February). doi:10.1063/5.0039425.

Han, S., Meng, Q., Chand, A., Wang, S., Li, X., Kang, H., & Liu, T. (2019). A comparative study of two graphene based elastomeric composite sensors. *Polymer Testing*, *80*(July), 106106–106112. doi:10.1016/j.polymertesting.2019.106106.

He, Y., Li, W., Yang, G., Liu, H., Lu, J., Zheng, T., & Li, X. (2017). A novel method for fabricating wearable, piezoresistive, and pressure sensors based on modified-graphite/polyurethane composite films. *Materials*, *10*(7), 684–699. doi:10.3390/ma10070684.

He, Z., Zhou, G., Byun, J. H., Lee, S. K., Um, M. K., Park, B., Kim, T., Lee, S., Chou, T. (2019). Highly stretchable multi-walled carbon nanotube/thermoplastic polyurethane composite fibers for ultrasensitive, wearable strain sensors. *Nanoscale*, *11*(13), 5884–5890. doi:10.1039/C9NR01005J.

Hong, S. Y., Lee, Y. H., Park, H., Jin, S. W., Jeong, Y. R., Yun, J., You, I., Zi, G., Ha, J. (2016). Stretchable active matrix temperature sensor array of polyaniline nanofibers for electronic skin. *Advanced Materials*, *28*(5), 930–935. doi:10.1002/adma.201504659.

Hudec, R., Matuska, S., Kamencay, P., & Hudecova, L. (2020). Concept of a wearable temperature sensor for intelligent textile. *Advances in Electrical and Electronic Engineering*, *18*(2), 92–98. doi:10.15598/aeee.v18i2.3610.

Jang, S., Kim, H., Ha, H., Jung, S. H., Lee, M., & Kim, J. (2017). Simple approach to high-performance stretchable heaters based on kirigami patterning of conductive paper for wearable thermotherapy applications. *ACS Applied Materials and Interfaces*, *9*(23), 19612–19621. doi:10.1021/acsami.7b03474.

Jeon, J., Lee, H., & Bao, Z. (2013). Flexible wireless temperature sensors based on Ni microparticle-filled binary polymer composites. *Advanced Materials*, *25*(6), 850–855. doi:10.1002/adma.201204082.

Ji, Y. H., Liu, Y., Li, Y. Q., Xiao, H. M., Du, S. Sen, Zhang, J., Hu, N., Fu, S. (2016). Significantly enhanced electrical conductivity of silver nanowire/polyurethane composites via graphene oxide as novel dispersant. *Composites Science and Technology*, *132*, 57–67. doi:10.1016/j.compscitech.2016.07.002.

Jia, Y., Yue, X., Wang, Y., Yan, C., Zheng, G., Dai, K., Shen, C. (2020). Multifunctional stretchable strain sensor based on polydopamine/ reduced graphene oxide/ electrospun thermoplastic polyurethane fibrous mats for human motion detection and environment monitoring. *Composites Part B: Engineering*, *183*(December 2019), 107696. doi:10.1016/j.compositesb.2019.107696.

Jing, X., Mi, Y., Salick, M. R., Peng, F., & Turng, S. (2014). Preparation of thermoplastic polyurethane/graphene oxide composite scaffolds by thermally induced phase separation. *Polymer Composites*, *35*(7), 1408–1417. doi:10.1002/pc.22793.

Kantoch, E., Augustyniak, P., Markiewicz, M., & Prusak, D. (2014). Monitoring activities of daily living based on wearable wireless body sensor network. *2014 36th Annual International Conference of the IEEE Engineering in Medicine and Biology Society, EMBC 2014*, 586–589. doi:10.1109/EMBC.2014.6943659.

Khalifa, M., Ekbote, G. S., Anandhan, S., Wuzella, G., Lammer, H., & Mahendran, A. R. (2020). Physicochemical characteristics of bio-based thermoplastic polyurethane/ graphene nanocomposite for piezoresistive strain sensor. *Journal of Applied Polymer Science*, *137*(44), 0–11. doi:10.1002/app.49364.

Khalifa, M., Wuzella, G., Lammer, H., & Mahendran, A. R. (2020). Smart paper from graphene coated cellulose for high-performance humidity and piezoresistive force sensor. *Synthetic Metals*, *266*(May), 116420–116426. doi:10.1016/j.synthmet.2020.116420.

Kim, H., Seo, M., Kim, W., Kwon, D. K., Choi, E., Kim, W., & Myoung, M. (2019). Highly stretchable and wearable thermotherapy pad with micropatterned thermochromic display based on ag nanowire–single-walled carbon nanotube composite. *Advanced Functional Materials*, *29*(24), 1–11. doi:10.1002/adfm.201901061.

Lai, C., Mei, J., Jia, Y., Li, C., You, X., & Bao, Z. (2016). A stiff and healable polymer based on dynamic-covalent boroxine bonds. *Advanced Materials*, *28*(37), 8277–8282. doi:10.1002/adma.201602332.

Lee, J., Kim, S., Lee, J., Yang, D., Park, B. C., Ryu, S., & Park, I. (2014). A stretchable strain sensor based on a metal nanoparticle thin film for human motion detection. *Nanoscale*, *6*(20), 11932–11939. doi:10.1039/c4nr03295k.

Li, C. H., Wang, C., Keplinger, C., Zuo, J. L., Jin, L., Sun, Y., Zheng, P., Cao, Y., Lissel, F., Linder, C., You, Z., Bao, Z. (2016). A highly stretchable autonomous self-healing elastomer. *Nature Chemistry*, *8*(6), 618–624. doi:10.1038/nchem.2492.

Li, H., Zhang, J., Jiang, C., Liu, Q., Zhang, M., & Zhou, J. (2020). Muscle temperature sensing and control with a wearable device for hand rehabilitation of people after stroke. *International Conference on Sensing, Measurement and Data Analytics in the Era of Artificial Intelligence, ICSMD 2020- Proceedings*, 94–99. doi:10.1109/ ICSMD50554.2020.9261634.

Li, J., Fang, L., Sun, B., Li, X., & Kang, S. H. (2020). Review—Recent progress in flexible and stretchable piezoresistive sensors and their applications. *Journal of The Electrochemical Society*, *167*(3), 037561. doi:10.1149/1945-7111/ab6828.

Liang, R., Luo, A., Zhang, Z., Li, Z., Han, C., & Wu, W. (2020). Research progress of graphene-based flexible humidity sensor. *Sensors (Switzerland)*, *20*(19), 1–17. doi:10.3390/ s20195601.

Liu, C., Han, S., Du, Z., Liu, Y., & Liu, C. (2018). Highly sensitive wearable strain sensors using copper nanowires and elastomers. *2018 International Conference on Electronics Packaging and IMAPS All Asia Conference, ICEP-IAAC 2018*, 333–338. doi:10.23919/ICEP.2018.8374317.

Liu, H., Li, Q., Zhang, S., Yin, R., Liu, X., He, Y., … Guo, Z. (2018). Electrically conductive polymer composites for smart flexible strain sensors: A critical review. *Journal of Materials Chemistry C*, 6(45), 12121–12141. doi:10.1039/C8TC04079F.

Liu, H., Li, Y., Dai, K., Zheng, G., Liu, C., Shen, C., … Guo, Z. (2015). Electrically conductive thermoplastic elastomer nanocomposites at ultralow graphene loading levels for strain sensor applications. *Journal of Materials Chemistry C*, 4(1), 157–166. doi:10.1039/c5tc02751a.

Liu, Q., Tian, B., Liang, J., & Wu, W. (2021). Recent advances in printed flexible heaters for portable and wearable thermal management. *Materials Horizons*, 8(6), 1634–1656. doi:10.1039/d0mh01950j.

Liu, S., Lin, Y., Wei, Y., Chen, S., Zhu, J., & Liu, L. (2017). A high performance self-healing strain sensor with synergetic networks of poly(ε-caprolactone) microspheres, graphene and silver nanowires. *Composites Science and Technology*, 146, 110–118. doi:10.1016/j.compscitech.2017.03.044.

Lü, X., Yu, T., Meng, F., & Bao, W. (2021). Wide-range and high-stability flexible conductive graphene/thermoplastic polyurethane foam for piezoresistive sensor applications. *Advanced Materials Technologies*, 6(10), 1–13. doi:10.1002/admt.202100248.

Luan, H., Tien, H. N., Cuong, V., Kong, B. S., Chung, S., Kim, J., & Hur, H. (2012). Novel conductive epoxy composites composed of 2-D chemically reduced graphene and 1-D silver nanowire hybrid fillers. *Journal of Materials Chemistry*, 22(17), 8649–8653. doi:10.1039/c2jm16910j.

Ma, Z., Li, H., Jing, X., Liu, Y., & Mi, Y. (2021). Recent advancements in self-healing composite elastomers for flexible strain sensors: Materials, healing systems, and features. *Sensors and Actuators, A: Physical*, 329, 112800. doi:10.1016/j.sna.2021.112800.

Maiti, M., Bhattacharya, M., & Bhowmick, A. K. (2008). Elastomer nanocomposites. *Rubber Chemistry and Technology*, 81(3), 384–469. doi:10.5254/1.3548215.

Nardo, N. R. (1987). End use applications for thermoplastic polyurethane elastomers. *Journal of Elastomers and Plastics*, 19. doi:10.1177/009524438701900106.

Ni, N., & Zhang, L. (2017). Dielectric elastomer sensors. *Elastomers*. doi:10.5772/intechopen.68995.

Niu, D., Jiang, W., Ye, G., Wang, K., Yin, L., Shi, Y., … Liu, H. (2018). Graphene-elastomer nanocomposites based flexible piezoresistive sensors for strain and pressure detection. *Materials Research Bulletin*, 102, 92–99. doi:10.1016/j.materresbull.2018.02.005.

Noh, J. S. (2016). Conductive elastomers for stretchable electronics, sensors and energy harvesters. *Polymers*, 8(4). doi:10.3390/polym8040123.

Oh, J. Y., Jun, G. H., Jin, S., Ryu, H. J., & Hong, S. H. (2016). Enhanced electrical networks of stretchable conductors with small fraction of carbon nanotube/graphene hybrid fillers. *ACS Applied Materials and Interfaces*, 8(5), 3319–3325. doi:10.1021/acsami.5b11205.

Pang, H., Piao, Y. Y., Xu, L., Bao, Y., Cui, C. H., Fu, Q., & Li, Z. M. (2013). Tunable liquid sensing performance of conducting carbon nanotube-polyethylene composites with a porous segregated structure. *RSC Advances*, 3(43), 19802–19806. doi:10.1039/c3ra43375g.

Park, E. B., Yazdi, M., & Lee, H. (2020). Development of wearable temperature sensor based on peltier thermoelectric device to change human body temperature. *Sensors and Materials*, 32(9), 2959–2970. doi:10.18494/SAM.2020.2741.

Park, J., Jeon, Y., Kang, B. C., & Ha, J. (2021). Wearable temperature sensors based on lanthanum-doped aluminum-oxide dielectrics operating at low-voltage and high-frequency for healthcare monitoring systems. *Ceramics International*, 47(4), 4579–4586. doi:10.1016/j.ceramint.2020.10.023.

Payandehpeyman, J., Mazaheri, M., & Khamehchi, M. (2020). Prediction of electrical conductivity of polymer-graphene nanocomposites by developing an analytical model considering interphase, tunneling and geometry effects. *Composites Communications*, *21*(March), 100364. doi:10.1016/j.coco.2020.100364.

Pu, Z., Zou, C., Wang, R., Lai, X., Yu, H., Xu, K., & Li, D. (2016). A continuous glucose monitoring device by graphene modified electrochemical sensor in microfluidic system. *Biomicrofluidics*, *10*(1), 1–11. doi:10.1063/1.4942437.

Sadasivuni, K. K., Ponnamma, D., Kim, J., & Thomas, S. (2015). Graphene-based polymer nanocomposites in electronics. *Graphene-Based Polymer Nanocomposites in Electronics*, (April), 1–382. doi:10.1007/978-3-319-13875-6.

Sang, G., Xu, P., Yan, T., Murugadoss, V., Naik, N., Ding, Y., & Guo, Z. (2021). Interface engineered microcellular magnetic conductive polyurethane nanocomposite foams for electromagnetic interference shielding. *Nano-Micro Letters*, *13*(1), 1–16. doi:10.1007/s40820-021-00677-5.

Sedlačík, M. (2021). Advances in elastomers. *Materials*, *14*. doi:10.3390/ma14020348.

Senior, M. (2014). Novartis signs up for Google smart lens. *Nature Biotechnology*, *32*(9), 856. doi:10.1038/nbt0914–856.

Shyu, T. C., Damasceno, F., Dodd, M., Lamoureux, A., Xu, L., Shlian, M., Shtein, M., Glotzer, S., Kotov, N. A. (2015). A kirigami approach to engineering elasticity in nanocomposites through patterned defects. *Nature Materials*, *14*(8), 785–789. doi:10.1038/nmat4327.

Soni, M., Bhattacharjee, M., Ntagios, M., & Dahiya, R. (2020). Printed temperature sensor based on PEDOT: PSS-graphene oxide composite. *IEEE Sensors Journal*, *20*(14), 7525–7531. doi:10.1109/JSEN.2020.2969667.

Tallman, T. N. (2018). Damage detection in nanofiller-modified composites with external circuitry via resonant frequency shifts. *ASME 2018 Conference on Smart Materials, Adaptive Structures and Intelligent Systems, SMASIS 2018*, *2*(December). doi:10.1115/SMASIS2018–8008.

Tan, C., Dong, Z., Li, Y., Zhao, H., Huang, X., Zhou, Z., … Sun, B. (2020). A high performance wearable strain sensor with advanced thermal management for motion monitoring. *Nature Communications*, *11*(1), 1–10. doi:10.1038/s41467-020-17301-6.

Tee, C. K., Wang, C., Allen, R., & Bao, Z. (2012). An electrically and mechanically self-healing composite with pressure- and flexion-sensitive properties for electronic skin applications. *Nature Nanotechnology*, *7*(12), 825–832. doi:10.1038/nnano.2012.192.

Tien, T., Jeon, S., Kim, D. Il, Trung, T. Q., Jang, M., Hwang, B. U., … Park, J. J. (2014). A flexible bimodal sensor array for simultaneous sensing of pressure and temperature. *Advanced Materials*, *26*(5), 796–804. doi:10.1002/adma.201302869.

Tierney, M. J., Tamada, J. A., Potts, R. O., Jovanovic, L., & Garg, S. (2001). Clinical evaluation of the GlucoWatch® biographer: A continual, non-invasive glucose monitor for patients with diabetes. *Biosensors and Bioelectronics*, *16*(9–12), 621–629. doi:10.1016/S0956–5663(01)00189–0.

Toi, P. T., Trung, T. Q., Dang, T. M. L., Bae, C. W., & Lee, N. E. (2019). Highly electrocatalytic, durable, and stretchable nanohybrid fiber for on-body sweat glucose detection [Research-article]. *ACS Applied Materials and Interfaces*, *11*(11), 10707–10717. doi:10.1021/acsami.8b20583.

Trung, Q., Dang, L., Ramasundaram, S., Toi, T., Park, Y., & Lee, E. (2019). A stretchable strain-insensitive temperature sensor based on free-standing elastomeric composite fibers for on-body monitoring of skin temperature. *ACS Applied Materials and Interfaces*, *11*(2), 2317–2327. doi:10.1021/acsami.8b19425.

Trung, Q., Duy, T., Ramasundaram, S., & Lee, E. (2017). Transparent, stretchable, and rapid-response humidity sensor for body-attachable wearable electronics. *Nano Research*, *10*(6), 2021–2033. doi:10.1007/s12274-016-1389-y.

Tung, T., Nine, J., Krebsz, M., Pasinszki, T., Coghlan, J., Tran, H., & Losic, D. (2017). Recent advances in sensing applications of graphene assemblies and their composites. *Advanced Functional Materials*, *27*(46), 1–57. doi:10.1002/adfm.201702891.

Verma, M., Chauhan, S., Dhawan, K., & Choudhary, V. (2017). Graphene nanoplatelets/ carbon nanotubes/polyurethane composites as efficient shield against electromagnetic polluting radiations. *Composites Part B: Engineering*, *120*, 118–127. doi:10.1016/j. compositesb.2017.03.068.

Vo, T., Lee, J., Kim, Y., & Suk, W. (2020). Synergistic effect of graphene/silver nanowire hybrid fillers on highly stretchable strain sensors based on spandex composites. *Nanomaterials*, *10*(10), 1–14. doi:10.3390/nano10102063.

Vuorinen, T., Niittynen, J., Kankkunen, T., Kraft, T. M., & Mäntysalo, M. (2016). Inkjet-printed graphene/PEDOT:PSS temperature sensors on a skin-conformable polyurethane substrate. *Scientific Reports*, *6*(September), 1–8. doi:10.1038/srep35289.

Wang, Z., Gao, W., Zhang, Q., Zheng, K., Xu, J., Xu, W., … Liu, Y. (2019). 3D-printed graphene/polydimethylsiloxane composites for stretchable and strain-insensitive temperature sensors. *ACS Applied Materials and Interfaces*, *11*(1), 1344–1352. doi:10.1021/ acsami.8b16139.

Wu, J., Han, S., Yang, T., Li, Z., Wu, Z., Gui, X., … Huo, F. (2018). Highly stretchable and transparent thermistor based on self-healing double network hydrogel. *ACS Applied Materials and Interfaces*, *10*(22), 19097–19105. doi:10.1021/acsami.8b03524.

Wu, J., Wu, Z., Lu, X., Han, S., Yang, B. R., Gui, X., … Liu, C. (2019). Ultrastretchable and stable strain sensors based on antifreezing and self-healing ionic organohydrogels for human motion monitoring [Research-article]. *ACS Applied Materials and Interfaces*, *11*(9), 9405–9414. doi:10.1021/acsami.8b20267.

Wu, S., Ladani, B., Zhang, J., Ghorbani, K., Zhang, X., Mouritz, A. P., Kinloch, G., Wang, H. (2016). Strain sensors with adjustable sensitivity by tailoring the microstructure of graphene aerogel/PDMS nanocomposites. *ACS Applied Materials and Interfaces*, *8*(37), 24853–24861. doi:10.1021/acsami.6b06012.

Wu, T., & Chen, B. (2017). Facile fabrication of porous conductive thermoplastic polyurethane nanocomposite films via solution casting. *Scientific Reports*, *7*(1), 1–11. doi:10.1038/ s41598-017-17647-w.

Xing, W., Tang, M., Wu, J., Huang, G., Li, H., Lei, Z., Fu, X., Li, H. (2014). Multifunctional properties of graphene/rubber nanocomposites fabricated by a modified latex compounding method. *Composites Science and Technology*, *99*(July), 67–74. doi:10.1016/j. compscitech.2014.05.011.

Yamada, T., Hayamizu, Y., Yamamoto, Y., Yomogida, Y., Izadi-Najafabadi, A., Futaba, D. N., & Hata, K. (2011). A stretchable carbon nanotube strain sensor for human-motion detection. *Nature Nanotechnology*, *6*(5), 296–301. doi:10.1038/nnano.2011.36.

Yan, C., Wang, J., & Lee, S. (2015). Stretchable graphene thermistor with tunable thermal index. *ACS Nano*, *9*(2), 2130–2137. doi:10.1021/nn507441c.

Yang, J., Wei, D., Tang, L., Song, X., Luo, W., Chu, J., Goa, T., Shi, H., Du, C. (2015). Wearable temperature sensor based on graphene nanowalls. *RSC Advances*, *5*(32), 25609–25615. doi:10.1039/c5ra00871a.

Yao, H. Bin, Ge, J., Wang, C. F., Wang, X., Hu, W., Zheng, Z. J., Ni, J., Yu, S. H. (2013). A flexible and highly pressure-sensitive graphene-polyurethane sponge based on fractured microstructure design. *Advanced Materials*, *25*(46), 6692–6698. doi:10.1002/ adma.201303041.

Yu, Y., Peng, S., Blanloeuil, P., Wu, S., & Wang, C. H. (2020). Wearable temperature sensors with enhanced sensitivity by engineering microcrack morphology in PEDOT:PSS-PDMS sensors. *ACS Applied Materials and Interfaces*, *12*(32), 36578–36588. doi:10.1021/acsami.0c07649.

Zhang, M., Yang, L., & Wang, Y. (2019). Conductive graphene/polydimethylsiloxane nano-composites for flexible strain sensors. *Journal of Materials Science: Materials in Electronics, 30*(21), 19319–19324. doi:10.1007/s10854-019-02292-y.

Zhao, J., Lin, Y., Wu, J., Nyein, Y., Bariya, M., Tai, C., Chao, M., Zhang, G., Fan, Z., Javey, A. (2019). A fully integrated and self-powered smartwatch for continuous sweat glucose monitoring. *ACS Sensors, 4*(7), 1925–1933. doi:10.1021/acssensors.9b00891.

Zhong, W., Jiang, H., Yang, L., Yadav, A., Ding, X., Chen, Y., Li, M., Sun, G., Wang, D. (2019). Ultra-sensitive piezo-resistive sensors constructed with Reduced Graphene Oxide/Polyolefin Elastomer (RGO/POE) nanofiber aerogels. *Polymers, 11*(11), 1–12. doi:10.3390/polym11111883.

Zhou, G., Byun, J. H., Oh, Y., Jung, B. M., Cha, H. J., Seong, D., Um, M., Hyun, S., Chou, T. W. (2017). Highly sensitive wearable textile-based humidity sensor made of high-strength, single-walled carbon nanotube/poly(vinyl alcohol) filaments. *ACS Applied Materials and Interfaces, 9*(5), 4788–4797. doi:10.1021/acsami.6b12448.

Zhou, R., Li, P., Fan, Z., Du, D., & Ouyang, J. (2017). Stretchable heaters with composites of an intrinsically conductive polymer, reduced graphene oxide and an elastomer for wearable thermotherapy. *Journal of Materials Chemistry C, 5*(6), 1544–1551. doi:10.1039/c6tc04849h.

15 Thermally Conducting Graphene-Elastomer Nanocomposites
Preparation, Properties, and Applications

Pritam V. Dhawale
Institute of Chemical Technology

Deepthi Anna David
Cochin University of Science and Technology (CUSAT)

Ajish Babu
Indian Institute of Technology Patna, Bihar

Peter Samora Owuor
Rice University

Leonardo Dantas Machado
Universidade Federal do Rio Grande do Norte

Vijay Kumar Thakur
Biorefining and Advanced Materials Research Centre

Jinu Jacob George and Prasanth Raghavan
Cochin University of Science and Technology (CUSAT)

CONTENTS

DOI: 10.1201/9781003200444-15

15.1 INTRODUCTION

Carbon materials are composed of carbon atoms with diverse structures and extraordinary properties. Carbon is capable of forming allotropes predominantly graphite and diamond based on the arrangement of carbon atoms and hybridisation. Other allotropes of carbon include graphene, fullerene (also known as buckminsterfullerene or bucky balls), carbon nanotube, and carbon quantum dots. Figure 15.1 schematically illustrates various allotropes of carbon.

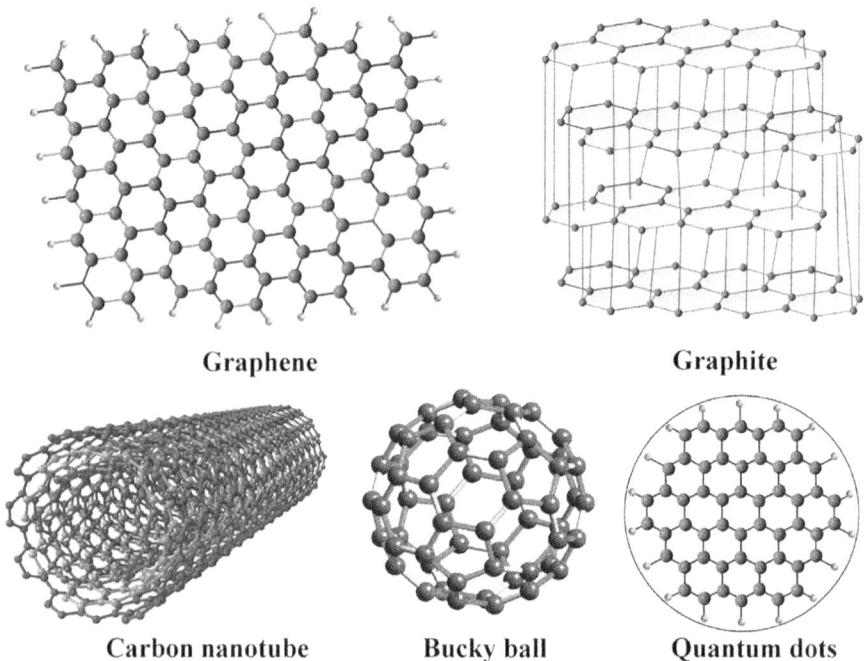

Graphene   Graphite

Carbon nanotube   Bucky ball   Quantum dots

FIGURE 15.1   Schematic illustration of different allotropes of carbon.

Graphene is one of the carbon allotropes with two-dimensional arrangement of sp^2 hybridised carbon atoms in single layer of hexagonal lattice. Figure 15.2 schematically illustrates the two-dimensional structure of graphene. Graphene is well-known in terms of excellent mechanical, gas barrier properties, electrical and thermal conductivity, while elastomers are widely accepted for their good elasticity, processability, and corrosion resistance. Moreover, graphene is the thinnest and strongest material reported. It is the most promising nanomaterials capable for wide range of commercial applications like transistors, batteries, supercapacitors, and solar cells. other vital properties of graphene include high elasticity, hardness, flexibility, and resistance.

The combination of these two materials to form graphene/elastomer composite gives rise to the synergistic effect of their individual properties for developing lightweight, flexible and conductive composites for versatile applications (Wang et al. 2017). High anisotropy of graphene enhances the heat flow through the polymer matrix by efficient heat transfer mechanism. Figure 15.3 schematically illustrates the heat flow in polymer, oriented graphene/polymer composite, and 3D graphene/polymer composite (Zhang and Zhang 2017).

Graphene can be synthesised by two broad approaches: top-down approach and bottom-up approach. The widely used method for the synthesis of graphene from a carbonaceous source in atomic level is the chemical vapour deposition (CVD) method in which the graphene is deposited on a transition metal catalytic substrate, mainly Ni, Cu, Ru, etc., or can be substrate-free. Other bottom-up approach methods include physical vapour deposition (PVD), epitaxial growth, and/or organic synthesis. In top-down approaches, exfoliation of precursors like graphite is the best method for the graphene synthesis. Exfoliation methods include mechanical exfoliation, liquid phase exfoliation (LPE), electrochemical exfoliation, and/or exfoliation of graphite intercalation compounds (GICs) (George and Bhowmick 2008a). Other top-down approaches include unzipping of carbon nanotubes (CNTs), arc discharge,

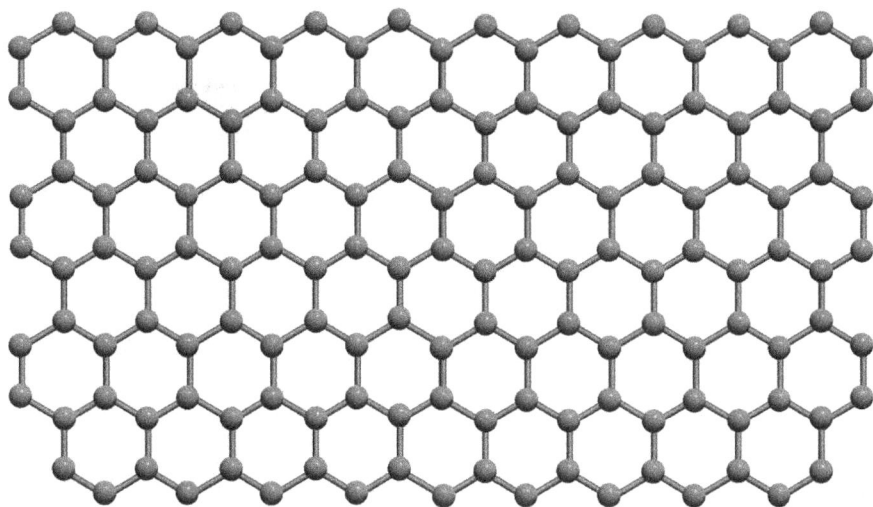

FIGURE 15.2 Schematic illustration of two-dimensional structure of graphene nanosheet.

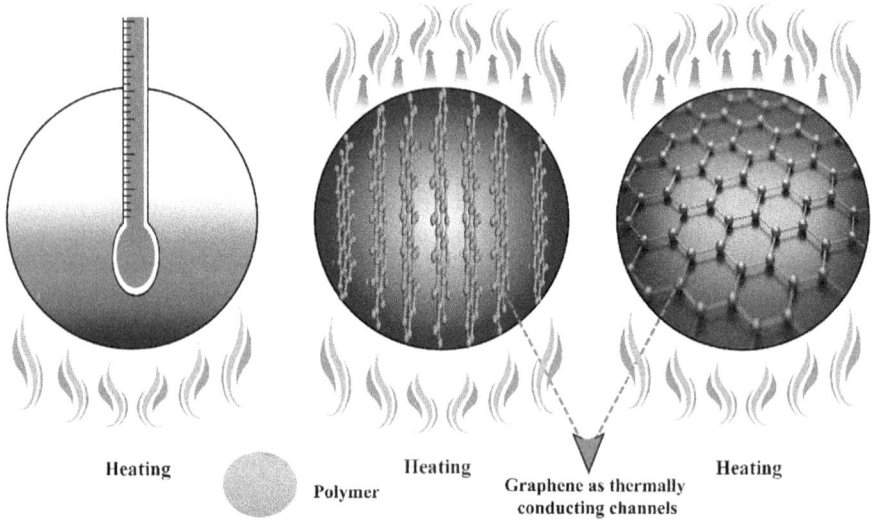

FIGURE 15.3 Schematic of thermal conductance in a polymer, oriented graphene/polymer composite, and 3D graphene/polymer composite.

and oxidation-exfoliation-reduction of graphite oxide. It is well known that phonons are the key basis for thermal transmission in polymers which is shown in Figure 15.4 (Yue, Yang, and Liao 2019).

Due to high phonon scattering, polymers have low thermal conductivity in contrast to metals. However, the incorporation of thermally conductive fillers in polymers could make sufficient thermally conductive pathways for achieving satisfying thermal conductivity. Moreover, their easy processability, electrical insulation, low coefficient of thermal expansion, high breakdown strength, and corrosion resistance make them quite relevant in many applications. In general, the fillers used for thermally conductive

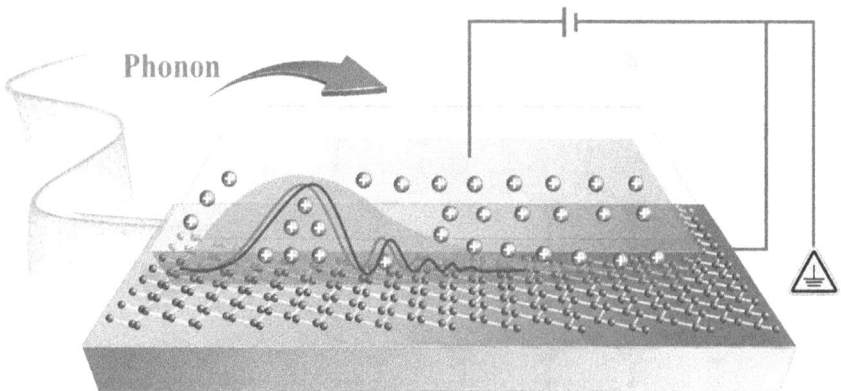

FIGURE 15.4 The concept of controlling phonon transport in 2D materials via externally induced phonon-electron scattering (Yue, Yang, and Liao 2019).

polymeric composites have relied on the mode of applications. For instance, in applications where electrical insulation is required along with high thermal conductivity (TC), fillers like boron nitride, aluminium nitride, and aluminium oxide are used. Graphite-derived and metal-based fillers are used where both thermal and electrical conductivity are required (Huang, Jiang, and Tanaka 2011; Sengupta et al. 2011). Even though polymers have a lower TC, the factors like orientation and structure of the polymer chains, crystallinity, liquid crystal domain size, and distribution and interactions of intermolecular bonds play a key role. For fillers, their type, morphology, size, shape, and electrical and thermal properties are important. Furthermore, the filler-filler and filler-polymer interfacial interactions are also highly significant for attaining better dispersibility to provide higher TC for polymer composites (Chen et al. 2016). Important factors to be considered for designing thermally conductive polymers are schematically shown in Figure 15.5 (Chen et al. 2016). However, developments in controlling the morphology of fillers, filler orientations, and synergistic effects of hybrid fillers are widely researched for fabricating enhanced thermal conductivity combining with other properties for polymer composites.

Unlike traditional fillers, graphene gained considerable attention due to its excellent properties among various thermally conductive filler particles. Despite its high thermal conductivity of 2,000–5,000 W m^{-1} K^{-1}, the adaptability of graphene into its derivatives, like graphene oxide (GO) and reduced GO (rGO), makes it appropriate for versatile applications (Balandin 2011). Moreover, the other notable advantages of graphene over other organic and inorganic fillers are as follows:

i. excellent thermoelectric property can be used for both thermal as well as electrically conductive applications;
ii. low filler loading, the better property even at low loadings;
iii. higher modulus and high strength, excellent mechanical properties;

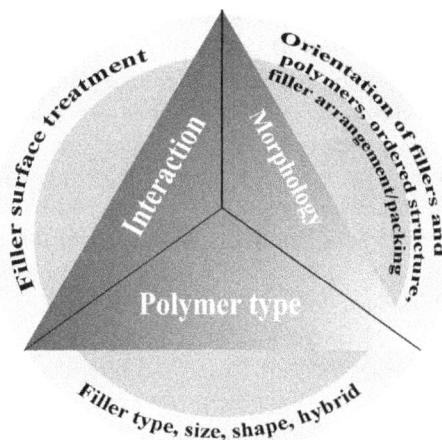

FIGURE 15.5 Schematic representation of factors to be considered while designing thermally conductive polymer composites.

iv. large specific surface area provides better interfacial interactions with matrix;
v. easily functionalised due to the presence of polar groups;
vi. environmental friendly dispersion methods (water-based) in rubber (Wang et al. 2017).

Composite materials with high thermal conductivity and electrically insulating property are exploited as thermal interface materials (TIMs) for heat dissipation in battery packs, electric vehicles, electronic devices, and solar cells to improve their safety (Niu et al. 2020). Graphene-based silicone rubber composites have been recently been used as TIMs (Zhang et al. 2019). The main functions of these composites such as high flexibility and stretchability are successfully applied in sensors (Han et al. 2019), electromagnetic shielding materials for radars and satellites (Al-Ghamdi et al. 2016), and drag reduction materials (Tian et al. 2017). Moreover, commercial rubber products like gloves, rubber hoses, sealants, and tyres, in a high-temperature environment, need a better heat transmission for prolonged service life (George et al. 2017).

In this chapter, a detailed study about thermally conductive graphene/elastomer composites is summarized. Starting from thermally conductive fillers used in elastomers, their limitations, and advantages of graphene as a filler in rubber matrix are discussed. Later, a critical study on various methods used for preparing graphene/elastomer composite and their specialities is summarised. Dedicated sections for thermally conductive composites reinforced with graphene and their comparison study with traditional fillers regarding thermal conductivity have been reviewed. Moreover, microstructural analysis of thermal transition mechanism through graphene interface matrix is discussed along with potential applications of graphene-elastomer composite.

15.2 THERMALLY CONDUCTIVE ELASTOMERIC COMPOSITES

The development of thermally conductive elastomeric composite materials has great significance especially for the design of electronic devices. Although the thermal conductivity of elastomers is quite low (0.15–0.4 W m^{-1} K^{-1}) (Kerschbaumer et al. 2019), the incorporation of fillers with high heat transmission ability is used to increase the thermal conductivity of elastomeric matrix. The role of conductive fillers is to reduce the thermal interface resistance between matrix and filler, as well as resistance in between fillers in composite through reducing the scattering of phonons or electrons (Ma et al. 2021). Early studies on polyurethane elastomer with carbon black, silica, oxides of aluminium, and zirconium fillers suggested a small improvement in thermal conductivity (Benli et al. 1998). More studies have been explored for designing thermally conductive elastomer composites using metallic, ceramic, carbon fillers and their hybrids (Liu et al. 2004; Sim et al. 2005; Zhou et al. 2007; Mu and Feng 2007).

Metals and metal oxides were widely used as a filler in elastomeric matrices due to their high heat transmission ability. Metals like Fe, Cu, brass, and Al have been reported as a filler in both particle and powder form with elastomers like styrene butadiene rubber (SBR), natural rubber (NR), nitrile rubber (NBR), and ethylene propylene diene monomer rubber (EPDM) (Shojaei, Fahimian, and Derakhshandeh 2007; Gwaily, Nasr, and Badawy 2001; Vinod et al. 2001; Vinod, Varghese, and Kuriakose 2004; Niu et al. 2012; Anuar, Mariatti, and Ismail 2007). Due to the incorporation and processing difficulties of metal particles-based elastomer composites, metal oxides like zinc oxide (ZnO) were also explored (Mu and Feng 2007). Recently, attempts were made to prepare soft thermally conductive elastomer composites without significantly reducing elastomer's elasticity. Incorporating liquid metal, like eutectic gallium–indium (EGa-In) microdroplets, in elastomers was observed to enable highly thermally conductive and stretchable elastomer composite (Bartlett et al. 2017). High TC unique properties of flexibility, biocompatibility, electrical conductivity, and low melting points of these composites have been explored widely in applications such as soft robotics, wearable computing, and biomedical therapy (Wang, Lu, and Rao 2021; Zhou et al. 2020).

For applications involving electrical insulation along with TC, ceramic fillers like alumina (with their different crystal forms α-Al_2O_3, γ-Al_2O_3, etc.), titania, zirconia, barium titanate, and strontium titanate are widely studied (Gwaily et al. 1995; Namitha and Sebastian 2017; Zhou et al. 2007). Boron nitride (BN) is well accepted regarding their high thermal conductivity and low coefficient of thermal expansion, which is suitable for TIM applications. Although the TC of BN-elastomer composite increases with BN composition, an optimum volume fraction of 40% was preferred (Zhou et al. 2007). Similarly, other ceramic fillers like aluminum nitride, silicon carbide, and silicon nitride were also explored (He, Chen, and Ma 2009; Ma et al. 2012; Cui et al. 2012). Figure 15.6 illustrates the heat flow in polymer nanocomposites with and without inter-filler resistance (Huang, Qian, and Yang 2018).

Carbon-based fillers have received a lot interests from researchers due to their multifunctional characteristics. With an exception of carbon black, other fillers like carbon fibres, graphite, diamond, CNT, and 1D fillers have found promising considerations for thermally conductive composite applications (Aguilar-Bolados, Yazdani-Pedram, and Verdejo 2020; Chen et al. 2016). Due to its large surface-to-volume ratio, graphite has been studied extensively for thermally conductive applications. At lower loadings, the formation of conductive networks results in enhanced TC of elastomeric

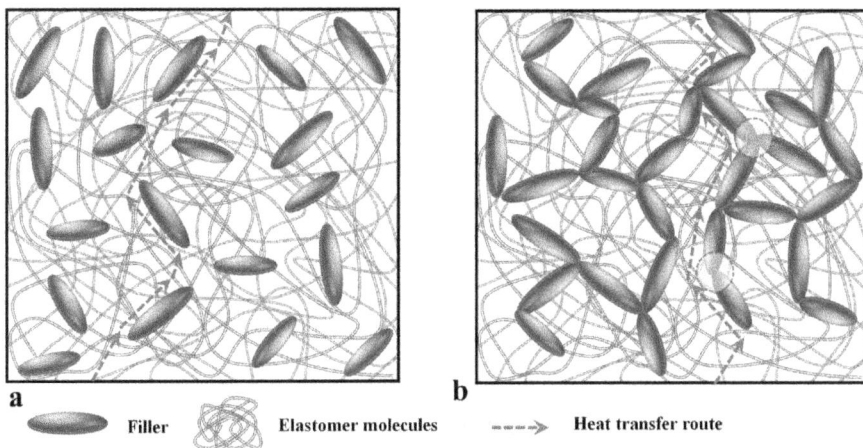

FIGURE 15.6 Schematic diagrams of polymer nanocomposites: (a) without inter-filler network and (b) with inter-filler networks. Thermally conductive pathway is identified with dashed lines (Huang, Qian, and Yang 2018).

composites (Mu and Feng 2007). CNT incorporation in elastomer has also been studied (Liu et al. 2004). Along with its high TC, other exceptional mechanical properties enable the design of CNT-filled composites for versatile applications. Recently, 2D nanomaterials like MXene have been studied for elastomeric composites and found to increase the TC even with small volume fractions (Aakyiir et al. 2020).

15.3 LIMITATIONS OF THERMALLY CONDUCTIVE ELASTOMERIC COMPOSITES

Even though progressive studies for the improvement of TC of elastomeric composites are underway with various filler types, their hybrids, optimal processing techniques, modification of filler and matrix for interfacial interactions, etc., many challenges are yet to be addressed. Some of the challenges that conventional, organic, or inorganic fillers face are the following: (i) simple mixing of ceramics in elastomer could not improve a satisfied TC, in contrast to the theoretically higher TC of nanomaterials like graphene; (ii) higher filler loadings are widely adapted mostly to increase TC, however, these could negatively impact the other desired properties as well as processing; (iii) good dispersion of fillers in rubber matrix is very necessary for improving interfacial interaction and is still a critical challenge; and (iv) high filler density, challenge for lightweight applications (Niu et al. 2020). However, it is believed that nanofillers like graphene can solve most of the above discussed challenges which enable the development of high thermally conductive elastomeric composites.

15.4 GRAPHENE ELASTOMERIC COMPOSITE PREPARATION

15.4.1 MELT MIXING

The widely used method for the preparation of graphene/elastomeric composite in which graphene is directly dispersed within the elastomeric matrix in its molten state, by applying shear force. Melt mixing is a cost-effective, faster, and eco-friendly method that can be used for mass production of composites and chemical modifications as well as compatibilisation. Furthermore, solvent is not required for the dispersion process since the procedure involves the melting of elastomeric matrix which can be done by injection moulding, extrusion, or even by internal mixing (Verma and Goh 2018). In contrast, poor dispersion of graphene, at higher loading, within the elastomeric matrix due to the enhanced viscosity, may affect the vital properties of the composite (Singh et al. 2011). Likewise, the composite may experience buckling/shortening of graphene due to strong mechanical shearing and elevated temperature may also favour the degradation process that will affect the thermal conductivity (Du and Cheng 2012).

15.4.2 TWO-ROLL MILLING/INTERNAL MIXING

A milling process in which the mechanical shear force is applied where whole material is made to pass through two horizontal rollers that rotate opposite to each other at different speeds with required nip gap determines the thickness of the graphene/elastomeric composite (Potts et al. 2013). Other compounding ingredients like softeners, accelerators, activators, and vulcanising agent can also be added for the vulcanisation of the elastomer counterpart. Likewise, the mixing of graphene elastomeric composite can also be done using internal mixer like Brabender having a chamber to which the compounding ingredients are added and the shear force generated between the two rotors gives graphene/elastomeric composites (Rodgers and Waddell 2013; Freakley and Wan Idris 1979). Composites prepared using elastomers like NR, nitrile rubber, and SBR can be done by this technique (Amutha Jeevakumari, Indhumathi, and Arun Prakash 2020; Potts et al. 2012). The temperature and speed of roller, mixing time, and nip gap should be closely monitored (Potts et al. 2012).

15.4.3 SOLUTION/LATEX STAGE MIXING

A solution stage mixing is a process in which graphene suspension is prepared and incorporated into the latex/elastomer solution by magnetic stirring or ultra-sonication. Later, the solvent is evaporated to get the graphene/elastomeric composite. This method is known for its better filler dispersion since the elastomeric matrix is enveloped over the graphene (Kuilla et al. 2010). The elastomers include SBR (Mao et al. 2013), silicone rubber (Xu et al. 2016), NR (Liu, Gao, and Hu 2019), and nitrile rubber (Zhang and Cho 2018). The evaporation of solvent is difficult as well as the solvent is of high cost which limits the process of the preparation of composites in large scale, but makes the laboratory research easy. The mixing time, composition, sonication, along with the solvent used may affect the properties of the graphene/ elastomeric composites dramatically. Furthermore, the co-coagulation of latex can be also done in order to prevent the agglomeration of graphene, thus promoting better dispersion of filler in the elastomeric matrix and enhancing the thermal conductivity (Wu et al. 2013; Zhan et al. 2012; Tang et al. 2012). The transmission electron microscopy (TEM) images of the graphene NR composite are prepared by ultrasonically assisted latex mixing followed by in situ reduction and two-roll milling and show that the former method promotes good dispersion of graphene in rubber matrix as shown in Figure 15.7 (Zhan et al. 2011). Schematic illustration of latex stage mixing of graphene/SBR composite is shown in Figure 15.8.

15.4.4 IN SITU POLYMERISATION

The elastomeric composites are prepared within the polymerisation mixture that involves the formation of covalent bond between the functionalised graphene and the elastomeric matrix by means of heat or radiation (Zheng, Lu, and Wong 2004). The reduction of graphene by the in situ polymerisation in elastomeric matrix also favoured the functionalisation of graphene, thus enhancing the interaction with elastomeric matrix (Liu, Kuang, and Guo 2015; Dong, Zhang, and Wu 2016). Figure 15.9 shows the schematic illustration of the in situ polymerisation of polymer-graphene nanocomposite (Gangarapu et al. 2019). In contrast, the enhanced viscosity along with the polymerisation reaction negatively affects the properties of the elastomeric composites (Park and Ruoff 2009).

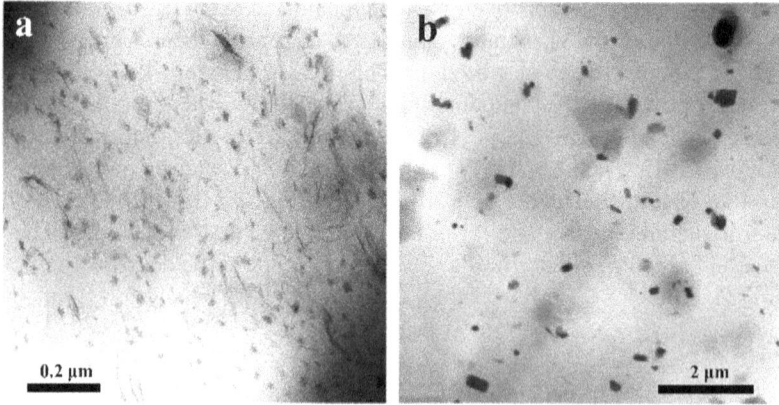

FIGURE 15.7 TEM images of graphene/natural rubber composite prepared by (a) ultrasonically assisted latex mixing and in situ reduction and (b) conventional two-roll milling (Zhan et al. 2011). (Adapted and reproduced with permission © copyright 2011, John Wiley and Sons.)

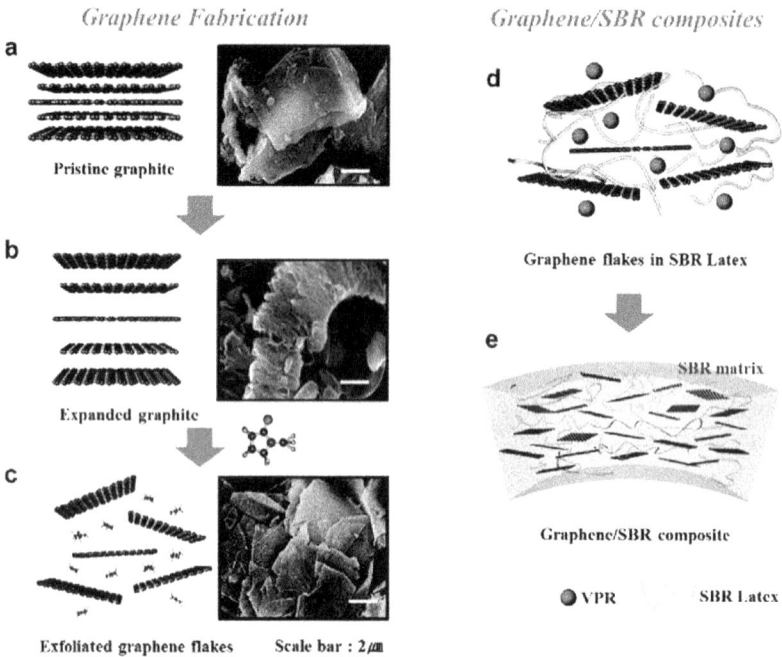

FIGURE 15.8 Schematic diagram showing the overall processing for preparing the *l*-GFs and *l*-GFs/SBR composites: (a) Schematic model and SEM images of graphite; (b) schematic model and SEM images of expanded graphite prepared by a conventional acid process combined with thermal shock method; (c) schematic model and SEM images of the prepared *l*-GFs by solvent exfoliation in NMP; (d) coagulation of the *l*-GFs aqueous solution and SBR latex. VPR: butadiene-styrene-vinyl-pyridine rubber, and (e) schematic model image of *l*-GFs/SBR composites (Song et al. 2015). Adapted and reproduced with permission @copyright 2015, The Royal Society of Chemistry.

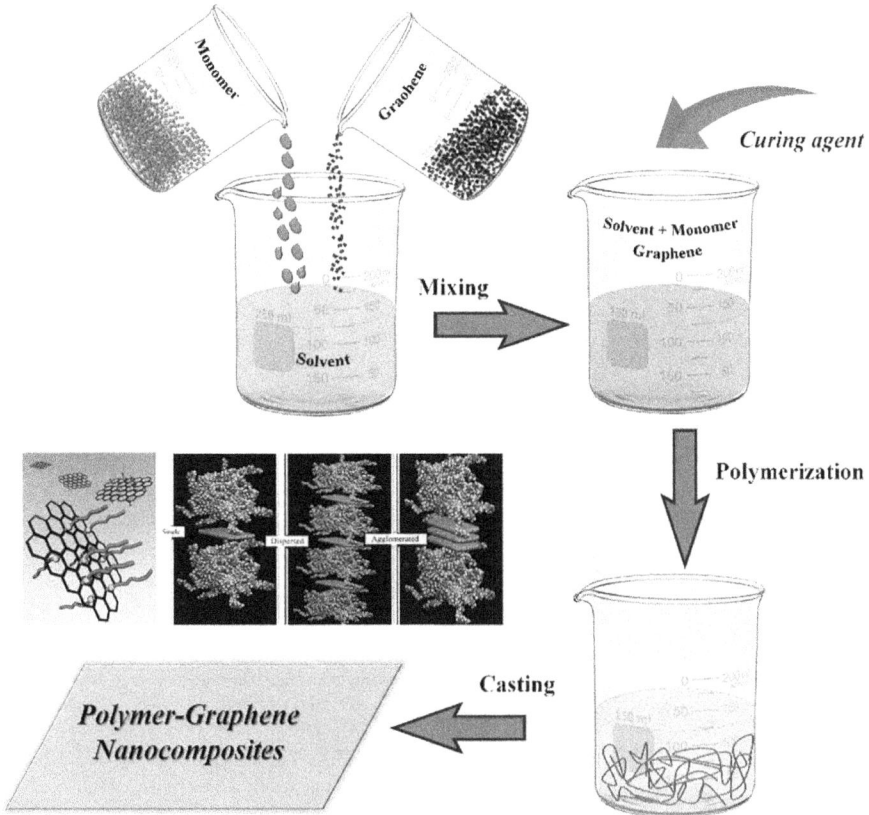

FIGURE 15.9 Schematic illustration of the in situ polymerisation of polymer-graphene nanocomposite.

15.4.5 ELECTROSPINNING

Electrospinning is an advanced powerful technique for the preparation of graphene/elastomeric composite which was very firstly experimented with NR latex (colloidal suspension) (Cacciotti et al. 2015). Suspension of graphene and the elastomer are prepared using suitable solvents like chloroform by sonication/magnetic stirring. The concentration and the viscosity of the prepared suspension are tailored for electrospinning. Further, the suspension is filled in the syringe pump with a needle connected to a spinneret. High voltage is applied through the graphene/elastomer suspension, and subsequently, electrostatic potential generated within the suspension repels each other that helps in overcoming the surface tension of the suspension and results in ejection of charged liquid jet into a fibrous which is collected by the metal collector (Cacciotti et al. 2015; Reddy et al. 2015; Zárate et al. 2020) and is illustrated in Figure 15.10. Good orientation of polymeric chains of aligned fibres was obtained.

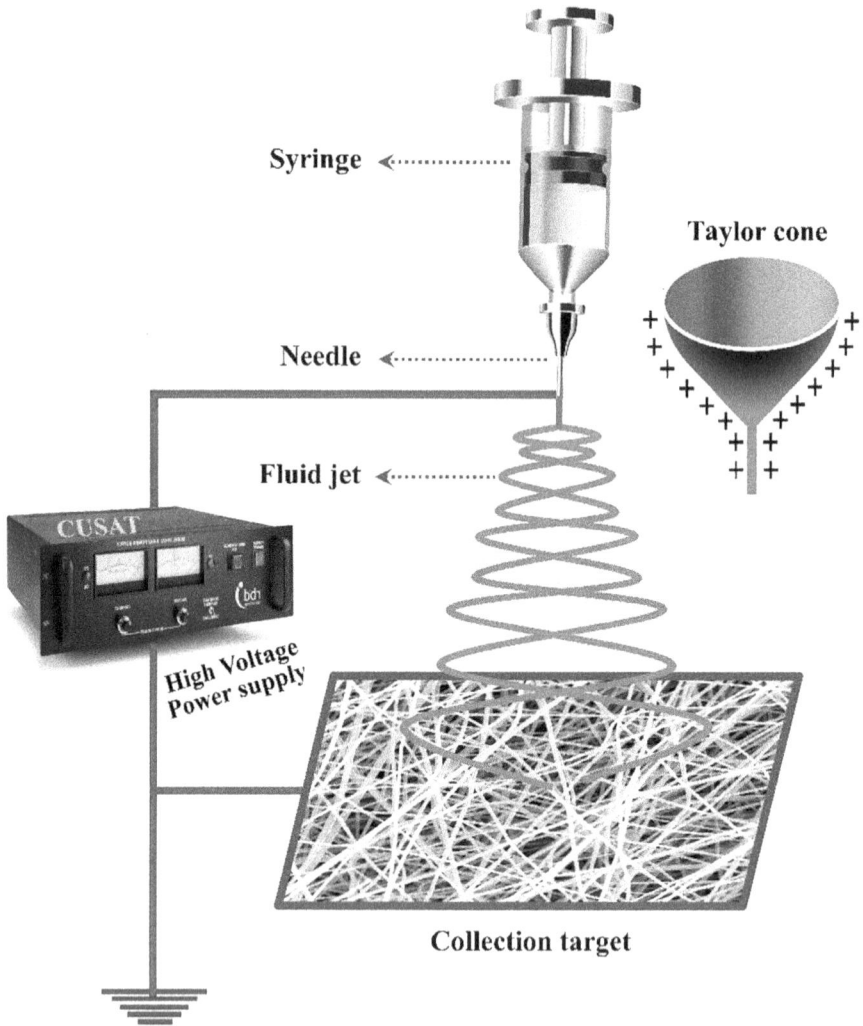

FIGURE 15.10 Schematic illustration of the electrospinning process.

15.5 THERMALLY CONDUCTING GRAPHENE/ ELASTOMERIC COMPOSITES

In recent years, there has been a surge in interest in elastomeric materials for a variety of applications, ranging from automotive to energy, aerospace to the manufacturing of everyday goods. The traditional rubber material could not keep up with the ever-increasing demand for applications, so developing innovative multifunctional rubber-based composites is critical. Since 2013, one of the hottest subjects in the graphene field has been the modification of rubber using graphene nanofillers with applications in car and bicycle tyres. With the combined effect of its typical structural characteristics and conductivity, a carbonaceous natured graphene is opening up new avenues for the creation of high performing multifunctional elastomer nanocomposites. Graphene has several benefits as a potential filler in rubber materials:

i. exceptional modulus of 1 TPa with elasticity and light weight, resulting in a superior mechanical reinforcing effect than traditional filler;
ii. exceptionally high electrical and thermal conductivity;
iii. greater gas barrier characteristics than clay;
iv. greater specific area, resulting in more interfacial interaction and a bigger influence on the molecular chain, and crystallisation, especially strain-induced crystallisation potential, as compared to clay and CNT;
v. excellent heat resistance; and
vi. some polar groups, such as hydroxyl, carboxyl, and epoxide, allow graphene to be readily functionalised by attaching small molecules or polymers.

Graphene may give an excellent chance for the revolution of conventional rubber manufacturing. Despite tremendous advances in scientific research and industry for rubber composites, several barriers have still to be addressed (Liu et al. 2018). Some of these concerns are as follows:

i. exfoliation and dispersion of graphene nanoplatelets;
ii. fabrication of well-defined graphene structures (i.e., segregated structure or layered structure and random dispersion);
iii. enhanced interfacial binding of 2D graphene with macromolecular chain.

NR is among the most extensively researched rubbers for real situations such as tyres, gloves, and footwear due to its biomaterial, high flexibility, cracking resistance, and other excellent mechanical properties (Ahmed et al. 2013). However, for practical applications, NR, like other polymers, must be supplemented with fillers to enhance mechanical and physical characteristics, as well as to provide flexibility and processability in product design and to minimise product manufacturing costs (Fahim, Elhaggar, and Elayat 2012). Tyres is an example: when alternate tension is applied to a rubber-filled tyre, it begins to create greater heat. Simultaneously, heat accumulation is evident due to the low thermal conductivity of rubber. Because of the heat accumulation, their usage in tyres is severely limited (Akutagawa, Hamatani, and Nashi 2015). Recently, the marvellous materials graphene and graphite have been utilised to modify the thermal characteristics of pure NR for a variety of applications. Figure 15.11 shows the thermal conductivity of the core-shell structured fillers-based graphene oxide and alumina-incorporated NR composite (Zhuang et al. 2021). Variation in the homologous surface temperature of the samples with heating and cooling time is shown in Figure 15.12 (Zhuang et al. 2021).

When attempting to enhance the thermal characteristics of rubber, we must minimise the thermal resistance at the filler-matrix contact as well as the heat generation caused by friction in between filler and matrix. To modify the fillers and enhance the filler-matrix interaction, a surface modifier is essential (Yang et al. 2016; Zheng et al. 2018; George and Bhowmick 2008b). Researchers have shown a lot of interest

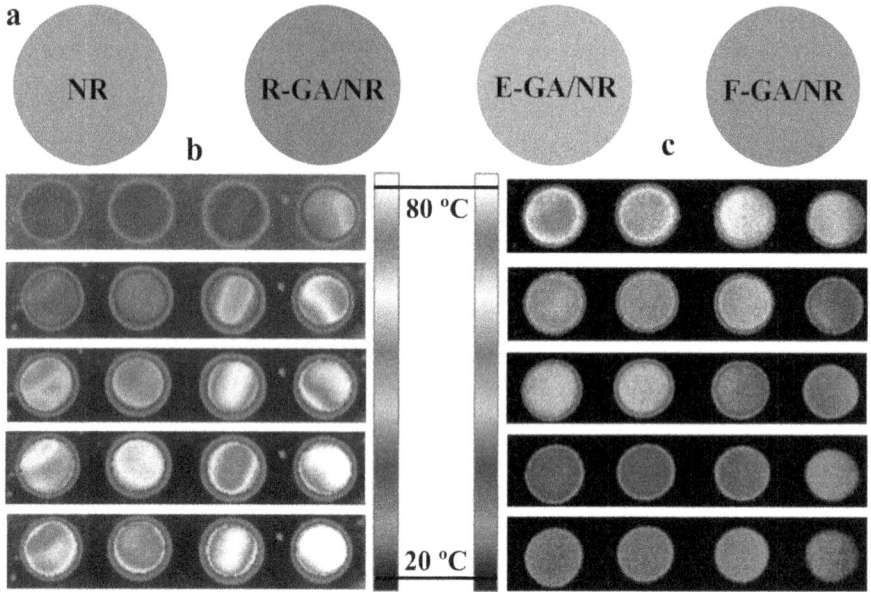

FIGURE 15.11 (a) Four diverse sample models based on natural rubber composites incorporated with core-shell structured fillers. IR thermal images of samples during (b) heating and (c) cooling process (Zhuang et al. 2021). (Adapted and reproduced with permission © copyright 2021, Elsevier.)

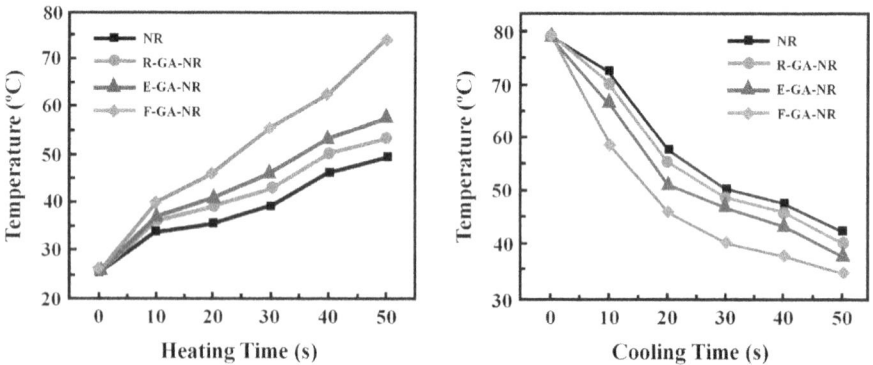

FIGURE 15.12 Variation in the homologous surface temperature of the samples with (a) heating and (b) cooling time (Zhuang et al. 2021). (Adapted and reproduced with permission © copyright 2021, Elsevier.)

in the usage of rubber additives as surface modifiers of nanometre-scale fillers so far. An effective, green, and simple approach was used to do simultaneous reduction and modification using 2-mecapto benzothiozol molecule (*M*) as shown in Figure 15.13 (Dong et al. 2018). Apart from the outstanding reduction capacity to GO, *M* molecules were covalently grafted onto the reduced GO surface in this study,

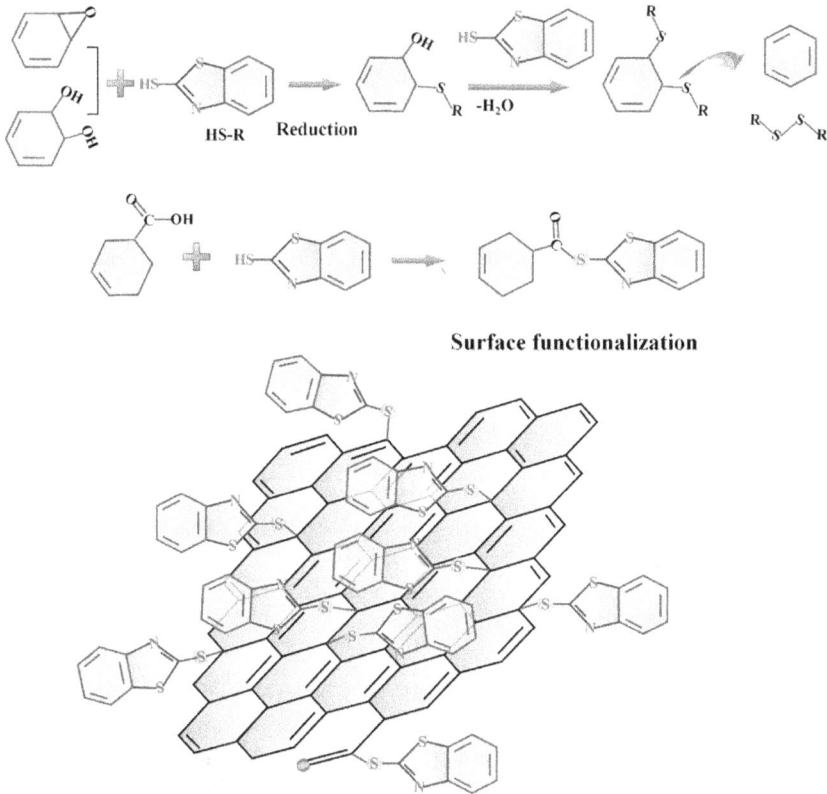

FIGURE 15.13 Mechanisms for reducing and surface functionalizing GO with vulcanisation accelerator (Dong et al. 2018).

endowing rubber composites featuring high mechanical strength and thermal conductivity.

In similar attempt, to shed light on the impact of GO on vulcanised rubber heat production and thermal conductivity, N-tert-butyl-2-benzothiazole sulfonamide (NS) (as a reducing agent and organic modifier) was utilised in a streamlined one-step process to concurrently reduce and modify GO (Cheng et al. 2021). This newly synthesised interfacial structure of reduced graphene oxide (rGO)/N-tert-butyl-2-benzothiazole sulfonamide (NS)/natural rubber (NR) (NR/NS-rGO) was then added to rubber using a latex co-precipitation technique, resulting in strong covalent bond linkages between the filler/matrix interfaces, as shown in Figure 15.14 (Cheng et al. 2021). Finally, when the filler loading of NS-rGO is 0.42 vol.%, not only does the thermal conductivity of the rubber composite increase to 0.237 W m^{-1} K^{-1}, which is 21.5% higher than that of pure NR; but also the internal heat generation decreases to 2.6°C, which is 45.8% lower than that of reduced graphene oxide (rGO)/ascorbic acid (VC)/natural rubber (NR) (NR/VC-rGO). The findings showed that covalent link connections significantly lowered interfacial heat resistance at the filler/matrix junction.

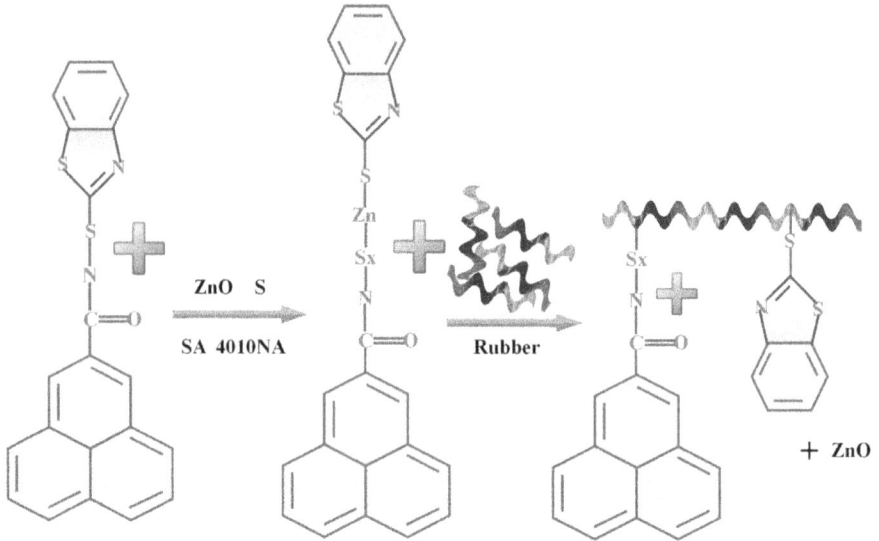

FIGURE 15.14 The chemical process between NS-rGO and rubber molecules is depicted schematically (Cheng et al. 2021).

In order to improve interfacial adhesion and boost compatibility of composite with a variety of polymers (Du and Cheng 2012) such as polyaniline, polylactic acid, polycaprolactone, and PEG, diversities in graphene enable NR are getting popular. In one such example, producing defect-free monolayer to few and multiple layers graphene, graphene oxide (GO), reduced GO (rGO), and functionalised graphene (Edwards and Coleman 2013) has been described in recent years. In one such recent work, a simple, scalable, and low-cost technique was used to produce few (2–5) layers graphene utilising sonication in the presence of anionic surfactant, resulting in a stable graphene-incorporated NR-latex dispersion. This modification not only showed enhanced thermal conductivity (by 480%–980%) but also increased tensile strength (by 40%) and electrical conductivity (60%) (George et al. 2017).

The right mixing procedure is another crucial aspect for maximising graphene's high conductivity in a polymer matrix. Ultrasound-assisted supercritical CO_2 (scCO_2) technique to create graphene infiltrated conductive NR is a new development to achieve high conductivity of graphene in a polymer matrix (Gao et al. 2+019). Since the entire procedure was carried out with only scCO_2 (one of the most ecologically friendly solvents), this approach is not only effective but it is also safe for the environment.

In addition to homogeneous filler dispersion, the selection of an appropriate polymer matrix is emerging as a significant problem in the field of polymer nanocomposite. In this regard, NR might be proposed as a promising choice for an appropriate polymer matrix for the manufacture of filler-reinforced nanocomposites. The benefits of utilising NR are numerous. NR has outstanding elasticity, flexibility, and antiviral permeability as a polymer matrix for nanocomposite production. Furthermore, NR is readily moulded and readily biodegradable, resulting in low environmental impact. Even with its unique properties, NR poses a major challenge when employed as a single polymer matrix in nanocomposite systems. The non-polar character of NR makes functionalised polar fillers less compatible with it. As a result, using NR as a single polymer matrix in a nanocomposite can never give a comprehensive cure. Epoxidised natural rubber (ENR) is one such modified form of NR in which the double bond of the NR is selectively oxidised utilising several types of peroxy agents.

SBR is the most versatile rubber in the synthetic elastomer family, and it is utilised for a variety of tyre and non-tyre applications. It is a random copolymer of styrene and butadiene that is made by polymerizing styrene and butadiene in a solution (anionic kind) or an emulsion (free-radical kind). As mentioned earlier, rubber reinforcements using nanofillers have been extensively studied to meet the rising demand in practical applications. Graphene oxide (GO), a derivative of graphene, has lately received a lot of attention in the rubber reinforcement industry due to its exceptional mechanical, thermal, and barrier characteristics, as well as its ease of production. In an investigation of the effects of the degree of GO oxidation on SBR/GO nanocomposites' curing characteristics, mechanical properties, thermal conductivity, thermal stability, and solvent resistance were comprehensively studied. In the thermal conductivity study of SBR/GO nanocomposites, a particular pattern was seen as the degree of oxidation of GO increased. The thermal conductivity of SG9 achieves its maximum (0.207 W m^{-1} K^{-1}), rising by 28.6% as compared to plain SBR. This is possibly due to the fact that phonons scattering or impedance mismatch on the interface between GO9 and SBR matrix is minimised resulting in improved compatibility of GO9 with SBR matrix. However, when the oxidation degree of GO grows, the interfacial adhesion between GO and SBR weakens, thus increasing the impedance mismatch and degrading the thermal conductivity of nanocomposites. Due to its unique mechanical, electrical, and thermal characteristics, graphene is an excellent nanocomposites filler. However, because graphene is difficult to functionalise, it is difficult to evenly distribute the vast proportion of graphene fillers into a polymer matrix. A new approach for introducing a substantial proportion of graphene into a SBR matrix is described in research.

In addition to the previously described problems connected with rubber-graphene composites, another sub-challenge is the production of stable colloid comprising few-layered graphene. Achieving higher thermal conductivity in graphene-polymer nanocomposites is often limited by mainly two parameters: (i) quality of graphene (isotopes, defects, impurities, or vacancies) and (2) the large interfacial thermal resistance between the polymer-mediated graphene boundaries. Consequently, previous studies employed high loading of graphene (up to 25 wt.%) into polymer matrix to achieve a considerable increase in thermal conductivity (Guo and Chen 2014). Conventionally, the target thermal conductivity values (>1 W mK^{-1}) can be achieved by dispersing high loadings (50–80 vol.%) of the thermally conductive micron-size fillers in thermally insulating polymers (TC < 0.2 W mK^{-1}) (Shtein et al. 2015). However, such high filler loadings resulted in high density and expensive composites with poor mechanical properties all of which combined to limit their practical applications. In addition, the physical mechanism that affects the thermal conductivity of polymers as well as polymer nanocomposite is shown in Figure 15.15 (Huang, Qian, and Yang 2018).

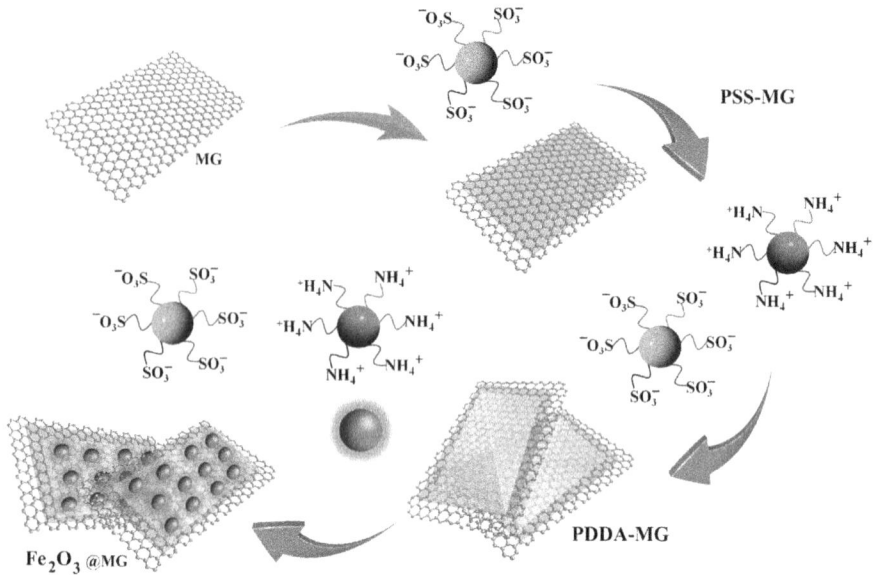

FIGURE 15.15   Schematic diagram of multilayer graphene functionalised with magnetite nanoparticles (Shi et al. 2019).

15.6   MECHANISMS OF THERMAL CONDUCTIVITY OF GRAPHENE/ELASTOMERIC COMPOSITES

Graphene exhibits thermal conductivity due to the flow of phonons and electrons in which phonons play the vital role (Pu et al. 2013; Balandin 2011). When some of the atoms come in contact with the heat source, they begin to vibrate. The vibrations fortified with strong covalent force and densely packed atoms within the hexagonal crystal lattice are transferred to the atoms in close proximity and thus further passed on, which is shown in Figure 15.16a–l (Burger et al. 2016). Rapid heat transfer is accomplished in the form of phonon waves (thermal energy carrier) resulting in collective excitation of atoms in the crystal lattice (Narula et al. 2012; Yao and Cao 2016, 2014). In contrast, the multilayer graphene shows difficulty in phonon transmission owing to the weak van der Waal's force that holds the layers together and thus shows anisotropy (Balandin et al. 2008; Ghosh et al. 2008). The notion of phonons can be extended to polymers, but it is quite complicated since a polymer is made of discrete molecular chains held together by weak attractive forces. Polymers experience slow diffusion of heat from one atom to another in the direction of molecular chain as shown in Figure 15.16m (Huang, Qian, and Yang 2018; Li, Zhang, and Zhang 2017; Burger et al. 2016), which is affected by the parameters like backbone, morphology, orientation, crystallinity, temperature, hydrogen bonding, cross-linking, and side chains (Rossinsky and Müller-Plathe 2009; Choy and Greig 1975; Choy, Chen, and Luk 1980; Zhang and Luo 2016; Wei and Luo 2019; Huang, Qian, and Yang 2018). Inharmonic vibrations accompanied by the rotation of the atoms result in phonon

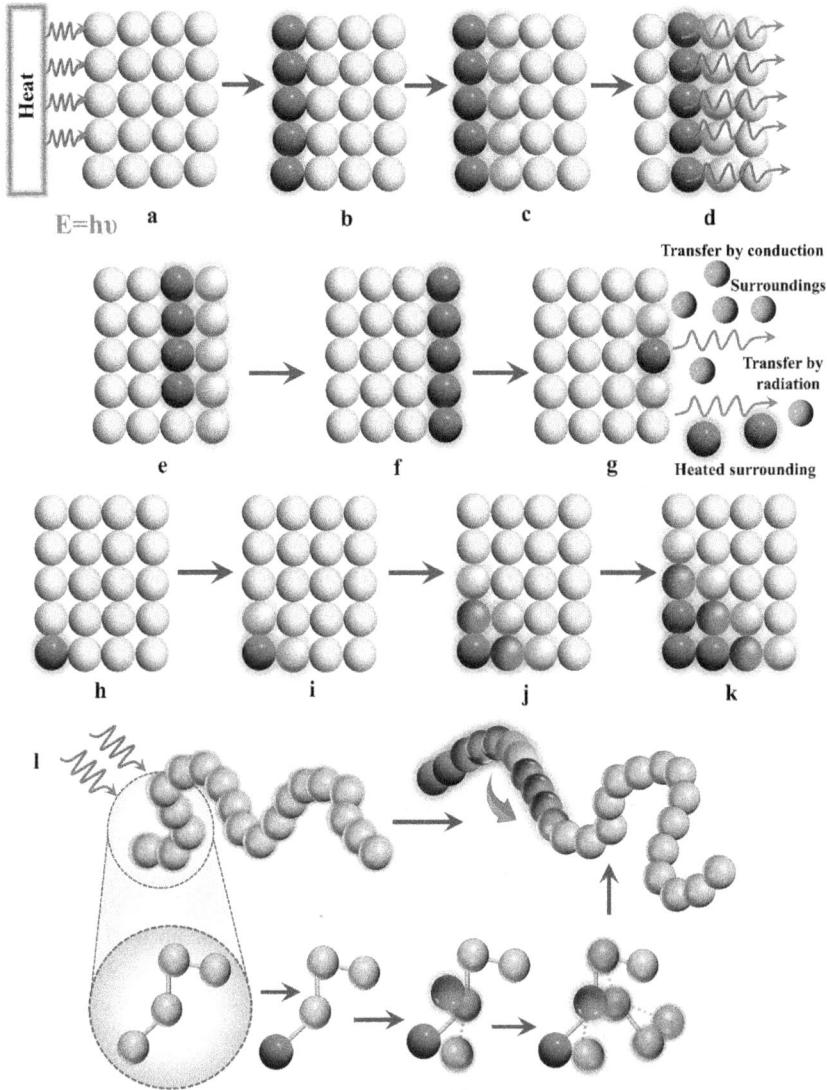

FIGURE 15.16 Schematic representation of phonon transition. (a–g) Thermal conductivity mechanism in a crystalline material; (h–k) illustration of the temperature gradient in a crystalline material; and (l) thermal conductivity mechanism in an elastomer (amorphous polymer) (Burger et al. 2016). (Adapted and reproduced with permission © copyright 2016, Elsevier.)

scattering that reduces thermal conductivity, usually seen in the case of elastomers which are amorphous polymers (Guo et al. 2020; Burger et al. 2016).

The trajectory is different when it comes with elastomeric composites that attribute to the better interaction between filler (here, graphene) and elastomer. Poor interaction or incongruity may lead to phonon scattering and thus creates high interfacial thermal resistance at the interface of the graphene-elastomer system that reduces the thermal conductivity of composites (Luo and Lloyd 2012; Liu et al. 2014). The mechanisms of thermal conductivity can be explained based on thermal percolation, thermal conduction path, and thermoelastic coefficient theories (Li, Zhang, and Zhang 2017; Soga et al. 2017; Zhang et al. 2019; Li et al. 2009). At lower loading, graphene gets isolated within the elastomeric matrix and thus interferes with the thermal conductivity pathway since the rate of diffusion of phonons within the graphene and elastomer is different (Zhang et al. 2019; Gu et al. 2017; Suh et al. 2016). Increasing and optimizing the volume fraction of graphene in elastomeric matrix may lead to enhancement in thermal conductivity, paving the basis for thermal percolation theory (Gu et al. 2014; Su, Li, and Weng 2018; Wu et al. 2019). Researches also shows that, as the thermal conductivity coefficient increases, thermoelastic coefficient also increases, thus reducing phonon scattering within the elastomeric composites which is a cumulative mechanism of classical vibration and elastic mechanics (Li et al. 2017; Bigg 1995, 1986). Surface modification can also enhance the graphene-polymer matrix interaction, thus contributing higher to the thermal conductivity (Kuila et al. 2011; Roy, Sengupta, and Bhowmick 2012; George, Bandyopadhyay, and Bhowmick 2008).

15.7 SURFACE MODIFICATION OF GRAPHENE VS. THERMAL CONDUCTIVITY

15.7.1 COMPOSITION OF GRAPHENE

Graphene's high thermal conductivity is effectively utilised in elastomers by incorporating it as a filler. As with any thermally conductive filler, the TC is observed to be increased as there is increment of graphene composition in the elastomeric matrix. Early Studies (Araby et al. 2013) on GnPs-SBR nanocomposites showed a 240% increase in TC compared with the unfilled SBR by incorporating 41.6% of GnPs into the rubber matrix. Similarly, in the graphene-SBR nanocomposite, a 15 vol.% increase in graphene content gave rise to 1,330% of TC compared to neat SBR (Li et al. 2017). Besides, for lower graphene loadings of up to 0.02 wt.%, an insignificant improvement of TC was observed (Xu et al. 2016). This was observed mainly due to lesser filler distribution in the elastomeric matrix. Likewise, many studies suggested a gradual increase in TC with the increase in graphene composition in composites, ascribed to a quick transfer of thermal energy through phonons (Lin et al. 2015; Wilk et al. 2020; Araby et al. 2014, 2013; Li et al. 2017; Tian et al. 2017). Beyond certain dosage of graphene, re-stacking of the filler happens and thereby hinders the regular arrangement of the rubber chains. This induces a limited compatibility of the filler with the rubber matrix. (Shen et al. 2013; Tian et al. 2017). This could adversely affect the mechanical properties and processing of composites which makes it difficult for the desired application.

The difference in thermal conductivity between the elastomeric matrix and graphene fillers, as well as phonon scattering at the elastomer-filler and filler-filler interfaces, resists the nanocomposites from achieving a satisfying level of thermal conductivity (Kim, Abdala, and Macosko 2010). Because of the strong van der Waals force between graphene sheets, uniform dispersion of inert graphene in rubber matrix is quite challenging. Besides, its hydrophobic properties lead to feeble interfacial bonding with the rubber matrix, which gives rise to poor stress transfer from rubber to graphene as well as limited interfacial thermal conductivity. Due to the abundant oxygen sites on basal planes and edges, GO is an alternatively suggested graphene-derivative for thermally conductive elastomeric applications. However, their hydrophilic behaviour towards nonpolar elastomers like SBR indicates its drawback for wide applications (Sadasivuni et al. 2014; Zhong et al. 2017). Thus, an enhanced TC with low graphene loadings, with better interfacial compatibility is preferred. In graphene-polymer nanocomposites, (i) alteration of the interfacial chemistries through surface modification of fillers; (ii) orientation of fillers in specific directions and creating 3D filler network structure to create an easier thermal conductive pathway; (iii) hybridisation of graphene with other conductive materials or fillers; and (iv) coating & decoration of fillers are the methods adopted to overcome these issues (Ma et al. 2021).

15.7.2 GRAPHENE ELASTOMER INTERFACE

Interfacial compatibility between elastomer and graphene fillers is an unavoidable factor for the thermal properties of elastomeric nanocomposites. Studies were conducted to understand the role of graphene's surface chemistry on interfacial bonding and dispersibility with SBR matrix, employing surface energy as a measurement (Tang et al. 2014). The results suggested that graphene with a lesser CO_x proportion, below a critical surface energy value of 0.2, can be dispersed quite effectively and exhibits stronger interfacial bonding with the rubber matrix. It is well known that pure graphene sheets are difficult to disperse in rubber matrix, for better dispersion with the matrix as well as to avoid agglomeration tendencies, functionalisation, and surface modifications of fillers are important. In general, surface functionalisation of fillers like graphene includes physical or non-covalent functionalisation and chemical or covalent functionalisation.

Physical functionalisation involves the attachment of functionalisation agents like surfactants, small molecules with benzene structures, ionic liquids, saccharides, and polymers onto the surface of the graphene mainly through π–π interactions, which effectively limits the interfacial thermal resistance by reducing the acoustic impedance mismatch (Shang et al. 2019). For instance, the impact of surface modifications can be observed in a study (Zhang et al. 2016) carried out on properties of polyvinyl pyrrolidone (PVP)-functionalised GO-NR composites (PGO-NR). As shown in Figure 15.17a, the poor dispersion (indicated by red arrow) of GO in NR matrix was observed, whereas in PGO-NR (Figure 15.17b), a superior and uniform dispersion of GO in rubber matrix without any aggregation to obtain an improved thermal conductivity was observed.

Yin et al. (2017) carried out a comparison study of pristine and ionic liquid (IL)-functionalised GO with SBR. The IL, functionalisation agent, 1-allyl-3-methylimidazolium chloride (AMICl), grafted onto GO surface through ultrasonic treatment of GO-IL suspension, demonstrated a combination of hydrogen bond and cation-π interaction with GO sheets, as shown in Figure 15.18 (Yin et al. 2017). It was found that the interaction of IL and GO effectively created a superior interfacial adhesion with the SBR matrix, resulting in approximately 14.5% higher thermal conductivity than the GO-SBR nanocomposite. In zinc dimethacrylate (ZDMA)-functionalised

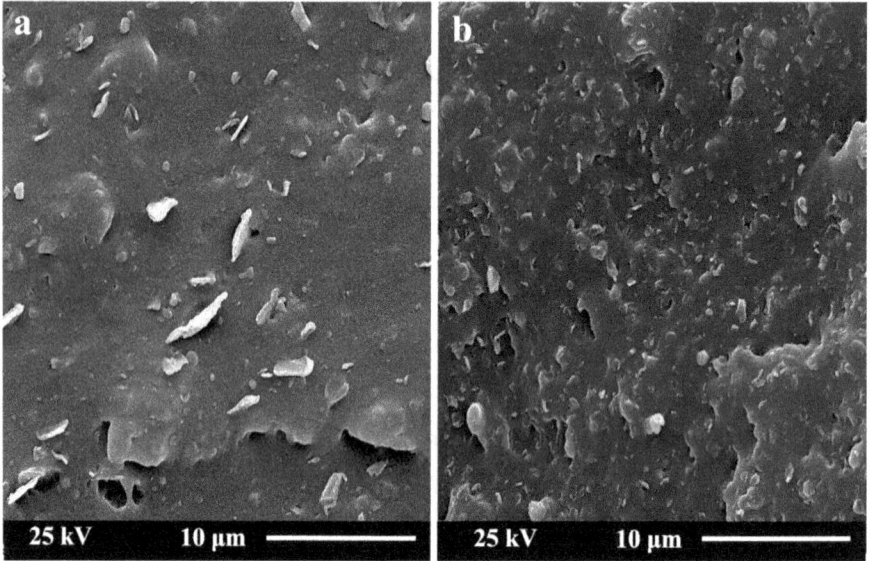

FIGURE 15.17 SEM images of (a) GO-NR and (b) polyvinyl pyrrolidone (PVP)-functionalised GO-NR composites (Zhang et al. 2016).

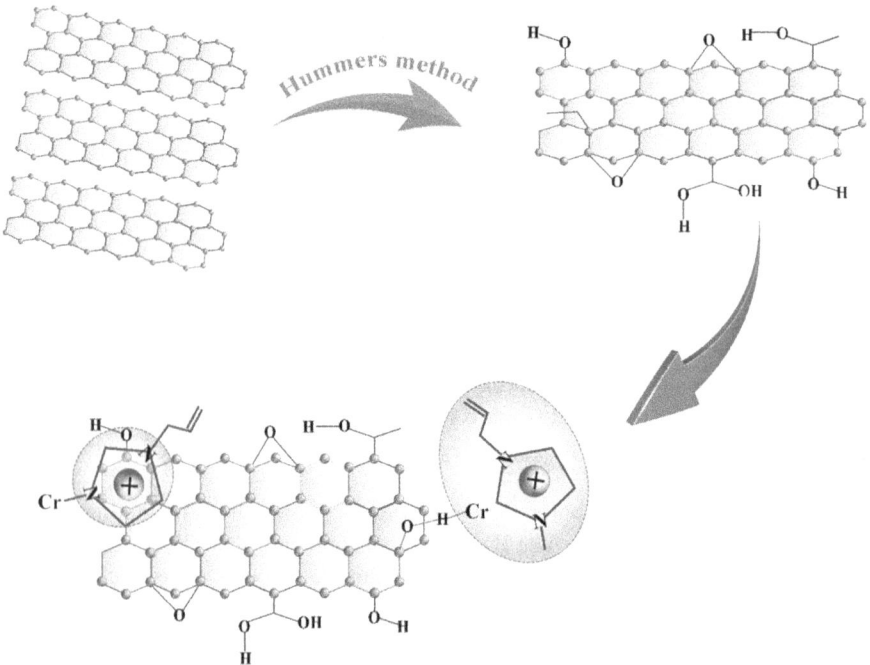

FIGURE 15.18 Schematic representation of the interaction between GO and IL (Yin et al. 2017).

graphene-NR composites, the "bridging" character of poly-ZDMA, which acts in between rubber matrix and graphene, is attributed to the efficient interfacial interaction and increase in TC by 29% (Lin et al. 2015). However, many studies using polymers like poly(ethyleneimine), amines, silanes, and ILs also suggest similar results (Xiong et al. 2013; Abd Razak et al. 2015; Lin et al. 2015; Tian et al. 2017).

Chemical interactions between filler and molecules are considered to be stronger than physical interactions. In chemically functionalised graphene, foreign molecules are covalently or chemically attached to its surface. These tightly grafted molecules provide a firm filler-matrix interfacial adhesion to improve the thermal conductivity. For instance, as shown in Figure 15.19 (Kuila et al. 2012), the preparation of amine-modified graphene, starting from oxidation of graphite sheets, followed by several

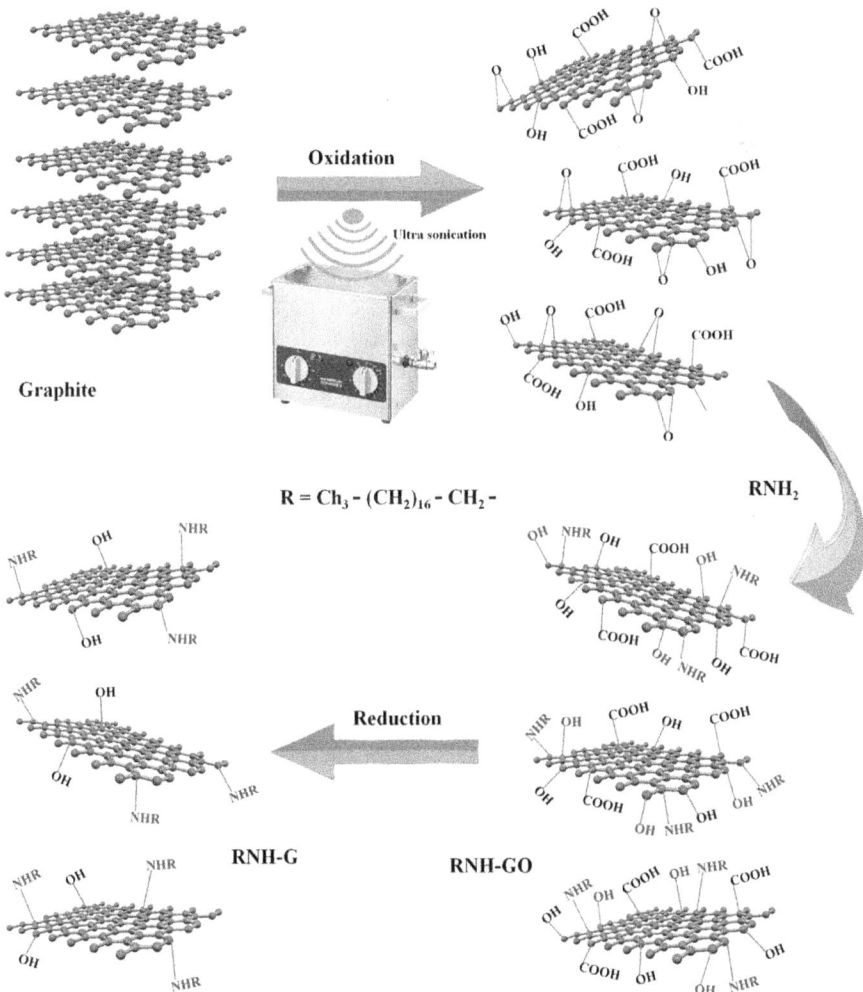

FIGURE 15.19 Schematic representation of the preparation of amine-modified graphene (Kuila et al. 2012).

processes such as sonication, treatment with alkylamine, and finally reduction of amine-modified GO (RNH-GO), gives a fair idea of chemical functionalisation of graphene (Kuila et al. 2012). On considering the thermal parameters, the increment in weight fraction of immobilised rubber layer (χ_{im}) and the depression in heat capacity jump (ΔC_{pn}) establish the interfacial bonding between filler and rubber matrix. Investigations on N-1,3-dimethylbutyl-N'-phenyl-p-phenylenediamine-functionalised GO-SBR composite show that a lower ΔC_{pn}, and a higher χ_{im} were observed, indicating a higher amount of immobilised rubber chains by the filler and a stronger interfacial interaction than the unfunctionalised GO-SBR composite (Zhong et al. 2017).

Studies on amino-functionalised graphene-silicone rubber nanocomposite show that the observed limited acoustic impedance mismatch at the interface for improved thermal conductivity was attributed to three reasons: (i) the wrapping of polymer chains by wrinkled surfaces of ultrathin graphene sheets; (ii) formation of hydrogen bond between the oxygen molecules of functionalised graphene and rubber matrix; and (iii) covalent bond formation for filler-matrix interface through amino groups (Shen et al. 2013). Antioxidants while used as a modifier for rGO act as capping agents to both prevent the stacking of graphene sheets in rubber matrix as well as direct the graphene sheets to form a thermally conductive pathway. These graphene-graphene pathways lower the thermal interface resistance and improve TC (Zhong et al. 2019). Apart from these, silane, polymers and molecules, rubber antioxidants and accelerators, etc., are also grafted with graphene for achieving an improved thermal conductivity in graphene-elastomer nanocomposite (Zhao et al. 2015; Wang et al. 2016; Wang et al. 2019; Dong et al. 2018; Cai et al. 2020; Zhong et al. 2017; Xu et al. 2016; Zhong et al. 2019). A comprehensive data chart (neglecting the type of functionalisation) involving the maximum thermal conductivity achieved for functionalised and non-functionalised graphene-rubber composite is shown in Table 15.1.

15.7.3 ORIENTATION OF GRAPHENE IN ELASTOMERIC MATRIX

How to increase the TC of polymer composites is quite an important question, which attracted the researcher's attention. Enhancing the composite filler content can generate more heat transfer pathways, but several other desired properties would be sacrificed. Increasing filler-matrix dispersion through the physical and chemical functionality of graphene was widely accepted. However, these methods were not ample to achieve a satisfying TC for elastomeric composites. Many investigators believed tailoring the orientation of graphene fillers is the best method to increase the TC of polymer composite materials. Aligned fillers in a particular direction can provide a facile heat transfer route across composite, in contrast to randomly oriented fillers. Numerous reports suggest the improvement of TC by specific graphene alignment in polymer composite materials, however, only a few reports are there in graphene-elastomers composites (Li, Zhang, and Zhang 2017).

Highly oriented graphene-silicone rubber composites were synthesised via the layer-by-layer method by Song et al. The well-packed, alternatively arranged graphene sheets and silicone rubber provided a consistent thermal path, resulting in a higher TC of about 2.03 W m^{-1} K^{-1}, even with small composition (2.53 wt.%) of graphene. Moreover, without any changes in TC up to 500 flexure cycles, the observed

TABLE 15.1
Thermal Conductivity for Functionalised and Non-Functionalised Graphene Rubber Composites

Rubber Matrix	Functionalising Agent	TC_{max} of Rubber Composite with Functionalised (FN) or Non-Functionalised (NFN) Graphene (GO) (W m^{-1} K^{-1})		Ref.
		NFN-GO	FN-GO	
SR	TEVS	0.300	0.380	Zhao et al. (2015)
NR	Polyvinyl pyrrolidone (PVP)	0.127	0.146	Xumin Zhang et al. (2016)
SR	Aminopropyltriethoxysilane (APTES)	0.232	0.250	Guangwu Zhang et al. (2016)
	Vinyltrimethoxysilane (VTMS)		0.264	
	Triton X-100		0.268	
SBR	Polyvinylpyrrolidone (PVP)	0.189	0.210	Yin et al. (2016)
SBR	1-allyl-3-methyl-imidazolium chloride (AMICl)	0.188	0.215	Yin et al. (2017)
SBR	N-1,3-dimethylbutyl-N0-phenyl-p-phenylenediamine	0.290	0.374	Zhong et al. (2017)
SBR	2-mercaptobenzothiazole	0.330	0.386	Dong et al. (2018)
SR	Γ-methacryloxypropyltrimethoxy silane	0.810	1.31	Chen and Liu (2018)
SBR	2-mercaptobenzimidazole	0.266	0.352	Zhong et al. (2019)
SBR	Dodecylamine	0.249	0.265	J. Park et al. (2020)
SBR	Pentaerythritol tetra(3-mercapto propionate) (PETMP)	0.176	0.185	Cai et al. (2020)
NR	N-tert-butyl-2-benzothiazole sulfonamide	0.325	0.40	Cheng et al. (2021)
XNBR	Polyethyleneimine (PEI)	0.149	0.194	Cai et al. (2021)

superior mechanical properties mitigated the drawbacks of methods discussed previously in this chapter (Song, Chen, and Zhang 2018). Studies for aligning Fe-induced magnetically functionalised graphene (Fe_3O_4@MG) in SR by the use of an external magnetic field were done by Shi and co-workers. A vertical chain-like alignment of graphene particles was observed in SR matrix, which induced a superior interaction between graphene particles for creating a heat transfer pathway. Figure 15.20 shows the SEM image of the aggregated chain-like structure aligned in the vertical direction (marked with dotted yellow and red elliptical shapes), which became more prominent with the increase in filler content. Compared to the randomly arranged, the vertically aligned Fe_3O_4@MG-SR composite shown an excellent 191% higher through-plane TC at a lower graphene loading (Shi et al. 2019). Similar research using silicon carbide nanowires, rGO, and cellulose nanofibre hybrid filler aligned in SR matrix also suggest comparable results (Song and Zhang 2020).

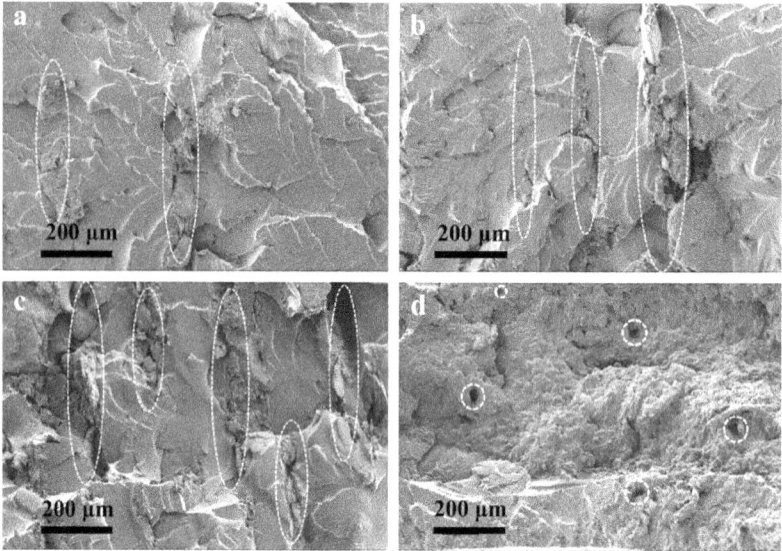

FIGURE 15.20   SEM images of different fillers loaded (a) 1, (b) 3, (c) 5, (d) 7 wt.%, vertically aligned Fe_3O_4@MG/SR composites (Shi et al. 2019). (Adapted and reproduced with permission © copyright 2019, John Wiley and Sons.)

New strategies involving the fabrication of 3D interconnected graphene networks are gaining more attention due to its rapid heat transfer capability, which enables the preparation of highly thermally conductive polymer composites. Fabrication of polydopamine-rGO (PDA-rGO) 3D networks using NaCl incorporated in liquid SR has been successfully adapted to prepare thermally conductive PDA-rGO/SR composites (Figure 15.21) (Song et al. 2019). Even at low filler loadings of 1.46 wt.%, these composites have obtained a TC of 1.50 W m^{-1} K^{-1}. Similar experiments on SBR, polysulfide rubber, and silicone rubber (Tao et al. 2021; Song et al. 2019; Zhang et al. 2019) provide a strong insight for further exploration of researches on highly thermally conductive graphene-elastomer nanocomposite.

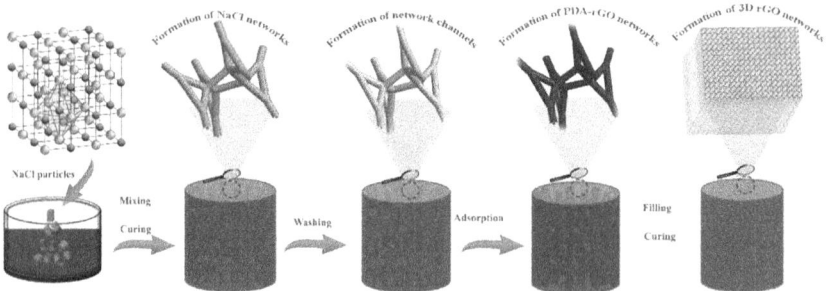

FIGURE 15.21   Schematic representation of the fabrication process of 3D interconnected graphene networked PDA-rGO/SR composite (Song et al. 2019). (Adapted and reproduced with permission © copyright 2009, Royal Society of Chemistry.)

15.8 APPLICATIONS

Thermally conductive graphene/elastomeric composites have wide range of applications in electronic packaging, thermal energy storage, sensing, aerospace, automobile, etc. (Verma and Goh 2018; Kumar et al. 2013). One of the key applications is as thermal interface materials (TIMs). In order to control excess heat developed within different components of an electronic device (integrated circuit, heat sink), a proper mechanism is required (Yu et al. 2007). Due to the high thermal conductivity of graphene, TIMs made of graphene-based composites are of prime interest, particularly in the field of electronic packaging. TIMs can improve the thermal coupling between two solid surfaces. Graphene-based silicone rubber or phenyl silicone rubber is found to be an effective TIM (Xu et al. 2016; Zong et al. 2015). The schematic diagram of TIMs is shown in Figure 15.22 (Lv et al. 2018; Shahil and Balandin 2012). The development of thermally conductive graphene-based elastomeric composites as phase change materials (PCMs) created jolt in the field of electronics that can provide cooling/heating effect on demand. And invigorated studies on the paraffin-based graphene/EPDM composites and even in silicone rubber for thermal energy storage application gained great attention of experimentalists (Ding et al. 2020; Deng et al. 2019; Feng et al. 2019). Likewise, thermally conductive graphene/elastomeric composites are used in wide range of applications by taking

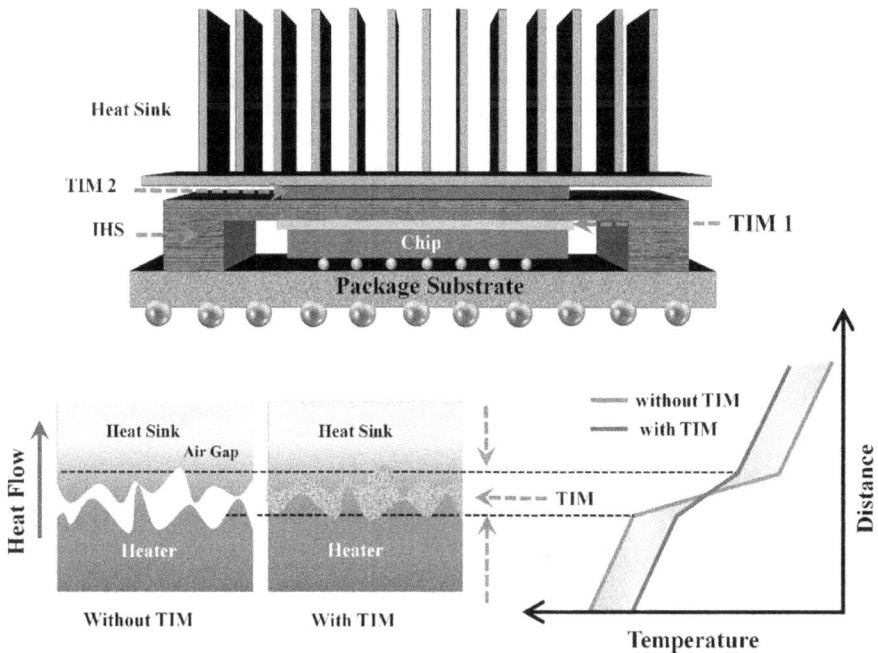

FIGURE 15.22 Schematic representation of thermal interface materials (TIMs) in electronic packaging. [TIM 1 is placed between chip and integrated heat spreader (IHS), whereas TIM 2 is placed between heat sink and IHS] (Lv et al. 2018), (Shahil and Balandin 2012).

into account the thermal conductivity coefficient, effective dispersion and alignment filler in elastomeric matrix, functionalisation of graphene, and other thermomechanical properties.

15.9 CONCLUSIONS

Graphene with its two-dimensional anisotropic structure is one of the best carbonaceous materials with high thermal conductivity. The heat flow in graphene is exhibited by the flow of phonons as well as electrons. Orientation and composition of graphene in the rubber as well as the graphene-rubber interface play a vital role in determining the thermal conductivity of graphene/elastomeric composites. Likewise, the large specific surface area of graphene results in better interfacial interactions with elastomer and the functionalisation of graphene is quite easy due to the presence of polar groups. Elastomeric composites fortified with graphene result in a synergistic impact of their unique features like better property even at low loadings, excellent mechanical properties, light-weight, flexible, and high thermal conductivity. Thus, thermally conductive graphene/elastomeric composites can be used for various applications including thermal energy storage, electronic packaging, sensing, aerospace, and automobile.

REFERENCES

Aakyiir, Mathias, Sherif Araby, Andrew Michelmore, Qingshi Meng, Yousef Amer, Yu Yao, Min Li, Xiaohui Wu, Liqun Zhang, and Jun Ma. 2020. "Elastomer Nanocomposites Containing MXene for Mechanical Robustness and Electrical and Thermal Conductivity." *Nanotechnology* 31 (31): 315715. doi:10.1088/1361-6528/ab88eb.

Abd Razak, Jeefferie, Sahrim Haji Ahmad, Chantara Thevy Ratnam, Mazlin Aida Mahamood, and Noraiham Mohamad. 2015. "Effects of Poly(Ethyleneimine) Adsorption on Graphene Nanoplatelets to the Properties of NR/EPDM Rubber Blend Nanocomposites." *Journal of Materials Science* 50 (19): 6365-81. doi:10.1007/s10853-015-9188-5.

Aguilar-Bolados, Héctor, Mehrdad Yazdani-Pedram, and Raquel Verdejo. 2020. "Thermal, Electrical, and Sensing Properties of Rubber Nanocomposites." In *High-Performance Elastomeric Materials Reinforced by Nano-Carbons*, 149-75. Elsevier. doi:10.1016/B978-0-12-816198-2.00007-4.

Ahmed, Khalil, Shaikh Nizami, Nudrat Raza, and Khalid Mahmood. 2013. "Effect of Micro Sized Marble Sludge on Physical Properties of Natural Rubber Composites." *Chemical Industry and Chemical Engineering Quarterly* 19 (2): 281-93. doi:10.2298/CICEQ111225062A.

Akutagawa, Keizo, Satoshi Hamatani, and Takayuki Nashi. 2015. "The New Interpretation for the Heat Build-up Phenomena of Rubbery Materials during Deformation." *Polymer* 66 (June): 201-9. doi:10.1016/j.polymer.2015.04.040.

Al-Ghamdi, Ahmed A., Attieh A. Al-Ghamdi, Yusuf Al-Turki, F. Yakuphanoglu, and Farid El-Tantawy. 2016. "Electromagnetic Shielding Properties of Graphene/Acrylonitrile Butadiene Rubber Nanocomposites for Portable and Flexible Electronic Devices." Composites *Part B: Engineering* 88: 212-19. doi:10.1016/j.compositesb.2015.11.010.

Amutha Jeevakumari, S. A., K. Indhumathi, and V. R. Arun Prakash. 2020. "Role of Cobalt Nanowire and Graphene Nanoplatelet on Microwave Shielding Behavior of Natural Rubber Composite in High Frequency Bands." *Polymer Composites.* doi:10.1002/pc.25718.

Anuar, J., M. Mariatti, and H. Ismail. 2007. "Study on Tensile, Electrical, and Thermal Properties of Aluminium Particle Filled Natural Rubber (NR) and Ethylene-Propylene-Diene Terpolymer (EPDM) Composites." *Polymer-Plastics Technology and Engineering* 46 (12): 1201–6. doi:10.1080/03602550701575920.

Araby, Sherif, Liqun Zhang, Hsu-Chiang Kuan, Jia-Bin Dai, Peter Majewski, and Jun Ma. 2013. "A Novel Approach to Electrically and Thermally Conductive Elastomers Using Graphene." *Polymer* 54 (14): 3663–70. doi:10.1016/j.polymer.2013.05.014.

Araby, Sherif, Qingshi Meng, Liqun Zhang, Hailan Kang, Peter Majewski, Youhong Tang, and Jun Ma. 2014. "Electrically and Thermally Conductive Elastomer/Graphene Nanocomposites by Solution Mixing." *Polymer* 55 (1): 201–10. doi:10.1016/j.polymer.2013.11.032.

Balandin, Alexander A. 2011. "Thermal Properties of Graphene and Nanostructured Carbon Materials." *Nature Materials.* doi:10.1038/nmat3064.

Balandin, Alexander A., Suchismita Ghosh, Wenzhong Bao, Irene Calizo, Desalegne Teweldebrhan, Feng Miao, and Chun Ning Lau. 2008. "Superior Thermal Conductivity of Single-Layer Graphene." *Nano Letters.* doi:10.1021/nl0731872.

Bartlett, Michael D., Navid Kazem, Matthew J. Powell-Palm, Xiaonan Huang, Wenhuan Sun, Jonathan A. Malen, and Carmel Majidi. 2017. "High Thermal Conductivity in Soft Elastomers with Elongated Liquid Metal Inclusions." *Proceedings of the National Academy of Sciences* 114 (9): 2143–48. doi:10.1073/pnas.1616377114.

Benli, Salih, Ülkü Yilmazer, Fikret Pekel, and Saim Özkar. 1998. "Effect of Fillers on Thermal and Mechanical Properties of Polyurethane Elastomer." *Journal of Applied Polymer Science* 68(7): 1057–65. doi:10.1002/(SICI)1097-4628(19980516)68:7<1057::AID-APP3>3.3.CO;2-E.

Bigg, D. M. 1986. "Thermally Conductive Polymer Compositions." *Polymer Composites.* doi:10.1002/pc.750070302.

Bigg, D. M. 1995. "Thermal Conductivity of Heterophase Polymer Compositions." *Advances in Polymer Science.* doi:10.1007/bfb0021279.

Burger, N., A. Laachachi, M. Ferriol, M. Lutz, V. Toniazzo, and D. Ruch. 2016. "Review of Thermal Conductivity in Composites: Mechanisms, Parameters and Theory." *Progress in Polymer Science.* doi:10.1016/j.progpolymsci.2016.05.001.

Cacciotti, Ilaria, John N. House, Claudia Mazzuca, Manlio Valentini, Francesco Madau, Antonio Palleschi, Paolo Straffi, and Francesca Nanni. 2015. "Neat and GNPs Loaded Natural Rubber Fibers by Electrospinning: Manufacturing and Characterization." *Materials and Design.* doi:10.1016/j.matdes.2015.09.054.

Cai, Fei, Yanlong Luo, Wei Yang, Xin Ye, Hao Zhang, Jing Zhu, and Sizhu Wu. 2021. "Study on the Thermal and Dielectric Properties of Covalently Modified GO/XNBR Composites." *Materials & Design* 198 (January): 109335. doi:10.1016/j.matdes.2020.109335.

Cai, Fei, Guohua You, Kaiqiang Luo, Hao Zhang, Xiuying Zhao, and Sizhu Wu. 2020. "Click Chemistry Modified Graphene Oxide/Styrene-Butadiene Rubber Composites and Molecular Simulation Study." *Composites Science and Technology* 190 (February): 108061. doi:10.1016/j.compscitech.2020.108061.

Chen, Hongyu, Valeriy V. Ginzburg, Jian Yang, Yunfeng Yang, Wei Liu, Yan Huang, Libo Du, and Bin Chen. 2016. "Thermal Conductivity of Polymer-Based Composites: Fundamentals and Applications." *Progress in Polymer Science* 59 (August): 41–85. doi:10.1016/j.progpolymsci.2016.03.001.

Cheng, Shuaishuai, Xiaoyuan Duan, Zhiyi Zhang, Dong An, Guizhe Zhao, and Yaqing Liu. 2021. "Preparation of a Natural Rubber with High Thermal Conductivity, Low Heat Generation and Strong Interfacial Interaction by Using NS-Modified Graphene Oxide." *Journal of Materials Science* 56 (5): 4034–50. doi:10.1007/s10853-020-05503-8.

Choy, C. L., F. C. Chen, and W. H. Luk. 1980. "Thermal Conductivity of Oriented Crystalline Polymers." *Journal of Polymer Science. Part A-2, Polymer Physics.* doi:10.1002/pol.1980.180180603.

Choy, C. L., and D. Greig. 1975. "The Low-Temperature Thermal Conductivity of a Semi-Crystalline Polymer, Polyethylene Terephthalate." *Journal of Physics C: Solid State Physics.* doi:10.1088/0022–3719/8/19/012.

Cui, Wei, Yuan Zhu, Xuanyi Yuan, Kexin Chen, and Feiyu Kang. 2012. "Gel-Cast-Foam-Assisted Combustion Synthesis of Elongated β-Si3N4 Crystals and Their Effects on Improving the Thermal Conductivity of Silicone Composites." *Journal of Alloys and Compounds* 540 (November): 165–69. doi:10.1016/j.jallcom.2012.04.123.

Deng, Hao, Yongli Guo, Fangfang He, Zhijian Yang, Jinghui Fan, Ren He, Kai Zhang, and Wenbin Yang. 2019. "Paraffin@graphene/Silicon Rubber Form-Stable Phase Change Materials for Thermal Energy Storage." *Fullerenes Nanotubes and Carbon Nanostructures.* doi:10.1080/1536383X.2019.1624539.

Ding, Ze, Fangfang He, Yongsheng Li, Zhuoni Jiang, Hongjian Yan, Ren He, Jinghui Fan, Kai Zhang, and Wenbin Yang. 2020. "Novel Shape-Stabilized Phase Change Materials Based on Paraffin/EPDM@Graphene with High Thermal Conductivity and Low Leakage Rate." *Energy and Fuels.* doi:10.1021/acs.energyfuels.9b04000.

Dong, Bin, Liqun Zhang, and Youping Wu. 2016. "Highly Conductive Natural Rubber–Graphene Hybrid Films Prepared by Solution Casting and in Situ Reduction for Solvent-Sensing Application." *Journal of Materials Science.* doi:10.1007/s10853-016-0276-y.

Dong, Huanhuan, Zhixin Jia, Yongjun Chen, Yuanfang Luo, Bangchao Zhong, and Demin Jia. 2018. "One-Pot Method to Reduce and Functionalize Graphene Oxide via Vulcanization Accelerator for Robust Elastomer Composites with High Thermal Conductivity." *Composites Science and Technology* 164 (May): 267–73. doi:10.1016/j.compscitech.2018.05.047.

Du, Jinhong, and Hui-Ming Cheng. 2012. "The Fabrication, Properties, and Uses of Graphene/Polymer Composites." *Macromolecular Chemistry and Physics* 213 (10–11): 1060–77. doi:10.1002/macp.201200029.

Edwards, Rebecca S., and Karl S. Coleman. 2013. "Graphene Synthesis: Relationship to Applications." *Nanoscale* 5 (1): 38–51. doi:10.1039/C2NR32629A.

Fahim, Irene S., Salah M. Elhaggar, and Hatem Elayat. 2012. "Experimental Investigation of Natural Fiber Reinforced Polymers." *Materials Sciences and Applications* 03 (02): 59–66. doi:10.4236/msa.2012.32009.

Feng, Jing, Zhan Jun Liu, Dong Qing Zhang, Zhao He, Ze Chao Tao, and Quan Gui Guo. 2019. "Phase Change Materials Coated with Modified Graphene-Oxide as Fillers for Silicone Rubber Used in Thermal Interface Applications." *Xinxing Tan Cailiao/New Carbon Materials.* doi:10.1016/S1872–5805(19)60011–9.

Freakley, P. K., and W. Y. Wan Idris. 1979. "Visualization of Flow During the Processing of Rubber in an Internal Mixer." *Rubber Chem Technol.* doi:10.5254/1.3535197.

Gangarapu, Satesh, Kiran Sunku, P. Suresh Babu, and Putla Sudarsanam. 2019. "Fabrication of Polymer-Graphene Nanocomposites BT - Handbook of Polymer and Ceramic Nanotechnology." In, edited by Chaudhery Mustansar Hussain and Sabu Thomas, 1–15. Cham: Springer International Publishing. doi:10.1007/978-3-030-10614-0_31-1.

Gao, Hanyang, Haijun Liu, Chengzhi Song, and Guoxin Hu. 2019. "Infusion of Graphene in Natural Rubber Matrix to Prepare Conductive Rubber by Ultrasound-Assisted Supercritical CO2 Method." *Chemical Engineering Journal* 368 (March): 1013–21. doi:10.1016/j.cej.2019.03.026.

George, Jinu Jacob, Abhijit Bandyopadhyay, and Anil K. Bhowmick. 2008. "New Generation Layered Nanocomposites Derived from Ethylene-Co-Vinyl Acetate and Naturally Occurring Graphite." *Journal of Applied Polymer Science.* doi:10.1002/app.25067.

George, Jinu Jacob, and Anil K. Bhowmick. 2008a. "Ethylene Vinyl Acetate/Expanded Graphite Nanocomposites by Solution Intercalation: Preparation, Characterization and Properties." In *Journal of Materials Science.* doi:10.1007/s10853-007-2193-6.

George, Jinu Jacob, and Anil K. Bhowmick. 2008b. "Fabrication and Properties of Ethylene Vinyl Acetate-Carbon Nanofiber Nanocomposites." *Nanoscale Research Letters.* doi:10.1007/s11671-008-9188-3.

George, Gejo, Suja Bhargavan Sisupal, Teenu Tomy, Bincy Akkoli Pottammal, Alaganandam Kumaran, Vemparthan Suvekbala, Rajmohan Gopimohan, Swaminathan Sivaram, and Lakshminarayanan Ragupathy. 2017. "Thermally Conductive Thin Films Derived from Defect Free Graphene-Natural Rubber Latex Nanocomposite: Preparation and Properties." *Carbon* 119 (August): 527–34. doi:10.1016/j.carbon.2017.04.068.

Ghosh, S., I. Calizo, D. Teweldebrhan, E. P. Pokatilov, D. L. Nika, A. A. Balandin, W. Bao, F. Miao, and C. N. Lau. 2008. "Extremely High Thermal Conductivity of Graphene: Prospects for Thermal Management Applications in Nanoelectronic Circuits." *Applied Physics Letters.* doi:10.1063/1.2907977.

Gu, Junwei, Chaobo Liang, Xiaomin Zhao, Bin Gan, Hua Qiu, Yonqiang Guo, Xutong Yang, Qiuyu Zhang, and De Yi Wang. 2017. "Highly Thermally Conductive Flame-Retardant Epoxy Nanocomposites with Reduced Ignitability and Excellent Electrical Conductivities." *Composites Science and Technology.* doi:10.1016/j.compscitech.2016.12.015.

Gu, Junwei, Chao Xie, Hailin Li, Jing Dang, Wangchang Geng, and Qiuyu Zhang. 2014. "Thermal Percolation Behavior of Graphene Nanoplatelets/Polyphenylene Sulfide Thermal Conductivity Composites." *Polymer Composites.* doi:10.1002/pc.22756.

Guo, Wenman, and Guohua Chen. 2014. "Fabrication of Graphene/Epoxy Resin Composites with Much Enhanced Thermal Conductivity via Ball Milling Technique." *Journal of Applied Polymer Science.* doi:10.1002/app.40565.

Guo, Yongqiang, Kunpeng Ruan, Xuetao Shi, Xutong Yang, and Junwei Gu. 2020. "Factors Affecting Thermal Conductivities of the Polymers and Polymer Composites: A Review." *Composites Science and Technology.* doi:10.1016/j.compscitech.2020.108134.

Gwaily, S. E., G. M. Nasr, and M. M. Badawy. 2001. "Thermal and Electrical Properties of Irradiated Styrene Butadiene Rubber-Metal Composites." *Egyptian Journal of Solids* 24 (2): 193–205. doi:10.21608/ejs.2001.151442.

Gwaily, S. E., G. M. Nasr, M. M. Badawy, and H. H. Hassan. 1995. "Thermal Properties of Ceramic-Loaded Conductive Butyl Rubber Composites." *Polymer Degradation and Stability* 47 (3): 391–95. doi:10.1016/0141-3910(95)00004-6.

Han, Sensen, Qingshi Meng, Aron Chand, Shuo Wang, Xiaodong Li, Hailan Kang, and Tianqing Liu. 2019. "A Comparative Study of Two Graphene Based Elastomeric Composite Sensors." *Polymer Testing* 80 (September): 106106. doi:10.1016/j.polymertesting.2019.106106.

He, Yan, Zhen Chao Chen, and Lian Xiang Ma. 2009. "Thermal Conductivity and Mechanical Properties of Silicone Rubber Filled with Different Particle Sized SiC." *Advanced Materials Research* 87–88 (December): 137–42. doi:10.4028/www.scientific.net/AMR.87-88.137.

Huang, Xingyi, Pingkai Jiang, and Toshikatsu Tanaka. 2011. "A Review of Dielectric Polymer Composites with High Thermal Conductivity." *IEEE Electrical Insulation Magazine* 27 (4): 8–16. doi:10.1109/MEI.2011.5954064.

Huang, Congliang, Xin Qian, and Ronggui Yang. 2018. "Thermal Conductivity of Polymers and Polymer Nanocomposites." *Materials Science and Engineering R: Reports.* doi:10.1016/j.mser.2018.06.002.

Kerschbaumer, R. C., S. Stieger, M. Gschwandl, T. Hutterer, M. Fasching, B. Lechner, L. Meinhart, et al. 2019. "Comparison of Steady-State and Transient Thermal Conductivity Testing Methods Using Different Industrial Rubber Compounds." *Polymer Testing* 80 (September): 106121. doi:10.1016/j.polymertesting.2019.106121.

Kim, Hyunwoo, Ahmed A. Abdala, and Christopher W. Macosko. 2010. "Graphene/Polymer Nanocomposites." *Macromolecules* 43 (16): 6515–30. doi:10.1021/ma100572e.

Kuilla, Tapas, Sambhu Bhadra, Dahu Yao, Nam Hoon Kim, Saswata Bose, and Joong Hee Lee. 2010. "Recent Advances in Graphene Based Polymer Composites." *Progress in Polymer Science (Oxford)*. doi:10.1016/j.progpolymsci.2010.07.005.

Kuila, Tapas, Saswata Bose, Chang Eui Hong, Md Elias Uddin, Partha Khanra, Nam Hoon Kim, and Joong Hee Lee. 2011. "Preparation of Functionalized Graphene/Linear Low Density Polyethylene Composites by a Solution Mixing Method." *Carbon*. doi:10.1016/j.carbon.2010.10.031.

Kuila, Tapas, Saswata Bose, Ananta Kumar Mishra, Partha Khanra, Nam Hoon Kim, and Joong Hee Lee. 2012. "Chemical Functionalization of Graphene and Its Applications." *Progress in Materials Science* 57 (7): 1061–1105. doi:10.1016/j.pmatsci.2012.03.002.

Kumar, Sadasivuni Kishor, Mickael Castro, Allisson Saiter, Laurent Delbreilh, Jean Francois Feller, Sabu Thomas, and Yves Grohens. 2013. "Development of Poly(Isobutylene-Co-Isoprene)/Reduced Graphene Oxide Nanocomposites for Barrier, Dielectric and Sensingapplications." *Materials Letters*. doi:10.1016/j.matlet.2013.01.036.

Li, Chenlin, Huili Guo, Xin Tian, and Xiaogeng Tian. 2017. "Transient Response for a Half-Space with Variable Thermal Conductivity and Diffusivity under Thermal and Chemical Shock." *Journal of Thermal Stresses*. doi:10.1080/01495739.2016.1218745.

Li, Bin, Yan Liu, Bin Sun, Min Pan, and Gance Dai. 2009. "Properties and Heat-Conduction Mechanism of Thermally Conductive Polymer Composites." *Huagong Xuebao/CIESC Journal*.

Li, Ying, Fan Xu, Zaishan Lin, Xianxian Sun, Qingyu Peng, Ye Yuan, Shasha Wang, Zhiyu Yang, Xiaodong He, and Yibin Li. 2017. "Electrically and Thermally Conductive Underwater Acoustically Absorptive Graphene/Rubber Nanocomposites for Multifunctional Applications." *Nanoscale* 9 (38): 14476–85. doi:10.1039/C7NR05189A.

Li, An, Cong Zhang, and Yang Fei Zhang. 2017. "Thermal Conductivity of Graphene-Polymer Composites: Mechanisms, Properties, and Applications." *Polymers*. doi:10.3390/polym9090437.

Liu, Haijun, Hanyang Gao, and Guoxin Hu. 2019. "Highly Sensitive Natural Rubber/Pristine Graphene Strain Sensor Prepared by a Simple Method." *Composites Part B: Engineering*. doi:10.1016/j.compositesb.2019.04.032.

Liu, C. H., H. Huang, Y. Wu, and S. S. Fan. 2004. "Thermal Conductivity Improvement of Silicone Elastomer with Carbon Nanotube Loading." *Applied Physics Letters* 84 (21): 4248–50. doi:10.1063/1.1756680.

Liu, Ying, Jingsong Huang, Bao Yang, Bobby G. Sumpter, and Rui Qiao. 2014. "Duality of the Interfacial Thermal Conductance in Graphene-Based Nanocomposites." *Carbon*. doi:10.1016/j.carbon.2014.03.050.

Liu, Xuan, Wenyi Kuang, and Baochun Guo. 2015. "Preparation of Rubber/Graphene Oxide Composites with in-Situ Interfacial Design." *Polymer*. doi:10.1016/j.polymer.2014.11.048.

Lin, Yong, Konghua Liu, Yizhong Chen, and Lan Liu. 2015. "Influence of Graphene Functionalized with Zinc Dimethacrylate on the Mechanical and Thermal Properties of Natural Rubber Nanocomposites." *Polymer Composites* 36 (10): 1775–85. doi:10.1002/pc.23021.

Liu, Xin, Le-Ying Wang, Li-Fen Zhao, Hai-Feng He, Xiao-Yu Shao, Guan-Biao Fang, Zhen-Gao Wan, and Rong-Chang Zeng. 2018. "Research Progress of Graphene-Based Rubber Nanocomposites." *Polymer Composites* 39 (4): 1006–22. doi:10.1002/pc.24072.

Luo, Tengfei, and John R. Lloyd. 2012. "Enhancement of Thermal Energy Transport across Graphene/Graphite and Polymer Interfaces: A Molecular Dynamics Study." *Advanced Functional Materials*. doi:10.1002/adfm.201103048.

Lv, Le, Wen Dai, Aijun Li, and Cheng Te Lin. 2018. "Graphene-Based Thermal Interface Materials: An Application-Oriented Perspective on Architecture Design." *Polymers*. doi:10.3390/polym10111201.

Ma, Haoqi, Bin Gao, Meiyu Wang, Zhenye Yuan, Jingbo Shen, Jingqi Zhao, and Yakai Feng. 2021. "Strategies for Enhancing Thermal Conductivity of Polymer-Based Thermal Interface Materials: A Review." *Journal of Materials Science* 56 (2): 1064–86. doi:10.1007/s10853-020-05279-x.

Ma, Lian Xiang, Jin Xu, Yan He, and Jia Na Ke. 2012. "Study on Thermal Conductivity of Filled Silicone Rubber." *Key Engineering Materials* 501 (January): 88–93. doi:10.4028/www.scientific.net/KEM.501.88.

Mao, Yingyan, Shipeng Wen, Yulong Chen, Fazhong Zhang, Pierre Panine, Tung W. Chan, Liqun Zhang, Yongri Liang, and Li Liu. 2013. "High Performance Graphene Oxide Based Rubber Composites." *Scientific Reports.* doi:10.1038/srep02508.

Mu, Qiuhong, and Shengyu Feng. 2007. "Thermal Conductivity of Graphite/Silicone Rubber Prepared by Solution Intercalation." *Thermochimica Acta* 462 (1–2): 70–75. doi:10.1016/j.tca.2007.06.006.

Namitha, L. K., and M. T. Sebastian. 2017. "High Permittivity Ceramics Loaded Silicone Elastomer Composites for Flexible Electronics Applications." *Ceramics International* 43 (3): 2994–3003. doi:10.1016/j.ceramint.2016.11.080.

Narula, Rohit, Nicola Bonini, Nicola Marzari, and Stephanie Reich. 2012. "Dominant Phonon Wave Vectors and Strain-Induced Splitting of the 2D Raman Mode of Graphene." *Physical Review B - Condensed Matter and Materials Physics.* doi:10.1103/PhysRevB.85.115451.

Niu, Hongyu, Yanjuan Ren, Haichang Guo, Katarzyna Małycha, Kazimierz Orzechowski, and Shu-Lin Bai. 2020. "Recent Progress on Thermally Conductive and Electrical Insulating Rubber Composites: Design, Processing and Applications." *Composites Communications* 22 (September): 100430. doi:10.1016/j.coco.2020.100430.

Niu, Hui Jun, Zhi Yi Zhang, Wei Guo, Yi Xue, and Zhen Xing Yao. 2012. "Mechanical, Morphological and Thermally Behaviors of Natural Rubber/Aluminum Powder Composites." *Key Engineering Materials* 501 (January): 289–93. doi:10.4028/www.scientific.net/KEM.501.289.

Park, Sungjin, and Rodney S. Ruoff. 2009. "Chemical Methods for the Production of Graphenes." *Nature Nanotechnology.* doi:10.1038/nnano.2009.58.

Park, Jaehyeung, Jaswinder Sharma, Kyle W. Monaghan, Harry M. Meyer, David A. Cullen, Andres M. Rossy, Jong K. Keum, David L. Wood, and Georgios Polizos. 2020. "Styrene-Based Elastomer Composites with Functionalized Graphene Oxide and Silica Nanofiber Fillers: Mechanical and Thermal Conductivity Properties." *Nanomaterials* 10 (9): 1682. doi:10.3390/nano10091682.

Potts, Jeffrey R., Om Shankar, Ling Du, and Rodney S. Ruoff. 2012. "Processing-Morphology-Property Relationships and Composite Theory Analysis of Reduced Graphene Oxide/Natural Rubber Nanocomposites." *Macromolecules.* doi:10.1021/ma300706k.

Potts, Jeffrey R., Om Shankar, Shanthi Murali, Ling Du, and Rodney S. Ruoff. 2013. "Latex and Two-Roll Mill Processing of Thermally-Exfoliated Graphite Oxide/Natural Rubber Nanocomposites." *Composites Science and Technology.* doi:10.1016/j.compscitech.2012.11.008.

Pu, H. H., S. H. Rhim, C. J. Hirschmugl, M. Gajdardziska-Josifovska, M. Weinert, and J. H. Chen. 2013. "Anisotropic Thermal Conductivity of Semiconducting Graphene Monoxide." *Applied Physics Letters.* doi:10.1063/1.4808448.

Reddy, A. B, G. Siva Mohan Reddy, V. Sivanjineyulu, J. Jayaramudu, K. Varaprasad, and Emmanuel Rotimi Sadiku. 2015. "Hydrophobic/Hydrophilic Nanostructured Polymer Blends." In *Design and Applications of Nanostructured Polymer Blends and Nanocomposite Systems.* doi:10.1016/B978-0-323-39408-6.00016-9.

Rodgers, Brendan, and Walter Waddell. 2013. "The Science of Rubber Compounding." In *The Science and Technology of Rubber.* doi:10.1016/B978-0-12-394584-6.00009-1.

Rossinsky, Eddie, and Florian Müller-Plathe. 2009. "Anisotropy of the Thermal Conductivity in a Crystalline Polymer: Reverse Nonequilibrium Molecular Dynamics Simulation of the δ Phase of Syndiotactic Polystyrene." *Journal of Chemical Physics*. doi:10.1063/1.3103890.

Roy, Nabarun, Rajatendu Sengupta, and Anil K. Bhowmick. 2012. "Modifications of Carbon for Polymer Composites and Nanocomposites." *Progress in Polymer Science*. doi:10.1016/j.progpolymsci.2012.02.002.

Sadasivuni, Kishor Kumar, Deepalekshmi Ponnamma, Sabu Thomas, and Yves Grohens. 2014. "Evolution from Graphite to Graphene Elastomer Composites." *Progress in Polymer Science*. 39 (4): 749–80. doi:10.1016/j.progpolymsci.2013.08.003.

Sengupta, Rajatendu, Mithun Bhattacharya, S. Bandyopadhyay, and Anil K. Bhowmick. 2011. "A Review on the Mechanical and Electrical Properties of Graphite and Modified Graphite Reinforced Polymer Composites." *Progress in Polymer Science (Oxford)*. doi:10.1016/j.progpolymsci.2010.11.003.

Shahil, Khan M. F., and Alexander A. Balandin. 2012. "Thermal Properties of Graphene and Multilayer Graphene: Applications in Thermal Interface Materials." *Solid State Communications*. doi:10.1016/j.ssc.2012.04.034.

Shang, Songmin, Lu Gan, Changtong Mei, Lijie Xu, Lin Tan, and Enling Hu. 2019. "Wet Functionalization of Graphene and Its Applications in Rubber Composites." In *Carbon-Based Nanofillers and Their Rubber Nanocomposites*, 285–322. doi:10.1016/B978-0-12-813248-7.00010-9.

Shen, Jianfeng, Tie Li, Yu Long, Na Li, and Mingxin Ye. 2013. "Comparison of Thermal Properties of Silicone Reinforced by Different Nanocarbon Materials." *Soft Materials* 11 (3): 326–33. doi:10.1080/1539445X.2012.654585.

Shi, Yangyang, Wenshi Ma, Li Wu, Dechao Hu, Jinpeng Mo, Bo Yang, Shuanghong Zhang, and Zhilin Zhang. 2019. "Magnetically Aligning Multilayer Graphene to Enhance Thermal Conductivity of Silicone Rubber Composites." *Journal of Applied Polymer Science* 136 (37): 47951. doi:10.1002/app.47951.

Shojaei, A., M. Fahimian, and B. Derakhshandeh. 2007. "Thermally Conductive Rubber-Based Composite Friction Materials for Railroad Brakes – Thermal Conduction Characteristics." *Composites Science and Technology* 67 (13): 2665–74. doi:10.1016/j.compscitech.2007.03.009.

Shtein, Michael, Roey Nadiv, Matat Buzaglo, Keren Kahil, and Oren Regev. 2015. "Thermally Conductive Graphene-Polymer Composites: Size, Percolation, and Synergy Effects." *Chemistry of Materials*. doi:10.1021/cm504550e.

Sim, L. C., S. R. Ramanan, H. Ismail, K. N. Seetharamu, and T. J. Goh. 2005. "Thermal Characterization of Al2O3 and ZnO Reinforced Silicone Rubber as Thermal Pads for Heat Dissipation Purposes." *Thermochimica Acta* 430 (1–2): 155–65. doi:10.1016/j.tca.2004.12.024.

Singh, Virendra, Daeha Joung, Lei Zhai, Soumen Das, Saiful I. Khondaker, and Sudipta Seal. 2011. "Graphene Based Materials: Past, Present and Future." *Progress in Materials Science*. doi:10.1016/j.pmatsci.2011.03.003.

Soga, Kosuke, Takushi Saito, Tatsuya Kawaguchi, and Isao Satoh. 2017. "Percolation Effect on Thermal Conductivity of Filler-Dispersed Polymer Composites." *Journal of Thermal Science and Technology*. doi:10.1299/jtst.2017jtst0013.

Song, Jianan, Caibao Chen, and Yong Zhang. 2018. "High Thermal Conductivity and Stretchability of Layer-by-Layer Assembled Silicone Rubber/Graphene Nanosheets Multilayered Films." *Composites Part A: Applied Science and Manufacturing* 105 (February): 1–8. doi:10.1016/j.compositesa.2017.11.001.

Song, Shiqiang, Jinyuan Wang, Cheng Liu, Jincheng Wang, and Yong Zhang. 2019. "A Facile Route to Fabricate Thermally Conductive and Electrically Insulating Polymer Composites with 3D Interconnected Graphene at an Ultralow Filler Loading." *Nanoscale* 11 (32): 15234–44. doi:10.1039/C9NR05153H.

Song, Jianan, and Yong Zhang. 2020. "Vertically Aligned Silicon Carbide Nanowires/Reduced Graphene Oxide Networks for Enhancing the Thermal Conductivity of Silicone Rubber Composites." *Composites Part A: Applied Science and Manufacturing* 133 (January): 105873. doi:10.1016/j.compositesa.2020.105873.

Song, Sung Ho, Jung Mo Kim, Kwang Hyun Park, Dong Ju Lee, O. Seok Kwon, Jin Kim, Hyewon Yoon, and Xianjue Chen. 2015. "High Performance Graphene Embedded Rubber Composites." *RSC Advances*. doi:10.1039/c5ra16446j.

Su, Yu, Jackie J. Li, and George J. Weng. 2018. "Theory of Thermal Conductivity of Graphene-Polymer Nanocomposites with Interfacial Kapitza Resistance and Graphene-Graphene Contact Resistance." *Carbon*. doi:10.1016/j.carbon.2018.05.033.

Suh, Daewoo, Choong Man Moon, Duckjong Kim, and Seunghyun Baik. 2016. "Ultrahigh Thermal Conductivity of Interface Materials by Silver-Functionalized Carbon Nanotube Phonon Conduits." *Advanced Materials*. doi:10.1002/adma.201600642.

Tang, Zhenghai, Xiaohui Wu, Baochun Guo, Liqun Zhang, and Demin Jia. 2012. "Preparation of Butadiene-Styrene-Vinyl Pyridine Rubber-Graphene Oxide Hybrids through Co-Coagulation Process and in Situ Interface Tailoring." *Journal of Materials Chemistry*. doi:10.1039/c2jm00084a.

Tang, Zhenghai, Liqun Zhang, Wenjiang Feng, Baochun Guo, Fang Liu, and Demin Jia. 2014. "Rational Design of Graphene Surface Chemistry for High-Performance Rubber/Graphene Composites." *Macromolecules* 47 (24): 8663–73. doi:10.1021/ma502201e.

Tao, Wenjie, Shaohua Zeng, Ying Xu, Wangyan Nie, Yifeng Zhou, Pengbo Qin, Songhua Wu, and Pengpeng Chen. 2021. "3D Graphene – Sponge Skeleton Reinforced Polysulfide Rubber Nanocomposites with Improved Electrical and Thermal Conductivity." *Composites Part A: Applied Science and Manufacturing* 143 (December 2020): 106293. doi:10.1016/j.compositesa.2021.106293.

Tian, Limei, E. Jin, Haoran Mei, Qingpeng Ke, Ziyuan Li, and Hailin Kui. 2017. "Bio-Inspired Graphene-Enhanced Thermally Conductive Elastic Silicone Rubber as Drag Reduction Material." *Journal of Bionic Engineering* 14 (1): 130–40. doi:10.1016/S1672-6529(16)60384-0.

Verma, Deepak, and Kheng Lim Goh. 2018. "Functionalized Graphene-Based Nanocomposites for Energy Applications." In *Functionalized Graphene Nanocomposites and Their Derivatives: Synthesis, Processing and Applications*. doi:10.1016/B978-0-12-814548-7.00011-8.

Vinod, V. S., Siby Varghese, Rosamma Alex, and Baby Kuriakose. 2001. "Effect of Aluminum Powder on Filled Natural Rubber Composites." *Rubber Chemistry and Technology* 74 (2): 236–48. doi:10.5254/1.3544947.

Vinod, V. S., Siby Varghese, and Baby Kuriakose. 2004. "Aluminum Powder Filled Nitrile Rubber Composites." *Journal of Applied Polymer Science* 91 (5): 3156–61. doi:10.1002/app.13472.

Wang, Fuzhong, Lawrence T. Drzal, Yan Qin, and Zhixiong Huang. 2016. "Enhancement of Fracture Toughness, Mechanical and Thermal Properties of Rubber/Epoxy Composites by Incorporation of Graphene Nanoplatelets." *Composites Part A: Applied Science and Manufacturing* 87 (August): 10–22. doi:10.1016/j.compositesa.2016.04.009.

Wang, Jian, Guoxia Fei, Yunqi Pan, Kaiye Zhang, Shuai Hao, Zhuo Zheng, and Hesheng Xia. 2019. "Simultaneous Reduction and Surface Functionalization of Graphene Oxide by Cystamine Dihydrochloride for Rubber Composites." *Composites Part A: Applied Science and Manufacturing* 122 (April): 18–26. doi:10.1016/j.compositesa.2019.04.018.

Wang, Xiaohong, Chennan Lu, and Wei Rao. 2021. "Liquid Metal-Based Thermal Interface Materials with a High Thermal Conductivity for Electronic Cooling and Bioheat-Transfer Applications." *Applied Thermal Engineering* 192 (April): 116937. doi:10.1016/j.applthermaleng.2021.116937.

Wang, Jian, Kaiye Zhang, Qiang Bu, Marino Lavorgna, and Hesheng Xia. 2017. "Graphene-Rubber Nanocomposites: Preparation, Structure, and Properties." In *Carbon-Related Materials in Recognition of Nobel Lectures by Prof. Akira Suzuki in ICCE*, edited by Satoru Kaneko, Paolo Mele, Tamio Endo, Tetsuo Tsuchiya, Katsuhisa Tanaka, Masahiro Yoshimura, and David Hui, 175–209. Cham: Springer International Publishing. doi:10.1007/978-3-319-61651-3_9.

Wei, Xingfei, and Tengfei Luo. 2019. "Chain Length Effect on Thermal Transport in Amorphous Polymers and a Structure-Thermal Conductivity Relation." *Physical Chemistry Chemical Physics*. doi:10.1039/c9cp02397f.

Wilk, Joanna, Robert Smusz, Ryszard Filip, Grzegorz Chmiel, and Tomasz Bednarczyk. 2020. "Experimental Investigations on Graphene Oxide/Rubber Composite Thermal Conductivity." *Scientific Reports* 10 (1): 15533. doi:10.1038/s41598-020-72633-z.

Wu, Siwu, Zhenghai Tang, Baochun Guo, Liqun Zhang, and Demin Jia. 2013. "Effects of Interfacial Interaction on Chain Dynamics of Rubber/Graphene Oxide Hybrids: A Dielectric Relaxation Spectroscopy Study." *RSC Advances*. doi:10.1039/c3ra41998c.

Wu, Zhaohong, Chuan Xu, Chaoqun Ma, Zhibo Liu, Hui Ming Cheng, and Wencai Ren. 2019. "Synergistic Effect of Aligned Graphene Nanosheets in Graphene Foam for High-Performance Thermally Conductive Composites." *Advanced Materials*. doi:10.1002/adma.201900199.

Xiong, Xiaogang, Jingyi Wang, Hongbing Jia, Eryuan Fang, and Lifeng Ding. 2013. "Structure, Thermal Conductivity, and Thermal Stability of Bromobutyl Rubber Nanocomposites with Ionic Liquid Modified Graphene Oxide." *Polymer Degradation and Stability* 98 (11): 2208–14. doi:10.1016/j.polymdegradstab.2013.08.022.

Xu, Yan, Qun Gao, Hongqin Liang, and Kangsheng Zheng. 2016. "Effects of Functional Graphene Oxide on the Properties of Phenyl Silicone Rubber Composites." *Polymer Testing*. doi:10.1016/j.polymertesting.2016.07.013.

Xu, Hui, Li Xiu Gong, Xu Wang, Li Zhao, Yong Bing Pei, Gang Wang, Ya Jun Liu, Lian Bin Wu, Jian Xiong Jiang, and Long Cheng Tang. 2016. "Influence of Processing Conditions on Dispersion, Electrical and Mechanical Properties of Graphene-Filled-Silicone Rubber Composites." *Composites Part A: Applied Science and Manufacturing*. doi:10.1016/j.compositesa.2016.09.011.

Yang, Zhijun, Jun Liu, Ruijuan Liao, Ganwei Yang, Xiaohui Wu, Zhenghai Tang, Baochun Guo, et al. 2016. "Rational Design of Covalent Interfaces for Graphene/Elastomer Nanocomposites." *Composites Science and Technology* 132 (August): 68–75. doi:10.1016/j.compscitech.2016.06.015.

Yao, Wen Jun, and Bing Yang Cao. 2014. "Thermal Wave Propagation in Graphene Studied by Molecular Dynamics Simulations." *Chinese Science Bulletin*. doi:10.1007/s11434-014-0472-6.

Yao, Wen Jun, and Bing Yang Cao. 2016. "Triggering Wave-Domain Heat Conduction in Graphene." *Physics Letters, Section A: General, Atomic and Solid State Physics*. doi:10.1016/j.physleta.2016.04.024.

Yin, Biao, Jingyi Wang, Hongbing Jia, Junkuan He, Xumin Zhang, and Zhaodong Xu. 2016. "Enhanced Mechanical Properties and Thermal Conductivity of Styrene–Butadiene Rubber Reinforced with Polyvinylpyrrolidone-Modified Graphene Oxide." *Journal of Materials Science* 51 (12): 5724–37. doi:10.1007/s10853-016-9874-y.

Yin, Biao, Xumin Zhang, Xun Zhang, Jingyi Wang, Yanwei Wen, Hongbing Jia, Qingmin Ji, and Lifeng Ding. 2017. "Ionic Liquid Functionalized Graphene Oxide for Enhancement of Styrene-Butadiene Rubber Nanocomposites." *Polymers for Advanced Technologies* 28 (3): 293–302. doi:10.1002/pat.3886.

Yu, Aiping, Palanisamy Ramesh, Mikhail E. Itkis, Elena Bekyarova, and Robert C. Haddon. 2007. "Graphite Nanoplatelet-Epoxy Composite Thermal Interface Materials." *Journal of Physical Chemistry C*. doi:10.1021/jp071761s.

Yue, Sheng Ying, Runqing Yang, and Bolin Liao. 2019. "Controlling Thermal Conductivity of Two-Dimensional Materials via Externally Induced Phonon-Electron Interaction." *Physical Review B*. doi:10.1103/PhysRevB.100.115408.

Zárate, Ignacio A., Héctor Aguilar-Bolados, Mehrdad Yazdani-Pedram, Guadalupe Del C. Pizarro, and Andrónico Neira-Carrillo. 2020. "In Vitro Hyperthermia Evaluation of Electrospun Polymer Composite Fibers Loaded with Reduced Graphene Oxide." *Polymers* 12 (11): 1–16. doi:10.3390/polym12112663.

Zhan, Yanhu, Marino Lavorgna, Giovanna Buonocore, and Hesheng Xia. 2012. "Enhancing Electrical Conductivity of Rubber Composites by Constructing Interconnected Network of Self-Assembled Graphene with Latex Mixing." *Journal of Materials Chemistry*. doi:10.1039/c2jm31293j.

Zhan, Yanhu, Jinkui Wu, Hesheng Xia, Ning Yan, Guoxia Fei, and Guiping Yuan. 2011. "Dispersion and Exfoliation of Graphene in Rubber by an Ultrasonically- Assisted Latex Mixing and in Situ Reduction Process." *Macromolecular Materials and Engineering*. doi:10.1002/mame.201000358.

Zhang, Yinhang, and Ur Ryong Cho. 2018. "Enhanced Thermo-Physical Properties of Nitrile-Butadiene Rubber Nanocomposites Filled with Simultaneously Reduced and Functionalized Graphene Oxide." *Polymer Composites*. doi:10.1002/pc.24335.

Zhang, Fei, Yiyu Feng, Mengmeng Qin, Long Gao, Zeyu Li, Fulai Zhao, Zhixing Zhang, Feng Lv, and Wei Feng. 2019. "Stress Controllability in Thermal and Electrical Conductivity of 3D Elastic Graphene-Crosslinked Carbon Nanotube Sponge/Polyimide Nanocomposite." *Advanced Functional Materials*. doi:10.1002/adfm.201901383.

Zhang, Wenya, Qing-Qiang Kong, Zechao Tao, Jiacheng Wei, Lijing Xie, Xiaoyu Cui, and Cheng-Meng Chen. 2019. "3D Thermally Cross-Linked Graphene Aerogel–Enhanced Silicone Rubber Elastomer as Thermal Interface Material." *Advanced Materials Interfaces* 6 (12): 1900147. doi:10.1002/admi.201900147.

Zhang, Teng, and Tengfei Luo. 2016. "Role of Chain Morphology and Stiffness in Thermal Conductivity of Amorphous Polymers." *Journal of Physical Chemistry B*. doi:10.1021/acs.jpcb.5b09955.

Zhang, Guangwu, Fuzhong Wang, Jing Dai, and Zhixiong Huang. 2016. "Effect of Functionalization of Graphene Nanoplatelets on the Mechanical and Thermal Properties of Silicone Rubber Composites." *Materials* 9 (2): 92. doi:10.3390/ma9020092.

Zhang, Xumin, Jingyi Wang, Hongbing Jia, Biao Yin, Lifeng Ding, Zhaodong Xu, and Qingmin Ji. 2016. "Polyvinyl Pyrrolidone Modified Graphene Oxide for Improving the Mechanical, Thermal Conductivity and Solvent Resistance Properties of Natural Rubber." *RSC Advances* 6 (60): 54668–78. doi:10.1039/C6RA11601A.

Zhang, Xiang, Kai Wu, Yuhang Liu, Bowen Yu, Qin Zhang, Feng Chen, and Qiang Fu. 2019. "Preparation of Highly Thermally Conductive but Electrically Insulating Composites by Constructing a Segregated Double Network in Polymer Composites." *Composites Science and Technology*. doi:10.1016/j.compscitech.2019.03.017.

Zhang, Gang, and Yong-Wei Zhang. 2017. "Thermal Properties of Two-Dimensional Materials." *Chinese Physics B* 26 (3): 34401. doi:10.1088/1674-1056/26/3/034401.

Zhao, Xiong-wei, Chong-guang Zang, Yu-quan Wen, and Qing-jie Jiao. 2015. "Thermal and Mechanical Properties of Liquid Silicone Rubber Composites Filled with Functionalized Graphene Oxide." *Journal of Applied Polymer Science* 132 (38): n/a-n/a. doi:10.1002/app.42582.

Zheng, Junchi, Dongli Han, Suhe Zhao, Xin Ye, Yiqing Wang, Youping Wu, Dong, Jun Liu, Xiaohui Wu, and Liqun Zhang. 2018. "Constructing a Multiple Covalent Interface and Isolating a Dispersed Structure in Silica/Rubber Nanocomposites with Excellent Dynamic Performance." *ACS Applied Materials & Interfaces* 10 (23): 19922–31. doi:10.1021/acsami.8b02358.

Zheng, Wenge, Xuehong Lu, and Shing Chung Wong. 2004. "Electrical and Mechanical Properties of Expanded Graphite-Reinforced High-Density Polyethylene." *Journal of Applied Polymer Science*. doi:10.1002/app.13460.

Zhong, Bangchao, Huanhuan Dong, Yuanfang Luo, Dongqiao Zhang, Zhixin Jia, Demin Jia, and Fang Liu. 2017. "Simultaneous Reduction and Functionalization of Graphene Oxide via Antioxidant for Highly Aging Resistant and Thermal Conductive Elastomer Composites." *Composites Science and Technology* 151 (October): 156–63. doi:10.1016/j.compscitech.2017.08.019.

Zhong, Bangchao, Yongyue Luo, Wanjuan Chen, Yuanfang Luo, Dechao Hu, Huanhuan Dong, Zhixin Jia, and Demin Jia. 2019. "Immobilization of Rubber Additive on Graphene for High-Performance Rubber Composites." *Journal of Colloid and Interface Science* 550 (August): 190–98. doi:10.1016/j.jcis.2019.05.006.

Zhou, Wenying, Shuhua Qi, Chunchao Tu, Hongzhen Zhao, Caifeng Wang, and Jingli Kou. 2007. "Effect of the Particle Size of Al2O3 on the Properties of Filled Heat-Conductive Silicone Rubber." *Journal of Applied Polymer Science* 104 (2): 1312–18. doi:10.1002/app.25789.

Zhou, Wen-Ying, Shu-Hua Qi, Hong-Zhen Zhao, and Nai-Liang Liu. 2007. "Thermally Conductive Silicone Rubber Reinforced with Boron Nitride Particle." *Polymer Composites* 28 (1): 23–28. doi:10.1002/pc.20296.

Zhou, Lu-yu, Jiang-hao Ye, Jian-zhong Fu, Qing Gao, and Yong He. 2020. "4D Printing of High-Performance Thermal-Responsive Liquid Metal Elastomers Driven by Embedded Microliquid Chambers." *ACS Applied Materials & Interfaces* 12 (10): 12068–74. doi:10.1021/acsami.9b22433.

Zhuang, Changchang, Rongyao Tao, Xiaoqing Liu, Lizhao Zhang, Yiwen Cui, Yaqing Liu, and Zhiyi Zhang. 2021. "Enhanced Thermal Conductivity and Mechanical Properties of Natural Rubber-Based Composites Co-Incorporated with Surface Treated Alumina and Reduced Graphene Oxide." *Diamond and Related Materials*. doi:10.1016/j.diamond.2021.108438.

Zong, Yangyang, Dayong Gui, Shibin Li, Guiming Tan, Weijian Xiong, and Jianhong Liu. 2015. "Preparation and Thermo-Mechanical Properties of Functionalized Graphene/Silicone Rubber Nanocomposites." In *16th International Conference on Electronic Packaging Technology, ICEPT 2015*. doi:10.1109/ICEPT.2015.7236538.

16 Graphene-Elastomer Composite for Energy Storage Applications

A. M. Shanmugharaj
Vels Institute of Science, Technology and Advanced Studies

Anil K. Bhowmick
The University of Houston

CONTENTS

DOI: 10.1201/9781003200444-16

16.1 INTRODUCTION

According to the "International Union of Pure and Applied Chemistry (IUPAC)" nomenclature, elastomers are macromolecular materials exhibiting viscoelastic rubbery-like behavior, weak intermolecular forces, lower Young's modulus and higher % strain (McNaught and Wilkinson, 1997). In addition, excellent heat resistance, ease of deformation at room temperature, outstanding flexibility and elongation at break (EB) make elastomers an ideal cost effective candidate for packaging, automotive and healthcare applications. According to the historical records, elastomers derived from natural rubber (NR) latex were first used by Mesoamericans in about 1600 B.C., however, profound use of elastomers grown up rapidly after the development of vulcanization process by Charles Goodyear in 1839 (Hosler et al., 1999). Since then, there is a tremendous growth of elastomer industry due to the development of specialty elastomeric materials, resulting in the steady global annual revenue, foreseen to be US $84 billion by 2026 (Moore, 2021).

Though vulcanized neat elastomeric system is considered as the cost effective material for specialty applications, the mechanical damages such as fatigue cracks, blistering and swelling are the serious drawbacks that limit their application aspects. High-performance elastomeric composites with notable increment in the properties can be achieved by incorporating particulate fillers including carbon black, silica nanoparticles, carbon-silica dual phase fillers, inorganic fibers and inorganic fillers such as clay materials (Leblanc, 2002; Koerner et al., 2004; Sengupta et al., 2011; Roy and Bhowmick, 2012; Kotal and Bhowmick, 2015; Shanmugharaj and Bhowmick, 2003; Ray and Bhowmick, 2003; Rajeev et al., 2003; Kumar et al., 2001; Sadhu and Bhowmick, 2004; Bandyopadhyay et al., 2005; Agarwal et al., 2006; George et al., 2008). The final properties of the high-performance elastomeric composites can be effectively tuned *via*. proper selection of filler with distinguishable structural and geometrical features as well as manipulating its dispersion states in the chosen elastomeric system. Surface functionalization of fillers resulting in enhanced filler-elastomer interactions is the widely employed technique in improving their dispersion states in elastomeric system. After the development of nylon-clay nanocomposites by the Toyota researchers (Kojima et al., 1993), continuous research progress has been made in making high-quality elastomeric nanocomposites by exploring various nanofillers including nanoclays, nano-metal oxides and carbon allotropes as outstanding property enhancement can be achieved even at lower loading amount of these nanofillers.

Preparation of elastomeric nanocomposites using various carbon allotropes have been the subject of interest both in academic as well as industrial sectors in last few decades as these nanomaterials render outstanding property enhancements, making these nanocomposites an ideal candidate for a variety of applications. The oxidation states of carbon resulted in various allotropic forms including three-dimensional non-conductive nanodiamond particles, conductive crystalline graphite particles, hollow spherical buckminsterfullerenes and conductive cylindrical carbon nanotubes (multi-walled and single-walled carbon nanotubes). Incorporation of nanodiamond has profound effect on the thermal conductivity of NR composites, thus making it an ideal candidate for thermal energy storage applications (Poikelispää et al., 2021). In contrast to nanodiamond-based elastomeric composites, loading of electrically conductive graphite in elastomers has significant impact on the energy storage applications especially electrochemical (supercapacitors (SC); batteries and fuel cells) as well as thermal energy storages (i.e., Phase change materials). High-performance flexible printable SCs with energy (~10.22 µWh cm^{-2}) and power densities (~11.15 mW cm^{-2}) (measured at 10 mA cm^{-2}) and outstanding cycle life (~15,000 cycles) can be easily fabricated using graphite/polyurethane (PU) (1:1 wt.%) composites, thanks to its outstanding electroactive surface area with surface roughness of ~97.6 nm and noteworthy electrical conductivity (~0.318 S cm^{-1}) (Manjakkal et al., 2019). Alternatively, graphite/poly(vinylidene fluoride) (PVDF) composite anodes (with 90%–95% graphite content) are for the fabrication of high-performance lithium-ion (Li-ion) batteries. Yamaguchi and coworkers studied the effect of PVDF molecular weight and its functionality on the electrochemical characteristics of graphite-based Li-ion battery anode and they claimed that the PVDF on the composite plays a significant role in the enhancement of Li-ion storage capacity (Yoo et al., 2004). In addition, composite membranes based on graphite fiber/sulfonic acid grafted PVDF system exhibit outstanding proton conductivity with approximately ~3.8-fold rise in peak power output in comparison to nafion membrane, thus making it a potential electrolyte system for hydrogen-air fuel cells (Vinothkannan et al., 2021). The membrane-less air cathode-based single chamber microbial fuel cells with significantly reduced resistance can be made using nitrogen-doped polymer-metal-carbon composites based on polyvinyl alcohol and poly methyl vinyl ether-alt-maleic anhydride facing anolyte side and poly(dimethyl siloxane) (PDMS) on the air-facing side (Modi et al., 2016). On the other hand, advanced elastomeric nanocomposites based on hollow cylindrical carbon sheets, viz., carbon nanotubes, have been explored widely in energy storage applications due to the outstanding enhancement in electrical, mechanical and electrochemical properties. Incorporation of carbon nanotubes has shown the promise in the development of composite materials, yielding high dielectric constants, when loaded in dielectric elastomers (DEs) such as PDMS, thus exploring its application aspect in microcapacitor applications (Fan et al., 2018). However, dielectric properties of the carbon nanotube composite materials prepared using single host polymer face several technological issues such as precise control on CNT dispersion states, continuous carbon network formation at higher CNT loading and deterioration of other properties. However, the drawback can be effectively mitigated using polymer blends resulting in the double percolation network, thus making it an ideal candidate for microcapacitor applications (Yuan et al., 2012; Mao et al., 2020).

16.2 GRAPHENE NANOFILLERS

Among the various carbon nanostructures, which have been explored widely in a variety of applications, planar monolayer carbon allotrope, graphene, consisting of two-dimensional -(2D) array of carbon atoms is considered as the mother of all graphitic carbons as it can lead to the formation of various carbon nanomaterials by wrapping (zero-dimensional (0D) buckyballs), rolling (one-dimensional (1D) carbon nanotubes) or by stacking (three-dimensional (3D) graphite) (Figure 16.1) (Ibrahim et al., 2021). The unique and the intriguing properties of this most exciting planar 2D carbon have attracted the interest of many scientific groups to explore its usage in a

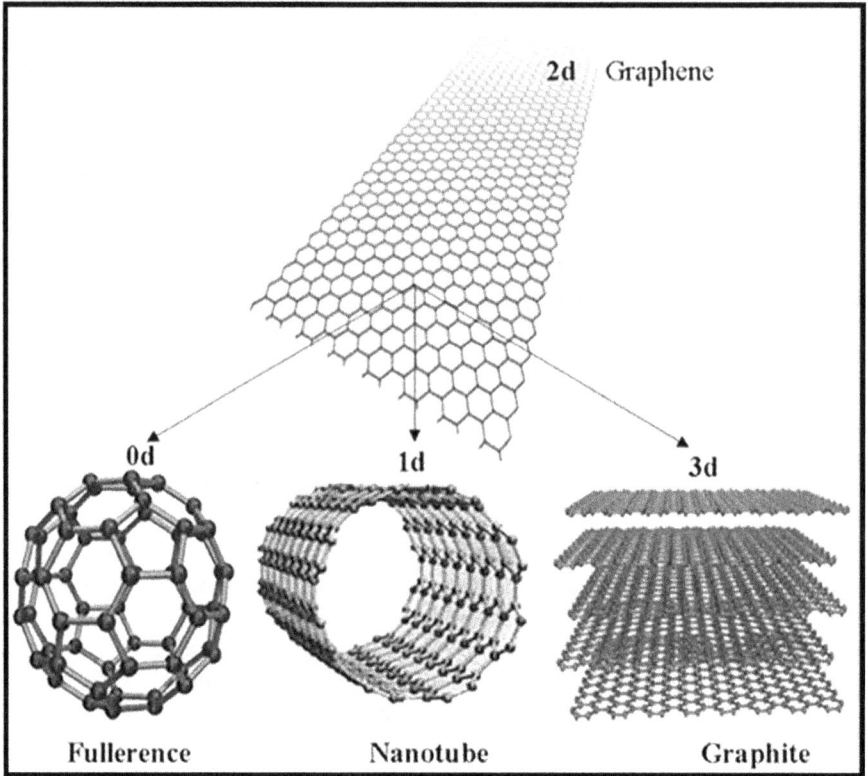

FIGURE 16.1   2D graphene derived carbon allotropes. (Reproduced from Ibrahim et al., 2021, Polymers, MDPI.)

variety of applications including electronic devices, energy storage materials, electronic devices, shielding and anti-static coatings. Though graphene is ideally a monolayer material consisting of honeycomb structures with two equivalent sub-lattices, due to the thermal fluctuations, wavy and rippled features observed in reality, thus promoting its ability to reinforce the polymeric materials. Alternatively, few-to-multilayered graphene layers can also be a potential candidate for the preparation of nanocomposite materials in relevance to the monolayer graphene material (Young et al., 2012). Owing to its nearly flat surface, high surface area and rumpled nature, graphene as nanofillers in polymeric system renders greater accessibility to the polymer molecules and induces strong interfacial interactions with the polymer matrix along with the decrement in the thermal interface resistance.

16.2.1 Preparation Methods

The demand for the production of high-quality, defect-free, low-cost graphene in bulk quantities has risen tremendously due to the endless innovations on its application aspects in various areas of technology such as composites, sensors, electronics,

SCs, quantum computations and medical applications. The market analysis by Reiss and coworkers revealed that the annual production capacity of graphene is likely to be 2,500 tons per annum since 2018 with the annual global demand of ~400 tons (Döscher et al., 2021). In the present scenario, the production supply of graphene is on the higher side than its demand and is attributed to the high production cost, which restricts the usage of graphene in certain applications that need low-cost raw materials. This can be offset by employing improved and low-cost production methods. Also, the structure, thickness and properties of the graphene nanomaterials can be effectively tuned by adopting a suitable preparation method. The most commonly employed synthetic approaches of graphene fall under two sub-categories, viz., (a) Bottom-up and (b) Top-down approach.

16.2.1.1   Bottom-Up Approach

Preparation of high-quality graphene using small and atomically precise building blocks is termed as "bottom-up" approach and the commonly employed techniques are classified as chemical vapor deposition (CVD) and epitaxy growth (Tour, 2013). CVD is the widely employed technique for the preparation of defect-free monolayer graphene by vapor deposition of gaseous hydrocarbon reactants decomposing to carbon radicals at the selected temperature on the metal surface such as nickel (Ni) substrate, which subsequently lead to the formation of graphene film (Yu et al., 2008; Zhang et al., 2013). The schematic representation depicting the deposition and nucleation of carbon atoms and its subsequent nucleation leading to the large domains is shown in Figure 16.2a. The corresponding field emission scanning electron microscopic (FE-SEM) images revealed the influence of annealing temperature on the growth of the graphene structures (Yan et al., 2012). Since its first development, several types of carbon feedstock's as well as metal substrates have been explored for the fabrication of graphene film (Gong et al., 2012; Ruan et al., 2011). Though CVD is an excellent technique to produce high-quality few-layer graphene with controlled grain size, one of the technological issues that need to be addressed is the proper exfoliation of the graphene from the metal substrate as its structure and properties depend on the successful exfoliation technique. In 2009, Ruoff and coworkers established a scale-up process in developing large-area defect-free graphene by roll-to-roll process, which could subsequently be transferred to a polymer film (Li et al., 2009). In the current status, graphene films prepared by CVD process find applications such as transparent electrodes, sensors, coatings or electronic devices. However, the mass production of graphene by CVD process is potentially hindered by high energy consumption process, and with the existing technologies, it is impossible to produce bulk quantities for nanocomposite preparation. However, recent attempts pave a way in preparing bulk production of graphene by CVD process (Kim et al., 2009).

Alternatively, in the epitaxy method, single crystalline silicon carbide (SiC) wafer is subjected to high temperature (~2,000°C) under argon or vacuum, generating highly crystalline graphene due to the sublimation of silicon atoms from the (0001) face of the crystal (Figure 16.2b). Low temperature growth and weaker interaction with the SiC substrate make it a most viable technique for the production of the graphene. Another key advantage of the epitaxy method is the generation of homogenous graphene film with the control over the number of layers just by adjusting the temperature. In addition, there is a formation of electrically inert buffer layer (shown as the broken line in Figure 16.2b) with honeycomb lattice structure having periodical arrangement of pentagon/heptagon superstructures (Norimatsu and Kusunoki, 2014; Qi et al., 2010). Earlier chapters have discussed these in details.

16.2.1.2 Top-Down Approach

The synthesis of graphene by top-down approach involves four major production methods including (i) mechanical exfoliation, (ii) chemical synthesis route, (iii) liquid phase exfoliation and (iv) electrochemical exfoliation.

   i. Mechanical exfoliation

      In 2004, Geim and coworkers first prepared the monolayer graphene using the scotch tape method, involving repeated peeling of graphene layers

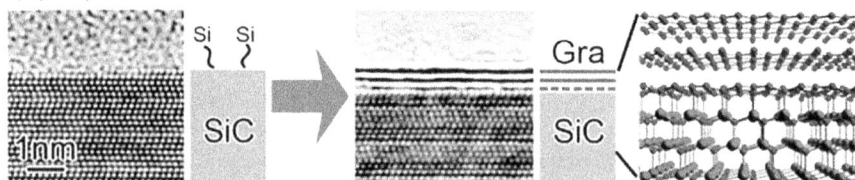

FIGURE 16.2 Preparation of graphene by (a) chemical vapor deposition and FE-SEM images of graphene films at different annealing temperatures. (Reproduced with permission from Yan et al., Copyright 2012, ACS Publications.); (b) epitaxial method, (Reproduced with permission from Norimatsu and Kusunoki, Copyright 2014, Royal Society of Chemistry.)

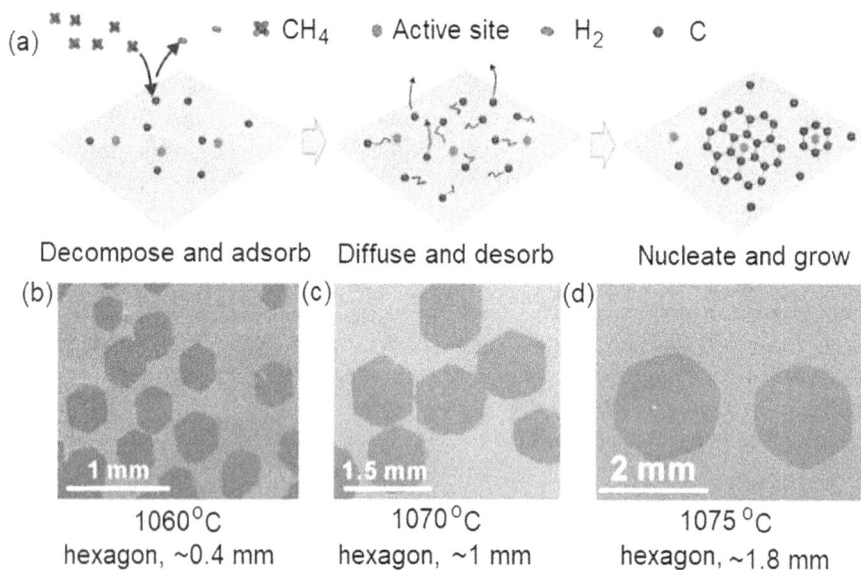

from highly oriented pyrolytic graphite or nature graphite which subsequently received wide attention among the scientific community (Novoselov et al., 2004). Through this mechanical exfoliation, the number of graphene layers is effectively controlled; on the other hand, dimensions of graphene solely depend on the grain size of the graphite, which is generally in the range of tens of microns. Though this scotch tape method yields high-quality graphene with excellent properties, bulk production of graphene is not feasible by this mechanical exfoliation (Gong et al., 2010).

ii. Liquid-phase exfoliation

The liquid-phase exfoliation method is one commonly employed effective technique in preparing graphene nanosheets by exposing graphite to various organic solvents such as tetrahydrofuran, N-methyl pyrrolidone and dimethylsulfoxide (Hernandez et al., 2008; Park et al., 2009).

Figure 16.3 shows the schematic representation of the solvent-induced graphite exfoliation, which involves following steps; (i) dispersion, (ii) exfoliation and (iii) purification (Ciesielski and Samori, 2014). By the proper choice of the organic solvent with surface tension equal to the graphene-graphene interaction energy, exfoliated monolayer graphene can be effectively prepared by simple sonication process, which is aided by the graphene-solvent interaction.

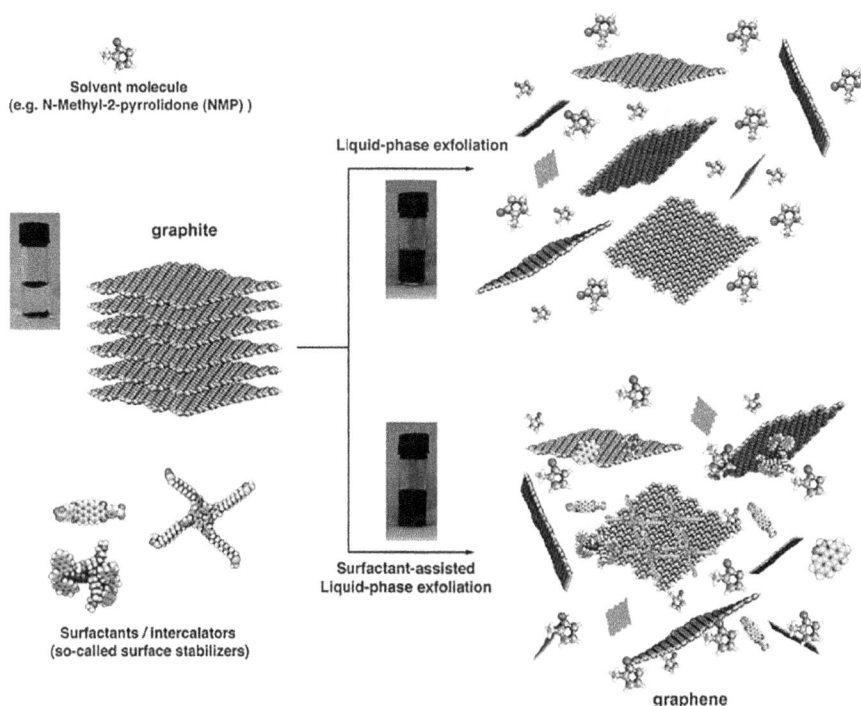

FIGURE 16.3 Graphite exfoliation under liquid phase in the presence and absence of surfactant. (Reproduced with permission from Ciesielski and Samori, Copyright 2014, Royal Society of Chemistry.)

Alternatively, aqueous-surfactant suspension is also demonstrated to be a most effective technique in preparing graphene sheets from the graphite (Lotya et al., 2009). Alternatively, high-quality few-layer graphene can be made using the aqueous dispersion of graphite in the presence of inorganic salts such as sodium chloride (NaCl) or copper chloride ($CuCl_2$) (Niu et al., 2013). Though liquid phase exfoliation is effective in preparing high-quality few-layer graphene, it cannot be scaled up to the industrial scale for the preparation of composite materials as the process is expensive and also not an eco-friendly procedure. In addition, electrical property of the graphene material prepared by liquid exfoliation is far inferior to the high-quality graphene prepared by mechanical exfoliation, CVD or epitaxial method.

iii. Thermal exfoliation

In contrast to the liquid-phase exfoliation technique, thermal shock treatment involving rapid microwave heating of acid-treated graphite is the effective technique in producing bulk quantities of 10-nm-thick graphite nanoplatelets (Xiang and Drzal, 2011). The transmission electron microscopic (TEM) results of thermally exfoliated graphite nanoplatelets with large aspect ratio are shown in Figure 16.4 (Xiang and Drzal, 2011).

Microwave treatment of acid intercalated graphite resulted in the rapid expansion of the graphite layers due to the vaporization of the intercalated acid molecules yielding pressures exfoliating the graphene sheets (McAllister et al., 2007). Alternatively, few-to-multilayered graphene with desired properties of single-layer graphene can also be produced by simple thermal shock treatment employing high temperature (~1,050°C), elevated pressure (~700 $m^2 g^{-1}$) and fast heating rates (> 2,000°C min^{-1}) (Zang et al., 2011).

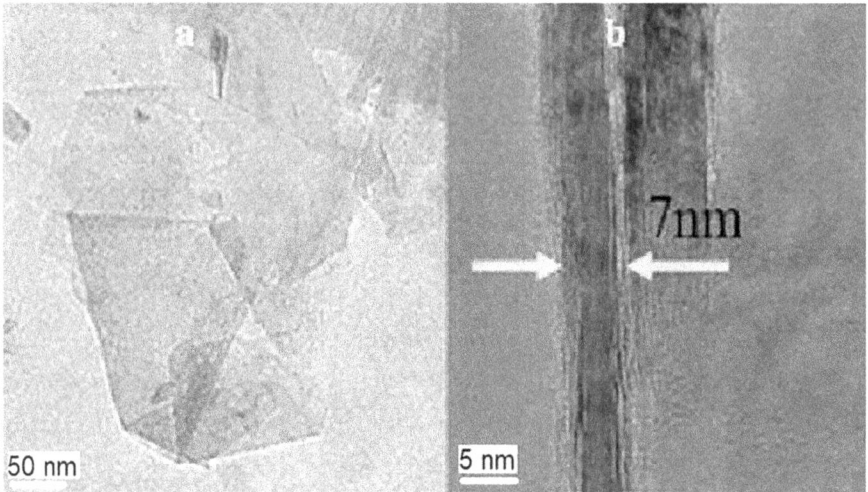

FIGURE 16.4 TEM results of the graphitic nanoplatelets prepared by microwave exfoliation. (Reproduced with permission from Xiang and Drzal, Copyright 2011, Elsevier.)

iv. Chemical exfoliation of graphite

Though thermal exfoliation of graphite can be an efficient technique in producing the bulk quantities of graphite nanoplatelets, synthetic technique used in the preparation of bulk quantity of the single-layered graphene is still a difficult process and needs a development of suitable processing technique. Alternatively, chemical exfoliation of graphite resulting in the formation of stacked layers of graphene oxide is a well-known concept, which received wide attention in recent years as it is one of the best techniques in producing bulk quantities. The first synthetic approach on chemical exfoliation of graphite was reported by Brodie 160 years ago by oxidizing the graphite using fuming nitric acid and potassium chlorate (Brodie, 1860). Later, the preparation steps were modified by using the mixture of sodium nitrate, concentrated sulfuric acid and potassium permanganate (Staudenmaier, 1898; Hummers Jr and Offeman, 1958). The main advantage of the chemical exfoliation technique is its potentiality in synthesizing monolayer graphene with the tuned interlayer spacing of graphite, ranging from 0.6 to 1 nm by introducing oxygen functionalities, which subsequently lead to the intercalation of the water molecules (Hummers Jr and Offeman, 1958). Also, the production of graphene oxide by chemical exfoliation technique can be scaled up to the industrial scale for the mass production of this material. However, poor electrical conductivity and reduced thermal stability are the major drawbacks associated with the graphene oxide prepared by chemical exfoliation technique. However, these demerits of graphene oxide can be offset by chemical or thermal reduction leading to a reduced graphene oxide with partial recovery of the graphene properties. The reducing agents that are often employed in the chemical restoration of graphene properties through the removal of oxygen functionalities includes L-ascorbic acid (Park and Ruoff, 2009), sodium borohydride (Zhang et al., 2010) and hydrazine hydrate (Shin et al., 2009). Alternatively, base hydrolysis using aqueous sodium hydroxide (NaOH) solution of graphene oxide is also a commonly employed technique in producing high-quality graphene with restoration of the electrical conductivity values (Rourke et al., 2011). Alternatively, thermal reduction is the widely employed technique in removing the oxygen functionalities of the graphene oxide and it is one of the best techniques in producing bulk quantities of graphene-like material for the nanocomposite preparation (Acik et al., 2011; Guerra et al., 2013; Guo et al., 2010; Kumar et al., 2016).

16.2.2 PHYSICOCHEMICAL PROPERTIES

16.2.2.1 Mechanical Properties

As discussed in the previous sections, synthesized graphene materials can be of monolayer or few-to-multilayered structures depending upon the method of preparation. Based on the nanoindentation measurements, Young's modulus of monolayer graphene prepared through bottom-up approach is ~1,000 GPa, whereas the accepted Young's modulus value of chemically prepared graphene-like materials, i.e. graphene oxide value is observed to be ~250 GPa (Lee et al., 2008; Gomez-Navarro et al., 2008).

The relatively lower value of the modulus of elasticity for GO is attributed to the structural disruption caused by the formation of sp^3 hybridized carbon due to chemical oxidation of graphite. In addition, chemical oxidation results in sheet thickening, which further be responsible for the reduction in Young's modulus values (Dikin et al., 2007). Raman spectroscopic characterization has also been employed in calculating the modulus of elasticity by inducing mechanical strains on the monolayer graphene and the obtained values are consistent with Young's modulus values obtained through the nanoindentation technique (Ferralis, 2010). Shen and coworkers employed Raman spectroscopy to determine the mechanical properties of the graphene oxide (GO) paper under strain conditions and the calculated effective Young's modulus (~230 GPa) is consistent with the indentation technique (Ni et al., 2008).

16.2.2.2 Thermal Conductivity

Owing to its 2D configuration, monolayer graphene has extraordinary thermal conductivity (~3,000 W mK^{-1}) at room temperature in comparison to its bulk graphite counterpart. Based on optothermal Raman studies, it has been further clarified that the thermal conductivity of the graphene decreases with increasing number of graphene layers and reaches the value of bulk graphite after stacking 3~4 layers of graphene (Balandin, 2011). The progressive decrement in the thermal conductivity is correlated to the phonon scattering, which substantially varies due to the increasing number of layers rendering more phase-space states for phonon scattering, thereby leading to the decrement in the thermal conductivity values. Contrastingly, graphene oxide prepared through chemical exfoliation has significantly lower thermal conductivity at room temperature depending on the oxidation level (~72 W mK^{-1} at the oxidation level 0.35; ~670 W mK^{-1} at the oxidation level 0.05) (Chen and Li, 2020).

16.2.2.3 Electrical Conductivity

The outstanding electrical properties (10^4~10^5 S m^{-1}) of the monolayer graphene are attributed to the zero overlap semi-metal characteristics consisting of both electrons and holes as the charge carriers with experimental carrier mobility value as high as 15,000 cm^2 V^{-1} S^{-1}. However, carrier mobility can be significantly improved on reducing the impurity level in the graphene (Geim and Novoselov, 2007). These overwhelming properties make the graphene as the ideal candidate for nanoelectronics applications. On the contrary, graphene oxide derived from chemical exfoliation is insulating in nature, which can be made potentially conductive by applying various reduction methods (Liu et al., 2013).

16.2.3 TERMINOLOGIES OF GRAPHENE-BASED MATERIALS

Rapid growth in the synthetic strategies and production methods of graphene and its derivatives has caused misunderstandings regarding the terminologies to be used for graphene-based materials. Zhang and coworkers provide some insights on the graphene family tree, eliminating these misunderstandings (Bianco et al., 2013). A report by Bianco and coworkers clearly demonstrates the classification framework of graphene-based materials, which are categorized through

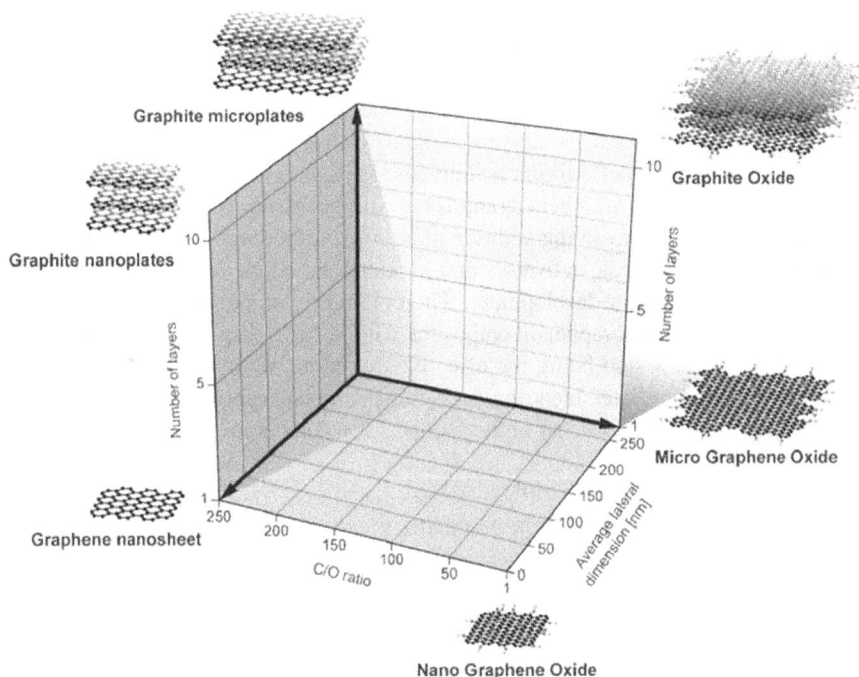

FIGURE 16.5 Fundamental classification of graphene-based materials based on number of layers, lateral dimensions and C/O ratio. (Reproduced with permission from Wick et al., Copyright 2014, John Wiley & Sons.)

three fundamental properties: (i) number of graphene layers, (ii) average lateral dimension and (iii) atomic carbon/oxygen ratio (Wick et al., 2014). Included in Figure 16.5 is the classification grid of graphene-based materials, distinguished through the abovementioned frameworks.

16.3 GRAPHENE-ELASTOMER COMPOSITES

16.3.1 PREPARATION METHODS

Owing to its high flexibility, affordable cost and good processing characteristics, elastomers are employed as lightweight multicomponent structural counterparts to metals in various engineering applications. Though the incorporation of conventional reinforcing fillers such as carbon black or silica has profound impact on the mechanical properties, viz., Young's modulus, hardness, tear strength, abrasion resistance and dynamic mechanical properties, loading amount of these fillers should be substantially higher to achieve the desired properties. On the other hand, loading of nanofillers including graphene has significant effect on the engineering properties of the elastomeric composites at significantly lower loading amount. Significant work has been done in recent past on the graphene-based elastomeric composites pertaining to various advanced applications including tires, automotive components, seals, biomedical, sensing devices and nanoelectronics. However, properties of the nanocomposites need to be effectively tuned in realizing its application aspects, which solely depend on the dispersion states of the nanofillers in elastomeric systems. Dispersion states of graphene filler in elastomers inducing elastomer-filler and filler-filler interaction can be altered by employing various processing methods such as latex blending, in situ polymerization, solution and melt mixing techniques (Papageorgiou et al., 2015). Since the preparation of elastomer/graphene composites with necessary property enhancement is the complex process involving addition of multiple ingredients including crosslinking agents, processing aids, activators and accelerators, single-step processing is not a viable technique for its development. The commonly employed technique in the fabrication of elastomer/graphene composites with desired properties is the two-step process with the first step being the effective dispersion of graphene nanofillers in elastomers by solution/latex blending, followed by the incorporation of curing ingredients using melt mixing process in the second step (Potts et al., 2012, 2013).

16.3.1.1 In situ Polymerization

An efficient synthetic approach in preparing graphene-based composites is the in situ polymerization, where the macromolecular chains can be successfully incorporated between the layered graphene structures leading to its complete exfoliation in the matrix (Paszkiewicz et al., 2014). The standard technique employed in the in situ synthesis of elastomer/graphene composites is the mixing of graphene nanofillers with the polymerizable monomer in a solvent followed by the in situ polymerization (Lee et al., 2009). Lu and coworkers synthesized waterborne polyurethane/graphene (PU/GNS) composites by in situ polymerization using fine dispersed graphene in the monomers of PU, viz., diphenylmethane diisocyanate (MDI) and poly (tetramethylene glycol) (Wang et al., 2011). Outstanding property enhancement especially mechanical and thermal properties can be achieved through this in situ polymerization method, thanks to the elastomer-graphene interaction inducing fine dispersion of graphene fillers in the elastomer matrix (Wang et al., 2011). Alternatively, Cui and coworkers developed a novel in situ emulsion polymerization strategy in synthesizing highly aligned graphene oxide (GO) nanofillers in the PU matrix (Figure 16.6a) (Sang et al., 2018). The morphological features of the GO, pre-polymerized PU and WPU-GO emulsions obtained from TEM characterization are included in Figure 16.6b–d. In contrast to GO, which showed 2D sheet morphology (Figure 16.6b), the pre-polymerized WPU showed irregular bubble patterns on the GO surface (Figure 16.6c); on the other hand, solid sphere like configurations are observed at the edges of GO in case of WPU-PU system. This morphological change is corroborated by the in situ polymerization and chain extending reactions of WPU, where the carbonyl and hydroxyl groups are present in the GO. The modulus and thermal stability of GO-filled waterborne polyurethane (WPU) are significantly improved, when compared to the unfilled WPU system.

For instance, loading 2 wt.% GO filler has increased the modulus to about 193% and a blue shift in thermal degradation temperature (~30°C) in comparison to neat WPU system. However, lower elastomer viscosity and inferior electrical conductivity values of synthesized elastomer/graphene nanocomposites are the limiting factors, which offset its usage in certain advanced applications.

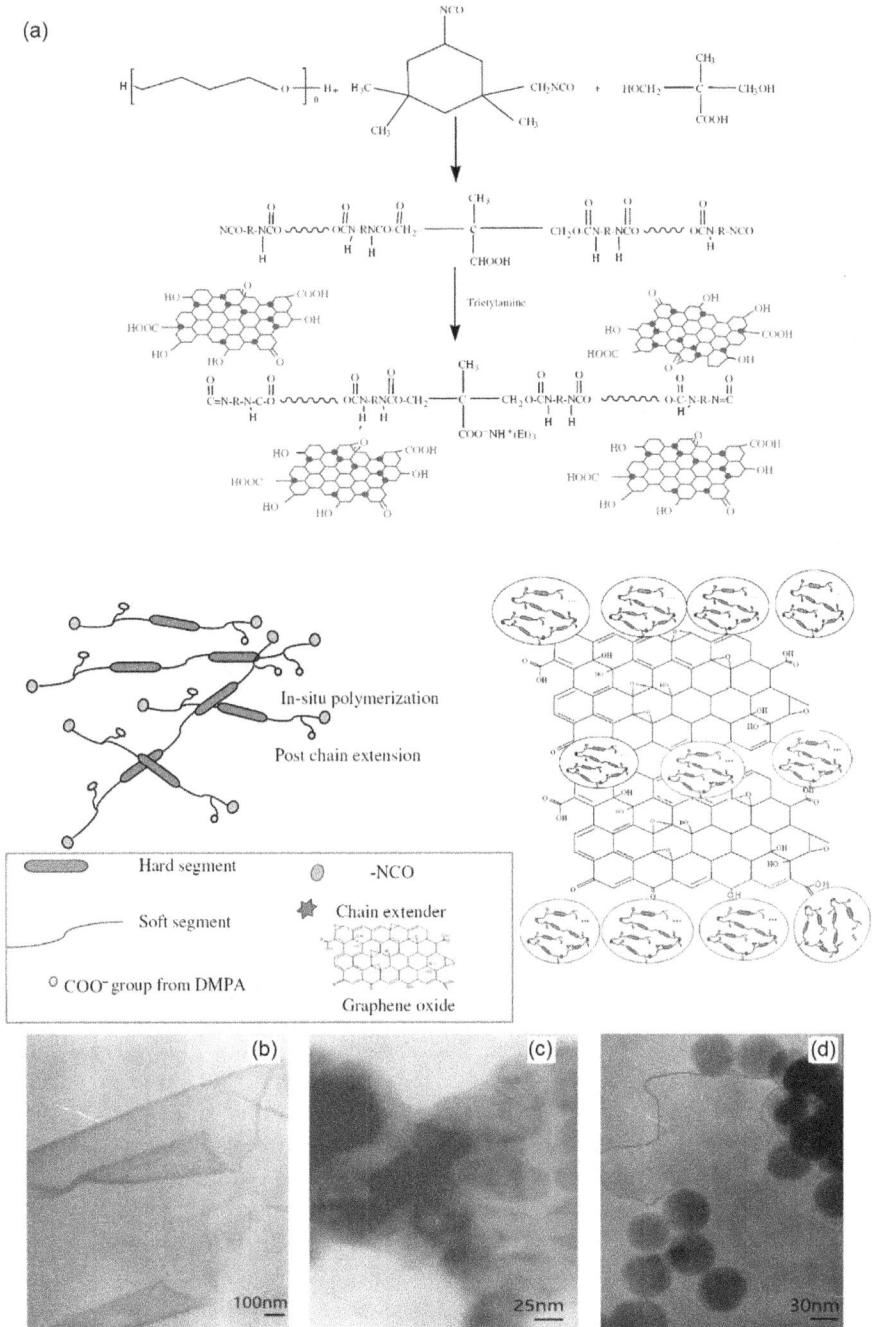

FIGURE 16.6 (a) Schematic representation of waterborne polyurethane (WPU)/graphene oxide (GO) nanocomposites. (b–d) TEM results of graphene oxide (GO). Pre-polymerized waterborne polyurethane (WPU) and WPU-GO emulsion. (Reproduced from Sang et al. 2017, De Gruyter.)

16.3.1.2 Solution/Latex Blending

The most frequently employed technique by the academic researchers in the preparation of elastomer/graphene nanocomposites is the solution/latex blending method, since the suspensions of the graphene can be incorporated into the elastomeric matrix without further processing (Potts et al., 2012, 2013; Wu et al., 2013; Araby et al., 2014; Zhan et al., 2011, 2012). Furthermore, fine dispersion of 2D sheet-like fillers in the elastomers can be easily achieved by solution mixing without any functionalization procedure. Elastomer/graphene composites can generally be prepared by dissolving the elastomer with the graphene dispersion in the same solvent or by dispersing the graphene filler in the elastomer solution (elastomer/solvent) by adopting high shear mixing or by ultrasonication. However, from the environmental aspect, the latex blending method is expected to be an industrially viable technique in preparing the elastomer/graphene composites as the starting precursor used for the preparation is in latex form. Significant number of preparation strategies have been proposed in recent past using latex precursors and functionalized graphene materials in making elastomer/graphene composites with improved graphene dispersion states (Potts et al., 2013; Zhan et al., 2012, 2013; Yan et al., 2013; Kim et al., 2011; Tian et al., 2014; Xing et al., 2014). The faster co-coagulation process induced by latex emulsions, prevents the aggregation of graphene fillers, thereby improving the quality of dispersion (Wei et al., 2014).

Nishi and coworkers prepared graphene oxide nanosheet (GONS) (~0.75 vol.%) loaded carboxylated nitrile rubber (XNBR) composites by latex blending technique (Tian et al., 2014). Figure 16.7 depicts the mechanism of GONS encapsulation on XNBR latex particles and the subsequent formation of segregated network. Ultrasonic-assisted latex mixing technique was employed in the preparation of GONS/XNMBR composites, which was subsequently dried by a spin flash drying process, resulting in the formation of GONS-encapsulated XNBR microspheres that are held together by the hydrogen-bonded interaction between carboxylated nitrile rubber and oxygen functionalities of GONS (Figure 16.17a and b).

Additionally, π–π interaction between C=N or C=C groups of nitrile rubber and C=C groups of GONS is equally responsible for the formation of segregated networks. The adoption of in situ vulcanization using hot pressing technique resulted in the formation of GONS-encapsulated XNBR networks (Figure 16.17c). Finally, the GONS segregated network is formed, since the crosslinked XNBR create an excluded volume that essentially distributes the GONS in the interstitial space between them (Figure 16.17d). These segregated networks are retained even after the cooling as the diffusion of GONS within the XNBR matrix due to the high viscosity of the crosslinked rubber particles. Thermal reduction at 190°C for 2 hours resulted in the formation of thermally reduced graphene (TRG)/XNBR composites with segregated networks (Figure 16.8a). As shown in Figure 16.8a, crosslinked GONS/XNBR composites has a lot of functional sites in the segregated GONS network, which is drastically reduced upon thermal treatment. The corresponding TEM images of TRG/XNBR composites at two different magnifications are given in in Figure 16.8b and c , which clearly indicates the formation of a segregated network as proposed in the scheme (Figure 16.8a). A notable increment in the dielectric constant value (at 100 Hz) is observed on loading 0.75 vol.% of TRG in XNBR composites (~5,542), when compared to XNBR composites (~23) (Tian et al., 2014).

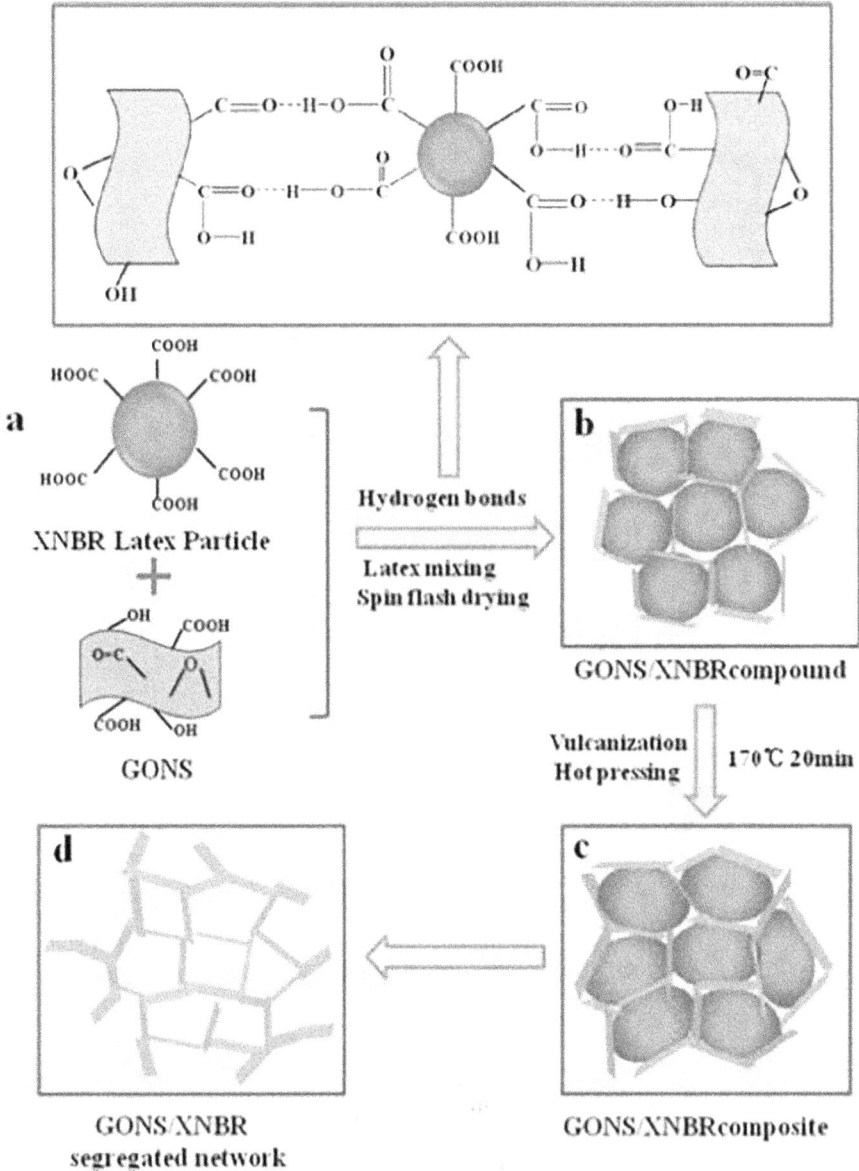

FIGURE 16.7 Schematics of the preparation XNBR/GONS composites by latex blending. (Reproduced with permission from Tian et al., Copyright 2014, Elsevier.)

16.3.1.3 Melt Mixing

Owing to its low cost and high speed, the melt mixing method has been considered as the industrially viable technique for the preparation of elastomer/graphene composites (Das et al., 2012; Lin et al., 2014). In contrast to solution/latex blending, dispersion of graphene nanofillers in the elastomeric matrix is achieved by applying high shear force under the molten state.

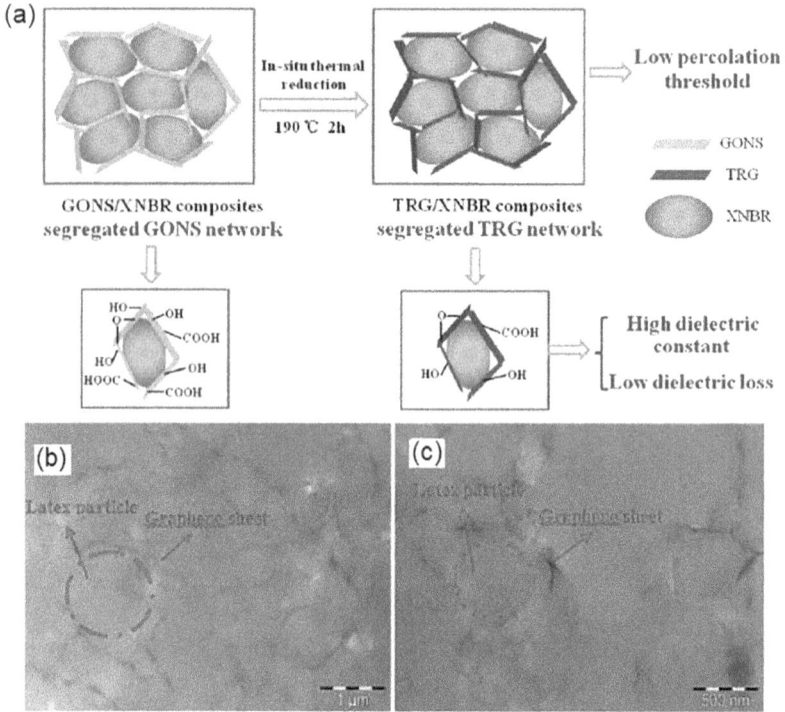

FIGURE 16.8 (a) Schematic representation of TRG/XNBR composites. (b, c) TEM results of TRG/XNBR composites at two different magnifications. (Reproduced with permission from Tian et al., Copyright 2014, Elsevier.)

However, the issues associated with the melt mixing method are the degradation of the elastomers during the high temperature processing. In addition, drastic rise in viscosity due to the addition of graphene fillers also hampers its dispersion states in the elastomer matrix (Zhan et al., 2011). Though the high shear forces induced in the melt mixing method are sufficient enough to overcome the viscosity of the matrix, and it can also lead to the breakage of the graphene or graphene oxide sheets. In contrast to other methods, melt mixing process generally resulted in poor dispersion of nanofillers in the elastomeric matrix (Kim et al., 2010). Figure 16.9 displayed the morphological features of graphite, graphene oxide (GO) and thermally reduced graphene oxide (TRG) in thermoplastic polyurethane (TPU), which clearly depict inhomogeneities in the graphene dispersion prepared by melt mixing in relevance to solution/latex blending or in situ polymerization.

16.3.2 Physicochemical Properties

16.3.2.1 Mechanical and Dynamic Mechanical Properties

Loading of graphene as the nanofiller in an elastomer matrix showed outstanding improvements in the mechanical properties such as Young's modulus and the tensile

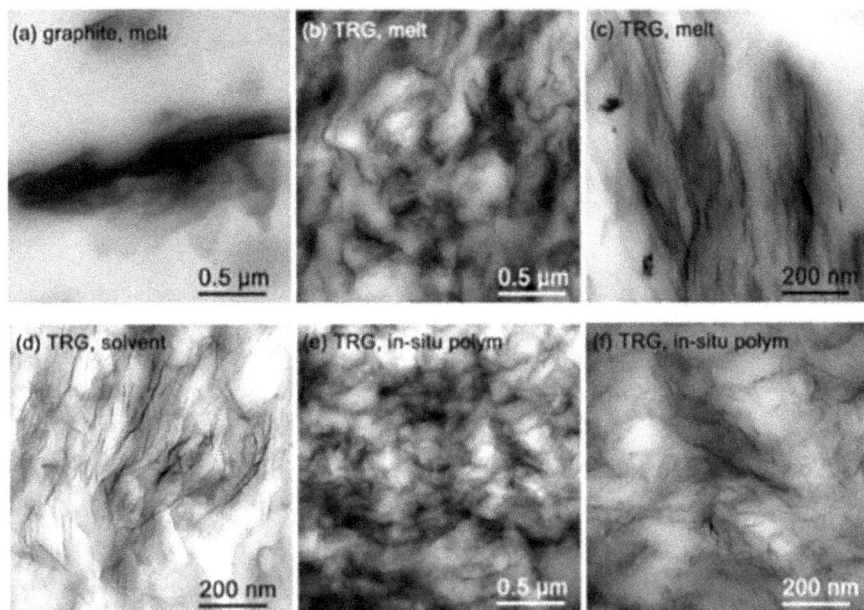

FIGURE 16.9 TEM results depicting graphene dispersions in the TPU matrix: (a) 5 wt.% (2.7 vol.%) graphite, (b, c) melt blended, 5 wt.% TRG, (d) solvent blended, TRG, (e, f) in situ polymerized 3 wt.% TRG. (Reproduced with permission from Kim et al., Copyright 2010, American Chemical Society.)

strength in comparison to the pure elastomer. In contrast, the variation in the EB values depends on the processing method and filler loading amount. For instance, notable improvement in EB is observed at lower graphene content, whereas the values decreased at higher concentration due to the formation of aggregates. Bhattacharya, Maiti, and Bhowmick (2008) reported first the effect of expanded graphite and various nanomaterials on the mechanical properties of NR and demonstrated improved modulus and hysteresis by using expanded graphite at low loading. They also reported the viscoelastic properties of many nanocomposites (Bhattacharya and Bhowmick, 2010). Later extensive research on the reinforcing ability of the reduced graphene oxide (RGO) in NR composites has been done by Ruoff and coworkers (Potts et al., 2012, 2013). According to their report, latex co-coagulation of NR/ RGO composites resulted in web-like morphology, whereas subsequent two-roll mill processing resulted in homogenous dispersion of the graphene nanofillers in the NR matrix. While the web-like morphology is highly beneficial for the stiffness improvement, it is detrimental to the EB values (Potts et al., 2013).

Included in Figure 16.10a and b are the stress-strain results of NR/RGO composites prepared by solution/latex mixing and mill mixing method. As evidenced from the figure, NR/RGO composites prepared by latex/solution mixing showed significantly higher modulus, whereas EB value is higher for milled NR/RGO composites. In addition to the mechanical properties, dynamic mechanical analysis of NR/RGO

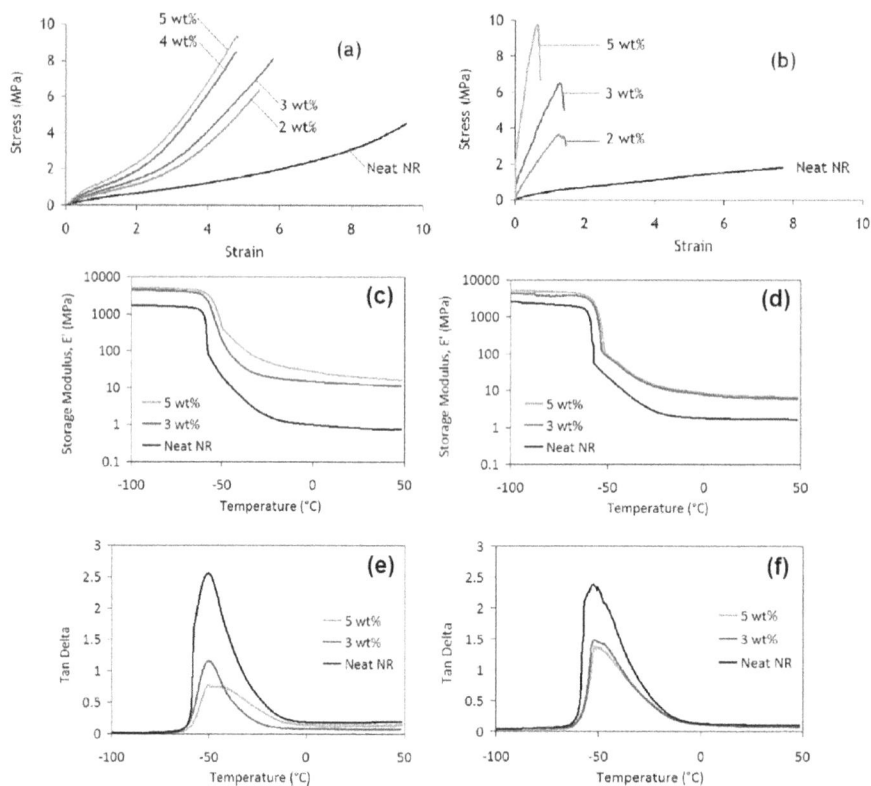

FIGURE 16.10 (a, b) Stress-strain results of solution and mill-processed NR/RGO composites. (c, d) Variation of storage modulus with temperature for solution and mill-processed NR/RGO composites. (e, f) Tan delta vs. temperature results of solution and mill-processed NR/NRGO composites. (Reproduced with permission from Potts et al., Copyright 2012, American Chemical Society.)

composites showed significant improvement in the storage modulus, E', over the range of temperature for both the solution as well as mill processed composite materials, when compared to neat NR samples. The variation of the tan delta with temperature range for both the NR/RGO composites is included in Figure 16.10e and f, which showed drastic reduction in the peak height and breadth, when compared to neat NR. Effective immobilization of NR chains near the elastomer-graphene interface is responsible for the variation in the tan delta values. In addition, the decrement in the tan delta peak amplitude is more pronounced for the solution-treated NR/RGO composites relative to the mill-processed NR/RGO composites.

In the subsequent study by Potts et al. (2012) on conventional melt processing, thermally exfoliated graphene oxide loaded NR composites (NR/TRGO) showed significant mechanical property enhancements. Mülhaupt and coworkers (Beckert et al., 2014) studied the reinforcing ability of chemically reduced GO and thermally reduced GO in styrene-butadiene rubber (SBR) system, which showed marked improvement in the mechanical properties such as modulus, stiffness and strain at break.

In general, graphene-filled elastomer composites enhance both the mechanical and dynamic mechanical properties, with the marked improvements being obtained with well-exfoliated, uniformly dispersed fillers. The greatest improvements in the properties are generally correlated to the synergistic contribution of modulus of graphene filler (up to 1,012 MPa), which is notably higher, when compared to the neat elastomers. In addition, by adopting suitable functionalization method, filler content and processing method, the properties of the resultant composite materials can be effectively tuned. The subject has been extensively reviewed by Bhowmick and coworkers (Maiti et al., 2008; Bhowmick, 2008).

16.3.2.2 Thermal Conductivity

The thermal conductivity of the elastomers showed drastic enhancement on suspending graphene in the amorphous elastomeric matrix, thanks to the sheet-like morphology of graphene rendering lower interfacial resistance, higher heat capacity and exceptional electrical conductivity, when compared to the host polymer. The strong interfacial bonding characteristics induced by the exfoliated graphene reduced phonon scattering or acoustic impedance mismatch, thereby enhancing the thermal conductivity in elastomeric composites. However, Kapitza resistance (interfacial resistance) induced by the graphene filler aggregates in the elastomer matrix as well as poor thermal coupling between the elastomer and the graphene is the serious limiting factor of the resultant composite materials (Kim et al., 2010).

Significant research has been done in recent past on the thermal conductivity of the elastomer/graphene composites (Potts et al., 2012, 2013; Zhan et al., 2011; Sherif et al., 2013; Xiong et al., 2013; George and Bhowmick, 2009). Thermal conductivity of the elastomer/graphene composites can be effectively tuned by employing a suitable processing method. For instance, thermal conductivity of NR/graphene composites is notably higher for the composites prepared by latex coagulation rather than the milling method due to the formation of interconnected morphologies in the former compared to the latter (Figure 16.11a). Another study by Ding and coworkers (Xiong et al., 2013) reported the role of ionic liquid functionalized graphene oxide (GO-IL) on the thermal conductivities of the bromobutyl (BIIR) rubber composites, which showed marked improvements with increasing wt.% of GO-IL (Figure 16.11b). Thermal conductivity of EVA–Expanded graphite at four parts was found to be in the range of 0.71–0.89 W mK^{-1} depending on the vinyl acetate content (higher value was obtained for higher VA content (George and Bhowmick, 2009).

The thermal conductivity of the elastomer/graphene composites is one of the desirable properties which upon tuning can lead to its application potential in power electronics, electric motors, etc.

16.3.2.3 Electrical Properties

Introduction of electrically conductive graphene and its derivatives as nanofillers in elastomers has profound impact on the electrical properties of the resultant composites, thanks to its high surface area and inherent electrical conductivity. Significant number of research publications have been made in recent times on the graphene nanofiller loaded elastomeric composites (George et al., 2010; Liu et al., 2013; Araby et al., 2014; Zhan et al., 2012; Kim et al., 2010, 2011; Luo et al., 2014;

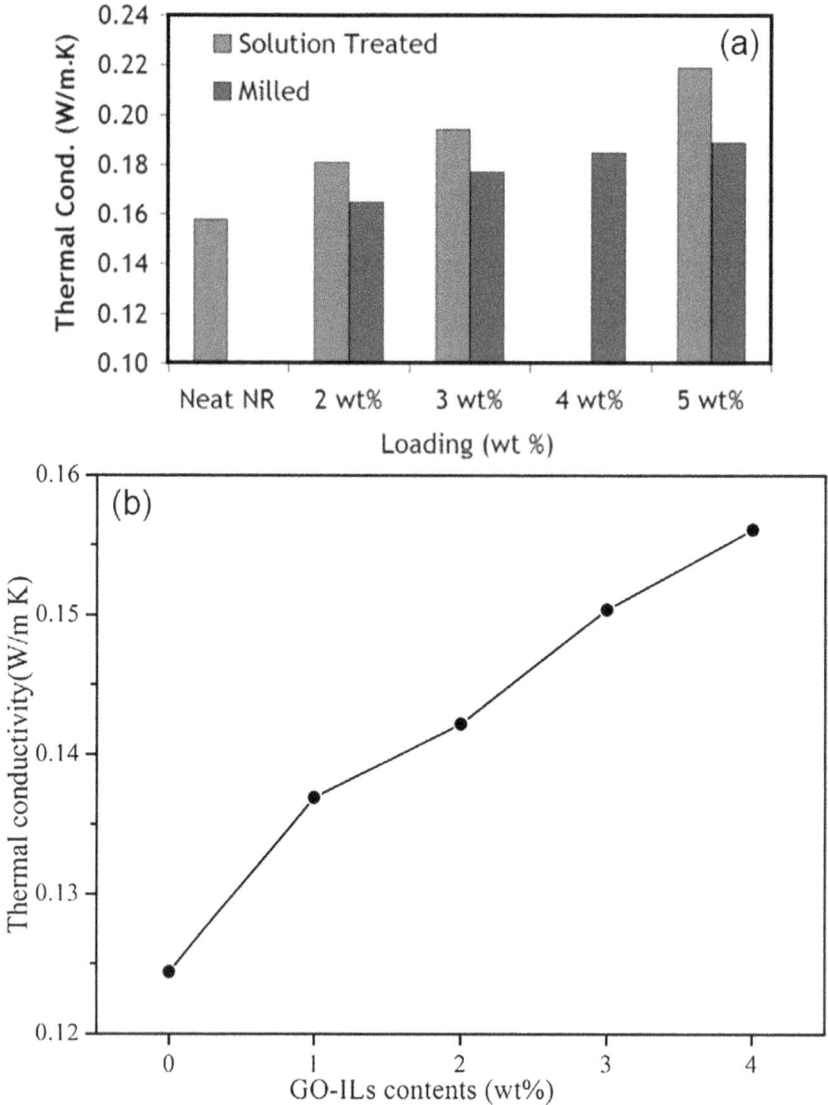

FIGURE 16.11 (a) Thermal conductivity results of NR/RGO composites. (Reproduced with permission from Potts et al., Copyright 2012, American Chemical Society.) (b) Variation of thermal conductivity of BIIR/GO-IL composites with varying filler content. (Reproduced with permission from Xiang et al., Copyright 2013, Elsevier.)

Yang et al., 2014; Liu et al., 2014), which clearly revealed the positive contribution of graphene and its derivatives in improving the electrical properties at significantly lower loading amount, when compared to the other carbon allotropes such as graphite, carbon black or carbon fibers. However, the dispersion states of graphene in the elastomer matrix is the influencing factor in designing composite materials with outstanding electrical property enhancement, which involves the formation of interconnected graphene networks surrounded by the elastomers. In contrast to other physicochemical properties, which demand fine dispersion states, electrical property enhancement depends on the partial segregation of graphene fillers in the elastomer composites.

Xia and coworkers (Zhan et al., 2012) highlighted the preparation strategy of graphene-loaded NR composites with interconnected morphological network (Figure 16.12a). The prepared composite materials exhibited maximum electrical conductivity (~0.03 S m^{-1}) at lower percolation threshold (~0.62 vol.%) and they corroborated this fact to the formation of segregated network in the resultant composite materials. Shown in Figure 16.12b is the variation of electrical conductivity with graphene content in the NR-based composite materials. In comparison to NR/graphene composites prepared by other processing methods, fivefold rise in the electrical conductivity is observed at the same graphene content for the latex-assisted method confirming the positive contribution of segregated network of fillers in the NR/graphene composites. George et al. reported electrical conductivity of EVA-based carbon nanocomposites and compared three different nanocomposites using expanded graphite, carbon nanofiber and carbon nanotube. CNF displays the lowest percolation threshold and highest conductivity due to high aspect ratio (George et al., 2010).

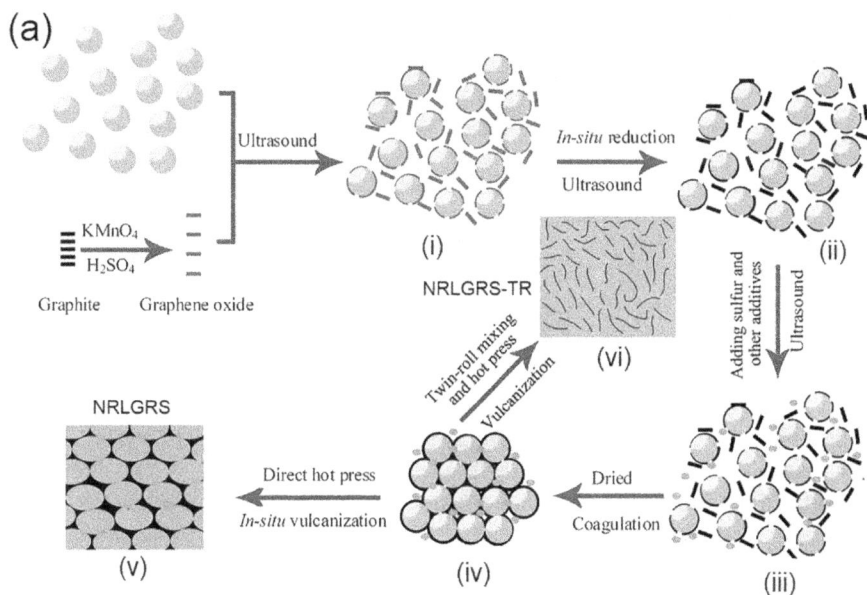

FIGURE 16.12 (a) Schematic representation depicting the preparation steps of NR/GR composites with segregated network by latex self-assembly followed by hot pressing: (i) Uniformly distributed GO in NR latex, (ii) In situ reduction of GO resulting in self-assembled GR platelets in NR latex, (iii) Step involving addition of curing ingredients, (iv) Latex coagulation resulting solid product (v) NR/GR composites with segregated network, (vi) NR/GR composites with fine dispersion of GR. (Reproduced with permission from Zhan et al., Copyright 2012, Royal Society of Chemistry.)

(*Continued*)

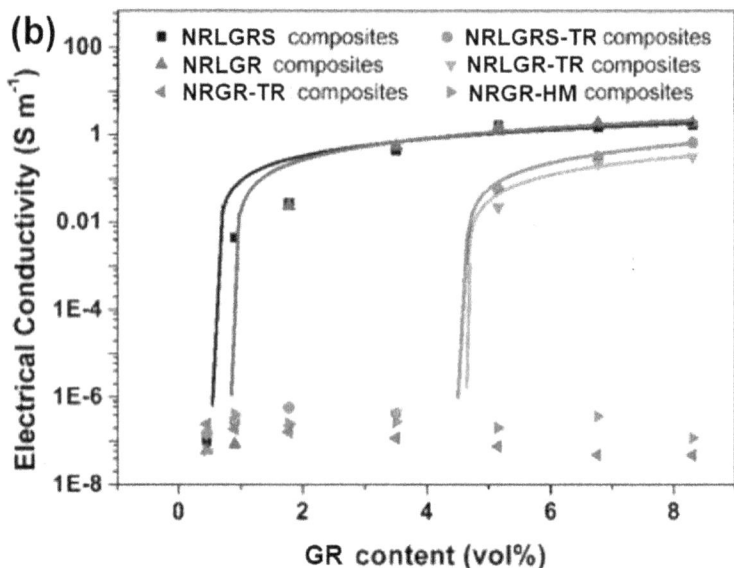

FIGURE 16.12 (*Continued*) (b) Electrical conductivity results of natural rubber (NR)/graphene (GR) composites prepared by different methods. NRLGRS: crosslinked NR/GR composites with the segregated network prepared by latex self-assembly followed by hot pressing. NRLGRS-TR: cross-linked NR/GR composites without the segregated network prepared by latex mixing followed by mill mixing. NRLGR: uncrosslinked NR/GR composites with a segregated network prepared by self-assembly in latex self-assembly followed by hot pressing. NRLGR-TR: uncrosslinked NR/GR composites without a segregated network prepared by latex mixing followed by mill mixing. NRGR-TR: composites prepared by mill mixing of GR powders and rubber. NRGR-HM: composites prepared by direct Haake mixing of GR powders and rubber. (Reproduced with permission from Zhan et al., Copyright 2012, Royal Society of Chemistry.)

16.3.2.4 Dielectric Properties

Dielectric property of the elastomeric composites is one of the key parameters in realizing its application potential in energy storage applications. Dielectric relaxation spectroscopy provides insights on the hindered dipolar rotation and intermolecular cooperative motion in the elastomeric systems, and hence, the technique has been widely explored by various academic researchers to reveal the dielectric characteristics of the graphene nanocomposites (Tian et al., 2014; Liu et al., 2014; Singh et al., 2012). High aspect ratio of the graphene, its adhesion with the elastomeric matrix and graphene loading are the influencing parameters in engineering the dielectric properties of the resultant nanocomposites. In addition, the thickness of the specimen used for dielectric measurements is also an influencing parameter, which critically affects the dielectric properties of the nanocomposites.

Ning and coworkers (Liu et al., 2014) reported noticeable increment in dielectric constant with low dielectric loss on loading TRG in the PU matrix. The role of interfacial bonding characteristics on the chain dynamics of butadiene-styrene-vinyl pyridine (VPR)/GO nanocomposites has been investigated by Jia and coworkers (Wu et al., 2013) (Figure 16.13). According to their study, the prepared elastomeric (VPR/GO) nanocomposites showed two distinct relaxation processes, viz., segmental and interfacial relaxations. However, engineering the interfacial bonding characteristics of VPR/GO nanocomposites by co-coagulation process using two different flocculants, viz., hydrochloric acid (HCl) and calcium chloride (CaCl$_2$), has profound influence on both the segmental as well as the interfacial relaxations. While the VPR/GO nanocomposites (HVPR) are prepared using HCl flocculant-induced hydrogen bonded interactions, nanocomposites (CaVPR) are prepared using CaCl$_2$-induced ionic interactions. These interfacial bonding (ionic and hydrogen) characteristics have profound influence on the segmental as well as the interfacial relaxations. For instance, HVPR nanocomposites exhibit faster segmental relaxation in comparison to the CaVPR nanocomposites. On the contrary, the interfacial relaxations are noticeably slower in CaVPR, when compared to HVPR. In addition, the dielectric strength is significantly higher for HVPR in comparison to CaVPR at a particular GO filler loading. George et al. reported that for all the composites and neat EVA, dielectric constant (ε') decreases with the increase in frequency. However, frequency dependency of ε' varies from system to system. For EVA-EG system, frequency dependency of ε' is increased with the increase in EG loading. They described the effect of loading using a power law equation and observed that the exponent was 0.78 for EVA-EG system (George et al., 2010).

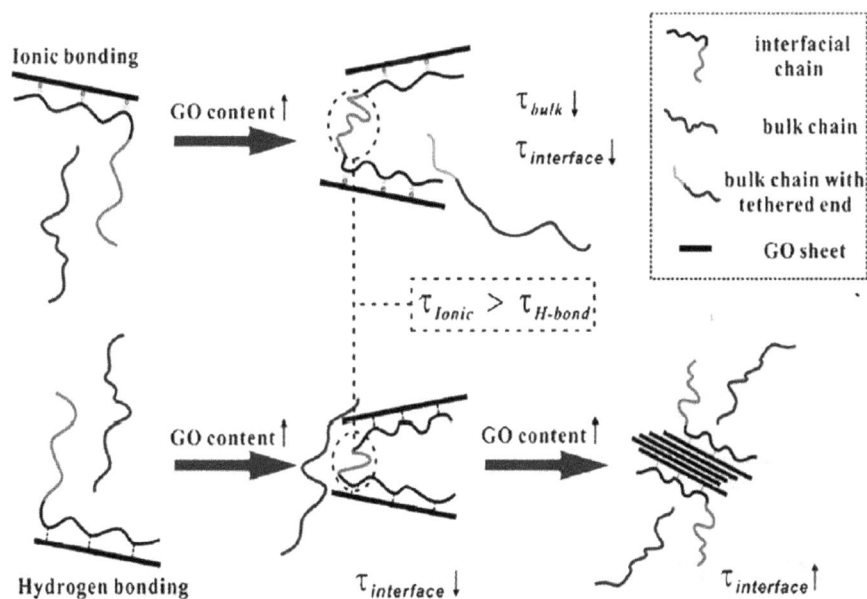

FIGURE 16.13 Schematics depicting chain dynamics in VPR/GO nanocomposites. (Reproduced with permission from Wu et al., Copyright 2013, Royal Society of Chemistry.)

16.4 ELASTOMER-GRAPHENE COMPOSITES FOR ENERGY STORAGE APPLICATIONS

Owing to the faster depletion of the natural resources such as coal and fossil fuels, development of new hybrid energy storage system is indeed needed in storing the energies generated through the intermittent renewable energy technologies for the sustainable use. Owing to its intrinsic career mobility, noteworthy specific surface areas, extraordinary optical transparency and exceptional mechanical properties make graphene and its polymer composites as the ideal candidate for energy-related applications such as rechargeable batteries, dielectric capacitors, flow batteries, SCs and thermal storage systems (Dai, 2012; Kim et al., 2017; Dai et al., 2020; Bellani et al., 2021; Chan et al., 2020; Veeman et al., 2021). Though significant work has been devoted to the fabrication of elastomer/graphene composites for automotive applications, very few works have been done on the energy storage applications that have been highlighted below.

16.4.1 DIELECTRIC CAPACITORS

Owing to the high energy density, large deformations, low cost and self-sensing abilities, DEs have received wide attention among the researchers to explore its application aspects in the fabrication of the flexible electronics, soft robotics, biomedical implantation and in aerospace (Yin et al., 2021). Devices fabricated using these highly stretchable DEs by sandwiching between the two compliant electrodes are termed as dielectric capacitors. Applying high voltage between the electrodes generates electrostatic forces, which subsequently result in the mechanical deformation along the transversal direction, due to the incompressibility of the DEs (Figure 16.14) (Singh and Singh, 2021).

Recently, dielectric elastomeric composites loaded with graphene nanofillers have received significant attraction by researchers, thanks to the interfacial polarization effects induced between the polymer matrices and the filler (Wu et al., 2017; Zhang et al., 2018; Panahi-Sarmad et al., 2019). Quite similar to the electrical properties of the elastomer/graphene composites, non-linear increment in the dielectric properties is observed with increasing graphene content. For instance, abrupt increase in dielectric loss as well as permittivity is observed with increasing loading amount of graphene.

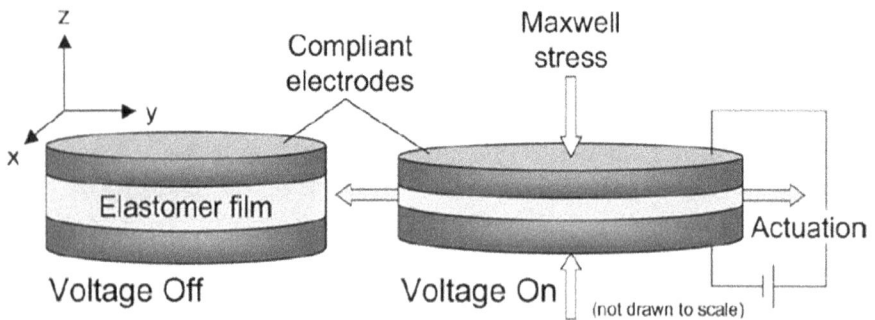

FIGURE 16.14 Schematic representation of dielectric capacitors under voltage on/off condition. (Reproduced from Singh and Singh, IOP publishing.)

However, in contrast to the electrical property variation of elastomer/graphene composites at the electrical percolation threshold, dielectric loss occurs below this threshold and this fact is correlated to the outflowing of the current from the aggregated conducting graphene network at the elastomer/graphene interface (Yin et al., 2021).

Panahi-Sarmad and Razzaghi-Kashani (2018) studied the effect of graphene oxide (GO) on the actuation behavior of the PDMS DE composites, which showed monotonic increment in the Maxwell stress with the volume fraction of the GO in the elastomer. Contrastingly, thermal reduction of GO has profound impact on the Maxwell stress and they have corroborated this fact to the decrement in the dielectric breakdown strength due to the formation of conductive network formation (electrical percolation threshold). For instance, PDMS composites prepared using 1 vol.% loading of reduced graphene oxide (rGO 400°C) showed considerable decrement in the Maxwell stress due to the larger drop in the dielectric breakdown strength (Figure 16.15).

They have used the ANOVA method to optimize the conditions in PDMS/RGO composites and they claimed that maximum dielectric permittivity with minimal dielectric loss and highest actuation strain (%) can be achievable for PDMS system loaded with 0.52 vol.% of RGO (reduced at 283°C), and subsequently, they validated the results using the experimental method, which showed highest actuation strain (13.32%) on loading 0.5 vol.% of graphene in the dielectric PDMS matrix (Panahi-Sarmad and Razzaghi-Kashani, 2018).

Similarly, a study by Li and coworkers (Chen et al., 2016) revealed that the incorporation of titanium dioxide (TiO_2)-functionalized graphene (RGO) has significant impact on the electric induced strain of dielectric PU/graphene (RGO) composites, due to the enhanced elastomer-filler interactions (Figure 16.16). Shown in Figure 16.16a is the schematics and pictorial representation of the PU/graphene (RGO) composites displaying electric field induced strain. The electric induced strain of PU/TiO_2-RGO composites is 72.4% at 38.7 V μm^{-1}, which is about ~1.8 and ~3.2 times higher, when compared to the PU/graphene (RGO) composites and pristine PU elastomer (Figure 16.16b).

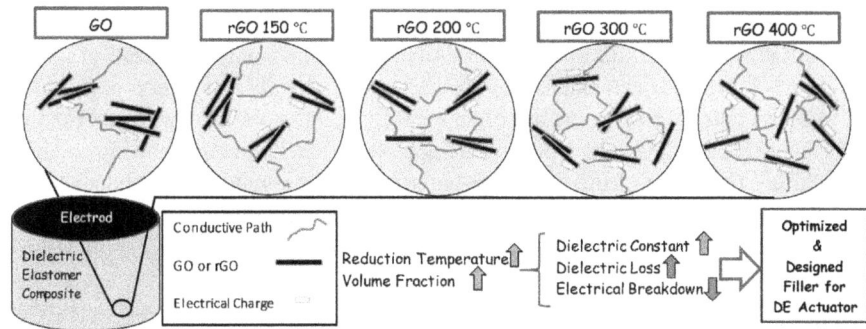

FIGURE 16.15 Effect of GO/RGO and their loading amount on the electrical properties of PDMS composites. (Reproduced with permission from Panahi-Sarmad and Razzaghi-Kashani, Copyright 2018, IOP publishing.)

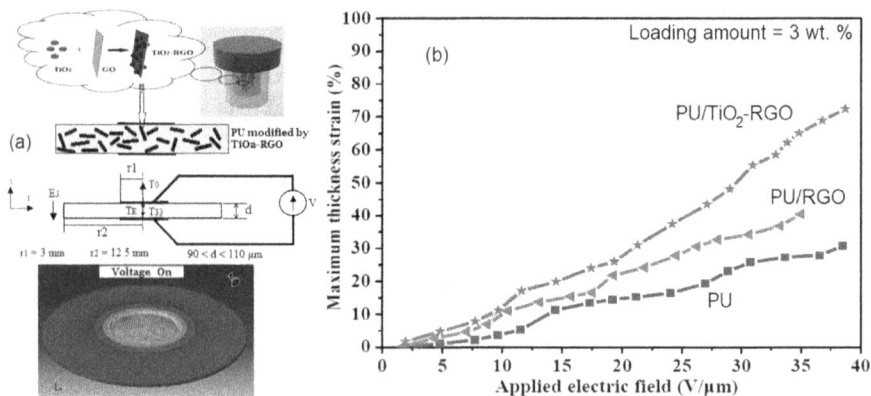

FIGURE 16.16 (a) Representation of dielectric capacitor fabricated using polyurethane composites. (b) Variation of electric strain with the applied electric field for polyurethane composites. (Reproduced with permission from Chen et al., Copyright 2019, Elsevier.)

16.4.2 SUPERCAPACITORS

SCs, which have tendency to temporarily store huge amount of electrical energy and release it as and when needed, have been considered as the next-generation portable energy storage device, thanks to its outstanding performance characteristics, i.e., high power density and long-term cycling stability in comparison to the conventional capacitors and lithium-ion (Li-ion) batteries. Owing to its high power density, SCs are currently being employed in the regenerative braking systems present in the automobiles with the positive scope of being implemented in the future electric vehicles as well as grid level storage systems. The electrodes in the SC are the key component in storing the energy as the double layer capacitance, which in turn depends on the mechanical and electrical properties of the material used. In contrast to the dielectric capacitors, which often use dielectric materials as the electrode, materials with porous configuration as well as high surface area are employed as the SC electrodes. Owing to its high specific surface area, excellent mechanical and electrical properties, graphene is considered as the ideal electrode candidate for SC applications (Kotal and Bhowmick, 2013; Ji et al., 2016). In addition, conducting polymer-based graphene composites have also been extensively studied in realizing its application potential as the SCs. In contrast, elastomers are often employed as the binders for the fabrication of the electrodes. In the recent past, some notable research attempts have been made in exploring the usage of the elastomer-based graphene composites as the electrode candidate for the SCs (Kaur et al., 2020; Suriani et al., 2015, 2017; Roy et al., 2021).

NR-based graphene nanocomposites have been explored for SC applications. For instance, Masrom and coworkers (Suriani et al., 2015) reported a novel one-step synthetic route in the preparation of natural rubber latex (NRL)/graphene oxide (GO) composites concurrently with the GO production. Included in Figure 16.17a and b is the schematics of the NR-based graphene nanocomposites prepared by one-step and two-step processes. In contrast to two-step processes, where the latex particles adsorbed to graphene platelets, one-step process involves coating of graphene platelets over the NR latex particles. Displayed in Figure 16.17c and d are the capacitance-voltage (C-V) results of NR/graphene composites prepared by one-step and two-step processes. NRL/GO composites by one-step process exhibited specific capacitance of 103.7 F g⁻¹, when compared to NRL/GO composites (specific capacitance, ~32.6 F g⁻¹) prepared by two-step process. In another study, surfactant-assisted preparation of NRL/reduced graphene oxide has been done by the same research group, with the specific capacitance value as high as 95 F g⁻¹ (Suriani et al., 2017).

Very recently, Bhowmick and coworkers (Roy et al. 2021) adopted pyrolytic route in the preparation of graphene from tannic acid and subsequently used the synthesized graphene (A11) in styrene butadiene rubber (SBR) latex, resulting in the formation of

(a) One-step method

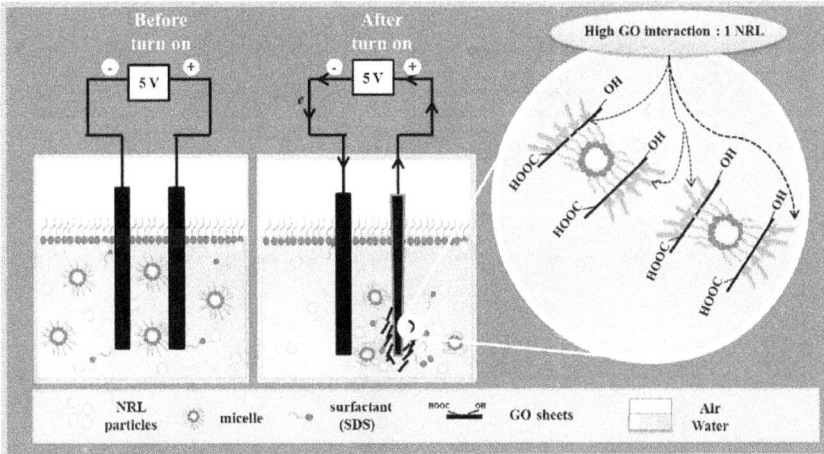

(b) Two-step method

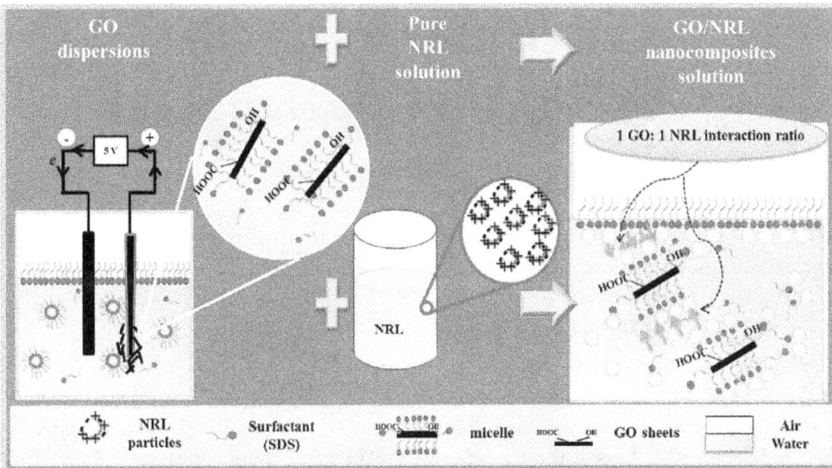

FIGURE 16.17 (a, b) Schematics depicting the preparation of NR/graphene nanocomposites prepared by one-step and two-step processes. (c, d) C-V results of NR/graphene composites prepared by one-step and two-step processes. (a–d: Reproduced with permission from Suriani et al., Copyright 2015, Elsevier.)

(Continued)

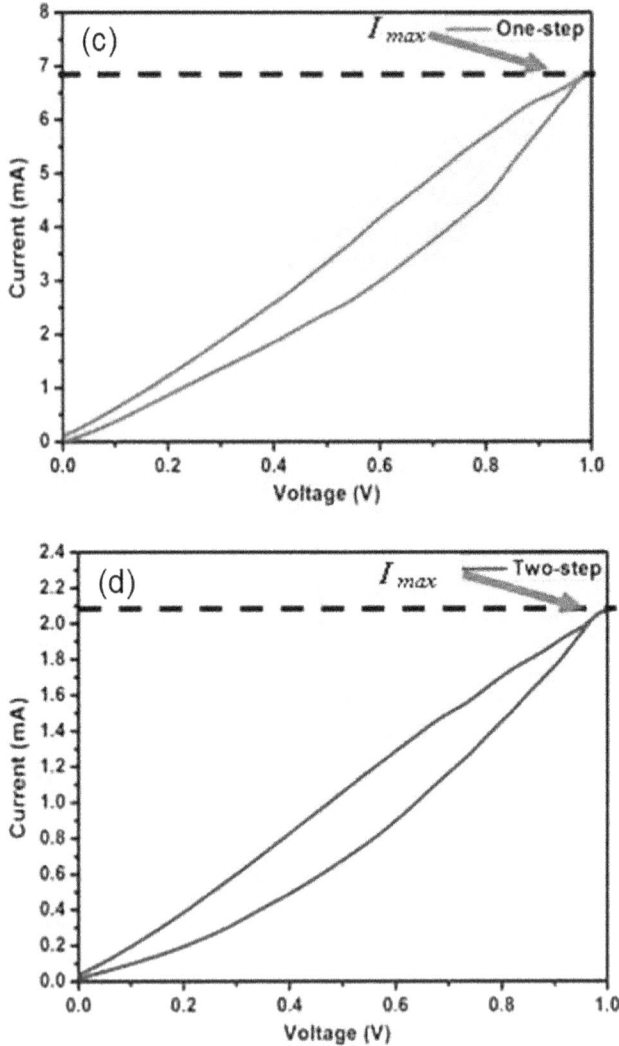

FIGURE 16.17 (*Continued*) (a, b) Schematics depicting the preparation of NR/graphene nanocomposites prepared by one-step and two-step processes. (c, d) C-V results of NR/graphene composites prepared by one-step and two-step processes. (a–d: Reproduced with permission from Suriani et al., Copyright 2015, Elsevier.)

SBR/graphene composites (SLA11) (Figure 16.18a). The dispersion states of the reduced graphene oxide (RGO) in the SBR latex are clearly corroborated using Atomic force microscopic (AFM) height image, which also revealed the formation of three-dimensional rubber-graphene network (Figure 16.18b). Finally, they also investigated the electrochemical properties of pristine graphene and SBR composites using cyclic voltametry (CV) and charge-discharge studies (Figure 16.18c and d). The specific capacitance values of pristine graphene and SBR/graphene composites measured using three electrode

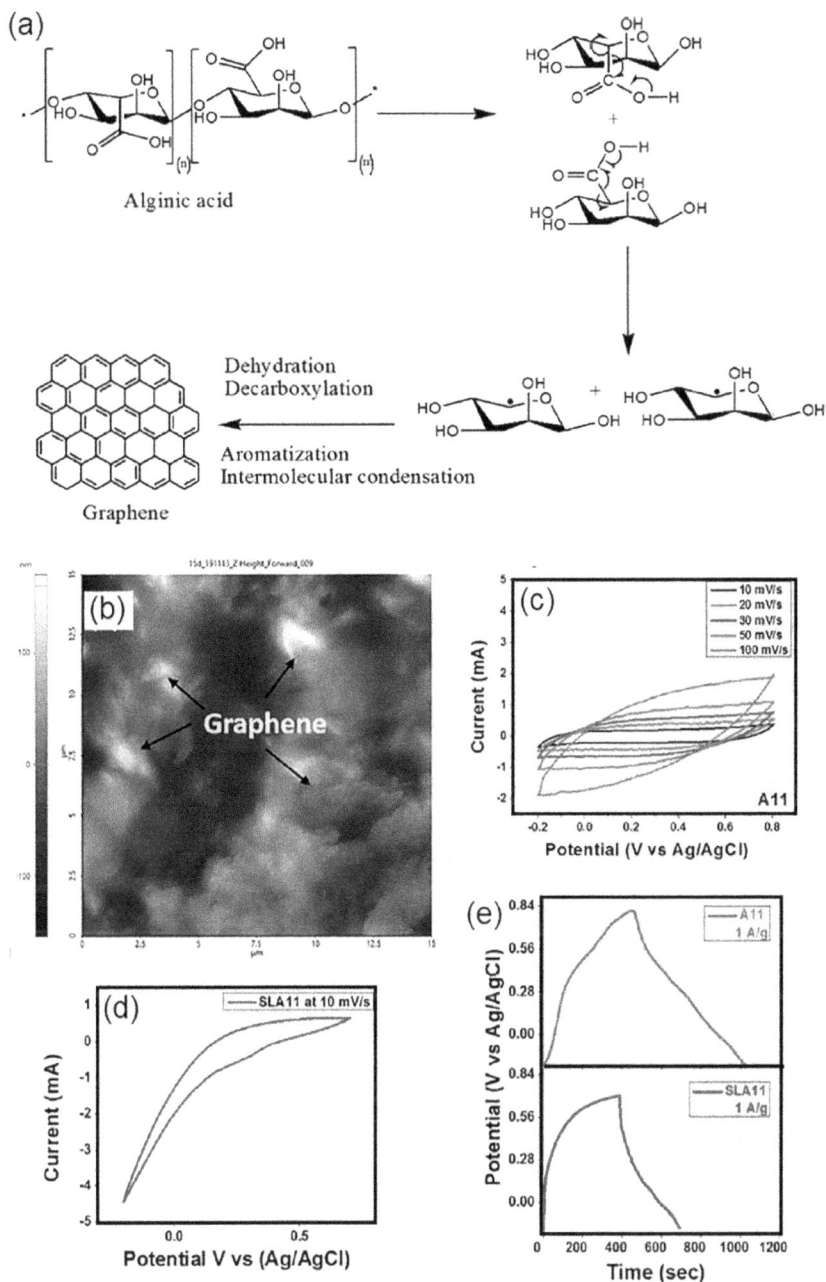

FIGURE 16.18 (a) Schematics of the steps involved in the synthesis of graphene from alginic acid. (b) Atomic force microscopic (AFM) height image of Styrene Butadiene rubber (SBR)/graphene (SLA11) nanocomposites. (c, d) Cyclic voltammograms of graphene (A11) and SBR/graphene (SLA11) nanocomposites (e) Charge-discharge results of graphene (A11) and SBR/graphene (SLA11) nanocomposites. (Reproduced from Roy et al., 2021, American Chemical Society.)

assembly are observed to be 315 and 137 F g^{-1}, respectively, revealing that the resultant elastomeric composites can be a suitable candidate for SC applications.

In the subsequent study, Bhowmick and coworkers (Roy et al., 2022) proposed a novel concept of preparing carbon black decorated graphene (GCB) by direct pyrolysis of molasses-carbon black mixture. Incorporation of the versatile GCB in SBR latex showed profound effect on the electrochemical characteristics, when measured in the three-electrode configuration. The CV results of bare GCB electrode and its latex nanocomposite film based GCB electrodes (SLGCB) displayed typical rectangular shaped structures corroborating electrochemical double layer capacitance behavior in both the electrodes. The specific capacitance of both the electrodes was measured using the chrono potentiometry and the results were observed to be 297 F g^{-1} (GCB) and 127 F g^{-1} (SLGCB), respectively. Finally, electrochemical impedance studies revealed SLGCB electrodes exhibited diffusional resistances with the knee frequency of 883 Hz, whereas Warburg type line was observed in bare GCB-based electrodes with the knee frequency of 1,024 Hz (Roy et al., 2022).

16.4.3 RECHARGEABLE BATTERIES

Owing to its low self-discharging, long cycle life and high energy density, rechargeable batteries (lithium-ion) have received significant interests in wide variety of applications, viz., consumer electronics and electric vehicles. In the current scenario, elastomers are employed as the binders, which hold the graphene-based electrodes tightly to the current collectors during the electrochemical cycling. The most commonly employed polymeric binder is poly(vinylidene difluoride) (PVDF). However, elastomers such as styrene butadiene rubber (SBR), carboxymethyl cellulose (CMC) and nitrile rubbers (NBR) have also been explored (Guerfi et al., 2007; Bresser et al., 2018; Shah et al, 2019).

Shown in Figure 16.19a and b is the schematic representation on the binding mechanism of graphite/PVDF and graphite/amorphous nitrile (NBR) composite electrodes. While semi-crystalline PVDF forms discrete coating on graphite exposing its underlying surfaces to electrolyte, homogenous coating is observed using nitrile rubber, protecting the underlying surfaces from direct contact with electrolyte. Two types of graphitic structures, viz., spherical (S-graphite) and flaky (F-graphite), were employed for the fabrication of the electrodes. Included in Figure 16.19c–h are the TEM and SEM images of the pure graphite and its composite electrodes prepared using PVDF and NBR. From the TEM images, it is clearly evident that spherical graphite has low-curvature surfaces, whereas flaky structures exhibit highly crystalline sheet morphologies with sharper edges (Figure 16.19c and f).

Displayed in Figure 16.19d, e, g, and h are the SEM morphologies of graphite electrode surfaces, which apparently look same irrespective of the binder used, thus revealing the fact that the electrochemical properties of the fabricated electrodes depend on their microscopic interfacial structure instead of its bulk electrode configuration (Shah et al., 2019). The long-term electrochemical cycling performances and the Coulombic efficiencies of the fabricated graphite electrodes are included in Figure 16.19i and j. From the results, it is quite evident that the graphite electrode fabricated using nitrile rubber binder has better electrochemical performance characteristics irrespective of the structure of the graphite. Thus, amorphous elastomers

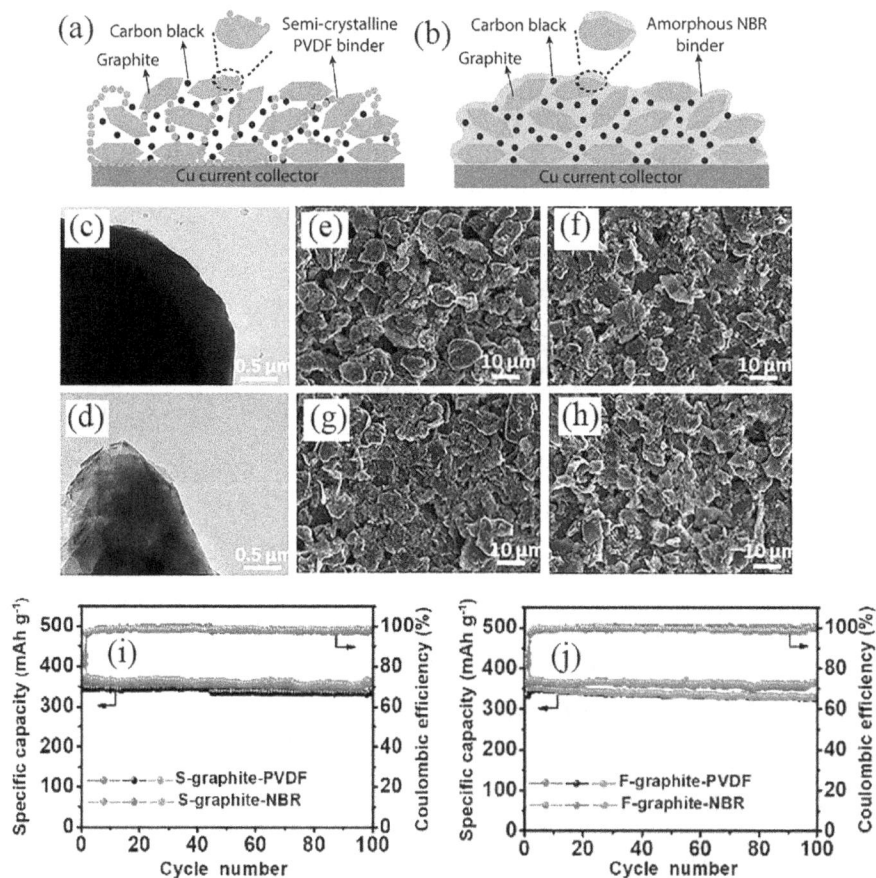

FIGURE 16.19 Schematics on binding morphologies of graphite electrodes: (a) PVDF binder, (b) nitrile rubber binder, (c, d) transmission electron micrographic (TEM) image of spherical (S-graphite) and flake graphite's (F-graphite). FE-SEM images of S-graphite/PVDF: (e), S-graphite/Nitrile (f), F-graphite/PVDF (g), F-graphite/Nitrile (h), electrochemical performances of S-graphite and F-graphite composites. (Reproduced from Shah et al., 2019, APCS, COME, PKU.)

have outstanding potential in improving the electrochemical performance characteristics, when used as a binder for battery fabrication. However, SBR/NBR elastomers consisting of C–C double bonds at its backbone oxidized at higher potential, thereby limiting its usage in high voltage applications (Yabuuchi et al., 2015).

Alternatively, elastomer/graphene composites with multiscale architecture design have greater potential for energy storage applications, thanks to its mechanical stability and high charge-transfer rate (Yuanzheng et al., 2021). Recently, a unique nanofiber enhanced elastomer/graphene composites mimicking nature's vein like textures with porous morphology, which subsequently loaded with sulfur particles have been explored as the electrode material for lithium-sulfur (Li-S) batteries. Included in Figure 16.20a is the schematics on the preparation steps of cellulose nanocrystals/

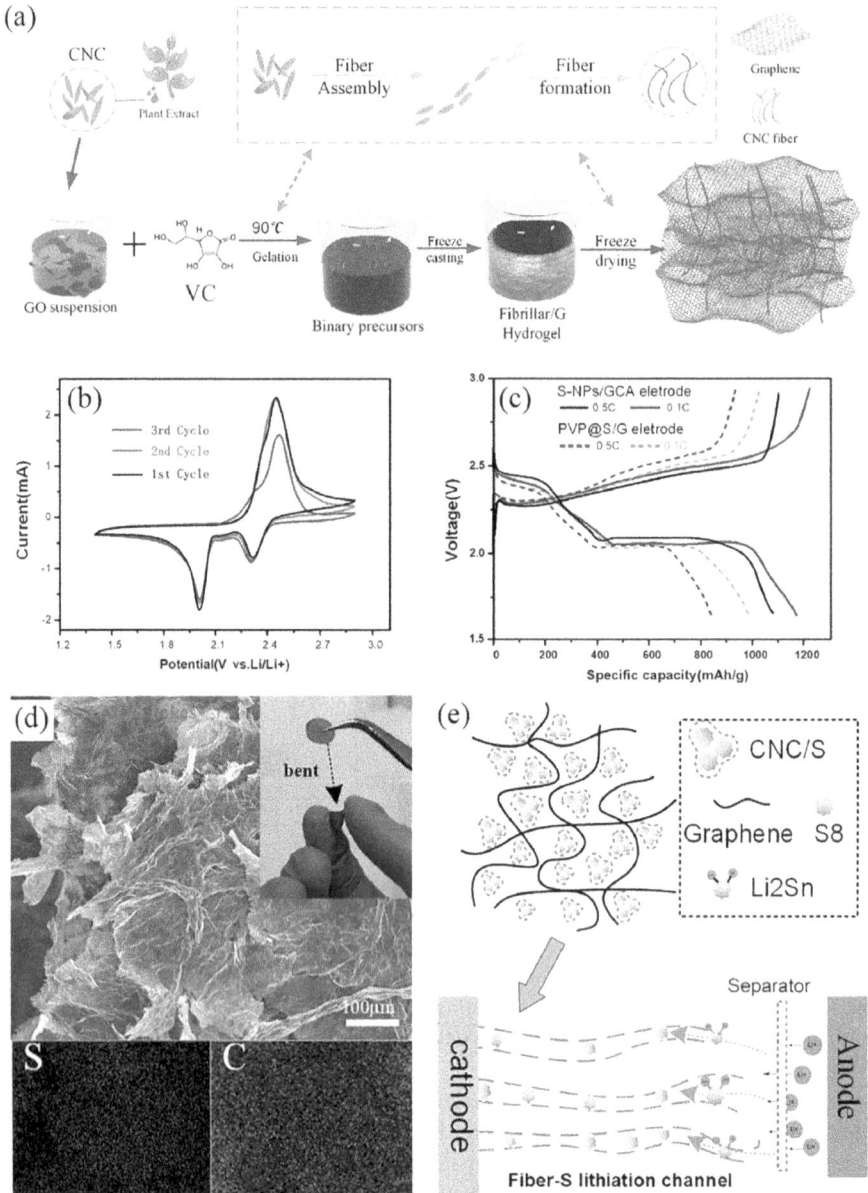

FIGURE 16.20 (a) Schematic illustration of the preparation steps of cellulose/graphene (GCA) aerogels, (b, c) CV and charge-discharge profiles of Li-S batteries fabricated using sulfur nanoparticle-loaded GCA aerogels, (d) SEM/EDX mapping of the GCA aerogels and (e) Underlying mechanism in prepared composite electrode. (Reproduced with permission from Yuanzheng.et.al., copyright 2021, Elsevier.)

graphene aerogel (GCA) monoliths interlinked with nanofibrous structures, mimicking vein-like textures. The prepared GCA aerogels (~107.62 m^2 g^{-1}) have relatively higher surface area in comparison to the graphene aerogels without cellulose nanocrystals (~54.32 m^2 g^{-1}). Subsequently, incorporation of sulfur particles resulted in the development of novel self-standing cathode material for lithium-sulfur batteries. Cyclic voltammetry (CV) results of the fabricated Li-S cell revealed the phase conversion of short-chain to long-chain sulfides, corroborating that the prepared electrodes have better reversibility, polarization and conductivity, when compared to the cyclic sulfur (cyclo-S8)-based cathode materials. Galvanostatic charge-discharge studies of Li-S cell prepared using GCA aerogels (S-NPs/GCA) displayed outstanding specific charge/discharge capacity values and these values are significantly higher, when compared to polyvinylpyrrolidone@sulfur/graphene (PVP@S/G) electrodes (Figure 16.20c). The prepared GCA aerogel-based sulfur electrode displayed outstanding reversible capacity of 880 mAh g^{-1} at 0.5°C rate after 300 cycles. The excellent rate capability of the prepared elastomer/graphene composite anode (S-NPs/GCA) is attributed to its flexible characteristics with relatively fine distribution of CNC/sulfur flakes.

This fact is further corroborated by the energy dispersive X-ray (EDX) mapping results depicting the fine dispersion states of sulfur and carbon in the fabricated electrode. The overwhelming electrochemical performances of Li-S batteries fabricated using elastomer/graphene electrode (S-NPs/GCA) is attributed to the restricted polysulfide shuttling as proposed in the scheme (Figure 16.20e). Nanofibers formed on the aerogels act as the physical barrier firmly immobilized with polysulfides on its surface, thereby enhancing cycling stability as well as sulfur utilization rate. In addition, reduced graphene oxide (RGO) renders synergistic contribution in controlling the polysulfide shuttling. Thus, the engineered elastomer/graphene composites can be an ideal candidate for battery applications.

16.4.4 Thermal Energy Storage Systems

The most encouraging and low-cost energy conservation technique is the thermal energy storage (TES) systems that are classified into two types, i.e., sensible heat thermal energy storage (SHTES) and latent heat thermal energy storage (LHTES) techniques. The most efficient and compact thermal energy storage technique with outstanding storage density and minimal temperature difference between the storage and release is the phase change material based LHTES. The heat absorbed by the LHTES material undergoes changes in the state of matter (Such as solid-liquid transition), which subsequently resulted in the release of heat during the inversion of state (i.e., transition from liquid to solid state).

The commonly employed phase change materials fall under two categories, i.e., Inorganic- and organic-based phase change materials (Pielichowska, and Pielichowski, 2014). However, the major drawback associated with the organic phase change materials (PCMs) is its low thermal conductivity. Loading of certain nanofiller with higher thermal conductivity can offset this problem. Extremely high thermal conductivity value of graphene (~4,000 W mK^{-1}) makes it a potential nanofiller in engineering the thermal conductivity characteristics of the organic PCMs (Hu, 2020). Also, leakage during the melting process is another critical issue in PCMs, which can be controlled by preparing microencapsulated PCMs.

Yang and coworkers (Deng et al., 2019) prepared the graphene-loaded paraffin-based microcapsules (MEPCMs), which are subsequently incorporated in the silicone rubber matrix at various loading amount resulting in the formation of stable phase change elements as proposed in the scheme (Figure 16.21a).

The phase change latent heat values, calculated from the differential scanning calorimetric (DSC) melting peaks of the pure MEPCMs, are observed to be 251.4 J g^{-1}, whereas the incorporation of MEPCMs in the silicone rubber matrix showed noticeable increment in the latent heat values from 9.97 to 53.39 J g^{-1} on increasing the loading amount of MEPCM from 5 to 30 phr (Figure 16.21b). The thermal conductivity of the prepared composite material determined using a transient plane source method showed tremendous improvement for the MEPCM-loaded silicone matrix in comparison to the neat silicone matrix. For instance, thermal conductivity of neat silicone rubber is 0.152 W mK^{-1}, which increased to about 198% on loading MEPCM in the silicone elastomer (0.453 W mK^{-1}).

Alternatively, long-chain alkylamine-grafted graphene-loaded elastomeric composites can also be a potent phase change element with improved thermal conductivity. For instance, Qian and coworkers (Zhang et al., 2020) prepared polyolefin/octadecylamine-functionalized graphene composites with stable thermal energy storage capacity over 200 cycles. Among the various organic PCMs, polyethylene glycol (PEG) has been considered as the most promising candidate due to its higher

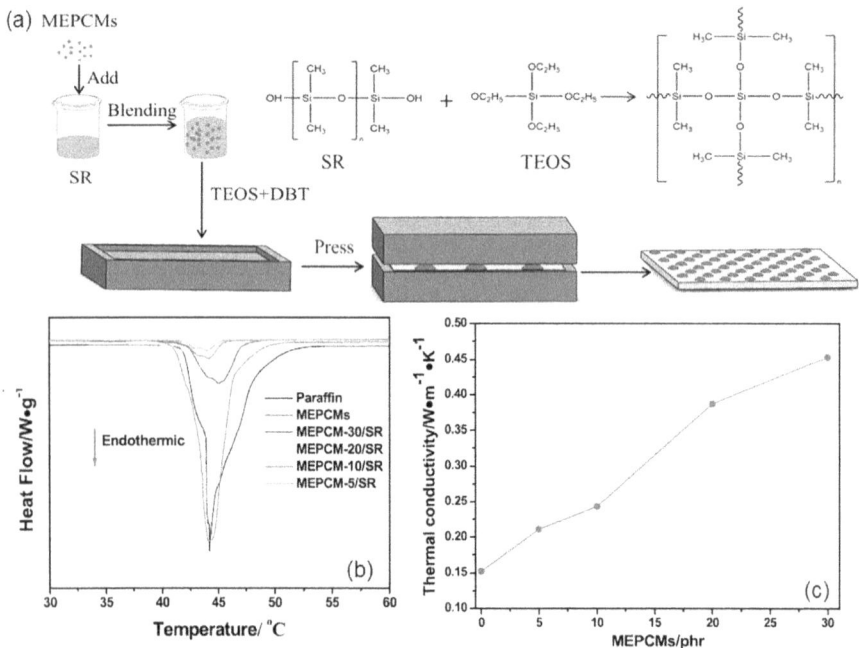

FIGURE 16.21 (a) Schematic representation of MEPCMs loaded silicon rubber vulcanizates; Differential scanning calorimetry (DSC) (b) and thermal conductivity (c) results of silicone/MEPCM's composites. (Reproduced with permission from Deng et al., Copyright 2019, Taylor & Francis.)

LHS and suitable phase change temperature, which can subsequently be tuned simply by varying its molecular weight. Yang and coworkers (Yang et al., 2016) studied the effect of hybrid graphene aerogels (graphene oxide (GO) + graphene nanoplatelets (GNP)) on the thermal conductivity and shape stabilization effect of PEG-based PCMs. Outstanding enhancement in the thermal conductivity is observed on loading hybrid graphene aerogels (0.45 wt.% GO + 1.8 wt.% graphene nanoplatelets) with the value as high as 1.43 mW mK^{-1} in comparison to the pure PEG-based system.

Finally, they have investigated the light-to-heat conversion and storage behavior of the prepared composite materials. Shown in the Figure 16.22a and b is the setup used for light-to-heat conversion, schematics showing light-to-heat conversion and storage process, temperature evolution curves as well as the energy storage efficiency with the varying weight ratios of GNP/GO (HGA1-1; HGA 2-2; HGA3-4; and HGA4-10) in the composites. On illumination of light, PCMs based on PEG/HGA absorbed the heat and resulted in profound rise in the temperature. Upon reaching the melting point of the composites, an inflection point aroused, indicating the phase change process with the storage of thermal energy. Once switching off the solar light, temperature of the PCM

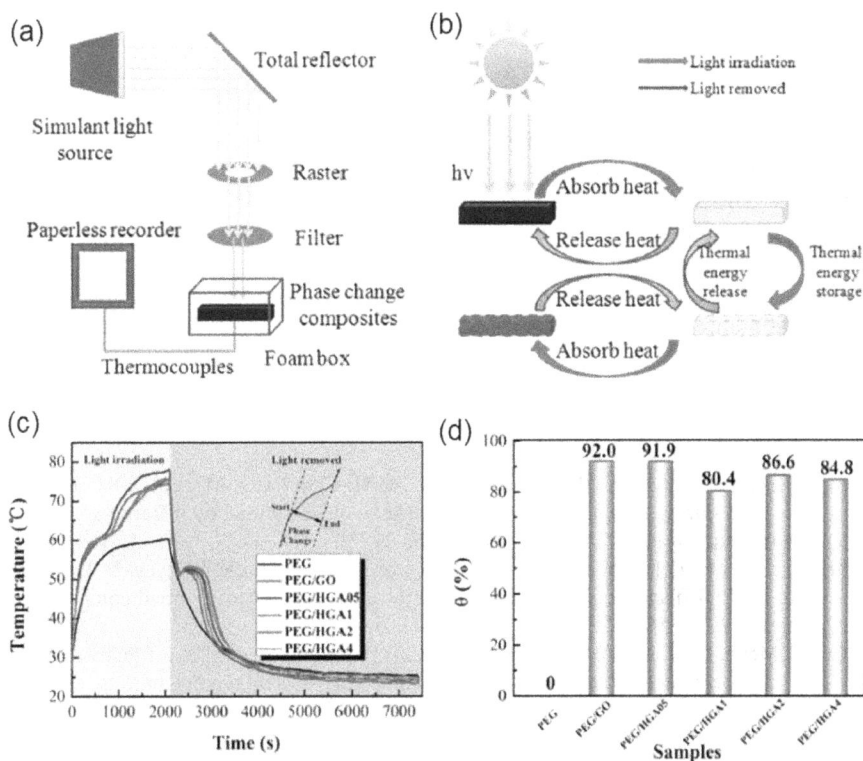

FIGURE 16.22 Storage and release of thermal energy (a) Set-up used for light-to-heat conversion. (b) Schematics depicting light-to-heat conversion and storage. (c) Temperature variation with light illumination for PEG, PEG/GO and PEG/HGA composites. (d) Energy storage efficiency (θ) of the composite PCM's. (Reproduced with permission from. Yang et al., Copyright 2016, Elsevier.)

decreased at slower pace till the PCMs get solidified, after which rapid decrement in thermal energy is observed (Figure 16.22c) for all the composite materials. Phase change efficiency (θ) decreased with increasing HGA content signifying better performances of light-to-thermal conversion as well as thermal storage efficiency (Figure 16.22d).

16.5 CONCLUSIONS

The preparation and properties of graphene, oxidized derivatives of graphene and elastomer/graphene nanocomposites have been reviewed. It has been proposed that the unique characteristics of the elastomer/graphene nanocomposites made it as the ideal candidate for wide range of applications with better reinforcing ability. In addition, various processing methods in preparing the resultant elastomer/graphene composites have also been discussed. For instance, processing approaches such as solution blending or in situ polymerization may better be suited for academic research purposes though higher degree of graphene dispersion can be achieved through these methods. On the contrary, melt mixing is the feasible method in scaling up to the industrial scale, though the degree of homogenization is not at the par to the other processing methods. Also, the bulk production of graphene is the determinant factor for the commercial preparation of elastomer/graphene nanocomposites. In this context, various synthetic processes, leading to the bulk production of the graphene and its derivatives, have also been reviewed in this chapter. The impact of graphene and its derivatives on the property enhancement of the elastomers as well as its role in the energy storage applications pertaining to dielectric capacitors, SCs, batteries and thermal energy storage systems has also been highlighted.

Surprisingly, little work has been done in the development of elastomer/graphene composite membranes for fuel cell applications (Nafion and PVDF are the extensively researched plastics).

REFERENCES

Acik, M. Lee, G. Mattevi, C. Pirkle, A. Wallace, R. M. Chhowalla, M. et al. 2011. The role of oxygen during thermal reduction of graphene oxide studied by infrared absorption spectroscopy. *J. Phys. Chem. C.*, **115**: 19761–19781.

Agarwal, S. L. Mandol, S. K. Mandal, N. Bandyopadhyay, S. Mukhopadhyay, R. Deuri, A. et al. 2006. Effect of corn powder as filler in radial passenger tire tread compound. *J. Mater. Sci.*, **41**: 5657–5665.

Araby, S. Meng, Q. Zhang, L. Kang, H. Majewski, P. Tang, Y. et al. 2014. Electrically and thermally conductive elastomer/graphene nanocomposites by solution mixing. *Polymer*, **55**: 201–210.

Balandin, A. A. 2011. Thermal properties of graphene and nanostructured carbon materials. *Nat. Mater.*, **10**: 569–581.

Bandyopadhyay, A. DeSarkar, M. and A. K. Bhowmick. 2005. Poly (vinyl alcohol)/silica hybrid composites by sol-gel technique synthesis and properties. *J. Mater. Sci.*, **40**: 5233–5241.

Beckert, F. Trenkle, S. Thomann, R. and R. Mülhaupt. 2014. Mechanochemical route to functionalized graphene and carbon nanofillers for graphene/SBR nanocomposites. *Macromol. Mater. Eng.*, **299**: 1513–1520.

Bellani, S. Najafi, L. Prato, M. Oropesa-Nuñez, R. Martin-Garcia, B. Gagliani, L. et al. 2021. Graphene based electrodes in vanadium redox flow battery produced by rapid low-pressure combined gas plasma treatments. *Chem. Mater.*, **33**: 4106–4121.

Bhattacharya, M. and A. K. Bhowmick. 2010. Correlation of vulcanization and viscoelastic properties of nanocomposites based on natural rubber and different nanofillers, with molecular and supramolecular structure. *Rubber Chem. Technol.*, **83** (1): 16–34.

Bhattacharya, M. Maiti, M. and A. K. Bhowmick. 2008. Influence of different nanofillers and their dispersion methods on the properties of natural rubber nanocomposites, *Rubber Chem. Technol.* **81** (5): 782–808.

Bhowmick, A. K., Ed. 2008, *Current topics of Elastomer Research, Chapters 2–4*, CRC Press, Taylor and Francis Group, Boca Raton, FL.

Bianco, A. Cheng, H. M. Enoki, T. Gogotsi, Y. Hurt, R. H. Koratkar, N. et al. 2013. All in the graphene family – A recommended nomenclature for two-dimensional carbon materials. *Carbon*, **65**: 1–6.

Bresser, D. Buchholz, D. Moretti, A. Varzi, A. and S. Passerini. 2018. Alternative binders for sustainable electrochemical energy storage – The transition to aqueous electrode processing and bio-derived polymers. *Energy Environ. Sci.*, **11**: 3096–3127.

Brodie, B. 1860. Sur le poids atomique du graphite. *Annales de Chimie et de Physique*, **59**: 466–472.

Chan, K. Y. Pham, D. Q. Demir, B. Yang, D. Mayes, E. D. H. Mouritz, A. P. et al. 2020. Graphene oxide thin film structural dielectric capacitors for aviation static electricity harvesting and storage. *Compos. Part B: Eng.*, **201**: 108375.

Chen, J. and L. Li. 2020. Effect of oxidation degree on the thermal properties of graphene oxide. *J. Mater. Res. Technol.*, **9**: 13740–13748.

Chen, T. Qiu, J. Zhu, K. and J. Li. 2016. Electro-mechanical performance of polyurethane dielectric elastomer flexible micro-actuator composite modified with titanium dioxide-graphene hybrid fillers. *Mater. Design*, **90**: 1069–1076.

Ciesielski, A. and P. Samori. 2014. Graphene via sonication assisted liquid-phase exfoliation. *Chem. Soc. Rev.*, **43**: 381–398.

Dai, L. 2012. Functionalization of graphene for efficient energy conversion and storage. *Acc. Chem. Res.*, **46**: 31–42.

Dai, C. Sun, G. Hu, G. Xiao, Y. Zhang, Z. and Q. Liu. 2020. Recent progress in graphene-based electrodes for flexible batteries. *InfoMat*, **2**: 509–526.

Das, A. Kasaliwal, G. R. Jurk, R. Boldt, R. Fischer, D. Stöckelhuber, K. W. et al. 2012. Rubber composites based on graphene nanoplatelets, expanded graphite, carbon nanotubes and their combination: A comparative study. *Compos. Sci. Technol.*, **72**: 1961–1967.

Deng, H. Guo, Y. He, F. Yang, Z. Fan, J. He, R. et al. 2019. Paraffin@graphene/silicon rubber form-stable phase change materials for thermal energy storage. *Fullerenes, Nanotube. Nanostruct.*, **27**: 626–631.

Dikin, D. A. Stankovich, S. Zimney, E. J. Piner, R. D. Dommett, G. H. B. Evmenenko, G. et al. 2007. Preparation and characterization of graphene oxide paper. *Nature*, **448**: 457–460.

Döscher, H. Schmaltz, T. Neef, C. Thielmann, A. and T. Reiss. 2021. Graphene roadmap briefs (No. 2): Industrialization status and prospects 2020. *2D Mater.*, 8: 022005.

Fan, B. Liu, Y. He, D. and J. Bai. 2018. Achieving polydimethylsiloxane/carbon nanotube (PDMS/CNT) composites with extremely low dielectric loss and adjustable dielectric constant by sandwich structure. *Appl. Phys. Lett.*, **112**: 052902.

Ferralis, N. 2010. Probing mechanical properties of graphene with Raman spectroscopy. *J. Mater. Sci.*, **45**: 5135–5149.

Geim, A. K. and K. S. Novoselov. 2007. The rise of graphene. *Nat. Mater.*, **6**: 183–191.

George, J. J. Bandyopadhyay, A. and A. K. Bhowmick. 2008. New generation layered nano-composites derived from ethylene-co-vinyl acetate and naturally occurring graphite. *J. Appl. Polym. Sci.*, **108**: 1603–1616.

George, J. J. Bhadra, S. and A. K. Bhowmick. 2010. Influence of carbon-based nanofillers on the electrical and dielectric properties of ethylene vinyl acetate nanocomposites. *Polymer Compos.*, 31 (2), 218–225.

George, J. J. and A. K. Bhowmick. 2009. Influence of matrix polarity on the properties of ethylene vinyl acetate–carbon nanofiller nanocomposites, *Nanoscale Res. Let.*, 4 (7), 655–664.

Gomez-Navarro, C. Burghard, M. and K. Kern. 2008. Elastic properties of chemically derived single graphene sheets. *Nano Letters*, **8**: 2045–2049.

Gong, L. Kinloch, I. A. Young, R. J. Riaz, I. Jalil, R. and K. S. Novoselov. 2010. Interfacial stress transfer in a graphene monolayer nanocomposite. *Adv. Mater.*, **22**: 2694–2697.

Gong, Y. Zhang, X. Liu, G. Wu, L. Geng, X. Long, M. et al. 2012. Layer controlled and wafer-scale synthesis of uniform and high quality graphene films on a polycrystalline nickel catalyst. *Adv. Funct. Mater.*, **22**: 3153–3159.

Guerfi, A. Kaneko, M. Petitclerc, M. Mori, M. and K. Zaghib. 2007. LiFePO4 water soluble binder electrode for Li-ion batteries. *J. Power Sources*, **163**: 1047–1052.

Guerra, E. L. Shanmugharaj, A. M. Choi, W. S. and S. H. Ryu. 2013. Thermally reduced graphene oxide-supported nickel catalyst for hydrogen production by steam reforming. *Appl. Catal. A: Gen.*, **468**: 467–474.

Guo, X. Jung, J. and S. Nagase. 2010. Hydrazine and thermal reduction of graphene oxide: Reaction mechanisms, product structures and reaction design. *J. Phys. Chem. C.*, **114**: 832–842.

Han, S. Pu, X. Li, X. Liu, M. and W. Hu. 2017. High areal capacity of Li-S batteries enabled by freestanding CNF/RGO electrode with high loading of lithium polysulfide. *Electrochim. Acta*, **241**: 406–413.

Hernandez, Y. Nicolosi, V. Lotya, M. Blighe, F. M. Sun, Z. De, S. et al. 2008. High-yield production of graphene by liquid-phase exfoliation of graphite. *Nat. Nanotechnol.*, **3**: 563–568.

Hosler, D., Burkett, S. L. and M. J. Tarkanian. 1999. Prehistoric polymers: Rubber processing in ancient Mesoamerica. *Science.* **284**: 1988–91.

Hu, H. 2020. Recent advances of polymeric phase change composites for flexible electronics and thermal energy storage system. *Compos. Part B*, **195**: 108094.

Hummers Jr, W. S. and R. E. Offeman. 1958. Preparation of graphitic oxide. *J. Am. Chem. Soc.*, **80**: 1339.

Ibrahim, A. Klopocinska, A. Horvat, A. and Z. A. Hamid. 2021. Graphene-based nanocom-posites: Synthesis, mechanical properties and characterizations. *Polymers*, **13**: 2869.

Ji, L. Meduri, P. Agubra, V. Xiao, X. and M. Alcoutlabi. 2016. Graphene-based nanocompos-ites for energy storage. *Adv. Energy Mater.*, **6**: 1502159.

Kaur, M. Twinkle. Kumar, S. and J. K. Gowsamy. 2020. Graphene-PVDF flexible nanohy-brids for supercapacitor application. *AIP Conf. Proceed.*, **2220**: 020197.

Kim, H. Abdala, A. A. and C. W. Macosko. 2010. Graphene/polymer nanocomposites. *Macromolecules*, **43**: 6515–6530.

Kim, M. Hwang, H. M. Park, G. H. and H. Lee. 2017. Graphene based composite electrodes for electrochemical energy storage devices: Recent progress and challenges. *FlatChem*, **6**: 48–76.

Kim, H. Miura, Y. and C. W. Macosko. 2010. Graphene/polyurethane nanocomposites for improved gas barrier and electrical conductivity. *Chem. Mater.*, **22**: 3441–3450.

Kim, J. S. Yun, J. H. Kim, I. and S. E. Shim. 2011. Electrical properties of graphene/SBR nanocomposite prepared by latex heterocoagulation process at room temperature. *J. Indust. Eng. Chem.*, **17**: 325–330.

Kim, K. S. Zhao, Y. Jang, H. Lee, S. Y. Kim, J. M. Kim, K. S. et al. 2009. Large-scale pattern growth of graphene films for stretchable transparent electrodes. *Nature*, **457**: 706–710.

Koerner, H. Price, G. Pearce, N. A. Alexander, M. and R. A. Vaia. 2004. Remotely actuated polymer nanocomposites--stress-recovery of carbon-nanotube-filled thermoplastic elastomers. *Nat. Mater.*, **3**: 115–120.

Kojima, Y. Usuki, A. Kawasumi, M. Okada, A. Kurauchi, T. and O. Kamigaito. 1993. Synthesis of nylon 6–clay hybrid by montmorillonite intercalated with ε-caprolactam. *J. Polym. Sci., Part A: Polym. Chem.*, **31**: 983–986.

Kotal, M. and A. K. Bhowmick, 2013. Multifunctional hybrid materials based on carbon nanotube chemically bonded to reduced graphene oxide. *J. Phys. Chem. C.*, **117**: 25865–25875.

Kotal, M. and A. K. Bhowmick. 2015. Polymer nanocomposites from modified clays: Recent advances and challenges. *Progress Polym. Sci.*, **51**: 127–187.

Kumar, P. V. Bardhan, N. M. Chen, G. Y. Li, Z. Belcher, A. M. and J. C. Grossman. 2016. New insights into the thermal reduction of graphene oxide: Impact of oxygen clustering. *Carbon*, **100**: 90–98.

Kumar, B. De, P. P. De, S. K. Peiffer, D. G. Majumdar, S. and A. K. Bhowmick. 2001. Influence of fillers and oil on mill processability of brominated isobutylene-co-paramethylstyrene and its blends with EPDM. *Polym. Eng. Sci.*, **41**: 2266–2280.

Leblanc, J. L. 2002. Rubber–filler interactions and rheological properties in filled compounds. *Progress Polym. Sci.*, **27**: 627–687.

Lee, Y. R. Raghu, A. V. Jeong, H. M. and B. K. Kim. 2009. Properties of waterborne polyurethane/functionalized graphene sheet nanocomposites prepared by an in-situ method. *Macromol. Chem. Phys.*, **210**: 1247–1254.

Lee, C. Wei, X. Kysar, J. W. and J. Hone. 2008. Measurement of the elastic properties and intrinsic strength of monolayer graphene. *Science*, **321**: 385–388.

Li, X. Cai, W. An, J. Kim, S. Nah, J. Yang, D. et al. 2009. Large-area synthesis of high-quality and uniform graphene films on copper foils. *Science*, **324**: 1312–1314.

Lin, Y. Liu, K. Chen, Y. and L. Liu. 2014. Influence of graphene functionalized with zinc dimethacrylate on the mechanical and thermal properties of natural rubber nanocomposites. *Polym. Compos.*, **36**: 1775–1785.

Liu, S. Tian, M. Yan, B. Yao, Y. Zhang, L. Nishi. et al. 2014. High performance dielectric elastomers by partially reduced graphene oxide and disruption of hydrogen bonding of polyurethanes. *Polymer*, **56**: 375–384.

Liu, H. Zhang, L. Guo, L. Cheng, Y. Yang, L. Jiang, L. et al. 2013. Reduction of graphene oxide to highly conductive graphene by lawesson's reagent and its electrical applications. *J. Mater. Chem. C*, **1**: 3104–3109.

Lotya, M. Hernandez, Y. King, P. J. Smith, R. J. Nicolosi, V. Karlsson, L. S. et al. 2009. Liquid phase production of graphene by exfoliation of graphite in surfactant/water solutions. *J. Am. Chem. Soc.*, **131**: 3611–3620.

Luo, Y. Zhao, P. Yang, Q. He, D. Kong, L. and Z. Peng. 2014. Fabrication of conductive elastic nanocomposites via framing intact interconnected graphene networks. *Compos. Sci. Technol.*, **100**: 143–151.

Maiti, M. Bhattacharya, M. and A. K. Bhowmick. 2008. Elastomer nanocomposites. *Rubber Chem. Technol.* **81** (3): 384–469.

Manjakkal, L. Navaraj, W. T. Núñez, C. G. and R. Dahiya. 2019. Graphene-graphite polyurethane composite based high-energy density flexible supercapacitors. *Adv. Sci.*, **6**: 1802251.

Mao, H. –J. Liu, D. –F., Zhang, N. Huang, T. KÜhnert, I. Yang, J. –H. et al. 2020. Constructing a Microcapacitor Network of Carbon Nanotubes in Polymer Blends via Crystallization-Induced Phase Separation Toward High Dielectric Constant and Low Loss. *ACS Appl. Mater. Interfaces*, **12**: 26444–26454.

McAllister, M. J. Li, J. –L. Adamson, D. H. Schniepp, H. C. Abdala, A. A. Liu, J. et al. 2007. Single sheet functionalized graphene by oxidation and thermal expansion of graphite. *Chem. Mater.*, **19**: 4396–4404.

McNaught, A. D. and A. Wilkinson. 1997. *Compendium of Chemical Terminology*. Blackwell Science, Oxford.

Modi, A. Singh, S. and N. Verma. 2016. Improved performance of a single chamber microbial fuel cell using nitrogen-doped polymer-metal-carbon nanocomposite-based air-cathode. *Int. J. Hyd. Energy*, **42**: 3271–3280.

Moore, A. 2021. Market study: Elastomers Market share and Industry growth revenue 2021 [cited: Available from https://www.theexpresswire.com/pressrelease/Elastomers-Market-Share-and-Industry-Growth-Revenue-2021-Regional-Overview-Latest-Technology-Business-Status-Top-Growing-Factors-and-Market-Dynamics-Forecast-to-2026-with-Impact-of-Covid-19_14315903].

Ni, Z. H. Yu, T. Lu, Y. H. Wang, Y. Y. Feng, Y. P. and Z. X. Shen. 2008. Uniaxial strain on graphene: Raman spectroscopy study and band-gap opening. *ACS Nano*, **2**: 2301–2305.

Niu, L. Li, M. Tao, X. Xie, Z. Zhou, X. Raju, A. P. A. et al. 2013. Salt-assisted direct exfoliation of graphite into high-quality, large-size, few-layer graphene sheets. *Nanoscale*, **5**: 7202–7208.

Norimatsu, W. and M. Kusunoki. 2014. Epitaxial graphene on SiC {0001}: Advances and perspectives. *Phys. Chem. Chem. Phys.*, **16**: 3501–3511.

Novoselov, K. S. Geim, A. K. Morozov, S. Jiang, D. Zhang, Y. Dubonos, S. et al. 2004. Electric field effect in atomically thin carbon films. *Science*. **306**: 666–669.

Panahi-Sarmad, M. and M. Razzaghi-Kashani. 2018. Actuation behavior of PDMS dielectric elastomer composites containing optimized graphene oxide. *Smart Mater. Struct.*, **27**: 085021.

Panahi-Sarmad, M. Zahiri, B. and M. Noroozi. 2019. Graphene based composite for dielectric elastomer actuator: A comprehensive review. *Sensors. Actuat. A: Phys.*, **293**: 222–241.

Papageorgiou, D. G. Kinloch, I. A. and R. J. Young. 2015. Graphene/elastomer nanocomposites. *Carbon*, **95**: 460–484.

Park, S. An, J. Jung, I. Piner, R. D. An, S. J. Li, X. et al. 2009. Colloidal suspensions of highly reduced graphene oxide in a wide variety of organic solvents. *Nano Letters*, **9**: 1593–1597.

Park, S. and R. S. Ruoff. 2009. Chemical methods for the production of graphene's. *Nat. Nanotechnol.*, **4**: 217–224.

Paszkiewicz, S. Szymczyk, A. Špitalský, Z. Mosnáček, J. Kwiatkowski, K. and Z. Rosłaniec. 2014. Structure and properties of nanocomposites based on PTT-block-PTMO copolymer and graphene oxide prepared by in-situ polymerization. *Euro. Polym. J.*, **50**: 69–77.

Pielichowska, K. and K. Pielichowski. 2014. Phase change materials for thermal energy storage. *Progress. Mater. Sci.*, **65**: 67e123.

Poikelispää, M. Honkanen, M. Vippola, M. and E. Sarlin. 2021. Effect of carbon nanotubes and nanodiamonds on the heat storage ability of natural rubber composites. *J. Elast. Plast.*, **53**: 311–322.

Potts, J. R. Shankar, O. Du, L. and R. S. Ruoff. 2012. Processing–morphology–property relationships and composite theory analysis of reduced graphene oxide/natural rubber nanocomposites. *Macromolecules*, **45**: 6045–6055.

Potts, J. R. Shankar, O. Murali, S. Du, L. and R. S. Ruoff. 2013. Latex and two-roll mill processing of thermally-exfoliated graphite oxide/natural rubber nanocomposites. *Compos. Sci. Technol.*, **74**: 166–172.

Qi, Y. Rhim, S. H. Sun, G. F. Weinert, M. and L. Li. 2010. Epitaxial growth of graphene on 6H-silicon carbide substrate by simulated annealing method. *Phys. Rev. Lett.*, **100**: 016602.

Rajeev, R. S. Bhowmick, A. K., De, S. K. and A. Bandyopadhyay. 2003. Short melamine fiber filled nitrile rubber composites. *J. Appl. Polym. Sci.*, **90**: 544–548.

Ray, S. and A. K. Bhowmick. 2003. Influence of untreated and novel electron-beam-modified surface coated silica filler on rheometric and mechanical properties of ethylene-octene copolymer. *J. Mater. Sci.*, **38**: 3199–3210.

Rourke, J. P. Pandey, P. A. Moore, J. J. Bates, M. Kinloch, I. A. Young, R. J. et al. 2011. The real graphene oxide revealed: Stripping the oxidative debris from the graphene-like sheets. *Angew. Chem. Inter. Ed.*, **50**: 3173–3177.

Roy, N. and A. K. Bhowmick. 2012. Modifications of carbon for polymer composites and nanocomposites. *Progress Polym. Sci.*, **37**: 781–819.

Roy, A. Kar, S. Ghosal, R. Mukhopadhyay, R. Naskar, K. and A. K. Bhowmick. 2022. Synthesis and characterization of graphene sheets decorated with carbon black by direct pyrolysis of a molasses –carbon black mixture as a potential versatile filler for rubber. *Rubber Chem. Technol.*, (In press; DOI: 10.5254/rct.21.79928).

Roy, A. Kar, S. Ghosal, R. Naskar, K. and A. K. Bhowmick. 2021. Facile synthesis and characterization of few layered multifunctional graphene from sustainable precursors by controlled pyrolysis, understanding of the graphitization pathway, and its potential application in polymer nanocomposites. *ACS Omega*, **6**: 1809–1822.

Ruan, G. Sun, Z. Peng, Z. and J. M. Tour. 2011. Growth of graphene from food, insects, and waste. *ACS Nano*, **5**: 7601–7607.

Sadhu, S. and A. K. Bhowmick. 2004. Preparation and properties of nanocomposites based on acrylonitrile butadiene rubber, styrene butadiene rubber and polybutadiene rubber. *J. Polym. Sci. Part B: Polym. Phys.*, **42**: 1573–1585.

Sang, L. Hao, W. Zhao, Y. Yao, L. and P. Cui. 2018. Highly aligned graphene oxide/waterborne polyurethane fabricated by in-situ polymerization at low temperature. *e-Polymers*, **18**: 75–84.

Sengupta, R. Bhattacharya, M. Bandyopadhyay, S. and A. K. Bhowmick. 2011. A review on the mechanical and electrical properties of graphite and modified graphite reinforced polymer composites. *Progress Polym. Sci.*, **36**: 638–670.

Shah, R. Alam, N. Razzaq, A. A. Cheng, Y. Yuje, C. Jiapeng, H. et al. 2019. Effect of binder conformity on the electrochemical behavior of graphite anodes with different particle shapes. *Acta Phys. Chim. Sin.*, **35**: 1382–1390.

Shanmugharaj, A. M. and A. K. Bhowmick. 2003. Influence of novel electron beam modified dual phase filler on the rheometric and mechanical properties of styrene butadiene rubber vulcanizates, *Rubber Chem. Technol.*, **76**: 299–317.

Sherif, A. Izzuddin, Z. Qingshi, M. Nobuyuki K, Andrew, M. Hsu-Chiang, K. et al. 2013. Melt compounding with graphene to develop functional, high-performance elastomers. *Nanotechnology*, **24**: 165601.

Shin, H. J. Kim, K. K. Benayad, A. Yoon, S. M. Park, H. K. Jung, I. S. et al. 2009. Efficient reduction of graphite oxide by sodium borohydride and its effect on electrical conductance. *Adv. Funct. Mater.*, **19**: 1987–1992.

Singh, V. K. Shukla, A. Patra, M. K. Saini, L. Jani, R. K. Vadera, S. R. et al. 2012. Microwave absorbing properties of a thermally reduced graphene oxide/nitrile butadiene rubber composite. *Carbon*, **50**: 2202–2208.

Singh, P. K. and P. K. Singh. 2021. A dielectric elastomer and graphene nanocomposites: A review. *IOP Conf.: Mater. Sci. Engg.*, **1116**: 012034.

Staudenmaier, L. 1898. Verfahren zur darstellung der graphitsäure. *Berichte der deutschen chemischen Gesellschaft*, **31**: 1481–1487.

Suriani, A. B. Nurhafizah, M. D., Mohamed, A. Masrom, A. K. Mamat, M. H. Malek, M. F. et al. 2017. Electrical enhancement of radiation vulcanized natural rubber latex added with reduced graphene oxide additives for supercapacitor electrodes. *J. Mater. Sci.*, **52**: 6611–6622.

Suriani, A. B. Nurhafizah, M. D., Mohamed, A. Zainol, I. and A. K. Masrom. 2015. A facile one-step method for graphene oxide/natural rubber latex nanocomposite production for supercapacitor applications. *Mater. Lett.*, **161**: 665–668.

Tian, M. Zhang, J. Zhang, L. Liu, S. Zan, X. Nishi, T. et al. 2014. Graphene encapsulated rubber latex composites with high dielectric constant, low dielectric loss and low percolation threshold. *J. Colloid Inter. Sci.*, **430**: 249–256.

Tour, J. M. 2013. Top-down versus bottom-up fabrication of graphene-based electronics. *Chem. Mater.*, **26**: 163–171.

Veeman, D. Sai, M. S. Rajkumar, V. Ravichandran, M. and S. Manivannan. 2021. Graphene for thermal storage applications: Characterization, simulation and modelling. *J. Electronics Mater.*, **50**: 5090–5105.

Vinothkannan, M. Kim, A. R. Ramakrishnan, S. Yu, Y. –T. and D. J. Yoo. 2021. Advanced nafion nanocomposite membrane embedded with unzipped and functionalized graphite nanofibers for high-temperature-air fuel cell system: The impact of filler on power density, chemical durability and hydrogen permeability of membrane. *Compos. Part B: Engg.*, **215**: 108828.

Wang, X. Hu, Y. Song, L. Yang, H. Xing, W. and H. Lu. 2011. In-situ polymerization of graphene nanosheets and polyurethane with enhanced mechanical and thermal properties. *J. Mater. Chem. A*, **21**: 4222–4227.

Wei, J. Jacob, S. and J. Qiu. 2014. Graphene oxide-integrated high-temperature durable fluoroelastomer for petroleum oil sealing. *Compos. Sci. Technol.*, **92**: 126–133.

Wick, P. Louw-Gaume, A. E. Kucki, M. Krug, H. F. Kostarelos, K. Fadeel, B. et al. 2014. Classification framework for graphene-based materials. *Angew. Chem. Inter. Ed.*, **53**: 7714–7718.

Wu, J. Huang, G. Li, H. Wu, S. Liu, Y. and J. Zheng. 2013. Enhanced mechanical and gas barrier properties of rubber nanocomposites with surface functionalized graphene oxide at low content. *Polymer*. **54**: 1930–1937.

Wu, S. Tang, Z. Guo, B. Zhang, L. and D. Jia. 2013. Effects of interfacial interaction on chain dynamics of rubber/graphene oxide hybrids: A dielectric relaxation spectroscopy study. *RSC Adv.*, **3**: 14549–14559.

Wu, S. Q. Wang, J. W. Shao, J. Wei, L. Yang, K. and H. Ren. 2017. Building a novel chemically modified polyaniline/thermally reduced graphene oxide hybrid through π-π Interaction for fabricating acrylic resin elastomer-based composites with enhanced dielectric property, *ACS Appl. Mater. Interfaces*, **9**: 28887–28901.

Xiang, J. L. and L. T. Drzal. 2011. Thermal conductivity of exfoliated graphite nanoplatelet paper. *Carbon*, **49**: 773–778.

Xing, W. Tang, M. Z. Wu, J. R. Huang, G. S. Li H, Lei, Z. Y. et al. 2014. Multifunctional properties of graphene/rubber nanocomposites fabricated by a modified latex compounding method. *Compos. Sci. Technol.*, **99**: 67–74.

Xiong, X. Wang, J. Jia, H. Fang, E. and L. Ding. 2013. Structure, thermal conductivity, and thermal stability of bromobutyl rubber nanocomposites with ionic liquid modified graphene oxide. *Polym. Degradat. Stab.*, **98**: 2208–2214.

Yabuuchi, N. Kinoshita, Y. Misaki, K. Matsuyama, T. and S. Komaba. 2015. Electrochemical properties of LiCoO$_2$ electrodes with latex binders on high-voltage exposure. *J. Electrochem. Soc.*, **162**: A538–A544.

Yan, Z. Lin, J. Peng, J. Sun, Z. Zhu, Y. Li, L. et al. 2012. Toward the synthesis of wafer-scale single-crystal graphene on copper foils. *ACS Nano*, **6**: 9110–9117.

Yan, N. Xia, H. Wu, J. Zhan, Y. Fei, G. and C. Chen. 2013. Compatibilization of natural rubber/high density polyethylene thermoplastic vulcanizate with graphene oxide through ultrasonically assisted latex mixing. *J. Appl. Polym. Sci.*, **127**: 933–941.

Yang, H. Liu, P. Zhang, T. Duan, Y. and J. Zhang. 2014. Fabrication of natural rubber nanocomposites with high graphene contents via vacuum-assisted self-assembly. *RSC Adv.*, **4**: 27687–27690.

Yang, J. Qi, C. Liu, Y. Bao, R. Liu, Z. Yang, W. et al. 2016. Hybrid graphene aerogels/phase change material composites: Thermal conductivity, shape-stabilization and light-to-thermal energy storage. *Carbon*, **100**: 693–702.

Yin, L. –J. Zhao, Y. Zhu, J. Yang, M. Zhao, H. Pei, J. –Y. et al. 2021. Soft, tough, and fast polyacrylate dielectric elastomer for non-magnetic motor, *Nat. Commun.*, **12**: 4517.

Yoo, M. Frank, C. W. Mori, S. and S. Yamaguchi. 2004. Interaction of poly (vinylidene fluoride) with graphite particles. 2. Effect of solvent evaporation kinetics and chemical properties of PVDF on the surface morphology of a composite film and its relation to electrochemical performance. *Chem. Mater.*, **16**: 1945–1953.

Young, R. J. Kinloch, I. A. Gong, L. and K. S. Novoselov. 2012. The mechanics of graphene nanocomposites: A review. *Compos. Sci. Technol.*, **72**: 1459–1476.

Yu, Q. Lian, J. Siriponglert, S. Li, H. Chen, Y. P. and S. -S. Pei. 2008. Graphene segregated on Ni surfaces and transferred to insulators. *Appl. Phys. Lett.*, **93**: 113101.

Yuan, J. K. Yao, S. H. Sylvestre, A. and J. Bai. 2012. Biphasic polymer blends containing carbon nanotubes: Heterogeneous nanotube distribution and its influence on the dielectric properties. *J. Phys. Chem. C*, **116**: 2051–2058.

Yuanzheng, L. Youqi, W. Jiang, H. and L. Buyin. 2021. Nanofiber enhanced graphene-elastomer biomimetic hierarchical structure for energy storage and pressure sensing. *Mater. Des.*, **203** (1–10): 109612.

Zhan, Y. Lavorgna, M. Buonocore, G. and H. Xia. 2012. Enhancing electrical conductivity of rubber composites by constructing interconnected network of self-assembled graphene with latex mixing. *J. Mater. Chem.*, **22**: 10464–10468.

Zhan, Y. Wu, J. Xia, H. Yan, N. Fei, G. and G. Yuan. 2011. Dispersion and exfoliation of graphene in rubber by an ultrasonically-assisted latex mixing and in-situ reduction process. *Macromol. Mater. Eng.*, **296**: 590–602.

Zhang, F. Li, T. and Y. Luo. 2018. A new low moduli dielectric elastomer nano-structured composite with high permittivity exhibiting large actuation strain induced by low electric field. *Compos. Sci. Technol.*, **156**: 151–157.

Zhang, H. Meng, Y. Cao, Y. Yao, Y. Fan, D. Yang, T. et al. 2020. Form-stable phase change materials based on polyolefin elastomer and octadecylamine-functionalized graphene for thermal energy storage. *Nanotechnol.*, **34**: 245402.

Zhang, H. –B. Wang, J. –W. Yan, Q. Zheng, W. –G. Chen, C. and Z. –Z. Yu. 2011. Vacuum-assisted synthesis of graphene from thermal exfoliation and reduction of graphite oxide. *J. Mater. Chem.*, **21**: 5392–5397.

Zhang, J. Yang, H. Shen, G. Cheng, P. Zhang, J. and S. Guo. 2010. Reduction of graphene oxide vial-ascorbic acid. *Chem. Commun.*, **46**: 1112–1114.

Zhang, Y. Zhang, L. and C. Zhou. 2013. Review of chemical vapor deposition of graphene and related applications. *Acc. Chem. Res.*, **46**: 2329–2339.

Zhao, Q. Zhao, K. Ji, G. Guo, X. Han, M. Wen, J. et al. 2019. High sulfur loading, RGO-linked and polymer binder-free cathodes based on RGO wrapped n, p-codoped mesoporous carbon as sulfur host for Li-S batteries. *Chem. Eng. J.* **361**: 1043–1052.

17 Graphene-Elastomer Composite for Biomedical Applications

Sanjoy Kumar Ghorai and Santanu Chattopadhyay
IIT Kharagpur

CONTENTS

DOI: 10.1201/9781003200444-17

17.1 INTRODUCTION

In recent years, the use of nanomaterials emerged with nanotechnology has made a revolutionary change in various fields of biomedical applications. The development of novel and efficient nanocomposites using suitable nanomaterials and matrices (elastomers) for biomedical applications has become a promising topic of interest in the recent era. Various nanomaterials such as titanium dioxide (TiO_2), bioglass (BG), tricalcium phosphate (TCP), layered silica (Si), nanohydroxyapatite (nHA), polyhedral oligomeric silsesquioxane (POSS), carbon nanotube (CNT), graphene oxide (GO), and graphene have been widely used along with suitable elastomer to formulate the nanocomposites with improved physico-mechanical and bioactive properties (Sadasivuni et al. 2014, Maiti, Bhattacharya, and Bhowmick 2008, Ghorai et al. 2019b). Among various nanocomposites, the carbonaceous nanomaterial-based nanocomposites impart significant improvement of numerous multifunctional attributes such as physico-mechanical properties, thermal properties, electrical conductivity, and strong near-infrared (NIR) absorbance as compared to other nanomaterials-based nanocomposites (Oliveira et al. 2015). Thus, carbonaceous nanomaterials-based nanocomposites are extensively used in different wings of the biomedical field such as drug delivery, gene delivery, cancer therapy, bone tissue engineering, nerve regeneration, cardiac tissue regeneration, biosensor, and imaging in the recent past (Shen et al. 2012).

Until the 1980s, only three allotropes of carbon were commonly known as graphite, diamond and amorphous carbon. But the discovery of graphene in 1904 by A. Geim and K. Novoselov made a revolutionary paradigm shift of application of graphene-based nanocomposites in biomedical applications (Novoselov and Geim 2007). Graphene is the mother of all allotropes of carbon. It is considered as the basic structure of graphite with a 2-dimensional layered structure where the sp^2 carbons are interconnected in a hexagonal honeycomb-like structure (Goenka, Sant, and Sant 2014). The multilayers are stacked together through weak van der Waals force of interaction with successive layer distance of 0.334 nm. Due to the presence of weak van der Waals force, various molecules and ions can easily be penetrated in between the successive layers of the graphene, resulting in the intercalation and exfoliation of graphene to produce graphene and its derivatives. Graphene is considered as the basic constructive unit of other carbon allotropes with different geometries (Liu, Cui, and Losic 2013). Figure 17.1 shows the single-layered graphene sheet with spherical geometry form '0' dimensional fullerene (Kroto et al. 1985), single or double or multilayer graphene sheet with a rolled structure formed 1D carbon nanotube (Iijima 1991), and multiple graphene sheets stacked together in 3D geometry generate graphite (Novoselov et al. 2004). However, graphene and its derivatives are usually referred to as sheet-like structures of graphene-based nanomaterials.

An elastomer is a group of polymers having viscoelastic characteristics with high elongation at break (EB) and low elastic modulus. The intermolecular force of interaction of elastomer is very weak. It shows a large deformation under the application of small stress and retains its original shape upon withdrawal of the applied stress. The use of elastomers started long back from ancient Mesoamerican people about 3,000 years ago (Papageorgiou, Kinloch, and Young 2015). The Mesoamerican

2-D (Graphene)

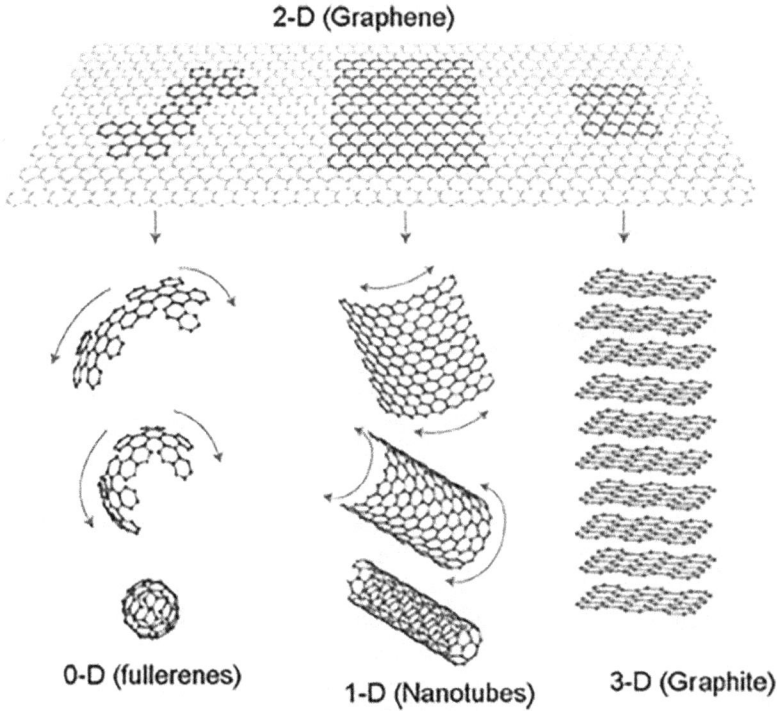

0-D (fullerenes)   1-D (Nanotubes)   3-D (Graphite)

FIGURE 17.1 Different allotropes of graphene generated from 2D graphene. (Reproduced with permission from Ref. (Liu, Cui, and Losic 2013). Copyright 2013, Elsevier.)

people harvested the natural latex from *Castilla elastica* and processed the latex using a natural liquid extracted from *Ipomoea alba* to prepare different artifacts such as hollow rubber figurines and rubber balls. The organic components present in *I. alba* purified the latex, increased the intermolecular force of attraction, and improved the mechanical properties of the processed latex. The processed latex was used to prepare the rubber balls which in later course become a central ritual element of the Mesoamerican ball game of the ancient Mesoamerican society (Hosler, Burkett, and Tarkanian 1999). However, the use of elastomer increased rapidly after the innovation of the vulcanization technique (Raue et al. 2014, Coleman, Shelton, and Koenig 1974). Through the use of vulcanization technique, the viscoelastic materials become more elastic with improved mechanical, thermal aging and resilience (Coran 1994, 2003).

It is interesting to amalgamate the nanostructure of the graphene layer and its various derivatives within the polymer/elastic matrix produced the graphene-elastomer nanocomposites with superior chemical, thermal, electrical, and bio-active properties as compared to the pristine elastomer. Moreover, the sheet-like nanostructure is difficult to process into various architecture. However, due to the low elastic characteristics of graphene, it is difficult to use graphene and graphene derivatives as solo materials. The incorporation of graphene and the derivatives of graphene with the elastomer matrix mitigates the brittleness and improves the elastic characteristics of the nanocomposites. On the other hand, the introduced graphene improves the mechanical, fatigue, thermal, electrical, bioactive, and recyclable properties of the polymer/elastomer nanocomposites. Thus, the amalgamation of nanographene with the elastomer experienced a synergistic effect and improved the various desirable multifunctional properties useful for various important applications like nanocomposites (Fan et al. 2010, Yang et al. 2010), capacitor (Stoller et al. 2008), energy storage (Yuanzheng et al. 2021), sensors (Lu et al. 2009), and flexible electronics (Xuan et al. 2008, Liang et al. 2010). Moreover, the non-toxic nature of graphene entices its significant application in the biomedical field which has been explored first in Liu et al. (2008). Chemically functionalized PEG on the surface of nanographene (NGO) produced an environment of effective nanocarriers of water-insoluble anticancer drug like SN-38 through π–π stacking and increased the efficiency of the drug (Liu et al. 2008). This breakthrough work recognizes graphene as a non-toxic, biocompatible material. Following this, numerous researches have been conducted in the last few years to explore significant uses of graphene-elastomer in various wings of nanomedicine (Sun et al. 2018, Jakus et al. 2015). In short, Patel et al. have shown that the sulfate functionalization of GO surface followed by grafting of PU on the surface produced elastomer nanocomposites with improved various properties exhibiting sustained release of the encapsulated drug as compared to the pure PU (Patel et al. 2017). Chen et al. have prepared the branched polyethyleneimine-functionalized GO for efficient target specific gene carrier for the delivery of plasmid DNA (pDNA) (Chen et al. 2011). Graphene-based highly elastic and biocompatible elastomeric scaffolds have been generated using polyacrylamide/calcium alginate hydrogel coupled with chitosan for the detection of high strain sensor for detection of human motion (Cai et al. 2019).

17.2   SYNTHESIS OF GRAPHENE

In the early 1970, single- or multilayered graphite synthesis has been tried through the decomposition of the carbonaceous substrate on the single-crystal Pt substrate (Eizenberg and Blakely 1979). But the lack of consistency of layer structure and the properties of the deposited graphite layer on different crystal planes of Pt limit its application. The lack of proper characterization and scarcity to identify the suitable application of the deposited graphite layer failed to get much interest at that point of time. In the middle of 1970s, single-layered or few-layered graphene nanosheets (GNSs) were prepared from graphite flakes by mechanical exfoliation technique (Scotch Tape method) though not published in the open forum until 2004. But in 2004, Novoselov et al. have synthesized the single-layered GNS through a chemical exfoliation technique from graphite (Novoselov et al. 2004). For this revolutionary invention of 2D graphene, K. Novoselov and A. Geim were awarded the most prestigious award of science, 'Noble Prize' in 2010. They have prepared graphene by exfoliation repeatedly with consistent properties. Afterward, various modifications have been adopted which enhance the large-scale production, reactivity, and stability of the GNS. However, 2D pristine GNS is unstable in aqueous media and lacks the

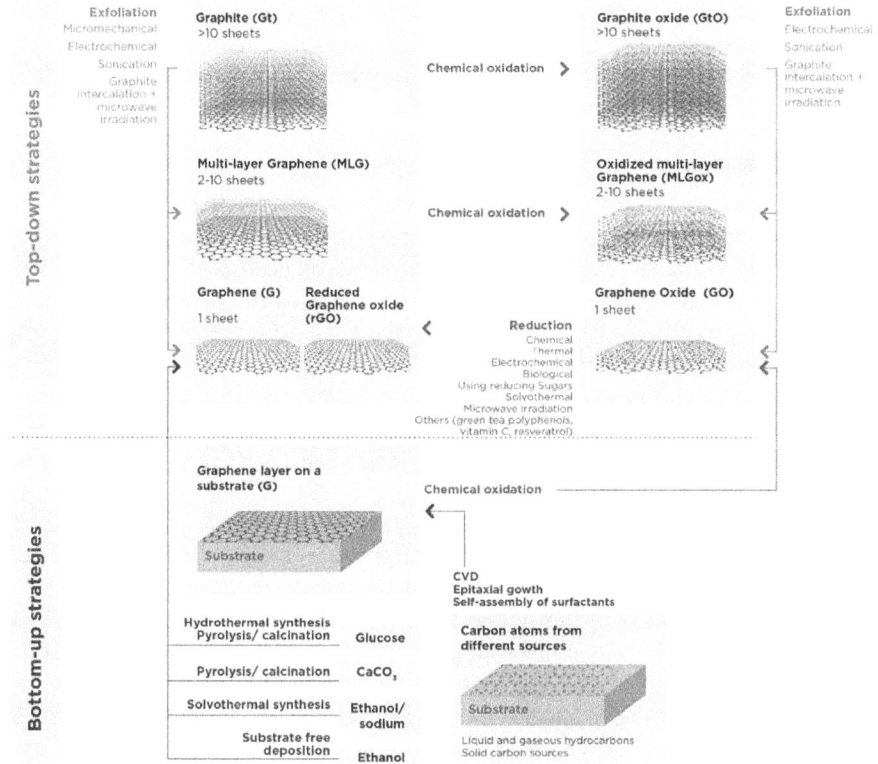

FIGURE 17.2 Different graphene-based materials and their synthesis by top-down and bottom-up methods. (Schematic representation of different synthesis strategies is reproduced with permission from Ref. (Henriques et al. 2018). Copyright 2018, Elsevier.)

active center as it is devoid of functional groups on the surface (Nwosu, Iliut, and Vijayaraghavan 2021). Thus, different modification processes have been employed to introduce the active functional group on the surface of the GNS which enhances the stability and reactivity of the synthesized GNS. In this regard, different graphene derivatives such as graphene oxide (GO), reduced graphene oxide (rGO), graphene nanoplatelets (GNPs), graphene quantum dots (GQDs), and graphene nanoribbons (GNRs) are synthesized by various modified techniques and widely utilized along with polymer matrix for different biomedical applications. Basically, the graphene and its derivatives are synthesized by two methods broadly known as the 'Top-down' method such as chemical exfoliation, mechanical exfoliation, and thermal exfoliation and the 'Bottom-up' method such as chemical vapor deposition (CVD) as shown in Figure 17.2 (Sadasivuni et al. 2014, Papageorgiou, Kinloch, and Young 2015).

17.2.1 'TOP-DOWN' METHOD

This process involves the synthesis of GNS and its derivative by the separation of an individual layer of graphite by physical and chemical approaches.

In a physical process, the layers are separated from the graphite nanoflakes by applying a mechanical force that destroys the interplanar van der Waals force. In a chemical process on contrary, the functionalization is followed by sonication of the graphite to generate graphene and its derivatives.

17.2.1.1 Mechanical Exfoliation

The mechanical exfoliation technique was first adopted by Novoselov et al. (2004) for the preparation of monolayer GNS which later on became a very interesting material for various applications. This method involved repeated peeling of successive layers of graphite manually using mechanical force, known as the Scotch Tape method (Novoselov et al. 2004). The dimension of the peeled graphene layers generally depends on the grain size of the graphite. This method produced a high quality of graphene with batch-to-batch consistency. However, this Scotch Tape method fails to draw a significant mark for the long run in the engineering community due to its low-scale production which is not enough for the preparation of polymer nanocomposites.

17.2.1.2 Chemical Exfoliation

Till date, the chemical exfoliation method of graphite is the most widely accepted technique for the preparation of GNSs and derivative of graphene due to its large-scale production and owing to batch-to-batch consistency in terms of quality (Park and Ruoff 2009). The method involved acids that react with the graphite, dispersed in different suitable solvents like dimethyl sulfoxide, tetrahydrofuran, dimethylformamide, and N-methyl-pyrrolidone followed by sonication. During sonication, the applied vibration energy to the graphene-solvent system plays a vital role in the separation of the individual layer of the graphite. Thus, the surface tension of the added solvent (solvation force) should circumvent the graphene-graphene interaction force. Generally, organic solvents and aqueous media are widely used for sonication. This acid intercalation method is also applicable for CVD and electrochemical processes to synthesize the nanographene sheet. The treated acid reacts with the graphite and generates different active functional groups such as hydroxyl, ketone, and epoxide groups on the surface of the sheets which under sonication exfoliated and generated single-layer graphene derivatives like GO. Different oxidants like $KMnO_4$, H_3PO_4, $NaNO_2$, and $KClO_3$ are widely used to oxidize the nanosheet to produce GO in the presence of acids like HNO_3, H_2SO_4, or mixed acid according to Staudenmaier, Hummers, Brodie, and modified methods (Hummers Jr and Offeman 1958, Marcano et al. 2010, Brodie 1855, Staudenmaier 1898). The generated functional groups increased the interplanar distance from 0.335 nm of graphite to 0.6–1.2 nm of the GO and minimized the van der Waals force between the layers which facilitates the exfoliation process. Moreover, the generated surface functional groups on the GO surface interact with polymer matrix by covalent, or ionic or electrostatic forces which help to produce a strong polymer-nanomaterials interface within the polymer composites. The graphite oxide (GtO) and graphene oxide have similar chemical properties but GtO has 3D structure with less exfoliation of graphene layers whereas GO has 2D structure with minimum size and high exfoliation. Both the GtO and the GO can be reduced to graphene or in true sense to be reduced to graphene oxide (rGO) by heat treatment or by chemical reduction using chemical reductants such as sodium borohydride and hydrazine. The reduction of GO to rGO increased the unsaturation of the graphene sheet which increased the electrical conductivity of rGO as compared to GO.

17.2.1.3 Thermal Exfoliation

Thermal exfoliation is another important technique to generate the graphene sheets in bulk quantities for the preparation of elastomer-based composites. The acid-intercalated graphite material is heated to high temperature (~1,100°C) at a very high rate of heating about 2,000°C min^{-1} and high pressure which leads to the evaporation of surface functional groups of the intercalated graphite oxide (Xiang and Drzal 2011). Due to rapid heating, the evaporated gaseous materials could not diffuse through the graphite layers rather generated internal pressure between the layers which can overcome the weak van der Waals force to exfoliate the graphene layers. Single and few layers of graphene are generated through this technique with satisfactory chemical properties as compared to single-layer graphene.

17.2.2 BOTTOM-UP PROCESS

In the bottom-up process, the carbonaceous elements are assembled on the specified surface of a substrate to generate the graphene sheet. Different techniques have been exploited for the synthesis of GNSs such as CVD, arc discharge, chemical conversion of hydrocarbons, and epitaxial conversion on SiC (Xiang and Drzal 2011, Fan et al. 2010, Li et al. 2010). In CVD, the gaseous hydrocarbons are deposited on the surface of polycrystalline Ni substrate and segregated as graphene at ambient temperature. In this process, the hydrocarbon feedstock was passed through the hot channel to the reactor where the gaseous hydrocarbon reacts culminating to decompose carbon radical deposited on the Ni substrate as a materials film (Eizenberg and Blakely 1979). Controlling the reaction temperature, the cooling rate of the deposited material and the proper bypassing of the disposal gases influence the quality of the graphene materials. Currently, Cu substrate for deposition of different types of carbon containing substrate as feed-stock is in practice for the preparation of graphene in CVD method. Though single- and double-layer graphene with satisfactory quality is produced by CVD, there exist certain limitations of using CVD. As the graphene is generated on the surface of the metal substrates, during collection there is a chance of changing of properties and defects may cause deterioration of quality of the product. CVD involves high energy consumption and replacement of the metal substrate in certain intervals causes the extra burden of cost to the production of graphene (Papageorgiou, Kinloch, and Young 2015).

17.3 PREPARATION OF GRAPHENE-ELASTOMER NANOCOMPOSITES

In order to achieve the improved properties such as physico-mechanical properties, thermal, electrical properties of elastomer, graphene and graphene-based nanomaterials are incorporated within the elastomer matrix. The extent of improvement of these desired properties is related to the degree of dispersion of the nanomaterials within the elastomer which depends on the filler-filler and polymer-filler interaction. The higher polymer-filler interaction compared to filler-filler interaction assures better dispersion probability of the nanomaterials. Various techniques such as melt mixing, solution mixing, and in situ polymerization have been adopted for homogeneous dispersion of nanomaterials within the matrix. Sometimes combination of two or more techniques are employed in the preparation of graphene-elastomer nanocomposites. For the preparation of elastomer nanocomposites, different ingredients like curing agents, crosslinkers, processing aids, stabilizers, and antiaging chemicals are added at different stages of mixing. The rubber-based composites are prepared using two roll and internal mixers to reduce the viscosity of rubber through mastication. The rubber latex based nanocomposites are generally prepared using a magnetic stirrer, mechanical stirrer, bath sonicator, etc. whereas, Brabender mixture, injection, and extrusion molding machines are largely used commercially for the preparation of thermoplastic elastomer by the melt mixing process.

17.3.1 MELT MIXING PROCESS

Melt mixing is one of the most favorable industrial processes because of its high productivity and cost-effectiveness. The added fillers are incorporated within the elastomer matrix in molten conditions using mechanical forces. This process is also environment friendly as it does not involve any toxic solvent and is useful for both the polar and non-polar elastomers. Sometimes compatibilizers are used for the better dispersion of filler through improving polymer-filler interaction. Despite having several advantages, the melt mixing process also has some disadvantages. Applying high energy for melting of the elastomer may lead to the degradation of other ingredients. The high viscosity of the elastomer matrix and the high loading of filler may cause the aggregation of filler. To improve the dispersion of the filler uniformly within the matrix, high shearing force is to be applied which breaks the structure of graphene layers. Kim et al. have prepared the thermoplastic polyurethane (TPU)-based polymer nanocomposites using functionalized graphene or graphene oxide (GO) via melt mixing process to improve the electrical and gas barrier properties of the TPU/GO nanocomposite (Kim, Miura, and Macosko 2010). Araby et al. have incorporated graphene nanoplates (GnPs) into the ethylene propylene diene monomer (EPDM) rubber by the melt mixing process. The incorporated GnPs improved the YM and TS by 710% and 404%, respectively, at 26.7% vol.% of the filler loading (Araby et al. 2013). Paran et al. have prepared the GnPs-incorporated thermoplastic vulcanizates (TPVs) based on reclaim rubber and linear low-density polyethylene via melt mixing process. The morphological study by FESEM and HRTEM indicates the uniform dispersion of GnPs into the rubber matrix (Paran et al. 2018). Yang et al. have prepared the acrylonitrile butadiene rubber (NBR) composites using graphite powder via melt mixing process. The incorporated graphite particles improved the TS and wear resistance of the NBR/graphite nanocomposites (Yang et al. 2006). Qiu et al. have shown the significant improvement of mechanical and viscoelastic properties of the surface-grafted GNS-incorporated thermoplastic polyester elastomer (TPEE) nanocomposites by melt mixing process (Qiu et al. 2016). Ke et al. have prepared branched carbon nanostructure (CNS) and GnPs-incorporated flexible pressure-sensitive TPU nanocomposites by melt mixing process using an internal mixture. The optimum level of CNS and GnPs with the mass ratio of 3:1 showed the highest pressure sensitivity of 2.05 MPa^{-1} for 0–1.2 MPa pressure as compared to 0.18 MPa^{-1} for pristine TPU. A combination of interconnected CNS along with GnPs with high dielectric constant has shown maximum pressure sensitivity of the TPU elastomer nanocomposites which is suitable for prosthetic limbs and wearable electronics (Ke et al. 2019).

17.3.2 SOLUTION MIXING

Though large-scale melt mixing process is widely accepted in the industry for the preparation of graphene-elastomer nanocomposites, the high viscosity of the elastomer matrix leads to inferior dispersion and reduces the chances of breakage of graphene structure at a high shear rate alarming an unavoidable issue. Solution mixing or latex blending, owing to its ease of dispersion of graphene in appropriate solvents, favors this technique for the preparation of graphene-based elastomer nanocomposites especially for lab-scale preparation in the academic implications (Sadasivuni et al. 2013, Kumar et al. 2013, Zhan et al. 2011, Mensah et al. 2014, Yan et al. 2013, Wu et al. 2013, Chen and Lu 2012, Nawaz et al. 2012, Xiong et al. 2013). The graphene nanomaterials are incorporated within the elastomer either by a common solvent or by mixing the dispersion of graphene with the elastomer matrix dissolved in another solvent. Proper mechanical stirring, sonication, or magnetic stirring ensures homogeneous dispersion and distribution of the nanomaterials within the elastomer leading to improve various desired properties of the nanocomposites (Mu and Feng 2007). Latex blending is similar to solution mixing where the elastomer is mixed with the nanomaterials in the form of latex. After incorporation of graphene nanomaterials within the latex, sonication and co-coagulation ensure the uniform dispersion of the nanomaterials into the elastomer nanocomposites (Zhan et al. 2012, Tang et al. 2012, Tian et al. 2014, Cao et al. 2019, Kang et al. 2014). Bahrami et al. have prepared the polyurethane (PU)/graphene elastomeric nanocomposites for tissue engineering application by solution casting technique (Bahrami et al. 2019). Zhan et al. have incorporated exfoliated graphene (GE) nanomaterials into the NR latex by ultrasonication process followed by coagulation to produce NR/GE masterbatch. The incorporated GE within the rubber matrix improves the mechanical properties of the NR significantly (Zhan et al. 2011). Frasca et al. have incorporated a very low concentration of multilayer graphene (MLG) by solution mixing process into the chlorine-isobutylene-isoprene rubber (CIIR), nitrile butadiene rubber (NBR), natural rubber (NR), and styrene-butadiene rubber (SBR). SEM study indicates the uniform dispersion of MLG in the rubber matrices which improved the various multifunctional properties of the graphene-elastomer nanocomposites as compared to the pristine rubber (Frasca et al. 2015). Niu et al. have fabricated the graphene nanoplatelets polydimethylsiloxane (GNPs/PDMS)-based elastomeric flexible piezoresistive sensors by solution process. The GNPs/PDMS piezoresistive sensors can sense the subtle bending and accurately detect the finger motion which is useful for artificial skin and wearable sensors (Niu et al. 2018). Ho et al. have prepared the stretchable and multimodal electronic skin using PDMS solution and graphene derivative through layer-by-layer laminating technique for humidity, thermal, and pressure sensor applications (Ho et al. 2016). Liu et al. have incorporated GO/zwitterionic chitin nanocrystals hybrids (GC) into the NR latex solution which generated a molecular bridge between the functional groups of GC and NR particles to produce NR composites with high TS and large elongation properties. The GO-incorporated NR composites have been used as good biocompatible, recyclable gloves for biomedical applications without using conventional sulfur vulcanization (Liu et al. 2021).

Even though there exist numerous advantages of the solution mixing process for the preparation of graphene-elastomer nanocomposites, some notable issues cannot be avoided such as complete removal of the solvents from the end products and proper disposal of toxic solvents. The use of costly solvents, removal, and disposal of the solvent at the end makes the process lengthy and costlier in large-scale industrial applications.

17.3.3 IN SITU PROCESS

In situ polymerization process is another important method for the preparation of graphene-elastomer nanocomposites used in the recent past (Paszkiewicz et al. 2014, Kim, Miura, and Macosko 2010, Lee et al. 2009). Generally, in situ process involves the addition of nanomaterials within the monomer or the solution containing monomer followed by the polymerization of the monomer producing nanomaterials-incorporated polymer nanocomposites. The added monomer may attach to the surface of the added nanomaterials increasing the layer gap of the successive layers during the growth of the polymer chains that result in interaction or exfoliation of the nanomaterials. In the solution process, a high level of dispersion of the graphene nanomaterials can be achieved without prior exfoliation of graphene (Shioyama 2000, Sadasivuni et al. 2014). Moreover, the polymers are generated on the surface of the graphene layers that help to form the strong polymer-nanomaterials interface of the nanocomposites. This strong interface easily transfers the applied load from matrix to nanofiller without failure improving the mechanical properties of the graphene-elastomer nanocomposites (Papageorgiou, Kinloch, and Young 2015). Bahena et al. have prepared the polymercene- and polyfarnesene-based bioelastomer nanocomposites by in situ polymerization technique using amine-grafted graphene. The alkyl chain length of the grafted amine influenced the intercalation distance between the graphene layers (Bahena et al. 2020). Ghorai et al. have synthesized the nanohydroxyapatite (nHA) decorated GO-incorporated polyurethane-urea (PUU)-based thermoplastic elastomer (PUU/nHA-GO) nanohybrid composites by in situ solution polymerization technique. Incorporation of optimum amount of GO within the PUU matrix increased the tensile strength (TS) and modulus value of 74% and 34%, respectively, of the nanohybrid composites as compared to the pristine PUU. Moreover, decorated GO-incorporated nanohybrid composites have exhibited superior % cell viability and the proliferation of the osteoblast cell which indicated the better bone regeneration capability of the GO added nanohybrid composites as compared to the pristine scaffold (Ghorai et al. 2019a). Zhang et al. have prepared the disulfide-bonded self-healable liquid crystalline elastomer (SHLCE)-based nanocomposites by introducing GnPs through in situ polymerization technique.

GnPs (20 wt.%)-incorporated SHLCE/GnPs exhibit good thermal conductivity and excellent reprocessability with satisfactory recycle properties compared to the pristine polymer. Instead, lot of merits, there are some limitations of in situ nanocomposites preparation technique. Sometimes this process allows preparing a low viscous elastomer which is not suitable for specific applications (Zhang et al. 2021). This technique suffers limitations in terms of the separation of the unreacted monomers and other reagents after the preparation of nanocomposites deteriorating the desired properties of the nanocomposites. This may also exert health hazards limiting its viability specifically in biomedical applications.

17.4 SURFACE MODIFICATION AND FUNCTIONALIZATION OF GRAPHENE

The graphene has a nanosheet-like structure with 2D geometry having no functional groups on the surface. Thus, it is devoid of any function groups in the true sense. Hence, graphene is hydrophobic in nature and very unstable in an aqueous medium which leads to poor dispersion. However, GO is dispersible in aqueous media but in physiological media, in the presence of solvated salts, it tends to aggregate due to the charge screening effect. The dispersibility and the binding efficiency of the graphene as well its derivative primarily depend on the chemical nature of the surface. Therefore, surface modification and functionalization of graphene and its derivative play a vital role in carrying of different therapeutic particles. Mainly two types of modification have been performed namely covalent modification and non-covalent modification. In a covalent modification, the graphene sheet reacts with different chemicals and forms some active functional groups at the expense of unsaturated bond structure on the surface of the GNS. The active site of the generated functional groups acts as binding sites for different therapeutic particles increasing their efficacy. Whereas, in the case of non-covalent modification, no structural change of the nanosheet happens and the added particle interacts through weak van der Waals force or by electrostatic interaction with the nanosheet. Moreover, the surface-modified graphene nanomaterials facilitate the interaction with the polymer matrix (elastomer) which forms polymer nanocomposites with improved physico-mechanical properties.

17.5 USE OF GRAPHENE-ELASTOMER IN VARIOUS FIELDS OF BIOMEDICAL APPLICATIONS

In the current scenario, many researchers have engaged to find out a promising strategy to explore the new function and application of nanomaterials including inorganic materials, polymer micelles, noble metals, and carbon nanomaterials in various fields of biomedical applications. The unique characteristics of nanomaterials including the high surface area to volume ratio, good stability in physiological condition, ease of functionalization, and biocompatibility make them an attractive counterpart in biomedical application. Among these nanomaterials, 2D graphene and its derivative have created a distinguished field of interest in the area of biomedical application. 2D graphene nanomaterials possess several unique properties such as high Young's modulus (YM, 1 TPa), remarkable TS (125 GPa), good electrical

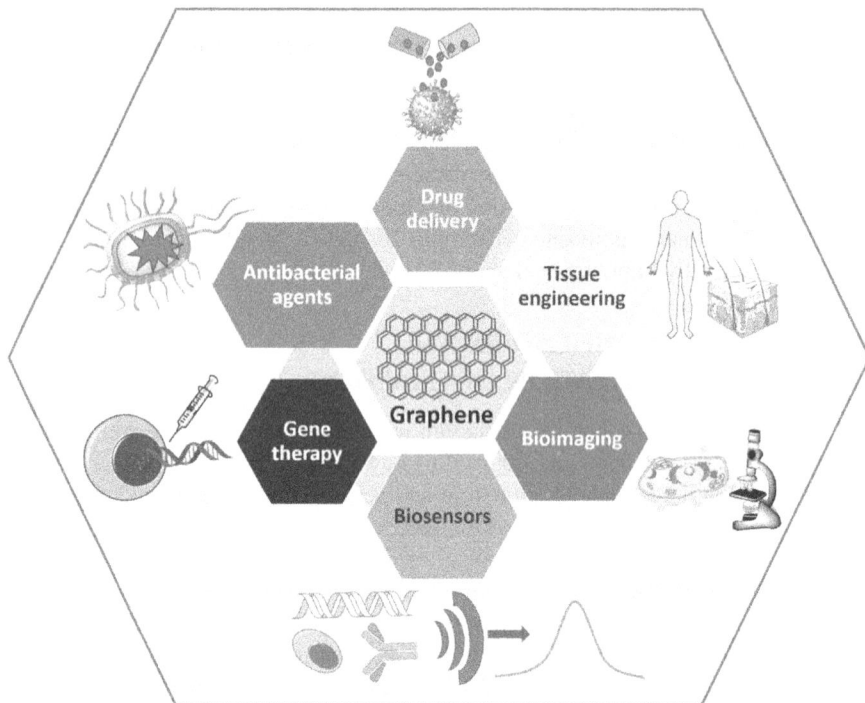

FIGURE 17.3 Schematic representation of application of graphene and graphene-based materials in various fields of biomedical applications.

conductivity (6000 S/cm), high thermal conductivity (4800 W/mK), and nanoscale size stability (Lee et al. 2008, Sadasivuni et al. 2014). For these numerous good merits, graphene nanomaterials are widely used in nanocomposite (Fan et al. 2010), energy storage (Stoller et al. 2008, Wang, Lee, et al. 2009), nanoelectronics (Liu et al. 2010), sensors (He et al. 2012), etc. But beyond the aforementioned traditional usage, the graphene material was first introduced by Liu et al. in 2008 in the biomedical field for the delivery of water-insoluble anticancer drugs (Liu et al. 2008). After that, a new all-round horizon had opened up of using graphene and its various derivative in numerous fields of biomedical application from drug delivery to imaging as well as antibacterial materials in scaffolds for tissue engineering as shown schematically in Figure 17.3. Various fascinating properties of graphene-based materials such as high surface area, good physico-mechanical properties, electrical conductivity, biocompatibility in physiological condition, size stability, scalable production, and tailorable surface modification or chemical functionalization make it a promising new generation nanomaterials in biomedical applications (Shen et al. 2012).

17.5.1 NANOCOMPOSITES FOR BIOMEDICAL APPLICATION

A 2-dimensional thin layer-like structure of graphene with uniform dispersion within the elastomer matrix improves the physico-mechanical properties of graphene-elastomer nanocomposite. Graphene-elastomer nanocomposites-based bio-compatible items such as catheters, surgical gloves, and heart valves are widely used in biomedical applications. In this regard, George et al. have prepared the scalable, defect-free graphene by using natural curcumin as a solid-phase exfoliation agent. The incorporation of exfoliated graphene within the NR latex displayed improved TS with low modulus and minimized the % loss of EB. Good *in vitro* and *in vivo* bio-compatibility of the nanocomposites indicates that graphene/NR latex can be useful as surgical gloves and catheters (George et al. 2018). Liu et al. have prepared the GO/zwitterionic chitin nanocrystal hybrids for the improvement of mechanical proper-ties of the NR latex to prepare the recyclable gloves. The chitin nanocrystal improved the dispersibility of GO in NR latex as well as formed an intermediate bond between the NR and the GO via amine and carboxylic groups that omit conventional sulfur vulcanization process (Liu et al. 2021).

17.5.2 GRAPHENE AND GRAPHENE-BASED ELASTOMERIC NANOCOMPOSITES AS NANOCARRIER IN THERAPEUTIC APPLICATION

The advancement of nanoscience and nanotechnology has empowered to synthe-size new and attractive nanomaterials with improved therapeutic profiles leading to boost the efficiencies of therapeutic agents in modern medicine. In the 21st century, the wonder material, graphene and its derivative, with a lot of interesting unique properties like high surface area, size stability, high mechanical property, electri-cal and thermal properties, and good biocompatibility in physiological conditions enables them as suitable components in modern therapeutic applications. Thin layer graphene with SP^2 hybridized carbon and a six-member л-conjugated aromatic mac-romolecular structure can bind with various kinds of substances like biomolecules, drugs, proteins, DNA, antibodies, and cells. Therefore, graphene and graphene-based derivative has acquired a great interest in modern nanomedicine and biomedi-cal applications as potential nanocarriers. Since the application of GO as an efficient nanocarrier of hydrophobic anticancer drugs in 2008 by Liu et al. (2008), a new area of utilization of graphene and its derivative has been explored in the area of nano-medicine field as nanocarriers as illustrated schematically in Figure 17.4.

17.5.2.1 Graphene and Graphene-Based Elastomer for Drug Delivery

For the last few years, graphene and graphene-based nanomaterials are explored extensively for drug delivery applications as nanocarriers. The π-electron charges of graphene surfaces and the surface functional groups of the GO play a significant role in the binding of drug carrying and release which increased the efficiency of the drugs. The high surface area with available surfaces of the graphene layer enables it to load a higher amount of drug as compared to other nanocarriers (Yang et al. 2013). Additionally, modifying the surface or conjugating site-specific ligand on the surface of graphene improved the cell uptake efficiency and specific accumulation of drugs at the site of the disease which increase the therapeutic efficiency of the drug. Surface-modified graphene or the GO may act as a binding site for a variety of chemotherapy drugs such as Doxorubicin (DOX), SN-38, camptothecin (CPT), β-lapachone, and ellagic acid with either chemical or physical adsorption. In 2008

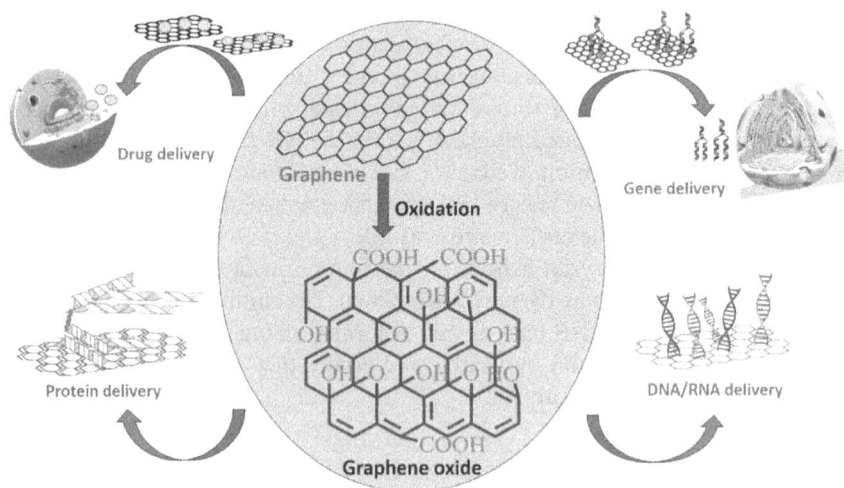

FIGURE 17.4 Scheme of application of graphene and derivative of graphene as nanocarriers for different therapeutic agents.

for the first time, Liu et al. (2008) have utilized the pegylated nGO by conjugating the PEG molecule on the surface of the nGO and used it as a nanocarrier of water-insoluble cancer drugs SN-38. Pegylated nGO shows excellent stability in physiological conditions and increases the drug loading capacity of aromatic SN-38 drug by $\pi-\pi$ interaction with the nGO surface. SN-38 loaded pegylated nGO exhibits superior cancer cell killing efficiency as compared to the nGO-PEG and the water-soluble pro-generic drug CPT as depicted in Figure 17.5.

Targeted drug delivery to specific diseases site is another advantageous strategy for efficient delivery of drugs to improve the therapeutic application. The site-specific delivery of the cancerous drug increases the cancer-killing efficiency of the drug. Zhang et al. have functionalized sulfonic acid on the surface of the nGO followed by conjugating the folic acid (FA), which improved the stability of the nGO in physiological conditions and augmented the specific release to human breast cancer cells, MCF-7 cells. It demonstrates that the control loading of two anticancer drugs DOX and CPT on the nGO-FA surface and more specific release of the FA conjugated nGO to the MCF-7 cells increased the cytotoxicity of the drug-loaded nGO-FA than drug-loaded nGO (Zhang et al. 2010). Depan et al. have loaded the DOX into GO by $\pi-\pi$ stacking interaction between DOX and GO followed by the incorporation of DOX-loaded GO into FA-conjugated chitosan. High DOX-loaded GO with hydrophilic character and cationic nature of chitosan increase the stability of the nanocarrier in the physiological media. The FA-conjugated GO-chitosan nanocarrier exhibits target-specific controlled release of DOX and increase the efficacy of anticancer drugs (Depan, Shah, and Misra 2011). Pan et al. have prepared the poly(*N*-isopropylacrylamide) (PNIPAM)-grafted GO by click chemistry. The PNIPAM-GO nanocarrier shows good stability in physiological media and experiences a hydrophilic to hydrophobic transition at 33°C below the lower critical temperature (LCST) of PNIPAM 37.8°C. The PNIPAM-GO nanocarriers bind high capacity (18.5 wt.%)

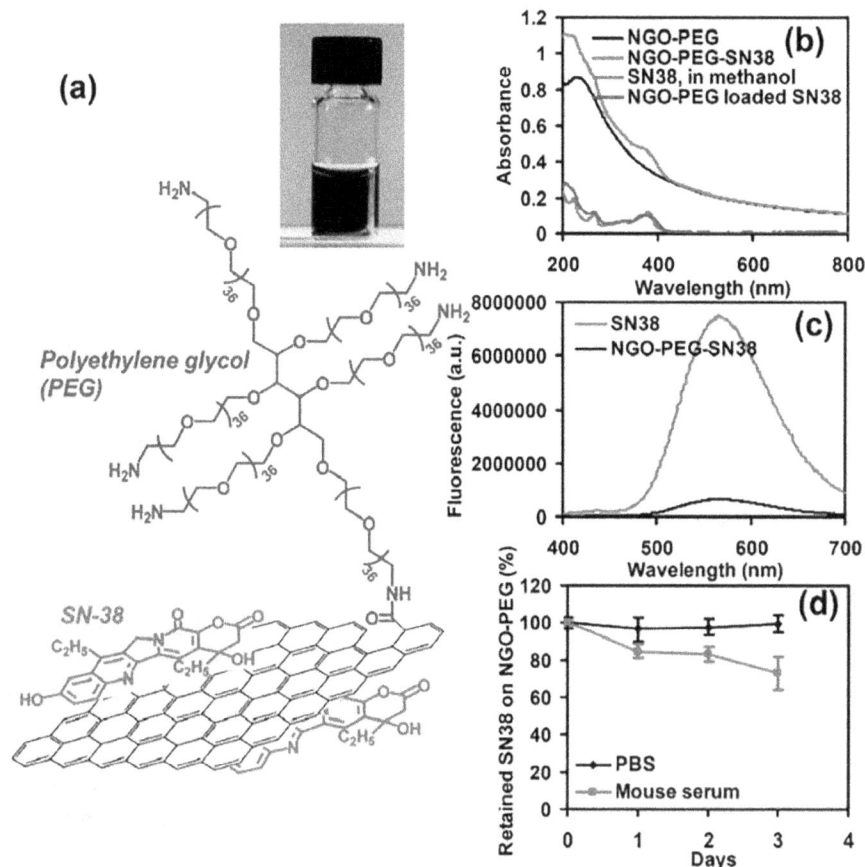

FIGURE 17.5 (a) Schematic representation of SN-38 loaded nGO-PEG in aqueous media, (b) UV-vis spectra of nGO-PEG, nGO-PEG-SN-38, and SN-38 in methanol and different spectrum of SN-38 loaded nGO-PEG. The SN-38 loading capacity of nGO-PEG was measured at the wavelength of 380 nm, (c) fluorescence quenching spectra of SN-38 on the GO nanosheet at the drug loading concentration of 1 μM, and (d) amount of retained SN-38 loaded on nGO-PEG in PBS and serum. The loaded SN-38 is stable in PBS, but slowly releases in serum. (Reprinted with permission from Ref. (Liu et al. 2008). Copyright 2008, American Chemical Society.)

of water-insoluble anticancer drug, CPT, by hydrophobic and π–π interaction. The unique amphiphilic nature of PNIPAM-GO with good dispersion stability and control release of the drug makes a promising strategy for the delivery of anticancer drugs without exhibiting any cytotoxicity (Pan et al. 2011).

Recently, stimuli-responsive GO-based nanocarriers have attracted much attention for the development of nano vehicles for the delivery of drugs in a controlled way at specific sites. Tao et al. have prepared the GO-based hydrogel using low concentration metformin as a model drug. GO-metformin produces a supramolecular assembly gel by H-bonding electrostatic interaction. Drug release kinetics shows a higher amount of drug release efficacy (74%) of the capsule in strongly acidic media

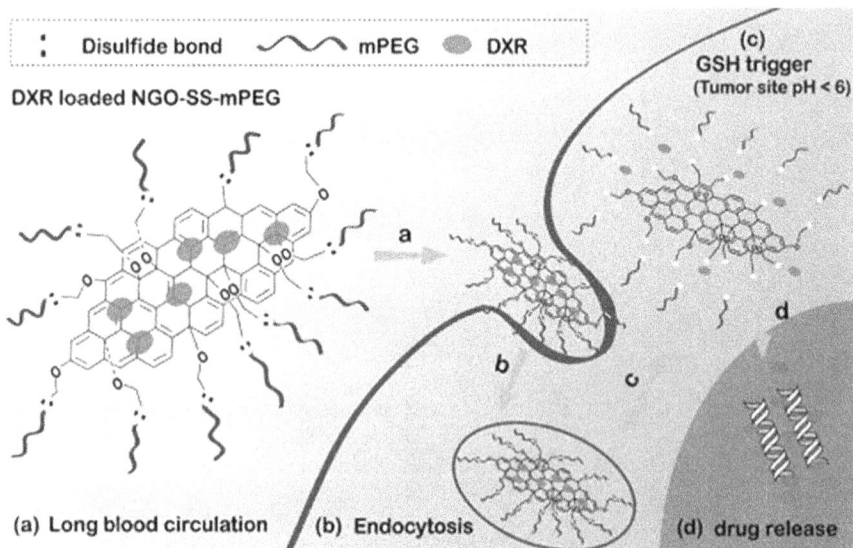

FIGURE 17.6 Schematic representation of drug release mechanism of DXR-loaded nGO-PEG nanocarriers: (a) PEG shielded nGO with loaded DOX ensures prolonged blood circulation, (b) endocytosis of DXR-loaded nanocarriers, (c) cleavage of S–S bond at high GLH level at the tumor site, and (d) release of the loaded drug at the tumor site. (Reproduced with permission from Ref. (Liu, Cui, and Losic 2013). Copyright 2013, Elsevier.)

as compared to the neutral media (50%). The pH-sensitive release of GO-metformin makes it a promising nanocarrier for the release of drugs in the acidic environment of the stomach (Tao et al. 2012). Wen et al. have grafted PEG on the surface of nano GO via disulfide (S–S) bond which is easily cleavable at the reducing environment. Doxorubicin hydrochloride (DXR)-loaded nGO-PEG shows good nanoscale stability in physiological media and releases the drug drastically in a reducing environment improving the therapeutic activity of the loaded drug. The red-ox responsive DXR-loaded PEGylated nanographene oxide (nGO-mPEG) nanocarriers for the rerelease of DXR at the cancerous cell at tumor related high glutathione (GSH) level and improved the site-specific activity of the drug as depicted in Figure 17.6 (Wen et al. 2012).

17.5.2.2 DNA/RNA Delivery

Surface functionalized graphene sheets increase the target-specific release of the therapeutic agents which improves the efficacy of the therapeutic agents. Delivery of free RNA is very difficult due to the negative charge on the surface of RNA which exerts cytotoxicity (Aigner 2006). Zhang et al. have attached the polyethylenimine (PEI) on the surface of the GO sheet through the formation of an amide bond using EDC chemical agent to generate GO-PEI as a nanocarrier. The positively charged GO-PEI surface can bind with the negatively charged siRNA showing efficient delivery of siRNA without showing any cytotoxicity (Zhang et al. 2011). Hence, graphene can be efficiently exploited for DNA/RNA fishing and their delivery.

17.5.2.3 Gene Delivery

In therapeutic application, gene therapy to cure various gene-related diseases is wisely explored for practical clinical application for the last few years (Whitehead, Langer, and Anderson 2009). The main difficulty of gene therapy is to formulate an efficient and safe vehicle that will deliver the target-specific gene in order to minimize the related side effect (Liu et al. 2007). Various nanomaterials are widely used for gene therapy; still the development of a highly efficient and non-toxic nano-vector with high gene transfection is highly desirable (Hom et al. 2010, Meng et al. 2013, Chertok, David, and Yang 2010). GO has lots of carboxyl group on the surface which can bind with different polymer matrices such as PEI and chitosan via electrostatic interaction or by π–π stacking for the delivery of plasmid DNA (pDNA) or RNA. In this regard, PEI is considered as a gold standard polymer for the binding and delivery of genes, owing to specific binding to the nucleic acids and high uptake by the cell. But high molecular weight and toxic nature of bare PEI limits its clinical applications (Figure 17.7). To overcome these problems, Feng et al. have conjugated GO with PEI by electrostatic interaction and generated stable GO-PEI nanocarriers with the positive charge on the surface which enables to bind the gene on its surface. The GO-PEI nano vehicle exhibits high gene transfection efficacy without showing any cytotoxicity, which even uses the high molecular weight of PEI (10 kDa) as compared to the bare PEI (Feng, Zhang, and Liu 2011). Yang et al. have functionalized GO with PEG followed by conjugated FA for specific delivery of siRNA. The loading of siRNA on the surface of GO has been enhanced by 1-pyrenemethyamine hydrochloride through π–π stacking. The PEG functionalization enhanced the stability of GO and conjugated FA release of the siRNA to the specific cell site increases the therapeutic efficacy (Yang et al. 2012). The π–π interaction helps to bind different drugs and biomolecules to generate the nanosupramolecular assembly on the surface of graphene and acts as a suitable nanocarrier for therapeutic agents.

17.5.3 TISSUE ENGINEERING

Graphene and different derivatives of graphene owing to high TS, thermal, electrical properties along with suitable elastomer form bioactive nanocomposites which are suitable for different fields of tissue engineering applications such as skin, bone, nerve, and cardiac tissue regeneration (Ivanoska-Dacikj et al. 2020). The graphene and surface-functionalized graphene interact strongly with the elastomer matrix and amplify the physicochemical properties of the nanocomposites as compared to the pristine elastomer. The high physico-mechanical properties of the graphene-incorporated nanocomposites help to carry the body load which is highly recommended for hard tissue regeneration. Moreover, the bioactive nature of graphene and its derivative make it a promising new-generation engineering material for the fabrication of scaffolds for the repair of a wide variety of tissues from soft tissue to hard tissue.

17.5.3.1 Bone Regeneration

The suitable biocompatibility with various superior physico-mechanical properties of graphene-based elastomer nanocomposites is widely used in various fields

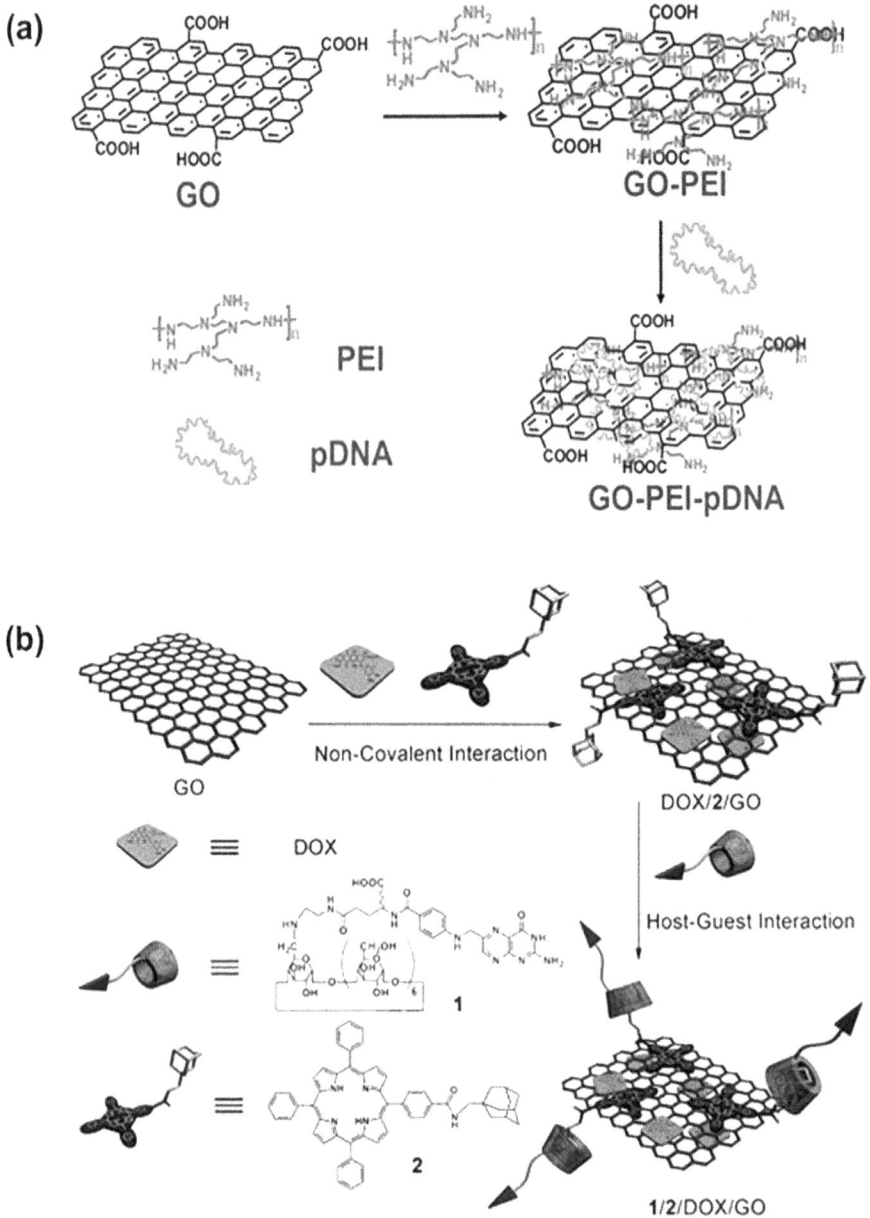

FIGURE 17.7 (a) Schematic representation of produced GO-PEI-pDNA complex through layer-by-layer (LBL) assembly. First, the positively charged PEI is immobilized on the GO surface that produces a positive charge on the GO surface followed by attachment of pDNA through π–π interaction and produced GO-PEI-pDNA complex. (b) Synthesis of DOX-loaded 1/2/DOX/GO from DOX, GO, adamantane-modified porphyrin, and FA-modified cyclodextrin. (Reproduced with permission from Ref. (Liu, Cui, and Losic 2013). Copyright 2013, Elsevier.)

of tissue engineering applications (Kim et al. 2019, Zhang et al. 2012). The incorporation of graphene-based nanomaterials within the polymer matrix improved the physico-mechanical properties in terms of TS, EB, nanohardness of the prepared nanocomposites which is favorable for the application of hard tissue regeneration like bone regeneration. Girase et al. have prepared the silicone/GO-based nanocomposite with superior physicochemical properties such as high TS, high EB, and the more hydrophilic surface of the GO-reinforced silica elastomer as compared to the pristine silica elastomer. Biocompatibility study with osteoblast MC3T3-E1 cells shows superior osteogenic bioactivity in terms of cell attachment, proliferation, and detection of actin protein, fibronectin, and osteoinductive signaling of GO-added silicone/GO elastomer compared to the virgin silicone, indicating a suitable bioelastomer for joint reconstruction (Girase, Shah, and Misra 2012). Jo et al. have synthesized the highly flexible nano GO-incorporated polyurethane scaffolds for skeletal tissue regeneration. The introduced nGO increased the surface hydrophilicity, elasticity, and stress relaxation capacity of the scaffolds. The PUU/GO elastomer scaffolds exhibit good cells adhesion, proliferation with significant up-regulated myogenic mRNA and myocin level which augmented the regeneration of skeletal tissue (Jo et al. 2020). Liao et al. (2015) have introduced the GO nanoparticle into the hybrid porous scaffold (CSMA/PECA/GO) generated from poly(ethylene glycol) methyl ether-ε-caprolactone-acryloyl chloride (PECA) and methacrylate chondroitin sulfate (CSMA) for cartilage regeneration.

Recently, our group also worked on the use of carbonaceous nanomaterial for the preparation of thermoplastic elastomer based nanocomposites using biomolecule-incorporated PUU as a polymer matrix in bone tissue regeneration application extensively (Figure 17.8). GO and osteoconductive nanomaterial (nHA) decorated GO hybrid nanomaterial (GO/nHA) were synthesized by the solvothermal process. The hybrid nanomaterials were incorporated within the newly synthesized thermoplastic elastomer PUU by in situ technique. The hybrid nanomaterial-incorporated PUU/GO-nHA hybrid nanocomposites experienced a significant amount of improvement of TS (72%), EB (48%), and hardness (152%) as compared to the pristine PUU. The nHA decorated GO showed good biocompatibility and promoted osteogenic gene expression of the nanocomposites which accelerated bone tissue regeneration (Ghorai et al. 2019a). In another article, a very low concentration (0.15%) of carboxyl acid functionalized carbon nanotube (CCNT) was incorporated within the PUU matrix along with nHA which improved the TS and hardness of the hybrid nanocomposites by 94.5% and 173.6%, respectively. RT-PCR study showed that the CCNT-incorporated PUU elastomer hybrid nanocomposites exhibit positive expression of osteogenic genes such as alkaline phosphatase, osteocalcin, and collagen I. Moreover, *in vivo* study in rat tibia and skull defects site indicates the excellent new bone regeneration efficacy of the nanohybrid composite based fabricated scaffolds as compared to the pristine PUU elastomer without showing any organ toxicity. In a judicious way, with the incorporation of a very low concentration of carbonaceous nanomaterials within the PUU-based elastomer can be used as an alternative potential platform for new bone regeneration applications (Ghorai et al. 2022).

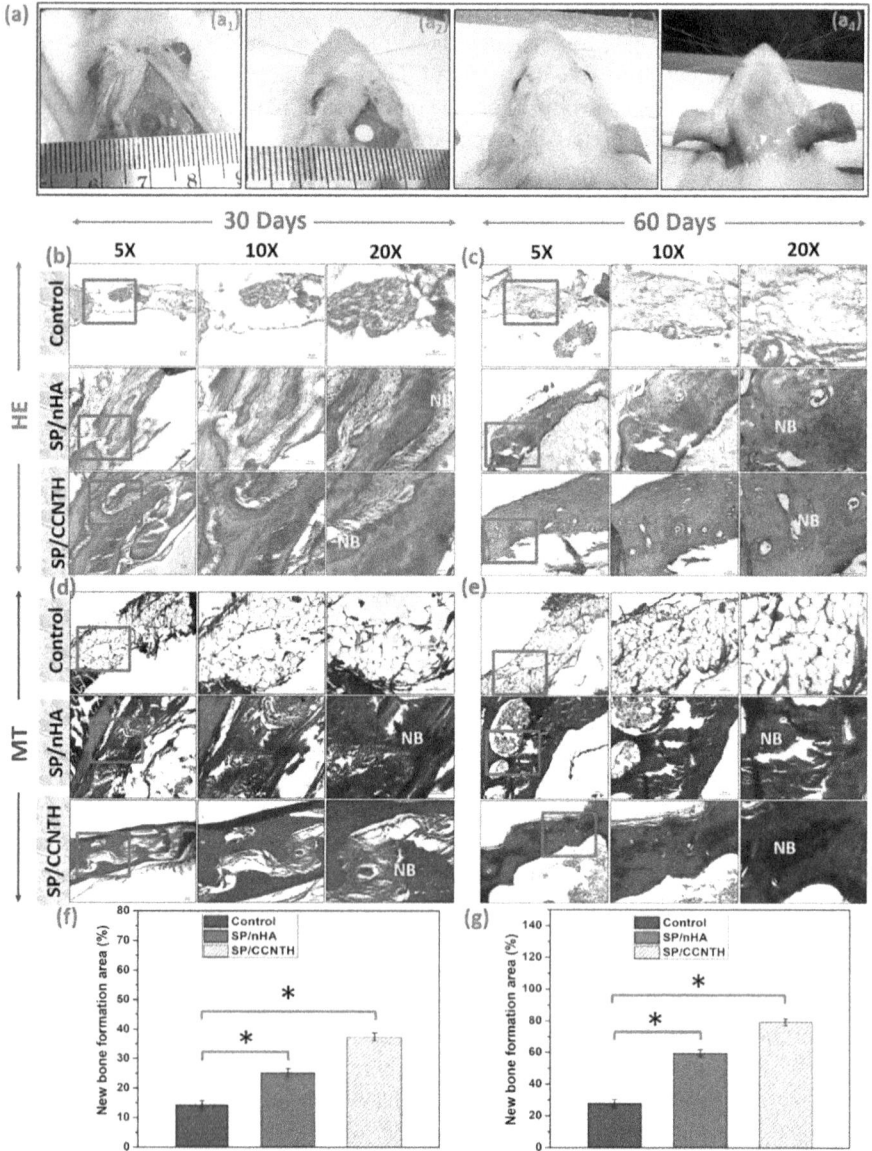

FIGURE 17.8 Evaluation of *in vivo* bone regeneration of functionalized CNT-incorporated thermoplastic elastomer polyurethane-urea-based nanohybrid composites: (a₁-a₄) represent the macroscopic images of the sample implantation and peripheral healing of the operated site without any inflammation and infection after 60 days of operation, (b, c) and (d, e) represent the hematoxylin & eosin and MT-stained bone section that indicates the superior bone regeneration efficacy of CCNT-incorporated PUU scaffolds as compared to the control, respectively, (f, g) represents the calculation of new bone formation area after 30 and 60 days of implantation, respectively. (This is reproduced with permission from Ref. (Ghorai et al. 2022). Copyright 2022, Elsevier.)

17.5.3.2 Nerve Tissue Regeneration

For the development of nerve tissue engineering, materials should have good electrical conductivity which augmented the nerve tissue regeneration and conveyed the signals from one neuron to another neuron (Li et al. 2013, Talebi et al. 2021). Owing to high electrical conductivity, thermal stability, large surface area, and good nanoscale stability, graphene and its derivatives serve as a promising candidate for nerve tissue regeneration (Bei et al. 2019, Aydin et al. 2018). In the recent past, graphene-incorporated elastomer based on a numerous number of biocompatible polymers is widely explored in nerve tissue regeneration. Ginestra et al. have fabricated the nanofibrous scaffolds from graphene suspended polycaprolactone-cyclopentanone solution by electrospinning technique. The hierarchical nanofibrous scaffolds mimic the extracellular matrix (ECM) and influence the morphology of adhered cells and proliferation. The morphology and fiber diameter of the electrospun mats depend on the content of graphene which guides the morphology and differentiation into neurons of the seeded rat mesenchymal stem cells (rMSCs) (Ginestra 2019). Graphene incorporation within the elastomer increases the neuron regeneration efficacy as reported by Lee et al. The incorporation of 5% graphene within flexible PU nanocomposites (PU-G5) exhibits better neuron stem cells activity and upregulated neuron related gene expression such as GFAP, VE-Cadherin, and KDR. They conducted *in vivo* study by implantation of PU-G5 in rat subcutaneous site which showed a higher rate of regeneration of peripheral nerve of rat sciatic nerve. The elastomeric conduits showed a significant amount of regeneration of neuron number and new blood vessels in the regenerated tissue of about 72% and 50%, respectively, as compared to the pristine PU (Lee, Yen, and Hsu 2019). The incorporation of graphene increases the electrical conductivity (0.1 to 3.9 ± 0.3 S m^{-1}) and mechanical properties of polycaprolactone/gelatine/polypyrrole (PCL/gelatin/PPy) and polycaprolactone/polyglycerol-sebacate/polypyrrole (PCL/PGS/PPy) fibrous scaffolds and exhibits good biocompatibility to the human fibroblast cells (Talebi et al. 2021). Zhang et al. have shown promoted proliferation of Schwann cells through the coating of GO on the electrospun poly-l-lactide (PLLA) scaffolds. The coated GO improved the surface hydrophilicity of the scaffolds and also exhibits promoted proliferation of rat pheochromocytoma 12 (PC12) cells and accelerates neurogenic differentiation to generate new neurite (Zhang et al. 2016).

17.5.3.3 Cardiac and Vascular Tissue Engineering

Myocardial infarction (MI) is the major cause of cardiac tissue damage and heart failure leading to the morbidity of the patients. The limited self-regeneration capability of new primary human cardiomyocytes (CMs) urge to develop a new approach that accelerates the regeneration of cardiac tissue. New generation fabricated substitutes having high electrical conductivity, high-order anisotropic structural and mechanical properties are widely applicable for the regeneration of cardiac tissue. Graphene and its derivatives based elastic nanocomposites having good mechanical properties and excellent electrical conductivity are suitable for cardiac tissue regeneration applications (Li et al. 2021, Hsiao et al. 2013). Ghasemi et al. have fabricated GO-incorporated polyethylene terephthalate electrospun scaffold for the electroconductive cardiac patch. With the introduction of 0.05% GO within the core-shell electrospun scaffolds, the electrical conductivity of the scaffolds is increased from 0.7×10^{-6} to 1.3×10^{-6} S cm^{-1} which is highly comparable to cardiac electro-activity value. *In vitro* cell culture study of human umbilical vein and endothelial cells shows good biocompatibility and the spread CM morphology indicates potential application as a cardiac patch (Ghasemi et al. 2019). Park et al. (2015) used GO flakes as cellular adhesives which adsorbed the ECM proteins and prevented the death of myocardium tissue by scavenging reactive oxygen species (ROS). Saravan et al. (2018) have shown good ventricular contractility of rat MI model using gold nanoparticle (Au) decorated GO-incorporated biodegradable chitosan scaffolds. Carbonaceous nanomaterials-incorporated polymer gels with good water dispersibility and electrical conductivity are widely used as cardiac tissue engineering patches (Shin et al. 2013, Ramón-Azcón et al. 2013).

17.5.4 Antibacterial Agent

The pathogenic infection caused by bacteria, fungi, protozoa, amoebas, and other different vectors of disease is a key challenging issue in biomedical applications (Giraud, Tourrette, and Flahaut 2021). Carbonaceous nanomaterials having intrinsic antibacterial activity have generated enormous interest to use graphene and various graphene-based derivatives as antibacterial agents (Szunerits and Boukherroub 2016, Mondal, Bhowmick, and Krishnamoorti 2012). For the last few decades, many researchers are engrossed to formulate carbonaceous nano-materials amalgamated polymer nanocomposites to fullfill the target. The polymer material has many advantages including homogeneous dispersion of the antibacterial materials and other decorative agents and finally reforms as a use-able artifact (Liu, Li, and Liu 2014). Antibacterial materials may directly kill the pathogens or resist their proliferation by prohibiting the adhesion of microbes on the surface of the agents (Huang et al. 2016). The intrinsic activity of graphene and graphene derivatives to kill the pathogens as well as doping of carbonaceous nanomaterials by other metal nanomaterials such as Ag, Cu, and Zn showed the improved efficacy of the antibacterial agents. The incorporation of graphene and graphene derivative within the polymer matrix exerts the good antibacterial activity as compared to the pristine polymer matrix. The introduced layered structure graphene flattened the bacterial cells and induced stress on the cell membrane, eventually losing the intracellular matrix and causing cell death. The sharp edge of the graphene sheet also acted as a cutter within the elastomer which also helps to rapture the cell membrane and kill the pathogens (Ghorai et al. 2019a). Liu et al. have stated that GO produced the superoxide anion (O_2^-) which acts as ROS and oxidized different intracellular components like proteins, lipid, and DNA by producing a conducive bridge over the lipid bilayer caused bacterial cell death as shown in Figure 17.9 (Liu, Zeng, et al. 2011). Moreover, the surface modification with other antibacterial agents and the introduction of trace amount of metal ions along with the graphene exhibit many-fold increase in antibacterial activity of the graphene-elastomer (Wei et al. 2017, Cheng et al. 2017, Nicosia et al. 2020, Nuñez-Figueredo et al. 2020).

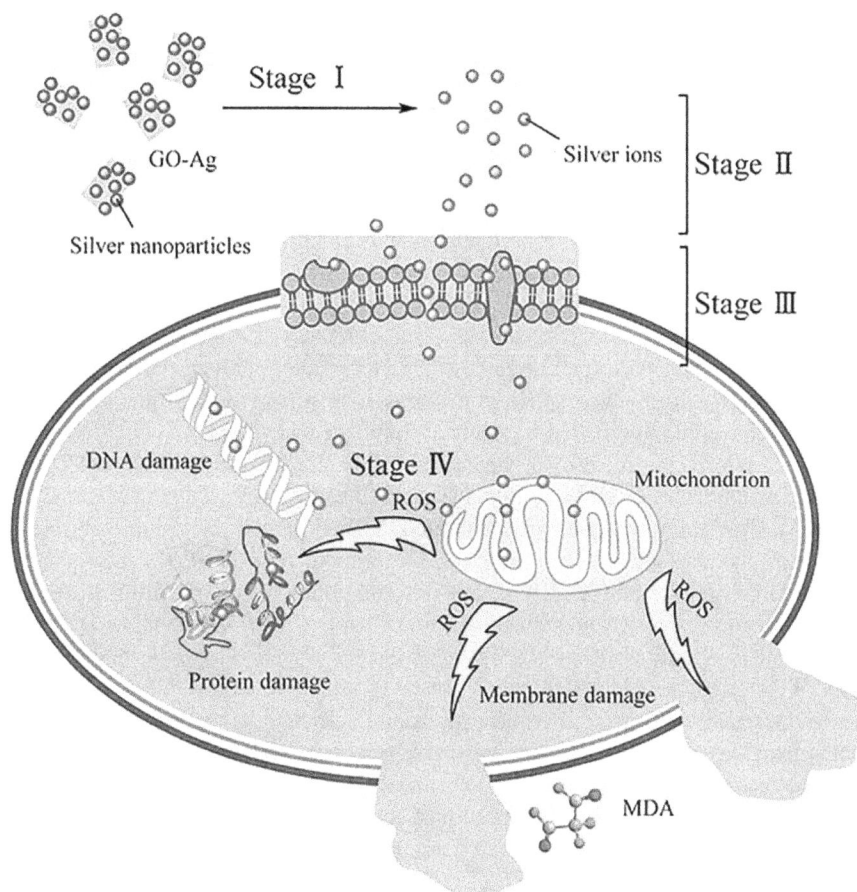

FIGURE 17.9 Schematic representation of the mechanism of antibacterial activity of GO-Ag. In stage I, the GO-Ag dissolved and Ag nanoparticles are generated in the aqueous media; in stage II, the positively charged Ag ion interacts with the negatively charged lipid layer of the bacterial cell membrane and is attached to the bacterial cells; in stage III, the Ag ion destroys the protein and phospholipid bilayer and enters within the bacterial cells; in stage IV, ROS is generated by mitochondrial oxidation chain reaction and interacts with the protein, lipid, DNA. The ROS destroys the cell membrane, generates malonaldehyde (MDA), and destroys the intracellular components. (The scheme is reproduced with permission from Ref. (Song et al. 2016). Copyright 2016, Elsevier.)

17.5.5 BIOIMAGING

The intrinsic optical, magnetic, and electrical characteristics of graphene enable to use it for the formulation of bioimaging devices. GO has been used as a bioimaging agent using intrinsic fluorescence activity to detect the cell uptake of anticancer drugs. Biocompatible and biodegradable polymer or elastomer are widely used as the binding fluorescent agent for imaging. Sun et al. have studied on cell uptake activity of anticancer drug DOX using intrinsic NIR active nGO. The PEG-functionalized

GO surface binds the water-soluble anticancer drug and is monitored by fluorescence imaging (Sun et al. 2008). Liu et al. have loaded anticancer drug DOX on gelatine-grafted GNS and tracked the cell uptake efficiency through bioimaging study (Liu, Zhang, et al. 2011). Recently, graphene-based quantum dots (GQDs) generated by bottom-up technique are extensively used for bioimaging techniques in nanomedi-cine fields. Alkyl amine-functionalized GQDs show excellent fluorescent activity which helps to track the cell uptake process of different therapeutic agents (Mei et al. 2010, Tian et al. 2016). Moreover, GQDs show high biocompatibility, physiological stability, and upconversion fluorescence properties which help to capture bioimage *in vitro* and *in vivo* in NIR (Zhu et al. 2011, Shen et al. 2011). In NIR region, gra-phene possesses high photothermal conversion efficiency. The photons or NIR light can interact strongly with the graphene and GO surface by forced resonance vibra-tion process which involves in the conversion of kinetic vibration energy to heat energy, resulting in hyperthermia effects of graphene or GO (Acik et al. 2010). This hyperthermia effect is utilized in various fields of nanomedicine like control drug delivery, actuators, and self-healing patches (Patil et al. 2021). Shuai et al. have for-mulated a minute amount of GO-incorporated PNIPAM nanocomposites using N,N'-methylene-bis-acrylamide as a crosslinker. The prepared GO-PNIPAM showed good photothermal efficacy and leads to shape deformation which is useful for the deliv-ery of target-specific therapeutic agents in a controlled manner (Shuai et al. 2013). Additionally, high dimensional stability, thermal stability, and high thermal conduc-tivity of the graphene and different derivatives of graphene-incorporated composites due to the negative value of thermal expansion coefficient values are widely appli-cable in various biomedical applications. rGO-incorporated nanocomposites exhibit high heat generation efficacy as compared to the GO upon irradiation of NIR. The generated heat increased the temperature of the nanocomposites which reduce cell viability of the cancerous cells. This phenomenon is widely used as chemo-photo-thermal therapy for cancer treatment. Lima-Sousa et al. have prepared the GO- and rGO-incorporated thermoresponsive chitosan-agarose hydrogel by in situ technique. rGO introduced hydrogel shows 3.8 times higher temperature increment upon NIR irradiation as compared to the GO-based hydrogel. Dox- and Ibuprofen-loaded gels show a diminished cancer cell viability of value 34% and antibacterial activity in the presence of NIR irradiation (Lima-Sousa et al. 2020).

17.5.6 BIOSENSORS

Based on numerous interesting intrinsic properties such as good thermal, electrical, and optical properties, graphene, GO, and rGO decorated graphene are widely used in the detection of various biomolecules in nanomedicine. Graphene-based biosen-sors are used for the detection of ATP, amino acid, strain, dopamine, oligonucle-otide, etc. (Chang et al. 2010, Tang et al. 2011, He et al. 2010, Wang, Li, et al. 2009). High elastic graphene nanocomposites with good electrical conductivity are used as a good strain sensor. The elastic sensors generally work by changing the electrical resistance of the nanocomposites in response to the change in length of the nanocom-posites. Boland et al. have prepared the exfoliated graphene incorporated conducting NR-based elastic nanocomposites which shows good dynamic sensitivity with gauge factor up to 35 at high strain and strain rate (Boland et al. 2014). Coskun et al. have prepared the graphene-based cellular elastomer as a flexible sensor by freeze-casting process for the detection of a sound wave with a wide window of the frequency range of 300–20,000 Hz. The elastomer shows excellent sensing capability of subtle vibration with a wide range, suitable for flexible electronics devices for biomedical application (Coskun et al. 2017). Berger et al. have prepared ultra-thin graphene-PMMA based highly sensitive captive pressure sensors up to the pressure of 25 kPa (Berger et al. 2017). Highly sensitive graphene elastomer is fabricated by freeze-casing technique for wearable electrocardiogram (ECG) by Asadi et al. The highly conductive, flexible sensor exhibits very low skin impedance compared to other conventional sensors and functions without using any intermediate gel layer up to high mechanical pressure which is useful for long-term pressure sensing utilized for artificial skin (Asadi et al. 2020). Dong et al. have synthesized rGO-based NR composites using gelatine as a stabilizer. The incorporated rGO increased the YM and dynamic storage modulus of the rGO/NR nanocomposites manifolds as compared to the pristine NR with high electrical conductivity at a low percolation threshold. The rGO/NR elastomer exhibits a high sensitivity of strain and repeatability which is very useful to sense the cyclic movement of human joints (Dong et al. 2016). Shaun et al. also have prepared the GO/waterborne polyurethane (GO/WPU)-based highly elastic nanocomposites as a highly elastic intelligent smart fiber for strain sensor for biomedical applications (Shaun et al. 2020). The electrical conductivity of the GO and rGO makes it suitable for the preparation of graphene-based elastomer with high electrical conductive nanocomposites that enable us to detect the high strain and can be used in the monitoring of movement of the finger (Costa et al. 2019, Liu et al. 2018, Huang et al. 2019, Jian et al. 2017). Graphene-incorporated NR-based conductive composites film shows variable electrical resistance for different organic solvents which is useful for the detection of different organic solvents (Dong, Zhang, and Wu 2016).

17.6 CONCLUSIONS AND FUTURE PROSPECTS

The current chapter summarized the recent progress of graphene-elastomer nanocomposites and their various derivatives in different important wings of biomedical applications. In the beginning, different important routes of the synthesis, and surface modification of graphene have been described, which is followed briefly by the preparation of graphene-based elastomer-nanocomposites in this chapter. Functionalization of graphene by different active functional groups and the surface modification of graphene increased the binding efficiency of different therapeutic molecules that improved therapeutic efficacy. Different biomolecules like drugs, protein, siRNA, and pDNA can bind on the surface of graphene-elastomer used for specific drug release and gene transfection. Moreover, the chemical conjugation with various ligands or active biomolecules such as FA conjugation and disulfide bonding on the surface of the graphene increases the site-specific, stimuli-responsive release of the branded drugs. The high surface area with numerous number of acid-functionalized groups of graphene helps to bind the positively charged imine type of polymers which effectively bind the negatively charged DNA and effectively release the DNA without showing toxicity. The inherent optical and NIR activity of graphene plays a significant role in bioimaging and photothermal tumor therapy (PTT). Capping of radioactive nucleotide and fluorescent dye on the surface-modified graphene is useful for optical and nuclear imaging. Inherent bioimaging characteristics along with plugging of different active molecules on the surface of graphene are widely applicable for tracking cell uptake of different therapeutic molecules. In addition, the inherent antibacterial activity and doping with metal ions on the surface of the graphene exhibit good antibacterial activity against different pathogens.

Previous studies show that graphene-based materials demonstrate a potential application in various biomedical fields. For unique structural, chemical properties and inherent characteristics of graphene and its derivatives, they are widely used in numerous fields of biomedical applications for the last few years. Although the preclinical studies encourage the significant use of graphene-based materials in various biomedical fields, there are some major unavoidable concerns like product stability, long-term toxicity, and accumulation of carbon burden in the body which are to be resolved for practical clinical applications. It is known that layer size, the number of layers, and the chemical modification of the graphene not only play a vital role in therapeutic efficacy but also have a major role in long-term toxicity. In the present scenario, the synthesis of nanographene and its derivatives with the exact required size and number of layers is still unknown for safe biomedical application. As a nanocarrier, the graphene and its derivatives based materials function through cellular uptake mechanism where long-term toxicity and the fate of the carbonaceous materials within the biological system should be evaluated by long-term *in vitro* and *in vivo* study. Moreover, the biological response of cells in contact with the used elastomer matrix or the degraded matrix is an important matter of concern.

Till date, graphene and derivatives of graphene-based elastomers have many advantageous properties compared to other applicable nanomaterials-based elastomers in biomedical fields. The ease of synthesis, high surface area, ease of encapsulating of different therapeutic molecules by simple chemical modification, and low cost indicate the high-level application of graphene-based nanocomposites. Moreover, the intrinsic optical, electrical, and NIR absorbance activity of the graphene-based nanocomposites serves as effective bioimaging and cancer therapeutic nano-platform. Thus, using of graphene and derivatives of graphene-based elastomeric nanocomposites with in-depth futuristics studies and long-term *in vitro* and *in vivo* evaluation would be efficient therapeutic agents in near future.

REFERENCES

Acik, M, Geunsik Lee, C Mattevi, Manishkumar Chhowalla, K Cho, and YJ Chabal. 2010. "Unusual infrared-absorption mechanism in thermally reduced graphene oxide." *Nature Materials* 9 (10):840–845.

Aigner, Achim. 2006. "Gene silencing through RNA interference (RNAi) in vivo: Strategies based on the direct application of siRNAs." *Journal of Biotechnology* 124 (1):12–25.

Araby, Sherif, Izzuddin Zaman, Qingshi Meng, Nobuyuki Kawashima, Andrew Michelmore, Hsu-Chiang Kuan, Peter Majewski, Jun Ma, and Liqun Zhang. 2013. "Melt compounding with graphene to develop functional, high-performance elastomers." *Nanotechnology* 24 (16):165601–165615.

Asadi, Sajjad, Zijun He, Fatemeh Heydari, Dan Li, Mehmet R Yuce, and Tuncay Alan. 2020. "Graphene elastomer electrodes for medical sensing applications: Combining high sensitivity, low noise and excellent skin compatibility enabling continuous medical monitoring." *IEEE Sensors Journal* 21 (13):13967–13975.

Aydin, Tugce, Cansu Gurcan, Hadiseh Taheri, and Açelya Yilmazer. 2018. "Graphene based materials in neural tissue regeneration." *Cell Biology and Translational Medicine* 3:129–142.

Bahena, Arely, Ilse Magaña, Héctor Ricardo López González, Rishab Handa, Francisco Javier Enríquez-Medrano, Sugam Kumar, Ricardo Mendoza Carrizales, Salvador Fernandez, Luis Valencia, and Ramón Enrique Díaz de León Gómez. 2020. "Bio-elastomer nanocomposites reinforced with surface-modified graphene oxide prepared via in situ coordination polymerization." *RSC Advances* 10 (60):36531–36538.

Bahrami, Saeid, Atefeh Solouk, Hamid Mirzadeh, and Alexander M Seifalian. 2019. "Electroconductive polyurethane/graphene nanocomposite for biomedical applications." *Composites Part B: Engineering* 168:421–431.

Bei, Ho Pan, Yuhe Yang, Qiang Zhang, Yu Tian, Xiaoming Luo, Mo Yang, and Xin Zhao. 2019. "Graphene-based nanocomposites for neural tissue engineering." *Molecules* 24 (4):658–672.

Berger, Christian, Rory Phillips, Alba Centeno, Amaia Zurutuza, and Aravind Vijayaraghavan. 2017. "Capacitive pressure sensing with suspended graphene–polymer heterostructure membranes." *Nanoscale* 9 (44):17439–17449.

Boland, Conor S, Umar Khan, Claudia Backes, Arlene O'Neill, Joe McCauley, Shane Duane, Ravi Shanker, Yang Liu, Izabela Jurewicz, and Alan B Dalton. 2014. "Sensitive, high-strain, high-rate bodily motion sensors based on graphene–rubber composites." *ACS Nano* 8 (9):8819–8830.

Brodie, B. 1855. "Note sur un nouveau procédé pour la purification et la désagrégation du graphite." *Annales de Chimie et de Physique* 45:351–353.

Cai, Yuting, Jinbao Qin, Weimin Li, Abhishek Tyagi, Zhenjing Liu, Md Delowar Hossain, Haomin Chen, Jang-Kyo Kim, Hongwei Liu, and Minghao Zhuang. 2019. "A stretchable, conformable, and biocompatible graphene strain sensor based on a structured hydrogel for clinical application." *Journal of Materials Chemistry A* 7 (47):27099–27109.

Cao, Lan, Tridib K Sinha, Lei Tao, Huan Li, Chengzhong Zong, and Jin Kuk Kim. 2019. "Synergistic reinforcement of silanized silica-graphene oxide hybrid in natural rubber for tire-tread fabrication: A latex based facile approach." *Composites Part B: Engineering* 161:667–676.

Chang, Haixin, Longhua Tang, Ying Wang, Jianhui Jiang, and Jinghong Li. 2010. "Graphene fluorescence resonance energy transfer aptasensor for the thrombin detection." *Analytical Chemistry* 82 (6):2341–2346.

Chen, Biao, Min Liu, Liming Zhang, Jie Huang, Jianlin Yao, and Zhijun Zhang. 2011. "Polyethylenimine-functionalized graphene oxide as an efficient gene delivery vector." *Journal of Materials Chemistry* 21 (21):7736–7741.

Chen, Zhongxin, and Hongbin Lu. 2012. "Constructing sacrificial bonds and hidden lengths for ductile graphene/polyurethane elastomers with improved strength and toughness." *Journal of Materials Chemistry* 22 (25):12479–12490.

Cheng, Chong, Shuang Li, Arne Thomas, Nicholas A Kotov, and Rainer Haag. 2017. "Functional graphene nanomaterials based architectures: Biointeractions, fabrications, and emerging biological applications." *Chemical Reviews* 117 (3):1826–1914.

Chertok, Beata, Allan E David, and Victor C Yang. 2010. "Polyethyleneimine-modified iron oxide nanoparticles for brain tumor drug delivery using magnetic targeting and intra-carotid administration." *Biomaterials* 31 (24):6317–6324.

Coleman, Michael M, J Reid Shelton, and Jack L Koenig. 1974. "Sulfur vulcanization of hydrocarbon diene elastomers." *Industrial & Engineering Chemistry Product Research and Development* 13 (3):154–166.

Coran, Aubert Y. 1994. "Vulcanization." In Science and Technology of Rubber, Editor(s): James E. Mark, Burak Erman, Frederick R. Eirich, 339–385. Elsevier.

Coran, AY. 2003. "Chemistry of the vulcanization and protection of elastomers: A review of the achievements." *Journal of Applied Polymer Science* 87 (1):24–30.

Coskun, M Bulut, Ling Qiu, Md Shamsul Arefin, Adrian Neild, Mehmet Yuce, Dan Li, and Tuncay Alan. 2017. "Detecting subtle vibrations using graphene-based cellular elastomers." *ACS Applied Materials & Interfaces* 9 (13):11345–11349.

Costa, P, S Gonçalves, H Mora, SAC Carabineiro, JC Viana, and S Lanceros-Mendez. 2019. "Highly sensitive piezoresistive graphene-based stretchable composites for sensing applications." *ACS Applied Materials & Interfaces* 11 (49):46286–46295.

Depan, D, J Shah, and RDK Misra. 2011. "Controlled release of drug from folate-decorated and graphene mediated drug delivery system: Synthesis, loading efficiency, and drug release response." *Materials Science and Engineering: C* 31 (7):1305–1312.

Dong, Bin, Sizhu Wu, Liqun Zhang, and Youping Wu. 2016. "High performance natural rubber composites with well-organized interconnected graphene networks for strain-sensing application." *Industrial & Engineering Chemistry Research* 55 (17):4919–4929.

Dong, Bin, Liqun Zhang, and Youping Wu. 2016. "Highly conductive natural rubber–graphene hybrid films prepared by solution casting and in situ reduction for solvent-sensing application." *Journal of Materials Science* 51 (23):10561–10573.

Eizenberg, M, and JM Blakely. 1979. "Carbon monolayer phase condensation on Ni (111)." *Surface Science* 82 (1):228–236.

Fan, Hailong, Lili Wang, Keke Zhao, Nan Li, Zujin Shi, Zigang Ge, and Zhaoxia Jin. 2010. "Fabrication, mechanical properties, and biocompatibility of graphene-reinforced chitosan composites." *Biomacromolecules* 11 (9):2345–2351.

Feng, Liangzhu, Shuai Zhang, and Zhuang Liu. 2011. "Graphene based gene transfection." *Nanoscale* 3 (3):1252–1257.

Frasca, Daniele, Dietmar Schulze, Volker Wachtendorf, Christian Huth, and Bernhard Schartel. 2015. "Multifunctional multilayer graphene/elastomer nanocomposites." *European Polymer Journal* 71:99–113.

George, Gejo, Suja Bhargavan Sisupal, Teenu Tomy, Alaganandam Kumaran, Prabha Vadivelu, Vemparthan Suvekbala, Swaminathan Sivaram, and Lakshminarayanan Ragupathy. 2018. "Facile, environmentally benign and scalable approach to produce pristine few layers graphene suitable for preparing biocompatible polymer nanocomposites." *Scientific Reports* 8 (1):1–14.

Ghasemi, Azin, Rana Imani, Maryam Yousefzadeh, Shahin Bonakdar, Atefeh Solouk, and Hossein Fakhrzadeh. 2019. "Studying the potential application of electrospun polyethylene terephthalate/graphene oxide nanofibers as electroconductive cardiac patch." *Macromolecular Materials and Engineering* 304 (8):1900187–1900199.

Ghorai, Sanjoy Kumar, Somnath Maji, Bhuvaneshwaran Subramanian, Tapas Kumar Maiti, and Santanu Chattopadhyay. 2019a. "Coining attributes of ultra-low concentration graphene oxide and spermine: An approach for high strength, anti-microbial and osteoconductive nanohybrid scaffold for bone tissue regeneration." *Carbon* 141:370–389.

Ghorai, Sanjoy Kumar, Somnath Maji, Bhuvaneshwaran Subramanian, Tapas Kumar Maiti, and Santanu Chattopadhyay. 2019b. "Promoted Osteoconduction of Polyurethane–Urea Based 3D Nanohybrid Scaffold through Nanohydroxyapatite Adorned Hierarchical Titanium Phosphate." *ACS Applied Bio Materials* 2 (9):3907–3925.

Ghorai, Sanjoy Kumar, Trina Roy, Somnath Maji, Preetam Guha Ray, Kajal Sarkar, Abir Dutta, Amiyangshu De, Sharba Bandyopadhyay, Santanu Dhara, and Santanu Chattopadhyay. 2022. "A judicious approach of exploiting polyurethane-urea based electrospun nanofibrous scaffold for stimulated bone tissue regeneration through functionally nobbled nanohydroxyapatite." *Chemical Engineering Journal* 429:132179–132201.

Ginestra, Paola. 2019. "Manufacturing of polycaprolactone-Graphene fibers for nerve tissue engineering." *Journal of the Mechanical Behavior of Biomedical Materials* 100:103387–103393.

Girase, Bhupendra, Jinesh S Shah, and R Devesh K Misra. 2012. "Cellular mechanics of modulated osteoblasts functions in graphene oxide reinforced elastomers." *Advanced Engineering Materials* 14 (4):B101–B111.

Giraud, Laure, Audrey Tourrette, and Emmanuel Flahaut. 2021. "Carbon nanomaterials-based polymer-matrix nanocomposites for antimicrobial applications: A review." *Carbon* 182:463–483.

Goenka, Sumit, Vinayak Sant, and Shilpa Sant. 2014. "Graphene-based nanomaterials for drug delivery and tissue engineering." *Journal of Controlled Release* 173:75–88.

He, Qiyuan, Herry Gunadi Sudibya, Zongyou Yin, Shixin Wu, Hai Li, Freddy Boey, Wei Huang, Peng Chen, and Hua Zhang. 2010. "Centimeter-long and large-scale micropatterns of reduced graphene oxide films: Fabrication and sensing applications." *Acs Nano* 4 (6):3201–3208.

He, Qiyuan, Shixin Wu, Zongyou Yin, and Hua Zhang. 2012. "Graphene-based electronic sensors." *Chemical Science* 3 (6):1764–1772.

Henriques, Patricia C, Ines Borges, Artur M Pinto, Fernao D Magalhaes, and Ines C Goncalves. 2018. "Fabrication and antimicrobial performance of surfaces integrating graphene-based materials." *Carbon* 132:709–732.

Ho, Dong Hae, Qijun Sun, So Young Kim, Joong Tark Han, Do Hwan Kim, and Jeong Ho Cho. 2016. "Stretchable and multimodal all graphene electronic skin." *Advanced Materials* 28 (13):2601–2608.

Hom, Christopher, Jie Lu, Monty Liong, Hanzhi Luo, Zongxi Li, Jeffrey I Zink, and Fuyuhiko Tamanoi. 2010. "Mesoporous silica nanoparticles facilitate delivery of siRNA to shutdown signaling pathways in mammalian cells." *Small (Weinheim an der Bergstrasse, Germany)* 6 (11):1185–1190.

Hosler, Dorothy, Sandra L Burkett, and Michael J Tarkanian. 1999. "Prehistoric polymers: Rubber processing in ancient Mesoamerica." *Science* 284 (5422):1988–1991.

Hsiao, Chun-Wen, Meng-Yi Bai, Yen Chang, Min-Fan Chung, Ting-Yin Lee, Cheng-Tse Wu, Barnali Maiti, Zi-Xian Liao, Ren-Ke Li, and Hsing-Wen Sung. 2013. "Electrical coupling of isolated cardiomyocyte clusters grown on aligned conductive nanofibrous meshes for their synchronized beating." *Biomaterials* 34 (4):1063–1072.

Huang, Kai, Shaoming Dong, Jinshan Yang, Jingyi Yan, Yudong Xue, Xiao You, Jianbao Hu, Le Gao, Xiangyu Zhang, and Yusheng Ding. 2019. "Three-dimensional printing of a tunable graphene-based elastomer for strain sensors with ultrahigh sensitivity." *Carbon* 143:63–72.

Huang, Keng-Shiang, Chih-Hui Yang, Shu-Ling Huang, Cheng-You Chen, Yuan-Yi Lu, and Yung-Sheng Lin. 2016. "Recent advances in antimicrobial polymers: A mini-review." *International Journal of Molecular Sciences* 17 (9):1578–1591.

Hummers Jr, William S, and Richard E Offeman. 1958. "Preparation of graphitic oxide." *Journal of the American Chemical Society* 80 (6):1339–1339.

Iijima, Sumio. 1991. "Helical microtubules of graphitic carbon." *Nature* 354 (6348):56–58.

Ivanoska-Dacikj, Aleksandra, Gordana Bogoeva-Gaceva, Andres Krumme, Elvira Tarasova, Chiara Scalera, Velimir Stojkovski, Icko Gjorgoski, and Trpe Ristoski. 2020. "Biodegradable polyurethane/graphene oxide scaffolds for soft tissue engineering: In vivo behavior assessment." *International Journal of Polymeric Materials and Polymeric Biomaterials* 69 (17):1101–1111.

Jakus, Adam E, Ethan B Secor, Alexandra L Rutz, Sumanas W Jordan, Mark C Hersam, and Ramille N Shah. 2015. "Three-dimensional printing of high-content graphene scaffolds for electronic and biomedical applications." *ACS Nano* 9 (4):4636–4648.

Jian, Muqiang, Chunya Wang, Qi Wang, Huimin Wang, Kailun Xia, Zhe Yin, Mingchao Zhang, Xiaoping Liang, and Yingying Zhang. 2017. "Advanced carbon materials for flexible and wearable sensors." *Science China Materials* 60 (11):1026–1062.

Jo, Seung Bin, Uyanga Erdenebileg, Khandmaa Dashnyam, Guang-Zhen Jin, Jae-Ryung Cha, Ahmed El-Fiqi, Jonathan C Knowles, Kapil Dev Patel, Hae-Hyoung Lee, and Jung-Hwan Lee. 2020. "Nano-graphene oxide/polyurethane nanofibers: Mechanically flexible and myogenic stimulating matrix for skeletal tissue engineering." *Journal of Tissue Engineering* 11:2041731419900424–2041731419900433.

Kang, Hailan, Kanghua Zuo, Zhao Wang, Liqun Zhang, Li Liu, and Baochun Guo. 2014. "Using a green method to develop graphene oxide/elastomers nanocomposites with combination of high barrier and mechanical performance." *Composites Science and Technology* 92:1–8.

Ke, Kai, Michael McMaster, William Christopherson, Kenneth D Singer, and Ica Manas-Zloczower. 2019. "Highly sensitive capacitive pressure sensors based on elastomer composites with carbon filler hybrids." *Composites Part A: Applied Science and Manufacturing* 126:105614–105619.

Kim, Junghoon, Juyoung Leem, Hong Nam Kim, Pilgyu Kang, Jonghyun Choi, Md Farhadul Haque, Daeshik Kang, and SungWoo Nam. 2019. "Uniaxially crumpled graphene as a platform for guided myotube formation." *Microsystems & Nanoengineering* 5 (1):1–10.

Kim, Hyunwoo, Yutaka Miura, and Christopher W Macosko. 2010. "Graphene/polyurethane nanocomposites for improved gas barrier and electrical conductivity." *Chemistry of Materials* 22 (11):3441–3450.

Kroto, Harold W, James R Heath, Sean C O'Brien, Robert F Curl, and Richard E Smalley. 1985. "C 60: Buckminsterfullerene." *Nature* 318 (6042):162–163.

Kumar, Sadasivuni Kishor, Mickael Castro, Allisson Saiter, Laurent Delbreilh, Jean Francois Feller, Sabu Thomas, and Yves Grohens. 2013. "Development of poly (isobutylene-co-isoprene)/reduced graphene oxide nanocomposites for barrier, dielectric and sensingapplications." *Materials Letters* 96:109–112.

Lee, Changgu, Xiaoding Wei, Jeffrey W Kysar, and James Hone. 2008. "Measurement of the elastic properties and intrinsic strength of monolayer graphene." *Science* 321 (5887):385–388.

Lee, Tsung-Han, Chen-Tung Yen, and Shan-Hui Hsu. 2019. "Preparation of polyurethane-graphene nanocomposite and evaluation of neurovascular regeneration." *ACS Biomaterials Science & Engineering* 6 (1):597–609.

Lee, Yu Rok, Anjanapura V Raghu, Han Mo Jeong, and Byung Kyu Kim. 2009. "Properties of waterborne polyurethane/functionalized graphene sheet nanocomposites prepared by an in situ method." *Macromolecular Chemistry and Physics* 210 (15):1247–1254.

Li, Xiao-Pei, Kai-Yun Qu, Bin Zhou, Feng Zhang, Yin-Ying Wang, Oluwatosin David Abodunrin, Zhen Zhu, and Ning-Ping Huang. 2021. "Electrical stimulation of neonatal rat cardiomyocytes using conductive polydopamine-reduced graphene oxide-hybrid hydrogels for constructing cardiac microtissues." *Colloids and Surfaces B: Biointerfaces* 205:111844–111854.

Li, Nan, Zhiyong Wang, Keke Zhao, Zujin Shi, Zhennan Gu, and Shukun Xu. 2010. "Large scale synthesis of N-doped multi-layered graphene sheets by simple arc-discharge method." *Carbon* 48 (1):255–259.

Li, Ning, Qi Zhang, Song Gao, Qin Song, Rong Huang, Long Wang, Liwei Liu, Jianwu Dai, Mingliang Tang, and Guosheng Cheng. 2013. "Three-dimensional graphene foam as a biocompatible and conductive scaffold for neural stem cells." *Scientific Reports* 3 (1):1–6.

Liang, Jiajie, Yongsheng Chen, Yanfei Xu, Zhibo Liu, Long Zhang, Xin Zhao, Xiaoliang Zhang, Jianguo Tian, Yi Huang, and Yanfeng Ma. 2010. "Toward all-carbon electronics: Fabrication of graphene-based flexible electronic circuits and memory cards using maskless laser direct writing." *ACS Applied Materials & Interfaces* 2 (11):3310–3317.

Liao, JinFeng, Ying Qu, BingYang Chu, XiaoNing Zhang, and ZhiYong Qian. 2015. "Biodegradable CSMA/PECA/graphene porous hybrid scaffold for cartilage tissue engineering." *Scientific Reports* 5 (1):1–16.

Lima-Sousa, Rita, Duarte de Melo-Diogo, Cátia G Alves, Cátia SD Cabral, Sónia P Miguel, António G Mendonça, and Ilídio J Correia. 2020. "Injectable in situ forming thermoresponsive graphene based hydrogels for cancer chemo-photothermal therapy and NIR light-enhanced antibacterial applications." *Materials Science and Engineering: C* 117:111294–111301.

Liu, Chang, Subbiah Alwarappan, Zhongfang Chen, Xiangxing Kong, and Chen-Zhong Li. 2010. "Membraneless enzymatic biofuel cells based on graphene nanosheets." *Biosensors and Bioelectronics* 25 (7):1829–1833.

Liu, Jingquan, Liang Cui, and Dusan Losic. 2013. "Graphene and graphene oxide as new nanocarriers for drug delivery applications." *Acta Biomaterialia* 9 (12):9243–9257.

Liu, Chengshun, Shasha Huang, Jiarui Hou, Wei Zhang, Jinheng Wang, Hongsheng Yang, and Jianming Zhang. 2021. "Natural rubber latex reinforced by graphene oxide/zwitterionic chitin nanocrystal hybrids for high-performance elastomers without sulfur vulcanization." *ACS Sustainable Chemistry & Engineering* 9 (18):6470–6478.

Liu, Lingyun, Wenchen Li, and Qingsheng Liu. 2014. "Recent development of antifouling polymers: Structure, evaluation, and biomedical applications in nano/micro-structures." *Wiley Interdisciplinary Reviews: Nanomedicine and Nanobiotechnology* 6 (6):599–614.

Liu, Xu, Dan Liu, Jeng-hun Lee, Qingbin Zheng, Xiaohan Du, Xinyue Zhang, Hongru Xu, Zhenyu Wang, Ying Wu, and Xi Shen. 2018. "Spider-web-inspired stretchable graphene woven fabric for highly sensitive, transparent, wearable strain sensors." *ACS Applied Materials & Interfaces* 11 (2):2282–2294.

Liu, Zhuang, Joshua T Robinson, Xiaoming Sun, and Hongjie Dai. 2008. "PEGylated nanographene oxide for delivery of water-insoluble cancer drugs." *Journal of the American Chemical Society* 130 (33):10876–10877.

Liu, Zhuang, Mark Winters, Mark Holodniy, and Hongjie Dai. 2007. "siRNA delivery into human T cells and primary cells with carbon-nanotube transporters." *Angewandte Chemie International Edition* 46 (12):2023–2027.

Liu, Shaobin, Tingying Helen Zeng, Mario Hofmann, Ehdi Burcombe, Jun Wei, Rongrong Jiang, Jing Kong, and Yuan Chen. 2011. "Antibacterial activity of graphite, graphite oxide, graphene oxide, and reduced graphene oxide: Membrane and oxidative stress." *ACS Nano* 5 (9):6971–6980.

Liu, Kunping, Jing-Jing Zhang, Fang-Fang Cheng, Ting-Ting Zheng, Chunming Wang, and Jun-Jie Zhu. 2011. "Green and facile synthesis of highly biocompatible graphene nanosheets and its application for cellular imaging and drug delivery." *Journal of Materials Chemistry* 21 (32):12034–12040.

Lu, Chun-Hua, Huang-Hao Yang, Chun-Ling Zhu, Xi Chen, and Guo-Nan Chen. 2009. "A graphene platform for sensing biomolecules." *Angewandte Chemie* 121 (26):4879–4881.

Maiti, Madhuchhanda, Mithun Bhattacharya, and Anil K Bhowmick. 2008. "Elastomer nanocomposites." *Rubber Chemistry and Technology* 81 (3):384–469.

Marcano, Daniela C, Dmitry V Kosynkin, Jacob M Berlin, Alexander Sinitskii, Zhengzong Sun, Alexander Slesarev, Lawrence B Alemany, Wei Lu, and James M Tour. 2010. "Improved synthesis of graphene oxide." *ACS Nano* 4 (8):4806–4814.

Mei, Qingsong, Kui Zhang, Guijian Guan, Bianhua Liu, Suhua Wang, and Zhongping Zhang. 2010. "Highly efficient photoluminescent graphene oxide with tunable surface properties." *Chemical Communications* 46 (39):7319–7321.

Meng, Huan, Wilson X Mai, Haiyuan Zhang, Min Xue, Tian Xia, Sijie Lin, Xiang Wang, Yang Zhao, Zhaoxia Ji, and Jeffrey I Zink. 2013. "Codelivery of an optimal drug/siRNA combination using mesoporous silica nanoparticles to overcome drug resistance in breast cancer in vitro and in vivo." *ACS Nano* 7 (2):994–1005.

Mensah, Bismark, Sungjin Kim, Sivaram Arepalli, and Changwoon Nah. 2014. "A study of graphene oxide-reinforced rubber nanocomposite." *Journal of Applied Polymer Science* 131 (16):40640–40648.

Mondal, Titash, Anil K Bhowmick, and Ramanan Krishnamoorti. 2012. "Chlorophenyl pendant decorated graphene sheet as a potential antimicrobial agent: Synthesis and characterization." *Journal of Materials Chemistry* 22 (42):22481–22487.

Mu, Qiuhong, and Shengyu Feng. 2007. "Thermal conductivity of graphite/silicone rubber prepared by solution intercalation." *Thermochimica Acta* 462 (1–2):70–75.

Nawaz, Khalid, Umar Khan, Noaman Ul-Haq, Peter May, Arlene O'Neill, and Jonathan N Coleman. 2012. "Observation of mechanical percolation in functionalized graphene oxide/elastomer composites." *Carbon* 50 (12):4489–4494.

Nicosia, Angelo, Fabiana Vento, Anna Lucia Pellegrino, Vaclav Ranc, Anna Piperno, Antonino Mazzaglia, and Placido Mineo. 2020. "Polymer-based graphene derivatives and microwave-assisted silver nanoparticles decoration as a potential antibacterial agent." *Nanomaterials* 10 (11):2269–2285.

Niu, Dong, Weitao Jiang, Guoyong Ye, Kun Wang, Lei Yin, Yongsheng Shi, Bangdao Chen, Feng Luo, and Hongzhong Liu. 2018. "Graphene-elastomer nanocomposites based flexible piezoresistive sensors for strain and pressure detection." *Materials Research Bulletin* 102:92–99.

Novoselov, Konstantin S, and AK Geim. 2007. "The rise of graphene." *Nature Materials* 6 (3):183–191.

Novoselov, Kostya S, Andre K Geim, Sergei V Morozov, De-eng Jiang, Yanshui Zhang, Sergey V Dubonos, Irina V Grigorieva, and Alexandr A Firsov. 2004. "Electric field effect in atomically thin carbon films." *Science* 306 (5696):666–669.

Nuñez-Figueredo, Yuresis, Saul Sánchez-Valdes, Eduardo Ramírez-Vargas, Luis F Ramos-deValle, Jorge Albite-Ortega, Oliverio S Rodriguez-Fernandez, Mario Valera-Zaragoza, Antonio S Ledezma-Pérez, Alberto A Rodríguez-González, and Ana B Morales-Cepeda. 2020. "Influence of ionic liquid on graphite/silver nanoparticles dispersion and antibacterial properties against Escherichia coli of PP/EPDM composite coatings." *Journal of Applied Polymer Science* 137 (21):48714–48723.

Nwosu, Christian, Maria Iliut, and Aravind Vijayaraghavan. 2021. "Graphene and water-based elastomer nanocomposites–a review." *Nanoscale.* 13:9505–9540

Oliveira, Sabrina F, Gili Bisker, Naveed A Bakh, Stephen L Gibbs, Markita P Landry, and Michael S Strano. 2015. "Protein functionalized carbon nanomaterials for biomedical applications." *Carbon* 95:767–779.

Pan, Yongzheng, Hongqian Bao, Nanda Gopal Sahoo, Tongfei Wu, and Lin Li. 2011. "Water-soluble poly (N-isopropylacrylamide)–graphene sheets synthesized via click chemistry for drug delivery." *Advanced Functional Materials* 21 (14):2754–2763.

Papageorgiou, Dimitrios G, Ian A Kinloch, and Robert J Young. 2015. "Graphene/elastomer nanocomposites." *Carbon* 95:460–484.

Paran, SMR, G Naderi, MHR Ghoreishy, and A Heydari. 2018. "Enhancement of mechanical, thermal and morphological properties of compatibilized graphene reinforced dynamically vulcanized thermoplastic elastomer vulcanizates based on polyethylene and reclaimed rubber." *Composites Science and Technology* 161:57–65.

Park, Jooyeon, Bokyoung Kim, Jin Han, Jaewon Oh, Subeom Park, Seungmi Ryu, Subin Jung, Jung-Youn Shin, Beom Seob Lee, and Byung Hee Hong. 2015. "Graphene oxide flakes as a cellular adhesive: Prevention of reactive oxygen species mediated death of implanted cells for cardiac repair." *ACS Nano* 9 (5):4987–4999.

Park, Sungjin, and Rodney S Ruoff. 2009. "Chemical methods for the production of graphenes." *Nature Nanotechnology* 4 (4):217–224.

Paszkiewicz, Sandra, Anna Szymczyk, Zdenko Špitalský, Jaroslav Mosnáček, Konrad Kwiatkowski, and Zbigniew Rosłaniec. 2014. "Structure and properties of nanocomposites based on PTT-block-PTMO copolymer and graphene oxide prepared by in situ polymerization." *European Polymer Journal* 50:69–77.

Patel, Dinesh K, Sudipta Senapati, Punita Mourya, Madan M Singh, Vinod K Aswal, Biswajit Ray, and Pralay Maiti. 2017. "Functionalized graphene tagged polyurethanes for corrosion inhibitor and sustained drug delivery." *ACS Biomaterials Science & Engineering* 3 (12):3351–3363.

Patil, Tejal V, Dinesh K Patel, Sayan Deb Dutta, Keya Ganguly, and Ki-Taek Lim. 2021. "Graphene Oxide-Based Stimuli-Responsive Platforms for Biomedical Applications." *Molecules* 26 (9):2797–2815.

Qiu, Yaxin, Jun Wang, Defeng Wu, Zhifeng Wang, Ming Zhang, Ye Yao, and Nengxin Wei. 2016. "Thermoplastic polyester elastomer nanocomposites filled with graphene: Mechanical and viscoelastic properties." *Composites Science and Technology* 132:108–115.

Ramón-Azcón, Javier, Samad Ahadian, Mehdi Estili, Xiaobin Liang, Serge Ostrovidov, Hirokazu Kaji, Hitoshi Shiku, Murugan Ramalingam, Ken Nakajima, and Yoshio Sakka. 2013. "Dielectrophoretically aligned carbon nanotubes to control electrical and mechanical properties of hydrogels to fabricate contractile muscle myofibers." *Advanced Materials* 25 (29):4028–4034.

Raue, Markus, M Wambach, S Glöggler, Dana Grefen, R Kaufmann, C Abetz, P Georgopanos, UA Handge, Thomas Mang, and B Blümich. 2014. "Investigation of historical hard rubber ornaments of Charles Goodyear." *Macromolecular Chemistry and Physics* 215 (3):245–254.

Sadasivuni, Kishor Kumar, Deepalekshmi Ponnamma, Sabu Thomas, and Yves Grohens. 2014. "Evolution from graphite to graphene elastomer composites." *Progress in Polymer Science* 39 (4):749–780.

Sadasivuni, Kishor Kumar, Allisson Saiter, Nicolas Gautier, Sabu Thomas, and Yves Grohens. 2013. "Effect of molecular interactions on the performance of poly (isobutylene-co-isoprene)/graphene and clay nanocomposites." *Colloid and Polymer Science* 291 (7):1729–1740.

Saravanan, Sekaran, Niketa Sareen, Ejlal Abu-El-Rub, Hend Ashour, Glen Lester Sequiera, Hania I Ammar, Venkatraman Gopinath, Ashraf Ali Shamaa, Safinaz Salah Eldin Sayed, and Meenal Moudgil. 2018. "Graphene oxide-gold nanosheets containing chitosan scaffold improves ventricular contractility and function after implantation into infarcted heart." *Scientific Reports* 8 (1):1–13.

Shaun, Bayazid Bustami, Runxuan Cai, Xiaojiang Sun, Chaokun Huang, Shuguang Bi, and Jianhua Ran. 2020. "Strain sensing properties of graphene/elastic fabric." *IOP Conference Series: Materials Science and Engineering* 774(1):012122–012129.

Shen, He, Liming Zhang, Min Liu, and Zhijun Zhang. 2012. "Biomedical applications of graphene." *Theranostics* 2 (3):283–294.

Shen, Jianhua, Yihua Zhu, Cheng Chen, Xiaoling Yang, and Chunzhong Li. 2011. "Facile preparation and upconversion luminescence of graphene quantum dots." *Chemical Communications* 47 (9):2580–2582.

Shin, Su Ryon, Sung Mi Jung, Momen Zalabany, Keekyoung Kim, Pinar Zorlutuna, Sang bok Kim, Mehdi Nikkhah, Masoud Khabiry, Mohamed Azize, and Jing Kong. 2013. "Carbon-nanotube-embedded hydrogel sheets for engineering cardiac constructs and bioactuators." *ACS Nano* 7 (3):2369–2380.

Shioyama, Hiroshi. 2000. "The interactions of two chemical species in the interlayer spacing of graphite." *Synthetic Metals* 114 (1):1–15.

Shuai, Hung-Hsun, Chung-Yao Yang, I Hans, Chen Harn, Roger L York, Tzu-Chun Liao, Wen-Shiang Chen, J Andrew Yeh, and Chao-Min Cheng. 2013. "Using surfaces to modulate the morphology and structure of attached cells–a case of cancer cells on chitosan membranes." *Chemical Science* 4 (8):3058–3067.

Song, Biao, Chang Zhang, Guangming Zeng, Jilai Gong, Yingna Chang, and Yan Jiang. 2016. "Antibacterial properties and mechanism of graphene oxide-silver nanocomposites as bactericidal agents for water disinfection." *Archives of Biochemistry and Biophysics* 604:167–176.

Staudenmaier, L. 1898. "Verfahren zur darstellung der graphitsäure." *Berichte der deutschen chemischen Gesellschaft* 31 (2):1481–1487.

Stoller, Meryl D, Sungjin Park, Yanwu Zhu, Jinho An, and Rodney S Ruoff. 2008. "Graphene-based ultracapacitors." *Nano Letters* 8 (10):3498–3502.

Sun, Xiaoming, Zhuang Liu, Kevin Welsher, Joshua Tucker Robinson, Andrew Goodwin, Sasa Zaric, and Hongjie Dai. 2008. "Nano-graphene oxide for cellular imaging and drug delivery." *Nano Research* 1 (3):203–212.

Sun, Bohan, Richard N McCay, Shivam Goswami, Yadong Xu, Cheng Zhang, Yun Ling, Jian Lin, and Zheng Yan. 2018. "Gas-permeable, multifunctional on-skin electronics based on laser-induced porous graphene and sugar-templated elastomer sponges." *Advanced Materials* 30 (50):1804327–1804334.

Szunerits, Sabine, and Rabah Boukherroub. 2016. "Antibacterial activity of graphene-based materials." *Journal of Materials Chemistry B* 4 (43):6892–6912.

Talebi, Alireza, Sheyda Labbaf, Mehdi Atari, and Maryam Parhizkar. 2021. "Polymeric nanocomposite structures based on functionalized graphene with tunable properties for nervous tissue replacement." *ACS Biomaterials Science & Engineering* 7 (9):4591–4601.

Tang, Longhua, Ying Wang, Yang Liu, and Jinghong Li. 2011. "DNA-directed self-assembly of graphene oxide with applications to ultrasensitive oligonucleotide assay." *ACS Nano* 5 (5):3817–3822.

Tang, Zhenghai, Xiaohui Wu, Baochun Guo, Liqun Zhang, and Demin Jia. 2012. "Preparation of butadiene–styrene–vinyl pyridine rubber–graphene oxide hybrids through co-coagulation process and in situ interface tailoring." *Journal of Materials Chemistry* 22 (15):7492–7501.

Tao, Cheng-an, Jianfang Wang, Shiqiao Qin, Yanan Lv, Yin Long, Hui Zhu, and Zhenhua Jiang. 2012. "Fabrication of pH-sensitive graphene oxide–drug supramolecular hydrogels as controlled release systems." *Journal of Materials Chemistry* 22 (47):24856–24861.

Tian, Xin, Bei-Bei Xiao, Anqing Wu, Lan Yu, Jundong Zhou, Yu Wang, Nan Wang, Hua Guan, and Zeng-Fu Shang. 2016. "Hydroxylated-graphene quantum dots induce cells senescence in both p53-dependent and-independent manner." *Toxicology Research* 5 (6):1639–1648.

Tian, Ming, Jing Zhang, Liqun Zhang, Suting Liu, Xiaoqing Zan, Toshio Nishi, and Nanying Ning. 2014. "Graphene encapsulated rubber latex composites with high dielectric constant, low dielectric loss and low percolation threshold." *Journal of Colloid and Interface Science* 430:249–256.

Wang, Lu, Kyuho Lee, Yi-Yang Sun, Michael Lucking, Zhongfang Chen, Ji Jun Zhao, and Shengbai B Zhang. 2009. "Graphene oxide as an ideal substrate for hydrogen storage." *ACS Nano* 3 (10):2995–3000.

Wang, Ying, Yueming Li, Longhua Tang, Jin Lu, and Jinghong Li. 2009. "Application of graphene-modified electrode for selective detection of dopamine." *Electrochemistry Communications* 11 (4):889–892.

Wei, Ting, Zengchao Tang, Qian Yu, and Hong Chen. 2017. "Smart antibacterial surfaces with switchable bacteria-killing and bacteria-releasing capabilities." *ACS Applied Materials & Interfaces* 9 (43):37511–37523.

Wen, Huiyun, Chunyan Dong, Haiqing Dong, Aijun Shen, Wenjuan Xia, Xiaojun Cai, Yanyan Song, Xuequan Li, Yongyong Li, and Donglu Shi. 2012. "Engineered redox-responsive PEG detachment mechanism in PEGylated nano-graphene oxide for intracellular drug delivery." *Small* 8 (5):760–769.

Whitehead, Kathryn A, Robert Langer, and Daniel G Anderson. 2009. "Knocking down barriers: Advances in siRNA delivery." *Nature Reviews Drug Discovery* 8 (2):129–138.

Wu, Jinrong, Wang Xing, Guangsu Huang, Hui Li, Maozhu Tang, Siduo Wu, and Yufeng Liu. 2013. "Vulcanization kinetics of graphene/natural rubber nanocomposites." *Polymer* 54 (13):3314–3323.

Xiang, Jinglei, and Lawrence T Drzal. 2011. "Thermal conductivity of exfoliated graphite nanoplatelet paper." *Carbon* 49 (3):773–778.

Xiong, Xiaogang, Jingyi Wang, Hongbing Jia, Eryuan Fang, and Lifeng Ding. 2013. "Structure, thermal conductivity, and thermal stability of bromobutyl rubber nanocomposites with ionic liquid modified graphene oxide." *Polymer Degradation and Stability* 98 (11):2208–2214.

Xuan, Yi, YQ Wu, T Shen, Minghao Qi, Michael A Capano, James A Cooper, and PD Ye. 2008. "Atomic-layer-deposited nanostructures for graphene-based nanoelectronics." *Applied Physics Letters* 92 (1):013101–013104.

Yan, Ning, Hesheng Xia, Jinkui Wu, Yanhu Zhan, Guoxia Fei, and Chen. 2013. "Compatibilization of natural rubber/high density polyethylene thermoplastic vulcanizate with graphene oxide through ultrasonically assisted latex mixing." *Journal of Applied Polymer Science* 127 (2):933–941.

Yang, Kai, Liangzhu Feng, Xiaoze Shi, and Zhuang Liu. 2013. "Nano-graphene in biomedicine: Theranostic applications." *Chemical Society Reviews* 42 (2):530–547.

Yang, Xiaoying, Gaoli Niu, Xiufen Cao, Yuku Wen, Rong Xiang, Hongquan Duan, and Yongsheng Chen. 2012. "The preparation of functionalized graphene oxide for targeted intracellular delivery of siRNA." *Journal of Materials Chemistry* 22 (14):6649–6654.

Yang, Jian, Ming Tian, Qing-Xiu Jia, Li-Qun Zhang, and Xiao-Lin Li. 2006. "Influence of graphite particle size and shape on the properties of NBR." *Journal of Applied Polymer Science* 102 (4):4007–4015.

Yang, Xiaoming, Yingfeng Tu, Liang Li, Songmin Shang, and Xiao-ming Tao. 2010. "Well-dispersed chitosan/graphene oxide nanocomposites." *ACS Applied Materials & Interfaces* 2 (6):1707–1713.

Yuanzheng, Luo, Wan Youqi, Huang Jiang, and Li Buyin. 2021. "Nanofiber enhanced graphene–elastomer with unique biomimetic hierarchical structure for energy storage and pressure sensing." *Materials & Design* 203:109612–109621.

Zhan, Yanhu, Marino Lavorgna, Giovanna Buonocore, and Hesheng Xia. 2012. "Enhancing electrical conductivity of rubber composites by constructing interconnected network of self-assembled graphene with latex mixing." *Journal of Materials Chemistry* 22 (21):10464–10468.

Zhan, Yanhu, Jinkui Wu, Hesheng Xia, Ning Yan, Guoxia Fei, and Guiping Yuan. 2011. "Dispersion and exfoliation of graphene in rubber by an ultrasonically-assisted latex mixing and in situ reduction process." *Macromolecular Materials and Engineering* 296 (7):590–602.

Zhang, Qian, Guokang Chen, Kun Wu, Jun Shi, Liyan Liang, and Mangeng Lu. 2021. "Self-healable and reprocessible liquid crystalline elastomer and its highly thermal conductive composites by incorporating graphene via in-situ polymerization." *Journal of Applied Polymer Science* 138 (4):49748–49759.

Zhang, Liming, Zhuoxuan Lu, Qinghuan Zhao, Jie Huang, He Shen, and Zhijun Zhang. 2011. "Enhanced chemotherapy efficacy by sequential delivery of siRNA and anticancer drugs using PEI-grafted graphene oxide." *Small* 7 (4):460–464.

Zhang, Yin, Tapas R Nayak, Hao Hong, and Weibo Cai. 2012. "Graphene: A versatile nano-platform for biomedical applications." *Nanoscale* 4 (13):3833–3842.

Zhang, Liming, Jingguang Xia, Qinghuan Zhao, Liwei Liu, and Zhijun Zhang. 2010. "Functional graphene oxide as a nanocarrier for controlled loading and targeted delivery of mixed anticancer drugs." *Small* 6 (4):537–544.

Zhang, Kuihua, Honghao Zheng, Su Liang, and Changyou Gao. 2016. "Aligned PLLA nano-fibrous scaffolds coated with graphene oxide for promoting neural cell growth." *Acta Biomaterialia* 37:131–142.

Zhu, Shoujun, Junhu Zhang, Chunyan Qiao, Shijia Tang, Yunfeng Li, Wenjing Yuan, Bo Li, Lu Tian, Fang Liu, and Rui Hu. 2011. "Strongly green-photoluminescent graphene quantum dots for bioimaging applications." *Chemical Communications* 47 (24):6858–6860.

18 Graphene-Elastomer Nanocomposites for Electromagnetic Interference (EMI) Shielding Applications

Shalmali Hui
Hijli College affiliated to Vidyasagar University

Narayan Chandra Das
IIT Kharagpur

CONTENTS

DOI: 1201/9781003200444-18

18.1   INTRODUCTION

In recent years, the electromagnetic interference (EMI) is an emerging area of research in the field of digital electronics and telecommunications. This phenomenon arises due to the extensive exploitation of commercial and scientific smart electronic devices which finally leads to electromagnetic (EM) pollution. The accumulation of EM radiation in space results in an undesirable harmful outcome on the performance of the electronic and telecommunication appliances, environment, and on the health of the living organisms as well. Therefore, nowadays it has become a serious concern of the modern civilization all over the world to prevent this EM pollution, and finally to protect the electronic systems, environment and public health from its adverse effect. In this connection, the multifunctional conductive polymer nanocomposites (PNCs) have gained a considerable attention to suppress the EM noises. These innovative class of materials have substituted the metal and metal alloy-based EMI shielding materials owing to their unique combination of electrical, dielectric, thermal, magnetic and/or mechanical properties. Furthermore, the booming of various carbon-based nanomaterials especially 2-dimensional (2D) graphene significantly endorses the development of the EM protection technology. Graphene is one of the most important and pioneering new generation carbon-based nanomaterials for science and technology. A combination of unique electrical, thermal and mechanical properties of graphene along with its high electron mobility and large specific surface area meets all the criteria of effective EMI shielding. Therefore, the development of conductive graphene-polymer nanocomposites will exhibit new and/or improved properties and provide outstanding prospects for designing a new kind of shielding material with a high EMI shielding effectiveness (SE). This chapter is an evaluation of the conductive graphene-based elastomer nanocomposites and it emphasizes their applications in EMI shielding. First, a brief overview of the EMI shielding phenomenon and the PNCs, a comprehensive description of the developing role of graphene as microwave shield and the synthesis methods of graphene-elastomer nanocomposites for EMI shielding applications are outlined. Then, the current research progress of different graphene-based elastomer nanocomposites in the context of their applications in EMI shielding till date in this era of electronic information technology are explored in this chapter. Finally, the current challenges in EMI shielding material research and perspectives for the future development are also highlighted in this chapter which will help to give valuable insights into new research directions.

18.2   EMI SHIELDING PHENOMENON

The EM radiation is a self-propagating diagonal oscillating wave in which electric field (E) and magnetic field (H) are perpendicular to each other and these two elements differ instantaneously. The wave propagates at a right angle to the plane containing these two components (Figure 18.1). It has the properties of both wave and particulate. The EM radiation spectrum is shown in Figure 18.2. Here, the EM waves are categorized on the basis of their wavelength and frequency, and the frequencies ranging from 10^4 to 10^{12} Hz are liable to EMI.

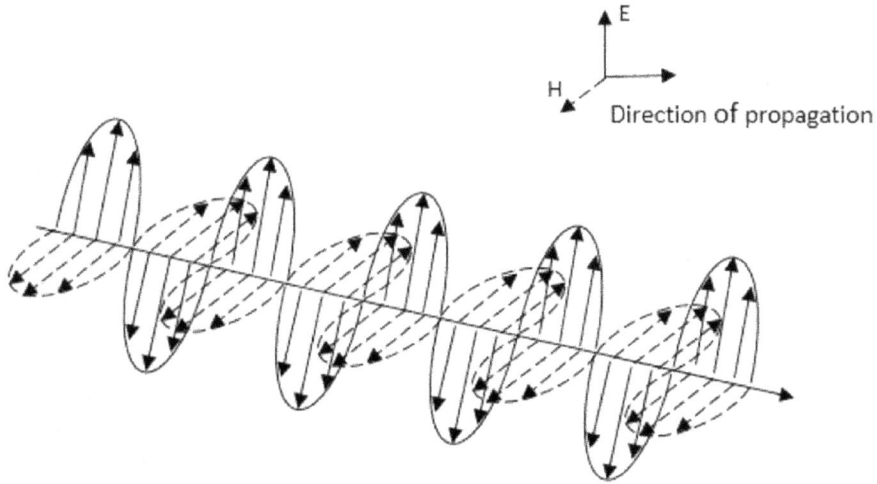

E: Electric field
H: Magnetic field

FIGURE 18.1 Schematic illustration of an electromagnetic transverse oscillating wave. (Adapted from Cezar Afilipoaei and Horatiu Teodorescu-Draghicescu, "A review over electromagnetic shielding effectiveness of composite materials," *Proceedings* 63, no. 1 (2020), 23–31). (Open access article.)

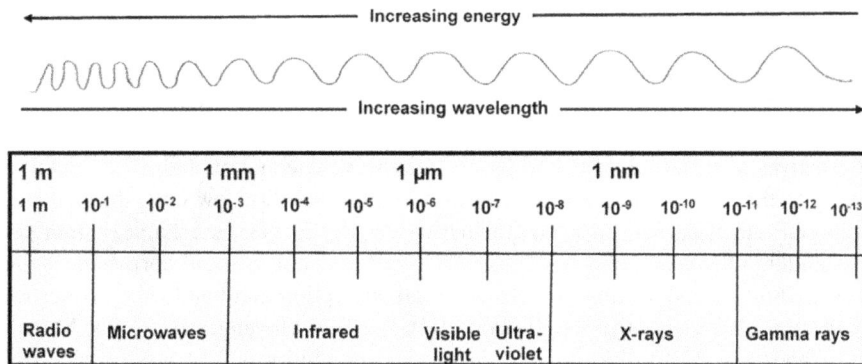

FIGURE 18.2 The electromagnetic (EM) spectrum.

The EM radiations emitted by modern electronic appliances may couple and interfere with the electronics which ultimately result in detrimental effects both on their performance and lifetime. This type of coupling and interference is defined as EMI (Saini and Arora 2012; Chung 2000). An undesirable and uncontrolled adverse effect on the environment, ecosystem and living organisms is also observed over prolonged exposure to high-energy EM radiations in the gigahertz (GHz) range. It causes severe health hazards after penetration into the human body such as psychological disorder, headaches, cancer, infertility, tissue damage and other mild or severe illnesses

(Raagulan et al. 2020; Kawamura et al. 2012; Deruelle 2020; Kostoff et al. 2020; Schüz et al. 2011). Natural phenomenon like solar flare, thunder and electrostatic discharge (ESD) can also generate EMI.

Here lies the importance of EMI shielding – the history of which was reported in the early 1830s when Faraday cage was discovered. This Faraday cage is a surrounded conductive frame along with zero electric fields which is used as a protective barrier against EM radiation (Chung 2001). EMI shielding is a phenomenon of reflection, absorption and/or multiple reflections of EM waves by either conductive- or magnetic-based shielding materials (Sankaran et al. 2018; Joshi and Datar 2015). In this scenario, designing of lightweight, cost-effective and multifunctional EMI shielding materials is of paramount importance both for radiation sources and electronics to minimize the undesirable effect of EM waves. This is the key technology to improve the performance and longevity of all electronic devices which are prone to EM radiation, to confirm their uninterrupted functioning even in the presence of external EM noises, and finally, to shield the living beings and environment as well. A good shielding material can form a Faraday cage during encountering EM waves. A large variety of EMI shielding materials like metals, carbons, ceramics, cement, polymers, hybrids and composites have grown rapidly over the last two decades (Chung 2000). Among these, conventional metal-based materials possess better EMI SE due to their excellent electrical conductivity (Das et al. 2001; Jagatheesan et al. 2015). Galvanized steel and aluminum are the most cost-effective metals out of various other metals (Zhang et al. 2018). However, metals exhibit shielding mostly via reflection mechanism and the reflected radiations are prone to interact with other electronic devices which ultimately cause severe EM pollution. Further, some other demerits of metals like high cost, heavy weight, high density, rigidity, poor processability, low chemical stability, susceptibility to corrosion and oxidation restrict their applications in contemporary devices (Zeng et al. 2016; Thomassin et al. 2013). In this situation, nowadays conductive PNC materials are considered as the promising alternatives to metal-based materials for microwave attenuation and EMI shielding owing to their unique properties like lightweight, flexibility, low cost, thermal stability, fair electrical and thermal conductivity, high SE, good mechanical strength, outstanding chemical stability, robust resistance to chemicals and corrosion, facile processability, good sealing properties, broad absorption and bandwidth properties (Han et al. 2021; Jia et al. 2020; Yousef et al. 2014; Ouyang et al. 2015; Modak and Nandanwa 2015). Specifically, their shielding ability of EM waves primarily through absorption process makes them suitable for military or stealth technology. These are generally prepared via the inclusion of conductive fillers into an insulating/conducting polymer matrix. In the recent past, carbon-based nanofillers stimulated a significant attention worldwide owing to their good conductivity and good absorption ability of EM radiation over a wide range of frequency. The addition of different types of nanoscale carbonaceous conductive fillers like carbon nanotubes (CNTs) (Zeng et al. 2016; Lin et al. 2016), graphite flakes (Jiang et al. 2015; Wu et al. 2016), reduced graphene oxide (rGO) (Wu et al. 2016; Yan et al. 2015) and graphene (Yousef et al. 2014; Modak et al. 2015; Song et al. 2014; Ha et al. 2019; Xia et al. 2017; Shen et al. 2016, 2017; Al-Ghamdi et al. 2016; Santhosi et al. 2020; Andersona et al. 2021; Khan et al. 2020; Xu et al. 2020; Pradhan et al. 2020; Barani et al. 2020; Hu et al. 2019) into the polymer provides a great connectivity in the composite materials which is one of the most important conditions for EMI shielding. The fine dispersion of these nanofillers within the polymer matrix enhances the electrical conductivity. It also improves the absorption as well as reflection of radiation, and fine tunes the electrical and electromagnetic properties of the resulting hybrid PNCs for effective EMI shielding applications. Furthermore, the intrinsic conductivity, dielectric constant and aspect ratio of a nanofiller contribute to the high EMI SE of a composite material (Bryning et al. 2005). Finally, these conductive PNCs provide unique prospects for environmental and energy related advanced applications such as electronics, portable and wearable electronic devices, electro-medical devices, sensors, information and communication technologies, automobiles, aircraft, defense safety systems, aerospace, coatings, military applications or stealth technology, automated industries, optical integrated circuits, electrical and magnetic shields.

18.3 EMI SHIELDING PROCESS AND MECHANISMS

18.3.1 TYPES OF EMI SHIELDING MECHANISMS

Usually the term shielding refers to an enclosure that completely/partially covers an electronic product/particular portion of the electronic product for its right functioning. Thus, it limits the incoming EM radiation from external source via penetration into the circuit, and on the other hand, prevents leaking out of the internal radiation from the device's electronic circuit into the external environment and interfering with adjacent electronics. A certain level of attenuation of electric and magnetic parts of EM waves is extended by this shielding process. To design a highly effective shielding material, it is utmost necessary to understand the response of bound and free electrons within a material to EM wave. Their interactions can be represented as follows (Balanis 2012):

$$D = \varepsilon E \tag{18.1}$$

$$B = \mu H \tag{18.2}$$

$$J = \sigma E \tag{18.3}$$

where E and H denote the electric and magnetic field intensity vectors, respectively. D, B and J represent the electric flux density, magnetic flux density and electric current density vectors, respectively. The magnetic permeability (measure of a material's magnetic polarizability) and the electric conductivity (measure of a material's ability to conduct electricity due to the presence of free electrons) are denoted by μ and σ, respectively. The permittivity is denoted by ε which is a measure of the electric polarizability of a dielectric. These three parameters (μ, σ, ε) are the main factors to understand the EM response of materials.

The shielding mechanism of the EM waves primarily depends on the nature of the EM waves and the properties of the shielding material (Jagatheesan et al. 2014).

When an EM wave encounters the shield, three different mechanisms namely reflection, absorption and multiple internal reflections contribute to the total shielding mechanism (Jaroszewski et al. 2019) which is schematically shown in Figure 18.3.

The primary mechanism of EMI shielding is reflection. The reflection of EM radiations depends on the presence of mobile charge carriers (electrons/holes) in the shielding material (Chung 2000; Joshi and Datar 2015). In this case, electrically conducting materials having at least a fair σ value of 1 S cm^{-1} act as the good shields owing to the presence of free electrons, e.g., metals. However, since conduction involves connectivity in the conduction path, electrical conductivity is not the scientific criterion for shielding. In this situation, another important criterion is developing a good connectivity within the conduction path (Dhakate et al. 2015).

Alternatively, the absorption is attributed as the secondary mechanism of EMI shielding. The absorption of radiations can be achieved due to the interaction of the material's electric and magnetic dipoles with the EM waves (Chung 2000; Joshi and Datar 2015). In this case, materials with a high permittivity, e.g., BaTiO$_3$, are the sources of the electric dipoles and those with a high value of the permeability, e.g., Fe$_3$O$_4$, provide the magnetic dipoles (Biter et al. 1994). However, from the

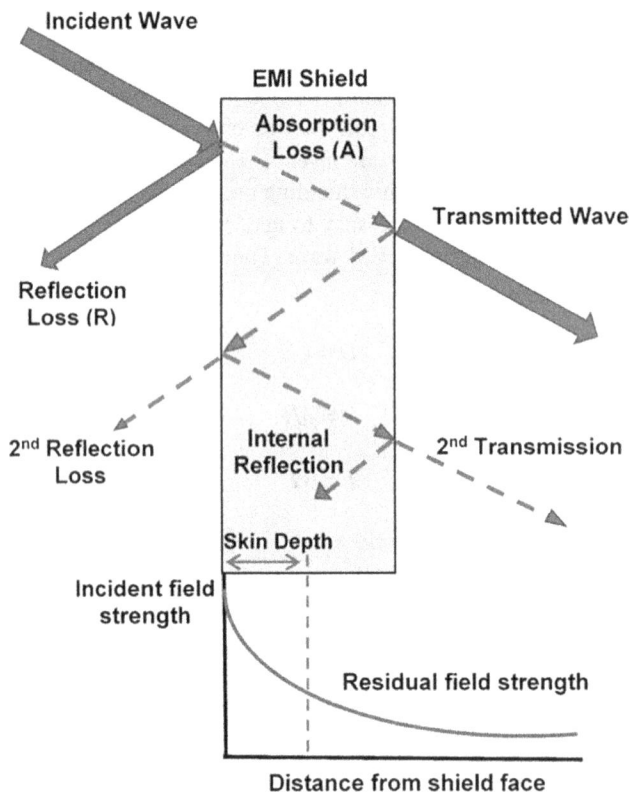

FIGURE 18.3 Schematic representation of the EMI shielding mechanisms for the thin plate shield.

safety lookout, the absorption mechanism is preferred over the reflection for shielding purpose.

The third shielding mechanism is the multiple reflections. It refers to the reflections at various surfaces or interfaces that take place within the shielding material. For this type of mechanism, the shielding material requires a high specific surface area (foam or porous materials) or large phase interface area (composites constituted of fillers with high specific surface areas) (Singh et al. 2020).

In a nutshell, for effective EMI shielding purpose, a shield should possess high conductivity and good electromagnetic attributes such as optimized permeability and permittivity with strong absorption and weak secondary reflection as well as a suitable physical structure and geometry. In addition to these, specific surface/interface area of the shielding material, radiation frequency, temperature, distance between the transmitter and the receiver, thickness of the shield, etc., control the overall effectiveness of a shield and its resulting EMI. Besides these, the other factors like size, shape, aspect ratio and morphology of the fillers, their aggregating tendency and synthesis conditions also have a potential impact on the effectiveness of a shield (Joshi and Datar 2015). Over 10^3 Hz, the reflection efficiency increases with the increasing conductivity of the shield and decreases with the increasing shield permeability and frequency of the EM radiation as well. Alternatively, the absorption efficiency increases with the increasing frequency of the EM radiation, with the increasing thickness and the permeability of the shielding material (Kruzelak et al. 2020). The SE of a shielding material, e.g., polymer matrix, increases with the increasing amount of the conductive filler because of the formation of the interconnected filler network path within it which finally leads to a sharp increase in the conductivity.

Skin effect is the propensity of the EM radiation at the higher frequency range to distribute into the near surface regions of the electric conductors (Kunkel 2020). The depth at which the electric field drops to $1/e$ (e is Euler's number and $1/e = 37\%$) of the incident value is called the skin depth (δ). It is represented as

$$\delta = 1/\sqrt{\pi f \mu \sigma} \tag{18.4}$$

where f is the frequency of the EM wave and μ is the magnetic permeability of the shield, where $\mu = \mu_0 \mu_r$ with μ_r being the relative magnetic permeability and $\mu_0 = 4\pi \times 10^{-7}$ H m^{-1}. σ is the electrical conductivity of the shield in Ω^{-1}m^{-1}. Therefore, it can be inferred that δ decreases with the increase in f, σ or μ. Jaroszewski et al. (2019) reported that the unit size of the filler should be less or comparable to the skin depth. Therefore, the composites containing conductive fillers with nano dimension are more effective shielding materials than the fillers having a large unit size.

18.3.2 THEORY OF EMI SHIELDING

The EM radiative region can be divided into three subclasses relative to the total wavelength λ of the EM wave according to the distance r between the source and the shield (Figure 18.4).

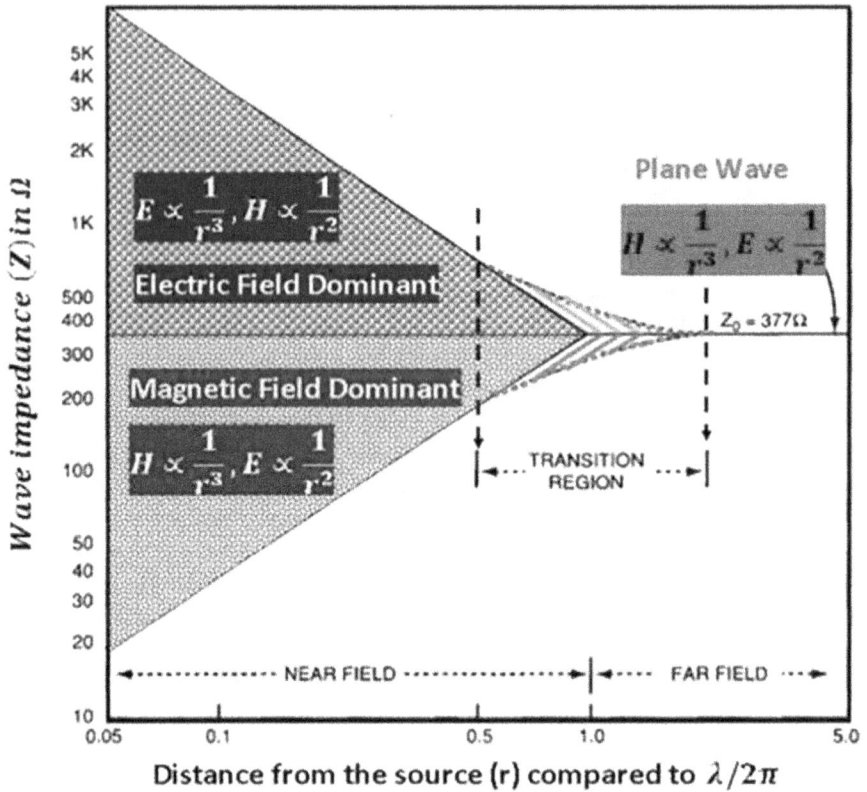

FIGURE 18.4 Dependence of wave impedance on distance from source normalized to $\lambda/2\pi$. (Adapted from Parveen Saini and Manju Arora, *New Polymers for Special Applications*, IntechOpen, (2012), 71–112). (Open access article.)

The EM plane wave theory is used for EMI shielding in the *far-field region* when $r > \lambda/2\pi$. In this region, the electric and magnetic vectors lie in an equal ratio in the plane waves, and they are in phase and orthogonal to each other (Ganguly et al. 2018; Geetha et al. 2009). This region is the most concern in measuring SE. Alternatively, the theory based on the contribution of the electric and magnetic dipoles is used for EMI shielding in the *near-field region* when $r < \lambda/2\pi$ (Geetha et al. 2009). When $r \approx \lambda/2\pi$, then the region is named as *transition region* which lies between the near-field and far-field regions (Al-Saleh and Sundararaj 2009).

The power loss due to shielding is referred to as the shielding effectiveness (SE) or the total shielding effectiveness (SE_T). It is a measurement of an attenuation of the EM signal after introducing a shield. It is expressed in decibels (*dB*) unit. Typically, the SE is defined as the ratio of the magnitude of the incident electric/magnetic field on the barrier to the magnitude of the transmitted electric/magnetic field through the barrier. Mathematically, it can be expressed as a function of the logarithm of the ratio of the incident and the transmitted electric, magnetic or plane-wave field intensities (Saini et al. 2009, 2011; Colaneri and Schacklette 1992; Schulz et al. 1988):

$$\mathrm{SE}_T\,(dB) = 10\,\log_{10}\,(P_T/P_I) = 20\,\log_{10}\,(E_T/E_I) = 20\,\log_{10}\,(H_T/H_I) \quad (18.5)$$

where P, E and H represent the plane-wave field intensity, electric and magnetic field intensities, respectively. Subscript I denotes the incident wave, while T represents the transmitted wave. SE is measured as a function of frequency (Geetha et al. 2009). The shielding material reflects as well as absorbs part of the incident radiation when the EM radiation impacts the surface of the conductive shield (Figure 18.3). The rear surface of the shield reflects another part of the radiation to the front where it can aid or hinder the effectiveness of the shield depending on its phase relationship with the incident wave, as shown in Eq. 18.6. Therefore, according to the Schelkunoffs theory, the total SE of a material is determined by the summation of contributions from absorption (A), reflection (R) and multiple reflections (M) (Paul 2004; Ott 1988; Schulz et al. 1988; Schelkunoff 1943; Ashokkumar et al. 2013). It can be mathematically represented as follows:

$$\mathrm{SE}_T\,(dB) = \mathrm{SE}_A + \mathrm{SE}_R + \mathrm{SE}_M \quad (18.6)$$

where SE_A and SE_R indicate the absorption factor and the reflection factor, respectively. SE_M is the correction factor to account for multiple reflections.

Out of several methods, the most commonly used SE measurement methods are (i) Open field or free space method, (ii) Shielded box method, (iii) Coaxial transmission line method and (iv) Shielded room method (Wanasinghe et al. 2020).

18.3.2.1   Absorption Loss (SE$_A$)

Absorption denotes the EM energy that is absorbed into the shield and dissipated as heat. Impedance matching of the incident wave and the shielding material causes absorption. SE_A is a function of the physical characteristics of the shield and is independent of the type of the source field. Hence, this term is same for all the three waves. As shown in Figure 18.3, the amplitude of an EM wave decreases exponentially during passage through a medium because of the induction of currents in the medium that generates ohmic losses and heating of the material, where E_T and H_T can be expressed as $E_T = E_I\,e^{-t/\delta}$ and $H_T = H_I{}^{e-t/\delta}$ (Ott 1988).

Therefore, the magnitude of SE_A in dB can be expressed by the following equation:

$$\mathrm{SE}_A\,(dB) = -20\,(t/\delta)\log_{10}e = -8.69\,(t/\delta) = -131t\sqrt{f\mu_r\sigma_r} \quad (18.7)$$

where t is the shield thickness in mm; f is the frequency in MHz. The parameters μ_r and σ_r represent the relative permeability and the conductivity, respectively, relative to copper.

Eq. 18.7 reveals that SE_A is directly proportional to the square root of the product of μ_r and σ_r of the shield (Saini et al. 2009). Further its magnitude increases with increase in the frequency. Therefore, in order to achieve the required number of skin depths at the lowest frequency of concern, the absorbing material must have high conductivity, high permeability and appropriate thickness as well. Usually, Superpermalloy and Mumetal exhibit very high absorption due to having high magnetic permeability.

18.3.2.2 Reflection Loss (SE$_R$)

SE$_R$ is a phenomenon related to the comparative mismatch between the surface intrinsic impedance of the shield and the propagating wave. The magnitude of SE$_R$ for the three principle fields can be represented as (Saini et al. 2011):

$$SE_R = C + 10\log_{10}\left(\sigma_r/\mu_r\right)\left(1/f^n r^m\right) \tag{18.8}$$

where σ is the relative conductivity of copper; f is the frequency in Hz; μ is the relative permeability of free space; r is the distance from the source to shielding in meter; C, n and m are constants for calculating reflection losses (in dB) for the plane waves, electric fields and magnetic fields, respectively.

Therefore, the magnitude of reflection (SE$_R$) for a plane wave radiation can be expressed as (Saini et al. 2011):

$$SE_R(dB) = -10\,\log_{10}\left(\sigma_T/16 f\varepsilon_0\mu_r\right) \tag{18.9}$$

where σ_T is the total electrical conductivity (in S cm^{-1}) relative to copper and μ_r is the relative magnetic permeability relative to free space. Eq. 18.9 clearly indicates that SE$_R$ is a function of the ratio of σ_T and μ_r of the shielding material and its magnitude decreases with increase in the frequency. In general, metals like silver, gold and copper possess a high conductivity and exhibit superior reflection property.

18.3.2.3 Multiple Reflections (SE$_M$)

In case of a thin shield, the reflected wave gets re-reflected and bounces through reflecting surfaces several times (Figure 18.3). The amplitude of the incident radiation is attenuated by this process within the shielding material. SE$_M$ can be expressed mathematically as follows (Saini et al. 2009):

$$SE_M = 20\,\log_{10}\left(1 - e^{-2t/\delta}\right) \tag{18.10}$$

where t is the thickness of the shielding material.

SE$_M$ has a potential impact on certain composite materials and porous structures. The chiral polymer composites where the dielectric matrix contains chiral fillers are generally useful for microwave applications (Varadan and Varadan 1990). The loss due to multiple reflections can be ignored when the distance between the reflecting surfaces/interfaces is large in comparison with δ (Chung 2001). SE$_M$ is only important when the shield is thin and the frequencies are low (<20 kHz), whereas SE$_M$ becomes insignificant in the case of a thick shielding material because of high absorption loss (SE$_A \geq$ 10 dB) which generally occurs at very high frequencies (~GHz or even higher) (Saini et al. 2009, 2011). SE$_M$ can also be irrelevant for electric fields and plane waves in practical practice. Mathematically, this factor is a positive or negative quantity, however, practically it is always negative.

Overall, to design and fabricate an efficient shielding material, it is of paramount importance to focus on both intrinsic and extrinsic parameters on which SE depends, and various theoretical and empirical relations associated with the above parameters with contributing reflection, absorption and multiple reflection components.

Extensive effort is devoted to the development of EMI shielding materials. In this context, high-performance electroconductive polymer composites and PNCs play an important role.

18.4   EMI SHIELDING MATERIALS

18.4.1   POLYMER COMPOSITES FOR EMI SHIELDING

Since discovery in 1960, polymer-based composites are continuously produced for scientific and commercial requirements due to their unique attributes, e.g., ease of production, lightweight, high strength, high durability, high flexibility and ductility. In polymer composites, heterophasic morphology is present where the polymer matrix acts as the binder between the reinforcing inclusions. The addition of conductive fillers into the polymer matrix makes them suitable for EMI shielding applications owing to their low density and reasonable conductivity. The overall processability of these composite materials depends on the types of polymer used (insulating/conducting) and on the nature of the filler. Intrinsically insulating polymers such as polystyrene (PS), polydimethylsiloxane (PDMS), natural rubber (NR), styrene-butadiene rubber (SBR) and polyurethanes (PU) can conduct electricity when loaded with conductive fillers or metals. Alternatively, intrinsically conducting polymers (ICPs) or conducting polymers such as polyaniline (PANI), polythiophene (PTh), polyacetylene (PA) and polypyrrole (PPy) possess high inherent electrical conductivity in the microwave and radio wave frequency regime (100 MHz-20 GHz) and serve as good EMI shielding materials. In order to satisfy all the shielding aspects including thickness, volume and absorption coefficient, these ICPs (as hosts) are combined with organic/inorganic fillers (as a guest) or ICPs (as a guest) added to the insulating matrices (as hosts) due to their inherent electrical conductivity and dielectric properties (Sankaran et al. 2018). In order to get an effective shielding, the minimum conductivity of the composite materials should be $> 10^{-2}$ S m^{-1} (Sudha et al. 2010).

18.4.2   POLYMER NANOCOMPOSITES (PNCs) FOR EMI SHIELDING

The interfacial incompatibility between organic polymer and added large volume of fillers sometimes leads to interfacial failures associated with the deterioration in some properties like tensile strength, tear strength and elongation at break. One approach to improve the interfacial interaction between the polymer matrix and the filler is to reduce the size of inclusions to nanoscale and to improve the interfacial adhesion. This approach can also control the electrical and EM properties of the resulting composites. It can also accomplish all the criteria for effective EMI shielding applications. Here lies the importance of emerging *Nanotechnology* which is now recognized as one of the most promising fields of research in the 21st century. PNCs are polymers which are reinforced with rigid inorganic/organic particles having at least one dimension in the nanometer size-range. These nanofillers include CNTs, layered silicates (e.g., montmorillonite, saponite), graphene, nanoparticles of metals (e.g., Au, Ag), metal oxides (e.g., TiO_2, Al_2O_3), polyhedral oligomeric silsesquioxanes (POSS) and semiconductors (e.g., PbS, CdS). The interfacial interaction between polymer and nanofiller strongly affects the mechanical, thermal, electrical and other properties of the nanocomposites. The importance of this PNC technology stems from providing value-added properties not present in the neat polymer, without sacrificing the polymer's inherent processibility, low density and mechanical properties. Any composite can be divided into three parts: the matrix, the reinforcing component and the so-called interfacial region. The interfacial region is responsible for communication between the matrix and the reinforcing component (filler) and its conventionally ascribed properties differ from the bulk matrix because of their proximity to the surface of the filler (Wagner and Vaia 2004). The high specific surface area of nanofillers (due to lower size scale and higher aspect ratio) is one of the reasons why the nature of reinforcement is different in nanocomposites. As compared to the conventional macroscopic conductive fillers, the nanofillers having a high aspect ratio are able to form excellent electrically conductive network within the polymer matrix which facilitates delocalization of the charge carriers resulting in improved conductivity. PNCs are one such potential area which can be utilized for EMI shielding owing to their commercial feasibility, lower weight, flexibility and ease of processing. Till now, a large number of PNCs have been reported to be useful for effective EMI shielding applications (Modak and Nandanwar 2015; Lin et al. 2016; Jiang et al. 2015; Wu et al. 2016; Yan et al. 2015; Song et al. 2014; Ha et al. 2019; Xia et al. 2017; Shen et al. 2016, 2017; Al-Ghamdi et al. 2016; Santhosi et al. 2020; Andersona et al. 2021; Khan et al. 2020; Xu et al. 2020; Pradhan et al. 2020; Barani et al. 2020; Hu et al. 2019).

18.5 GRAPHENE: ELECTRONIC STRUCTURE AND ELECTRICAL PROPERTIES

Among various nanoscale inclusions, graphene and graphene-based materials have stimulated a significant attention worldwide as versatile, environmentally friendly and available carbon nanomaterials for future technology advancement. Graphene – a monolayer of sp^2-hybridized carbon atoms arranged in a 2D crystal lattice structure like honeycomb – exhibits fascinating thermal, mechanical, electrical, physical and chemical properties which are significantly better than the other inorganic filler materials (Novoselov et al. 2004; Potts et al. 2011; Geimand and Novoselov 2007; Compton and Nguyen 2010). This type of close pack arrangement makes it a thinnest, strongest and stiffest material in the world. The structure of graphene is shown in Figure 18.5.

Here, the carbon atoms are bonded together through covalent bonding in the same plane and the length between two carbon atoms is around 0.14 nm (Neto et al. 2009). It has free conjugated π-electrons carriers, bounded in the plane drifted along the hexagonal chain of carbon atoms. It exhibits extremely high electrical conductivity of 6,000 S cm^{-1}, high thermal conductivity in the range of 1,500–5,000 W m^{-1} K^{-1} and a high carrier mobility ($\approx$ 2 00,000 cm^2 V^{-1} s^{-1}) at room temperature, strength of 130 GPa, Young's modulus of 1 TPa (Neto et al. 2009; Das et al. 2020; Lee et al. 2008). It can exist both in bilayer and few layer configurations as well. The conductivity decreases with increasing graphene layer, and finally, approaches to the conductivity of graphite (Nirmalraj et al. 2011; Powell and Childs 1972). It is defined as a

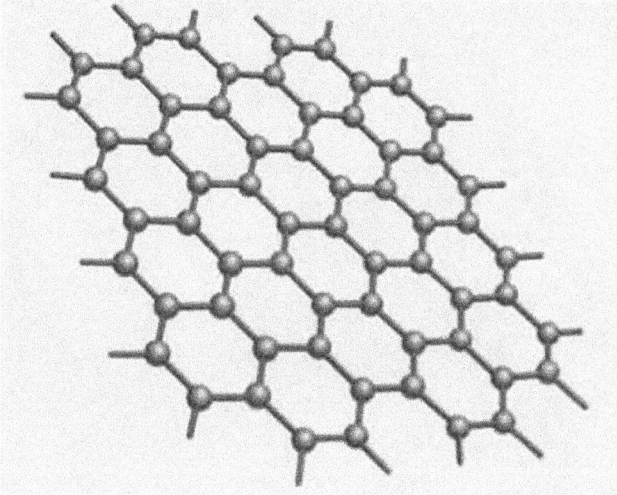

FIGURE 18.5 Structure of graphene. (Adapted from A.H. Castro Neto et al., "The electronic properties of graphene," *Reviews of Modern Physics* 81, (2009): 109–164, with permission from American Physical Society.)

zero-gap semiconductor on the basis of the lattice energy spectrum obtained through the tight binding (TB) model (Neto et al. 2009), and π and π^* energy band contacted together in Dirac points (K and K') which is the intersection of the two Dirac cones (Figure 18.6).

In comparison with other carbon nanomaterials, graphene-based materials have a unique 2D structure, ultrahigh surface area ($2{,}630\,m^2g^{-1}$), high aspect ratio, low density, good mechanical strength, ballistic electron transport properties, zero energy band gap, exceptional electronic and thermal conductivities, excellent optical transmittance, low cost, and so on. Owing to such attributes, these materials possess good shielding capabilities, not only in the microwave region but also in the other regions of the EM spectrum. The EMI SE of flexible graphite is as high as $130\,dB$ (Kim et al. 2014). These can be used as effective and electrically conductive nano fillers to combine with polymers in order to fabricate thinner EMI shielding PNC materials with improved SE. These developed nanocomposites have attained a good popularity for their inimitable properties such as lightweight, high aspect ratio, easy processability, flexibility, resistance to corrosion, outstanding electrical, mechanical, thermal and barrier properties and economic feasibility (Novoselov et al. 2004; Potts et al. 2011; Geimand and Novoselov 2007; Compton and Nguyen 2010; Xia et al. 2017; Jan et al. 2017; Chen et al. 2017; Shen et al. 2013). Some important types of graphene include graphene oxide (GO), rGO, multilayer graphene (MLG), few-layer graphene (FLG), graphene nanosheet (GNS), graphene nanoribbon (GNR), graphene nanoplatelets (GnPs) and graphene quantum dots (GQDs). GNSs provide a greater capacitive potential to the percolative polymer composites. This capacitive potential controls the dielectric characteristics which eventually boosts the EM energy attenuation capability of the material (Jan et al. 2012; Rao et al. 2015). Generally, agglomerated nanocomposites have a poor non-uniform dispersion of fillers which leads to a low

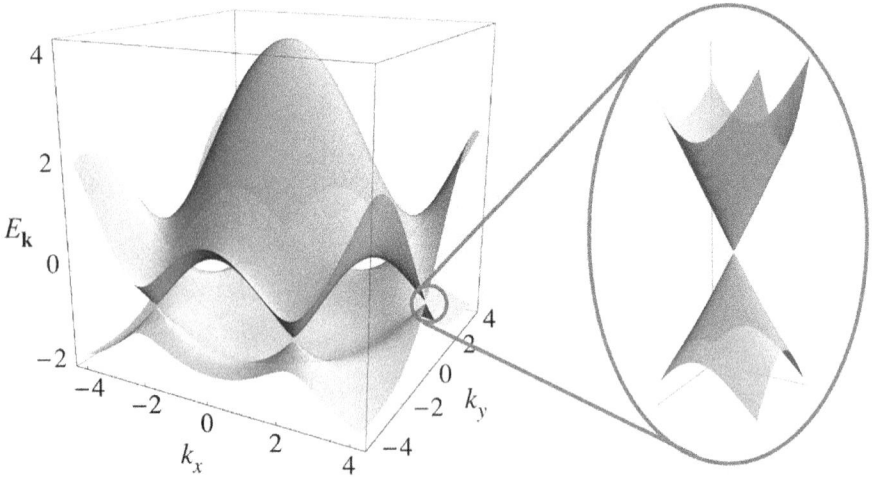

FIGURE 18.6 (a) Energy spectrum (in units of t) for finite values of t and t', with $t = 2.7\,\text{eV}$ and $t' = 0.2\,t$. (b) Zoom-in of the energy bands close to one of the Dirac points. t is the nearest neighbor hopping energy (hopping between different sublattices) and t' is the next nearest neighbor hopping energy (hopping in the same sublattice). (Adapted from A.H. Castro Neto et al., "The electronic properties of graphene," *Reviews of Modern Physics* 81, (2009): 109–164, with permission from American Physical Society.)

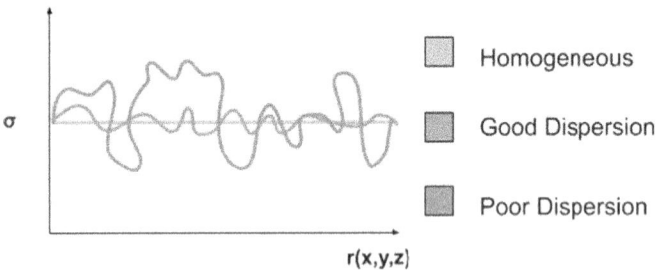

FIGURE 18.7 Schematic illustration of the difference in conductivity between homogeneous, well dispersed and poorly dispersed materials. (Adapted from Liam Anderson et al., "Modelling, fabrication and characterization of graphene/polymer nanocomposites for electromagnetic interference shielding applications," *Carbon Trends* 4 (2021): 100047–100066). (Open access article.)

conductivity (Figure 18.7). GO – a most accessible precursor for graphene – provides a more uniform dispersion throughout the polymer matrix which is produced by oxidizing graphite. This is due to the presence of various oxygenated functional groups on the surface of GO like carbonyl, epoxide, carboxylate and alcohols (Szabó et al. 2005) which results in a reduction in aggregation/restacking, higher processability, good dispersibility in aqueous media and/or polar organic solvents, and an enhancement in the interfacial interactions between GO and the polymer matrix as well. These functional groups as well as the defects enhance the polarization loss.

However, the oxygenated groups interrupt the conjugated carbon backbone of GO resulting in a poor electrical conductivity. But interestingly, conductivity can be re-established after reducing it to rGO via chemical or physical reduction (Park and Ruoff 2009). The heterogeneous structure of rGO consists of graphene-like basal plane, which is decorated with additional structural defects. The oxidized chemical groups and heteroatom are present within the rGO sheets (Tarcan et al. 2020). The propagation paths for the EM waves are improved to a great extent for a thin, flexible and corrugated graphene, whereas thick graphene increases the conductive paths which finally enhances the conduction loss (Cao et al. 2015).

18.6 SYNTHESIS OF GRAPHENE

The synthetic pathways are mainly divided into two classes – (i) "top-down" (destruction) and (ii) "bottom-up" (construction) approaches as illustrated in Figure 18.8.

In the top-down approach, a stack of graphene layers of graphite precursor breaks followed by separation into individual sheets of nano-sized graphene via a multitude of thermal, mechanical and chemical processes (Edwards and Coleman 2013). Some common top-down approaches are chemical exfoliation (Marcano et al. 2010; Choi and Lee 2012), chemical synthesis (Choi and Lee 2012), mechanical/micromechanical exfoliation/Scotch-tape method (Choi and Lee 2012; Geimand and Novoselov 2007; Emiru and Ayele 2017), arc discharge (Wu et al. 2016; Madurani et al. 2020), liquid-phase exfoliation (LPE) (Guler et al. 2016), sonication (Bourlinos et al. 2009), electrochemical method (Parvez et al. 2013), laser ablation (Kazemizadeh and Malekfar 2018), unzipping of CNTs (Dimiev et al. 2018), etc. GO, rGO and GnPs are achievable through these top-down approaches (Andersona et al. 2021). GO is synthesized via using Hummer's, Brodie's, Staudenmaier's, or similar methods (Marcano et al. 2010), whereas rGO is produced by thermal (Apollo et al. 2015) or chemical reduction (Apollo et al. 2018) processes. On the other hand, in the bottom-up approach, smaller carbon molecules which act as the building blocks are built up to match a material's specifications. Here, the atomic-sized carbon precursors are obtained from alternative sources other than graphite. GNRs and graphene dots (so-called nanoflakes) in large quantities are obtained by this process (Warner et al. 2012). This approach mainly includes epitaxial growth (Sprinkle et al. 2009), chemical vapor deposition (CVD) (Ao et al. 2020), substrate-free gas-phase (SFGP) synthesis (Dato and Michael 2010), template route (Yang et al. 2015) and pyrolysis (Choucair et al. 2008). These are covered in earlier chapters.

The top-down method produces highly pure and high-quality single-layered graphene (SLG) along with a high reproducibility where no substrate transfer is required and large-scale production can be achieved. At present, this method is less complex, cheaper and more scalable than the bottom-up method. However, it has some drawbacks such as less controllable, less environmentally friendly, low yield, tedious procedure and irregular properties of developed graphene (Tour 2014). On the other hand, defect-free graphene products with large surface area along with a good control of resolution are produced by the bottom-up methods. However, some issues including low quality, low purity, scalability, high cost, low yield and sophisticated experimental set-up remain in this process. Some typical approaches about synthetic methods,

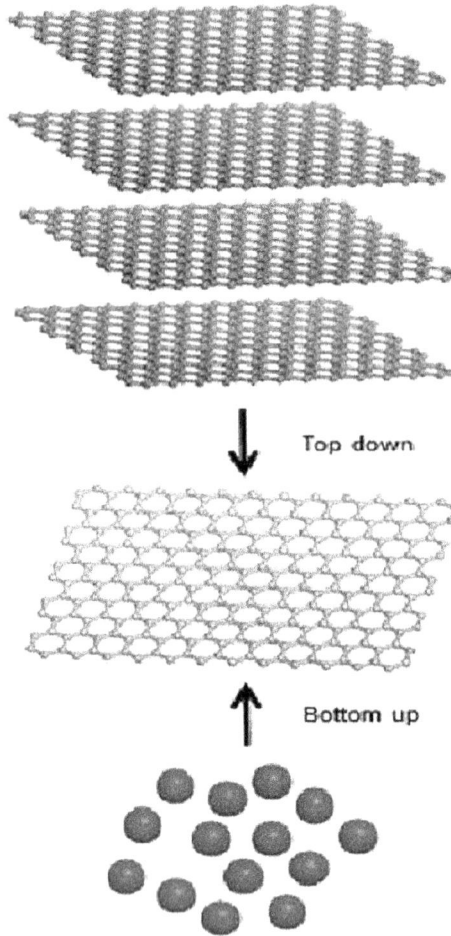

FIGURE 18.8 Synthesis methods of graphene. (Adapted from S. Saqib Shams, Ruoyu Zhang and Jin Zhu, "Graphene synthesis: a review," *Materials Science-Poland* 33, no. 3 (2015): 566–578). (Open access article.)

advantages, disadvantages and applications of graphene and its derivatives are summarized in Table 18.1. During the last few years, a large number of methodologies of synthesis of graphene of diverse nature were discussed in several articles (Tour 2014; Zhang et al. 2019; Madurani et al. 2020; Lee et al. 2019). To summarize, new advances/modifications are urgently needed for a large-scale production of high-quality graphene with a high yield to overcome the limitations of different approaches and to satisfy all the industrial needs. Covalent (Salavagione et al. 2011) and non-covalent methods (Shen et al. 2011) are generally used to modify the surface of graphene for interaction with various matrices, to provide uniform and wide dispersion of graphene sheets within the polymer matrix and to improve the EMI shielding properties of graphene as well. Graphene modification has been covered in Chapter 4 of this book.

TABLE 18.1

Advantages, Disadvantages and Potential Applications of Different Typical Synthetic Methods of Graphene and Its Derivatives

Methods	Subclassification	Typical Dimension	Advantage	Disadvantage	Applications	Ref.
Top-Down	Chemical exfoliation	Few layer (~1.1 nm)	Large-scale production	1. Low yield 2. Structural defects	Membranes, TEM grids sensitive device fabrication	Marcano et al. (2010)
	Micromechanical exfoliation	Few-layer (μm to cm)	1. Large size 2. Unmodified graphene sheets	Very small scale production	Fabrication of graphene-based composite materials electric batteries, sensors	Geimand Novoselov (2007)
	Arc Discharge	100–300 nm	1. Cost effective 2. Fast 3. Environmentally friendly 4. Large-scale production of graphene sheets	Low yield	Anode material for lithium-ion batteries, industrial synthesis of graphene, appropriate choice for an electric charger used for conducting composite materials	Wu et al. (2016)

(Continued)

TABLE 18.1 (*Continued*)

Advantages, Disadvantages and Potential Applications of Different Typical Synthetic Methods of Graphene and Its Derivatives

Methods	Subclassification	Typical Dimension	Advantage	Disadvantage	Applications	Ref.
	Electrochemical	Mono layer (~0.86 nm) Bi-layer (~1.5nm)	1. Cost effective 2. Fast 3. Simple step & Simple purification 4. High electrical conductivity of functionalized graphene 5. High quality 6. Environmentally friendly 7. High yield 8. Excellent electronic properties	Ionic liquids are costly	Fabrication of both rigid and flexible organic electronic devices	Parvez et al. (2013)
	Sonication	Few layer (μm to cm) Single- and multilayer (μm or sub μm)	1. Unmodified graphene sheets 2. Unmodified graphene 3. Low cost	1. Small production 2. Separation needed	Facilitates solution processing of several polymer composites	Bourlinos et al. (2009)

(*Continued*)

TABLE 18.1 (*Continued*)

Advantages, Disadvantages and Potential Applications of Different Typical Synthetic Methods of Graphene and Its Derivatives

Methods	Subclassification	Typical Dimension	Advantage	Disadvantage	Applications	Ref.
	Super acid dissolution of graphite	Mostly single layer	1. Unmodified graphene 2. Measurable	1. Utilization of hazardous chemical chlorosulfonic acid 2. Removal of acid is costly	Flexible electronics and multifunctional fibers	Behabtu et al. (2010)
Bottom-Up	CVD	Multilayer (~10 layers)	1. High quality graphene 2. Large size	1. Low yield 2. Expensive	High-performance EMI shielding, significant increase in mechanical performance	Ao et al. (2020)
	Confined self-assembly	Single layer (100 mm)	Thickness can be controlled	Existence of defects	Electronics, optoelectronics electrocatalysis/photocatalysis, environment, energy storage and conversion.	Zhang et al. (2009)
	Epitaxial growth on SiC	Few layers (up to cm size)	1. Clean, ordered, high-quality graphene 2. Controllable thickness	Low yield	Graphene electronics	Sprinkle et al. (2009)

Source: Modified from Madurani et al., "Progress in Graphene Synthesis and its Application: History, Challenge and the Future Outlook for Research and Industry," *ECS Journal of Solid State Science and Technology* 9, no. 9 (2020): 093013. (Open access article).

18.7 PREPARATION METHODS OF GRAPHENE-ELASTOMER NANOCOMPOSITES FOR EMI SHIELDING APPLICATIONS

The importance of graphene-based nanofillers over other nanoscale fillers in their capability to impart several functional properties to a rubber matrix was discussed in detail by Hamed and Zhang et al. (Hamed 2000; Zhang and Jia 2004). The three most commonly used methods to prepare graphene-elastomer nanocomposites are as follows:

i. Melt intercalation

It is the most scalable, environmentally friendly, simple and economically feasible method to fabricate graphene-elastomer nanocomposites for industrial applications (Prud'homme et al. 2009; Al-solamy et al. 2012). In this method, the polymers are allowed to melt first to produce viscous liquids followed by applying a high shear force to disperse the nanofillers into the polymer matrix. Thus, it does not require any solvent to disperse filler and polymer. This method can be applied to both polar and non-polar elastomers. However, the low dispersion ability of the graphene is the major drawback of this method (Li and Zhai 2015).

ii. Solution dispersion

It is an effective fabrication technique which has a promising potential to overcome the problem of non-uniform dispersion of graphene and its derivatives in the polymer matrix. In this method, the nanofiller is dispersed into the suitable solvent via ultrasonication process to produce a colloidal suspension which is then mixed with the elastomer-solvent mixture by stirring or shear mixing. Finally, the nanocomposite films are obtained by evaporating this solvent via thermal treatment or precipitation or distillation. Various graphene-elastomer nanocomposites were reported to be synthesized by using this technique (Kim et al. 2011; Lian et al. 2011). However, this method requires a large number of solvents which are very tough to evaporate.

iii. In situ polymerization

In this method, the filler and the monomers are initially mixed which is prospered by the incorporation of initiator, followed by in situ polymerization. As a result, a uniform and homogeneous dispersion of fillers are achieved throughout the polymer matrix without a prior exfoliation step. Various graphene-elastomer nanocomposites were prepared by using this technique (Mu and Feng 2007; Kim et al. 2010). Thermally unstable and insoluble elastomers which are difficult to process via melt or solution blending can be easily prepared by this method. However, the requirement of monomers as well as a lot of reagents limit the applicability of this process for naturally existing polymers.

18.8 RECENT PROGRESS IN GRAPHENE-ELASTOMER NANOCOMPOSITES FOR EMI SHIELDING

In recent years, one area that has popped up for intense inspection in the field of nanoscience and nanotechnology is the evolution of new generation multifunctional conductive graphene-based elastomer nanocomposites to overcome the limited permittivity and the lower environmental stability of the pure elastomers for shielding applications. An elastomer cannot form a shield by itself unless it is filled with a conductive filler having a small unit size, a high conductivity and a high aspect ratio (Chung 2001). In addition, low concentration of the filler is desirable to maintain the resilience, strength, ductility and processability of the elastomer, and overall cost and weight of the composite materials as well. In this context, the development of conductive graphene-elastomer nanocomposites aiming at achieving a high aspect ratio, lightweight, super electrical and thermal conductivity, microwave absorption, dielectric performance, efficient reinforcement, multifunctional properties and economic feasibility has grabbed a significant attention from the scientific community. It will provide promising applications in effective EMI shielding where the dual advantages of flexibility of the elastomer and the superior electrical conductivity of the filler are combined together (Singh et al. 2012). The dielectric and electrical properties of these composites can be further improved by proper functionalization/surface modification of graphene. The EMI shielding properties of some graphene-elastomer nanocomposites are summarized in Table 18.2.

An electrically conductive and crosslinked graphene/NR composite containing a segregated graphene network (GN) was reported to be prepared by using an innovative method – self-assembly in latex and static vulcanization (Zhan et al. 2012). At 1.78 vol.% graphene loading, this composite showed a percolation threshold of ~0.62 vol.% and a conductivity of 0.03 S m^{-1} which was about five orders of magnitude higher than that of the composites prepared by the conventional procedures. The resulting composites can be used in stretchable conductors, conductive seals, electromagnetic shielding and package materials. Interestingly, thickness dependent shielding was observed for MLG/ethylene vinyl acetate (EVA) composite (G-E composite) which was fabricated via wet casting method (Song et al. 2014). These G-E films contained 10, 30, 50 and 60 vol.% MLG loadings and their thickness were in the order of 40–60 μ. These films remained strong and flexible at 60 vol.% loading and showed the maximum conductivity of 2.5×10^2 S m^{-1}. These films were further used to form a wax/PVA/G-E film/PVA/wax sandwich structure for the measurement of EMI shielding. At 60 vol.% MLG loading, the range of δ was 290.72–335.70 μm and the overall EMI SE of G-E composite films were 15.19 –16.25 dB for 9–12 GHz frequency range, and a maximum absorption efficiency (70.6%) of the sandwich structure was achieved as well. Here, the reflection mechanism was the dominant mechanism and the contribution of absorption shielding increased with increase in the shielding thickness. Ultralightweight and compressible PU/graphene composite foams having an ultralow density (~0.027–0.030 g cm^{-3}) were prepared by solution dip-coating of graphene on commercial PU sponges (Shen et al. 2016). The developed foams exhibited an exceptional adjustable EMI shielding performance combined with an absorption-dominant shielding mechanism. These foams containing ~5 wt.% rGO loading showed the EMI SE of 12.4, 24.6, and 34.7 dB at thickness of ~2, ~4, and ~6 cm, respectively, in the 8–12 GHz frequency range. The SE was improved with further increase in the filler loading (up to ~10 wt.%), reached 19.9, 41.6 and 57.7 dB, respectively, due to the increased electrical conductivity. Although the average SE values were irreversibly reduced during the successive compression

TABLE 18.2

EMI Shielding Properties of Graphene-Based Elastomer Composites

Polymer Matrix	Filler	Filler Concentration (wt.% or vol.%)	Sample Thickness (mm)	σ(S cm^{-1})	EMI SE (dB)	Band Frequency (GHz)	Applications	Ref.
PDMS	Graphene	0.8 wt.%	~1	2	20	8–12	Aircraft, space-craft, automobiles and next-generation portable electronics	Chen et al. (2013)
PDMS	Graphene	3.6 vol.%	0.5	1.1	~25	1–3	EMI shielding in electric vehicles and the aviation industry, and in de-icing units	Ha et al. (2019)
PDMS	Graphene	0.42 wt.%	-	~0.005	~65	8.2–12.4	Civil military application	Gao et al. (2020)
PDMS	Graphene Network (GN)	1.2 wt.%	0.75	61	~90	8.2–12.4	High-performance EMI shielding applications	(Ao et al 2020)
EVA	MLG	60 vol.%	0.04–0.06	2.5	15.1–16.5	9.12	Lightweight EMI shielding coating materials	Song et al. (2014)
PU	GO	28.6 wt.%	10	-	35	8–12	EMI shielding product suited to military and civilian settings	Hu et al. (2019)
PA	Graphene	6 wt.%	2	~1.9	66	8.2–12.4	Fabrication of an ultra-efficient EMI shielding material	Li et al. (2016)

(Continued)

TABLE 18.2 (Continued)
EMI Shielding Properties of Graphene-Based Elastomer Composites

Polymer Matrix	Filler	Filler Concentration (wt.% or vol.%)	Sample Thickness (mm)	σ(S cm⁻¹)	EMI SE (dB)	Band Frequency (GHz)	Applications	Ref.
EMA	GO	5 wt.%	-	8.2×10^{-5}	30.6	8.2–12.4	Techno-commercial applications	Bhawal et al. (2018)
PDMS	rGO-Gnp	18.1 wt.%	2	>10	~86	8–11	EMI-shielding materials for applications in telecommunications aerospaces and portable electronics	Li et al. (2021)
PDMS	cMF-Au-G-IO	-	2	0.81.3	30.5	8.2–12.4	Promising engineering system for EMI shielding	Sun et al. (2018)
Silicone	Microwires graphene	0.059 wt.%	2	-	18	8.2–12.4	EMI shielding and microwave applications	Xu et al. (2020)
PDMS	GF/h Fe304	12 wt.%	2	84 ± 8.385	70.37	8.2–12.4	EMI shielding and thermal interface materials for chip cooling	Fang et al. (2020)
PDMS	CCA@rGO	3.05 wt.%	-	0.75	51	8.2–12.4	Lightweight, flexible electromagnetic shielding composites, portable and wearable electronic devices	Song et al. (2021)

cycles, their shielding performance remained constant and it could be adjusted due to the excellent compressibility and structural integrity of the 3D conductive GN. This technique could potentially promote the large-scale production of light-weight foam materials for EMI shielding.

At present, facile and scale development of ultrathin graphene EMI shielding material with special architecture still remains challenging. Chen and co-workers prepared a highly flexible graphene/PDMS foam composite (density: $< 0.06\,g\,cm^{-3}$) with seamlessly interconnected 3D network by using CVD technique (Chen et al. 2013). This composite showed a conductivity of 2 S cm^{-1} at $\sim 0.8\,wt.\%$ graphene loading. Its EMI SE was as high as 30 dB in the 30 MHz–1.5 GHz frequency range that could reach $\sim 500\,dB\,cm^{-3}g^{-1}$ and $\sim 20\,dB$. The conductivity and SE remained almost unchanged after repeatedly bending to a radius of $\sim 2.5\,mm$ for 10,000 times. Further, the SE was increased in stacked graphene/PDMS multiple foams (more than 33 dB) and the shielding mechanism was dominated by absorption rather than reflection in the X-band frequency range. In another study, nacre-mimetic graphene/PDMS composites with a very low graphene loading for high-performance EMI shielding were fabricated (Gao et al.2020). A unique bidirectional freezing technique was used to prepare nacre-mimetic 3D conductive GN with biaxial aligned lamellar structure along with an ultralow percolation threshold. The as-prepared biomimetic composites showed anisotropic conductivities, good mechanical properties and outstanding EMI SE at an extremely low graphene loading. Only at 0.42 wt.%, graphene loading an improved EMI SE of 65 dB was observed after annealing the graphene aerogels at 2,500°C, which was very close to the copper foil. Furthermore, in comparison with the metal foils and solid materials containing a high conductive filler loading, the specific SE of these low-density composites was even very high. Another composite film made of graphene-PDMS (G-PDMS) was fabricated by using a three-roll mill technique where 1.1 vol.% (low, i.e., LG), 2.3 vol.% (middle, i.e., MG) and 3.6 vol.% (high, i.e., HG) of graphene content were used, respectively (Ha et al. 2019). The uniform homogeneous dispersion of graphene in the PDMS matrix (Figure 18.9) resulted in an increase in the conductivity of the composites to 110 S m^{-1}. The EMI SE values of the LG-PDMS, MG-PDMS and HG-PDMS composites were 15, 21 and ~25 dB, respectively. Figure 18.10 exhibits the EMI shielding properties of this composite. Due to the high conductivity, the composites could be heated rapidly from room temperature up to 200°C in 50 seconds by electrical heating with low electric power to enable fast de-icing.

The potential of graphene sheets/Polyacrylate (GS/PA) composites for ultra-efficient EMI shielding was demonstrated in another study (Li et al. 2016). A sustainable and environmentally friendly solvent-free latex blending technology was used to prepare this composite for construction of a tunable architecture of GS in the polymer matrix. The maximum conductivity of ~190 S m^{-1} and a noteworthy low percolation threshold of ~0.11 mass percent were attributed to the highly intrinsic conductivity of GS having a high aspect ratio and the isolated architecture of GS in the PA matrix. The unexpectedly high complex permittivity evoked by the induction of a strong Maxwell–Wagner–Sillars (MWS) polarization at the conductive GS/non-conductive PA interfaces was reported here. The highest real polarization loss (ε') and imaginary electric loss (ε'') values of these composites containing 6 wt.% GS were in the ranges

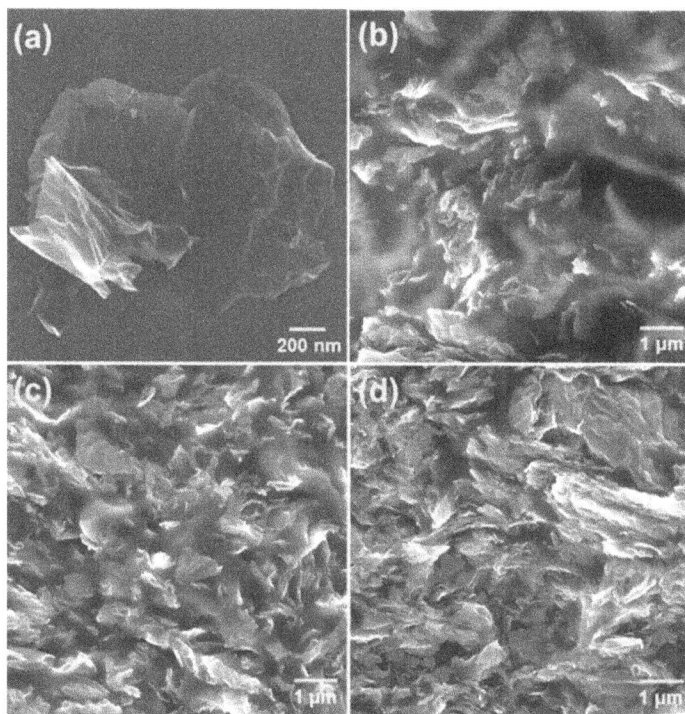

FIGURE 18.9 SEM photomicrograph of (a) graphene. Cross-section SEM view of (b) LG-PDMS, (c) MG-PDMS and (d) HG-PDMS. (Adapted from Ji-Hwan Ha et al., "Development of multi-functional graphene polymer composites having electromagnetic interference shielding and de-icing properties," *Polymers* 11 (2019): 2101–2110). (Open access article.)

of 21.4–26.1 and 17.9–18.3, respectively. The significant enhancement of ε' resulted from the improved aspect ratio along with the lattice defects on the GS. Alternatively, the improved conductivity of these composites was responsible for the enhancement of ε'' of these composites. This composite with 6 wt.% loading exhibited a high EMI SE of ~66 dB over the X-band frequency range which resulted from the pronounced conduction loss, dielectric relaxation and multi-scattering. Hu and co-workers synthesized a self-assembled graphene/PU sponge composite via a two-step hydrothermal reductive chemical reaction method where PU sponge worked as a robust scaffold for graphene to shape its 3D structure (Hu et al. 2019). This 3D conductive GN exhibited an absorption-dominant EMI shielding for PUG foams. This foam composite (density:0.11 mg cm^{-3}) showed an EMI SE of 35 dB in the X-band frequency region and a high specific EMI SE of 969–1,578 dB cm^{2}g^{-1}. The EMI SE of 23–26 and 18–23 dB of this composite were observed after 200 and 1,000 compression cycles, respectively. The superconductivity of GO and highly porous structure of the PU/GO sponge contributed to this excellent shielding performance of this composite. In another study, a highly conductive 3D GN was fabricated by CVD technique on a 3D nickel fiber network, followed by an etching process. Subsequently, a lightweight

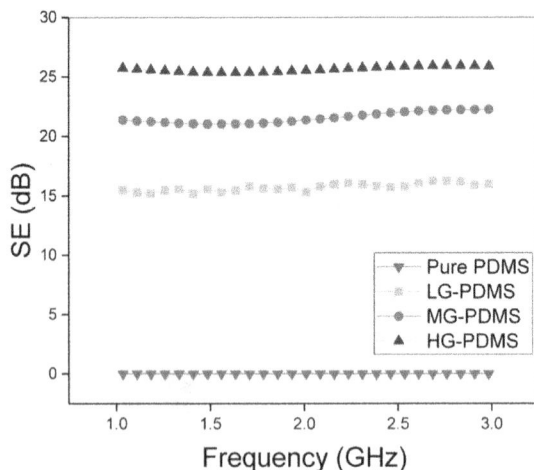

FIGURE 18.10 EMI SE of G-PDMS composites with various filler contents (1.1, 2.3, 3 vol.%, and pure PDMS) over the frequency range of 1.0–3.0 GHz. (Adapted from Ji-Hwan Ha et al., "Development of multi-functional graphene polymer composites having electromagnetic interference shielding and de-icing properties," *Polymers* 11 (2019): 2101–2110.) (Open access article.)

and flexible PDMS/GN composite (density: 0.8 g cm^{-3}) was synthesized by using this GN as a template via a vacuum infiltration method (Ao et al. 2020). The huge potentiality of this composite in the application of EMI shielding was evident from its superior conductivity of 6,100 S m^{-1} and an improved mechanical performance (256% increase in tensile strength in the PDMS/GN composite as compared to pristine PDMS at 1.2 wt.% graphene loading). This composite was highly stable and remained intact up to 100 stretching and release cycles as confirmed from mechanical properties, conductivity and EMI SE results (Figure 18.11). With this superior conductivity, superb EMI SE of about 40, 60 and 90 dB within the X-band frequency range with a thickness of 0.25, 0.5 and 0.75 mm, respectively, could be achieved by this composite (Figure 18.12).

Bhawal et al. (2018) prepared in situ rGO (IRGO) via in situ melt blending of ethylene methyl acrylate (EMA) and GO at 210°C in order to achieve an outstanding EMI SE with adjustable electro-mechanical properties of EMA in situ RGO (EIRGO) nanocomposites. At optimum filler loading (5 wt.%), IRGO were oriented selectively in a 3D conductive segregated network fashion into the matrix resulting in a high interfacial interaction between IRGO and EMA. As a result of this multifaceted segregated structure and dipolar interaction of nanofillers, a significant improvement in conductivity and SE as well as a constant enhancement in mechanical (27.71% and 30.0% improvement in tensile strength and 300% modulus) and thermal properties were observed. The conductivity was improved at a very lower percolation of 1.98% RGO than the conventional material. At 7 wt.% loading, the mechanical and thermal properties were reduced due to the re-aggregation of IRGO platelets but the conductivity and the SE were enhanced. All the composites exhibited an increased EMI

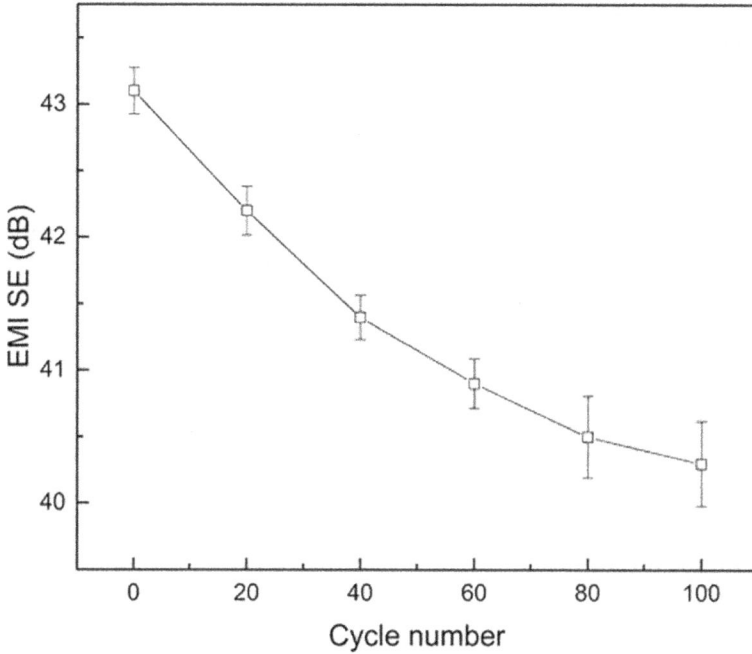

FIGURE 18.11 EMI SE of the 0.25 mm composite at 11.5 GHz after stretching and release cycles for 0–100 cycles. (Adapted from Dongyi Ao et al., "Highly conductive PDMS composite mechanically enhanced with 3D-graphene network for high-performance EMI shielding application," Nanomaterials 10 (2020): 768–779). (Open access article.)

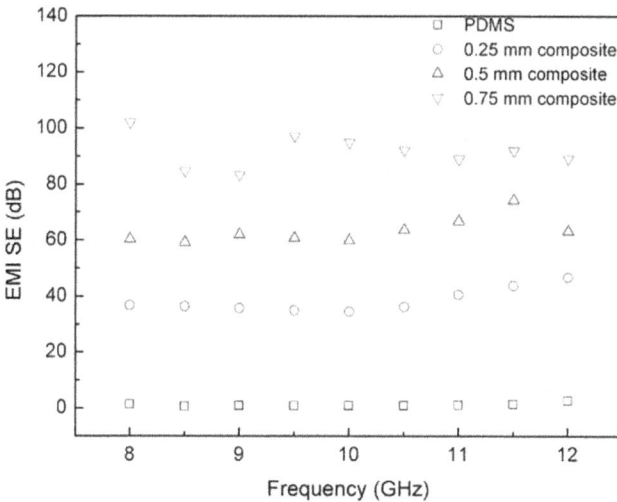

FIGURE 18.12 EMI SE of the PDMS/GN composite with different thickness in X-band. (Adapted from Dongyi Ao et al., "Highly conductive PDMS composite mechanically enhanced with 3D-graphene network for high-performance EMI shielding application," *Nanomaterials* 10 (2020): 768–779). (Open access article.)

SE which reached a maximum value of 34.9 dB at 7 wt.% loading, governed by an absorption mechanism. It was demonstrated that the composite had a most effective EMI shielding with optimized EM strength at 5 wt.% loading. Recently, the superiority of the bubble-templated 3D GN in EMI shielding was explored by Li et al. (2021). They prepared lightweight, thermally conductive and EMI shielding PDMS rubber composites filled with bubble-templated 3D GN (rGO-GnP foams) by using a novel and facile foaming method. Here, GO was used to support assembly of GnP into a 3D porous framework followed by a reduction process to produce rGO-GnP hydrogel which was further annealed at 1,500°C. This interconnected 3D conductive network and the closed pore structure with abundant internal wave-absorbing interfaces contributed to high conductivity and outstanding EMI shielding properties of this composite. The EMI-shielding performance increased with the increase in graphene loading from 9.7 to 18.1 wt.% and reached to a maximum EMI SE value of ~94 dB at ~4 mm thickness. Here, both absorption and reflection played a significant role on the overall shielding performance of this composite. This composite containing only 18.1 wt.% graphene content exhibited an EMI SE of about 90 dB and a thermal conductivity of >3 W mK^{-1}. The effect of graphene loading, absorber thickness and the frequency domain on the SE of acrylonitrile butadiene rubber/graphene nanocomposites prepared via conventional rubber roll milling technique was reported (Al-Ghamdi et al. 2016). The EMI SE of the developed nanocomposites varied from 22 to 48 and 38 to 77 dB at 1.0 and 4.0 wt.% graphene loading, respectively. The homogeneous nature of the conductive network of graphene was evident from the linearity nature of SE as a function of frequency. The addition of graphene into NBR matrix significantly improved the conductivity and the mechanical properties of the nanocomposites.

Recently, nanostructured elastomer composites filled with hybrid inorganic and carbon-based graphene and/or its derivatives afford a great prospect to fine tune permeability, thermal and electrical conductivity in order to obtain improved EMI performance along with outstanding physical-mechanical properties. Sun and co-workers made a 3D hierarchically porous structure in carbonized melamine foam (cMF) by tailored incorporation of Au nanoparticles (Au NPs), graphene, Fe$_3$O$_4$ (IO), and PDMS to construct a multifunctional composite (Sun et al. 2018). The resulting cMF-Au-G-IO/PDMS composite exhibited outstanding physical properties such as large specific surface area (708 m^{2}g^{-1}), moderate compressive strength (110 KPa) and low density (116 mg cm^{-3}). The relatively high conductivity of 81.3 S m^{-1} and super magnetization (Ms = 22.6 emu g^{-1}) endowed the porous cMF-Au-G-IO foam with excellent EMI shielding properties. The total EMI SE reached up to 30.5 dB in the X band with SSE of 263 dB cm^{3}g^{-1}. The EMI SE$_T$ reached to 30.5, 36.2, 40.6, 46.8 and 52.5 dB under the film thickness of 2, 4, 6, 8 and 10 mm, respectively. The superior EMI shielding performance could be attributed to the synergistic effect of Fe$_3$O$_4$/ graphene hybrids imposed on hierarchical structure of cMF-Au-G-IO. A multifunctional, hybrid shielding system consisting of microwire (M)/graphene(G)/silicone rubber was fabricated in another study (Xu et al. 2020). The soft magnetic microwires were used in order to get a better impedance matching confirming most of the EM wave entering into the shielding material without being secondarily reflected. In this nanocomposite, an improvement in absorption efficiency, impedance matching and polarization was prompted by diverse arrangements and distribution of both the integrating fillers which resulted in an outstanding SE as compared to the individual filler. The maximum SE reached to 18 dB (98.4% attenuation) for both periodic and random configurations with 0.059 wt.% filler loading at the frequency of 11.5 GHz. Here, the optimized arrangement permitted effective wave attenuation by the dielectric and magnetic properties of the microwires. Among the randomly dispersed arrays, the highest EMI SE value reached up to 6 dB for equal amount of G and M due to polarization effects at the interface between the two regions. Furthermore, the SE could be further increased by tailoring topological and structural factors of both fillers. Fang et al. reported a novel method to in situ grown hollow Fe_3O_4 sphere onto 3D graphene foam (GF) surface and then filled it with PDMS to prepare hybrid nanocomposites (Fang et al. 2020). It exhibited a high thermal conductivity (28.12 ± 1.212 W m^{-1} K^{-1}) at room temperature and an EMI SE of 70.37 dB in the X band due to the highly conductive interconnected GF networks and intensive interaction between h-Fe_3O_4 and graphene. Song et al. wrapped rGO on the surface of cellulose carbon aerogel (CCA) to form CCA@rGO having a 3D double-layer conductive network skin-core structure followed by prepared CCA@rGO/PDMS EMI shielding composites via backfilling with PDMS (Figure 18.13) (Song et al. 2021). This composite exhibited an outstanding EMI SE of 51 dB at 3.05 wt.% loading of CCA@rGO. At the same loading, it showed excellent thermal stability (T_{HRI} of 178.3°C), good thermal conductivity coefficient of 0.65 W m^{-1} K^{-1} and excellent mechanical properties. The variation of EMI SE of CCA@rGO/PDMS composite with the loading of CCA@rGO is shown in Figure 18.14. Zhan et al. (2018) fabricated flexible NR/magnetic $Fe_3O_4@$ rGO (NRMG) composites with segregated structure via a self-assembly method in latex where Fe_3O_4 enhanced the EMI SE of this composite. This composite showed

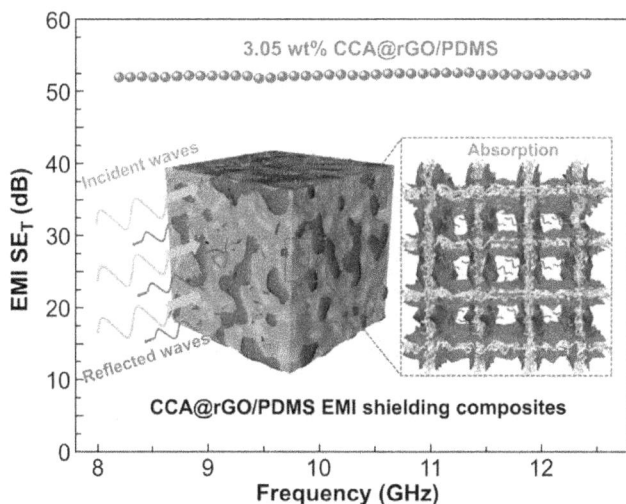

FIGURE 18.13 CCA@rGO/PDMS EMI shielding composites. (Adapted from Ping Song et al., "Lightweight, flexible cellulose-derived carbon aerogel@reduced graphene oxide/ PDMS composites with outstanding EMI shielding performances and excellent thermal conductivities," *Nano-Micro Letters* 13 (2021): 91–108). (Open access article.)

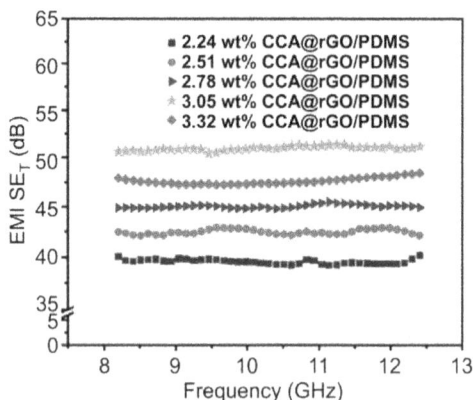

FIGURE 18.14 EMI SE_T of the CCA@rGO/PDMS shielding composites. (Adapted from Ping Song et al., "Lightweight, flexible cellulose-derived carbon aerogel@reduced graphene oxide/PDMS composites with outstanding EMI shielding performances and excellent thermal conductivities," *Nano-Micro Letters* 13 (2021): 91–108). (Open access article.)

an EMI SE value of 42.4 dB in the X band and the specific EMI SE value of 26.4 dB mm^{-1} which were attributed to the excellent magnetic property, perfect segregated network together with good electrical conductivity. This value was reduced to only 3.5% even after 2,000 bending-release cycles.

Sometimes, the shield is needed to be nonmagnetic in order to avoid interference of magnetic materials and to guard sophisticated precise electronic instruments as well. Pradhan et al. prepared GnP and α-MnO$_2$ decorated GnP filled EVA-45 composites by the dual mixing technique, i.e., solution followed by melt mixing where MnO$_2$ further improved the shielding properties of pure graphene (Pradhan et al. 2020). The mechanical properties, thermal stability, permeability, conductivity and EMI shielding performance were observed to be improved with increasing GnP loading. The EMI SE value of this composite was 22 dB at 8 phr α-MnO$_2$ loading which reached to 28 dB at 15 phr loading in the X band.

18.9 CONCLUSIONS AND OUTLOOKS

Nowadays, research on EMI shielding materials has attracted pivotal importance in the era of modern electronic information technology and revolution. The recent progress in the multifunctional graphene-based elastomer nanocomposites has led to substantial improvements in the properties of EMI shielding materials where a synergistic combination of properties of individual material can be achieved while eliminating all the disadvantages of typically used metallic shields. This chapter highlights the journey of conductive graphene-based elastomer nanocomposites designed for EMI shielding applications till date. Besides the applications, a brief overview of the EMI shielding phenomenon and the PNCs, a comprehensive description of the developing role of graphene as microwave shield and the synthesis methods of graphene-elastomer nanocomposites have also been discussed. It provides a detailed insight into graphene-based nanomaterials and their elastomeric composites. As compared to all other organic and inorganic fillers, the unique 2D structure, ultrahigh surface area, high aspect ratio and exceptional electronic and thermal conductivities of these reinforcement graphene-based materials can improve the electrical, piezoresistive, dielectric, thermal, mechanical and gas barrier properties of the elastomers even at extremely low loadings. Here, graphene is used in the form of ultrathin flexible films, papers, laminates, microcellular foams, sheets, etc. These enhanced properties along with lightweight, ease of processability, flexibility, good resistance to corrosion, super electrical and thermal conductivity, microwave absorption, efficient reinforcement, and economic feasibility have made these graphene-elastomer nanocomposites both economically and functionally promising for specific high-performance EMI shielding applications. A huge variety of graphene materials, types of elastomer, and processing conditions have a significant impact on the EMI shielding characteristics and mechanisms of resulting nanocomposites. By taking the advantages of improved EMI SE, low microwave reflectivity, low filler loading, high degree of design freedom and multifunctionality, these proposed composites can be considered as exciting replacements for the existing nanomaterials for a variety of applications.

However, although graphene and its derivatives can assimilate multiple functions into one system, their full potential in this field has yet to be known because of some critical issues and limitations. Nowadays, it has become a great challenge to the researchers to develop graphene-based elastomer nanocomposites as EMI shielding materials with lightweight, broadband, high efficiency, as well as stability at elevated temperature and broad frequency ranges as well. Future research should focus on the fabrication and optimization of these proposed materials with various architectures where permittivity, permeability, conductivity, frequency and thickness are suitably correlated to obtain a higher SE. From the prospect of environmental protection, special consideration should be given on green application which requires a strong absorption toward penetrating EM wave and weak secondary reflection from surface. It appears that significant efforts should be expanded to explore the diverse aspects of graphene aerogel as an EMI shielding material to control the EM pollution which is still very limited in EMI shielding industries. More tremendous efforts are required to realize the potential of utilization of defect-free pristine graphene as reinforcing fillers in polymer composites which is still in the infancy to date. So far, how to manage a suitable exfoliation and inclusion of graphene in a rubber matrix still remains a difficult challenge and needs to be further explored thoroughly. Therefore, it is also necessary to investigate controlled synthesis of graphene-elastomer composites, aggregation effects and interparticle interactions of these conductive fillers. More productive investigations are required to develop new manufacturing procedures that can tailor distribution and spatial orientation of 2D graphene sheets and fine-tune their properties for EMI shielding applications as well. Considerable work is needed to fabricate PNCs shielding materials with hybrid conductive fillers based on suitable combination of graphene and/or its derivatives, metal and dielectric/magnetic nanoparticles in a proper content and ratio. This strategy may be a future prospect for applications in the broadband frequency spectrum. Therefore, to meet the above-mentioned challenges, more scientific investigations need to be undertaken urgently for a systematic and long-term study on these proposed nanocomposites. It will develop more efficient EMI shielding materials that will satisfy all the techno-commercial specifications and will also check on the commercial viability. It is of paramount importance to implement such promising nanocomposites in many other wide ranges of multifunctional scientific disciplines including automotive, next-generation flexible and portable electronics, telecommunications, packaging, aerospace, environment, materials science, biomedical and shape memory applications. In the light of current scenario, these graphene-elastomer nanocomposites will have a great contribution in the EMI shielding in near future that are not achievable with the existing EMI shields. Finally, this chapter is expected to open up new avenues for graphene-elastomer based materials to be applied in wider fields of research where the aforementioned current challenges and perspectives for the future development will help to give valuable insights into new research directions.

REFERENCES

Afilipoaei, C. and H. Teodorescu-Draghicescu 2020. A review over electromagnetic shielding effectiveness of composite materials. *Proceedings* 63, no. 1: 23–31.

Al-Ghamdi, A.A., A.A. Al-Ghamdi, Y. Al-Turki, et al. 2016. Electromagnetic shielding properties of graphene/acrylonitrile butadiene rubber nanocomposites for portable and flexible electronic devices, *Composites Part B: Engineering* 88: 212–219.

Al-Saleh, M. H. and U. Sundararaj. 2009. Electromagnetic interference shielding mechanisms of CNT/polymer composites. *Carbon* 47: 1738–1746.

Al-Solamy, F.R., A.A. Al-Ghamdi, W.E. Mahmoud. 2012. Piezoresistive behavior of graphite nanoplatelets based rubber nanocomposites. *Polymers for Advanced Technologies* 23: 478–482.

Andersona, L., P. Govindaraj, A. Anga, et al. 2021. Modelling, fabrication and characterization of graphene/polymer nanocomposites for electromagnetic interference shielding applications. *Carbon Trends* 4: 100047–100066.

Ao, D., Y. Tang, X. Xu, et al. 2020. Highly conductive PDMS composite mechanically enhanced with 3D-graphene network for high-performance EMI shielding application. *Nanomaterials* 10, no. 4: 768–779.

Apollo, N.V., J. Jiang, W. Cheung, et al. 2018. Development and characterization of a sucrose microneedle neural electrode delivery system. *Advanced Biosystems* 2, no. 2: 1700187.

Apollo, N.V., M.I. Maturana, W. Tong, et al. 2015. Soft, flexible freestanding neural stimulation and recording electrodes fabricated from reduced graphene oxide. *Advanced Functional Materials* 25, no. 23: 3551–3559.

Ashokkumar, M, N.T. Narayanan, B.K. Gupta, et al. 2013. Conversion of industrial bio-waste into useful nanomaterials. *ACS Sustainable Chemistry & Engineering* 1, no. 6: 619–626.

Balanis C.A. 2012. *Advanced Engineering Electromagnetics*. Hoboken, NJ: John Wiley & Sons.

Barani, Z., F. Kargar, K. Godziszewski, et al. 2020. Graphene epoxy-based composites as efficient electromagnetic absorbers in the extremely high frequency band. *ACS Applied Materials & Interfaces* 12: 28635–62844.

Behabtu, N., J.R. Lomeda, M.J. Green, et al. 2010. Spontaneous high-concentration dispersions and liquid crystals of graphene. *Nature Nanotechnology* 5: 406–411.

Bhawal, P., S. Ganguly, T.K. Das, et al. 2018. Superior electromagnetic interference shielding effectiveness and electro-mechanical properties of EMA-IRGO nanocomposites through the in-situ reduction of GO from melt blended EMA-GO composites. *Composites Part B: Engineering* 134: 46–60.

Biter, W., P. Jamnicky, W. Coburn. 1994. Shielding improvement by use of thin multilayer films. *International Sampe Electronics Conference, Society for the Advancement of Material and Process Engineering* 234–234.

Bourlinos, A. B., V. Georgakilas, R. Zboril, et al. 2009. Liquid-phase exfoliation of graphite towards solubilized graphenes. *Small* 5, no. 16: 1841–1845.

Bryning, M.B., M.F. Islam, J.M. Kikkawa, A.G. Yodh. 2005. Very low conductivity threshold in bulk isotropic single-walled carbon nanotube-epoxy composites. *Advanced Materials* 17, no. 9: 1186–1191.

Cao, M.S., X.X. Wang, W.Q. Cao, J. Yuan. 2015. Ultrathin graphene: Electrical properties and highly efficient electromagnetic interference shielding. *Journal of Materials Chemistry C* 3: 6589–6599.

Castro Neto, A H., F. Guinea, N.M. R. Peres, et al. 2009 The electronic properties of graphene. *Reviews of Modern Physics* 81: 109–164.

Chen, C.C., W.F. Liang, Y.H. Nien, et al. 2017. Microwave absorbing properties of flake-shaped carbonyl iron/reduced graphene oxide/epoxy composites. *Materials Research Bulletin* 96: 81–85.

Chen, Z., C. Xu, C. Ma, et al. 2013. Lightweight and flexible graphene foam composites for high- performance electromagnetic interference shielding. *Advanced Materials* 25, no. 9: 1296–1300.

Choi, W. and J.W. Lee. 2012. *Graphene: Synthesis and Applications*. Boca Raton, FL: CRC Press, Taylor & Francis Group.

Choucair, M., P. Thordarson, J.A. Stride. 2008. Gram-scale production of graphene based on solvothermal synthesis and sonication. *Nature Nanotechnology* 4: 30–33.

Chung, D.D.L. 2000. Materials for electromagnetic interference shielding. *Journal of Materials Engineering and Performance* 9, no. 3: 350–354.

Chung, D.D.L. 2001. Electromagnetic interference shielding effectiveness of carbon materials. *Carbon* 39, no. 2: 279–285.

Colaneri, N. F. and L. Schacklette. 1992. EMI shielding measurements of conductive polymer blends. *IEEE Transactions on Instrumentation and Measurement* 41, no. 2: 291–97.

Compton, O. C. and S. T. Nguyen. 2010. Graphene oxide, highly reduced graphene oxide, and graphene: Versatile building blocks for carbon-based materials. *Small* 6, no. 6: 711–723.

Das, N. C., T. K. Chaki, D. Khastgir, A. Chakraborty. 2001. Electromagnetic interference shielding effectiveness of ethylene vinyl acetate based conductive composites containing carbon fillers. *Journal of Applied Polymer Science* 80: 1601–1608.

Das, P., A. B. Deoghareb, S.R. Maity. 2020. Exploring the potential of graphene as an EMI shielding material– an overview. *Materials Today: Proceedings* 22: 1737–1744.

Dato, Albert and F. Michael. 2010. Substrate-free microwave synthesis of graphene: Experimental conditions and hydrocarbon precursors. *New Journal of Physics* 12: 125013–125037.

Deruelle, F. 2020. The different sources of electromagnetic fields: Dangers are not limited to physical health. *Electromagnetic Biology and Medicine* 39, no. 2: 166–175.

Dhakate, S.R., K.M. Subhedar, B.P. Singh. 2015. Polymer nanocomposite foam filled with carbon nanomaterials as an efficient electromagnetic interference shielding material. *RSC Advances* 5: 43036–43057.

Dimiev, A.M., A. Khannanov, I. Vakhitov, et al. 2018. Revisiting the mechanism of oxidative unzipping of multiwall carbon nanotubes to graphene nanoribbons. *ACS Nano* 12: 3985–3993.

Edwards, R.S. and K.S. Coleman. 2013. Graphene synthesis: Relationship to applications. *Nanoscale* 5: 38–51.

Emiru, T.F. and D.W. Ayele. 2017. Controlled synthesis, characterization and reduction of graphene oxide: A convenient method for large scale production. *Egyptian Journal of Basic and Applied Sciences* 4: 74–79.

Fang, H., H. Guo, Y. Hu, et al. 2020. In-situ grown hollow Fe_3O_4 onto graphene foam nanocomposites with high EMI shielding effectiveness and thermal conductivity. *Composites Science and Technology* 188: 107975.

Ganguly, S., P. Bhawal, R. Ravindren, N. C. Das. 2018. Polymer nanocomposites for electromagnetic interference shielding: A review. *Journal of Nanoscience and Nanotechnology* 18: 7641–7669.

Gao, W., N. Zhao, T. Yu, et al. 2020. High-efficiency electromagnetic interference shielding realized in nacre-mimetic graphene/polymer composite with extremely low graphene loading. *Carbon* 157: 570–577.

Geetha, S., K. K. Satheesh Kumar, C. R. K. Rao, et al. 2009. EMI shielding: Methods and materials – A review. *Journal of Applied Polymer Science* 112: 2073–2086.

Geimand, A.K. and K.S. Novoselov. 2007. The rise of graphene. *Nature Materials* 6, no. 3: 183–191.

Guler, O., S.H. Guler, V. Selen, et al. 2016. Production of graphene layer by liquid-phase exfoliation with low sonication power and sonication time from synthesized expanded graphite. *Fullerenes, Nanotubes and Carbon Nanostructures* 24: 123–7.

Ha, J.H., S.K. Hong, J.K. Ryu, et al. 2019. Development of multi-functional graphene polymer composites having electromagnetic interference shielding and de-icing properties. *Polymers* 11: 2101–2110.

Hamed, G.R. 2000. Reinforcement of rubber. *Rubber Chemistry and Technology* 73: 524–533.

Han, G., Z. Ma, B. Zhou, et al. 2021. Cellulose based Ni-decorated graphene magnetic film for electromagnetic interference shielding. *Journal of Colloid and Interface Science* 583: 571–578.

Hu, Z., X. Ji, B. Li, Y. Luo. 2019. A self-assembled graphene/ polyurethane sponge for excellent electromagnetic interference shielding performance. *RSC Advances* 9, no. 44: 25829–25835.

Jagatheesan, K., A. Ramasamy, A. Das, A. Basu. 2014. Electromagnetic shielding behaviour of conductive filler composites and conductive fabrics–A review. *Indian Journal of Fibre & Textile Research* 39, no. 3: 329–342.

Jagatheesan, K., A. Ramasamy, A. Das, A. Basu. 2015. Fabrics and their composites for electromagnetic shielding applications. *Textile Progress* 47: 87–161.

Jan, R., A. Habib, M.A. Akram, et al. 2017. Flexible, thin flms of graphene—polymer composites for EMI shielding. *Materials Research Express* 4, no. 3: 35605–35614.

Jan, R., M. B. Khan, Z. M. Khan. 2012. Synthesis and electrical characterization of 'carbon particles reinforced epoxy-nanocomposite' in Ku-band. *Materials Letters* 70: 155–159.

Jaroszewski, M., S. Thomas, A. V. Rane. 2019. *Advanced Materials for Electromagnetic Shielding. Fundamentals, Properties and Applications.* Hoboken, NJ: John Wiley & Sons.

Jia, Y., T.D. Ajayi, B.H. Wahls, et al. 2020. Multifunctional ceramic composite system for simultaneous thermal protection and electromagnetic interference shielding for carbon fiber-reinforced polymer composites. *ACS Applied Materials & Interfaces* 12, no. 52: 58005–58017.

Jiang, X., D.-X. Yan, Y. Bao, et al. 2015. Facile, green and affordable strategy for structuring natural graphite/polymer composite with efficient electromagnetic interference shielding. *RSC Advances* 5, no. 29: 22587–22592.

Joshi, A. and S. Datar. 2015. Carbon nanostructure composite for electromagnetic interference shielding. *Pramana* 84: 1099–1116.

Kawamura, Y., T. Hikage, T. Nojima, et al. 2012. Experimental estimation of EMI from electronic article surveillance on implantable cardiac pacemakers and implantable cardioverter defibrillators: Interference distance and clinical estimation. *Transactions of Japanese Society for Medical and Biological Engineering* 50: 289–298.

Kazemizadeh, F. and R. Malekfar. 2018. One step synthesis of porous graphene by laser ablation: A new and facile approach. *Physica B: Condensed Matter* 530: 236–241.

Khan, R., Z. M. Khan, H. BinAqeel, et al. 2020. 2D nanosheets and composites for EMI shielding analysis. *Scientifc Reports* 10: 21550–21557.

Kim, H., Y. Miura, C.W. Macosko. 2010. Graphene/polyurethane nanocomposites for improved gas barrier and electrical conductivity. *Chemistry of Materials* 22: 3441–3450.

Kim. S., J.S. Oh, M.G. Kim, et al. 2014. Electromagnetic interference (EMI) transparent shielding of reduced graphene oxide (RGO) interleaved structure fabricated by electrophoretic deposition. *ACS Applied Materials and Interfaces* 6, no. 20: 17647–17653.

Kim, J.S., J.H. Yun, I. Kim, S.E. Shim. 2011. Electrical properties of graphene/SBR nanocomposite prepared by latex heterocoagulation process at room temperature. *Journal of Industrial and Engineering Chemistry* 17: 325–330.

Kostoff, R. N., P. Heroux, M. Aschner, A. Tsatsakis. 2020. Adverse Health Effects of 5G Mobile Networking Technology under Real-Life Conditions. *Toxicology Letters* 323: 35–40.

Kruzelak, J., A. Kvasnicakova, K. Hlozekova, I. Hudec. 2021. Progress in polymers and polymer composites used as efficient materials for EMI shielding. *Nanoscale Advances* 3: 123–172.

Kunkel, G. M. 2020. *Penetration of Electromagnetic Wave through Shielding Barrier, Shielding of Electromagnetic Waves*. 13–18. Cham: Springer.

Lee, X.J., H. Billie Yan Zhang, L. Kar Chiew, et al. 2019. Review on graphene and its derivatives: Synthesis methods and potential industrial implementation. *Journal of the Taiwan Institute of Chemical Engineers* 98: 163–180.

Lee, C., X. Wei, J.W. Kysar, J. Hone. 2008. Measurement of the elastic properties and intrinsic strength of monolayer graphene. *Science* 321: 385–388.

Li, Y. and W. Zhai. 2015. Graphene nanocomposites for electromagnetic induction shielding. In *Graphene-Based Polymer Nanocomposites in Electronics*, eds. K.K. Sadasivuni, D. Ponnamma, J. Kim, and S. Thomas, 345–372. Switzerland: Springer International Publishing.

Li, J., X. Zhao, W. Wu, et al. 2021. Bubble-templated rGO-graphene nanoplatelet foams encapsulated in silicon rubber for electromagnetic interference shielding and high thermal conductivity. *Chemical Engineering Journal* 415: 129054.

Li, Y., S. Zhang, Y. Ni. 2016. Graphene sheets stacked polyacrylate latex composites for ultra-efficient electromagnetic shielding. *Materials Research Express* 3, no. 7: 075012–075024.

Lian, H., S. Li, K. Liu, et al. 2011. Study on modified graphene/butyl rubber nanocomposites. I. Preparation and characterization. *Polymer Engineering & Science* 51: 2254–2260.

Lin, J.-H., Z.-I. Lin, Y.-J. Pan, et al. 2016. Polymer composites made of multi-walled carbon nanotubes and graphene nano-sheets: Effects of sandwich structures on their electromagnetic interference shielding effectiveness. *Composites Part B: Engineering* 89: 424–431.

Madurani, K.A., S. Suprapto, N.I. Machrita, et al. 2020. Progress in graphene synthesis and its application: History, challenge and the future outlook for research and industry. *ECS Journal of Solid State Science and Technology* 9: 093013–093125.

Marcano, D.C., D.V. Kosynkin, J.M. Berlin, et al. 2010. Improved synthesis of graphene oxide. *ACS Nano* 4, no. 8: 4806–4814.

Modak, P. and D.V. Nandanwar. 2015. A review on graphene and its derivatives based polymer nanocomposites for electromagnetic interference shielding. *International Journal of Advances in Science Engineering and Technology* 1: 212–214.

Mu, Q. and S. Feng. 2007. Thermal conductivity of graphite/silicone rubber prepared by solution intercalation. *Thermochimica Acta* 462: 70–75.

Nirmalraj, P. N., T. Lutz, S. Kumar, et al. 2011. Nanoscale mapping of electrical resistivity and connectivity in graphene strips and networks. *Nano Letters* 11, no. 1: 16–22.

Novoselov, K.S., A.K. Geim, S.V. Morozov, et al. 2004. Electric field effect in atomically thin carbon films. *Science* 306: 666–669.

Ott, H.W. 1988. *Noise Reduction Techniques in Electronic Systems*, 426–427. New York: John Wiley & Sons.

Ouyang, X., W. Huang, E. Cabrera, et al. 2015. Graphene-graphene oxide-graphene hybrid nanopapers with superior mechanical, gas barrier and electrical properties. *AIP Advances* 5, no. 1: (2015) 017135–017143.

Park, S. and R.S. Ruoff. 2009. Chemical methods for the production of graphenes. *Nature Nanotechnology* 4: 217–224.

Parvez, K., R. Li, S. R. Puniredd, et al. 2013. Electrochemically exfoliated graphene as solution-processable, highly conductive electrodes for organic electronics. *ACS Nano* 7, no. 4: 3598–3606.

Paul, C.R. 2004. *Electromagnetics for Engineers: With Applications to Digital Systems and Electromagnetic Interference*. Hoboken, NJ: John Wiley & Sons.

Potts, J. R., D. R. Dreyer, C. W. Bielawski, R. S. Ruoff. 2011. Graphene-based polymer nanocomposites. *Polymer* 52, no. 1: 5–25.

Powell, R. L. and G. E. Childs. 1972. *American Institute of Physics Handbook*, 3rd Edition, 142–160. New York: McGraw-Hill.

Pradhan, S., D. Goswami, D. Ganguly, et al. 2020. Graphene and MnO_2 decorated graphene filled composites for electromagnetic shielding applications having excellent dielectric properties. *Polymer Testing* 90: 106716.

Raagulan, K., B.M Kim, K.Y. Chai. 2020. Recent advancement of electromagnetic interference (EMI) shielding of two dimensional (2D) MXene and graphene aerogel composites. *Nanomaterials* 10, no.4: 702–724.

Rao, B.V.B., P. Yadav, R. Aepuru, et al. 2015. Single-layer graphene-assembled 3D porous carbon composite with PVA and Fe_3O_4 nano-fillers: An interface-mediated superior dielectric and EMI shielding performance. *Physical Chemistry Chemical Physics* 17: 18353–18363.

Saini, P., and M. Arora. 2012. Microwave absorption and EMI shielding behavior of nanocomposites based on intrinsically conducting polymers, graphene and carbon nanotubes. In *New Polymers for Special Applications*, ed. A.D.S. Gomes, 71–112. IntechOpen. doi:10.5772/48779.

Saini, P., V. Choudhary, B.P. Singh, et al. 2009. Polyaniline-MWCNT nanocomposites for microwave absorption and EMI shielding. *Materials Chemistry and Physics* 113: 919–926.

Saini, P., V. Choudhary, B.P., Singh, et al. 2011. Enhanced microwave absorption behavior of polyaniline-CNT/polyatyrene blend in 12.4–18.0GHz range. *Synthetic Metals* 161: 1522–1526.

Saini, P., V. Choudhary, K. Sood, S. Dhawan. 2009. Electromagnetic interference shielding behavior of polyaniline/graphite composites prepared by in situ emulsion pathway. *Journal of Applied Polymer Science* 113, no. 5: 3146–3155.

Salavagione, H.J., G. Martinez, G. Ellis. 2011. Recent advances in the covalent modification of graphene with polymers. *Macromolecular Rapid Communications* 32: 1771–1789.

Sankaran, S., K. Deshmukh, M. Basheer Ahamed, S.K. Khadheer Pasha. 2018. Recent advances in electromagnetic interference shielding properties of metal and carbon filler reinforced flexible polymer composites: A review. *Composites: Part A Applied Science and Manufacturing* 114: 49–71.

Santhosi, B., K. Ramji, N.M. Rao. 2020. Design and development of polymeric nanocomposite reinforced with graphene for effective EMI shielding in X-band. *Physica B: Condensed Matter* 586: 412144–412153.

Schelkunoff, S.A. 1943. *Electromagnetic Waves*. Princeton, NJ: Van Nostrand.

Schulz, R. B., V. C. Plantz, D.R. Brush. 1988. Shielding theory and practice. *IEEE Transactions on Electromagnetic Compatibility* 30, no. 3: 187–201.

Schüz, J., P. Elliott, A. Auvinen, et al. 2011. An international prospective cohort study of mobile phone users and health (Cosmos): Design considerations and enrolment. *Cancer Epidemiology* 35: 37–43.

Shen, B., Y. Li, D. Yi, et al. 2017. Strong flexible polymer/graphene composite films with 3D saw-tooth folding for enhanced and tunable electromagnetic shielding. *Carbon* 113: 55–62.

Shen B., Y. Li, W. Zhai, W. Zheng. 2016. Compressible graphene-coated polymer foams with ultralow density for adjustable electromagnetic interference (EMI) shielding. *ACS Applied Materials & Interfaces* 8, no. 12: 8050–8057.

Shen, B., W. Zhai, C. Chen, et al. 2011. Melt blending in situ enhances the interaction between polystyrene and graphene through p–p stacking. *ACS Applied Materials & Interfaces* 3: 3103–3109.

Shen, B., W. Zhai, M. Tao, et al. 2013. Lightweight multifunctional polyetherimide/graphene@Fe_3O_4 composite foams for shielding of electromagnetic pollution. *ACS Applied Materials & Interfaces* 5: 11383–11391.

Singh, K., A. Ohlan, S.K. Dhawan. 2012. Polymer-graphene nanocomposites: Preparation, characterization, properties, and applications. In *Nanocomposites-New Trends and Developments*, ed. F. Ebrahimi, 38–71. IntechOpen. doi:10.5772/50408.

Singh, A. K., A. Shishkin, T. Koppel, N. Gupta. 2020. Porous materials for EMI shielding. In *Materials for Potential EMI Shielding Applications*, eds. J. Kuruvilla, W. Runcy, G. Gejo, 287–314. Cambridge, MA: Elsevier Inc.

Song, P., B. Liu, C. Liang, et al. 2021. Lightweight, flexible cellulose-derived carbon aerogel@ reduced graphene oxide/PDMS composites with outstanding EMI shielding performances and excellent thermal conductivities. *Nano-Micro Letters* 13: 91–108.

Song, W.L., M.S. Cao, M.M. Lu, et al. 2014. Flexible graphene/polymer composite films in sandwich structures for effective electromagnetic interference shielding. *Carbon* 66: 67–76.

Sprinkle, M., P. Soukiassian, W. A. De Heer, et al. 2009. Epitaxial graphene: The material for graphene electronics. *Physica Status Solidi (RRL)- Rapid Research Letters* 3, no. 6: A91–A94.

Sudha, J. D., S. Sivakala, K. Patel, P.R. Nair. 2010. Development of electromagnetic shielding materials from the conductive blends of polystyrene polyaniline-clay nanocomposite. *Composites Part A: Applied Science and Manufacturing* 41, no. 11: 1647–1652.

Sun, Y., S. Luo, H. Sun, et al. 2018. Engineering closed-cell structure in lightweight and flexible carbon foam composite for high-efficient electromagnetic interference shielding. *Carbon* 136: 299–308.

Szabó, T, O. Berkesi, I. Dékány. 2005. DRIFT study of deuterium-exchanged graphite oxide. *Carbon* 43, no. 15: 3186–3189.

Tarcan, R. O., I. Todor-Boer, C. Petrovai, et al. 2020. Reduced graphene oxide today. *Journal of Materials Chemistry C* 8, no. 4: 1198–1224.

Thomassin, J.-M., C. Jérome, T. Pardoen, et al. 2013. Polymer/carbon based composites as electromagnetic interference (EMI) shielding materials. *Materials Science and Engineering R: Reports* 74: 211–232.

Tour, J.M. 2014.Top-down versus bottom-up fabrication of graphene-based electronics. *Chemistry of Materials* 26, no. 1: 163–171.

Varadan, V.K. and V.V. Varadan. 1990. Chiral polymer coatings for microwave applications. *Proceedings: Electro-Optical Materials for Switches, Coatings, Sensor Optics and Detectors* 1307: 122–129.

Wagner, D.H. and R.A. Vaia. 2004. Nanocomposites: Issues at the Interface. *Materials Today* 7, 38–42.

Wanasinghe, D., F. Aslani, G. Ma, D. Habibi. 2020. Review of polymer composites with diverse nanofillers for electromagnetic interference shielding. *Nanomaterials* 10, no. 3: 541–587.

Warner, J.H., F. Schaffel, A. Bachmatiuk, M.H. Rummeli. 2012. *Graphene: Fundamentals and Emergent Applications.* Oxford: Elsevier.

Wu, J., J. Chen, Y. Zhao, et al. 2016. Effect of electrophoretic condition on the electromagnetic interference shielding performance of reduced graphene oxide carbon fiber/epoxy resin composites. *Composites Part B: Engineering* 105: 167–171.

Wu, X., Y. Liu, H. Yang, Z. Shi. 2016. Large scale synthesis of high-quality graphene sheets by an improved alternating current arc-discharge method. *RSC Advances* 6: 93119–93124.

Xia, X., A.D. Mazzeo, Z. Zhong, G.J. Weng. 2017. An X-band theory of electromagnetic interference shielding for graphene-polymer nanocomposites. *Journal of Applied Physics* 122, no. 2: 025104–025115.

Xu, Y.L., A. Uddin, D. Estevez, et al. 2020. Lightweight microwire/graphene/silicone rubber composites for efficient electromagnetic interference shielding and low microwave reflectivity *Composites Science and Technology* 189 (2020): 108022–108031.

Yan, D.-X., H. Pang, B. Li, et al. 2015. Structured reduced graphene oxide/polymer composites for ultra-efficient electromagnetic interference shielding. *Advanced Functional Materials* 25, no. 4: 559–566.

Yang, Y., R. Liu, J. Wu, et al. 2015. Bottom -up fabrication of graphene on silicon/silica substarte via a facile soft-hard template approach. *Scientific Reports* 5: 13480–13487.

Yousef, N., X. Sun, X. Lin, et al. 2014. Highly aligned graphene/polymer nanocomposites with excellent dielectric properties for high-performance electromagnetic interference shielding. *Advanced Materials* 26, no. 31: 5480–5487.

Zeng, Z., M. Chen, H. Jin, et al. 2016. Thin and flexible multi-walled carbon nanotube/waterborne polyurethane composites with high-performance electromagnetic interference shielding. *Carbon* 96: 768–777.

Zhan, Y., M. Lavorgna, G. Buonocoreb, H. Xia. 2012. Enhancing electrical conductivity of rubber composites by constructing interconnected network of self-assembled graphene with latex mixing. *Journal of Materials Chemistry* 22: 10464–10468.

Zhan, Y., J. Wang, K. Zhang, et al. 2018. Fabrication of a flexible electromagnetic interference shielding Fe_3O_4@reduced graphene oxide/natural rubber composite with segregated network. *Chemical Engineering Journal* 344: 184–193.

Zhang. L., S. Bi, M. Liu. 2018. Lightweight electromagnetic interference shielding materials and their mechanisms. In *Electromagnetic Materials and Devices*, ed. M.G. Han, Gomes, 1–20. IntechOpen. doi:10.5772/intechopen.82270.

Zhang, W., J. Cui, C. Tao, et al. 2009. A strategy for producing pure single-layer graphene sheets based on a confined self-assembly approach. *Angewandte Chemie* 121, no. 32: 5978–5982.

Zhang, Z., A. Fraser, G. Merle, J. Barralet. 2019. Top-down bottom-up graphene synthesis. *Nano Futures* 3, no. 4: 042003.

Index

For Product Safety Concerns and Information please contact our EU
representative GPSR@taylorandfrancis.com
Taylor & Francis Verlag GmbH, Kaufingerstraße 24, 80331 München, Germany